AF767608

Surface and Interface Analysis

Surface and Interface Analysis

Principles and Applications

Seong H. Kim
The Pennsylvania State University, Pennsylvania, USA

WILEY

Published by John Wiley & Sons, Inc., Hoboken, New Jersey.
Published simultaneously in Canada.

Library of Congress Cataloging-in-Publication Data Applied for:

Paperback ISBN: 9781394218349

Cover Design: Wiley
Cover Image: © a-r-t-i-s-t/Getty Images

Set in 9.5/12.5pt STIXTwoText by Straive, Pondicherry, India

Printed and bound by CPI Group (UK) Ltd, Croydon, CR0 4YY

C9781394218349_021224

Contents

Preface

This textbook is written to help graduate students or novice researchers to expand upon what they have learned in undergraduate chemistry, physics, and math classes to comprehend fundamental principles of surface and interface characterization techniques that are widely used in materials-related research fields. Surface and interface characterization is essential for the advancement of almost all materials research areas including renewable materials for sustainable energy; catalytic materials for chemical synthesis; electronic materials for quantum computing, sensing, and communications; biological materials to uncover the rules of life and produce food; and support of manufacturing through quality control and failure analysis. In recent decades, advances in computer-aided equipment control and visualization software have made the operation and data collection of complicated characterization techniques highly user-friendly. Also, the pace of invention of new materials and the need to test them in mission-specific applications has become faster. This has shifted the focus of materials characterization training from *learning fundamental characterization principles sufficiently* to *routinely and efficiently operating user-friendly instruments*.

When we look at typical undergraduate curricula in the STEM fields, students take many basic science courses in sequence over several years. After finishing all these courses independently, most students do not see the necessity to integrate individually studied concepts and construct their own deeper knowledge that can be applied to materials characterization. We have witnessed that many students take or rely on short training sessions for instrument operation and data collection, then interpret or analyze the data by comparison to the literature (often online) without having deep knowledge of the working principles. Although this approach could be viewed as "pragmatic" for data generation for manuscript submission, it has been the root cause for the spread of incorrect information and misinterpretations in the literature. Moreover, without deep knowledge of the fundamental principles, it is difficult for students to design new experiments that have not been described in the literature, which is an educational necessity for future generations of research scientists.

Although comprehending deep knowledge of multiple techniques sounds like a big challenge, I believe that all students can find their own ways to deal with this challenge – mastering the characterization principles and utilizing them efficiently – once they see how to integrate what they have learned so far to construct a proper foundation and gain expertise to step into a new research field. There are many excellent technical books that introduce specific topics in depth; but I have often found that those books start from a very high level and go too fast into advanced applications, frequently leading to students feeling overwhelmed and discouraged. When I, too, studied those books in my early career, I felt a high "activation barrier" to finishing even the first chapter.

This textbook is intended to show how to "bridge" the gap between the knowledge gained from college education and the knowledge needed to confidently and independently study advanced technical books and papers. I hope this book will lower the barrier for students to start digging into deeper-level books and literatures and utilize what they already know properly to grasp advanced theoretical concepts needed for more appropriate and nuanced application of modern characterization techniques.

Please note that basic fundamentals that students are expected to know from their undergraduate chemistry, physics, and math classes could not be described in detail in this book. I know that not all readers have the same educational background as mine. Without going far outside, I see students in my research group have very different backgrounds and took different classes. To some students, certain physical concepts or terms might be completely new. Even for students who took relevant classes, a large fraction of previous course materials may have already faded from memory. I remind them that what they have not learned or what they may need to brush up on can typically be found easily through searching the plethora of resources available on the internet. By simply typing a few key words in the internet, most information can be found that is necessary to digest the materials covered in this book.

I might be one of the last generations who physically searched and used encyclopedia and chemical abstract numbers in the library. In the past, finding relevant information was very time-consuming. Nowadays, most information that students need can be found on the internet. I encourage students to use the internet as frequently as needed. At the same time, I urge students to use the internet as a starting point, not as a sole source. Although obtaining the results of a web search is now exceedingly fast, truly processing the information found can be slow and time-consuming. The ease of finding information does not necessarily mean that digesting it is also easy. It takes time and effort to understand the information at the level at which you can use it efficiently when you need it. So, don't be discouraged if learning those concepts is not as quick as the click of a mouse button. This book is designed to help students to connect the information found on the internet to the working principle of state-of-the-art characterization methods, especially focusing on surface and interface analysis of materials.

The book is organized in the following order. In Chapter 1, a list is compiled of various characterization methods used for surface and interface analyses. Many of these analyses are conducted in ultrahigh vacuum (UHV) conditions. So, we first will study the reason for that by reviewing the kinetic theory of gases, such as inter-particle collisions in the gas phase and gas-surface collisions. We will also consider these aspects from the thermodynamics point of view – i.e., surface energy. Then, we will move onto experimental methods for elemental analysis of surfaces. One of the most widely used methods is x-ray photoelectron spectroscopy (XPS). This ubiquity is probably because XPS can be applied to almost all surfaces as long as the sample can be introduced into UHV, and because it can provide both qualitative and quantitative information for about 90 elements in the periodic table (except hydrogen, helium, and unstable elements that are heavier than uranium). Moreover, thanks to technical advancements made by instrumentation industries, XPS operation has become relatively easy to learn. After several hours of training sessions, new users can reach the "skillful" level for independent operation and data collection. In addition, the availability of a plethora of XPS data in the literature has made this technique the first-to-try for elemental surface analysis. Although XPS can be somewhat expensive because most users may have to pay fees to get data, relatively easy access to the instrument and data sets has made XPS very popular. However, at the same time, this ease has made XPS analysis vulnerable to unintended errors in analysis. So, I have chosen XPS as the first technique to be studied in depth. Chapter 2 starts with brushing up on the atomic orbital theory and the surface-sensitivity principle.

To my own students, I often say that I am wary of the data collected by someone who does not know how the instrument works and why the experiment is conducted in that specific way. That is because, without knowing those principles, it is easy to make mistakes in sample preparation and instrument operation, and we end up wasting a lot of time by trying to interpret the data collected improperly or incorrectly. It would be "fortunate" if we find errors in measurement or data collection and re-do the experiment, instead of publishing inaccurate data and conclusions. Here I direct readers to read a perspective article by G. H. Major, et al. published in *Journal of Vacuum Science & Technology A* 2023, *41*(3):038501. For these reasons, the instrumentation principles of typical XPS systems are covered in Chapter 2. Then, the chapter delves into qualitative and quantitative analysis as well as depth profiling using XPS. This chapter also touches x-ray absorption spectroscopy such as near-edge x-ray absorption fine structure (NEXAF, also known as x-ray absorption near-edge spectroscopy, XANES) and extended x-ray absorption fine structure (EXAFS), but not as in-depth as XPS.

In Chapter 3, elemental analysis with electron irradiation – specifically, Auger electron spectroscopy (AES) – is covered. Energy-disperse x-ray spectroscopy (EDX or EDS) and electron probe micro-analysis (EPMA) are somewhat relevant, but not discussed in depth because they provide (to my view) bulk information (since the probe depth is large). Since the basic principle of AES somewhat overlaps with XPS, the theoretical coverage is kept light. In Chapter 4, secondary ion mass spectrometry (SIMS) is described, which is an elemental analysis with ion bombardment. Again, the basic principle of SIMS is somewhat touched in the depth profiling of XPS; so, the theoretical discussion on the ion–surface interaction is kept minimal. Another technique based on ion bombardment is ion scattering spectroscopy (ISS, also known as low-energy ion scattering [LEIS] spectroscopy). The probe depth of ISS is the shallowest among the ones mentioned above; but its usage is somewhat narrow or at least not as versatile as XPS, AES, and SIMS. So, LEIS is not included in this book.

Till this point, the elemental analyses based on photon-in/electron-out (XPS), electron-in/electron-out (AES), electron-in/photon-out (EDX, EPMA), and ion-in/ion-out (SIMS) are covered. The focus then shifts to molecular analysis. For this, we will study optical spectroscopy based on photon-in/photon-out. Chapter 5 describes how light propagates through a matter and is reflected at the interface of two matters with different refractive indexes. This lays out the foundation for almost all spectroscopy techniques – Snell's law and Fresnel coefficients. In this book, we will focus on vibrational spectroscopy techniques since they provide molecular information via identification of specific functional groups at the interface.

Chapter 6 shows how the Snell's law and Fresnel coefficients are used to conduct attenuated total reflection infrared spectroscopy (ATR-IR), specular reflection infrared spectroscopy (SR-IR), reflection absorption infrared spectroscopy (RAIRS), polarization-modulation RARIS (PM-RAIRS), Brewster-angle transmission IR spectroscopy (BAT-IR), and diffuse reflectance infrared spectroscopy (DRIFT). It will also touch infrared spectroscopic imaging that is coupled with atomic force microscopy (AFM). Going one-step deeper beyond the linear spectroscopy methods listed above, a nonlinear spectroscopy called sum frequency generation (SFG) vibrational spectroscopy is discussed in Chapter 7. This nonlinear spectroscopy is particularly useful and powerful for analysis of buried interfaces. Then, Chapter 8 discusses how spectroscopic data can be processed using multivariate analysis principles. For simplicity, only the basic principles of factor analysis and principal component analysis/regression (PCA/PCR) are covered.

Chapters 9–12 are about metrology. Reflectometry and ellipsometry are explained in Chapter 9, again utilizing the Snell's law and Fresnel coefficients for determination of thickness and optical constants of films on flat substrates. An extension of the same principle to the measurement of surface topography is optical profilometry, which is covered in Chapter 10. Then, we shift gears to topography analysis using a physical probe – i.e., atomic force microscopy (AFM) – in Chapter 11. An extension of the physical probe method is nanoindentation and micro-indentation (Chapter 12), which provide mechanical property information of the surface.

Note that this book is not intended to cover all the details of each technique to the expertise level. Also, it is not intended to cover a large number of techniques superficially with emphases on advanced or recent applications, as in typical survey-level courses or workshops. The book uses some examples from literature to show how key concepts are utilized or what kind of errors or misinformation prevails in the literature due to the lack of sufficient understanding of basic principles. The coverage of such examples is kept minimal to avoid diverting the focus from the fundamental principles. Reviewing various applications in the literature can be done independently or as a homework problem in a short summary report format.

Finally, I'd like to express my deep appreciation to the people who have influenced me one way or another. The biggest thanks go to my own students who conducted research with me and the students who took my surface characterization course (ChE/MatSE 510) at Penn State. They showed me what physical concepts they missed previously and why they had difficulties in remembering and connecting them; from that, I learned what I need to teach to them so they can build their own foundations by themselves from now on. I would like to express my special thanks to the technical staff at the Penn State Materials Characterization Lab, especially Vince Bojan and Jeff Shallenberger, for their dedication and assistance in helping our students and for my class. I acknowledge numerous supports from the National Science Foundation, the Department of Energy, the Air Force Office of Scientific Research, and several companies; many of the data shown in this book were obtained while conducting the research supported by them. I also acknowledge the support from the Brain-Pool program of the National Research Foundation of Korea during my sabbatical stay at the Korea Institute of Science and Technology (KIST). A big portion of this book was written during my sabbatical at KIST.

July 15, 2024

Seong H. Kim
The Pennsylvania State University, Pennsylvania, USA

About the Companion Website

This book is accompanied by a companion website:

www.wiley.com/go/Kim

The website contains answers to the practice problems at the end of each chapter.

1

Introduction

1.1 Types of Surface Analysis

Surface science deals with the chemical and physical phenomena occurring at the boundary of two bulk phases, which include solid/gas, solid/liquid, liquid/liquid, and solid/solid interfaces. As the size of materials gets smaller, the fraction of atoms exposed at the surface with respect to the number of atoms in the bulk (i.e., surface-to-volume ratio) increases; thus, knowledge of the interfacial composition and structure and fundamental understanding of their roles in governing materials behaviors and properties become more vital. Although "surface" is mathematically defined as a two-dimensional (2D) plane, in practice, "surface and interface science" involves not only studying the topmost atoms or molecules exposed at the phase boundary but also investigating the atoms and chemical bonds beneath the surface as they typically affect the structure and reactivity of the surface. Conversely, the topmost atoms and molecules interacting with environments can also influence the structure and chemistry of subsurface atoms and functional groups. Thus, the system of interest in surface science is not confined to the 2D plane only, but it involves the depth dimension, making it a three-dimensional (3D) problem.

In order to characterize the surface, one must send a "probe" which can be photons, electrons, ions, or a physical object to the surface and detect a "messenger" which the surface is sending out upon interactions with the probe. Such interactions vary depending on the type of probe sent to the surface, the surface–probe interactions, and the type of messenger (which we can call "signal") to be detected. Different surface analysis methods work with different physical principles with different length scales. For that reason, it is important to know the effective probe depth as well as types of information that each analytical technique can probe. Table 1.1 compiles a list of characterization techniques that are widely used in interfacial science study and research.

1.2 Why Is Surface Analysis Done in Vacuum?

In Table 1.1, one can see that many surface analysis techniques operate in vacuum conditions, especially when elemental analysis is done through detection of charged particles (electrons or ions) scattered or emitted from the surface. Why is vacuum required for such analyses? For fundamental surface science studies, one must be able to prepare a surface with well-defined structure and composition reproducibly. For surface characterization, one must be able to maintain the pristine surface for a sufficiently long time for experimental investigations. Having the sample in vacuum helps in these aspects. Then, what kind of vacuum is necessary? Would rough vacuum with a mechanical pump be sufficient? Or should it be an ultrahigh vacuum condition that can be reached with a turbo pump? To answer these questions, one can consider solid–gas collisions.

Let us start with the kinetic theory of ideal gas. Gas consists of many particles traveling in random trajectories (Figure 1.1). The size of gas molecules is much smaller than the average inter-molecular separation. Under the ideal gas condition, gas molecules are assumed to interact with other particles through elastic collisions only. In this situation, gas pressure results from their collisions with a wall. When a molecule with mass m ($= M_w/N_A$, where M_w is the molecular weight and N_A is the Avogadro's number) travels at a velocity v_x along the x-direction in a Cartesian coordinate and collides with a surface at an arbitrary location in x, its momentum changes from mv_x to $-mv_x$; thus, the total momentum change is $2mv_x$. For molecules

Surface and Interface Analysis: Principles and Applications, First Edition. Seong H. Kim.

Companion website: www.wiley.com/go/Kim

Table 1.1 Compilation of characterization techniques used as *surface-sensitive* or *surface analysis* methods.

Technique	Main information	Type of sample	Environment
Scanning electron microscopy (SEM)	Topography projected to 2D plane	Solid, conductive (or with conductive coating)	Vacuum (or limited control of environmental)
Tunneling electron microscopy (TEM)	Images and structures	Solid, should be thinner than 200 nm	Vacuum
Energy dispersive X-ray spectroscopy (EDX or EDS); electron probe microanalysis (EMPA)	Elemental composition of ∼1 μm depth	Solid, conductive	Vacuum
Low-energy electron diffraction (LEED)	Two-dimensional periodic structure with probe depth < 0.5 nm	Solid, crystalline, conductive	Vacuum
X-ray photoelectron spectroscopy (XPS)	Elemental composition and limited chemical state for top 3–10 nm	Solid, any type	Vacuum (or near-ambient pressure if a differentially pumped analyzer is used)
Near-edge X-ray absorption fine structure (NEXAFS); X-ray absorption near-edge structure (XANES)	Elemental and oxidation information; top 2–5 nm in electron yield detection; bulk in X-ray fluorescence detection	Solid, any type	Vacuum for electron detection; ambient for X-ray detection
Extended X-ray absorption fine structure (EXAFS)	Local order (nearest and next nearest atoms); typically averaged through the sample	Solid, any type	Vacuum to ambient
UV photoelectron spectroscopy (UPS)	Elemental composition and electronic band structure for top 1–2 nm	Solid	Vacuum
Auger electron spectroscopy (AES)	Elemental composition and limited chemical state for top 2–4 nm	Solid, conductive	Vacuum
Ion scattering spectroscopy (ISS); low-energy ion scattering spectroscopy (LEIS)	Elemental composition of topmost surface	Solid, flat	Vacuum
Secondary ion mass spectrometry (SIMS)	Elemental compositions with varying depth resolution depending on operation mode	Solid, flat (at least visually)	Vacuum
Atomic probe tomography (APT)	Three-dimensional (3D) compositional mapping	Sharp tip prepared by ion milling	Vacuum
High-resolution electron energy loss spectroscopy (HREELS)	Organic functional groups at the topmost surface	Solid, flat, conductive	Vacuum
Reflection–absorption infrared spectroscopy (RAIRS)	Functional groups on top of substrates (up to 10's nm thick)	Solid, flat	Vacuum to ambient
Diffuse reflectance infrared Fourier transform spectroscopy (DRIFT)	Functional groups on/in particulates	Solid, fine powder	Ambient
Attenuated total reflection infrared spectroscopy (ATR-IR)	Functional groups on top of substrates up to a few μm from the surface	Solid, flat	Ambient
Specular reflection infrared spectroscopy (SR-IR)	Functional groups up to a few μm from the surface	Solid, flat	Ambient
Raman spectroscopy	Functional groups up to a few μm from the surface	Solid, liquid	Vacuum to ambient
Sum frequency generation (SFG) vibrational spectroscopy	Functional groups at topmost surface or within field gradient (up to several μm)	Solid, liquid	Vacuum to ambient
Total internal reflection (TIR) UV–Vis spectroscopy	Functional groups up to a few 100's nm	Liquid	Ambient

Table 1.1 (Continued)

Technique	Main information	Type of sample	Environment
Photoluminescence	Chemical states up to a few μm	Solid, usually semiconductors	
Spectroscopic ellipsometry	Optical constants of overlayers and bulk and thicknesses of overlayers	Solid, flat liquid	Vacuum to ambient
Atomic force microscopy (AFM)	3D topographic images (up to a few μm) and other properties	Solid, any type	Vacuum to ambient
Scanning tunneling microscopy (STM)	3D images (up to a few Å) convoluted with electronic structure	Solid, flat, conductive	Vacuum to ambient
Optical microscopy	Topography projected to 2D plane	Any type	Any environment
Optical profilometry	3D images of topography	Any type	Ambient
Nanoindentation	Mechanical properties of top 100's nm region (up to several μm)	Solid, flat	Any environment

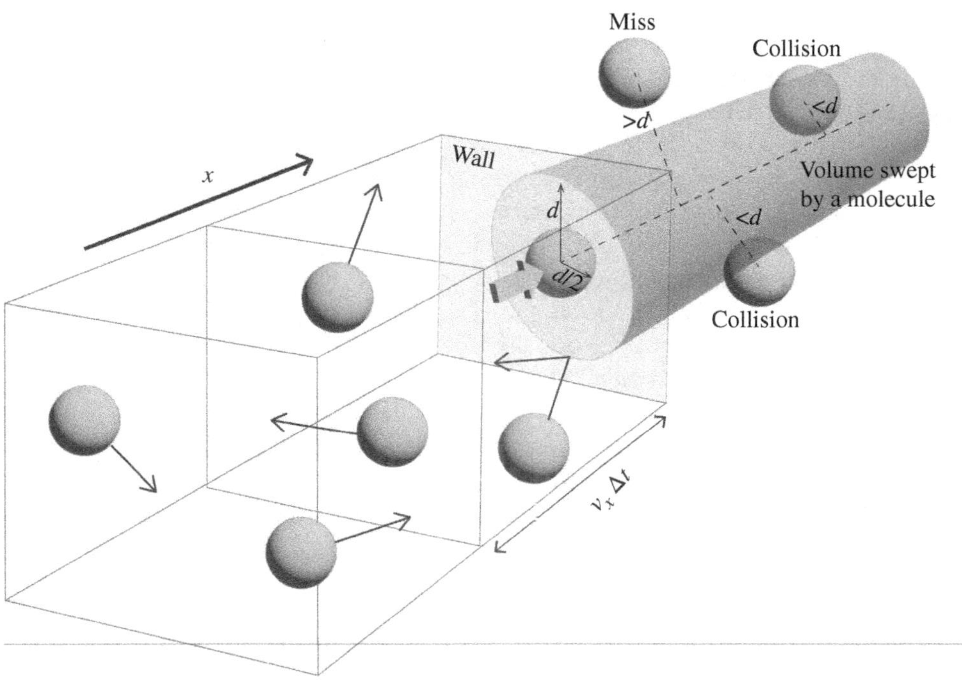

Figure 1.1 Collision of molecules with a wall and the volume swept by a molecule starveling along the x-direction.

traveling with average velocity $\langle v \rangle$, the distance they can travel in time interval Δt is $\langle v \rangle \Delta t$. If they move toward a wall of area A and the number density is n (= N/V, where N is the total number of molecules and V is the total volume), the number of molecules that strike the wall in time Δt is $nA\langle v \rangle \Delta t$. In the random trajectory condition, half of the molecules are traveling toward the wall and the other half are traveling away from the wall. So, the total number of collisions for one side of the wall is $\frac{1}{2} \times nA\langle v \rangle \Delta t$. Then, the total momentum change can be calculated as number of collisions times momentum change per collision, which gives $\frac{1}{2} nA\langle v_x \rangle \Delta t \times 2m\langle v_x \rangle = n\ m\ A\ \langle v_x \rangle^2 \Delta t$ along the x-direction. Since force is the time derivative of momentum, the force exerted by molecules to this surface is $F_x = n\ m\ A\ \langle v_x \rangle^2$. The pressure is the force divided by the area, thus the pressure in the x-direction is $P_x = n\ m\ \langle v_x \rangle^2$. Within a given volume with molecules traveling in random directions, the average velocity is $\langle v \rangle = \langle v_x \rangle + \langle v_y \rangle + \langle v_z \rangle = 3\langle v_x \rangle$. Thus, the pressure can be expressed as:

$$P = \frac{1}{3} n\, m \left\langle v^2 \right\rangle \tag{1.1}$$

Now, let us consider collisions among molecules in a volume V with a pressure P (Figure 1.1). A molecule of diameter d sweeps out a cylindrical volume with cross-sectional area $\sigma = \pi d^2/4$ and length $L = \langle v \rangle \Delta t$ during a time period Δt. For two colliding objects, we need to take into account their relative speed v_{rel}. The relative velocity of two particles can be described as $\vec{v}_{rel} = \vec{v}_1 - \vec{v}_2$. Then, the average of the velocities of particles moving in random directions is the square root of the scalar products: $\langle v_{rel} \rangle = \sqrt{\vec{v}_{rel} \cdot \vec{v}_{rel}} = \sqrt{\langle v_1 \rangle^2 + \langle v_2 \rangle^2} = \sqrt{2}\langle v \rangle$. Thus, the collision frequency per unit time per molecule in a given volume is:

$$z = \sqrt{2}\, n\, \sigma \, \langle v \rangle \tag{1.2}$$

Then, $1/z$ is the time a molecule spends between collisions. The distance it has traveled in this time is called the mean free path: $\lambda = \langle v \rangle / z = \dfrac{1}{\sqrt{2}\, n\, \sigma}$. Here the number density term n can be replaced with pressure P and temperature T using the ideal gas law, which gives:

$$\lambda = \frac{1}{\sqrt{2}\sigma} \frac{k_B T}{P} \tag{1.3}$$

where k_B is the Boltzmann constant. For nitrogen (N_2), σ is 0.43 nm^2. Then, at 1 atm and 300 K, λ = 70 nm.

Using eq. (1.3), one can estimate the pressure range in which the surface characterization experiment should be conducted. Let us assume that we send the probe beam to the surface and the surface emits particles which we need to detect. In an environment filled with gas at 1 atm, the emitted particles will be able to travel only ~70 nm before colliding with other gas molecules. If our detector is 1 m away from the surface, that means the particle would not be able to arrive at the detector. The charged particles (electrons and ions) are highly reactive; so, they get altered upon collision with gas phase molecules. If one needs to measure the kinetic energy of particles emitted from the surface, it should be done in the condition where they pass through the energy analyzer and arrive at the detector without any collisions in the gas phase. Otherwise, the particle energy will be fully equilibrated with the thermal energy of surrounding gas molecules. So, in order to allow the emitted particle to arrive to the detector without encountering any gas molecules in its trajectory, we need to increase the mean free path (λ) at least up to 1 m, which requires lowering the environment pressure down to $\sim 10^{-7}$ atm (7.6×10^{-5} Torr). This is the theoretical minimum of λ, in practice, the longer λ is, the lower the collision probability is, thus better for detection of the surface signal without any alteration.

Next, let us consider the collision with the gas phase molecules for the topmost atoms at the surface (Figure 1.1). This time, we are going to use a slightly different approach. The number of molecules crossing a plane per unit area per unit time can be calculated from the number density and velocity of the molecules, which can be expressed as $n \cdot \int_0^\infty v_x f(v_x) dv_x$, where $f(v_x)$ represents the velocity distribution of molecules. Here we integrate the velocity from zero to infinity, because the molecules traveling away from the surface (thus, having the negative velocity with respect to the reference plane) will not collide with the surface. Using the Maxwell–Boltzmann distribution of molecular speeds, one gets $\int_0^\infty v_x f(v_x) dv_x = \sqrt{\frac{k_B T}{2\pi m}}$. The ideal gas law says $n = \frac{N}{V} = P/k_B T$. Thus, the collision frequency with the wall per unit area becomes:

$$Z_A = P/\sqrt{2\pi m k_B T} \tag{1.4}$$

If the gas phase is filled with N_2 at 1 atm, then $Z_A = \sim 3 \times 10^{23}\ cm^{-1}\ s^{-1}$. In N_2 gas with $P = 1.315 \times 10^{-9}$ atm ($=10^{-6}$ Torr), the collision rate on a surface is $\sim 3.8 \times 10^{14}\ cm^{-1}\ s^{-1}$.

At this point, one may get a better sense of what this collision rate means if it is compared with the number of atoms at the surface. If we assume a typical interatomic distance for a solid surface is 3 Å, then the number of surface atoms is:

$$n_s = \frac{1}{(3 \times 10^{-10}\, m)^2} \approx 1 \times 10^{15}\ cm^{-2} \tag{1.5}$$

If every molecule that strikes this surface remains adsorbed (i.e., sticking probability = 1), the time it takes to form a full monolayer is:

$$\tau = n_s / Z_A \tag{1.6}$$

For nitrogen at 300 K, τ is 2 ns at 1 atm (760 Torr), 3 ms at 1.3×10^{-6} atm (10^{-3} Torr), 3 s at 1.3×10^{-9} atm (10^{-6} Torr), and 3000 s at 1.3×10^{-12} atm (10^{-9} Torr). This means that if a clean surface is exposed to atmospheric air, all surface atoms are bombarded by gas phase molecules every 2 ns. If the surface characterization takes several minutes per measurement and you need to repeat the measurement several times to check the reproducibility, then you need to be in ultra-high vacuum ($P < 1.3 \times 10^{-12}$ atm or 10^{-9} Torr) to keep the surface composition unaltered for the entire duration of experiment. This kind of simple estimation gives you a sense of what kind of vacuum is necessary to do surface characterization reliably without worrying about the surface being contaminated during the analysis, especially if the sample is highly reactive or easily get contaminated.

An example showing the importance of proper sample preparation is shown in Figure 1.2 [1]. In this example, iron nanoparticles covered with a thin layer of oxide were analyzed. The XPS principle will be discussed in Chapter 2; so, let us not focus on the detail of spectral interpretation, but we can still see the importance of proper sample handling. When the sample was removed from their initial packaging in a glove-bag pursed with dry N_2 which was connected to the sample mounting port of an XPS instrument and analyzed with XPS, the spectrum showed a relatively strong Fe^0 peak at 707 eV in addition to an oxide peak at 710–711 eV. When the sample was exposed to air for less than 5 minutes, the Fe^0 peak was significantly decreased relative to the oxide peak. This is because the sample surface was oxidized by O_2 gas and H_2O vapor impinging from the ambient air.

Another example is shown in Figure 1.3 [2]. When a friction test of hydrogenated diamond-like carbon (HDLC) was performed in a dry Ar or dry H_2 environment, the coefficient of friction (COF) was found to be ultralow (around 0.01–0.02). However, the COF of the same material was found to be an order of magnitude higher in dry O_2 or in Ar with n-propanol vapor with a partial pressure 40% of its saturation (P/P_{sat}). When the friction test was done in Ar with 40% relative humidity (RH), then COF was unsteady and varied over a large range (Figure 1.3A). Since the same material was tested, one could hypothesize that the difference in COF must be related to the surface chemistry during the friction test which is governed by reactions with molecules impinging from the gas phase. With a goal to identify the surface chemistry difference among the surfaces tested in different environments, *ex situ* XPS analysis was carried out; however, the high-resolution C 1s spectra of these samples did not show much difference. This is basically due to the post-experiment oxidation of the surface upon exposure to air during the sample transfer. The same study found that even the exposure of the surface that already reached the ultralow COF state to an ultrahigh purity Ar gas for 5 minutes, the COF increased momentarily and then decreased in repeated cycles of frictional sliding [2]. In a later study, it was found that the oxidation probability of the HDLC material shown in Figure 1.3B was on the order of 10^{-5}–10^{-4} $Torr^{-1}$ s^{-1} [3]. Note that the collision frequency of a gas molecule at the

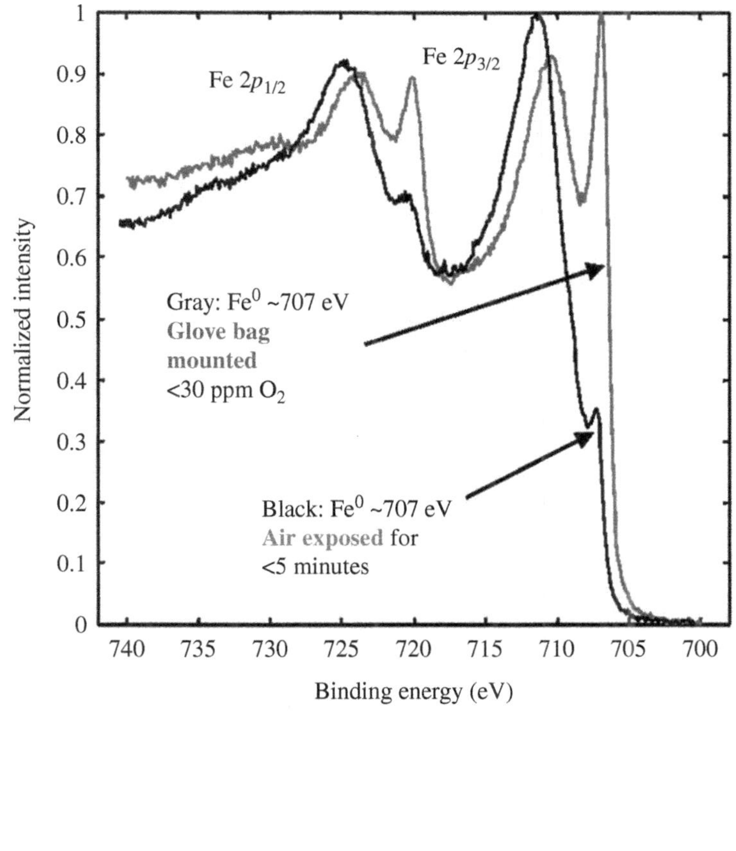

Figure 1.2 XPS spectra of iron nanoparticles prepared in two different ways. The gray spectrum is the data of the sample removed from its initial package in a N_2-purged glove-bag connected to the sample mounting port of the XPS system. The black spectrum is the data of the sample exposed to air for 5 minutes during the sample transfer from the package to the XPS chamber. Baer et al. [1]/Elsevier.

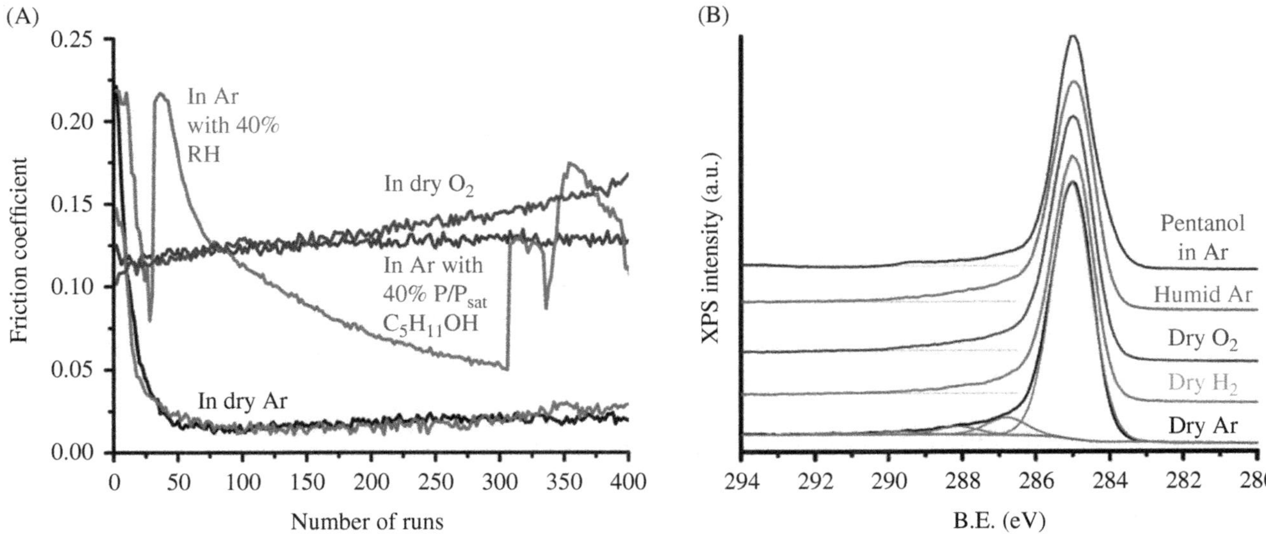

Figure 1.3 (A) Friction coefficient measured for HDLC in various environments. (B) C 1s XPS spectra of the friction-tested HDLC surface. The sample was exposed to air during the transfer from the environment-controlled tribo-testing unit to the XPS system. Marino et al. [2]/American Chemical Society.

surface is $\sim 7 \times 10^9\,s^{-1}$ at 760 Torr at room temperature (calculated for N_2; the values for O_2 and H_2O can be obtained by multiplying their concentrations in ambient air). Although the absolute reaction probability per collision is quite low, the number of collisions is extremely high. One may get a better sense of how reactive this surface is if one realizes that typical reaction probabilities of reactive gases on catalytically active surfaces are on the order of $10^{-3}\,Torr^{-1}\,s^{-1}$ [3].

1.3 Surface Energy

The gas collision frequency of the surface calculated from the kinetic theory is an extrinsic factor (not intrinsic to the surface) since it describes how frequently the surface atom interacts with molecules present in the environment. Even if the gas-surface collision frequency is extremely high, it may not be a serious concern if the surface of interest is chemically inert. There are many factors that govern how easily the surface gets contaminated by or reacts with molecules impinging from the gas phase. Such factors vary depending on the types of materials under consideration as well as the types of molecules arriving at the material surface, which invites discussion of the thermodynamic properties of the surface. This discussion can start from the surface energy argument. A material with a higher surface energy will have a higher propensity to hold molecules upon their landing on the surface as compared to a material with a lower surface energy. Table 1.2 lists some examples of the surface energy of several materials. Note that the surface energy and surface tension have the same value but different units. The former has the unit of J/m^2, and the latter has the unit of mN/m. So, their dimensions are the same. For that reason, they are often used interchangeably; both are represented with the same symbol (γ). But, depending on the application, the proper term should be used with the correct unit.

At the surface, chemical bonding with surrounding atoms is disrupted. Thus, the presence of the surface makes the system unstable, increasing the Gibbs free energy (G). The surface contribution to the Gibbs free energy is described as:

$$dG = -SdT + VdP + \gamma dA \tag{1.7}$$

where S and V are the entropy and molar volume of the material, respectively, and A is the surface area. So, at constant T and P, the surface energy can be defined as excess energy per unit area:

$$\gamma \equiv \left(\frac{dG}{dA}\right)_{T,P} \tag{1.8}$$

Table 1.2 Surface energy (γ) of selected solids and liquids [4].

Materials	Surface energy (J/m^2)	Surface tension (mN/m)	Temperature (°C)
He (liquid)	3.08×10^{-4}	0.308	−270.5
N_2 (liquid)	9.71×10^{-3}	9.71	−195
O_2 (liquid)	15.7×10^{-3}	15.7	−193
n-Octane	0.02180	21.80	20
Benzene	0.02888	28.88	20
Carbon tetrachloride	0.02695	26.95	20
Ethanol	0.02275	22.75	20
Acetic acid	0.0278	27.8	20
Water	0.07275	72.75	20
KCl	0.110	110	25
NaCl	0.227	227	25
LiF	0.340	340	−195
CaF_2	0.450	450	−195
MgO	1.2	1200	25
Glass [5]	3.5–5.3	3500–5300	25
Al_2O_3 (sapphire)	6.0–7.3	6000–7300	25
Hg (liquid)	0.487	487	16.5
Ag (liquid)	0.879	879	1100
Ag (solid)	1.14	1140	907
Au (solid)	1.41	1410	1027
Cu (liquid)	1.30	1300	1535
Cu (solid)	1.67	1670	1047
Fe (liquid)	1.88	1880	1535
Fe (solid)	2.15	2150	1400
Pt	2.34	2340	1311
W	2.90	2900	1727
C (diamond) [6]	5.03	5030 (111 facet)	25
C (graphite) [7]	0.065–0.088 4.8–6.0	65–88 (basal plane) 4800–6000 (edges)	25
C (graphene) [8]	~0.050	~50 (fresh)	25

Source: Adapted from Somorjai [4].

In typical introductory thermodynamics class, we deal with the first two terms on the right-hand side of eq. (1.7), but not the surface energy term. It is simply because the undergraduate course deals mostly with the bulk properties. When the size of system is large, the surface-to-volume ratio is negligible; thus, the surface energy contribution to the Gibbs free energy can be ignored.

Equation (1.8) tells us that γ is literally the change in free energy upon change of surface area. Since γ is positive, it means that at a given T and P, an increase in A leads to an increase in G. This is the reason that a given system always tries to assume the configuration with the smallest surface area. For a given volume, the sphere has the smallest surface area as compared to other geometric shapes. Thus, the liquid droplet suspended in air is always spherical when its diameter is smaller than the capillary length, $l_c = \sqrt{\gamma/(\Delta\rho g)}$ where $\Delta\rho$ is the density difference between liquid and air and g is the gravitational acceleration constant.

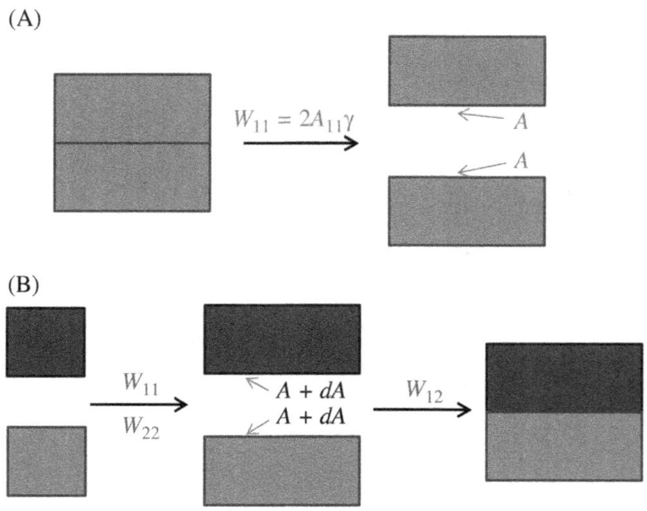

Figure 1.4 (A) Work of cohesion and (B) work of adhesion.

If a solid material is split into two pieces, the surface area of the system increases (Figure 1.4A). Then, the surface energy can be related to the work of cohesion (W_{11}), which is the free energy change (reversible work done) to cleave or separate a material from contact to infinity in vacuum, via:

$$W_{11} = \gamma \times 2A_{11} \tag{1.9}$$

where A_{11} is the total area of the newly creased surface. The work of cohesion is isotropic for amorphous materials, but it can be anisotropic for crystalline materials because fracture can occur through different crystallographic planes.

The work of adhesion for dissimilar materials can be considered with two hypothetical steps: (i) generating new surfaces for materials 1 and 2 in vacuum and (ii) bringing two new surfaces into contact (Figure 1.4B). The first step can be described with the work of cohesion per unit area for each material as $W_1 = \gamma_1$ and $W_2 = \gamma_2$. In the second step, the two new surfaces will get stabilized by interacting with each other; this energy per unit area can be expressed as γ_{12}. Then, the work of adhesion per unit area can be written as:

$$W_{12} = \gamma_1 + \gamma_2 - \gamma_{12} \tag{1.10}$$

This is the Dupré equation describing the work of adhesion for dissimilar materials. If materials 1 and 2 are the same, then, γ_{12} is zero and the, eq. (1.10) becomes the same as eq. (1.9).

1.4 Relationship Between Surface Energy and Physical Properties

For a brittle material, the surface energy (γ) can be related to the critical stress required to create a new crack surface (σ_f) that eventually leads to fracture of the material:

$$\sigma_f = \sqrt{\frac{2E\gamma}{\pi a}} \tag{1.11}$$

where E is the elastic modulus and a is a critical crack length. This is called the Griffith equation [9]. For silicate glasses, Young's modulus is typically ~70 GPa. If one sets $\gamma = 4\,\text{J/m}^2$ and assumes a is 0.2 nm which corresponds to the typical value of a single chemical bond length, then σ_f is estimated to be ~30 GPa. This can be viewed as the theoretical tensile strength of silicate glass. Practically, no glass exhibits this strength due to the presence of surface defects. For a compositionally and structurally homogeneous glass, there is no grain boundary, but a surface defect can be made during manufacturing or handling. If a tensile stress is applied to a glass object, then the stress is effectively concentrated at the surface defect. If the glass surface has a micron size defect, then σ_f becomes on the order of 100 MPa, so the glass breaks way before the applied stress reaches the theoretical strength. This argument also leads to the relationship between the fracture toughness in the crack opening mode (K_{Ic}) and the surface energy (γ):

$$K_{Ic} = \sqrt{2E\gamma} \tag{1.12}$$

which was used by Wiederhorn to estimate the surface energy of glass from the K_{Ic} measurement [5]. A typical K_{Ic} of a silicate glass is ~0.7 MPa·m$^{1/2}$.

The surface energy can also be related to the heat of sublimation. Suppose that the bond energy of a pair of atoms is E_b, the number of bonds or nearest neighbors is z_b for an atom in the bulk and z_s for an atom at the surface. When a crystal is cut, the number of bonds to be broken is $z_b - z_s$ Then the energy required to create the surface can be estimated to be:

$$E = A\, n_s\, (z_b - z_s)\, E_b \tag{1.13}$$

where A is the surface area and n_s is the number of atoms per unit surface area. Then, eq. (1.9) gives:

$$\gamma = \frac{E}{2A} = \frac{1}{2} n_s\, (z_b - z_s)\, E_b \tag{1.14}$$

Sublimation involves breaking all bonds. In order to isolate N_A number of atoms, the total number of bonds to break is $\frac{1}{2} z_b N_A$. Then, the heat of sublimation is $\Delta H_{subl} = \frac{1}{2} z_b N_A E_b$. Substituting this to eq. (1.14) gives:

$$\gamma = \frac{z_b - z_s}{z_b}\, n_s \frac{\Delta H_{subl}}{N_A} \tag{1.15}$$

Let us use this equation to estimate the surface energy of gold. Gold has a face-centered cubic (FCC) crystal in which each atom has 12 nearest neighbors. An atom on the (100) surface has eight nearest neighbors. The atomic number density of the (100) surface is $1.2 \times 10^{19}\ \mathrm{m^{-2}}$. If ΔH_{subl} is 368 kJ/mol and N_A is Avogadro's number (6.23×10^{23}/mol), then eq. (1.15) estimates the surface energy of Au(100) surface to be 2.4 J/m^2. The experimental value is ~1.4 J/m^2 near the melting temperature (Table 1.2). The discrepancy is due to the relaxation of the surface structure. As mentioned before, the creation of a new surface increases the free energy. In order to lower the free energy, surface atoms can be rearranged or restructured searching for a lower energy state [4]. This reconstruction process brings the surface energy lower than the value estimated from the heat of sublimation of bulk atoms.

For the sake of getting a physical sense of γ, one may practice estimating the order of surface energy for different crystallographic facets. There are many freeware programs that help you to visualize the surface structures of different crystal structures. Figure 1.5 is the (100), (110), and (111) facets of the FCC crystal visualized with the program Balsac (Build and Analyze Lattices, Surfaces, and Clusters; © K. Hermann, FHI 1991–2022) [10]. One can count that z_s is 8 for the (100)

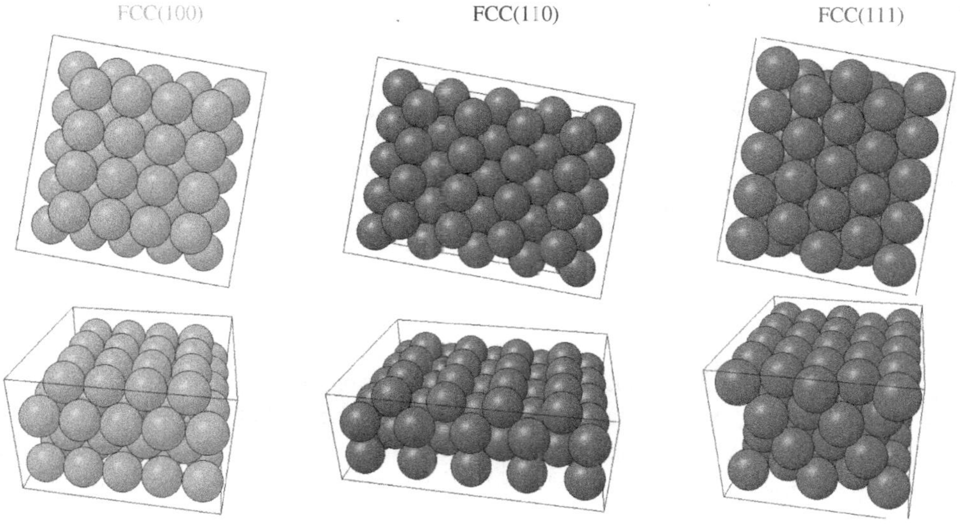

Figure 1.5 Structures of FCC(100), FCC(110), and FCC(111) facets. The images are created using the Balsac program. Adapted from Hermann [10].

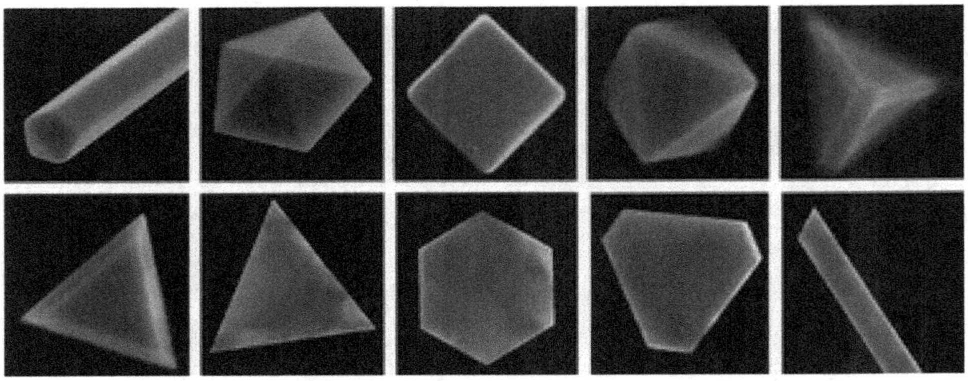

Figure 1.6 Images of nonspherical shapes found for Au and Ag nanoparticles. Reproduced from Walters and Parkin [11]/with permission of Royal Society of Chemistry.

surface, 7 for the (110) surface, and 9 for the (111) surface. Thus, the number of bonds ($z_b - z_s$) to break to produce (100), (110), and (111) surface is 4, 5, and 3, respectively. From this calculation, one can predict $\gamma_{(110)} > \gamma_{(100)} > \gamma_{(111)}$. This tells us that the polycrystalline sample of the FCC metal will have the (111) facets predominantly, which is indeed true. Note that the exact value γ can vary significantly from the value estimated by counting the number of bonds to break, because the surfaces will undergo structural relaxation whenever possible. For elemental crystals, the database for surface energies of various facets in vacuum is available [6].

In gas or liquid environments, the surface energy of a solid can be affected by chemical interactions of molecules adsorbed on the surface. For example, when a table salt (NaCl) is dissolved in water and recrystallized by evaporating water, one will see cubic crystals. The facet of this cube has the (100) plane of the NaCl crystal since it has the lowest surface energy among all possible crystallographic facets. When the same salt is dissolved in aqueous solution containing a certain amount of urea or citric acid and recrystallized, then pyramidal-shape crystals are precipitated. This is because the (111) plane of the NaCl crystal gets stabilized more in the presence of urea or citric acid. Similarly, the shape of nanoparticles can be controlled by using additives. For example, the shapes of very small gold nanocrystals (<20 nm) are inclined thermodynamically to the symmetric FCC types, but kinetic factors of growth and nucleation can vary depending on concentrations of surfactants or capping agents, which can be used to control or direct shape during nanoparticle synthesis as shown in Figure 1.6 [11].

1.5 Measuring Surface Energy

For a solid material that can react upon contact or should break chemical bonds for separation, a double cantilever beam (DCB) method is commonly used to measure the surface energy. In this method, two plates with smooth surfaces are brought into contact with a wedge in one corner. Then, the two plates will bend inward narrowing the distance from the gap to the junction (crack) point until the establishment of equilibrium between the elasticity of plates and the bonding interactions responsible for the adhesion of the plates (Figure 1.7A). At the equilibrium, the critical strain energy is equal to the work of adhesion and can be calculated from the crack length and the linear elastic fracture mechanics for a symmetric pair of beams [12]:

$$\gamma = \frac{3}{16} \frac{E\, t^3\, \delta^2}{(1-\nu)\, L^4} \tag{1.16}$$

where E and ν are Young's modulus and Poisson's ratio of the isotropic material of the beam, respectively, and t is the beam thickness, δ is the wedge thickness, and L is the crack length. Figure 1.7B shows an example of surface energy determined with the DCB method. Silicon wafers with thermal oxide layers were contacted in air and then annealed for 10 minutes in N_2 at various temperatures between 100 and 1400 °C and the measured γ is plotted as a function of annealing temperature [12]. Upon dehydration (desorption of physiosorbed water) at 100 °C, γ of the silicon oxide surface was ~0.1 J/m^2 (=100 erg/cm^2). As the temperature increases, silanol groups (Si-OH) at the surface undergoes dehydroxylation and condensation reactions between two contacting surfaces (Si-OH + HO-Si ➔ Si-O-Si + H_2O), and accordingly γ is augmented gradually to

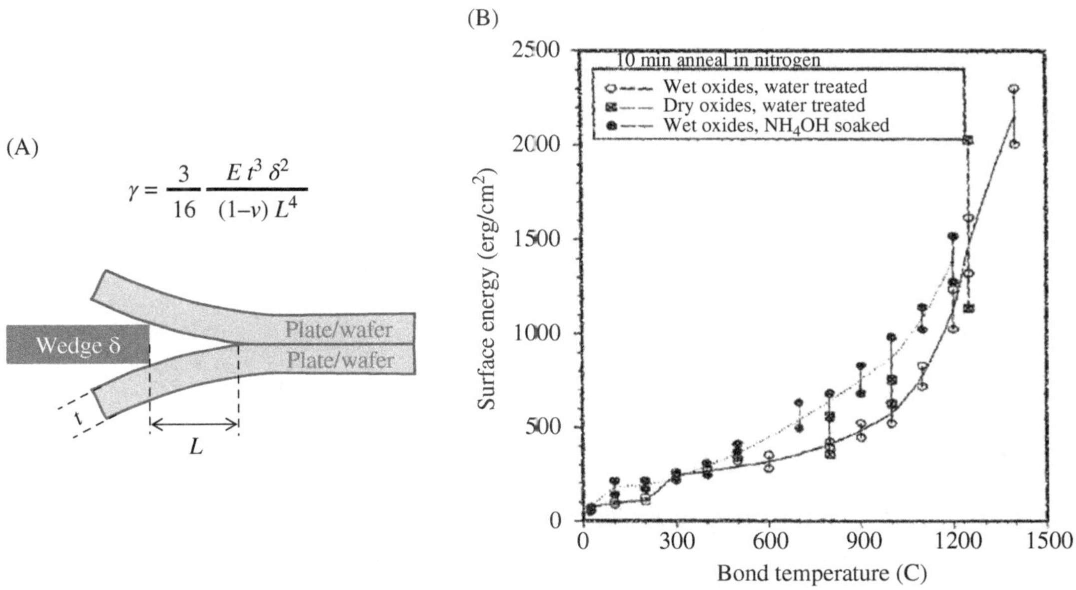

Figure 1.7 (A) Schematic of the DCB method measuring the surface energy of solid materials. E and ν are the elastic modulus and Poison's ratio of the material. (B) Surface energy of silicon oxide grown on silicon wafers bonded at different temperatures. Maszara et al. [12]/with permission of AIP Publishing.

$\sim 1.5\ \mathrm{J/m^2}$ at 1200 °C. The surface energy of the bonded wafer is similar to the dry-oxidized and steam-oxidized samples. The sample treated with basic solution (NH_4OH) exhibited a bit higher energy than the water-treated sample.

The example shown in Figure 1.7 is for the surface energy involving chemical bonds between two solid surfaces. In the case of physical interactions, the sessile drop method is commonly used. It involves applying a drop of a test liquid over the sample surface and measuring the contact angle formed at the three-phase interface (solid–liquid–vapor). Figure 1.8 depicts a liquid droplet placed on a solid surface. The force balance along the normal direction of the three-phase contact line can be written as:

$$\gamma_{SL} + \gamma_{LV}\cos\theta_c = \gamma_{SV} \tag{1.17}$$

where γ_{SL}, γ_{LG}, and γ_{SV} are the surface tensions of solid/liquid, liquid/vapor, and solid/vapor interfaces, respectively, and θ_c is the contact angle of the solid–liquid–gas equilibrium line. This is called Young's equation. Note that eq. (1.17) is for a homogeneous and flat surface; if the surface is rough, then the measured θ_c will vary accordingly [13, 14]. Although this equation is conceptually simple, it is not easy to use. Normally, the surface tension of liquid (which is γ_{LV}) is known, and the contact angle (θ_c) can be measured accurately. But there are still two unknowns in eq. (1.17); so, without knowing γ_{SL}, the determination of γ_{SV} is not possible. In addition, γ_{SV} corresponds to the interface of solid and saturated vapor, which is, strictly speaking, not the same as the surface energy of solid (γ) that we normally use for experiments in ambient air. The free energy change upon immersion of the solid in the saturated vapor phase is relatively small, so we conveniently replace the solid–liquid–vapor contact boundary with the solid–liquid-air boundary (i.e., $\gamma_{SV} \approx \gamma$).

Figure 1.8 Contact angle measurement of liquid on solid in the sessile drop method.

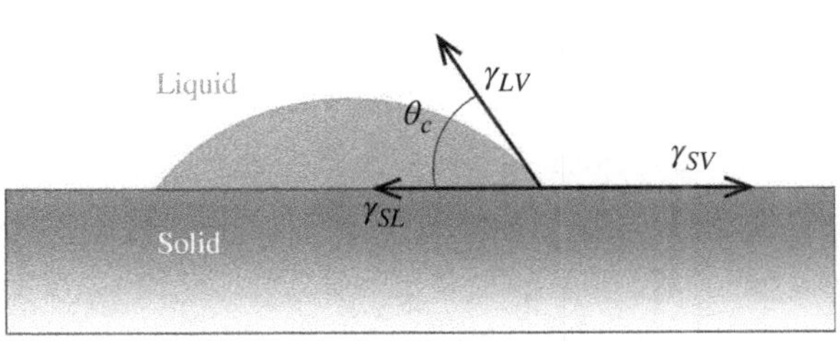

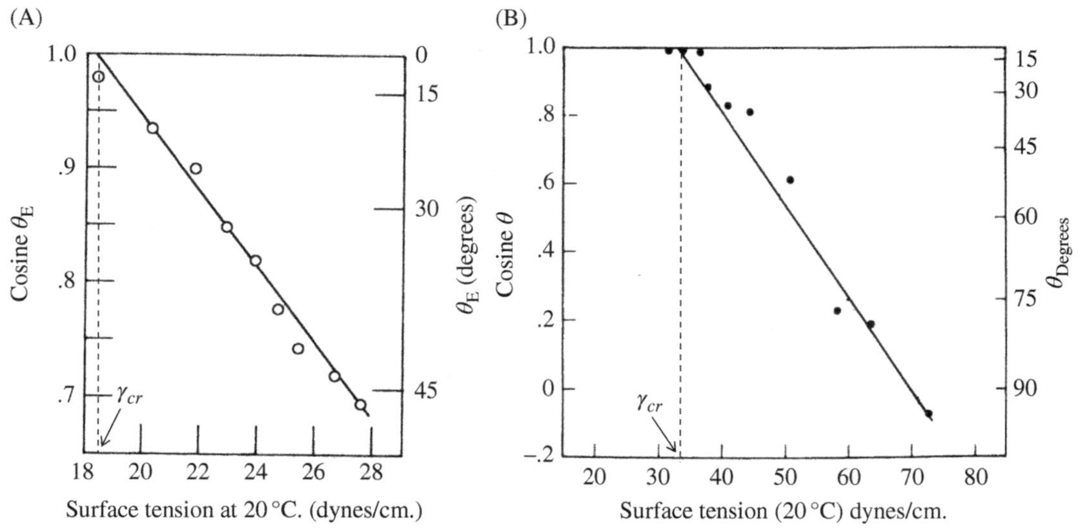

Figure 1.9 Wettability of (A) polytetrafluoroethylene by n-alkanes and (B) polyethylene by various liquids. Zisman [15]/with permission of American Chemical Society.

Zisman empirically found a rectilinear relation between $\cos\theta_c$ and the liquid surface tension (γ_L) for homologous series of organic liquids on solid surfaces with relatively low γ (<0.1 J/m^2) [15]:

$$\cos\theta_c = 1 - b(\gamma_L - \gamma_{cr}) \tag{1.18}$$

He then defined the intercept of the straight trend line at $\cos\theta_c = 1$ as the critical surface tension (γ_{cr}) (Figure 1.9). Any liquid with $\gamma_L < \gamma_{cr}$ will completely wet the surface. Based on the widespread occurrence of the rectilinear relationship between $\cos\theta_c$ and γ_L for many organic solid surfaces, Zisman showed that γ_{cr} can be used as the surface energy of the solid (γ).

The Zisman theory treats the surface energy as one single parameter, rather than accounting for the fact that polar interactions are much stronger than dispersive interactions. Then, Fowkes introduced a new term called the London dispersion force contribution (γ^D) and nondispersive contribution (γ^{nD}), that is, $\gamma = \gamma^D + \gamma^{nD}$, and the interactions between the solid and surface were interpreted as the geometric mean of these terms [16]:

$$\gamma_{SL} = \gamma_S + \gamma_L - 2\left(\sqrt{\gamma_S^D \gamma_L^D} + \sqrt{\gamma_S^{nD} \gamma_L^{nD}}\right) \tag{1.19}$$

Then, inserting eq. (1.19) into eq. (1.17) gives:

$$\gamma_L(\cos\theta_c + 1) = 2\left(\sqrt{\gamma_S^D \gamma_L^D} + \sqrt{\gamma_S^{nD} \gamma_L^{nD}}\right) \tag{1.20}$$

Since this equation has two unknowns, γ_S^D and γ_S^{nD}, it can be solved by measuring the contact angles of two liquids with known γ_L^D and γ_L^{nD}. Frequently used liquids for the Fowkes method are diiodomethane (γ_L^D = 50.8 mN/m and γ_L^{nD} = 0) and water (γ_L^D = 26.4 mN/m and γ_L^{nD} = 46.4 mN/m).

The nondispersive component of the Fowles method, γ^{nD}, is further defined as the polar component, γ^P, in the Owens–Wendt–Rabel–Kaelble (OWRK) method [17–19]. In the OWRK method, contact angles are measured with a series of homologous liquids with a wide range of γ_L^D and γ_L^P values, and the data are plotted in a linearized form: $\gamma_L(\cos\theta_C + 1)/\left(2\sqrt{\gamma_L^D}\right)$ vs. $\sqrt{\gamma_L^P}/\sqrt{\gamma_L^D}$. Then, the γ_S^D and γ_S^P are calculated from the intercept and slope, respectively (Figure 1.10). These two-term models describing γ as the sum of γ^D and γ^P are typically applied to surfaces with moderate polarity such as polycarbonates, polyesters, polyacrylates, polyurethanes, and polyvinylchloride.

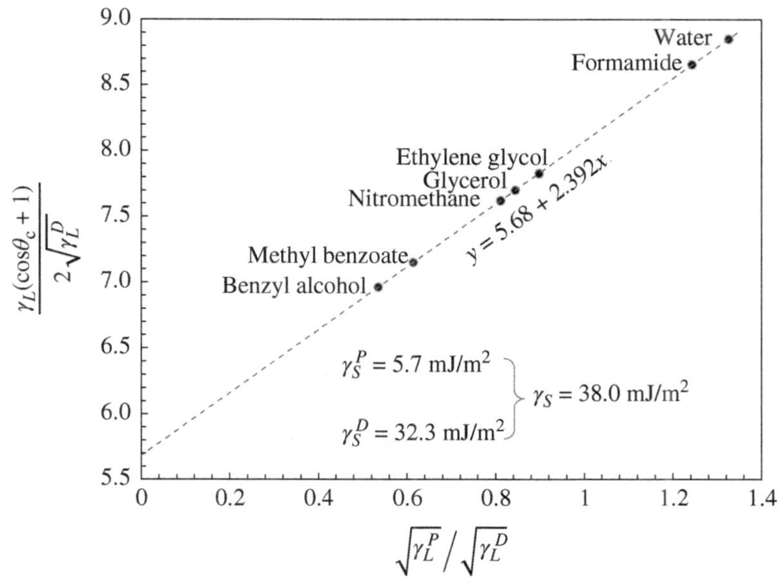

Figure 1.10 Determining the surface energy of PMMA using the WORK method.

In the Oss–Good method [20], the polar component (γ^P) is further specified as the sum of the acidic interactions (electron-donating, γ^+) and the basic interactions (electron-accepting, γ^-). Then, eq. (1.20) is written as:

$$\gamma_L(\cos\theta_c + 1) = 2\left(\sqrt{\gamma_S^D\gamma_L^D} + \sqrt{\gamma_S^+\gamma_L^-} + \sqrt{\gamma_S^-\gamma_L^+}\right) \tag{1.21}$$

To use this method, one needs to measure contact angles of at least three liquids – one with dispersive components only and two others with different acidic and basic contributions. Table 1.3 shows the surface tension values frequently used liquids for this purpose. This method is particularly useful for inorganic surfaces containing ions.

Note that the surface energy (J/m^2) and surface tension (N/m) of a liquid are a function of temperature [21]:

$$\gamma_L {V_m}^{2/3} = K(T_c - T) \tag{1.22}$$

where V_m and T_c are the molar volume and critical temperature of the liquid and K is a proportionality constant. Figure 1.11 shows the surface tension of water as a function of temperature along with empirical equation describing its temperature dependence [22]. In the case of liquid mixture, the surface tension varies nonlinearly with the composition [23] and its temperature dependence is even more complicated [24]. And, the surface tension also varies as the system pressure changes [25].

Table 1.3 Surface tension values of widely used liquids for the Oss–Good method.

Liquid	γ (mN/m)	γ^D (mN/m)	γ^+ (mN/m)	γ^- (mN/m)
Cyclohexane	25.5	25.5	0	0
Chloroform	27.1	23.3	3.8	0
Tetrahydrofuran	27.4	12.4	0	15
Diiodomethane	50.8	50.8	0	0
Water	72.8	26.4	23.2	23.2

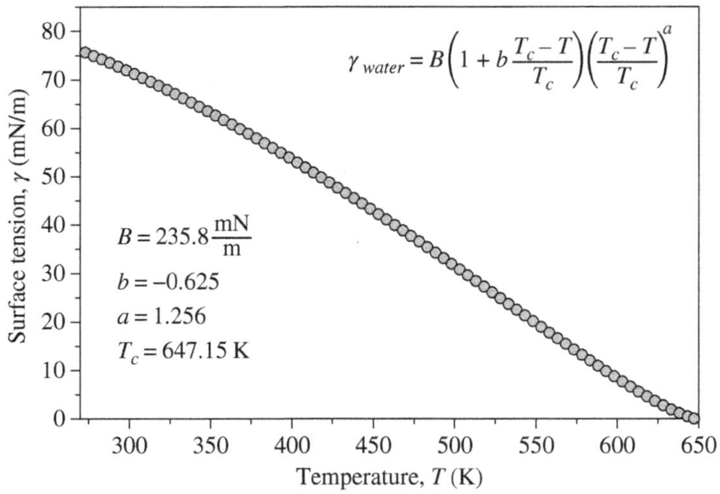

Figure 1.11 Surface tension of water as a function of temperature plotted using the data in [22]. The equation describes the empirical correlation between surface tension (mN/m) and absolute temperature (K). Adapted from [22].

1.6 Bulk Properties Affected by Surface Energy

Before we delve into detailed principles of surface characterization methods, let us ponder a brain teaser here. Since the surface energy is relevant to the surface phenomenon, one may think that it would only affect the surface properties and not the bulk properties. It is generally true if the material of interest is large enough, then the surface-to-volume ratio is essentially zero. That is the main reason the γdA term in eq. (1.7) is dropped out in most introductory books used in undergraduate thermodynamics courses. As the dimension (=characteristic length, L) of a material decreases, the bulk property governed by volume decreases in proportion to L^3, while the surface-related property decreases in proportion to L^2. Thus, at a smaller scale, the surface energy contribution to the total energy becomes larger than the bulk contribution. This is one of the reasons that nanomaterials often exhibit properties quite different from bulk materials of the same composition. There is a plethora of such examples in scientific literature; let us take a look at a few examples.

Figure 1.12 shows how the density (ρ) of solution-grown polyethylene crystals vary with the lamella thickness (l) of the crystal [26]. As the lamella thickness (l) decreases, the density of the crystal decreases in proportion to l^{-1}. To understand this phenomenon, one needs to recall how the surface energy is estimated from the bulk energy, eq. (1.13). Since atoms or molecules at the surface have less interatomic or intermolecular interactions than those in the bulk, their interatomic or intermolecular spacing is different from that of an infinite crystal where the surface effect can be ignored. As a result, there is a surface relaxation that manifests itself as surface stress (σ_s). This generates a strain which involves changes in the in-plane

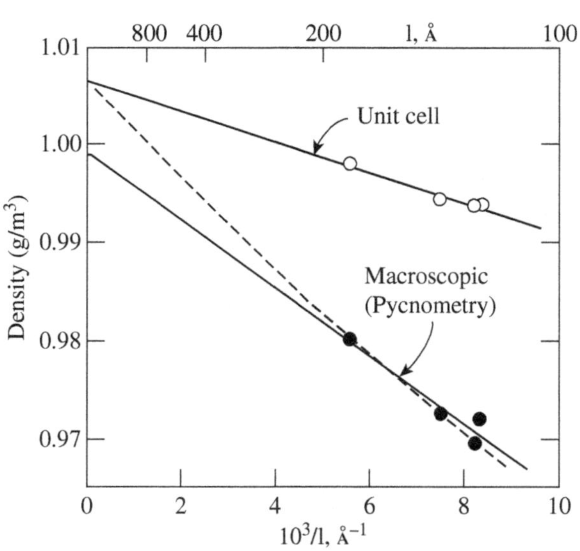

Figure 1.12 Density of solution-grown polyethylene crystal vs. reciprocal of lamella thickness (l) at 30 °C. Davis et al. [26]/with permission of AIP Publishing.

lattice parameters relative to those of a large crystal. For the unit cell, the dimension of a crystal is an equilibrium constant at a given temperature and pressure, which will vary only with the composition change. But, due to the surface strain, the unit cell dimension of extremely small materials can change for the same composition at the constant temperature and pressure. The induced strain (ϵ) can be related to σ_s [27]:

$$\epsilon = -\frac{2\sigma_s}{Br} \tag{1.23}$$

where B is the bulk modulus and r is the size of crystal. For very fine-grained materials (i.e., r is on the order of a few nm's), the induced strain can be quite large, of the order of 1% of the unit cell dimension. Depending on the sign of σ_s (i.e., tensile or compressive stress), the unit cell volume can be enlarged or reduced.

Figure 1.13A displays the effective elastic modulus determined by measuring the force required to bend a nanowire freely standing between two posts using atomic force microscopy (AFM) [28]. The measured modulus of silver and lead nanowires was significantly larger for the ones with smaller diameters than the larger ones. This increase of the apparent modulus could be attributed to the surface tension effect. In that study, the surface tension of the tested nanowires was obtained from the regression of the stiffness data with a model and found to be ~3.1 N/m for Ag and ~1.0 N/m for Pd. It is noted that the surface tension of Ag nanowire here is significantly larger than the value shown in Table 1.2. A portion of this large surface tension could be attributed to the difference in experimental conditions. The surface tension of nanowires was obtained from the experiment conducted at room temperature, while the value shown in Table 1.2 pertains to the high-temperature case. Also, the distribution of the crystallographic facets might be different between the nanowire and the bulk material.

In eq. (1.15), we have seen the relationship between the surface energy and the heat of sublimation. Sublimation is an equilibrium phase transition process occurring when the chemical potential of atoms at the surface is the same as that in the gas phase. Figure 1.13B is the plot of the onset temperature (T_{onset}) at which the size of Ag nanoparticles started to decrease (measured with tunneling electron microscopy [TEM]) vs. the reciprocal of the particle diameter ($1/d$) [29]. The decrease in size in the vacuum condition occurs due to evaporation of atoms from the surface. It is known that the vapor pressure of surface atoms or molecules is a function of the surface curvature. Thermodynamically, the curvature dependence of vapor pressure can be described by the Kelvin equation. However, the size dependence of T_{onset} of Ag nanoparticles was found to be much larger than predicted by the Kelvin equation. The data in Figure 1.13B suggests that the surface energy of Ag nanoparticles would be significantly larger (~7.4 J/m^2) than the bulk solid (~1.14 J/m^2) in a similar temperature range. This could be due to the increase in inward cohesive energy (something similar to the surface stress effect mentioned above) as the particle size decreases (so the surface-to-volume ratio increases), or the presence of different crystallographic facets at the nanoparticle surfaces under the TEM experiment condition.

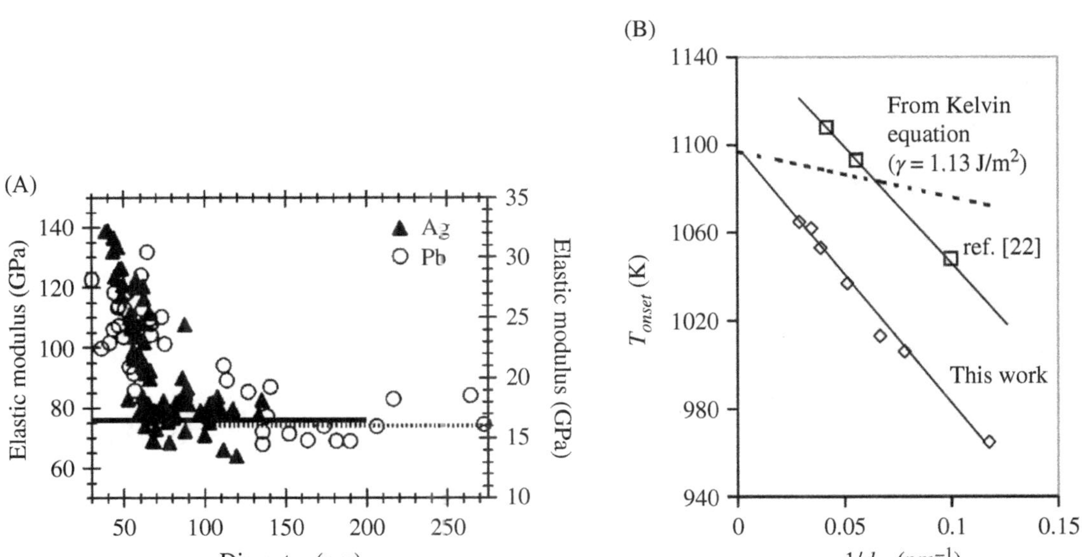

Figure 1.13 (A) Variation of the apparent elastic modulus measured for Ag (left scale) and Pb (right scale) nanowires as a function of diameter. Reprinted with permission from [28]. Copyright 2004 American Physical Society. (B) Onset temperature of evaporation of Ag vs. the inverse of nanoparticle size. Nanda et al. [29]/with permission of American Physical Society.

These are only a few examples demonstrating that surface phenomena are quite complicated and could not be fully explained by simple extrapolation of the principles governing bulk properties. There are many intriguing examples where surface energy plays a critical role in determining or altering the material's properties not only in physical conditions but also in biological systems.

1.7 Further Reading for Surface Contamination and Vacuum

Going back to the reason that many surface analysis methods listed in Table 1.1 require the vacuum condition, we have considered both kinetic and thermodynamic aspects. There are many other factors that affect the vacuum requirement, which can be associated with reliable operation of an instrument. The vacuum condition with a pressure of 10^{-12} atm or 10^{-9} Torr or lower is called "ultra-high vacuum" (UHV). Keeping the surface clean by lowering the gas collision probability is simply from a kinetic point of view. In humid air, any materials with a surface energy higher than that of water (72.8 mJ/m^2 at room temperature) will be covered with adsorbed water as soon as the surface is generated from the bulk. We know that polyethylene (PE) is relatively inert. The surface energy of PE is ~34 mJ/m^2 (Figure 1.9B). If the PE surface is exposed to air containing alcohol or hydrocarbon vapors, it will still be covered with physiosorbed molecules. The thickness or surface coverage of the adsorbed layer is determined by the equilibrium with the partial pressure of gas-phase molecules with respect to their saturation pressure (P/P_{sat}) at a given temperature. So, low-surface energy materials with low chemical reactivity can be made adsorbate-free by keeping them in a clean dry nitrogen environment. Thus, from thermodynamic point of view, if the surface energy of a material of interest is lower than that of the condensed phase of the potential interferent or contaminant molecules impinging from the gas phase (see Table 1.2), it is possible to keep the surface contamination-free even in the atmospheric pressure condition. Nonetheless, "surface-sensitive" elemental analyses are still required to be conducted in UHV conditions. That is because they rely on detecting electrons and ions emitted from the sample or bombarding the sample with electrons and ions. These charged particles have very high reaction probabilities or cross-sections for reaction upon collision with gas phase molecules, thus they can be easily annihilated if the pressure is not low enough. For these reasons, most surface-sensitive elemental analyses are carried out in UHV conditions.

There are many online materials that one can find and access freely to get deeper understanding of vacuum techniques and gas–solid interactions. A few examples are listed below. Please note that these are just suggested readings. The website links are shown as suggested examples where students can find additional information. Of course, there is a plethora of scientific articles and review papers from which students can find much deeper and thorough information.

Al-Ghouti MA, Da'ana DA. Guidelines for the use and interpretation of adsorption isotherm models: a review. Journal of Hazardous Materials. 2020;393:122383. https://doi.org/10.1016/j.jhazmat.2020.122383

Biolin Scientific. Glossary. https://www.biolinscientific.com/knowledge/glossary

Biolin Scientific. Measurements. https://www.biolinscientific.com/measurements

Foo KY, Hameed BH. Insights into the modeling of adsorption isotherm systems. Chemical Engineering Journal. 2010;156(1):2–10. https://doi.org/10.1016/j.cej.2009.09.013

Hoffman DM, Singh B, Thomas III JH, Thomas III JH, editors. Handbook of Vacuum Science and Technology. Amsterdam, Netherlands: Elsevier; 1998. ISBN 978-0-12-352 065-4; https://doi.org/10.1016/B978-0-12-352065-4.X5040-8

KRÜSS. Glossary. https://www.kruss-scientific.com/de-DE/know-how/glossar

Lafferty JM, editor. Foundations of Vacuum Science and Technology. Wiley; 1998. ISBN: 978-0-471-17 593-3

Marguardt N. Introduction to the Principles of Vacuum Physics. https://cds.cern.ch/record/582156/files/p1.pdf

ramé-hart. Glossary; 2024. http://www.ramehart.com/glossary.htm

Wang J., Guo X. Adsorption isotherm models: classification, physical meaning, application and solving method. Chemosphere. 2020;258:127 279. https://doi.org/10.1016/j.chemosphere.2020.127279

Practice Problems

1.1 Plot the pressure dependence of the mean free path (λ) and the wall collision frequency (Z_A) for N_2 at 1300 K.

1.2 Estimate the decreasing order of surface energies for (100), (110), and (111) facets of body-centered cubic (BCC) crystals. The crystal structure can be drawn with the Balsac program [10].

1.3 Determine surface energy of polyethylene (PE) and polymethylmethacrylate (PMMA) using the data shown in Table 1.4

Table 1.4 Surface tension (mN/m) and contact angle of liquid on PE and PMMA.

Solid: PE			Solid: PMMA		
Liquid	**γ (mN/m)**	**θ_c(deg)**	**Liquid**	**γ (mN/m)**	**θ_c(deg)**
n-Pentane	15.4	0	Cyclohexane	25.5	0
n-Hexane	18.4	0	n-Tetradecane	26.4	0
n-Heptane	19.9	0	Toluene	28.4	0
n-Octane	21.3	0	Nitromethane	36.5	16.5
n-Decane	23.8	18.5	Methyl benzoate	37.2	3.9
Cyclohexane	25.5	28.2	Benzyl alcohol	39	15.1
n-Tetradecane	26.4	32.1	Ethylene glycol	47.7	46.7
Toluene	28.4	38.7	Formamide	57	64.1
Benzyl alcohol	39	63.7	Glycerol	63.4	61.5
Ethylene glycol	47.7	81.1	Water	72.8	75.6

Source: Adapted from Krüss.

1.4 The contact angle of diiodomethane and water on PMMA is 53.6° and 75.6°, respectively. Calculate the surface energy of PMMA using the Fowkes method.

1.5 The dispersive (γ_L^D) and polar (γ_L^P) components of surface tension of the liquid used to test the surface energy of PMMA in Table 1.4 are given. Calculate the dispersive and polar components of PMMA surface energy using the OWRK method.

Liquid	**γ_L^D (mN/m)**	**γ_L^P (mN/m)**
Cyclohexane	25.5	0
n-Tetradecane	26.4	0
Toluene	26.1	2.3
Nitromethane	22	14.5
Methyl benzoate	27	10.2
Benzyl alcohol	30.3	8.7
Ethylene glycol	26.4	21.3
Formamide	22.4	34.6
Glycerol	37	26.4
Water	26.4	46.4

References

1 Baer DR, Artyushkova K, Richard Brundle C, Castle JE, Engelhard MH, Gaskell KJ, et al. Practical guides for x-ray photoelectron spectroscopy: first steps in planning, conducting, and reporting XPS measurements. Journal of Vacuum Science & Technology A. 2019;37(3):031401.

2 Marino MJ, Hsiao E, Chen Y, Eryilmaz OL, Erdemir A, Kim SH. Understanding run-in behavior of diamond-like carbon friction and preventing diamond-like carbon wear in humid air. Langmuir. 2011;27(20):12702–8.

3 Jang S, Chen Z, Kim SH. Environmental effects on superlubricity of hydrogenated diamond-like carbon: understanding tribochemical kinetics in O_2 and H_2O environments. Applied Surface Science. 2022;580:152299.

4 Somorjai GA. Introduction to Surface Chemistry and Catalysis. New York: John Wiley & Sons, Inc.; 1994.

5 Wiederhorn SM. Fracture surface energy of glass. Journal of the American Ceramic Society. 1969;52(2):99–105.

6 Tran R, Xu Z, Radhakrishnan B, Winston D, Sun W, Persson KA, et al. Surface energies of elemental crystals. Scientific Data. 2016;3(1):160080.

7 Abrahamson J. The surface energies of graphite. Carbon. 1973;11(4):337–62.
8 Kozbial A, Li Z, Conaway C, McGinley R, Dhingra S, Vahdat V, et al. Study on the surface energy of graphene by contact angle measurements. Langmuir. 2014;30(28):8598–606.
9 Griffith AA, Taylor VI GI. The phenomena of rupture and flow in solids. Philosophical Transactions of the Royal Society of London Series A, Containing Papers of a Mathematical or Physical Character. 1921;221(582–593):163–98.
10 Hermann K. Build and Analyze Lattices, Surfaces, And Clusters [Available from: https://www.fhi.mpg.de/1012355/Balsac].
11 Walters G, Parkin IP. The incorporation of noble metal nanoparticles into host matrix thin films: synthesis, characterisation and applications. Journal of Materials Chemistry. 2009;19(5):574–90.
12 Maszara WP, Goetz G, Caviglia A, McKitterick JB. Bonding of silicon wafers for silicon-on-insulator. Journal of Applied Physics. 1988;64(10):4943–50.
13 Wenzel RN. Resistance of solid surfaces to wetting by water. Industrial & Engineering Chemistry. 1936;28(8):988–94.
14 Cassie ABD, Baxter S. Wettability of porous surfaces. Transactions of the Faraday Society. 1944;40:546–51.
15 Zisman WA. Relation of the equilibrium contact angle to liquid and solid constitution. In: Frederick MF, editor. Contact Angle, Wettability, and Adhesion. Washington, D.C.: American Chemical Society; 1964. p. 1–51.
16 Fowkes FM. Attractive forces at interfaces. Industrial & Engineering Chemistry. 1964;56(12):40–52.
17 Owens DK, Wendt RC. Estimation of the surface free energy of polymers. Journal of Applied Polymer Science. 1969;13(8):1741–7.
18 Rabel W. Einige Aspekte der Benetzungstheorie und ihre Anwendung auf die Untersuchung und Veränderung der Oberflächeneigenschaften von Polymeren. Farbe und Lack. 1971;77(10):997–1005.
19 Kaelble DH. Dispersion-polar surface tension properties of organic solids. The Journal of Adhesion. 1970;2(2):66–81.
20 Van Oss CJ, Chaudhury MK, Good RJ. Interfacial Lifshitz-van der Waals and polar interactions in macroscopic systems. Chemical Reviews. 1988;88(6):927–41.
21 Shereshefsky JL. Surface tension of saturated vapors and the equation of Eötvös. The Journal of Physical Chemistry. 1931;35(6):1712–20.
22 Vargaftik NB, Volkov BN, Voljak LD. International tables of the surface tension of water. Journal of Physical and Chemical Reference Data. 1983;12(3):817–20.
23 Zhang Z, Wang W, Korpacz AN, Dufour CR, Weiland ZJ, Lambert CR, et al. Binary liquid mixture contact-angle measurements for precise estimation of surface free energy. Langmuir. 2019;35(38):12317–25.
24 Villers D, Platten JK. Temperature dependence of the interfacial tension between water and long-chain alcohols. The Journal of Physical Chemistry. 1988;92(14):4023–4.
25 Hinton ZR, Alvarez NJ. Surface tensions at elevated pressure depend strongly on bulk phase saturation. Journal of Colloid and Interface Science. 2021;594:681–9.
26 Davis GT, Weeks JJ, Martin GM, Eby RK. Cell dimensions of hydrocarbon crystals: surface effects. Journal of Applied Physics. 2003;45(10):4175–81.
27 Cammarata RC, Eby RK. Effects and measurement of internal surface stresses in materials with ultrafine microstructures. Journal of Materials Research. 1991;6(5):888–90.
28 Cuenot S, Frétigny C, Demoustier-Champagne S, Nysten B. Surface tension effect on the mechanical properties of nanomaterials measured by atomic force microscopy. Physical Review B. 2004;69(16):165410.
29 Nanda KK, Maisels A, Kruis FE, Fissan H, Stappert S. Higher surface energy of free nanoparticles. Physical Review Letters. 2003;91(10):106102.

2

Elemental Analysis via X-ray Irradiation

2.1 Electron Emission upon X-ray Absorption

2.1.1 Atomic Orbital and Photoemission

X-ray photoelectron spectroscopy (XPS) is one of the most widely used characterization techniques to find the elemental composition of the surface of various types of materials, and its use and appearance in the literature have been increasing significantly and this trend is expected to continue [1]. That is because the constituents of a surface are directly related to its functions and properties and, in principle, XPS can detect nearly all elements found in the periodic table (except ^{1}H, ^{2}He, and unstable elements that are heavier than ^{92}U). What specific aspect of a material does XPS probe? How can XPS identify what elements are present in the surface of a sample? The answers to these questions come from the periodic table of elements. For that, let us brush up the atomic orbital theory that we have learned in undergraduate general chemistry. The periodic table was originally constructed based on the correlation between the atomic numbers and approximate recurrences of evident properties; but the key essence of the table is the occupancy of electrons in atomic orbitals. An atomic orbital is the probability description of where an electron can be found as a function of distance from the nucleus (center of atom). The four basic types of orbitals are designated as *s*, *p*, *d*, and *f*, which have designated angular momentum quantum numbers of $l = 0, 1, 2,$ and 3, respectively. The probability density and phase of the hydrogen-like atomic orbitals are shown in Figure 2.1A [2]. The *s*-orbital is spherically symmetric. The *p* orbitals consist of three degenerated components with

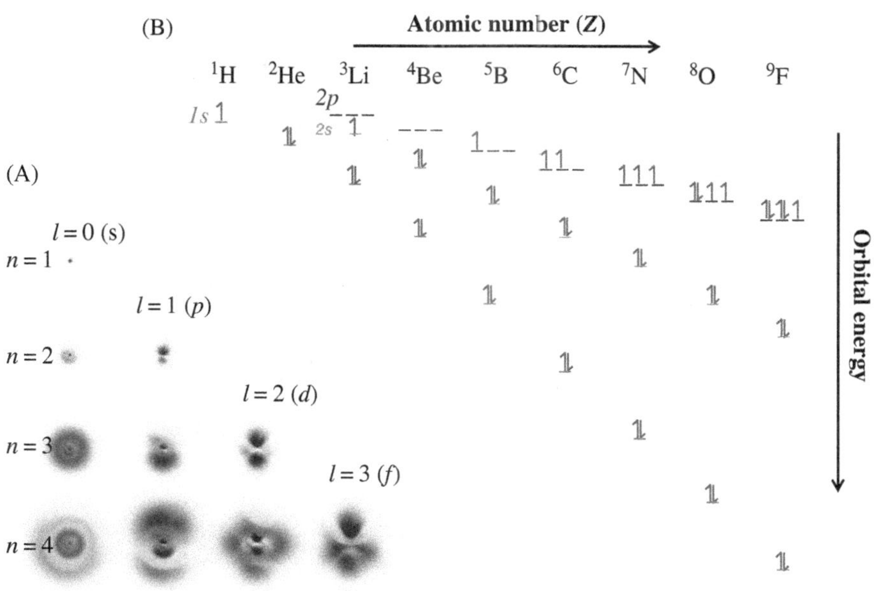

Figure 2.1 (A) Schematic views of hydrogen-like atomic orbitals showing probability density and phase. Atomic Orbital [2]/Wikipedia/Public domain. (B) Graphic illustration of changes in electron occupancy and orbital energy for the elements in the first two rows of the periodic table. The up and down half arrows represent the spin-up ($s = +1/2$) and spin-down ($s = -1/2$) quantum states of electron, respectively.

Surface and Interface Analysis: Principles and Applications, First Edition. Seong H. Kim.

Companion website: www.wiley.com/go/Kim

different magnetic components which are orthogonal to each other, for example, as in *x*, *y*, and *z* directions in the cartesian coordinate. The *d* orbitals consist of five sub-units which can be described as *xy*, *yz*, *zx*, $x^2 - y^2$, and z^2. The *f* orbitals have a total of seven sub-units with different magnetic quantum states. These angular orbital names, together with the principal quantum number *n*, are used to describe the electron configurations of atoms. The principal quantum number *n* is also called shell: *K*-shell for $n = 1$, *L*-shell for $n = 2$, *M*-shell for $n = 3$, *L*-shell for $n = 4$, and so on. Electrons occupy orbitals from the lowest energy state first, as shown in Figure 2.1B. Each of these orbitals can be occupied by a maximum of two electrons, each with its own spin (*s*) states: that is, spin-up (+½) or spin-down (−½). As the atomic number (*Z*) increases, the number of electrons in the atom increases accordingly. At the same time, the energy level of each orbital decreases with increasing *Z*. Once the lowest energy orbital is fully occupied by two electrons, the next available orbital for additional electron is the one with the next lowest energy orbital. Electrons fill the empty orbitals first before they become paired with the opposite spin electron.

Figure 2.2 depicts the energy levels of inner-shell orbitals of atoms [3]. The total and individual orbital energies calculated from quantum mechanics can be found in the NIST Standard Reference Database [4]. The atomic orbital energy level is fixed for a given element. Note that the energy scale is negative. The reference point can be typically taken as a vacuum level (E_{vac}); or, for a solid material, it can be set to the Fermi level (E_F) which is the highest energy level that an electron can occupy at the absolute zero temperature (0 K). Thus, the negative value in energy means that electrons occupying atomic orbitals are bound to that specific atom. When a sample is irradiated with an x-ray beam that has an energy higher than the orbital energy, a fraction of x-ray photons can be absorbed by the sample and electrons in atomic orbitals lying between the vacuum level and the photon energy can be ejected to the vacuum level. This process is called photoelectron emission. In typical commercial XPS systems, the x-ray source is made with Al or Mg. The x-ray energy from the Al source is 1486.6 eV. Then, the electrons in the shaded area in Figure 2.2 cannot be ejected by irradiating the sample with this x-ray source, because the x-ray energy is smaller than the binding energy of electrons in the shaded area. However, the electrons in

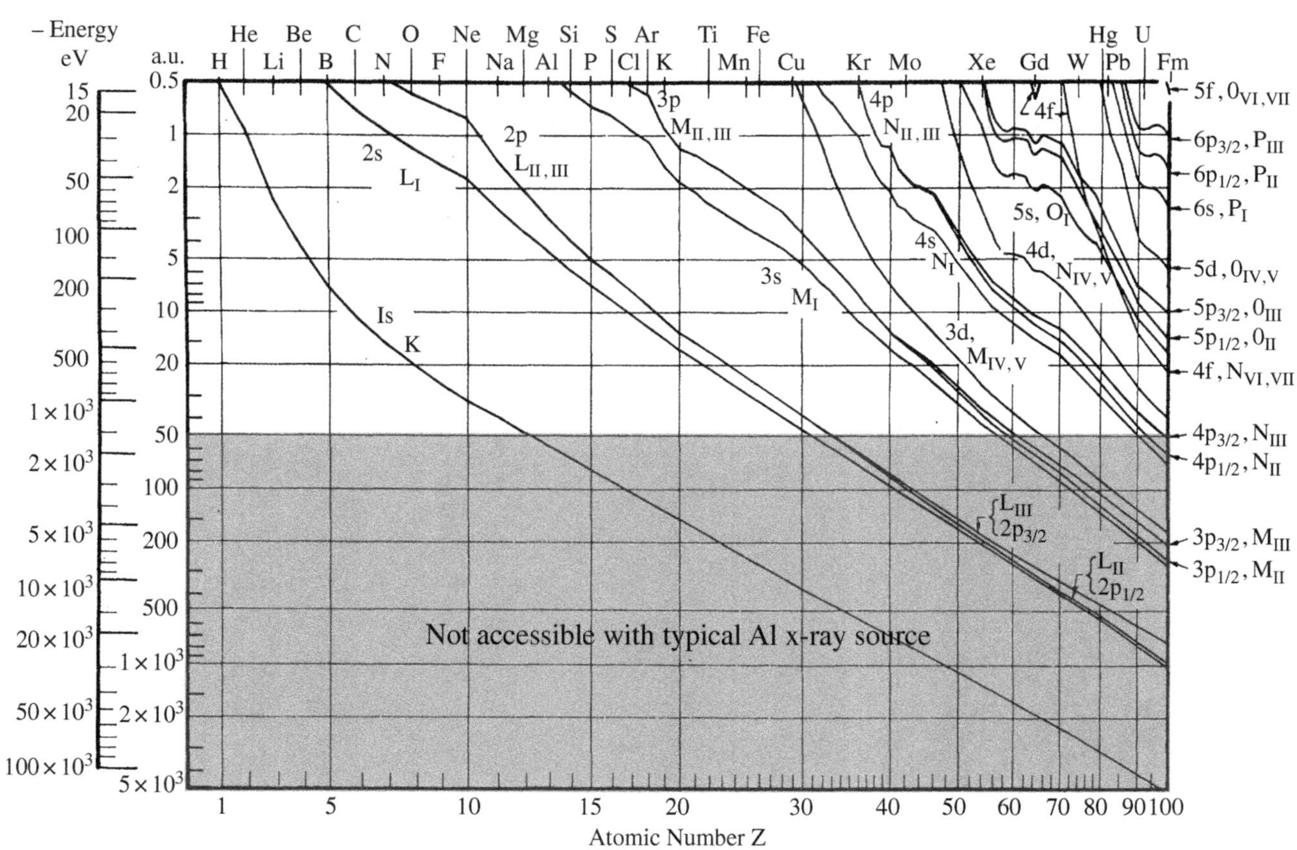

Figure 2.2 Energy levels of inner shells of atoms. Barat and Lichten [3]/with permission of American Physical Society.

the orbitals above −1486.6 eV can be emitted from the sample when irradiated with the x-ray from the Al source. In XPS, the kinetic energy (KE) and intensity of these photoemitted electrons are measured.

The first thorough investigation and publication of the photoemission effect was done by Heinrich Hertz in 1887 [5]. At that time, the electromagnetic radiation was considered to behave like a wave and the energy transferred by a wave in a given time was called intensity. Based on that idea, it was expected that a higher intensity light would give more energy to the photoelectron. However, the KE of photoelectron is independent of the intensity of the light; instead, it depends linearly on the frequency of the light. This puzzle was solved by Albert Einstein who published the theory describing the photoemission effect in terms of light quanta in 1905. He proposed that the energy of light is not transferred continuously as in a classical wave, but only in small "packets" or quanta, and the size of these packets, which was later named "photons," was to be defined by the Planck's "energy element." In 1914, Robert Millikan measured that the constant of proportionality between the frequency (ν) of incident light and the KE of photoelectrons was equal to the Planck constant ($h = 6.626 \times 10^{-34}$ J/s $= 4.136 \times 10^{-15}$ eV/s). That proof earned Einstein the Nobel Prize in 1921 [5]. And, Millikan was awarded the Nobel Prize in 1923. Then, Kai Siegbahn significantly contributed to the development of electron spectroscopy for chemical analysis (ESCA), now known as XPS, and was awarded the Nobel Prize in physics for this contribution in 1981.

Bringing these together, the photoelectron process can be described with the energy conservation principle. When a solid material is irradiated with x-ray whose energy ($h\nu$) is larger than the binging energy (BE) of electrons (here, the BE is defined as a negative energy with respect to the E_F of the solid), then electrons can be exited above the vacuum level (E_{vac}). The excess energy can contribute to the KE of photoelectrons emitted. There is another term that reduces the KE of the photoelectron, which is a work function (Φ_w). This is the energy difference between E_{vac} and E_F, which is conceptually similar to the ionization potential of elements in the gas phase. Balancing these terms, one can write the following equation for the KE of photoelectron emitted from the sample upon irradiation of x-ray:

$$KE = h\nu - (BE + \Phi_w) \tag{2.1}$$

This principle is schematically illustrated in Figure 2.3. Here the BE of atomic orbital is drawn with respect to the Fermi level (E_F) and the KE of photoelectron is measured with respect to the vacuum level (E_{vac}). The atomic orbitals are grouped into core levels and valence band. In organic species, the valence band consists of the molecular orbitals that are involved in hybridization to form covalent bonds, and the core levels are the inner-shell orbitals that are affected relatively less from the covalent bond formation. In the case of carbon, the core-level is 1s orbital and the valence band is the sp^3 or sp^2 hybrid orbitals formed from the 2s and 2p atomic orbitals. In the case of inorganic compounds, core-level electrons are the ones occupied in the inner orbitals localized to individual atoms, and balance electrons are the ones involved in the overlapping orbitals localized in the bond region. In the case of metals, the valence electrons are the ones in the conduction band formed by resonance of electron clouds between each atom and its neighbors. The atomic orbitals are quantized, and thus its state can be viewed as a delta function. The x-ray source has a finite energy width, so the KE of photoelectrons will have a finite width. The lowest-energy orbital from which photoelectron can be emitted upon x-ray irradiation is determined by the x-ray photon energy (see Figure 2.2). Figure 2.4 schematically illustrates how the XPS spectrum can be correlated to Bohr's electronic structure model [6]. There are other sub-atomic and energy-loss processes that affect the peak position, intensity, width, and the background during the photoelectron emission process.

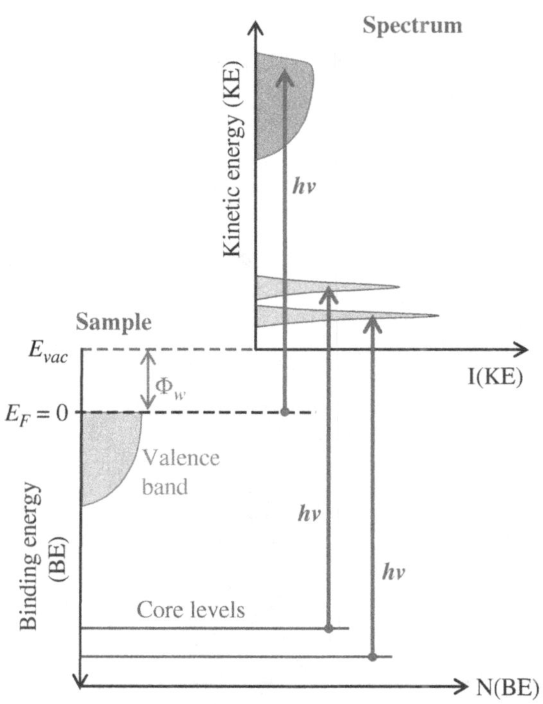

Figure 2.3 Schematic of the photoemission process in a single particle picture.

In quantum mechanics, the atomic orbitals are the "allowed" electronic functions subjected to a potential that confines electrons within the atomic structure. The consequence of this confinement is that the energy spectrum of a set of bound states is discrete, unlike free electrons which have a continuous spectrum. That is represented as horizontal lines for core electrons in Figure 2.3 and discrete rings in Bohr's

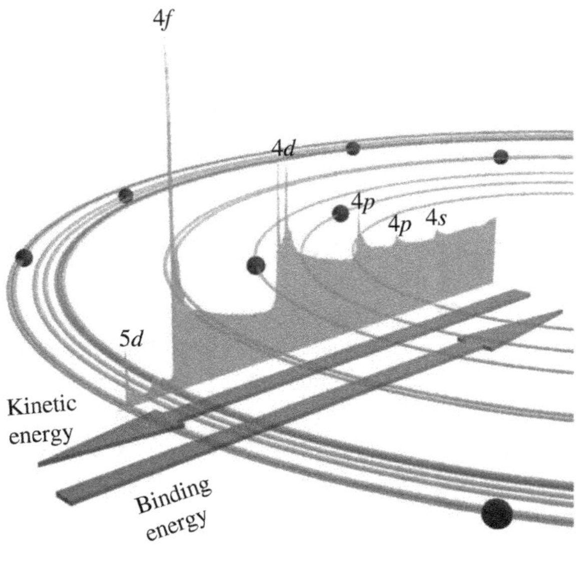

Figure 2.4 XPS spectrum of Pb superimposed on Bohr's electronic structure model to illustrate how each orbital gives rise to photoelectron peak. Watts and Wolstenholme [6]/with permission of John Wiley & Sons.

model in Figure 2.4. The sharp peaks in XPS spectrum correspond to these allowed electronic functions. However, the XPS spectrum shown in Figure 2.4 has finite and continuous intensities in the energy regions between sharp peaks where stable electronic functions are "forbidden." These regions are called background. Then, why are there background signals detected in the forbidden energy regions between specific orbital peaks which are the only allowed quantum states within a given atom? That is due to inelastic scattering of photoelectrons during their escape to vacuum. X-ray can penetrate deep into the sample and photoelectrons are generated below the surface. Photoelectrons produced inside the sample can undergo collisions with other electrons in their way to the vacuum and such collisions will lower the KE of photoelectrons (Figure 2.5A). The ones generated deeper inside the sample will go through more of these inelastic collisions, losing a larger portion of their KE. Although minor, another energy loss path is the Bremsstrahlung process. The trajectory of electrons traveling through the sample can be deflected and decelerated due to the electric field of the nuclei of atoms (Figure 2.5A). The inelastic collision and Bremsstrahlung are a continuum process. Thus, the background signal is continuous in the lower KE side of each photoelectron peak (Figure 2.5B). Although the photoelectron signal is measured as a function of KE, the XPS spectrum is plotted in the BE scale. By convention, the BE is plotted from high to low in the *x*-axis of the XPS spectrum (which scales from low to high if the spectrum is plotted in the KE scale). The conversion of the KE scale to the BE scale is done by rearranging eq. (2.1):

$$BE = h\nu - KE - \Phi_w \tag{2.2}$$

Thus, the loss of KE through inelastic collisions and Bremsstrahlung processes during the escape of photoelectrons makes the background signal in the higher BE side of the photoelectron peak higher than that in the lower BE side (Figure 2.5B).

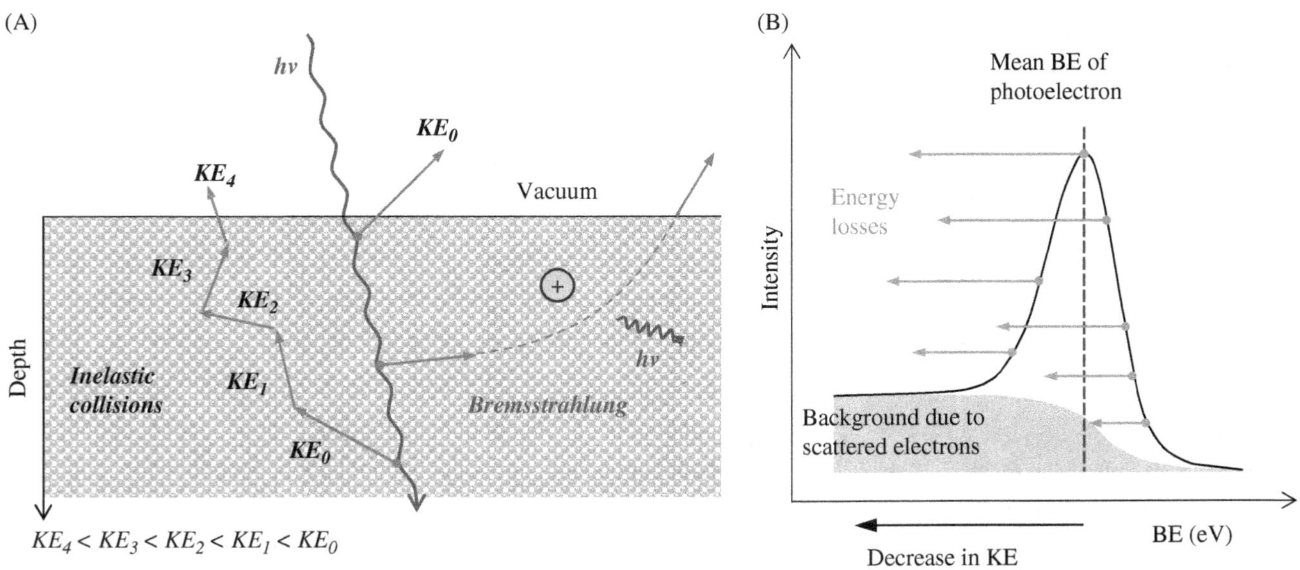

Figure 2.5 (A) Inelastic scattering and Bremsstrahlung processes, contributing to (B) background signal.

2.1.2 Relaxation During Photoemission

The photoelectron emission is accompanied by further relaxation of the electronic structure (Figure 2.6). When an electron is emitted from an inner shell orbital, the hole in that orbital can be refilled with another electron from a higher level (lower BE) orbital. This relaxation process can induce emission of another electron in the orbital whose BE is smaller than the energy difference between the inner shell orbital from which the photoelectron is emitted and the orbital from which the secondary electron leaves to fill the hole in the lower energy (higher BE) state. The electron emitted by this secondary relaxation process is called Auger electron and described with three alphabet letters. The first letter represents the orbital from which the photoelectron is emitted due to the x-ray absorption. The second letter represents the orbital from which the secondary electronic transition takes place. The third letter represents the orbital from which the Auger electron is emitted. For example, let us consider the C-KLL Auger electron. It means that a photoelectron is emitted from the carbon 1*s* orbital (*K*-shell) and one of electrons in the 2*s* or 2*p* orbitals (*L*-shell) fills the hole in the 1*s* orbital and the energy difference between these two states causes another electron emitted from the *L*-shell. If the Auger electron emission is caused by the ejection of a primary electron from a state in the *M*-shell ($n = 3$) which is accompanied by the transition of a secondary electron from the *N*-shell ($n = 4$) filling that hole and the emission of a third electron from another state in the *N*-shell, it will be called the MNN Auger electron. Note that the KE of Auger electron is not a function of the x-ray energy, but a function of energy differences of these three states. For example, the KE of a KLL Auger electron will be determined in the following form:

$$KE_{Auger} = (BE_K - BE_L) - BE_L \tag{2.3}$$

The primary electron ejection can occur not only by x-ray irradiation but also via irradiation of high-energy electron beam. The chemical analysis using the latter case is called Auger electron spectroscopy (AES), which is discussed in Chapter 3.

Another relaxation process accompanying the photoelectron process is x-ray emission (Figure 2.6C). The energy difference of the secondary electronic transition filling the hole in the lower energy state by an electron from the higher energy state can be emitted as a photon. If a carbon atom undergoes photoemission process, then it can emit a photon with an energy that is equivalent to the difference between the *K*- and *L*-shell:

$$h\nu_{emitted} = (BE_K - BE_L) \tag{2.4}$$

The energy of this photon happens to be in the x-ray range, so this process is called x-ray fluorescence and specific to the electronic states of the atom from which x-ray is emitted. As in the AES case, the primary electron emission can be induced by irradiation of higher-energy electron beam. By analyzing the energy of the emitted x-ray upon election irradiation, one can identify what elements are present in the sample. The technique using a solid-state crystal for energy dispersive detection is called energy-dispersive x-ray spectroscopy (EDX or EDS). The technique using a wavelength-dispersive spectrometer (WDS) is called electron probe microanalysis (EPMA). Examples of chemical analysis based on EDX/EDS and EPMA can be found in various review articles and books [7–9].

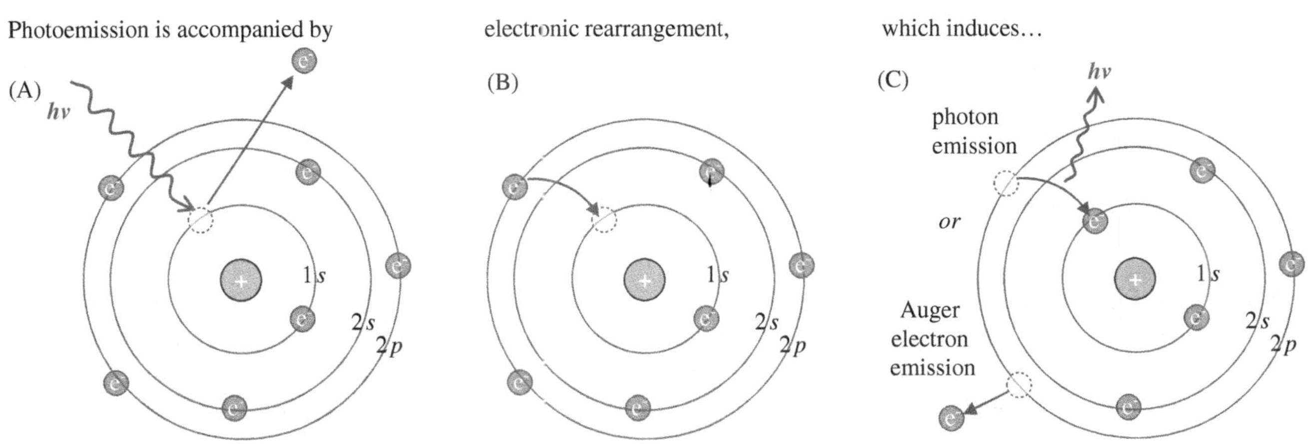

Figure 2.6 (A) Photoemission of a core-level electron. (B) Transition of an electron from a higher energy state filling the hole in the core level. (C) Emission of x-ray and Auger electron.

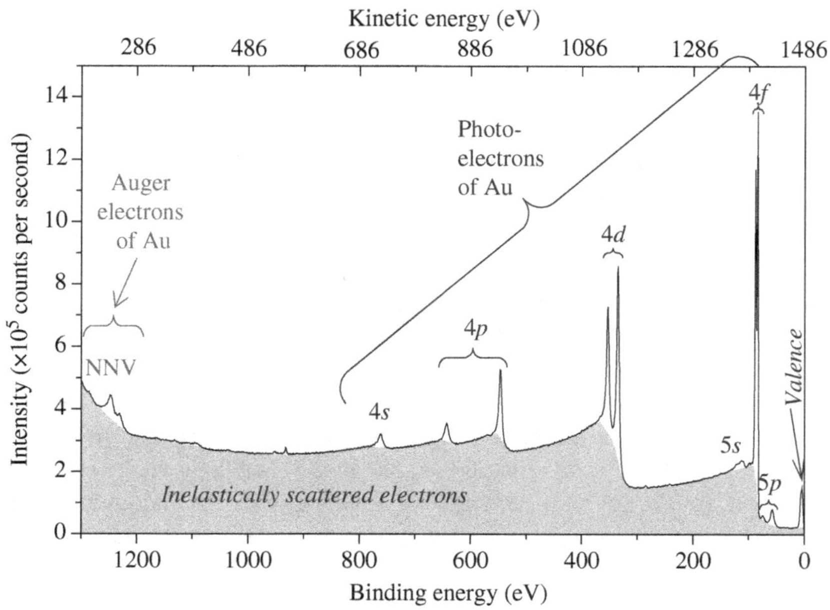

Figure 2.7 XPS spectrum of a gold foil produced with an Al x-ray source.

Figure 2.7 summaries what we have discussed so far using an XPS spectrum of gold. The photoelectron peaks are highlighted with the atomic orbitals of Au which have BE smaller than the photon energy from the Al x-ray source. The 1*s*, 2*s*, 2*p*, 3*s*, 3*p*, and 3*d* orbitals are inaccessible with the x-ray source used (see Figure 2.2, the atomic number of Au is 79). The baseline change is due to inelastic scattering of photoelectrons inside the sample (see Figure 2.5B). The small peaks centered at ~1247 eV in the BE scale (bottom axis) is the NNV Auger electron of gold (here V stands for the valence band). The KE of this Au-NNV Auger electron is ~240 eV (top axis), which can be calculated using eq. (2.3) and the atomic orbital energies of Au. When it is plotted in the BE scale with eq. (2.2), then it appears at ~1247 eV because the x-ray energy ($h\nu$) from the Al source is 1486.6 eV. Note that when a different x-ray source is used, the Auger electron signal is displayed at a different BE position. For example, if a Mg x-ray source ($h\nu$ = 1254 eV), then the Au-NNV peak appears at the 1014 eV in the XPS spectrum plotted in the BE scale. This is simply because the x-ray source energy in eq. (2.2) is changed, but the KE of the Auger electron does not change. The photoelectron signal does not change with the x-ray source since the KE of photoelectron varies with the x-ray source energy ($h\nu$) in eq. (2.2).

2.1.3 Spin–Orbit Splitting

In Figure 2.7, two peaks under the curly bracket are assigned to one atomic orbital, except the 4*s* peak. This brings up a question on how many peaks are detected for each orbital? How can we know they are from the same orbital? Upon photoemission from fully filled core orbitals, two energetically equivalent (degenerate) unfilled final states can be produced, that is, spin-up ($s = +½$) and spin-down ($s = -½$). The magnetic interaction between the orbital angular momentum (l) and the electron spin (s) can lead to a splitting of the energetically equivalent states into two components (Figure 2.8). These new states are described with a new quantum number (j):

$$j = l + s \text{ where } s \text{ can be } \pm ½ \tag{2.5}$$

The *s* orbital ($l = 0$) has spherical symmetry (see Figure 2.1A) and thus has no magnetic momentum. For that reason, the photoelectron signal from the *s* orbital is a singlet (i.e., no splitting). All other orbitals with $l \neq 0$ can split into a doublet. The doublet signal is described with the nomenclature nl_j. The ratio of their respective degeneracies is $2j + 1$, and this value determines the relative intensities of the spin–orbit split components. So, the photoelectron signal from the *p* orbital ($l = 1$) will have a doublet of $p_{1/2}$ and $p_{2/3}$ with relative intensity of 1:2 ($= (2 \times 0.5 + 1) : (2 \times 1.5 + 1)$). Similarly, the photoelectron signal from the *d* orbital ($l = 2$) will have a doublet of $d_{3/2}$ and $d_{5/2}$ with relative intensity of 2:3 ($=(2 \times 1.5 + 1) : (2 \times 2.5 + 1)$). The photoelectron peak of the *f* orbital ($l = 3$) will give a doublet of $f_{5/2}$ and $f_{7/2}$ with relative intensity of 3:4 ($= (2 \times 2.5 + 1) : (2 \times 3.5 + 1)$).

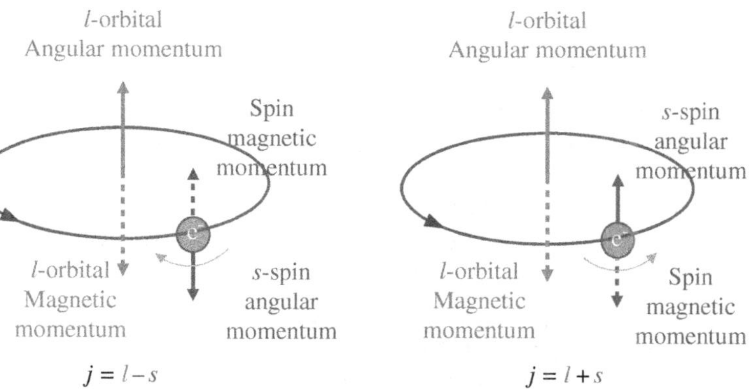

Figure 2.8 Schematic illustration of spin–orbing coupling.

The energy difference between the doublet peaks (Δ) is $p > d > f$ within a given atom. The smaller j state is always in the higher BE side than the larger j state, and their relative intensities follow the $(2j + 1)$ ratio.

In the XPS spectrum of Au shown in Figure 2.7, we can see that 4*s* peak has only one peak at 763 eV. The 4*p* peak is split to $4p_{1/2}$ at 643 eV and $4p_{3/2}$ at 547 eV ($\Delta_{4p} = 96$ eV) with 1:2 intensity ratio. The 4*d* peak is split to $4d_{3/2}$ at 353 eV and $4d_{5/2}$ at 335 eV ($\Delta_{4d} = 18$ eV) with 2:3 intensity ratio. The 4*f* peak is split to $4f_{5/2}$ at 88 eV and $4f_{7/2}$ at 84 eV ($\Delta_{4f} = 4$ eV) with 3:4 intensity ratio. Note that the spin–orbit split of the same type of orbital but in different shells is not the same. For examples, the $5p_{1/2}$ and $5p_{3/2}$ peaks are at 74 eV and 57 eV, respectively, giving $\Delta_{5p} = 17$ eV. This is because the electronic shielding from the nucleus field is different for electrons in $n = 4$ and $n = 5$ states.

The spin–orbit splitting becomes larger as the atomic number increases. An example is shown in Figure 2.9 for the 6*f* orbitals [10]. This is because the atomic magnetic field felt by orbiting electrons depends on the charge of the nucleus. The larger atomic magnetic field amplifies the spin–orbit splitting. In Figure 2.9, the higher BE peak is $f_{5/2}$ and the lower one is $f_{7/2}$; their relative intensity ratio may appear varying, but with proper background correction, their relative area ratio is 3:4 for every doublet of each element. Note that spin–orbit splitting is not always observed. For example, the ^{13}Al 2*p* and ^{14}Si 2*p* peaks often look like a singlet, while the spin–orbit splitting with the 1:2 intensity ratio is easily observed for ^{15}P 2*p* ($\Delta_{2p} = 0.84$ eV) and ^{16}S 2*p* ($\Delta_{2p} = 1.18$ eV) peaks. The singlet peak shape for the Al 2*p* and Si 2*p* orbitals does not mean the spin–orbit splitting does not occur. It is definitely present since the photoemission produces unpaired electrons, but the Δ_{2p} of these light elements is so small and unresolved unless the XPS system is operated with a very high energy resolution.

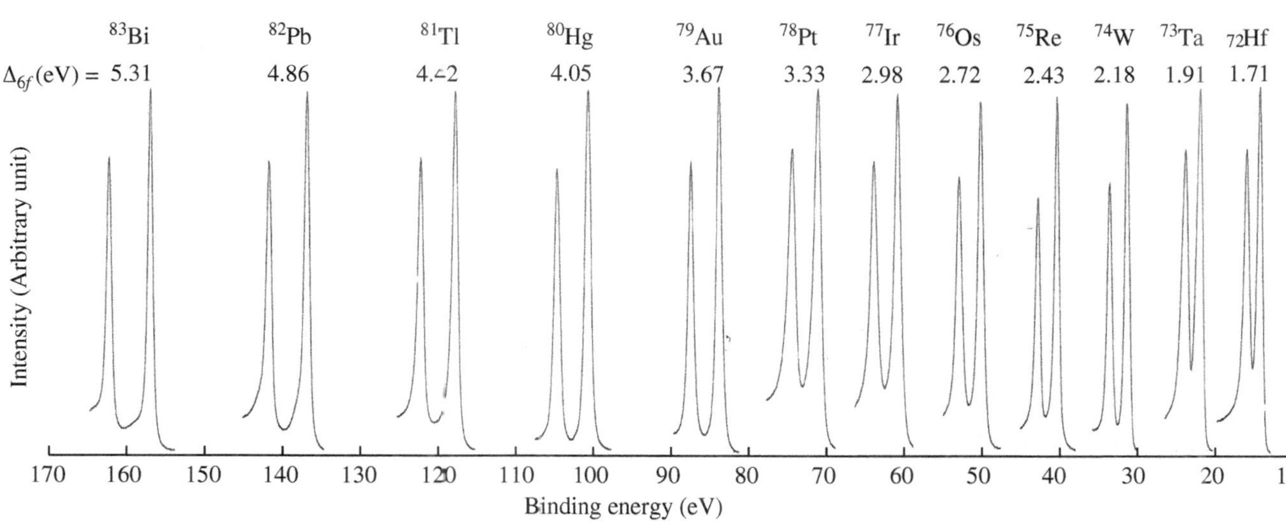

Figure 2.9 XPS spectra of *f*-orbitals of sixth period elements. Drawn using the data in the *Handbook of x-ray Photoelectron Spectroscopy*. Adapted from Moulder and Chastain [10]. The relative intensities among different elements cannot be compared in this plot. For each doublet, the higher BE component is the $6f_{5/2}$ peak and the lower BE component is the $6f_{7/2}$ peak of each element.

2.1.4 Surface Sensitivity

X-ray can penetrate the sample quite deeply. Recall that in hospital, x-ray transmission is used to diagnose bones in our body. Then, where does the surface sensitivity of XPS come from? It comes from the fact that XPS detects the KE of photoelectrons emitted from the sample. In Chapter 1, we studied why UHV is necessary to detect photoelectrons emitted from the sample without losing them through collisions with gas phase molecules. In Figure 2.5, we have studied that inelastic scattering of photoelectrons alter their KE. We do not know how many such collisions occurred before photoelectrons produced deep inside the sample escape from the sample surface. They contribute to the background rise in the higher BE side (so, lower KE). The shape of background can provide certain information about the sample structure, especially a layered distribution of specific element (see Figure 2.49 in Section 2.4.3). But it is not useful when it comes to elemental identification and quantification. So, it is essential to measure the pristine KE of photoelectrons before it is altered.

Intuitively, one can think that photoelectrons produced near the surface will have a higher probability of escaping from the surface without losing KE than those produced deeper in the sample. This escape probability decreases exponentially as the depth from the surface (z) increases (Figure 2.10A). The characteristic length is where the escape probability decreases to e^{-1}. This length defines the average distance an electron travels before it undergoes inelastic collisions and therefore loses energy; so, it is called inelastic mean free path (IMFP, λ). When this probability function is integrated from zero (surface) to 3λ (subsurface), it gives 0.95. This means that 95% of the photoelectrons detected at a given BE in an XPS spectrum comes from the region up to 3λ from the surface. So, the information depth (ID) is defined as 3λ. In the example shown in Figure 2.10A, IMFP (λ) is ~3.5 nm and the information depth (3λ) is ~10.5 nm.

If the sample has a layered structure as in Figure 2.10B, then, as the overlayer thickness (d) increases, the overlayer signal (I_O) increases while the substrate signal (I_S) decreases. When the emitted photoelectron signals are detected at the surface normal direction, their thickness dependences can be predicted with simple equations shown in Figure 2.10B. This also explains why the surface should be kept clean to get the proper XPS analysis. As adventitious carbon contaminants are deposited on the surface, the genuine sample signal will be attenuated accordingly.

The degree of signal attenuation or the magnitude of IMFP (λ) depends on the KE of photoelectron and the material in which it is traveling. Seah and Dench measured IMFPs for a large number of elemental solids as well as inorganic and organic compounds and observed recurring trends (Figure 2.11) [11]. The IMFP was found to decrease in proportion to the square of KE in the low KE regime and increase in proportion to the square root of KE in the high KE regime. Based on this observation, they derived the following empirical equations [11]:

$$\text{For elements: } \lambda(nm) = 538\,KE^{-2} + 0.41\sqrt{a \cdot KE} \tag{2.6}$$

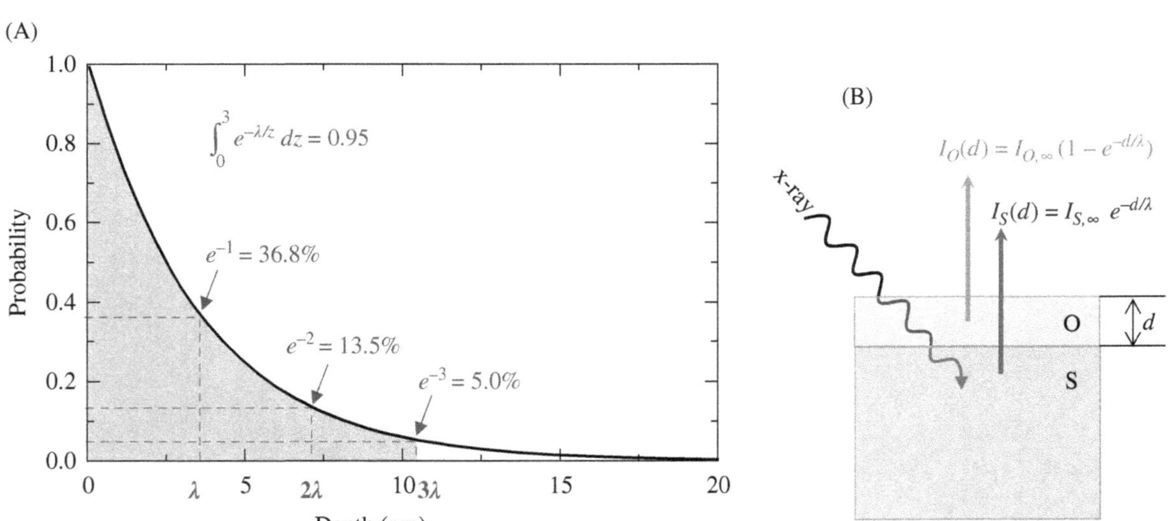

Figure 2.10 (A) Decay of escape probability of photoelectron as a function of depth (z) from the surface. (B) Overlayer thickness dependence of XPS signals detected at the surface normal direction. Here, $I_{O,\infty}$ and $I_{S,\infty}$ are the bulk signal of the overlayer and substrate, respectively.

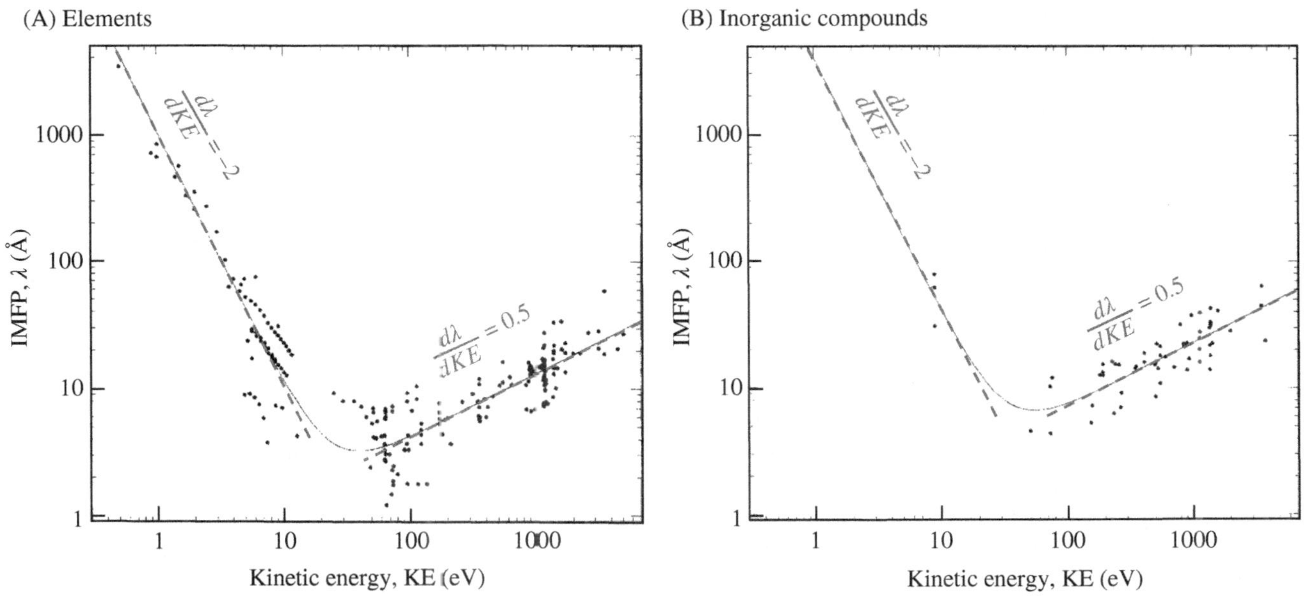

Figure 2.11 Universal curves of IMFP (λ) as a function of kinetic energy (KE) of electron for (A) elements and (B) inorganic compounds. Seah and Dench [11]/John Wiley & Sons.

$$\text{For inorganic compounds: } \lambda(nm) = 2170\,KE^{-2} + 0.72\sqrt{a \cdot KE} \tag{2.7}$$

$$\text{For organic compounds: } \lambda'(mg/m^2) = 49\,KE^{-2} + 0.11\sqrt{KE} \tag{2.8}$$

Here, a is the estimated monolayer thickness, $a = \left(\frac{A}{\rho n N} \times 10^{24}\right)^{1/3}$ where A is the atomic or molecular weight, n is the number of atoms in a molecule, ρ is the bulk density in kg/m^3, and N is Avogadro's number. In the case of organic compounds, the λ' value obtained from eq. (2.8) needs to be divided with the density to get λ in the nm unit.

In practice, information depth (ID) or effective attenuation length (EAL) are often used more conveniently. Note that while IMFP depends on the material and electron energy only, ID and EAL depend on not only IMFP but also the instrumental configuration, and elastic-scattering effects in the sample material. The EAL value can be defined differently for different XPS applications. Guidance on sources of IMFP and EAL data and on predictive equations can be found in a paper published by C. J. Powell [12].

2.2 Instrumentation for X-ray Photoelectron Spectroscopy (XPS)

Before we begin studying how to interpret XPS spectra and how to get qualitative and quantitative information of the sample, let us briefly go over instrumentation details. Knowing the instrument is important to properly design the experiment, optimally conduct the measurement, and accurately interpret the obtained data. In Section 2.1, we have studied how x-ray interacts with matter and produce photoelectrons from atomic orbitals, how many peaks are expected from each orbital, and how the surface sensitivity of XPS is attained. In order to induce all these processes, we need an x-ray source, which is the first subject to study in this section. X-ray induces electron emission, and the remaining surface will be positively charged unless the equal amount of electron is supplied to the surface. If the sample is conductive, this happens naturally as long as the sample is grounded properly. If the sample is not conductive, then charge neutralization of the surface should be done properly. Once photoelectron is emitted, then we need to measure its KE. So, the next subject is the energy analyzer. In Chapter 1, we studied why UHV is needed to measure KE of electrons without losing them on their way to the energy analyzer. For that reason, XPS experiments are typically done in UHV conditions; but it is also possible to conduct the XPS measurement while keeping the sample in a near-ambient pressure (NAP). This principle will be touched on as well.

Finally, we need to count the number of photoelectrons filtered as a function of their KE with a good signal-to-noise ratio to construct the XPS spectrum. The electron detection statistics will also be covered at the end.

2.2.1 X-ray Source

In most commercial XPS systems, x-ray is generated by bombarding a metallic anode with high-energy electrons. Such electrons are generated by passing high current through a filament (typically made of tungsten since it has a high melting temperature). Then, the electrons boiled off the hot filament are accelerated to the anode surface which is biased with a high positive voltage (typically 15–30 kV). Since the electron energy is higher than the binding energy of core electrons in the anode, it can induce emission of core electrons producing holes in the inner shell. Then, the relaxation processes discussed in Figure 2.6 take place, among which is x-ray emission. The energy of the emitted x-rays depends on the atomic orbital energy levels of the anode material, as depicted in eq. (2.4), and the beam intensity depends on the electron current striking the anode and its energy. Figure 2.12 schematically illustrates the x-ray emitted from an aluminum anode. When the x-ray is generated by electronic transition from higher orbitals (*L*-shell) to the hole generated in the *K*-shell, it is called Al-Kα. This line has two bands – Kα_1 and Kα_2 which correspond to the electron transition from the $2p_{3/2}$ and $2p_{1/2}$ states, respectively. Since the $2p_{3/2}$ state a lower BE than the $2p_{1/2}$ state, the Kα_1 line has a slightly higher energy than the Kα_2 line. Since the relative degeneracy of $2p_{3/2}$ and $2p_{1/2}$ is 2:1 (as discussed in Section 2.1.3), the relative intensity of Kα_1:Kα_2 is also 2:1.

Typical anode materials used in commercial XPS systems are Al and Mg. The Al Kα line is 1486.6 eV with a full-width half maximum (FWHM) of 0.9 eV. The Mg Kα line is 1253.6 eV with a FWHM of 0.7 eV. They are used because their energy is large enough to induce photoemission from most elements for identification of chemical species and their FWHM is narrow enough for high-resolution analysis for distinction of chemical states of elements such as oxidation states and chemical bonding to neighboring elements. Recently, the Ag Lα source is made commercially available. This one has a larger FWHM (2.6 eV), but its energy is much higher (2984.2 eV). The higher photon energy makes more accessible to higher BE orbitals (see Figure 2.2); but the gain from this aspect is relatively insignificant. Instead, as depicted by eq. (2.1), it makes the KE of photoelectrons significantly higher than the ones generated with the Al Kα or Mg Kα source. Thus, it makes IMFP significantly larger (see Figure 2.11), which allows probing deeper into the sample, as depicted by eqs. (2.6)–(2.8).

In addition to this characteristic main line, additional lines called satellites can exist. They can originate from the dipole-allowed de-excitation of multiply ionized or excited states. Other mechanisms are resonance emission, Auger effect, magnetic dipole, and electric quadrupole transitions.

Also, there is a broad continuous background. This is caused by the Bremsstrahlung process (Figure 2.12). While the electron ejected from the atomic orbital travels through the sample, its trajectory can be altered and decelerated via interactions with the nucleus of other atoms, which is accompanied by the emission of x-ray whose energy is equal to the change in KE of traveling electron. This Bremsstrahlung process is similar to the x-ray generation in a synchrotron facility. In synchrotrons, a pack of electrons is accelerated to a high energy and its trajectory is bent by magnetic fields, which causes emission of x-ray with a specific energy that vary depending on the electron energy and the applied magnetic field. While the synchrotron facility utilizes the Bremsstrahlung process to generate monochromatic x-ray with extremely high brightness at a specific energy, the Bremsstrahlung background in the commercial anode source is problematic since that can also induce photoemission process from the sample, adding to the background signal in the measured spectrum. So, the Bremsstrahlung background and the satellite bands need to be removed.

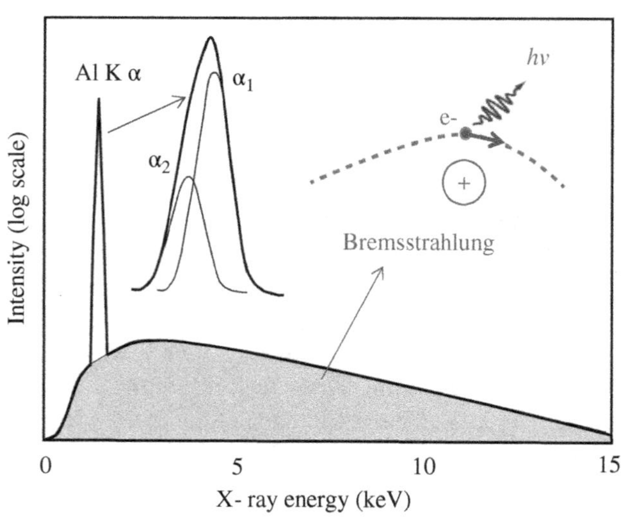

Figure 2.12 Schematic illustration of x-ray emitted from an anode bombarded with electron beam with an energy higher than 15 keV.

The monochromatic filtering or refining of the incident x-ray beam is done using the Bragg's principle which is the same as the underline principle of x-ray diffraction (XRD). In XRD, Bragg's law is used to find the distance between crystallographic planes with the same scattering density. In XPS, it is used to filter out satellite lines and backgrounds. The Bragg's law can be written as:

$$n\lambda = 2d \sin\theta \tag{2.9}$$

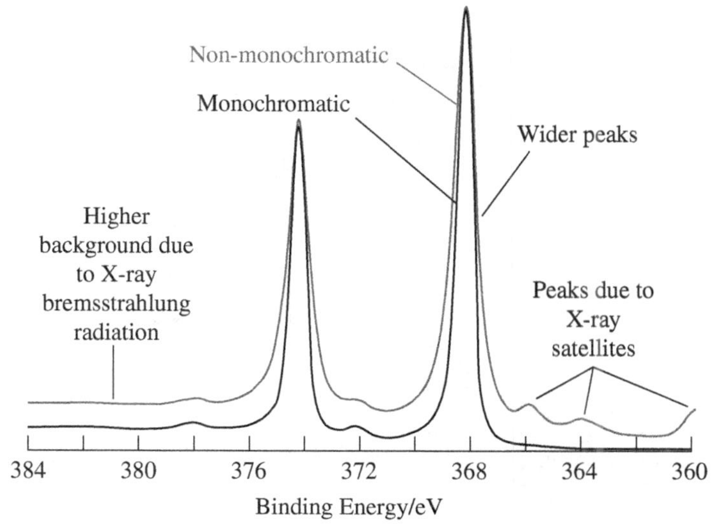

Figure 2.13 Ag 3*d* spectra obtained with nonmonochromatic and monochromatic X-ray sources. The intensities are normalized for comparison. Watts and Wolstenholme [13]/with permission of John Wiley & Sons.

where n is the diffraction order, λ is the x-ray wavelength, d is the lattice spacing, and θ is the incidence angle of x-ray with respect to the crystal plane used as a monochromator. For Al Kα line, the energy ($h\nu$) is 1486.6 eV, which corresponds to $\lambda = 8.34$ Å. The first-order diffraction from the (10$\bar{1}$0) plane of quartz ($d = 4.257$ Å) constructed with a Bragg angle of 78.5° is used for the monochromatic filtering of the Al Kα x-ray source. Figure 2.13 compares the Ag 3d spectra acquired with and without using the monochromator for the Al Kα x-ray source [13]. The data show the disappearance of the peaks due to x-ray satellites and the reduction of the background in the higher BE side due to the Bremsstrahlung continuum.

2.2.2 Energy Analyzer

In the case of photons in the visible wavelength, the energy filtering is done through a prism in a transmission mode or with a grating in a reflection mode. Prisms and gratings disperse the light as a function of wavelength; but they cannot be used to disperse electrons as a function of their kinetic energies since electrons are annihilated upon contact with these materials. Instead, an electrostatic analyzer is used to filter out electrons with energies that are different from the desired energy. Figure 2.14 illustrates the principle of electrostatic analyzer. Assume a pair of parallel electrodes is placed between the sample and the detector. The bottom electrode to the sample side has the entrance and exit holes through which electrons can pass and the top electrode is negatively biased. Photoelectrons ejected from the sample with a wide distribution of kinetic energies can enter the entrance hole. This parallel-plate analyzer will allow passing of electrons only with the proper KE whose trajectories are bent by the negative potential applied to the top plate and reach the exit hole in the bottom plate. If their kinetic energies are too low, they will hit the bottom electrode and get annihilated. If the KE is too high, electrons will hit the top electrode and get annihilated. This principle is used in various geometries to construct cylindrical mirror analyzer (CMA) or concentric hemispherical analyzer (CHA). The CMA has been widely used for the AES system and the CHA is more common in the XPS system, although both can be used in either analysis [13].

Let us look into the CHA more in detail (Figure 2.15) since it is more widely used for XPS. It consists of two hemispheres. The voltage difference between the inner (V_1) and outer (V_2) hemispheres and their radii (R_1 and R_2) define so called pass energy (PE):

$$PE = \frac{e(V_2 - V_1)}{R_2/R_1 - R_1/R_2} \quad (2.10)$$

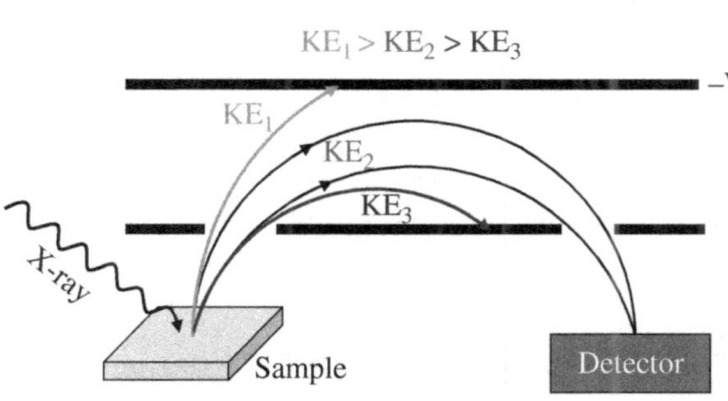

Figure 2.14 Kinetic energy filtering with two parallel plates with entrance and exit holes.

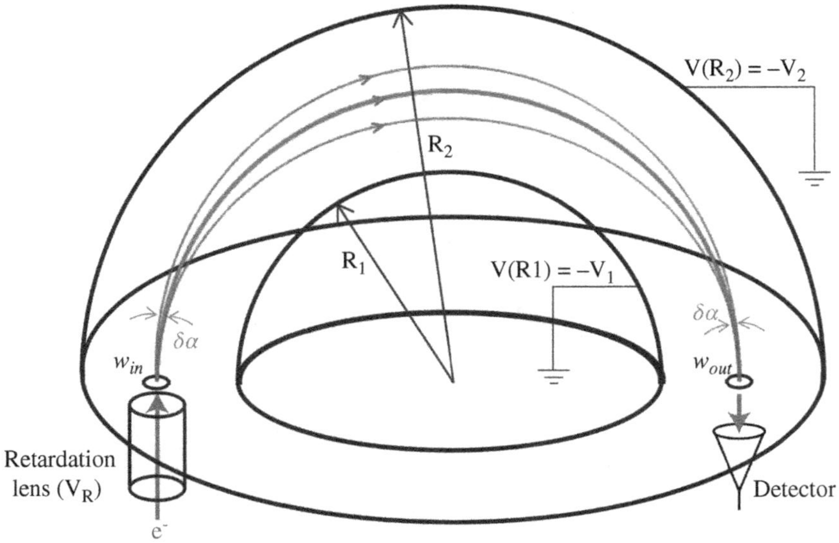

Figure 2.15 Common design of concentric hemispherical analyzer (CHA).

Then, the range of KE (ΔKE) of electrons that can pass through CHA is the function of physical dimensions of the analyzer and the pass energy:

$$\frac{\Delta KE}{PE} = \frac{w_{in} + w_{out}}{2(R_1 + R_2)} + \delta\alpha^2 \tag{2.11}$$

where w_{in} and w_{out} are the diameter of the inlet and outlet orifices, respectively, and $\delta\alpha$ is the acceptance angle of electron trajectory entering the CHA. Since the physical dimensions of the inner and outer hemispheres are fixed, the energy resolution (ΔKE) is controlled by changing the orifice sizes, tuning the acceptance angles with an electrostatic focusing lens, or adjusting the PE. For a given physical setting, the right-hand side of eq. (2.11) is constant. Thus, if PE is large, then ΔKE is also larger. In order to increase the energy resolution without changing the orifice size and the focusing lens condition, PE should be reduced.

The CHA is operated at a fixed PE while changing the retardation voltage (V_R) just before the entrance orifice is scanned stepwise (Figure 2.15). Then, the measured KE (KE_{meas}) of photoelectron passing through the analyzer is:

$$KE_{meas} = V_R + PE \tag{2.12}$$

Note that this is referenced to the vacuum level of the spectrometer (E_{vac}^{spectr}).

When the energy resolution is increased by decreasing PE, then the throughput (fraction of photoelectrons passing through CHA) is also decreased. The large PE (low-resolution) condition is typically used for survey analysis scanning over a large range of V_R with a relatively large step (ΔV_R) to find all elements present in the sample surface. The low PE (high-resolution) operation is used to determine chemical states of elements (such as oxidation states or bonding to other atoms) by scanning over a specific energy range with a small ΔV_R step. An example is shown in Figure 2.16.

Note that different instruments are configured with CHAs with different dimensions. So, using the same PE in operation does not mean the energy resolution is the same. Some systems allow changing the entrance orifice size. Then, even at the same PE, the energy resolution can be tuned differently. Even if all physical conditions (the FWHM of the incident x-ray beam, CHA dimensions, and PE) are the same, the FWHM of the peak in the XPS spectrum (i.e., ΔBE) can vary depending on the sample. Also, depending on the operation mode of CHA, that is, constant analyzer energy (CAE) mode vs. constant retard ratio (CRR) mode, the transmission efficiency and energy resolution may change for electrons with different KE [13]. This affects the relative intensities of photoelectron signals from different orbitals from the same element as well as the background shape. Typical XPS systems are operated in the CAE mode which keeps ΔKE constant over a wide range of KE.

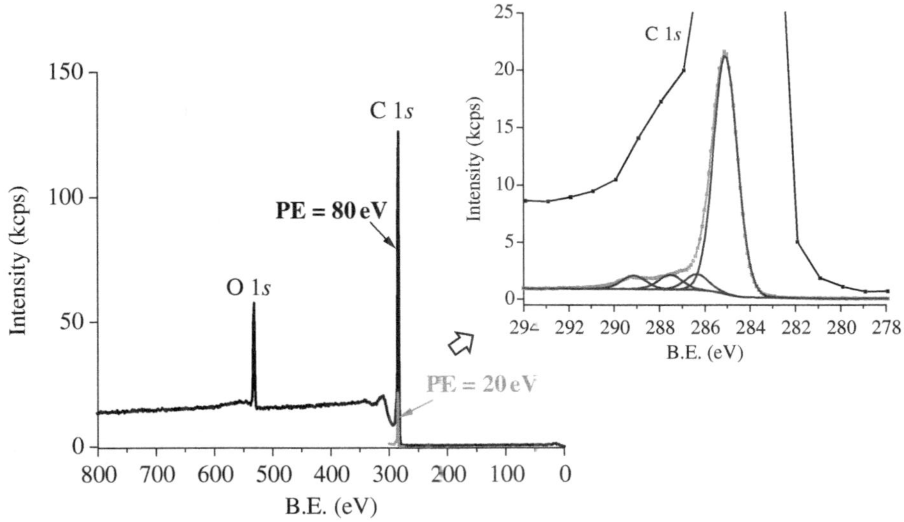

Figure 2.16 XPS spectra of amorphous carbon film (~1 μm thick) deposited on a silicon wafer collected with PE = 80 eV and 20 eV. The data were collected at every 1 eV in the PE = 80 eV operation and 0.1 eV in the PE = 20 eV operation.

2.2.3 Energy Reference and Charge Neutralization

All energies are measured with respect to a reference level. In the BE analysis, the reference level is the Fermi level (E_F) of the sample. For a conducting sample that is grounded to the spectrometer, the Fermi levels of the sample and the spectrometer are equilibrated to the ground level (Figure 2.17). Photoelectrons ejected from the sample will enter the spectrometer and travel through the spectrometer. Recall that the work function is the energy required to remove an electron from the highest occupied orbital to the vacuum level. In other words, the work function of the spectrometer (Φ_w^{spectr}) is the lowest energy that allows electrons traveling through the analyzer to stay in the vacuum. If the KE falls below Φ_w^{spectr}, then electrons will be attracted and annihilated at the spectrometer surface. So, by adding Φ_w^{spectr}, the measured KE of photoelectrons (KE_{meas}) passing through the analyzer can be referenced to the E_F of the sample that is properly grounded. Then, the BE of the photoelectron detected after passing the analyzer is calculated as:

$$BE = h\nu - \left(KE_{meas} + \Phi_w^{spectr}\right) = h\nu - \left(V_R + PE + \Phi_w^{spectr}\right) \tag{2.13}$$

This simplifies the BE determination in XPS [14]. For example, the work function of an unknown sample is unknown. If Φ_w^{sample} was to be used in eq. (2.13), one could not determine the BE of photoelectron from the unknown sample by running XPS only; another independent measurement would be needed to find Φ_w^{sample}. The energy diagram in Figure 2.17 shows that we do not need to know Φ_w^{sample}, since Φ_w^{spectr} is used to calculate the BE of photoelectron detected. The Φ_w^{spectr} value is constant for a given material used to construct the system.

When the sample of interest is not conductive, that is, an insulator, the situation becomes complicated because the Fermi level of the sample surface (E_F^{sample}) cannot be equilibrated to that of the spectrometer (E_F^{spectr}). So, the reference point is floated from the ground by an unknown value. In the case of conductive samples grounded to the spectrometer, the holes left in the sample upon emission of photoelectrons are refilled by electrons flowing from the ground. So, the sample surface remains uncharged. In the case of insulating samples, this electron transport path is essentially blocked, and the loss of electrons via photoemission causes the surface to be positively charged. As the positive charge builds up, the threshold energy for emitted electrons to overcome the work function and escape into vacuum increases. Thus, the KE of photoelectrons is reduced accordingly. Since the detected photoelectron signals are plotted in the BE scale, the decrease in KE_{meas} causes the signal to appear in a higher value in the BE scale than the true value. An example is shown in Figure 2.18.

Since the insulating samples cannot be neutralized via electron transport from the ground, charge neutralization must be done in an external way. This is done by supplying electrons from the vacuum side to compensate for the positive charge at the surface. For this, a low-energy electron flood gun is used. Typically, electron energy is less than 5 eV. It is difficult to determine the absolute energy and flux of electrons needed for charge neutralization because the degree of charging cannot

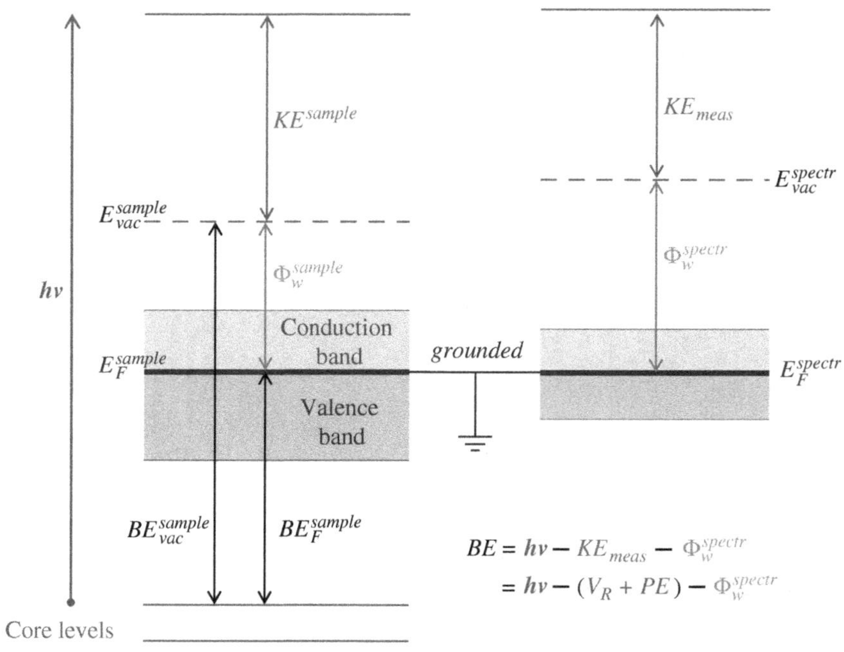

Figure 2.17 Energy level diagram for an electrically conducting sample that is grounded to the spectrometer.

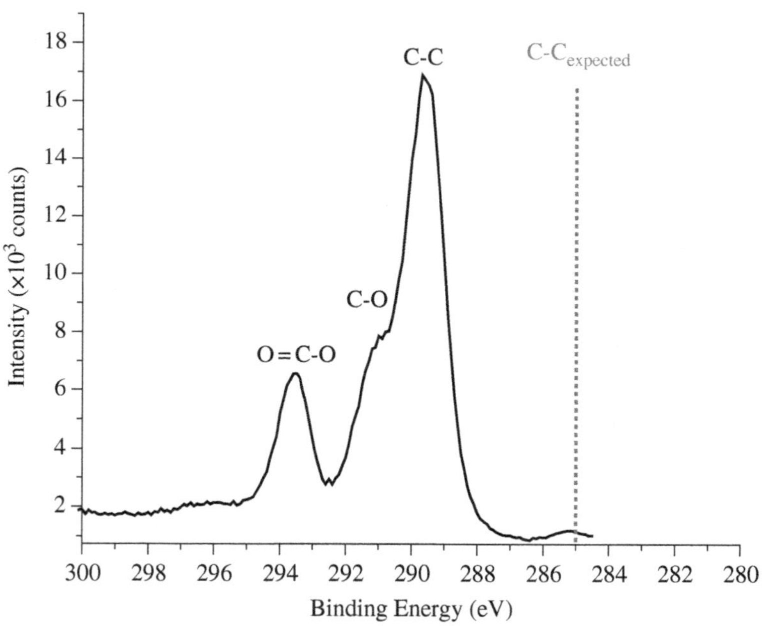

Figure 2.18 C 1*s* XPS spectrum of polyethylene terephthalate (PET) plastic collected without using the charge neutralization mode.

be determined independently. Practically, it is done by adjusting the flux and energy of the charge neutralizing electrons until a specific reference peak is shifted to the expected BE position [15]. The reference peak is usually chosen from one of the bulk signals, that is, element in the subsurface region so its BE is known based on the stoichiometry of the sample. In the case of example shown in Figure 2.18, the charger neutralizer operation condition would be tuned until the main C–C peak is shifted to the expected C 1*s* BE (which is 285 eV for the C–C species, see Section 2.3). For an unknown sample, the C 1*s* peak of adventitious carbon contaminant could be used as an approximation method [16]. Then, XPS spectra in other BE regions are collected while operating the charge neutralization with this reference peak chosen by the user.

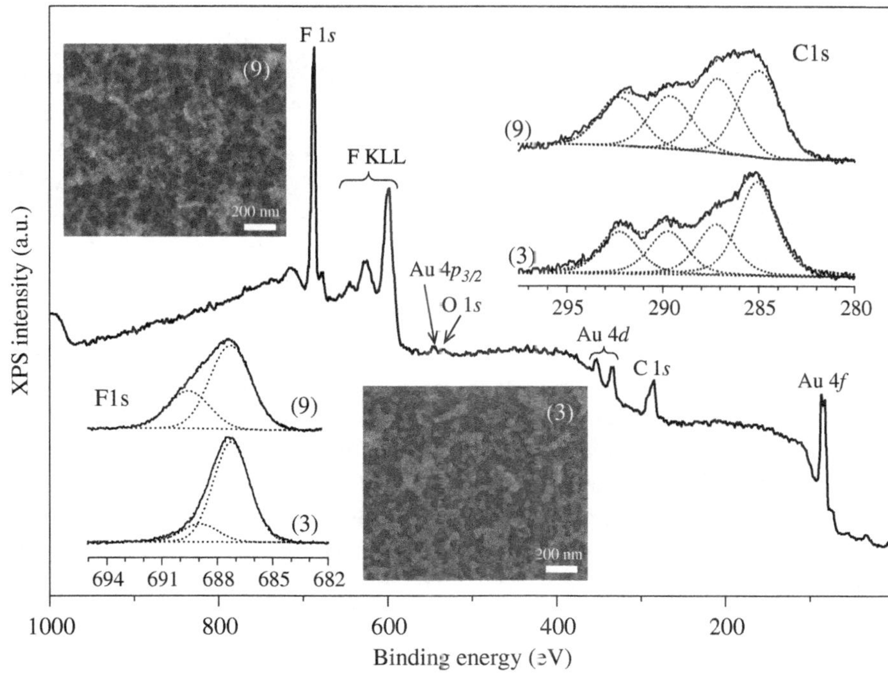

Figure 2.19 XPS spectra of superhydrophobic CF_x films deposited on an Au film using an atmospheric rf plasma method. The number in paratheses is the number of plasma treatments. The treatment cycle increases, the film gets thicker, and the surface roughness increases. Reproduced from Kim et al. [18]/with permission of American Chemical Society.

In most modern XPS systems, the electron flood gun operation is tuned automatically with a computer control; but it does not mean that it will be done properly and perfectly. Details of charge neutralization are much more complicated than what is described here [15]. The FWHM of the analyte peak can vary depending on the charge neutralization condition. If the sample is sensitive to redox reactions, the oxidation state of element of interest may vary during the XPS measurement under the electron flood condition [17]. If the charge neutralization is not done properly, all peaks will be shifted from the true values. Since the degree of shift is relatively constant for a given condition, the fine adjustment can be done mathematically during the data processing, again using the same principle – shifting the entire spectrum by the same amount until the BE of the known peak moves to the expected position.

There is one thing that the instrumental charge neutralization cannot correct. The charge neutralization is done assuming the degree of charging is the same over the entire sample. But, if the sample surface is rough, the degree of charging varies locally. Then, the KE of photoelectrons originating from different locations can experience different degrees of charging. This will broaden the measured KE since photoelectrons enter the energy analyzer with a broad distribution of KE due to differential charging. Figure 2.19 shows the XPS spectra of superhydrophobic CF_x films deposited on a gold film [18]. Since the superhydrophobicity requires amplifying the hydrophobicity originating from chemistry with surface roughness [19], the CF_x films was intentionally made rough. The high-resolution spectra show the C 1*s* and F 1*s* bands become broader as the surface roughness increases. In the F 1*s* region, the peak could be fitted nicely with two components. But this should not be interpreted as two types of chemical bonds for fluorine in the sample; such interpretation is not possible in the presence of differential charging. If the differential charging cannot be avoided, it becomes a nuisance in analysis. On the other hand, if it can be controlled intentionally, it could be an additional knob to play with to understand the sample structure better [20].

2.2.4 Transfer Lens and Near-Ambient Pressure Operation

As you can imagine from the schematic shown in Figure 2.15, the CHA is quite bulky. For that reason and other practical design issues, it is placed relatively far from the sample. If the analyzer orifice is too far from the sample, only a very small fraction of photoelectrons ejected from the sample will enter the analyzer. Increasing the orifice size is not ideal since it will reduce the energy resolution (see eq. (2.11)) and the actual increase in signal intensity would be minimal if the distance from the sample to the analyzer is large. To address this issue, a transfer lens is placed between the sample and the entrance orifice of the analyzer. This lens consists of several cylindrical tubes with electrical biases. The transfer lens entrance is placed close to the sample to maximize the collection angle to ensure high sensitivity. It also has the retardation lens that sets the cutoff energy (V_R) with which electrons are entering to the analyzer. In addition, it controls the acceptance angle ($\delta\alpha$) that affects the energy resolution (Figure 2.15).

For the reasons discussed in Section 1.2, XPS analysis is carried out in UHV conditions. Then, the sample surface is naturally placed in the UHV condition. In heterogenous catalysis research, most reactions, in particularly thermally-activated gas-solid reactions, are carried out in high pressures to ensure high flux of reactants colliding to the catalyst surface and/or increase the thermodynamic driving force to change the structure and reactivity of the catalyst. This incongruity between the UHV requirement for XPS analysis of catalyst surfaces and the high-pressure requirement for high reactivity is called "pressure gap" and has been a challenge in heterogeneous catalysis research [21]. To address this challenge, a breakthrough in XPS instrumentation was made in the late 1990s and early 2000s. The main idea was to differentially pump the transfer lens section, so that the sample can stay in near-ambient pressure (still low pressure, but at least not in UHV), while the CHA and the detector are in the UHV condition [22]. An example of a commercial near-ambient pressure XPS (NAP–XPS) system is shown in Figure 2.20. The development of this system makes it possible to conduct XPS analysis while the solid surface is being exposed to a few Torr or mbar of reactive gases or wet with liquid, enabling *operando* surface analysis.

In NAP–XPS, not only the molecules present at the sample surface but also the molecules in the gas phase can be excited by x-ray and emit photoelectrons. These electrons can enter the differentially pumped lens system and arrive at the analyzer entrance. For the molecules at the surface, the BE is referenced to the Fermi energy of the sample (see Figure 2.17). Gas phase molecules freely fly in the space between the sample surface and the nozzle of the differentially pumped lens and do not have any electrical contact with either of them. Thus, the reference level for the KE of photoelectrons emitted from the gas phase molecules is referenced to the vacuum level (E_{vac}) which is higher than the Fermi level (E_F) of the sample. Thus, photoelectrons of the gas phase molecules appear in the higher BE side of the same molecule adsorbed at the surface in NAP–XPS spectra. An example is shown in Figure 2.21A [22].

The maximum pressure that can be used in NAP–XPS operation is governed by the number of inelastic collisions that photoelectrons will have before they enter the collision-free zone, which will be a function of the distance from the sample to the entrance nozzle and the inelastic scattering cross section [22]. This cross section is a function of photoelectron KE, which varies depending on the type of molecules in the gas phase [23]. In typical KE range of photoelectrons encountered in XPS analysis, the dominant inelastic scattering process by gas-phase molecules is electronic excitation from an occupied low-energy state of a molecule to an unoccupied state at higher energy through the energy transfer during the collision with photoelectron. Thus, the inelastic scattering with the gas molecules in the photoelectron trajectory lowers the KE

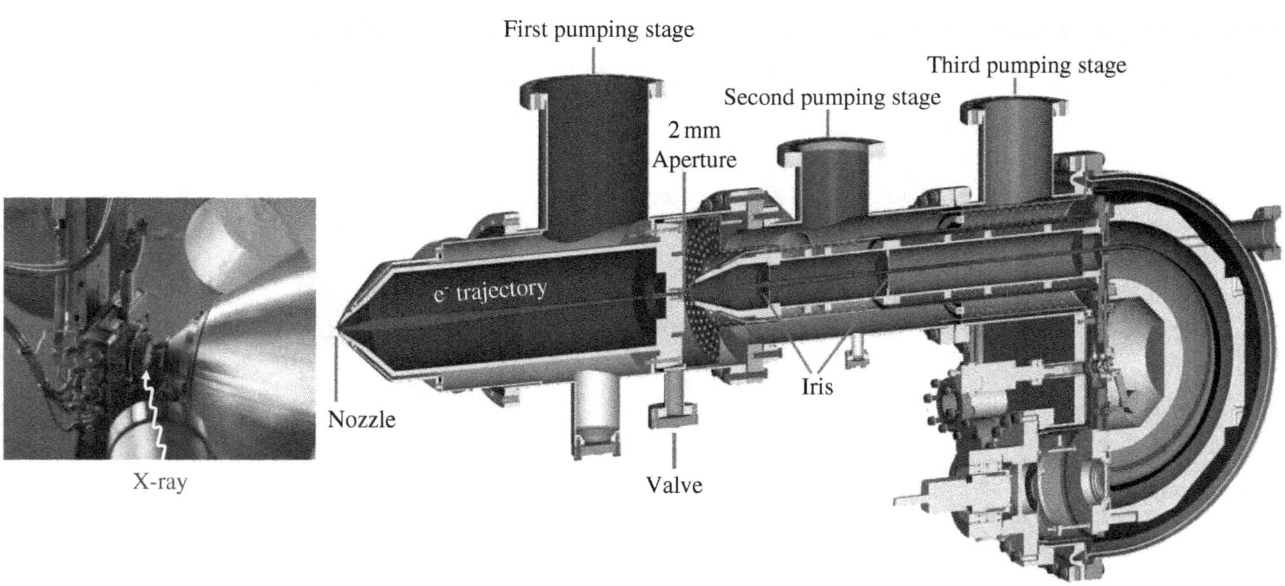

Figure 2.20 SPECS™ PHOIBOS 150 NAP analyzer equipped with a triply differentially pumped transfer lens system. The picture in the left shows the sample placed very close (typically a few to several hundred μm's) to the nozzle of the differentially pumped lens system. Reproduced with permission of SPECS GmbH.

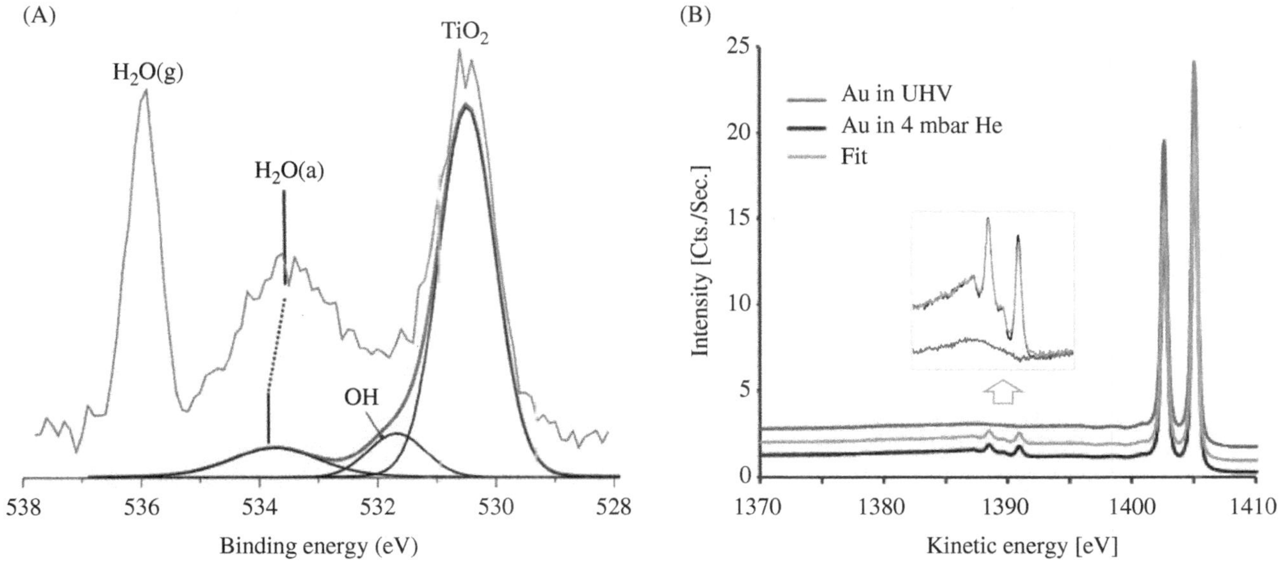

Figure 2.21 (A) XPS spectra in the O 1*s* region of a TiO_2 sample in the presence of 2.25 Pa of H_2O at 298 K (bottom curve, red) and in 133 Pa at 270 K (top curve, gray). Salmeron and Schlögl [22]/with permission of Elsevier. (B) Au 4*f* XPS spectra measured in UHV, in 4 mbar He, and the result of convolving the red spectrum with the energy-loss function of He. Pielsticker et al. [23]/permission of John Wiley & Sons.

of photoelectron, which creates new features in the higher BE side of the photoelectron of the sample in the NAP-XPS spectrum. An example is shown in Figure 2.21B [23].

2.2.5 Electron Detection

The last step is to detect and record the flux of electrons filtered as a function of KE. One may think "current" when considering the electron flux. The charge of electron is 1.6×10^{-19} C and the unit of ampere is C/s. So, the flux of one electron per second corresponds to 1.6×10^{-19} A. Commercially available ammeter can detect current down to pA, which means a flux of 10^7 electrons/sec. The flux of photoelectrons passing through the energy analyzer is much smaller than this, so the ammeter cannot be used for detection of photoelectron flux. Thus, an electron multiplier is needed for detection of energy-filtered photoelectrons. The most popular electron multiplier used nowadays is called channeltron, which has a channel shape (Figure 2.22). Most channeltrons are made with a glass tube coated with a conductive layer and a dielectric layer on top of that. There are many different shapes of channeltrons, but the basic operation principle is the same [24]. An electron striking the input face of the channeltron typically produces 2–3 secondary electrons. These secondary electrons are accelerated down the channel by a positive bias. These electrons strike the channel walls, producing additional secondary electrons. This process repeats along the tube until a pulse of 10^6–10^8 electrons is generated. This amplification gain depends on the channeltron geometry and the applied bias. The electron pulse emerging from the exit port of the channeltron is counted with a pulse counting circuit. So, the XPS signal intensity is typically recorded as "count per second (cps)." The

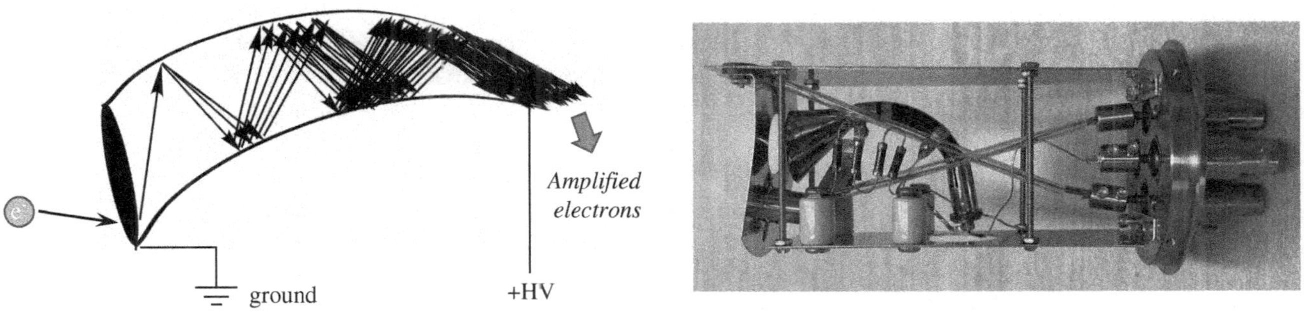

Figure 2.22 Schematic view of channeltron and a picture of a channeltron assembly.

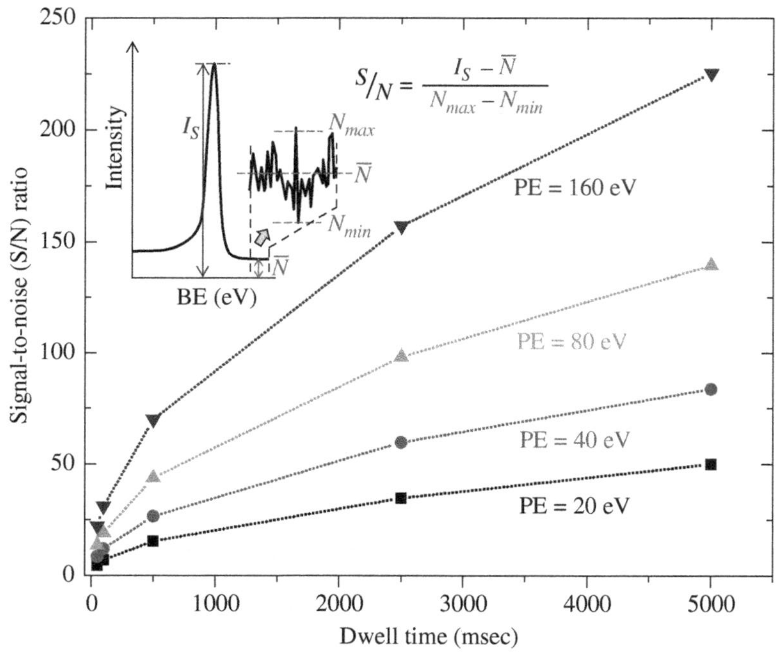

Figure 2.23 Plot of signal-to-noise (S/N) ratio of the O 1*s* peak from $HfSiO_4$ vs. dwell time per step scan. Data provided by Vince Bojan, Penn State Materials Characterization Lab. The inset shows who to calculate the S/N ratio. The low BE side background is magnified to show N_{min}, N_{max}, and $\overline{N}$.

typical pulse width is ~10 ns; so, if the pulse counting rate (cps) reaches $\sim 10^8$, then it becomes a continuous current and should be detected with an ammeter.

The probability for a single incident electron striking the multiplier surface with a certain gain to generate *n* secondary electrons follows the Poisson statistics. In the Poisson distribution, as the total number of secondary electrons increases to *n*, the spread of the probability increases by $\sqrt{n}$. In other words, the more counts are recorded, the larger standard deviation will be; but the relative standard deviation decreases as $\sqrt{n}/n = 1/\sqrt{n}$. This has an important consequence. When the signal intensity is low, one may increase the data accumulation time (often called dwell time per step) to increase the total signal intensity. Then, the signal count of the peak (I_S) will increase proportional to the dwell time, but at the same time the average ($\overline{N}$) and spread (i.e., N_{min} and N_{max}) of the noise level also increase. The spread of the noise signal will be proportional to $\sqrt{\overline{N}}$. Thus, if one increases the signal intensity by increasing the dwell time (t), the signal-to-noise (S/N) ratio of the resulting spectrum increases by $t^{1/2}$, while the signal intensity increases in proportion to t. An example showing this relationship is presented in Figure 2.23. The signal intensity can be increased by increasing the pass energy (at the expense of losing energy resolution) while keeping the dwell time; even in this case, the increase in S/N ratio is still proportional to the square root of the degree of signal count enhancement.

2.2.6 Small-Area Analysis and Imaging

Small-spot XPS analysis and elemental imaging have become increasingly popular. There are two strategies used for small spot analysis or imaging. One is to irradiate x-ray over a large area of the surface and use a small aperture of the electrostatic lens to allow photoelectrons from a small area of the sample only; then, the position of data collection is step-scanned for imaging. The other is to focus the x-ray into a small area and use a relatively large aperture to maximize the detection efficiency of the small amount of photoelectrons from that area; then, the focused x-ray beam is rastered for imaging. Each method has its own advantages and disadvantages. Typical lateral resolution is on the order of a few μm's (Figure 2.24) [25]; but this can vary depending on the x-ray/lens alignment of the system used, as well as the user's skill. If a 2D CCD camera coupled with a multichannel electron multiplier is used for imaging, elemental mapping with a higher lateral resolution is possible; in this case, the aberration of the analyzer can be the limiting factor. It should be noted that if the signal intensity is low, the small spot analysis is challenging. As the spot size decreases, the S/N ratio gets worse. Similarly, the lateral resolution in imaging depends on the signal intensity contrast of the elements. If the BEs of two species to be mapped are close to each other, the PE of the analyzer needs to be kept small, which makes the overall intensity and S/N

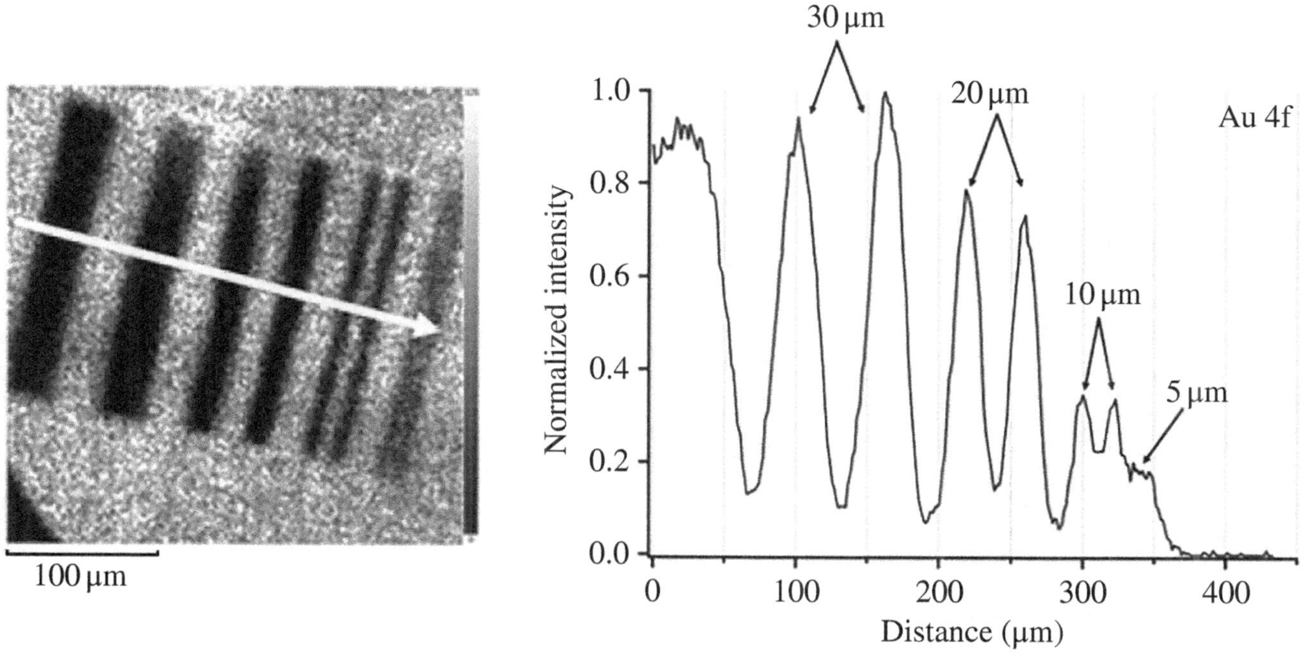

Figure 2.24 Test specimen consisting of consecutive parallel strips of Ti (black) having widths of 30, 20, 10, and 5 μm in on an Au substrate and line-scan displaying the Au 4*f* intensity across the test specimen (along the white arrow). Reproduced from Unger et al. [25]/with permission of American Vacuum Society.

ratio low. When the small spot analysis is used for a quantitative analysis of a small feature on the surface, the question arises as to what fraction of the signal originates from outside the analysis area. A proper strategy to control the intensity from outside the nominal analysis area should be implemented.

2.3 Qualitative Analysis with XPS

2.3.1 Identifying Elements Present in the Surface

The elemental identification is typically done with a so-called survey scan mode. The XPS spectrum is collected over a wide range of BE with a low resolution (i.e., high PE) with a relatively large step scan (e.g., every 1 eV). As an example, the XPS spectrum of a bottle glass is shown in Figure 2.25. We can see that photoelectron signals with no loss of KE (i.e., sharp peaks above the background) are only a few percents of the total signals, and most electrons detected in XPS analysis are inelastically scattered ones. The electrons escaped from the surface without losing their initial KEs appear above the background as peaks with varying widths. Those are the ones produced mostly within the information depth from the surface. In the high BE side, the KEs of photoelectron peaks are in the 1000–1400 eV range. From Figure 2.11B, we can see the IMFP is around ~3 nm in this KE energy range for most materials. So, those photoelectrons detected without inelastic scattering originate from the top ~9 nm region. The signals in the left end of the spectrum correspond to KE of 200–400 eV, which gives IMFP of ~1 nm; thus, the information depth in this high BE region is only ~3 nm. This tells you that the information depths of different orbitals of the same elements within a single survey spectrum vary depending on their BEs (more accurately, KEs). It is not uncommon to choose Auger signals of elements instead of their main photoelectron signals for quantitative analysis of all elements within a similar information depth from the surface. More details will be discussed in Section 2.4.

The peak identification is typically done through comparison with the database. A good collection of such data can be found at the NIST x-ray Photoelectron Spectroscopy Database [26]. *The Handbook of x-ray Photoelectron Spectroscopy* published by Perkin-Elmer (hereafter, called XPS Handbook) is another good resource [10]. Experienced users remember peak positions of frequently analyzed elements. Note that for a given peak position, multiple candidates can be found. It is very

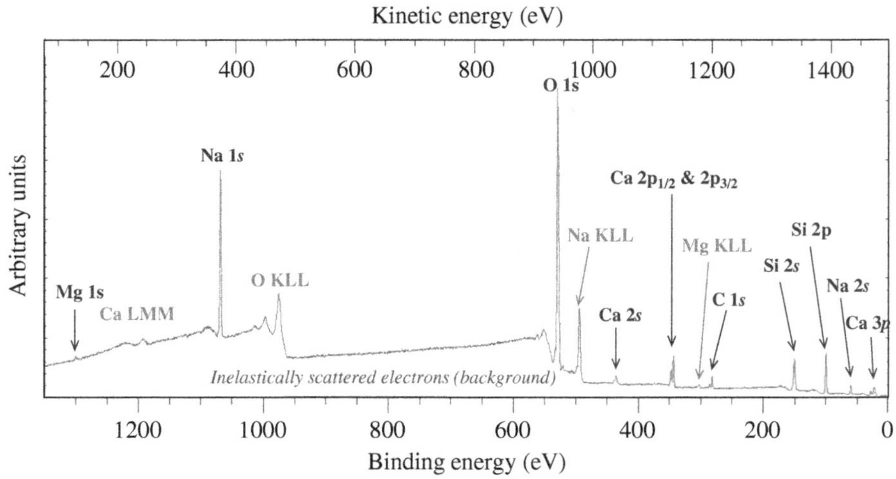

Figure 2.25 XPS spectrum of a green bottle glass. The sample was fractured in air prior to loading into the XPS chamber and the fractured surface was analyzed with charge neutralization.

rare to do the elemental assignment based on one single peak, unless the user knows the composition of the sample analyzed. For an unknown sample, all possible orbitals of a specific element should be identified with the relative intensity similar to the ones found in the XPS Handbook [10]. In Figure 2.25, the peak at ~498 eV is confidently assigned to Na KLL since we can confirm the presence of Na form its 1*s* peak at ~1072 eV. Then, some fine structures with smaller intensities at 524 eV, 536 eV, and 565 eV can be assigned to other minor Auger signals of the Na ion in the glass. Similarly, the confidence of assigning the small peak at ~439 eV to Ca 2*s* comes from the presence of the Ca $2p_{1/2}$ and $2p_{3/2}$ doublet at ~351 eV and ~347 eV. The peak positions of this doublet band are consistent with the spin–orbit split energy of Ca (Δ_{2p} = 3.6 eV) found in the XPS Handbook [10], and their relative intensities are 1:2 as expected from the degeneracy of the $p_{1/2}$ and $p_{3/2}$ states (see Section 2.1.3). For the Si 2*s* and Si 2*p* peaks, their peak positions are 154 eV and 102 eV, respectively, which are not the same as the 151 eV and 99 eV found in the XPS Handbook [10], but we are sure that they are from the Si 2*s* and 2*p* orbitals. This discrepancy is because the Si data in the XPS Handbook are from a clean silicon wafer, and the sample shown in Figure 2.25 is a silicate glass (i.e., Si atoms are oxidized). The discrepancy like this in BE is useful to find the chemical state of element (such degree of oxidation), which is explained in the next section. Such variations in BE of a specific photoelectron peak of an element can be seen in the NIST x-ray Photoelectron Spectroscopy Database [26]. Further identification of small bands in the lower BE side is also possible, although it is often not necessary if other major bands are assigned with confidence.

One of the most widely used software for XPS data processing is *CasaXPS* [27]. This is a quite versatile program that can handle XPS data from various instruments. This program shows rough positions of all orbitals of specific elements with their relative intensities. So, these suggested peak positions and relative intensities in the program become quite useful for elemental identification. More details are presented in Section 2.4 along with peak fitting and quantification.

Unless the surface energy of the sample is very low, there is a driving force for airborne organic molecules to adsorb on the surface. Thus, almost all samples that have been exposed to ambient air are typically covered with adventitious carbon species. So, if the C 1*s* signal is detected for inorganic samples in which the presence of carbon is not expected (as in the case shown in Figure 2.25), it is likely to be from these contaminants. Such contaminants can be removed prior to the measurement after the sample is loaded into the XPS chamber. A widely used method for sample cleaning is sputtering with an Ar^+ ion beam. But it should be noted that the sputter cleaning may alter the elemental composition of the sample. The reason for this phenomenon is discussed in Section 2.5 where depth profiling methods are discussed in depth. In shared-user facilities, it is possible that previous users may have left some residual contaminants in the system (unintentionally and unknowingly) due to outgassing from their samples. Some contaminants such as sulfur and fluorine containing vapors are notoriously difficult to pump out and reside in the UHV chamber for a long time. Unless the sample entry port and analysis chamber are properly cleaned subsequently, such contaminations will affect the analysis of the next user. The data analysis can still be done as long as contaminant signals do not overlap or interfere with the signals of interest or are clearly identified and removed through data processing. But it would be better to remove all complications as much as we can before the measurement.

2.3.2 Chemical Shift in Binding Engery (BE) Due to Oxidation State and Electronegativity

If the element of interest is initially ionized and possesses positive or negative charges, then the initial state energy of electrons is shifted from the neutral state energy because the screening of the nucleus charge is altered. The BE shift due to the oxidation state of atoms is called the "initial state effect." Table 2.1 lists typical binding energy shift (Δ_{BE}) of transition metal 2*p* orbitals upon formation of oxides. A general trend is an increase in Δ_{BE} as the oxidation state increases, *but not always*. The deviation from this trend is discussed in Section 2.3.5 and with an example in the Practice Problem at the end of the chapter.

Similarly, for covalently bonded atoms with different electronegativities, the overall electron density will be shifted depending on the electronegativity difference, and the energy of electronic states is altered accordingly. Figure 2.26 displays some of the original XPS spectra published by K. Siegbahn and colleagues [28]. In Figure 2.26A, two spectra of S 2*p* doublet bands from thiophene and carbonyl sulfide are superimposed. The BE of the S 2*p* doublet is higher when S is bonded to CO (higher electronegativity) than C (lower electronegativity). The spin–orbit splitting ($\Delta_{2p} = 1.2$ eV) and the 1:2 intensity ratio

Table 2.1 Typical binding energy shift (Δ_{BE}) of 2*p* orbital of first-row transition metal oxides with respect to the elemental metal value. The number in parentheses is the oxidation state of metal in the oxide form.

Element	Δ_{BE} (oxidation state)
^{22}Ti	~1 eV (+2), 4~5 eV (+4)
^{23}V	3~4 eV (varying)
^{24}Cr	1~2.5 eV (+3, +6)
^{25}Mn	2~3 eV (+2, +3)
^{26}Fe	~2.5 eV (+2), ~4 eV (+3)
^{27}Co	1.8 ~ 2 eV (+2, +3)
^{28}Ni	~1 eV (+2), 4~5 eV (+3)
^{29}Cu	~0 eV (+1), +1 eV (+2)
^{30}Zn	~0 eV (+2)

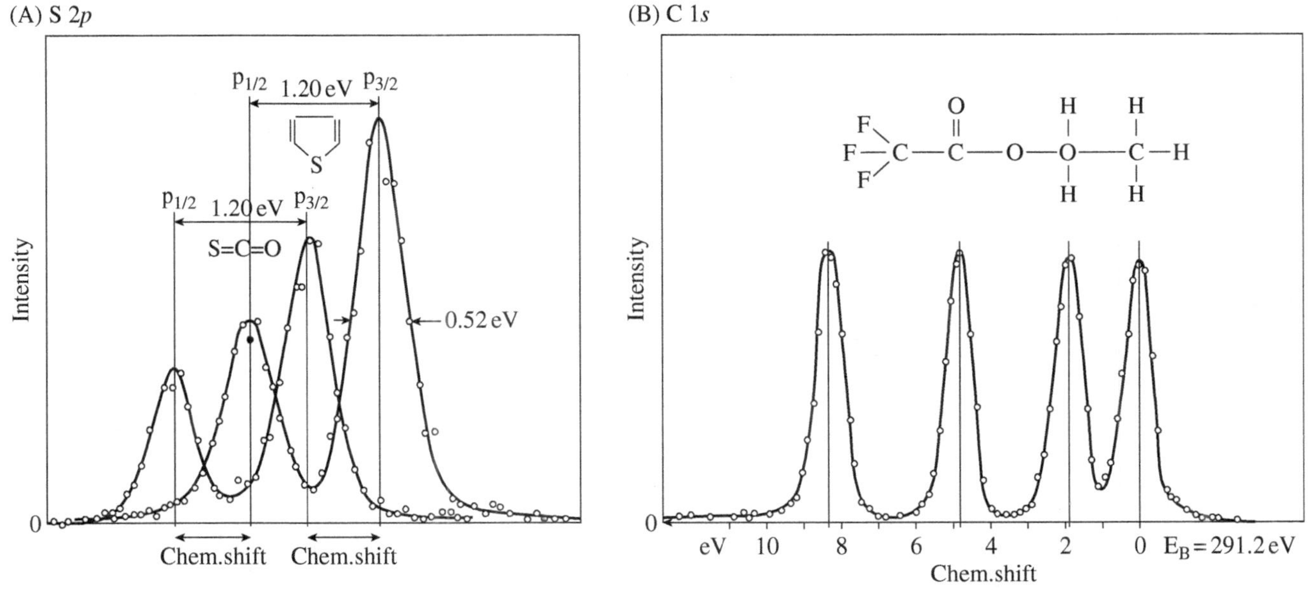

Figure 2.26 (A) Sulfur $2p_{1/2}$ and $2p_{3/2}$ lines from thiophene and carbonyl sulfide. (B) Carbon 1s electron lines in ethyl trifluoroacetate. Gelius et al. [28]/with permission of Elsevier.

do not change. Figure 2.26B exhibits the C 1*s* spectrum of ethyl trifluoroacetate molecule along with its molecular structure. Since this spectrum was collected for the gas phase molecule, the BE was referenced to the vacuum level. The BE of C 1*s* signal from the carbon attached to the oxygen appears at 1.8 eV higher than the carbon in the terminal CH_3 group. The C 1*s* peak of the O=C—O part appears at ~4.8 eV higher than the terminal CH_3 peak. The BE of the C 1*s* peak from the CF_3 group is ~8.2 eV higher compared to that of the CH_3 group.

Figure 2.27 displays the binding energy ranges for carbon, oxygen, silicon, and nickel in their compounds. The XPS Handbook [10] and the NIST XPS Database [26] have compilations of all elements analyzed and published. For carbon atoms in organic compounds, the degree of binding energy increase becomes larger as the electronegativity of covalently bonded atom (nearest atom) increases. As the number of higher electronegativity atoms bonded to C increases, the increase in BE of C 1*s* becomes larger. In the case of oxygen, the variance in Δ_{BE} is relatively small, but a general trend can still be found. For most metal oxides, the BE of O 1*s* is lower than 531 eV. In molecular anions, the BE of O 1*s* is typically around 531~533 eV. In SiO_2 and silicone, it is within 532.5~533 eV. In alumina, it can vary over a larger range near 531 eV. For Si 2*p*, the BE increases with increasing the electronegativity of nearest atoms to which Si is bonded. In silicate, the Si atoms are bonded to oxygen, the same as silica; but the Si 2*p* BE is lower in silicate than silica. This is due to the next nearest neighboring atom effect. In silicates, the next nearest atoms can be metallic ions which vary depending on their compositions. It is interesting to note that the Si 2*p* BE in silicide is slightly higher than that in silicon, even though the oxidation number of Si in silicide is −4. In the case of Ni, the 2*p* BE of Ni^{2+} in nickel sulfide is 853~853.5 eV, while that of Ni^{+} in nickel halides is 855.5~857.5 eV. These examples show that the general trend mentioned for the initial state effects – higher BE for higher oxidation state or with higher electronegativity atoms – should be used only as "an empirical rule of thumb," not as a rule, because exceptions are found frequently.

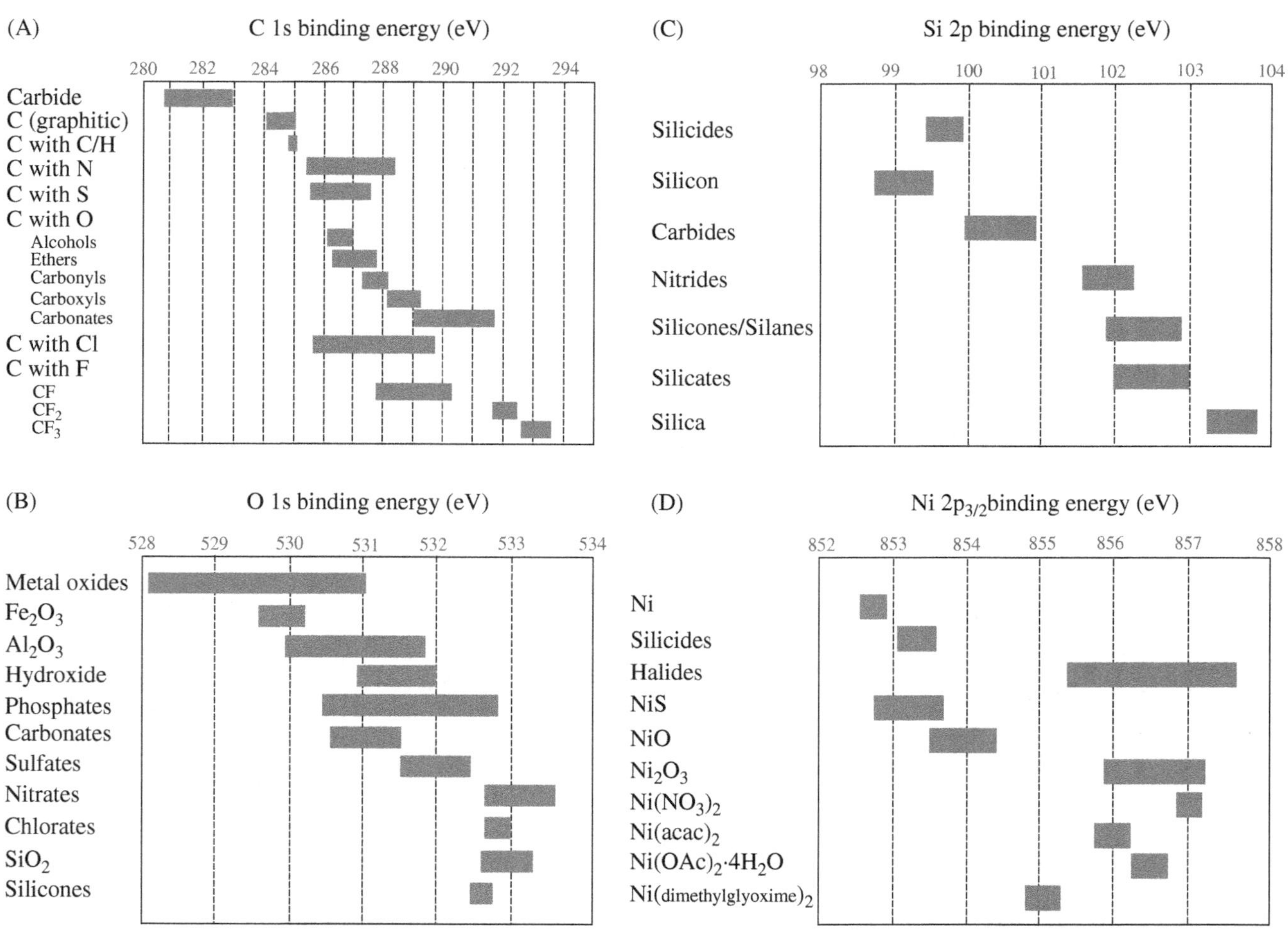

Figure 2.27 Ranges of BE observed various compounds of (A) carbon, (B) oxygen, (C) silicon, and (D) nickel. Redrawn using the data in the XPS Handbook. Adapted from Moulder and Chastain [10].

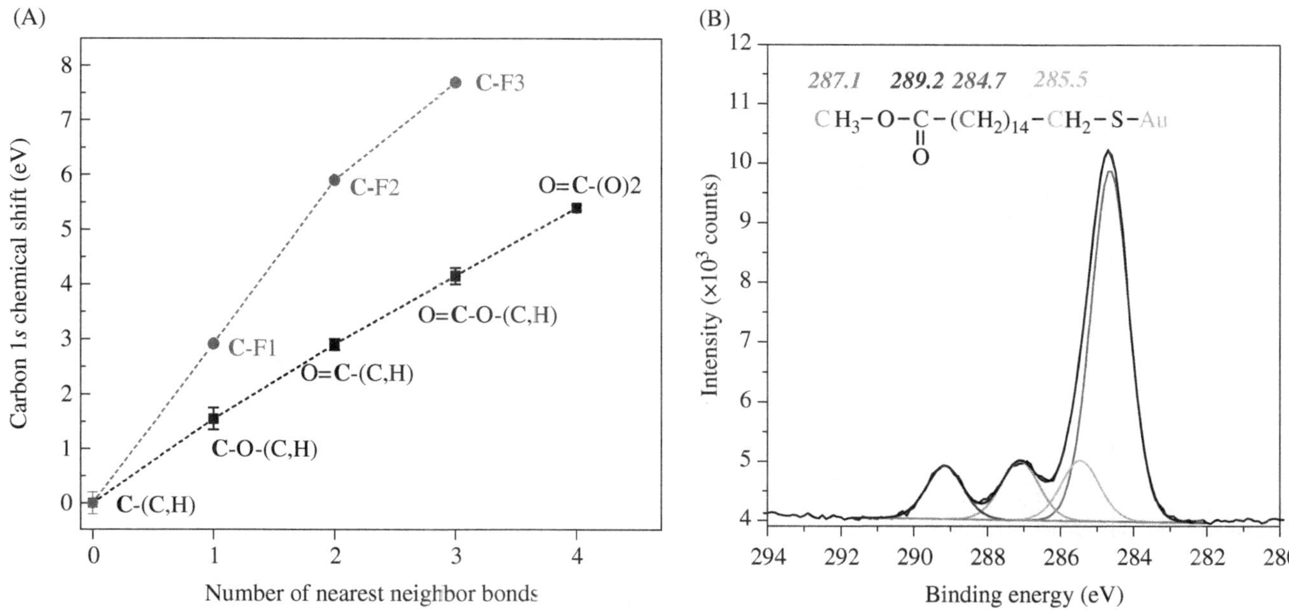

Figure 2.28 (A) Summary of binding energy shift (Δ_{BE}) measured for C 1*s* in organic compounds containing various oxygenated and fluorinated species. (B) High-resolution C 1*s* spectrum of a self-assembled monolayer of 16-mercaptohexadecanoic acid methyl ester on gold. Courtesy of Jeff Shallenberger, Penn State Materials Characterization Lab.

Figure 2.28A plots the degree in binding energy shift (Δ_{BE}) measured for C 1*s* with various organic compounds. As the number of nearest neighboring atoms with higher electronegativity increases, the magnitude of Δ_{BE} gets larger. This kind of empirically observed trend is very useful for data interpretation. For example, the C 1*s* spectra shown in the inset of Figure 2.19 could be fitted with multiple components. Comparing with the trend in Figure 2.28A, one can conclude that each component represents C atoms with different numbers of F atoms. Figure 2.28B plots the high-resolution C 1*s* spectrum of a self-assembled monolayer of 16-mercaptohexadecanoic acid methyl ester on gold. Using the chart in Figure 2.27A and the data in Figure 2.28A, one can assign each component identified in peak fitting to different carbon atoms in the molecule. In Figure 2.28B, the BE of the terminal CH_3-O is 287.1 eV, slightly higher than the typical BE of C–O species; this larger shift is due to the carbonyl group at the next nearest site.

2.3.3 Chemical Derivatization-Assisted Analysis

The binding energy shift charts shown in Figure 2.27 are very handy, but still somewhat limited. For example, the C 1*s* BE shifts for alcohol (C–OH) and ether (C–O–C) groups are almost identical (around 286.5 eV) and practically indistinguishable. Also, the C 1*s* BEs of carboxylic acid (–COOH) and ester (–COO–C) are difficult to distinguish. These are just a few examples where the identification of chemical groups cannot be made based on the core level BE shift only. The XPS-based chemical identification can be facilitated if one combines with chemical derivatization methods. In this approach, chemical reactions specific to the targeted functional group only are employed to differentiate that group from others. Figure 2.29 illustrates the basic concept of the chemical derivatization-assisted XPS analysis [29]. There are many derivatization reactions explored for this purpose. Table 2.2 lists such reactions that can be used to identify and quantify certain functional groups [30].

It should be noted that the choice of derivatization reactions and cross-contamination control are extremely important to use this method reliably. Ideal derivatization reactions are the ones that occur stoichiometrically with 100% reaction yields and high specificities to the target groups and leave elements in the reaction product group that have high detection sensitivity and do not interfere with other XPS peaks of the original surface. And the derivatization reagent should not swell or ingress into the sample and should not contaminate the XPS system during the measurement. If the elemental tag in the product is easily dissociated and desorbed by x-ray, then its signal will decrease continuously during the XPS analysis. Fluorine is well known for its high activity in x-ray-induced desorption [31]. It is also important to note that the derivatization reaction found in the literature for a different sample may not work for your sample to the same efficiency.

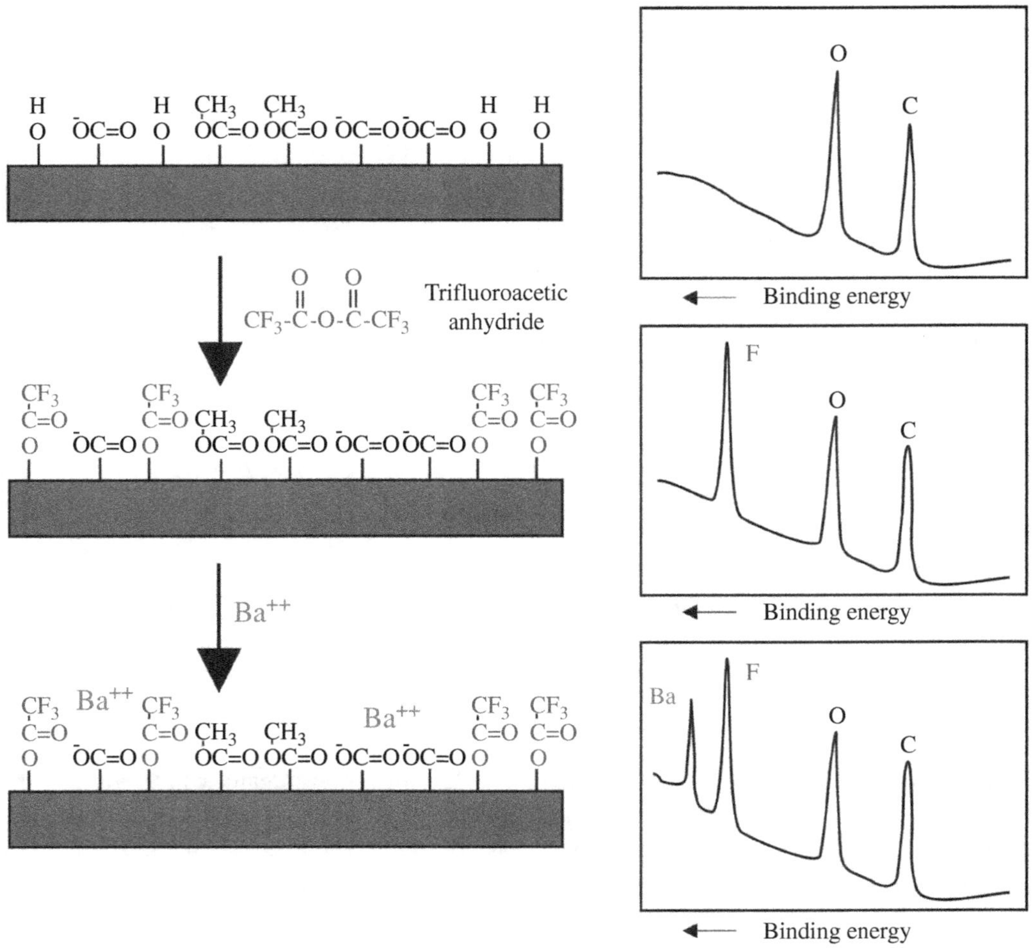

Figure 2.29 A strategy to differentiate hydroxyl, carboxylate, and ester groups at the surface. Here, trifluoroacetic anhydride is used to transform −OH to −OC(=O)CH_3 and Ba^{2+} is used to selectively bound to the −COO^- site. Then, XPS survey scan is used for quantification of F 1*s* and Ba 3*d* signals. Ratner and Castner [29]/with permission of John Wiley & Sons.

Table 2.2 Derivatization reactions used for selected chemical groups [30].

Surface functional group	Derivatization agent
C-OH (hydroxyl)	$(CF_3CO)_2O$, C_3F_7COCl, CBr_3COOH, ClC_6H_4NCO
C=O (carbonyl)	$CF_3CH_2NHNH_2$, $C_6F_5NHNH_2$
C(=O)OH (carboxyl)	$BaCl_2$, $Ba(OH)_2$, NaOH, $TlOC_2H_5$, CF_3CH_2OH, $C_6F_5CH_2Br$
C=C (alkene)	Br_2, OsO_4, $Hg(CH_3CO_2)_2$
$\overset{O}{—C—C—}$ (epoxy)	HCl, CH_3COCl, $(CF_3CO)_2O$
$-NH_2$ (amino groups)	C_6F_5CHO, CS_2, $C_2H_5SCOCF_3$
−SH (sulfide)	$AgNO_3$, $HgCl_2$

Source: Batich [30]/with permission of Elsevier.

2.3.4 Valence Band Analysis

In XPS analysis, core-level spectra are usually used to identify oxidation states of atoms and molecular functional groups in the surface region of the sample. Since the photoemission cross section of core-level electrons does not vary much with the chemical bonding or sample matrix, they are ideal for quantification of relative atomic concentrations (see Section 2.4). Valence electrons are directly involved in bond formation, forming molecular orbitals. Thus, the intensity and energy of the valence band peaks vary drastically depending on their bonding environment. But, since the spectral variance is too large to predict without detailed theoretical calculations, the valence bands are not extensively used for qualitative analysis. Nonetheless, the valence band can be useful to distinguish structural properties that cannot be resolved from the core orbital analysis.

Figure 2.30 compares the C 1*s* and valence band spectra of polyethylene (PE), polypropylene (PP), polystyrene (PS), polyethylene oxide (PEO), and polyvinyl alcohol (PVA) [32]. The C 1*s* spectra of PE and PP are practically identical and indistinguishable because both consist of C–C and C–H bonds only. In the valence band region, there is a unique spectral feature that allows distinction of those two polymers. In the valence band spectrum of PE, two broad bands in the 12–20 eV region are due to alterations of stabilizing interactions of the hyperconjugation character of C 2*s* orbitals in the extended chain conformation that ultimately lead to the large energetic dispersion of the states [33]. In the PP valence spectrum, there is an extra peak at 16.5 eV. This extra peak is due to the CH_3 side group which is not involved in the hyperconjugation along the extended chain. In the case of PS, the main C 1*s* peak position is expected to be ~284.8 eV, but in practice, it can be placed at ~285 eV during the charge neutralization operation. Although there is a small peak at 291.6 eV that suggests the sample contains aromatic functional groups (see Section 2.3.5 for interpretation of this peak), it is difficult to identify the molecular structure from the C 1*s* spectrum only. In this case, the valence band spectral features can be used as a fingerprint and compared with reference databases to confirm that the sample is PS. In the valence spectra of PEG and PVA, there is a relatively strong band at ~26.4 eV that mostly originates from the O 2*s* orbital [34]. These examples demonstrate the use of valence band spectra for molecular identification of polymeric samples [35]. These are relatively simple polymers with known structures; as the molecular composition and structure become more complicated, full spectral interpretations become extremely challenging and require molecular orbital theory calculations.

Other areas where valence band analysis is useful include analyses of electronic structures of metal clusters as a function of cluster size, electronic interactions in metallic alloys, etc. An example of valence band changes probed with XPS is shown in Figure 2.31 [36]. In the XPS spectra of NiCu and NiAu alloys, the peaks near E_F can be attributed to states of largely Ni 3*d* character, while the Cu 3*d* and Au 5*d* peaks are found at higher BE, near their positions in the pure elements. Owing to its

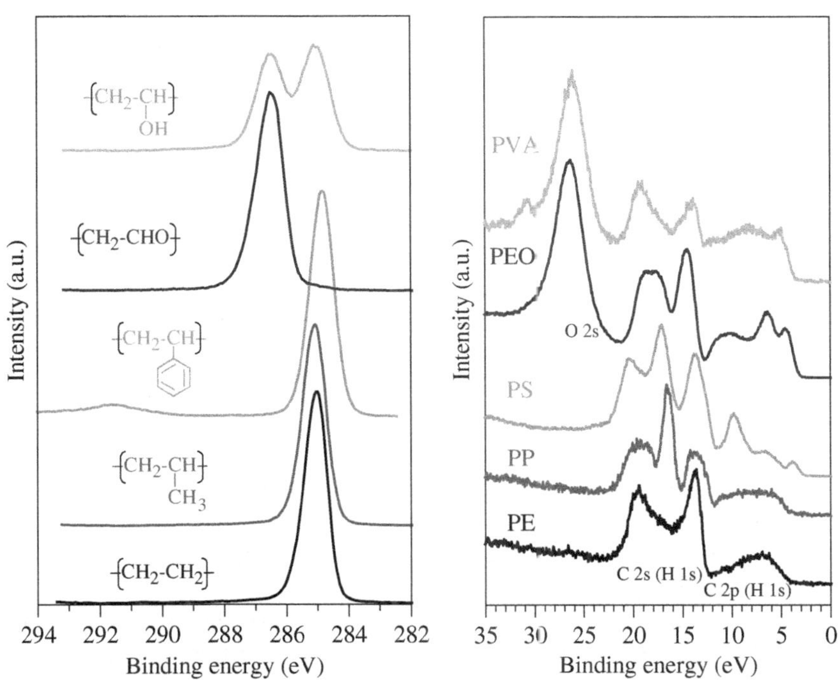

Figure 2.30 XPS spectra of polyethylene (PE), polypropylene (PP), polystyrene (PS), polyethylene oxide (PEO), and polyvinyl alcohol (PVA). Adapted from Beamson and Briggs [32].

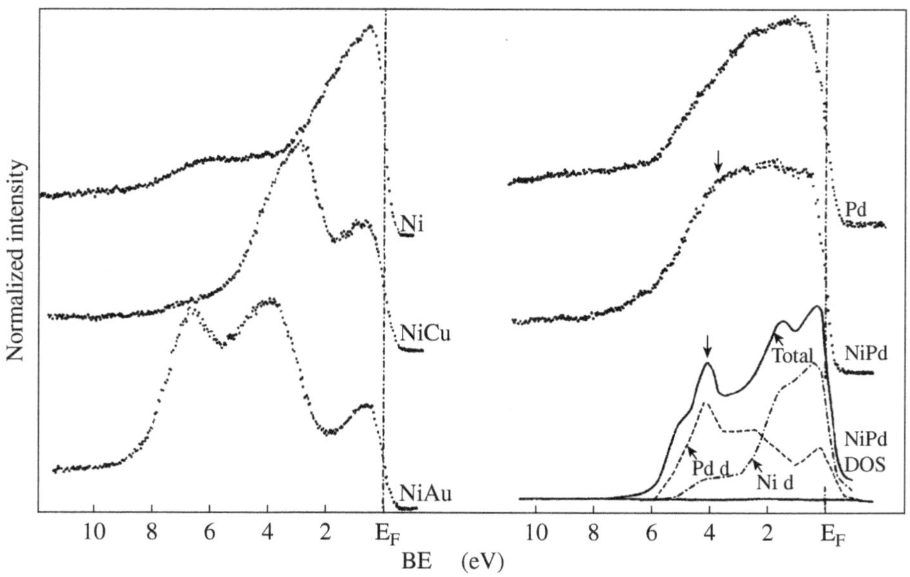

Figure 2.31 XPS valence band spectra from Ni, NiCu, NiAu, NiPd, and Pd. The calculated density of state (DOS) curves for NiPd are also shown. Fuggle et al. [36]/with permission of American Physical Society.

smaller photoelectron cross section, the Ni 3*d* signal is smaller than the Cu 3*d* and Au 5*d* signals. In NiPd, the valence band has a rather square appearance with an enhancement in the region marked with an arrow as compared to that of pure Pd. However, because the BEs of Ni 3*d* and Pd 4*d* levels are similar in the pure elements, it is not possible to separate *a priori* their contributions. The density of state (DOS) calculation showed that the Pd states are more concentrated towards the bottom of the valence band (marked with an arrow) while the Ni states are near the top close to E_F [36].

2.3.5 Final State Effects on BE

To describe the photoemission process, a schematic diagram like Figure 2.6A is widely used. If all other electrons are frozen during the photoemission of a core electron, the initial and final energy states of an atom can be calculated using quantum mechanics theory. If we use the vacuum level (E_{vac}) as a reference and set up an energy balance for the process shown in Figure 2.6A, it will be $h\nu - E_i = KE - E_f$, where E_i is the energy of the initial state and E_f is the energy of the final state with one hole in the core level. These energies can be calculated using the Hartree–Fock method. Rearranging this, we can arrive another expression defining the BE of photoelectron: $h\nu - KE = -(E_f - E_i) = BE$. This is the Koopman theorem. However, the BE calculated with this theorem is not sufficient to explain the experimentally observed BE values of compounds. This is because all electrons in the system are *rearranged during* the photoelectron process, which is accompanied by the relaxation of the system energy. Thus, photoemission of an electron from a core level orbital is a "many-electron" process, even though the x-ray absorption/transition operator may affect only one electron directly [37]. One may think the processes shown in Figure 2.6B,C may occur sequentially after the photoemission. But that view is not accurate; all processes shown in Figure 2.6 occur simultaneously. The spin–orbit splitting in Figure 2.8 is another consequence of the relaxation process of the final state effects. For the 2*p* orbital, $E_f(2p_{1/2})$ and $E_f(2p_{3/2})$ are different; thus, $\Delta_{2p} = E_f(2p_{3/2}) - E_f(2p_{1/2})$.

If the magnitude of the relaxation energy due to electronic rearrangements during the photoemission process varies significantly as the chemical environment of an atom is changed, the BE ranking that would be expected based on initial state considerations (such as oxidation state) can be altered. For example, the ranking of the Co $2p_{3/2}$ BE values is Co^0 (778.2 eV) < Co^{3+} (780 eV) < Co^{2+} (780.3 eV) [10]. Also, as shown in Table 2.1, the $2p_{3/2}$ BEs of Cu^0 and Cu^{1+} are indistinguishable from each other (932.5 eV) [10]. In the case of Ag, the $3d_{5/2}$ BE of Ag_2O is ~1 eV lower than that of Ag^0 [10]. The BE of Ca 2*p* is ~346.5 eV for Ca, CaO, and CaS, but ~348 eV for $CaCl_2$, CaF_2, and $CaSO_4$ [10].

When the oxidation state of an element cannot be distinguished from the BE of core electrons due to the atomic relaxation, it can be done by checking the KE of Auger electrons. In the Auger transition process, polarization effects are more important in determining chemical shifts than are changes in electron density on the atom in the ground state. Thus, the degrees of chemical shift in the BE of a core electron and the KE of the Auger electron involving the same core electron are different, and this phenomenon can be used to determine the chemical and physical state of solid compounds. An example is shown for copper compounds in Figure 2.32 [38]. Here, we can see that the Cu KLL KE varies from ~915 to ~918 eV among

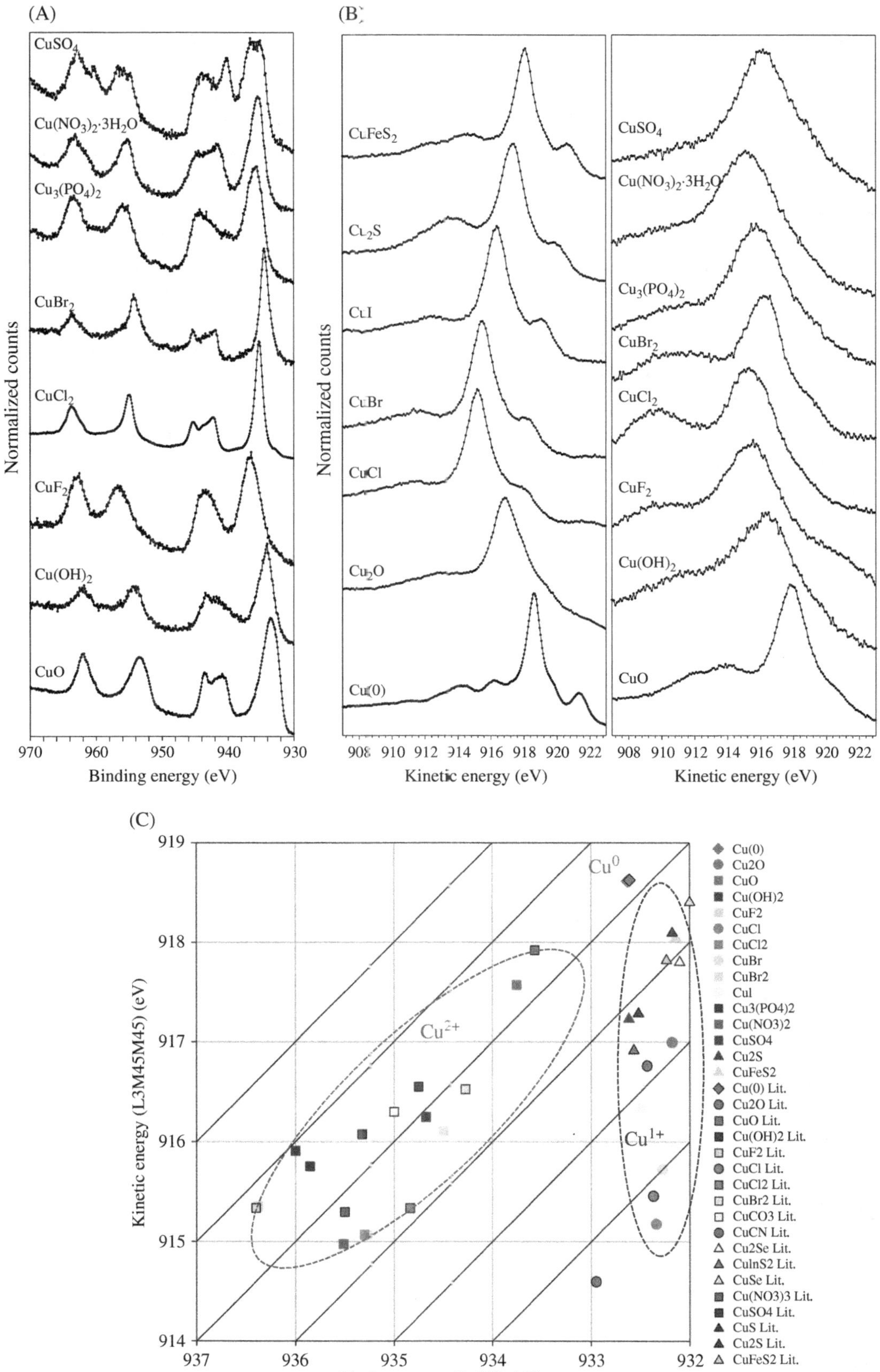

Figure 2.32 (A) Cu $2p_{3/2}$ spectra for Cu^{+2} species. (B) Cu LMM spectra for Cu^0, Cu^{1+}, and Cu^{2+} species. (C) Two-dimensional chemical state plot of Cu LMM KE vs. Cu $2p_{3/2}$ BE for various copper species. Biesinger [38]/with permission of John Wiley & Sons.

different Cu^{1+} compounds, while their Cu $2p_{3/2}$ BEs are almost invariant and identical to the Cu^0 value. In contrast, the Cu KLL KE and Cu $2p_{3/2}$ BE of Cu^{2+} species are along the diagonal direction of the two-dimensional correlation plot (a.k.a. Wagner plot).

The difference between the Auger KE and the photoelectron BE is called the Auger parameter [39]; a modified version of this is the sum of these two energies, which is readily available at the NIST database for various compounds [26]. In Figure 2.27B, we can see that the variance in O 1*s* BE is relatively small among various compounds. In this case, comparing both O 1*s* BE and O KLL KE (and Auger line shape as well) can be useful for the identification of surface chemical states [40]. The Auger parameters are independent of sample charging, as both peaks would shift by the same amount.

In Section 2.1.3, we learned that the *s*-orbital (with $l = 0$) does not exhibit spin–orbit splitting because the orbital magnetic momentum is zero. In transition metals containing a large number of unpaired electrons in the valence levels, there can be additional exchange splitting or electrostatic splitting of the core-level peaks due the high magnetic field induced collectively by those electrons in the final state. For example, ${}^{25}Mn^{2+}$ has the electronic structure of $1s^2\ 2s^2\ 2p^6\ 3s^2\ 3p^6\ 3d^5\ 4s^0$. The five unpaired electrons in the 3*d* orbitals have the same spins, thus, their spin state (S) corresponds to six in the initial state. Upon photoemission of a 3*s* electron, the remaining 3*s* electron is also unpaired. There are two possibilities – the spin of this electron in $3s^1$ can be parallel or antiparallel to those of the 3*d* electron, which leads to $S = 7$ or 5 in the final state. The $E_f(S = 7)$ and $E_f(S = 5)$ are different, thus, the Mn^{2+} 3*s* peak splits into two peaks. A process like this is called the "multiplet splitting." This effect is frequently observed for elements with high magnetic moments. The examples of Mn and Fe are shown in Figure 2.33 [41]. Similar effects are often observed for Cr 3*s*, Co $2p_{3/2}$, Ni $2p_{3/2}$, and the 4*s* levels of the rare earth elements.

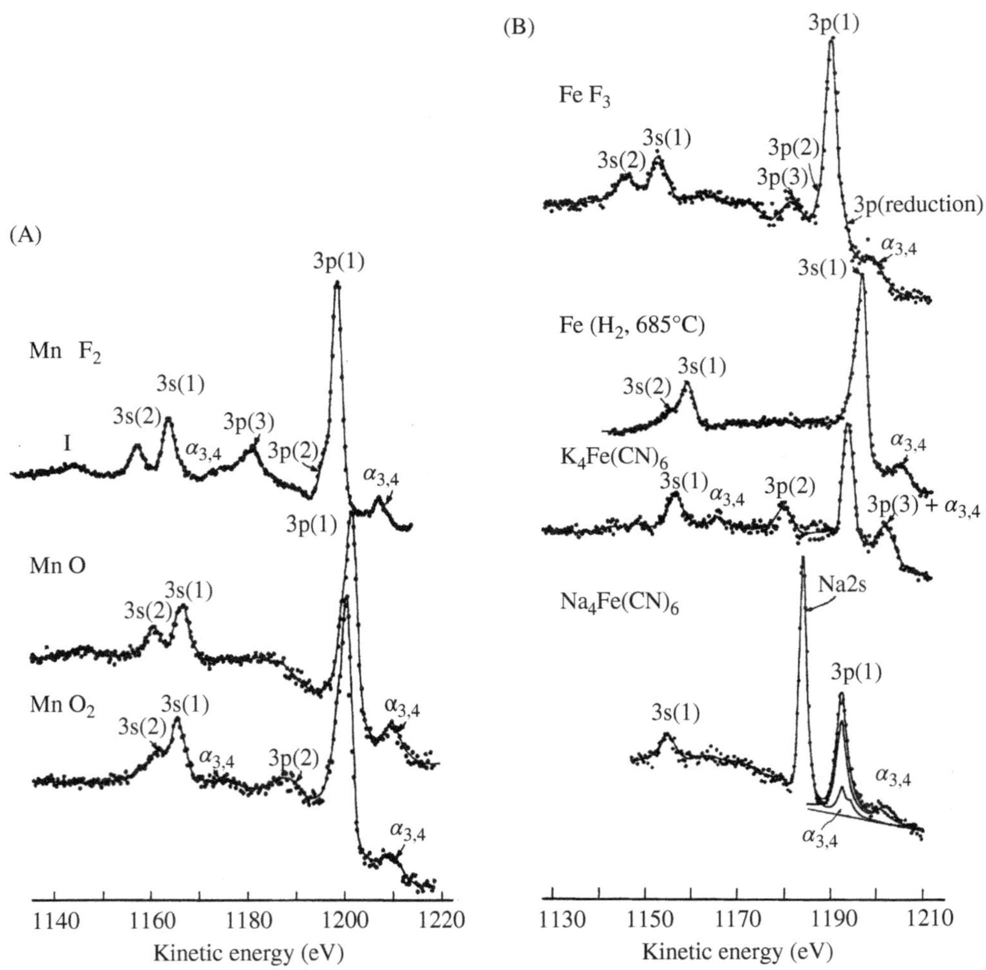

Figure 2.33 (A) Photoelectron spectra of Mn 3*s* and 3*p* electrons from MnF_2, MnO, and MnO_2. (B) Photoelectron spectra of Fe 3*s* and 3*p* electrons from FeF_3, Fe metal, $K_4Fe(CN)_6$, and $Na_4Fe(CN)_6$. The x-ray source was Mg Kα and the spectra were plotted in the KE scale. Fadley and Shirley [41]/with permission of American Physical Society.

Note that multiplet splitting is not expected to be seen always. For example, multiplet splitting occurs for Ni $2p_{3/2}$ of NiO, but not for $Ni(OH)_2$.

To the valence electrons, the loss of a core electron by photo-emission is equivalent to an increase in the nuclear charge. This perturbation gives rise to a substantial reorganization of the valence electrons which may lead to an electronically excited final state where a valence electron is promoted to a higher unfilled level. Since the final state energy will be higher in this *two-electron process* (i.e., photoemission of one core-level electron and electronic excitation of one valence electron), the KE of the detected photoelectron will be lower by the energy of the secondary excitation process. This is called the "shake-up" process. In Figure 2.30, we saw a small peak at 291.6 eV in the C 1*s* spectrum of PS. This ~6.5 eV difference corresponds to the electronic excitation of the π orbital of the aromatic side group ($-C_6H_5$ in PS) to the π^* anti-bonding orbital. Most aromatic hydrocarbons (and even heterocyclic compounds) exhibit this $\pi \rightarrow \pi^*$ shake-up peak in their C 1*s* spectra. In Figure 2.32A, the doublet-like features in 938–946 eV between the Cu $2p_{1/2}$ and $2p_{3/2}$ peaks and the broad feature in 960–966 eV above the Cu $2p_{1/2}$ peak are also shake-up features of Cu^{2+} species in different chemical environments. These shake-up satellites are not observed for Cu^{1+}.

If outgoing photoelectrons travel through a free-electron-like band of the material, free electrons in such a band can oscillate back-and-forth in response to the electric field generated by the photoelectrons. Such oscillations are called plasmons. Like the allowed energy states for a free electron confined in a potential box (which is one of the first subjects taught in undergraduate-level quantum mechanics course), plasmon energies are quantized. The outgoing photoelectron (and Auger electron as well) can excite not only the first excited state but also higher-order plasmonic states, and lose KE accordingly. Thus, as shown in Figure 2.34, the plasmonic excitation bands appear as a series of equally spaced peaks of decreasing intensity on the low KE (high BE) side of the main peak. There can be additional plasmons, called surface plasmons. The plasmon excitation bands are often observed for metals and semiconducting materials.

Note that the final state effects studied here – atomic relaxation, multiplet splitting, shake-up, and plasmon loss structure, etc. – make the 2*p* orbital analysis of first-row transition metal compounds quite complicated and challenging. Also, some of these effects overlap, making peak shapes asymmetric. In addition, the background change associated with the inelastic scattering of photoemission of 2*p* electrons of transition metals is quite large. All these make determination of oxidation states of transition metals (and quantification of different oxidation states) quite complicated (see Section 2.4). A few examples are shown in Figures 2.35 [42, 43]. The final state effects can be significantly less for different orbitals; in that case, analyzing the peaks with insignificant final state effects, although their intensities may be smaller than the most prominent peak, can be more amenable as long as they have sufficiently high S/N ratios [44]. Since recognizing these effects is somewhat complicated, unexperienced users are highly recommended to consult with XPS experts, rather than simply comparing the measured peak positions with literature values found for similar samples without knowing whether the chosen literature has any error in peak identification [45].

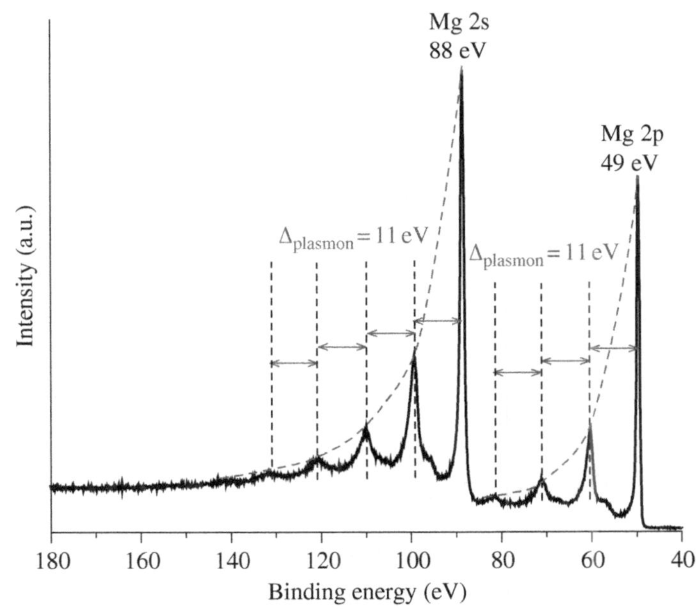

Figure 2.34 XPS spectrum of metallic Mg showing plasmon excitation bands.

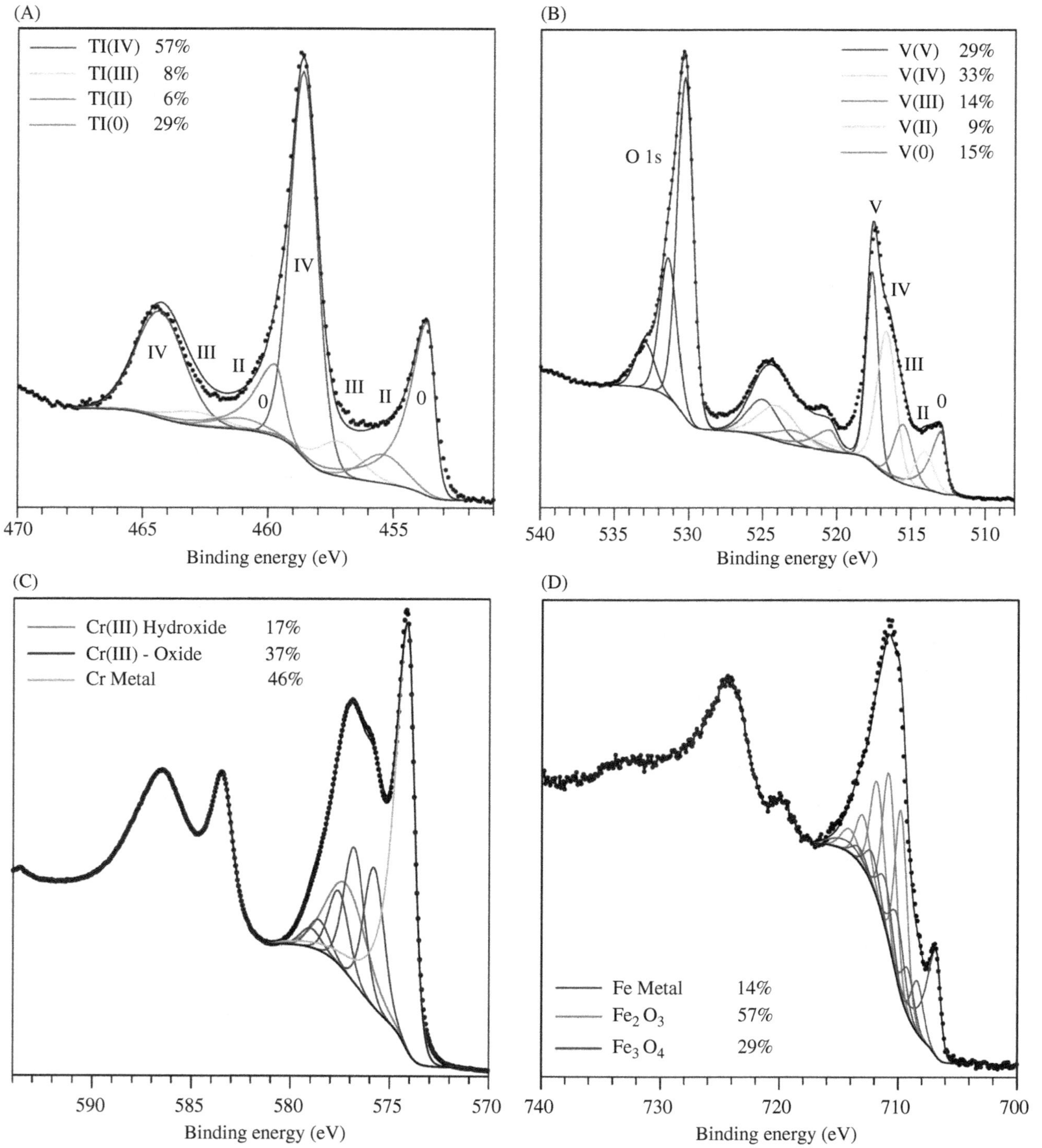

Figure 2.35 (A) Ti 2*p* spectrum of a heat-treated Ti-apatite composite using peak fittings. (B) V 2*p* (and O 1*s*) spectrum of a mixed vanadium oxide sample. Biesinger et al. [42]/with permission of Elsevier. (C) Curve-fitted Cr 2*p* spectrum from a sample of a vacuum sputter deposited decorative chromium-plated plastic. (D) Curve-fitted Fe 2*p* spectrum of a mixed metal/oxide system of Fe-based nanoparticles dispersed on a glass surface. Biesinger et al. [43]/with permission of Elsevier.

2.4 Quantitative Analysis with XPS

We now know how to identify elements present at/in the surface and their oxidation states and bonding to nearest atoms (Section 2.3). The next step is to do quantification of each element within the information depth from the surface (Section 2.1.4). The quantitative XPS analysis consists of two steps – the first is curve fitting to deconvolute peaks of interest with expected components and their areas and the second is to convert the relative ratios of the calculated areas to relative concentrations. The concentrations are typically presented in the atomic-% unit, as shown in Figure 2.35A,B; but it is also possible to calculate for relative abundance of model or expected chemical species, as shown in Figure 2.35C,D. Note that curve fitting is an *art*, meaning it is often not straightforward. The sole goal of peak fitting algorithms is to minimize the fit error; but the fitting program does not know the physics behind peak deconvolution. Thus, there is no guarantee that the fit result will be physically meaningful (although it may have mathematically the smallest error). So, in every single step, a judgment should be made on what is reasonable to do and what should not be done. Implementing proper fitting methods and constraints and judging if fit results are reasonable or not require fundamental knowledge of the photoelectron emission process (Section 2.1), how the emitted electrons are detected as a function of KE (Section 2.2), and how the detected signals can be interpreted (Section 2.3). There are many ways to fit the raw data shown in Figure 2.35; but not all of them would physically make sense. Getting the smallest fitting error should not be used as the criterion to judge the quality of fit result. This section describes the key principle of XPS data processing. Note that the material covered in this section is just a starting point for beginners to grasp the curve fitting process; further discussion with experts and practicing with a lot of real data are required to master it.

The content described in this section can be learned most efficiently by doing data processing, not by just reading and memorizing it. For practice, raw XPS data files of various materials can be downloaded from *XPSSurfA*, an open collaborative XPS data repository [46]. Note that every instrument has its own calibration parameters. Unless the proper calibration parameters are entered into the XPS data processing program, the accuracy of analysis result cannot be guaranteed. Although this sounds obvious, many users often do not pay attention to it and report inaccurate (sometime incorrect) analysis results.

Among various XPS data processing programs, *CasaXPS* software is widely used since it can read raw data from many commercial XPS systems and already has various mathematical functions and options embedded for data processing. So, the examples shown here are done with *CasaXPS*. The program can be downloaded for free trial from the *CasaXPS* website [27]. The website also provides rich information on how to do XPS peak fitting properly.

Other freeware designed for XPS data processing can be found on the Internet. Those programs work equally well as long as the user knows how the program works and what constraints can be used properly in data fitting. Not to mention, there are many websites delivering very useful tips and guides for curve fitting procedures and reference materials. One such site is *XPS Reference Pages* by the Surface Science Western lab [47], which provides curve-fitting tips and *CasaXPS* tips pertaining to frequently-asked element-specific questions.

2.4.1 Background Subtraction

The curve fitting starts with the background correction. The C 1*s* spectra shown in Figure 2.30 have little change in the background level before and after the peak. In that case, a simple straight line connecting the average levels of the baselines before and after the peak would be sufficient. In the examples shown in Figure 2.35, the background level before and after the peak varies substantially. Then, would it be sufficient to draw a linear interpolation between two points before and after the peak? That could be one way of doing it, but it is not the appropriate way of removing the background in the XPS data processing. Let us study how to do peak identification and background subtraction using the example shown in Figure 2.36. This is a screen shot of the *CasaXPS* program with the survey spectrum of a titanium reference sample.

The first step is to identify the elements present at/in the surface. Using the "Library" function in the *CasaXPS* program, all elemental lines are marked at their "expected" positions. In doing so, the program suggests three other possible peaks near the one marked as Ti 2*s* (562 eV); they are Pd 3$p_{1/2}$, Ta 4*s*, and Cu LMM (not shown here). Then, how do we know the peak at 562 eV is from Ti 2*s* only and not from others. In this case, other possible peaks from Ti, Pd, Ta, and Cu should be checked. Only Ti shows all peaks in the spectrum and no peaks from the other three elements can be found in the spectrum. So, the presence of Pd, Ta, and Cu can be ruled out. Other elements that we can identify in the spectrum are Ar, C, and O. Argon must be due to implantation of Ar^+ ion into the sample during the sputter cleaning process. Carbon is likely from the adventitious contaminants that were not fully removed by Ar-sputtering. Or, it can be due to re-adsorption of organic contaminants present in the chamber over time. Oxygen must be due to the native oxide layer remaining on the surface after sputtering. These could be completely removed upon further sputtering.

The next is to define the spectral ranges to analyze. In the case of Ti, there are five peaks identified, which are Ti LMM, Ti 2*s*, Ti 2*p* ($p_{1/2}$ and $p_{3/2}$), Ti 3*s*, and Ti 3*p*. For the quantification purpose, we choose only one peak for each element that is representative. Obviously, our choice is the one with the largest intensity. But, if the biggest peak overlaps with another element, then choosing an alternative one without the overlap would make sense. In this example, the gray background regions are the ranges chosen for quantification of each element. For the Ti 2*p* region, we have expanded the background to

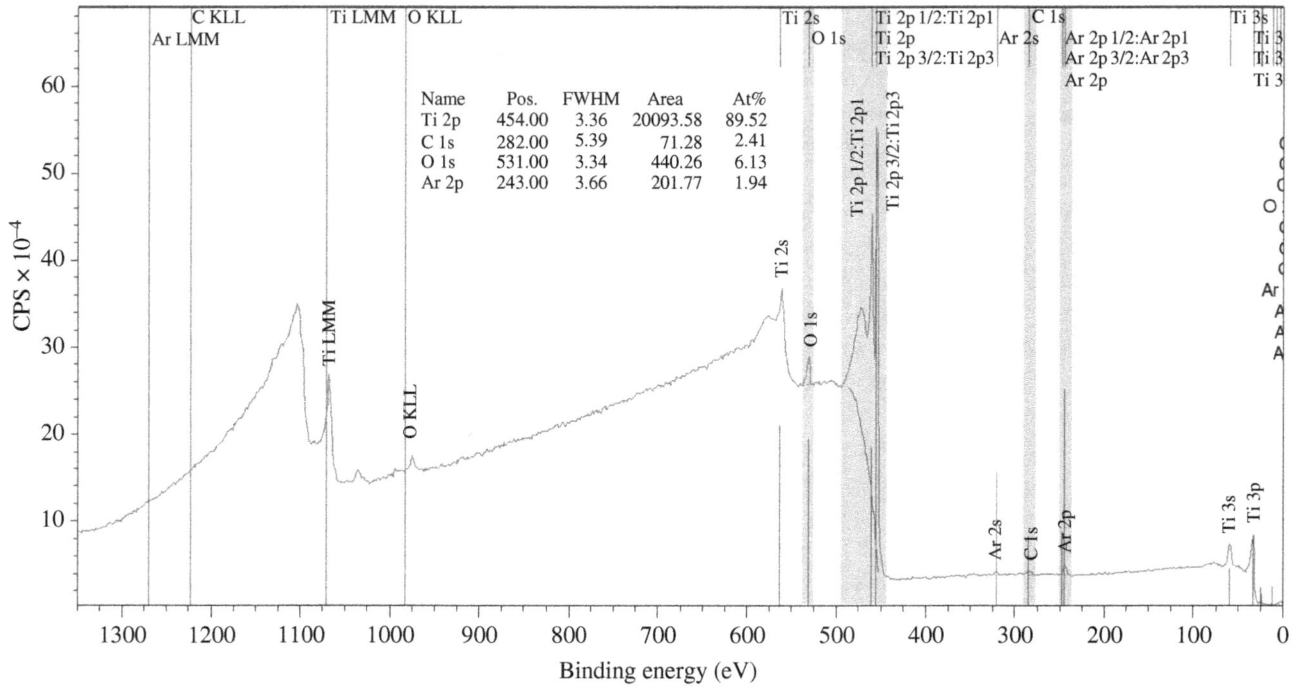

Figure 2.36 Survey spectrum of a titanium metal lightly sputtered with Ar^+ ion. Adapted from Jeff Shallenberger.

499 eV, so the broad component centered at ~470 eV is included. This is the loss peak originating from the photoemission of Ti 2*p* electrons. Since the loss peak is also part of the Ti element in the sample, it should be included. If it is excluded, then the peak area of the Ti 2*p* region decreases to one-third of the value shown in Figure 2.36.

Once the spectral ranges are defined for representative peaks for all elements identified, the baseline needs to be defined for each curve to calculate the peak area, which is discussed next in this section. Once the areas of the representative photoelectrons or Auger electron peaks of all elements are determined (Section 2.4.2), each area value is divided with the relative sensitivity factor (RSF) corresponding to each peak chosen for each element. This gives the relative concentration of each element in the atomic-% unit, as shown in Figure 2.36. In the *CasaXPS* program, the RSF data are saved in the "CasaXPS.lib" file. The default RSF values included in the zip file downloaded from the website [27] are generally good; but they are not tuned for individual instruments that are designed and operated differently. The RSF library file should be updated accordingly. The factors affecting RSF values are discussed in Section 2.4.3.

In the O 1*s*, C 1*s*, and Ar 2*p* regions, the baseline does not change much from the lower BE side to the higher BE side (i.e., from the higher KE side to the lower KE side) of the peak. Thus, we can simply draw a linear interpolation line and use it as the baseline. In the case of Ti 2*p* region, the baseline in the higher BE side is significantly higher than that in the lower BE side. Recall that the origin for the increase in background signal in the higher BE energy side is due to inelastic scattering of outgoing photoelectrons (Figure 2.5). Then, it would conceptually make sense to say that the degree of background signal increase at a given BE would be proportional to the amount of photoelectron signal at the higher KE side (i.e., lower BE side) of the point under consideration. With this concept, the Shirley algorithm iteratively computes the background contribution using a weighted average based on the peak area to the lower and higher KE sides of each point under consideration [48]:

$$S(E) = I_L + (I_H - I_L)\frac{\int_L^E (I(E') - S(E'))dE'}{\int_L^H (I(E') - S(E'))dE'} \tag{2.14}$$

where $S(E)$ and $I(E)$ are the background and signal intensity as a function of energy, respectively, and I_L and I_H are the background intensity at the lower (L) and upper (H) boundary of the region under consideration. An example of using the Shirley background is shown for the high-resolution spectrum of the Ti $2p_{1/2}$ and $2p_{3/2}$ region (without including the loss peak just for simplicity of display) in Figure 2.37. The Shirley background is relatively simple and thus widely used in practice.

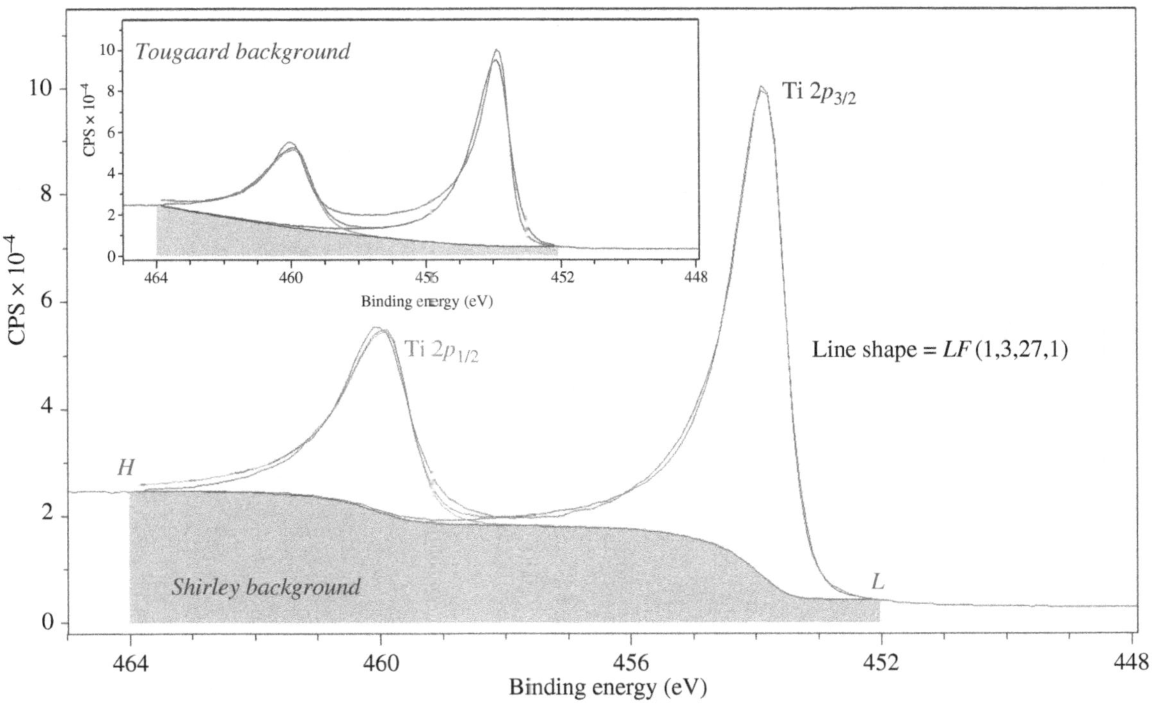

Figure 2.37 High-resolution spectrum of the Ti 2p region of the same sample as in Figure 2.36 with Shirley background correction and curve fitting. The Tougaard background correction is shown in the inset for comparison. The upper and lower limits of the region are marked with H and L. The line in the full range is the raw data, and the additional lines above the shaded region are fit results for the $2p_{1/2}$ and $2p_{3/2}$ components.

Another background model is the Tougaard background which is a physics-based model considering the energy loss cross section. The Tougaard background, $T(E)$, is also calculated from the measured spectrum with two fitting parameters (B, C) using the following equation [49]:

$$T(E) = I_L + (I_H - I_L)\int_E^{\infty} \frac{B\,(E' - E)}{\left(C + (E' - E)^2\right)^2} I(E')dE' \tag{2.15}$$

The merits of this method include the concept of calculating the probability for an electron at energy offset from the main peak to undergo a loss event and appear as a contribution to the background. But it has additional fitting parameters which complicates in practical use, particularly if there are overlapping peaks. The titanium reference example considered in Figure 2.36 does not work as good as the Shirley background; after doing curve fitting with the same parameters, the result with the Tougaard background (inset of Figure 2.37) is not as good as that with the Shirley background.

Many aspects of background signals are not fully understood. The Tougaard background based on the electron energy loss probability does not generate the Shirley background shape that works better, which suggests that the low KE electrons responsible for the Shirley background may not purely due to inelastic scattering of photoelectrons. Figure 2.38A shows that even photoelectrons with similar kinetic energies from the same element (so with similar IMFP lengths) can have different background shapes [50]. Then, which method is the best method? The rule of thumb is that the method that gives the atomic concentration or density closer to the theoretically expected value is the better one or at least the more appropriate one [51]. In other words, the background appropriate for composition calculations can be substantially different for different elements; an example is shown in Figure 2.38B [50]. In the case of NAP–XPS, photoelectron scattering in the gas phase also contributes to the background (in addition to extra peaks as shown in Figure 2.21B) [52].

2.4.2 Peak Shape and Width

Once the background is properly modeled, the next step is to fit the data to calculate the area under the peak. There are several factors contributing to the peak shape in the measured spectra. The first is the line shape of the incident x-ray. The

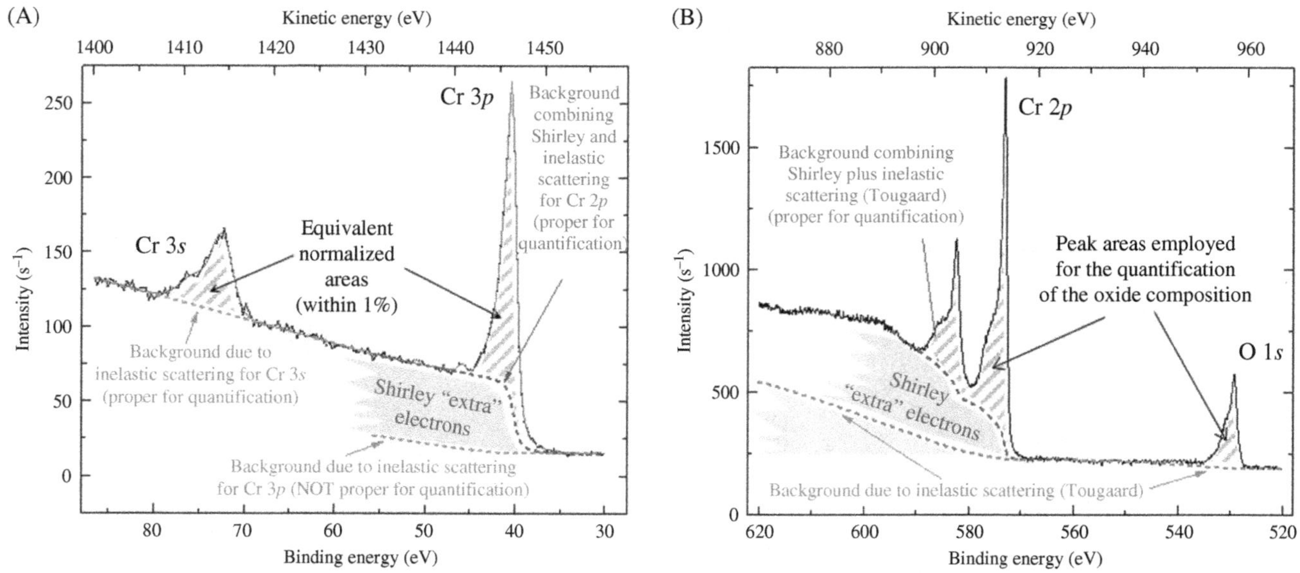

Figure 2.38 (A) Cr 3*s* and 3*p* spectrum from metallic chromium. (B) Cr 2*p* and O 1*s* spectrum from oxidized chromium. Herrera-Gomez et al. [50]/with permission of John Wiley & Sons.

natural line shape of x-ray is governed by the timescale of the electronic transition emitting x-ray photons (Figure 2.6C), which follows the Lorentzian profile [27]:

$$L(E; E_o, w) = \frac{1}{1 + 4\left(\frac{E - E_o}{w}\right)^2} \tag{2.16}$$

where E_o and w are the center and width (FWHM) of the Lorentzian profile (see Section 5.1.4 for the discussion of line shape and FWHM). The photoelectron emission process (Figure 2.6A) will also follow the Lorentzian profile. The emitted electrons are filtered with the energy analyzer as a function of KE. The energy filtering will impose the Gaussian line shape [27]:

$$G(E; E_o, w) = \exp\left[-4 \ln 2\left(\frac{E - E_o}{w}\right)^2\right] \tag{2.17}$$

For the same w (FWHM), the Lorentzian shape spreads more broadly and the Gaussian shape is centered more near E_o. The interplay between these two affects the actual line shape. Figure 2.39 shows how the line shape varies with the analyzer condition. When the pass energy (PE) is set high, then the energy resolution will be low (see eq. (2.11)). This means the peak shape of the survey spectrum will be mostly governed by the Lorentzian profile. In contrast, the peak shape of the high-resolution spectrum obtained with a low PE setting will be mostly governed by the analyzer, thus assuming the Gaussian profile. Thus, the line shape depends on not only intrinsic factors associated with photoemission but also extrinsic factors such as the spectrometer energy resolution.

In practice, the peak shape is often described by the sum (GLS) or product (GLP) of these two weighted by a mixing parameter (m) [53]:

$$GLS(E\!: E_o, w, m) = \exp\left[-4 \ln 2\left(\frac{E - E_o}{w}\right)^2\right](1 - m) + m\Big/\left[1 + 4\left(\frac{E - E_o}{w}\right)^2\right] \tag{2.18}$$

$$GLP(E\!: E_o, w, m) = \exp\left[-4 \ln 2\left(\frac{E - E_o}{w}\right)^2 (1 - m)\right]\Big/\left[1 + 4m\left(\frac{E - E_o}{w}\right)^2\right] \tag{2.19}$$

Mathematically, the GLS function represents the intermediate behaviors between the pure Gaussian and the pure Lorentzian. The GLS function is closer to the Voigt function which is a convolution of Gaussian and Lorentzian functions. For low values of m, the GLP function is a rather close approximation of the pure Gaussian shape. Because of the more compact nature of the Gaussian function, the GLP function has very little tails and a larger fraction of area weighs near E_o.

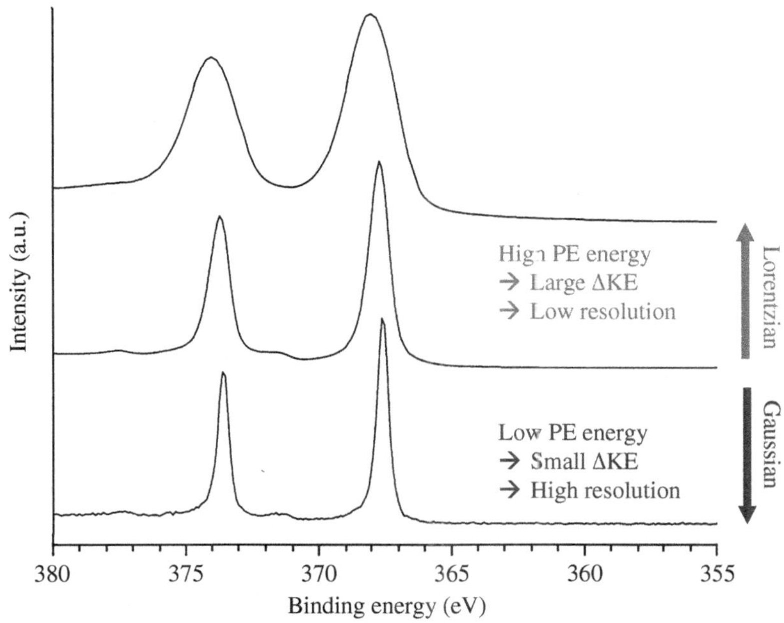

Figure 2.39 Ag 3*d* spectra collected with different CHA pass energies.

In the *CasaXPS* program, the GLS and GLP functions are defined with *SGL*(p) and *GL*(p), respectively, with $p = m \times 100\%$. So, *SGL*(100) and *GL*(100) are purely Lorentzian, and *SGL*(0) and *GL*(0) are purely Gaussian. A good strategy to find the optimum m value (p in *CasaXPS*) is to adjust it until the line shape matches well the low BE side tail of the peak [54]. This approach can be justified since the low BE tail (i.e., the high KE side) has the least influence from the inelastic scattering.

Let us try this method to process the XPS data of polymethylmethacrylate (PMMA). Figure 2.40 displays the survey spectrum of PMMA. Within the spectral range, we can see C 1*s*, O 1*s*, and O KLL peaks. In the valence spectrum, we can see four bands in the C 2*s* region (including a small one at ~10 eV). The O band predominantly contributed from O 2*s* (between 22 eV and 28 eV) appears to be doublet, which can be attributed to two different oxygens in the ester group. The atomic

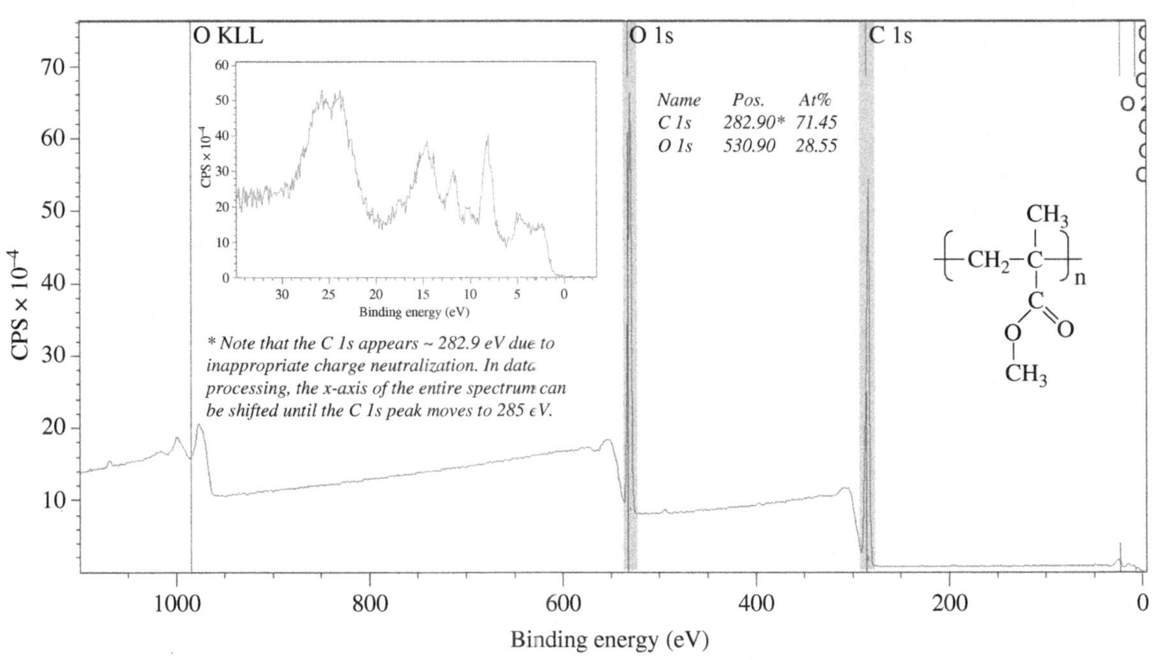

Figure 2.40 Survey and valence (inset) spectra of PMMA. Plotted using the Beamson and Briggs [32]/with permission of Elsevier.

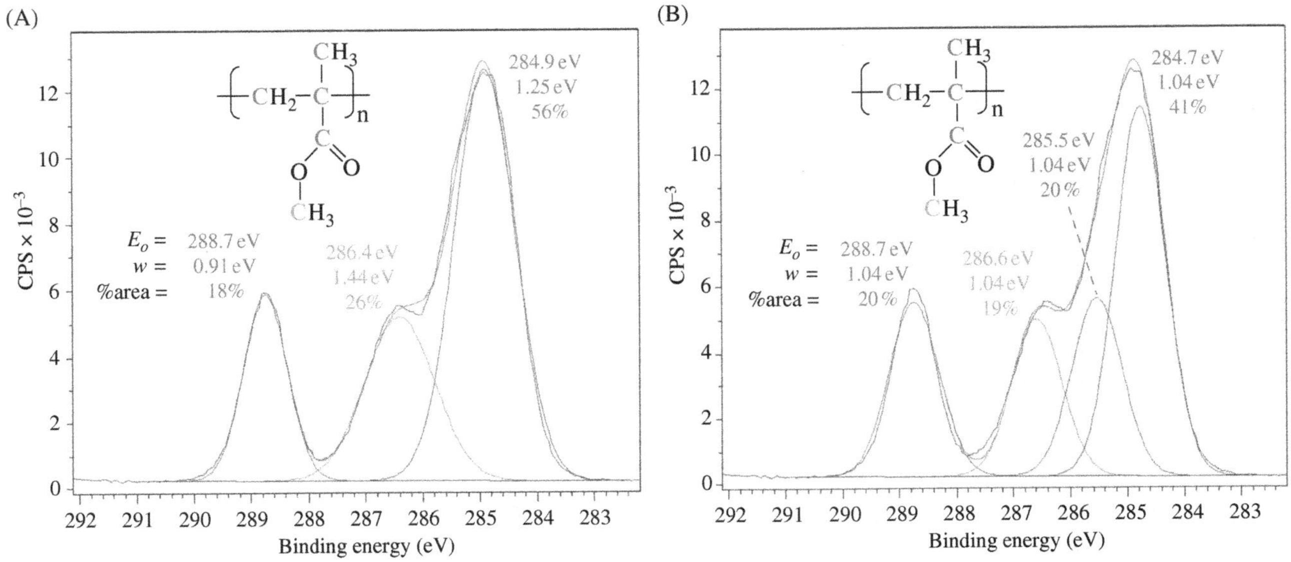

Figure 2.41 Curve fit results of C 1*s* spectrum of PMMA (A) without any constraints and (B) with a constant FWHM. The GL(10) was used for the line shape in both fitting.

concentration of C and O calculated from the C 1*s* and O 1*s* peak areas divided by their RSF values is 71.5% and 28.6%, which are consistent with the chemical composition of PMMA (C:O=5:2). In this example, choosing which baseline – linear, Shirley, or Tougaard – does not matter since the background change across the C 1*s* and O 1*s* peak is small.

Figure 2.41 compares the curve fitting results of the high-resolution C 1*s* spectrum of PMMA obtained with the *CasaXPS* program. The line shape was set to *GL*(10), which corresponds to the *GLP* function with $m = 0.1$. This line shape was chosen to match the data in the low BE (283–284.5 eV) region. Then, the same line shape was imposed on all other components. First, the data were fitted with three components without any constraints. The fit result in Figure 2.41A looks quite good from the error minimization point of view. But there are two problems we can see here. The FWHM of all three components are quite different and we cannot justify this large variance. Moreover, the relative area ratio of three components does not match with the molecular structure. The 284.9 eV peak is attributed to three carbons having the C–C bonds. The 286.4 eV and 288.7 eV peaks are from the O–CH_3 and O–C=O groups, respectively (see Figures 2.27A and 2.28A). Their atomic ratio should be 3:1:1, but the fit result (56%:26%:18%) does not match with this expected ratio. In Figure 2.41B, an additional component was added and the FWHM of all peaks were set to be the same. The fit result may not look as good as the one shown in Figure 2.41A, but this four-component fit result makes much more sense physically. The fourth component at 285.5 eV is the carbon atom in the backbone bonded to the ester carbon. The 0.8 eV shift in the BE from the other C–C carbon (284.7 eV) is due to the next-nearest atom effect (see Section 2.3.2). The relative intensity ratio of these four components closely matches with the molecular structure (color coded in Figure 2.41B). The fit of the 288.7 eV peak in Figure 2.41B is not as good as the one in Figure 2.41A. This suggests that the FWHM of the C 1*s* peak must be slightly smaller for the O–C=O group than the other three groups. If the FWHM of the 288.7 eV is reduced to 0.92 eV, then the curve fitting is improved; but the relative atomic ratio does not change (so, still good).

For the peaks that look symmetric, the line shapes based on the sum or product of Gaussian and Lorentzian functions work quite well. Since these functions are easy to visualize, they are widely used. However, photoelectron peaks do not always conform to these mathematical forms. For example, one can see that the Ti $2p_{1/2}$ and $2p_{3/2}$ peaks in Figure 2.37 have relatively long tails in the high BE side, that is, they are highly asymmetric. If those peaks are to be fitted with the *SGL*(*p*) or *GL*(*p*) options in *CasaXPS*, one must use two to three components for each spin–orbit split peak. One may argue the possibility of having Ti^{2+} and Ti^{4+} species to justify these additional components. But recall that the sample was a titanium metal. Based on the O:Ti ratio in the survey spectrum (Figure 2.36), the presence of large amount of oxidized titanium species cannot be justified (Figure 2.37). This means that the line shape of the Ti 2*p* peak of metal should be asymmetric. Such asymmetric shapes are caused by the final state effects.

To deal with asymmetric line shapes, an exponential decay function can be multiplied to the high BE side of the Gaussian–Lorentzian functions. Although this is convenient, it is an arbitrary modification. A more thorough line shape function was proposed considering many-electron interactions of metallic conduction electrons with accompanying deep holes in the final state [55]. But this requires the evaluation of the lifetime width of the initial core hole state, and the fit result is difficult to compare with parameters determined by other methods. An alternative approach is to use different powers for the left and right side of the peak. This method is called the Asymmetric Lorentzian, *LA*:

$$LA(E\colon \alpha, \beta) = \begin{cases} [L(E_o, w, m)]^{\alpha} \text{ for } E \le E_o \\ [L(E_o, w, m)]^{\beta} \text{for } E > E_o \end{cases} \tag{2.20}$$

By independently varying α and β, one can control how fast the fit curve decays separately in each side of the peak. Depending on the values chosen for α and β, the LA line shape can define the functional forms with infinite area. For this reason, a further modification is made to ensure that the line shape defines a finite area. This method is called the Finite Lorentzian function, $LF(E: \alpha, \beta, w', m)$, where the specified values of α and β are forced to increase to a constant value via a smooth function determined by the width parameter w'. Here, the last term m defines the width of Gaussian used to convolute the LA functional form.

The asymmetric line shape used to fit the Ti $2p_{1/2}$ and $2p_{3/2}$ peaks in Figure 2.37 is $LF(1, 3, 27, 1)$. Note that even for the same orbital of the same element, the degree of asymmetry can be different depending on the chemical status. For example, while the Ti $2p_{1/2}$ and $2p_{3/2}$ peaks of titanium metal are highly asymmetric, those of TiO_2 are much more symmetric. Another peculiar feature of the Ti $2p$ peak of TiO_2 is that the FWHM of the Ti $2p_{1/2}$ peak is larger than that of the Ti $2p_{3/2}$ peak in the TiO_2 spectrum. This is because the lifetimes of these two states are affected differently by the final state effects.

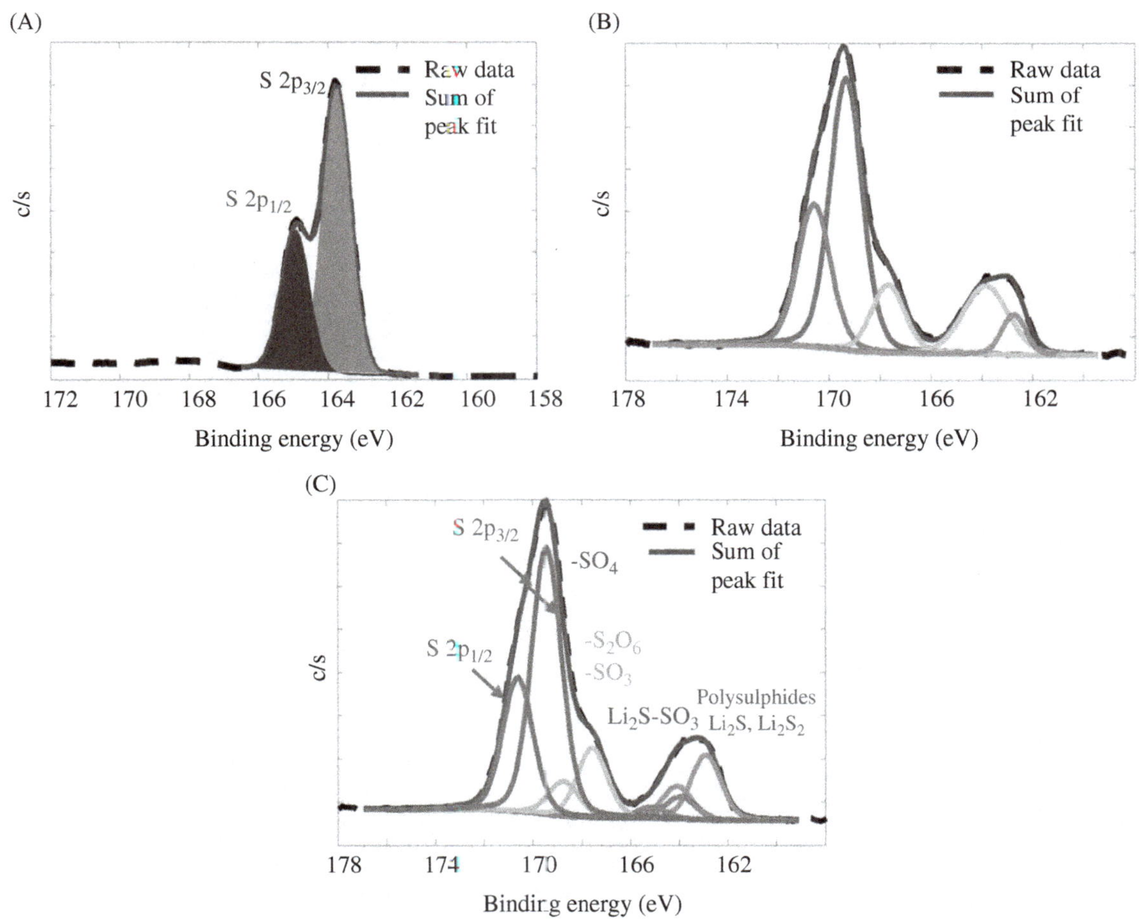

Figure 2.42 (A) Elemental sulfur spectrum showing the characteristic $2p_{1/2}$ and $2p_{3/2}$ doublet with $\Delta_{2p} \approx 1.2$ eV and a 2:1 peak ratio. (B) An example of a common error of mishandling of spin–orbit splitting during fitting that leads to a chemically and physically inappropriate analysis. (C) An appropriate fit to the spectrum from (B) accounting for the known peak splitting and peak intensity ratio. Baer et al. [1]/with permission of AIP Publishing.

All final state effects discussed in Section 2.3.5 can potentially affect the asymmetric line shape if the BE shift caused by those is smaller than the FWHM of the peak. For example, the energy of C–H stretching vibration modes is typically 2900 cm^{-1} mode, which corresponds to 0.36 eV. So, if the C 1*s* photoemission process causes vibrational excitations of organic molecules, then there will be a satellite peak at 0.36 eV lower than the main peak. In fact, multiple excitations of vibrational modes are possible. So, there can be a series of satellite peaks at $n \times 0.36$ eV with intensities decreasing as n increases. This is similar to the plasmon excitation peaks (Figure 2.34); but the energy shift is much smaller than the energy resolution of typical spectrometers, so the net effect is manifested as an asymmetric peak shape. The C 1*s* spectrum of PE in Figure 2.30 cannot be fitted with one single peak with a symmetric peak shape, but it can be fitted with a series of vibrationally excited peaks (up to $n = 4$) that have the same symmetric shape and width but are separated by 0.36 eV.

It is important to keep in mind that the spin–orbit splitting should not be ignored in peak fitting. As shown in Figure 2.42A, the S 2*p* peak always has the spin–orbit splitting with $\Delta_{2p} \approx 1.2$ eV. The peak fitting result shown in Figure 2.42B is the data extracted from a paper discussing a battery electrode [1]; here, the spin–orbit splitting was completely neglected, and five different sulfur species were claimed to exist. This kind of error can be prevented in curve fitting by setting tow components to be coupled with a fixed spin–orbit splitting and a fixed relative area ratio. In Figure 2.42C, the same data is fitted with properly accounting the spin–orbit splitting and using four components [1]. By setting the proper spin–orbit splitting parameters for the ~163 eV species, the presence of another species at ~164 eV with a minor amount could be found in this properly fitted result. Although rare, the spin–orbit splitting energy can vary depending on the chemical status. For example, the Δ_{2p} is 6.1 eV for titanium metal, 6.0 eV for titanium nitride, and 5.7 eV for titanium oxide. This augments complexity in fitting the Ti 2*p* region with multiple chemical states accurately.

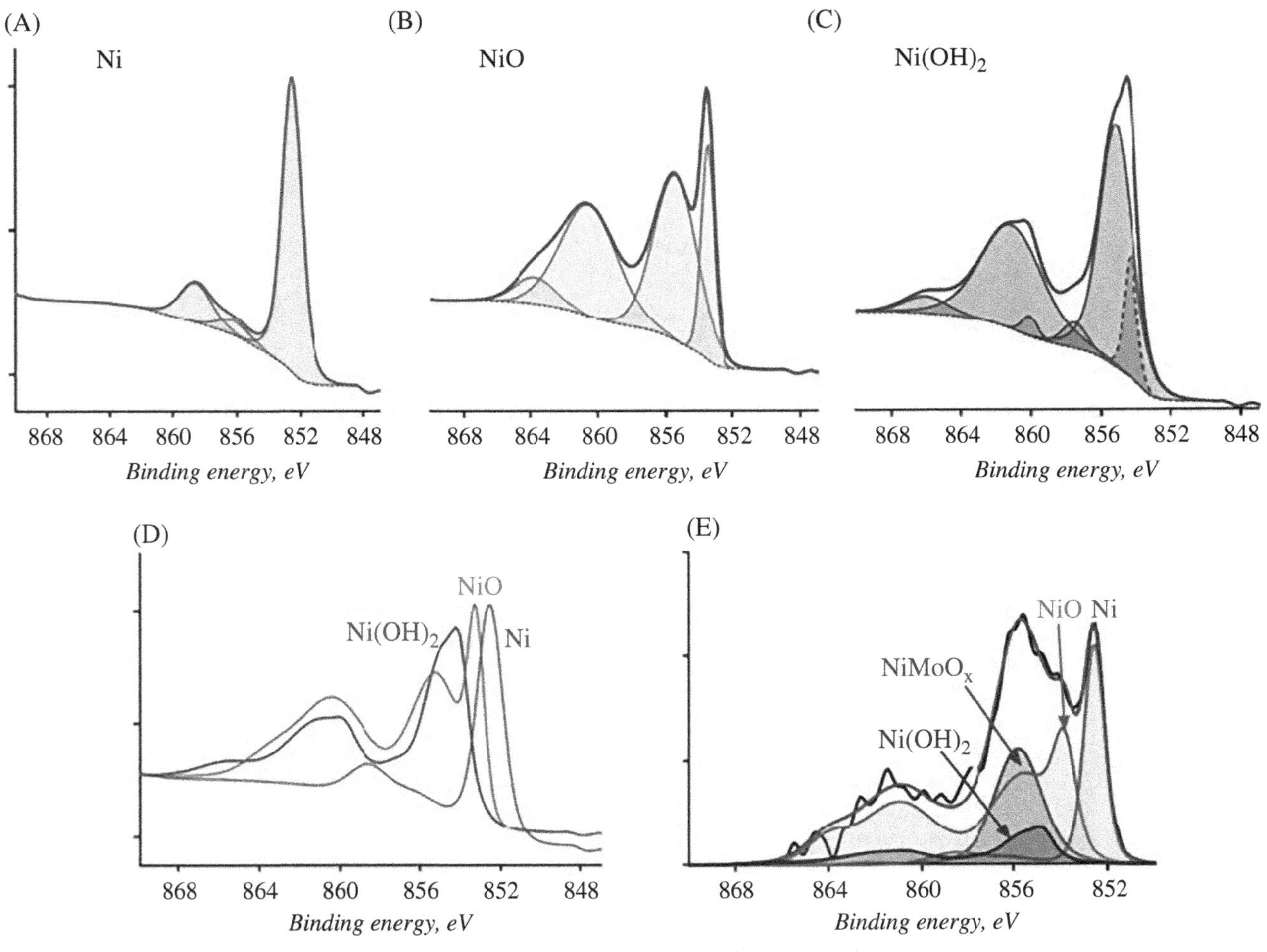

Figure 2.43 (A–C) Curve-fitted high-resolution Ni 2*p* region for three pure reference compounds. (D) Pure reference spectra overlayed. (E) Ni 2*p* for Ni–Mo oxides fitted using the peak models shown in (C). Major et al. [54]/with permission of AIP Publishing.

High-resolution spectra of transition metal compounds are among the most complex and challenging spectra to fit in XPS. The factors contributing to this complexity include final state effects, such as atomic relaxation, multiplet splitting, shake-up satellites, and plasmon loss structures, as well as the overlap of the various chemical states usually present in transition metal compounds. In addition, exposure to air during the sample mounting can cause oxidation which can be sub-stochiometric. If co-existing compounds have different electrical conductivities, it can cause a differential charging problem. When fitting transition metal peaks of complex compounds, an alternative method is to use curve-fitting parameters from reference compounds collected with the same instrument under the same data collection condition. An example is shown Figure 2.43 for a Ni compound [54]. In Figure 2.43A–C, a series of peaks are used to fit the peak envelope to create an overall shape of each reference sample (Figure 2.43D), and then these composite shapes are used in subsequent peak fitting of the data from a sample containing multiple chemical states (Figure 2.43E).

Even if the peak shape looks simple to naked eyes, peak fitting can be a delicate task that calls for deep knowledge and experience. For example, XPS has been used for speciation of oxygen atoms exposed at the surface because they play critical roles in determining the physicochemical properties of the oxide materials. A mathematical model can be used to find the concentration of surface hydroxyl groups from the intensity ratio of the OH to O^{2-} contributions to the O 1*s* XPS signal [56]. This requires reliable deconvolution of the O 1*s* peak. Figure 2.44 shows the O 1*s* region of the soda lime silica glass (sodium calcium silicate) that is similar to the one shown in Figure 2.25. The component at 530.2 eV is assigned to the non-bridging oxygen (NBO; $Si\text{-}O^-$) that is negatively charged and compensates the positive charges of Na^+ and Ca^{2+} ions in the glass [57]. Its position is within the typical O 1*s* range of metal oxides (see Figure 2.27B). The main band at 532.2 eV is the bridging oxygen (BO; Si–O–Si) connecting Si atoms in the glass network. In Figure 2.27B, we see that the typical position of O 1*s* in SiO_2 is ~533 eV. This difference is due to the next nearest atom effect. In pure silica, the O–Si–O units are fully connected; in silicate glass, the network connectivity is disrupted by the presence of modifier ions such as Na^+ and Ca^{2+}. The small peak at 536 eV in Figure 2.44 cannot be found in the O 1*s* reference chart (Figure 2.27B). We know that differential charging can cause the appearance of a peak at the position higher than theoretically expected (see Figure 2.18). Could it be due to differential charging of the sample surface, because glass is not conductive at all? In the absence of Na, one may interpret the 536 eV peak in that way. But we know sodium is present in the sample. As discussed in Figure 2.25, it is one of the minor peaks of the Na KLL Auger line [57]. In this example, the low BE side tail of the Na KLL peak overlaps with the high BE side tail of the O 1*s* peak where the Si–OH silanol contribution is expected. The contribution from the Na KLL peak at 536 eV should be subtracted from the O 1*s* peak area, and this can be done by using the intensity ratio between the 536 and 498 eV components of the Na KLL line which can be obtained using a reference sample independently.

When the 498 eV peak of Na KLL is analyzed for soda lime silica glass, it should be noted that if Sn is present in the glass, the Sn $3d_{3/2}$ peak can overlap with the Na KLL peak. If a flat glass panel of soda lime silica is made through the float process (such glass is called "float" glass), then one side of the glass contains Sn (because the glass melt is floated and processed on a

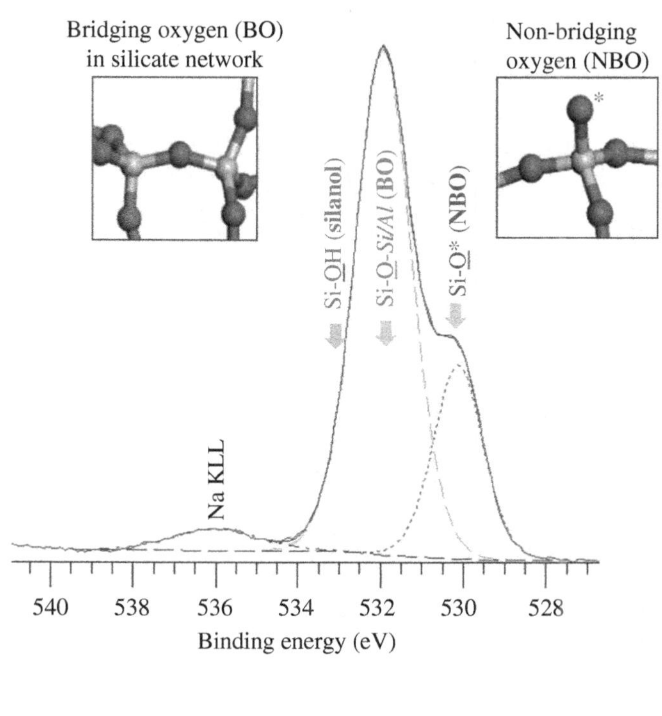

Figure 2.44 High-resolution XPS spectrum of soda lime silica glass in the O 1*s* region. Banerjee et al. [57]/with permission of John Wiley & Sons.

molten tin bath). If the float glass is rinsed with aqueous solutions during the manufacturing process, then the other side can also be contaminated with Sn. Based on the main component of the glass (which is reflected in the glass name, sodium calcium silicate), non-experts can miss the potential contribution of Sn $3d_{3/2}$ to the Na KLL peak area. Missing minor peaks is a common mistake found in XPS analyses reported in the literature.

Figure 2.45 shows another example demonstrating how delicate peak fitting can be. The data are the high-resolution O 1*s* spectrum of a calcium aluminosilicate glass with a small amount of magnesium (called E-glass) which had been treated with pH 1 solution at 90 °C for 24 hours [58]. In this treatment, the glass surface dissolves incongruently; all network modifier ions (Ca^{2+} and Mg^{2+}) are leached out and the aluminum in the network is also dissolved, which leads to the formation of a silica-rich surface layer. Since all cations are removed, there is no discernable peak at the NBO position in the O 1*s* spectrum. Upon the loss of cations to aqueous solution, the NBO groups should have been converted to the Si–OH groups by reaction with water. The O 1*s* peak of the Si–OH species is expected to be at ~533 eV. And the O 1*s* peak looks slightly asymmetric, with a higher intensity in the higher BE side. Thus, the presence of Si–OH species is reasonable.

To determine the relative abundance of Si–OH vs. Si–O–Si (BO), we can try to fit the O 1*s* peak with two components. Figure 2.45A is the result of fitting without any constraints. The fitting error appears very small, so one may think that the fit result looks reasonable. But the separation between the BEs of the silanol and BO components is smaller than what is expected from the literature. Also, the FWHMs of the two components are quite different, which does not make sense. In Figure 2.45B, the data is re-fitted with a constraint fixing the silanol BE is 0.55 eV above that of BO. The Si–OH concentration is reduced significantly compared to the previous case. The FWHMs of the two components are still quite different. Figure 2.45C shows the fit result obtained with a constraint keeping the FWHMs the same for both components. Now, the

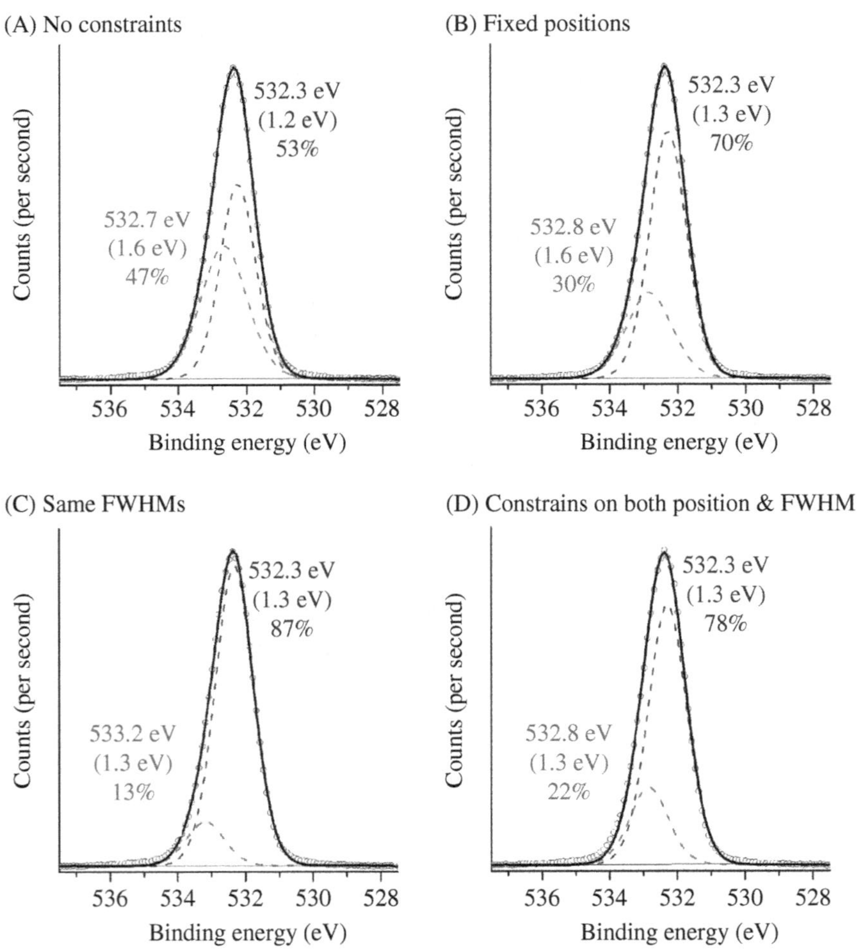

Figure 2.45 High-resolution XPS spectrum of a Na-leached E-glass surface with different fitting results: (A) No constraint is used. (B) Constraining the peak position of Si–OH to the BE of Si–O–S plus 0.55 eV. (C) Constraining the FHWM to be the same for both components. (D) Constraining both BE of Si–OH and FWHM. Banerjee et al. [58]/with permission of Elsevier.

BE separation of the two components is significantly larger than what is expected. Figure 2.45D displays the fit result obtained with two constraints – fixing the silanol BE to 0.55 eV above the BE of BO and keeping the FWHMs of both to be the same.

Now, the next question would be which one is the most reasonable fit result? Based on the overall elemental composition and charge balance of all elements, the silanol concentration was estimated to be ~15% in the acid-leached E-glass [58]. Then, one may say that Figure 2.45C is the best. Indeed, the 533.2 eV BE is acceptable for the Si–OH species. If there is any reason that the silanol BE should be at 0.55 eV higher than the BE of BO, then Figure 2.45D would be a more reasonable fit. Then, the charge balance model estimating the silanol concentration from the elemental composition may need to be revisited.

A flawed and erroneous fit will lead to unreliable results and, ultimately, invalid conclusions. Common mistakes found in the literature include overfitting (including too many components when they are not justified), not placing appropriate constraints on fit components, using inappropriate backgrounds and line shapes, incorrect assignments of chemical shifts by simply comparing the peak positions with a table or due to inappropriate charge correction, etc. [59]. Also, it is not uncommon to see the ratio of C–C/H vs. oxygenated species in the high-resolution C 1*s* fitting inconsistent with the C:O ratio from the survey spectrum analysis, or incorrect positioning of mixed graphitic/nongraphitic species in amorphous or partially graphitic carbon materials.

The only way to reduce the probability of making these mistakes is to practice curve fitting with real data and compare the results with others. XPS data of ~122 materials can be downloaded in the vms file format from *XPSSurfA* (1740 spectra as of May 2024) [46] which can be processed with *CasaXPS* software [27] or other freeware available online. XPS users who need to process their own data can practice with those data first and compare the fit results with known compositions and structures in the literature.

2.4.3 Relative Sensitivity Factor (RFS) to Conversion of Peak Area to Atomic-%

So far, we have learned how to identify elements present at/in the surface and how to calculate the peak areas of representative peaks of the detected elements. Now, let us study how to convert the calculated areas to the atomic concentrations. The factors affecting the photoelectron signal intensity include photoemission cross section (which is a function of the energy of the exciting radiation and the orbital from which the electron is ejected; Section 2.1.1), the escape depth of the photoelectron (which is a function of the KE of photoelectron and the nature of the sample; Section 2.1.4), the transmission function of the spectrometer (which varies with the pass energy of the analyzer and the lens operation mode; Sections 2.2.2 and 2.2.4), the efficiency of the electron multiplier (Section 2.2.5), the sample surface condition (area, roughness, etc.), and the concentration of element of interest at/in the surface. All these factors can be summarized into the following equation:

$$I_A = J\,\sigma_A\,W_A\,\frac{\Delta\Omega}{\pi}\,T\int_0^\infty exp\left[-\frac{z}{\lambda\cos\theta}\right]dz\,D\,S\,y_S\,N_A \tag{2.21}$$

where I_A is the signal intensity of element A, J is the photon flux, σ_A and W_A are the photoelectron cross section and angular symmetry parameter for the element and orbital under consideration, respectively, $\frac{\Delta\Omega}{\pi}$ is the angular acceptance of the spectrometer, T is the transmission function of the transfer lens and electron analyzer, z is the depth from the surface, λ is the IMFP for the KE of electron in the solid being analyzed, θ is the angle of emission from the surface normal direction (if the take-off angle from the surface plane is used, then cosθ is changed to sinθ; see Section 2.5), D is the electron multiplier amplification and counting efficiency, S is an effective surface area, y_S is the sample-specific parameter such as surface roughness, porosity, matrix effects, etc., and N_A is the atomic concentration of element under consideration.

Among all these parameters, many are typically fixed for a given instrument. Those include J, $\frac{\Delta\Omega}{\pi}$, θ, D, and S. For now, let us ignore the sample specific variance (y_S). Then, eq. (2.21) is reduced to:

$$I_A \propto (\sigma_A\,W_A\,T) \times \int_0^\infty exp\left[-\frac{z}{\lambda\cos\theta}\right]dz\,N_A \tag{2.22}$$

The first three terms in parentheses on the right-hand side determine the RSF. Here, σ_A is element specific, but W_A and T are instrument specific. This is the reason that each *CasaXPS* program has its own RSF library file to be used for analysis of the data collected with the intended instrument.

The probability of electron emission upon x-ray absorption can be described with quantum mechanics. The probability of sub-shell ionization is a function of initial and final state wavefunctions:

$$\sigma_{i,j} \propto A\left(BE_i, KE_j\right)\left|\left\langle \Psi_j|\mu|\Psi_i\right\rangle\right|^2 \tag{2.23}$$

where A depends on the BE of ionized core level and KE of emerging photoelectron, Ψ_i and Ψ_j are the wavefunctions of the initial and final states, respectively, and μ is the linear operator describing the time-dependent perturbation of electronic state under the electromagnetic field of photon. J. H. Scofield calculated the photoelectric cross sections, using the single-potential Hartree-Foch atomic model, for the Mg Kα (1254 eV) and Al Kα (1487 eV) lines [60]. Figure 2.46A plots the theoretically calculated photoemission cross sections as a function of binding energy, on a log–log scale, for major photoelectron signals of elements from ^{3}Li to ^{95}Am [61]. When normalized with the F 1*s* cross section, the theoretical cross sections of the strong photoemission lines fall on similar curves. Figure 2.46B plots the empirically determined sensitivity factors for the strongest lines of individual elements, which show qualitatively similar trends seen in the theoretical calculation plot [61].

The angular distribution of photoelectrons emitted without energy loss from a given subshell of uniformly distributed atoms in a homogeneous solid is not uniform. The anisotropic distribution of photoelectrons is governed by the dipole asymmetry parameter (β) and the angle (ϑ) between the photoelectron vector and the electric field vector of the x-ray. The β value is constant for the *s* orbital, but varies with atomic number (Z) for the *p*, *d*, and *f* orbitals [62]. In XPS experiment, photoelectrons emitted along the analyzer axis (transfer lens axis $\pm$ ΔΩ) are detected. Figure 2.47A–F schematically illustrates the relationship between the incidence angle and polarization of x-ray and the detection direction [63]. In the case of spherical emission or unpolarized x-ray, the ϑ dependence of the signal is negligible. But, in the case of polarized x-ray, the signal

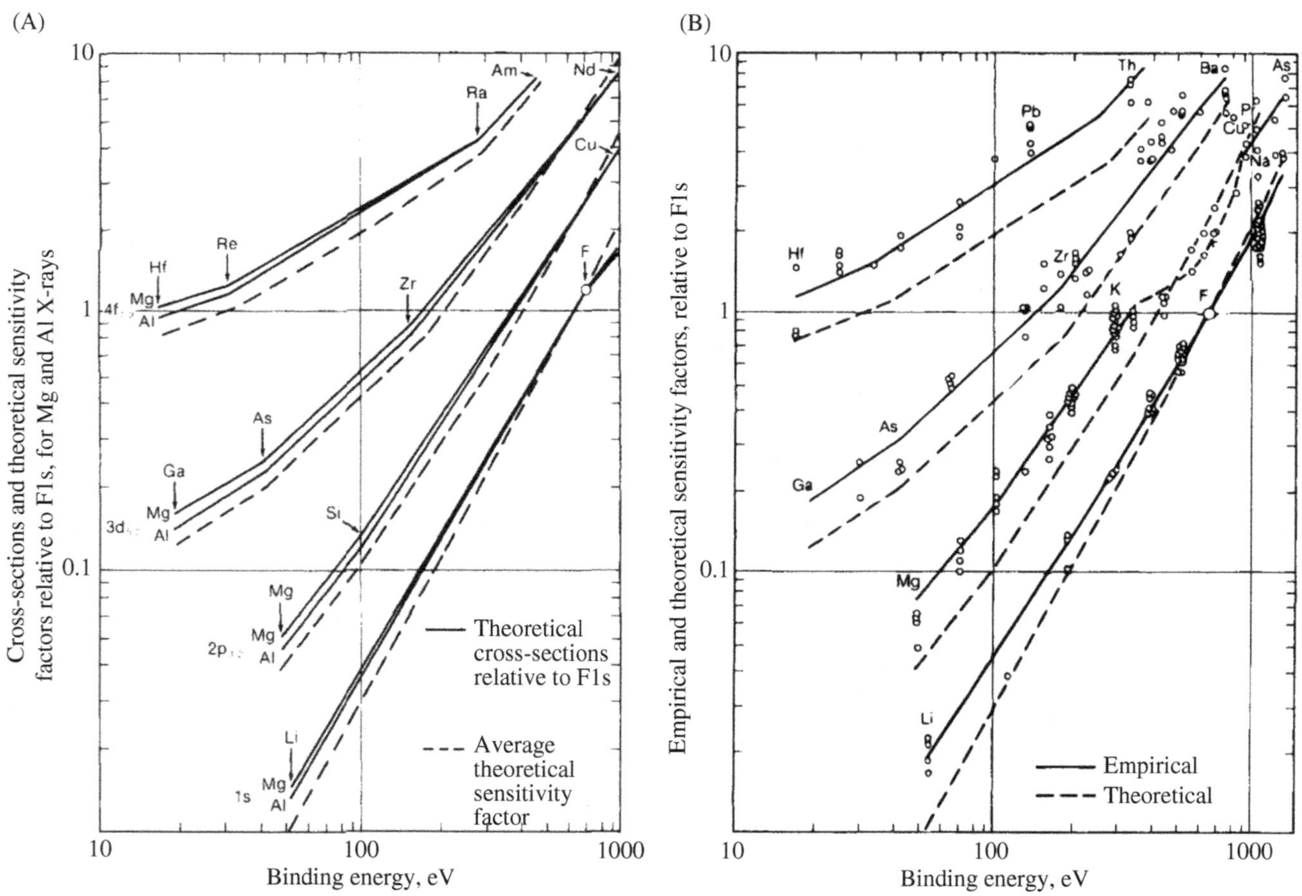

Figure 2.46 (A) Photoelectric cross sections, and derived theoretical sensitivity factors, of the strong lines, plotted against binding energy of the transitions. (B) Empirical sensitivity factors, normalized with the F 1*s* value, for 1*s* transitions of ^{3}Li through ^{11}Na, 2$p_{3/2}$ transitions of ^{12}Mg through ^{33}As, 3$d_{5/2}$ transitions of ^{31} through ^{59}Pr, and 4$f_{7/2}$ transitions of ^{72}Hf through ^{90}Th. Wagner et al. [61]/with permission of John Wiley & Sons.

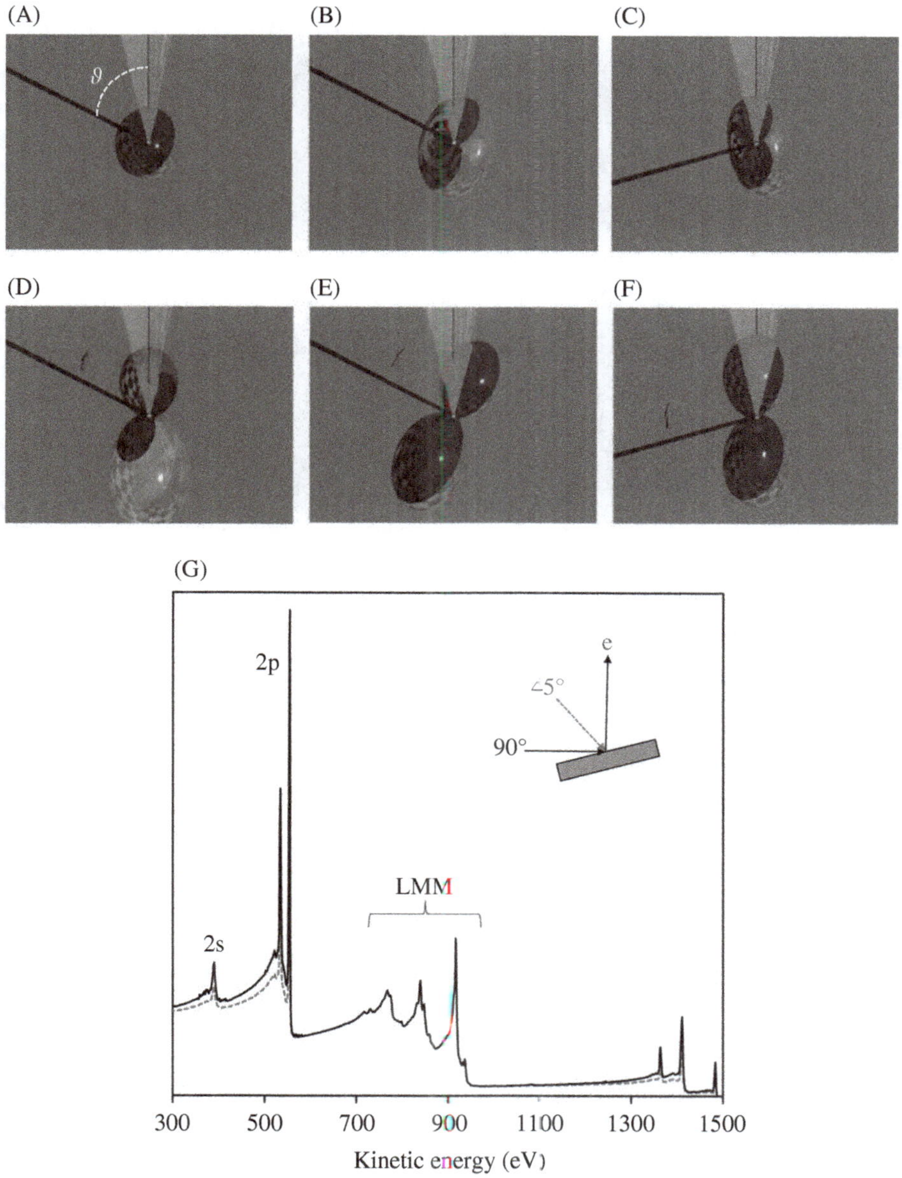

Figure 2.47 Illustration of the angular distribution of electron emission from an atom: (A) isotropic emission (typical of Auger electrons) and (B–F) examples of photoelectron emission with β = 1.5 for (B) unpolarized x-ray at ϑ = 55°, (C) unpolarized x-ray at ϑ = 90°, (D) polarized x-ray at ϑ = 55° with x-ray electric field 45° out of the plane of the x-ray and detection axes, (E) polarized x-ray at ϑ = 55° with in-plane electric field, and (F) polarized x-ray at ϑ = 90° with in-plane electric field. The red area is the overlap between the angular emission distribution and the analyzer acceptance angle. (G) XPS spectra of clean copper collected with an Al Kα at 45° (dashed) and 90° (solid) with respect to the detector axis. Reproduced from Shard [63]/with permission of American Institute of Physics.

intensity can vary drastically. This effect can be seen in the XPS spectra of copper shown in Figure 2.47G [63]. The absolute intensities of photoelectron peaks change by ~50% between $\vartheta = 45°$ and 90° cases, while the Auger electron peak intensities are unchanged. The peak intensity dependence on the angle between the incident x-ray and the detector axis disappears when ϑ is 54.7° (magic angle). For the system designed with ϑ different from this magic angle, the W_A parameter should be calculated based on the β of the orbital under consideration and the ϑ of the system used for XPS analysis [62].

Transmission function (T) is another important parameter that affects RSF. This is something that is empirically measured for each instrument and factored in the RSF calculation. Figure 2.48 shows examples of the transmission function data collected for low-density polyethylene (LDPE) and Au by four groups who participated in the round-robin analysis which

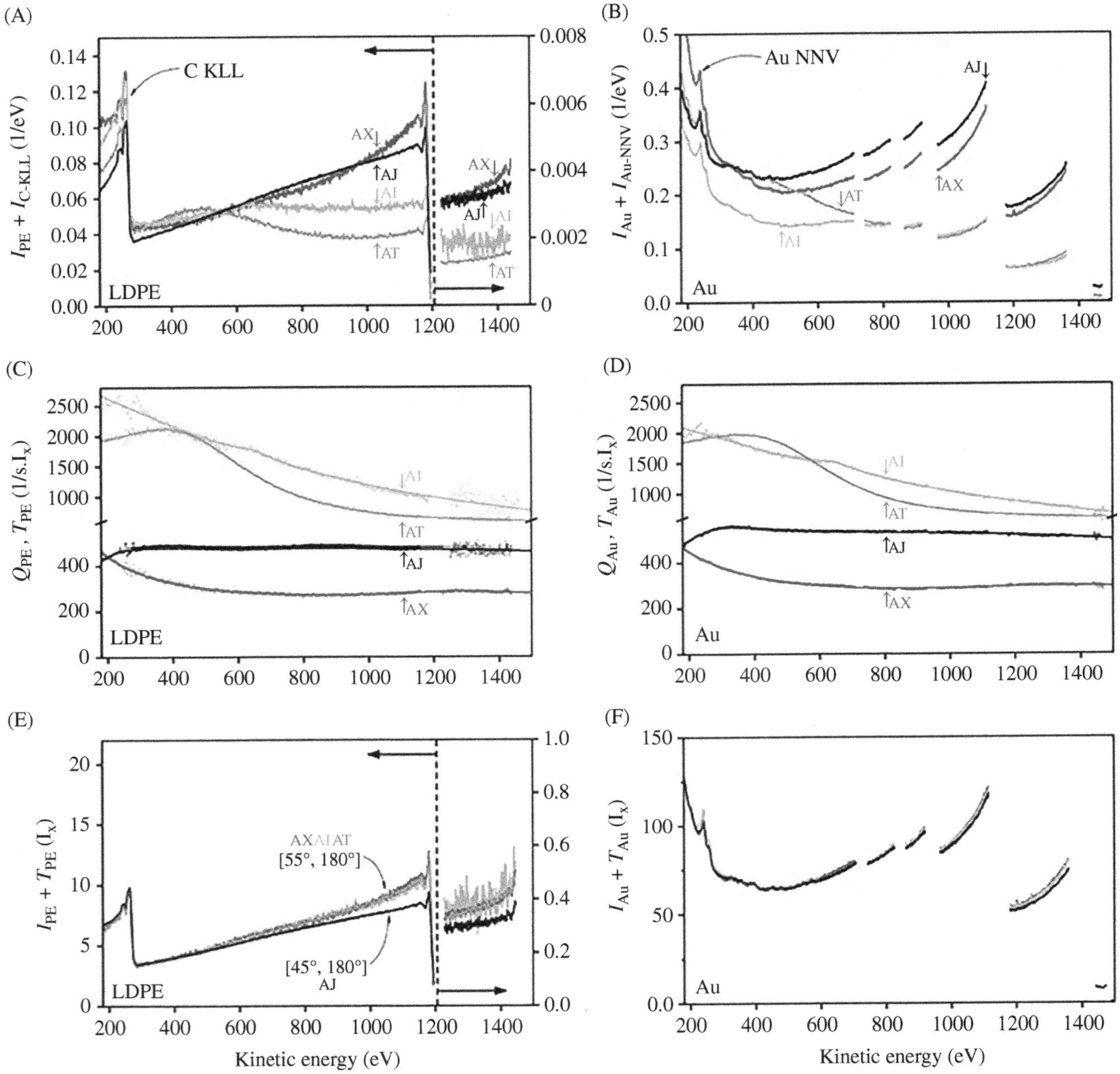

Figure 2.48 (A,B) Continuous background spectra of LDPE and Au normalized to the C KLL and Au NNV Auger peaks, respectively. (C,D) Transmission functions calculated by dividing the measured data with reference spectra of LDPE and Au, respectively. (E,F) Intensity-corrected LDPE and Au spectra using the transmission function of each sample, respectively. Reed et al. [64]/with permission of AIP Publishing.

was organized by Technical Working Area 2 (Surface Chemical Analysis) of the Versailles Project on Advanced Materials and Standards (VAMAS) organization in 2020 [64]. The significant variability of background shapes and intensities between different participants can be seen in Figure 2.48A,B, which visually demonstrates large variances among different instruments. The measured spectra are then divided by the corresponding reference spectrum and then fitted with some functional form to get the transmission function (*T*), which are shown in Figure 2.48C,D. When the measured intensities are corrected using the transmission function of each system, the resulting spectra are now very close to each other (Figure 2.48E,F). This example shows how important the transmission function correction is. The obtained transmission function is valid as long as the analyzer is stable; but, if a major maintenance or change is made to the instrument, it may vary.

Assuming the sample is homogeneous, we can integrate the photoelectron escape term, $\int_0^\infty exp\left[-\frac{z}{\lambda\cos\theta}\right]dz$, using a general trend of the IMFP (λ) value. M. P. Seah revisited the original data shown in Figure 2.11 and obtained an empirical equation for the KE range from 100 eV to 10000 eV:

$$\lambda(nm) = \left(0.73 + 0.0095\,KE^{0.872}\right)/Z^{0.3} \tag{2.24}$$

where Z is the atomic number [65]. When tested for ~40 elemental materials, this equation gives ~15% root-mean-square deviations. Another version, called Tanuma–Powell–Penn (TPP-2M) equation, is to calculate λ based on energy loss functions of 42 inorganic compounds [66]. It has a modified form of the relativistic Bethe equation. When compared with 40 compounds (excluding two outliers), the TPP-2M model gives an average RMS deviation of 8.7% in the KE range from 50 eV to 200 keV. In the absence of accurate data of λ from reference samples, these universal equations can be used as an approximation. Then, the atomic concentration (N_A) calculated from I_A using eq. (2.22) is the average value over the probe depth (3λ) that we would get if the sample were homogeneous.

When an overlayer is present, it can affect the photoelectron escape efficiency, which is a function of IMFP (λ). For example, the presence of adventitious carbon layer can make the atomic-% value calculated from a high BE (low KE) peak smaller compared to the value calculated from a low BE (high KE) peak of the same element [67]. We can model the attenuation effect due to the overlayer using the model illustrated in Figure 2.10. Considering that the detection axis may vary from the surface normal direction, the equations shown in Figure 2.10B are modified further using in the following forms:

$$I_O = \int_0^d I_{O,\infty}\, exp\left[-\frac{z}{\lambda\cos\theta_n}\right]dz = I_{O,\infty}\left(1 - exp\left[-\frac{d}{\lambda_O\cos\theta_n}\right]\right) \tag{2.25}$$

$$I_S = \int_d^\infty I_{S,\infty}\, exp\left[-\frac{z}{\lambda\cos\theta_n}\right]dz = I_{S,\infty}\, exp\left[-\frac{d}{\lambda_S\cos\theta_n}\right] \tag{2.26}$$

where d is the overlayer thickness, θ_n is the angle of the detection axis with respect to the surface normal direction, $I_{O,\infty}$ and $I_{S,\infty}$ are the pure bulk signal intensity of the overlayer and the substrate, respectively, and λ_O and λ_S are the IMFP of the overlayer and the substrate, respectively. These equations work only for the case where the overlayer thickness is homogeneous. If the overlayer is patchy or inhomogeneous, eqs. (2.25) and (2.26) should be modified further.

Now, let us bring up the y_s term in eq. (2.21). The photoelectron cross section discussed in Figure 2.46 describes the emission of an electron from an atom, but not all these electrons can escape into the vacuum. The ones contributing to the

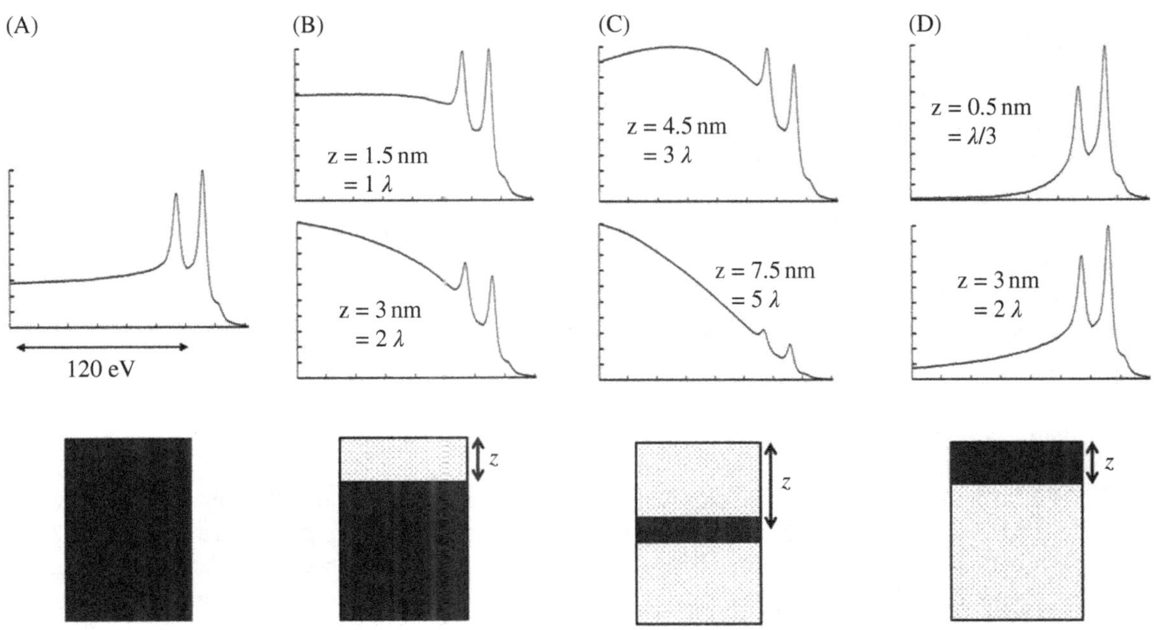

Figure 2.49 Model Au4d XPS for different distributions of Au atoms: (A) Bulk. (B) Buried under an overlayer. (C) Present as a thin layer below the surface. (D) Present as an overlayer. The host material has λ = 1.5 nm. Tougaard [68]/with permission of John Wiley & Sons.

background due to inelastic scattering (Section 2.4.2) are not included in the RSF value determination; so, if that fraction varies due to the structure of the sample (especially, layered materials), it can affect the atomic-% calculation determined using the pre-determined RSF value set.

Sometimes, the shape of the inelastic background can provide critical information about the overlayer structure. Figure 2.49 shows the model spectra for four different depth distributions of Au atoms in a fictitious sample [68]. Figure 2.49A is the case where photoelectrons are escaping from a homogeneous material. When there is an overlayer, the inelastic scattering loss of the substrate signal will be larger as compared to the homogeneous case; so, the ratio of the background signal to the pristine photoelectron signal will increase (Figure 2.49B). If photoelectrons originate from a buried layer within the probe depth, they also experience inelastic scattering in the overlayer (Figure 2.49D). If the signal originates from the top layer which is different from the substrate, then its photoelectrons mostly escape to the vacuum without inelastic scattering; thus, the background in the high BE side would be negligible (Figure 2.49D).

The loss peak is often ignored in quantitative analysis, but it can cause a relatively significant error. A systematic study example is shown in Figure 2.50 [69]. In this study, XPS measurements were conducted for a series of glass surfaces exposed by fracturing the samples inside the XPS chamber, and the measured intensities of all identifiable peaks were compared with the theoretical atomic-% predicted from the bulk composition. In the high BE regime (i.e., KE less than ~500 eV), the RSF values determined from glass samples were found to be quite different from the theoretically calculated ones as well as the default values in *CasaXPS* and the ones provided by the vender (Figure 2.50A). When the relative abundances of various elements in glass were calculated with different RSF values, they deviated from the expected values by ~10%; especially, the O/Si ratio deviated most significantly (Figure 2.50B). The magnitude of this discrepancy could not be explained by the transmission function (because it was already factored out in the calculation), effective attenuation length difference, or elastic scattering; instead, the satellite loss peak was found to be the main cause for this discrepancy [69]. The effect of energy loss process can vary in different materials [70]; thus, the y_s contribution in eq. (2.21) may vary with materials.

In summary, the RSF value of a specific element can vary drastically. The RSF values reported in the literature should be taken as an approximation only; the RSF value can vary substantially among different instruments. Even for the same system, it may vary over time as the instrument is aged or undergoes major repairs. The most accurate and reliable RSF values can be obtained only through careful control experiments with high-quality references samples. Moreover, depending on the material type, the fraction of the signal contributing to the final state effects (such as satellites, plasmons, and loss peaks) can vary. If these are not included properly in the peak area calculation, it can cause a significant error in quantification. That is the reason that the loss peak of the Ti 2*p* signal is included in the atomic-% calculation from the surface spectrum as shown in Figure 2.36.

2.4.4 A Demonstration of Quantitative Analysis

In this section, an example of using what we have studied so far is demonstrated. Figure 2.51 is the survey spectrum of an air-fractured surface of a MoS_2 crystal. Using the "Element Library" tool of *CasaXPS*, all possible peaks are marked. At first glance, the marked peaks are reasonable. In the NIST XPS database [26], we see the BE of Mo $3d_{3/2}$ and $3d_{5/2}$ is 232.5–232.8 eV and 228.8–229.5 eV, respectively, which are very close to the positions observed in the survey spectrum. However, the Mo:S ratio calculated with the Mo 3*d* (3.5177) and S 1*s* (0.6399) RSFs is only 1:1.56, instead of 1:2. Also, there are several peaks that do not belong to Mo and S. The automatic "Find Peaks" function suggests Be, B, N, Mg, Sc, Se, Br, and Rb as well; but those suggestions are simply based on the proximity of one of the observed peaks to their elemental peaks, instead of matching all possible peaks of those elements. So, their presence can be ruled out. In the case of B, there is only one peak (1*s*) that appears typically at 188–193 eV, which is close to the small peak at 185 eV that is not marked in the survey spectrum in Figure 2.51. We can still rule out the presence of B, since the observed BE is too low to be any of B species [10, 26], and the peak width is too broad to be the 1*s* peak. The non-stoichiometric atomic ratio and unassigned peaks suggest that something is missing here.

In close inspection (Figure 2.52), we can see that, for almost all Mo and S peaks, there are additional peaks spaced by ~24 eV with their intensities decreasing gradually as moving to higher BE sides. Based on this pattern (see Figure 2.34), we can tell that these are the plasmon excitation bands of MoS_2. This tells us that the sub-stoichiometric ratio found in the initial quantification trial is caused by ignoring the plasmon peaks in the calculation (see the final state effect discussion in Section 2.4.3). The high-resolution spectra of Mo 3*d* and S 2*p* regions (collected with much smaller PE setting of CHA) are shown as an inset in the top right corner of Figure 2.52. It can be seen that S 2*s* peak (227.2 eV) is very close to the Mo $3d_{5/2}$ peak. Since these two are not well resolved in the survey spectrum (due to the high PE setting of CHA), the Mo 3*d* area must

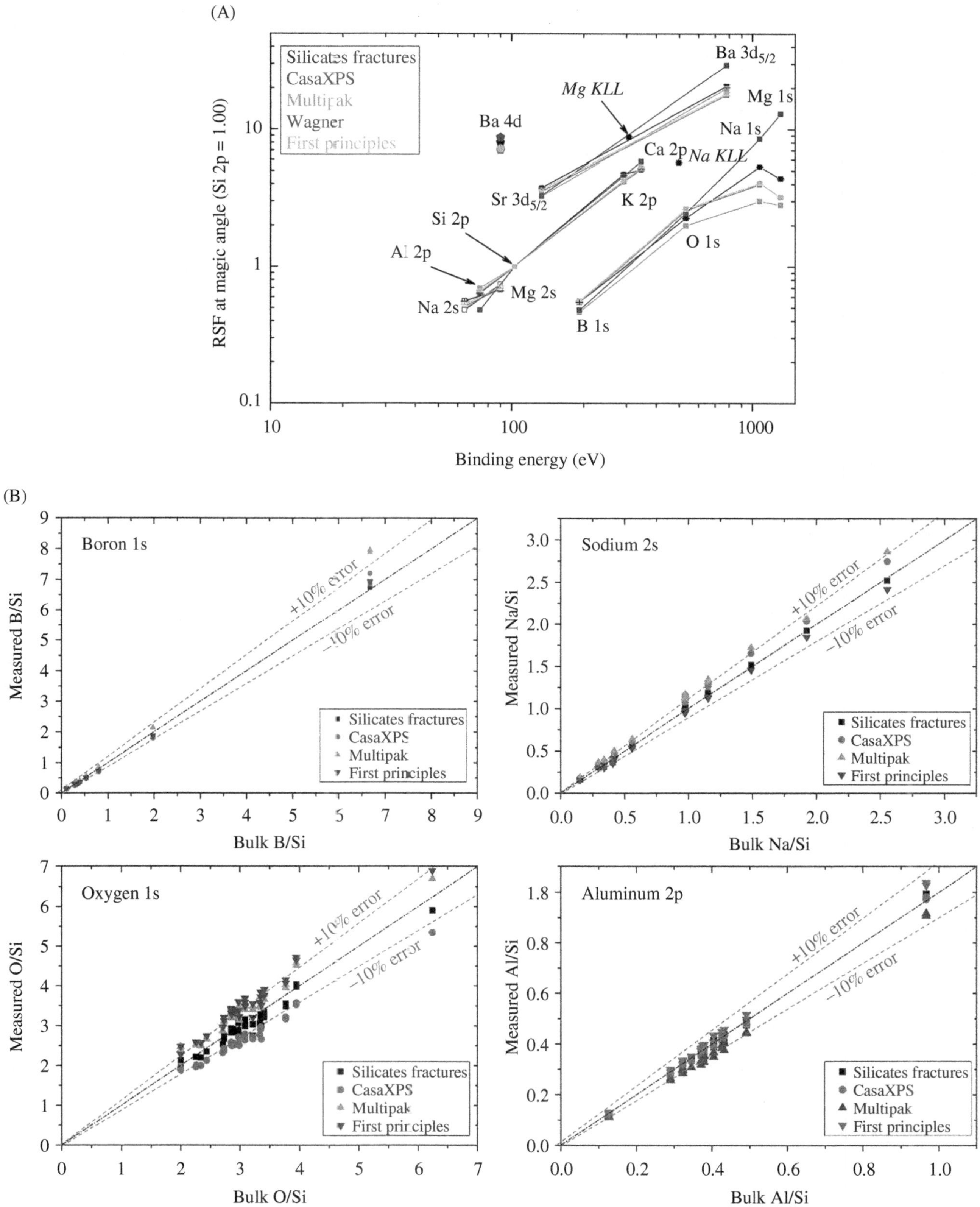

Figure 2.50 (A) Plot of RSF vs. binding energy for different RSF sources. (B) Error plots predicted from different RSFs. The silicate fracture values are the ones determined by comparing the measured intensity with the bulk composition. Other values are from the literature or the default set from venders. Shallenberger et al. [69]/with permission of John Wiley & Sons.

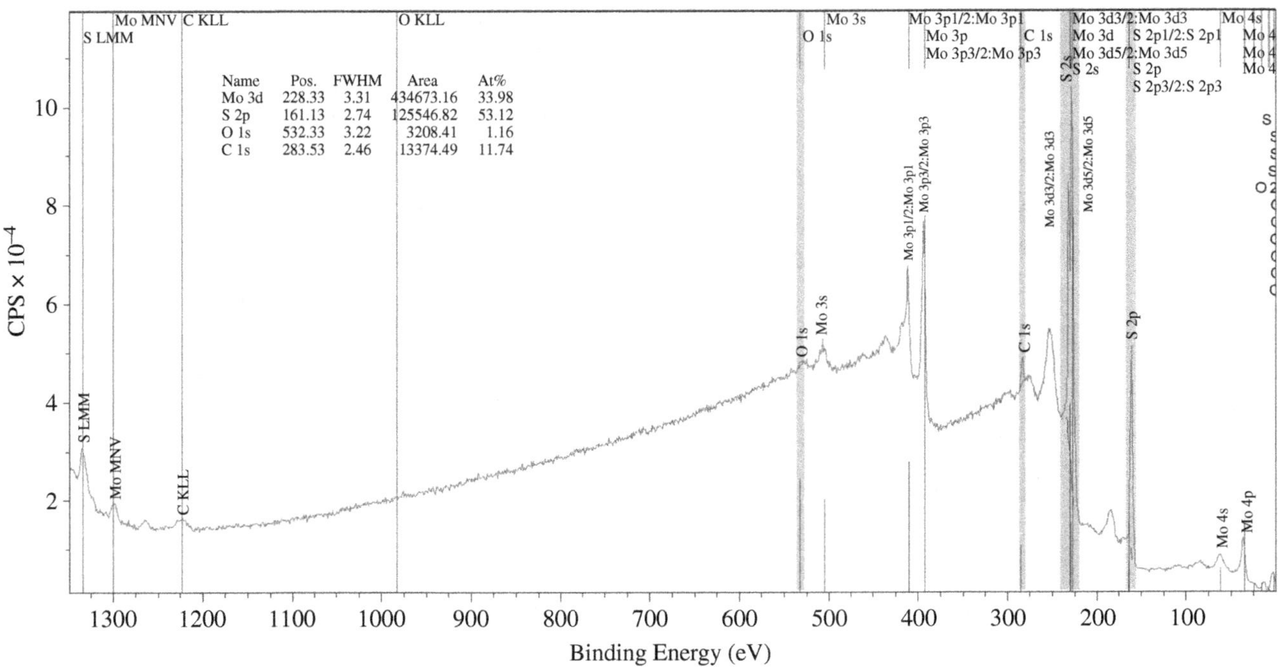

Figure 2.51 Survey spectrum of an air-fractured surface of MoS_2. The main photoelectron and Augers peaks are marked.

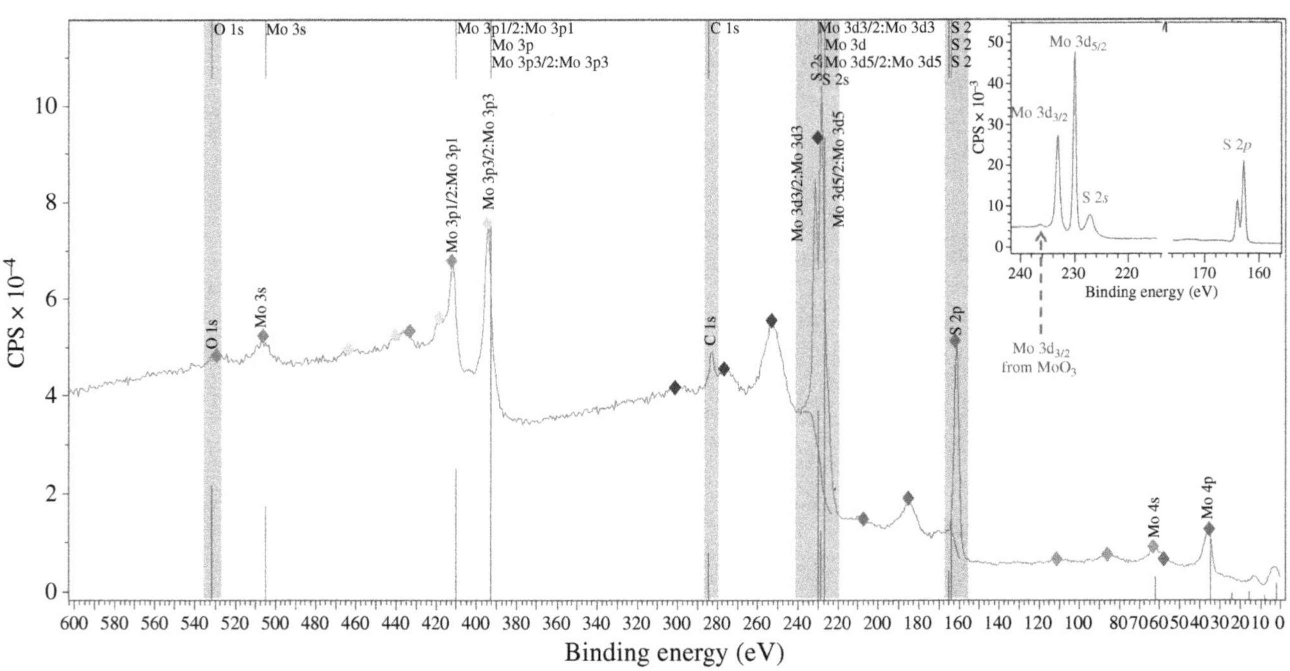

Figure 2.52 The low BE side of the survey spectrum is shown in Figure 2.50. The diamond marks are the peaks separated by ~24 eV. The inset in the top right corner shows the high-resolution spectra of Mo 3*d* and S 2*p* regions.

be affected by the overlapping S 2*s* peak. In addition, we can see that C 1*s* background is affected by the plasmon peak of the Mo 3*d* photoelectron. These overlaps also contribute to the error in the peak area calculation shown in Figure 2.51. The peak at 532 eV is not totally from O 1*s*; the plasmon peak of the Mo 3*d* photoelectron contributes to the area calculated for this peak in Figure 2.51. Then, how do we know if oxygen is present in the sample. We can see a tiny peak at 236.5 eV in the high-resolution Mo $3d$ region. That can be assigned to the Mo $3d_{3/2}$ of MoO_3. Then, we know that there should be the $3d_{5/2}$ peak at 233.4 eV (Δ_{3d} of Mo = 3.1 eV [10]); but this peak is buried under the Mo $3d_{3/2}$ of MoS_2.

Since the high-resolution spectra allow us to calculate the Mo $3d$ area only (without including the overlapping S $2s$ peak) and the S $2p$ area, one may think that using these spectra may work better for the quantification purpose. From the peak fitting (not shown here), we find that the areas of Mo $3d_{3/2}$ (without counting the contribution from $3d_{5/2}$ of MoO_3) and Mo $3d_{5/2}$ peaks are 20066.2 and 30099.3, respectively. And the areas of S $2p_{1/2}$ and $2p_{3/2}$ peaks are 6875.8 and 13751.5, respectively. Then, we get (50165.5/3.5177):(20527.3/0.6499) = 1:2.22, suggesting the sample is $MoS_{2.22}$. Does the sample really have excess sulfur? Here, we should recall that there are many plasmon peaks; as discussed in Section 2.4.3, these satellite peaks from the finite state effect should be included in the atomic-% calculation. So, the result from the high-resolution data suggesting the presence of excess sulfur is incorrect because the plasmon peaks are not included. We should re-do the calculation using the survey spectrum where we can see all plasmon excitation peaks.

Figure 2.53 shows the survey spectrum in the 340–140 eV region with the peak assignment and the area calculated for each region. In the S $2p$ region, the high BE side baseline is extended to 175 eV because there are two surface plasmon bands [71]. The next step is to subtract the S $2s$ contribution from the (Mo $3d$ & S $2s$) area. The S $2s$ contribution can be calculated from the S $2p$ area because their relative intensities are known from the RSF values for S $2s$ (0.4650) and $2p$ (0.6499). The total area of the S $2p$ peak and its plasmons is 194457. Then, the S $2s$ area is calculated to be: $194{,}457 \div 0.6499 \times 0.4650 = 139133$. Thus, the Mo $3d$ peak area in the 240–230 eV region is 297642. Then, the total Mo $3d$ peak area is $297642 + 160738 + 50018 + 15929 = 524327$. Then, we can calculate the atomic concentration ratio between Mo and S using the total areas of Mo $3d$ (+ its plasmons) and S $2p$ (+ its plasmons), which gives (524327/3.5177):(194457/0.6499) = 1:2.007. Now we get the stoichiometry that is consistent with the MoS_2 crystal.

As we can see in this example, the quantitative analysis of XPS data cannot be done reliably by simply taking the peak fitting results from *CasaXPS*. The user should be able to interpret the full spectral features and process the fit results, and often re-process the data with various constraints that physically make sense until all sources of error are eliminated. The only way to get to that level is through lots of practice. For that reason, readers are recommended to practice using the data of known materials. There are several online databases from which reference spectra and peak positions can be found. One of such sites is *XPS Reference Pages* by the Surface Science Western lab [47].

Another note that is important to remember is that XPS can be destructive, especially for the samples that are prone to damage by x-ray or secondary electrons. In such cases, the elemental composition determined with XPS can vary over time. Halide compounds are quite sensitive to such damage processes [31, 72]. The secondary electrons generated by x-ray can

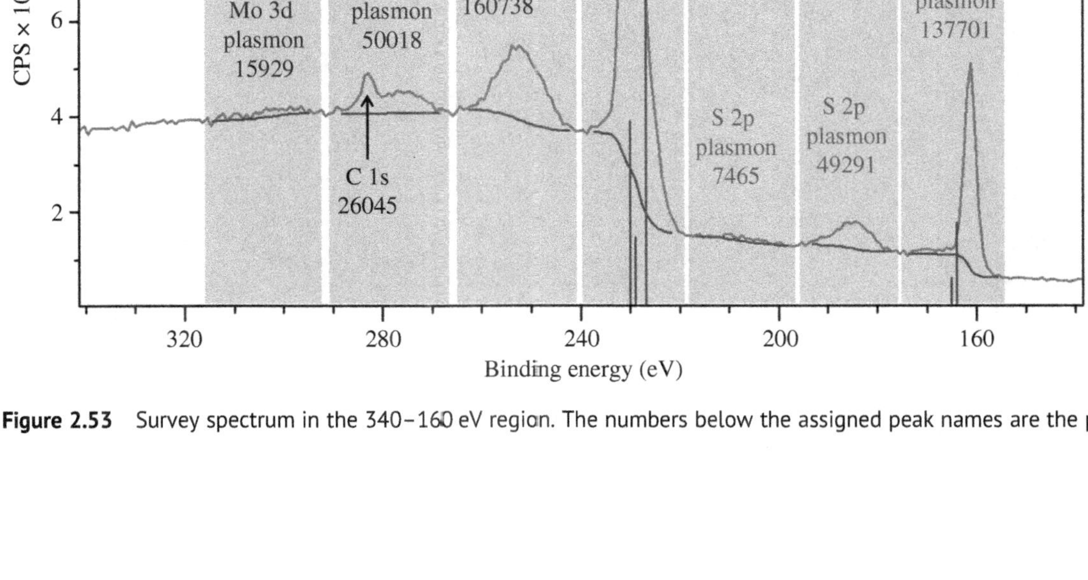

Figure 2.53 Survey spectrum in the 340–160 eV region. The numbers below the assigned peak names are the peak area.

cause electron-stimulated desorption of halide ions. Also, the low-energy electron flood gun used for charge neutralization of insulating samples can also cause changes in the oxidation state of transition metals in the sample [17]. If a long-time data accumulation is required for the sample prone to beam damages, frequently moving analysis spots to fresh locations is needed to minimize the measurement artifact.

2.5 Depth Profiling with XPS

2.5.1 Angle-Resolved XPS

If the sample composition varies within the XPS information depth, then depth profiling can be done by simply tilting the sample with respect to the analyzer. This is called angle-resolved (AR) XPS analysis. The principle of AR-XPS is schematically explained in Figure 2.54. As the sample is tilted, while keeping the x-ray source and detector in fixed positions, the effective information depth decreases by a factor of cosine of the emission angle (θ_n) from the surface normal direction. Conventionally, this relationship is described using the take-off angle, which is defined by the angle from the surface plane. Thus, the take-off angle (θ_s) is 90° minus the emission angle (θ_n). Using θ_s, the depth dependence of the XPS signals for a sample with an overlayer with a thickness of d is described as:

$$I_O = \int_0^d I_{O,\infty} \exp\left[-\frac{z}{\lambda \sin\theta_s}\right] dz = I_{O,\infty}\left(1 - \exp\left[-\frac{d}{\lambda_O \sin\theta_s}\right]\right) \quad (2.25a)$$

$$I_S = \int_d^\infty I_{S,\infty} \exp\left[-\frac{z}{\lambda \sin\theta_s}\right] dz = I_{S,\infty} \exp\left[-\frac{d}{\lambda_S \sin\theta_s}\right] \quad (2.26a)$$

where θ_s is the take-off angle from the surface plane (Figure 2.54). The only difference from eqs. (2.25) and (2.26) is the change of the cosine term for the emission angle (θ_n) to the sine of the take-off angle (θ_s). These equations assume that the overlayer is uniform.

If the sample is homogeneous, then the photoelectron signals with similar kinetic energies from different elements in the sample will have the same exponential dependence on the depth at which the electron is ejected from the atom. So, their relative ratio will have no dependence on the take-off angle. But, if an overlayer is present and its thickness is smaller than

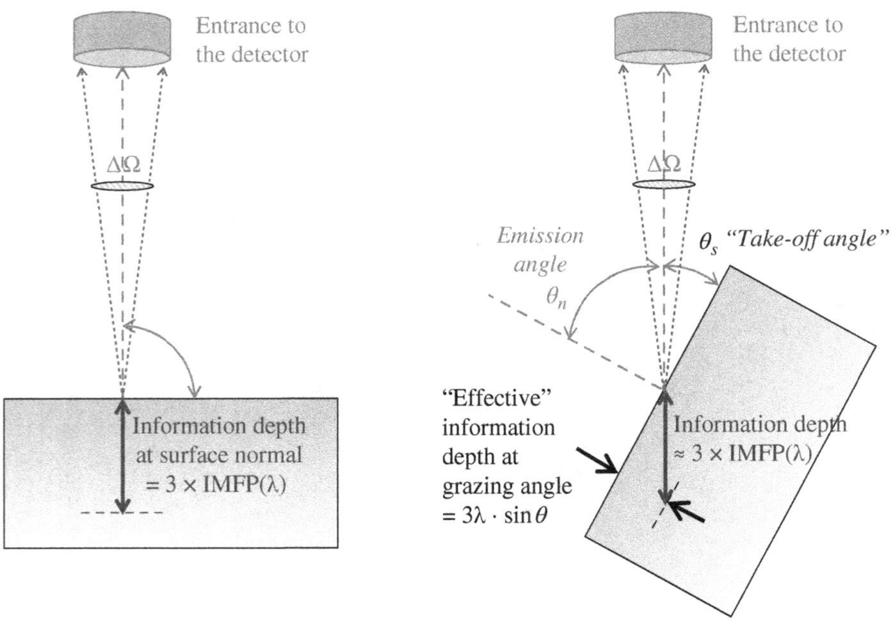

Figure 2.54 Change in the information depth as a function of take-off angle.

the information depth (3λ), then the ratio of the overlayer signal to the substrate signal will vary as a function of take-off angle. By dividing eq. (2.25a) with eq. (2.26a) and assuming $\lambda_O \approx \lambda_S$, we obtain:

$$\frac{I_O(\theta)}{I_S(\theta)} = \frac{I_O^{\infty}}{I_S^{\infty}}\left(\exp\left(\frac{d}{\lambda_O \sin\theta_s}\right) - 1\right) \tag{2.27}$$

Thus, by analyzing the overlayer and substrate signals as a function of take-off angle, one can determine the thickness of the overlayer. If the overlayer is not uniform, the equation can be modified to:

$$\frac{I_O(\theta)}{I_S(\theta)} = \frac{I_O^{\infty}}{I_S^{\infty}} \frac{\gamma\left(1 - e^{-d/\lambda_O \sin\theta_s}\right)}{(1-\gamma) + \gamma \cdot e^{-d/\lambda_O \sin\theta_s}} \tag{2.28}$$

where γ is the fractional coverage of the overlayer. If γ is known, then the overlayer thickness can be obtained from eq. (2.28). If both d and γ are unknown, then eq. (2.28) cannot be solved since there are two unknowns. If another photoelectron signal with drastically different binding energy (thus, with substantially different λ_O) is available, both d and γ can be determined by solving two of eq. (2.28) simultaneously [like Fowkes method in the surface energy calculation in Section 1.5]. Otherwise, the d value obtained from eq. (2.27) should be taken as an approximation calculated assuming the overlayer thickness is uniform.

Let us take a look at an example using AR-XPS to study the compositional change of the polymer surface as a function of depth. Figure 2.55A displays the C 1*s* spectra, measured at take-off angles at 90° and 15°, of a PVA film cast with a saturated aqueous solution on a silicon wafer [73]. As shown in Figure 2.30, we expect only two components (285 eV for C–C species and 286.5 eV for C–O species) for PVA, but we can see a small peak at 289.4 eV of the O–C=O species. Through peak fitting, the presence of C=O species at 287.8 eV is also identified. Considering the next nearest effect, the <u>R</u>COO component is added at 285.7 eV in the peak fitting. From the take-off angle dependence of each component (Figure 2.55B), it is clear that the O–C=O species are mostly present at the surface. By fitting the exponential decay of the COO component as a function of $\sin\theta_s$ and assuming IMFP = 2.7 nm, the thickness of the surface layer enriched with the COO species is estimated to be ~0.3 nm [73]. This suggests that acetate and carboxylate groups, which probably originate from impurities in the PVA sample used to prepare the aqueous solution, are preferentially segregated in the topmost layer of the film.

When the composition changes gradually over the information depth ($=3\lambda$), the degree of freedom increases and the equations to solve become under-defined. Also, the signal intensities of low concentration components and the ones in deeper regions are weak and noisy. This kind of situation can be dealt with the maximum-entropy method [74]. The key process is

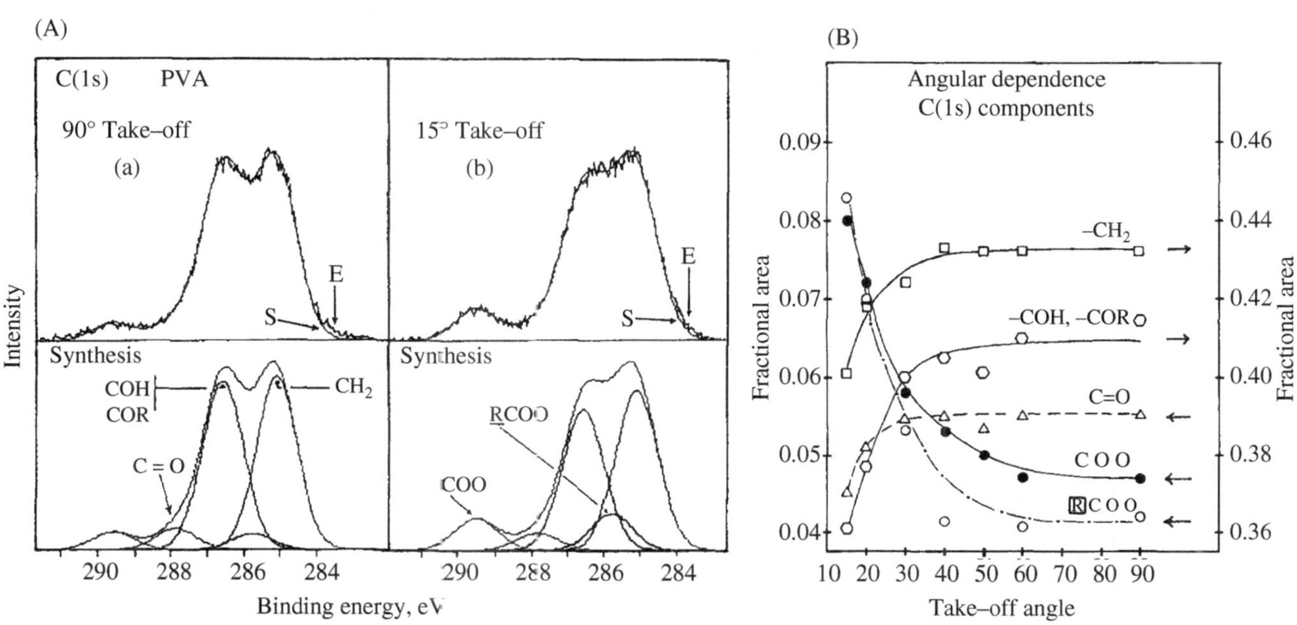

Figure 2.55 (A) C 1*s* spectra of PVA film taken at (a) 90° and (b) 15° take-off angles. (B) Angular dependence of C 1*s* components. Note the different *y*-axis scales marked with arrows. Akhter et al. [73]/with permission of Elsevier.

to iteratively find a layered model with compositional variances as a function of depth, described with the minimum amount of information, until it reproduces the experimental depth profile data with minimum artificial correlations within the data. It should be noted that, as in all peak fitting processes, the goodness of the result cannot exceed that of the data used. Any gross error entered to the algorithm can lead to physically unreasonable profiles. Also, the outcome varies depending on the probability density used which is not trivial to determine.

There is an important instrumental factor to consider in the AR-XPS experiment. The angle step size in AR-XPS is limited by the acceptance angle of the analyzer ($\Delta\Omega$). For the instrument with a blanket irradiation of x-ray over a larger area, the acceptance angle can be set to be small by adjusting the slit size or entrance orifice of the lens. In the instrument using a focused x-ray source for small spot or imaging analyses, the system is designed with a large acceptance angle to increase the detection sensitivity. In that case, the AR-XPS measurement with a small increment step in take-off angle cannot be done. To resolve this issue, a two-dimensional array detector is used to disperse photoelectrons entering the analyzer with different angles with respect to the analyzer axis and count them at different positions in the detector array.

2.5.2 Depth Profile with Ion Sputtering

When the depth of interest is larger than the XPS information depth, the sputter depth profiling method is used. Basically, this method employs a high energy ion gun to sputter atoms from the surface and then analyze the residual elements in the surface with XPS after a certain sputtering time [75]. The species sputtered off the surface during the sputtering can also be analyzed using a mass spectrometer; one of such methods is secondary ion mass spectrometry (SIMS), which is described in Chapter 4. An Ar^+ ion gun is widely used for this purpose. The same ion gun can be used for He, Ne, or Xe by simply changing the gas source for ion generation. Typical ion energy used is a few keV (ranging from 1 to 5 keV). Other ion guns include fullerene (C^{60}) and Ar-cluster sources.

Let us study the basic principle of sputter depth profiling using the data shown in Figure 2.56. The sample was a silicon wafer with a 20 nm thick SiO_2 film grown by thermal oxidation. The sample surface was sputtered with a 3 keV Ar^+ ion gun at a given ion flux for 30 second and the XPS measurement was conducted. This process was repeated in cycles. In Figure 2.56A, we can see that the Si 2*p* peak position is initially 103 eV and gradually shifting to 99 eV, which is accompanied by the decrease in O 1*s* signal. In Figure 2.56B, the atomic-% of Si and O is plotted as a function of total sputter time.

The next step is to convert the sputter time to the depth removed by sputtering. If the total depth sputtered can be measured using an independent method (such as atomic force microscopy or optical profilometry), then the sputter depth at each sputter time can be estimated assuming that the sputter rate is constant. In Figure 2.54B, the sputter depth scale converted from this calibration process is shown at the top. But, if a series of chemically and structurally different layers are present, each layer may sputter at a different rate. Different ions have different sputter efficiencies, which are also a function of the

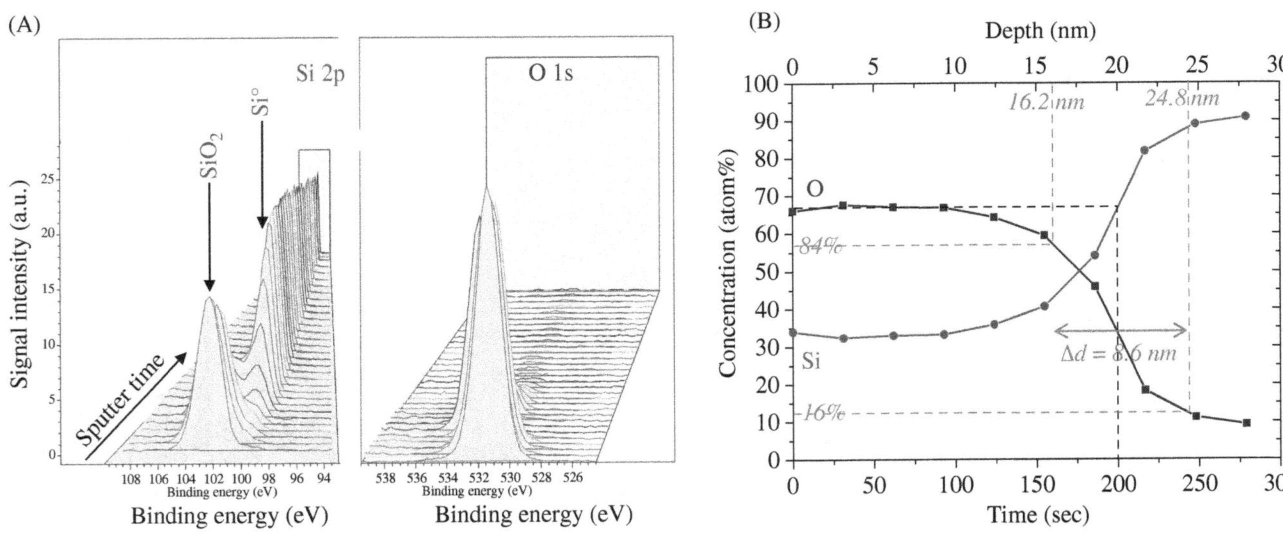

Figure 2.56 Depth profiling of a SiO_2 (20 nm thick) layer on a Si wafer. (A) Si 2p and O 1s XPS spectra. (B) Changes in concentration (atomic-%) of Si and O measured as a function of sputter time. Adapted from Jeff Shallenberger.

energy of impinging ion. Thus, the sputter rate calculated for one ion gun setting cannot be transferred to different operation conditions.

The depth resolution in the ion sputtering process can be estimated with an assumption that the ion concentration detected with XPS has the Gaussian profile. When a Gaussian distribution function is integrated, we get an error function. When we set the max of the error function to 100%, one standard deviation point from the mean position in the Gaussian distribution corresponds to 15.8% and 84.1% of the error function. Thus, we can read the depth points at which the post-sputter signal intensity falls to 16% and 84% of the maximum change. In Figure 2.56B, the O concentration of 66.7 atomic-% can be taken as the maximum signal. Then, the 84% and 16% of that concentration corresponds to 16.2 nm and 24.8 nm, respectively. So, the depth resolution (Δd) of this sputter process can be said to be 8.6 nm.

Here one may notice that the O atomic-% does not decrease to zero even after the sputter depth exceeds the thickness of the original SiO_2 layer. This is due to the atomic layer mixing during the sputtering process. Because of very high KE, the impinging ions can penetrate deep into the sample before they lose all KE through collisions and are stopped. The trajectories of Ar^+ ions entering the SiO_2 surface at the normal incidence with a KE of 1 and 10 keV are shown in Figure 2.57 [76]. The energy transfer to surrounding atoms during these collisions causes subsurface atoms to be displaced from their original positions, which eventually leads to atomic mixing. Also, the species removed by sputtering is not just from the topmost layer, but they can also come from subsurface regions. The dynamics of interlayer mixing and sputtering process can be seen in the snapshot of molecular dynamics simulations for Ga^+ ion and $C_{60}{}^+$ ions impinging onto a Ag(111) surface (Figure 2.58) [77]. Animations for Ar-cluster ions can be seen in the supporting information of a paper by B. Garrison [78]. For these reasons, the depth resolution (Δd) of typical Ar^+ ion sputtering process is around 10 nm.

How can one increase the depth resolution (i.e., decrease Δd)? One way is to minimize the atomic mixing caused by impinging ions. Decreasing the primary ion energy would be the easiest way to reduce the mixing process (Figure 2.57); the lower sputtering efficiency at lower energy could be compensated by increasing the ion flux. Another option is to use a heavier ion; at the same ion energy, the stopping range is much shorter for the heavier ion. Using the same principle of AR-XPS (Figure 2.54), one can increase the Ar^+ ion incidence angle. Then, the depth affected along the ion incidence axis would be the same, but the effective depth affected along the surface normal direction will decrease in proportion to the cosine of the ion incidence angle. Alternatively, different ion guns can be used. For this purpose, $C_{60}{}^+$ and Ar-cluster guns were developed [78–81]. This principle is described further in Chapter 4.

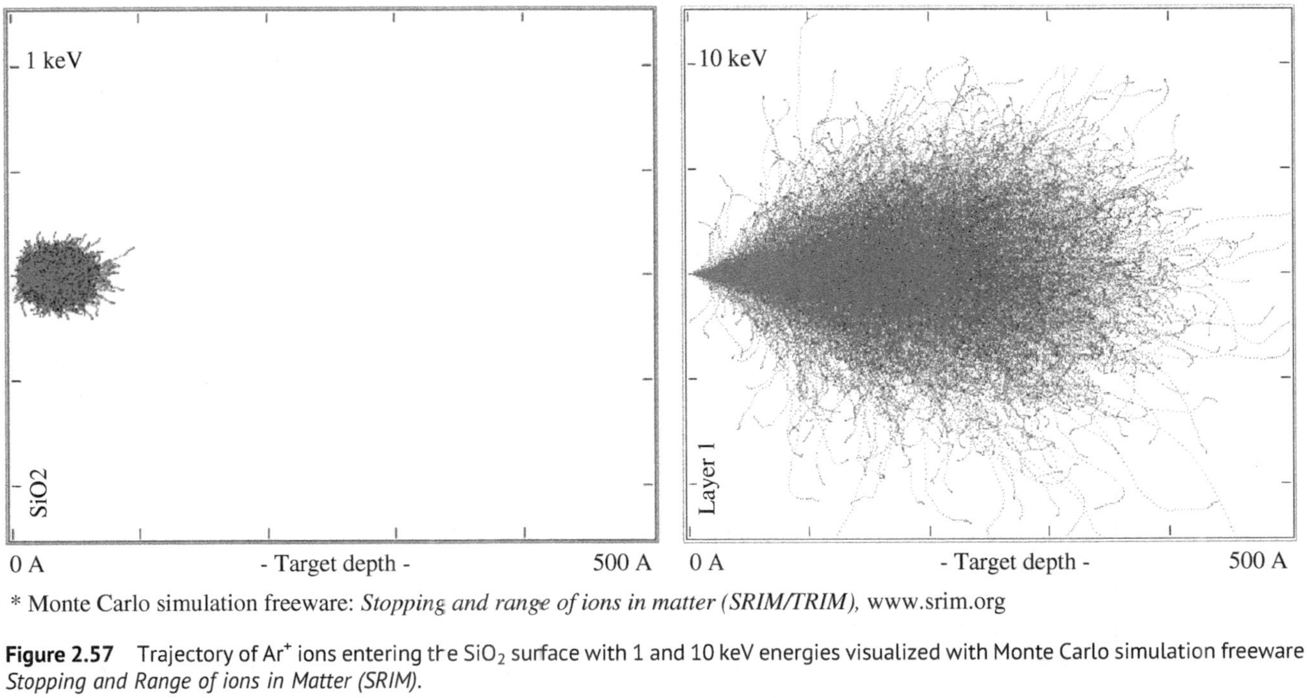

Figure 2.57 Trajectory of Ar^+ ions entering the SiO_2 surface with 1 and 10 keV energies visualized with Monte Carlo simulation freeware *Stopping and Range of ions in Matter (SRIM)*.

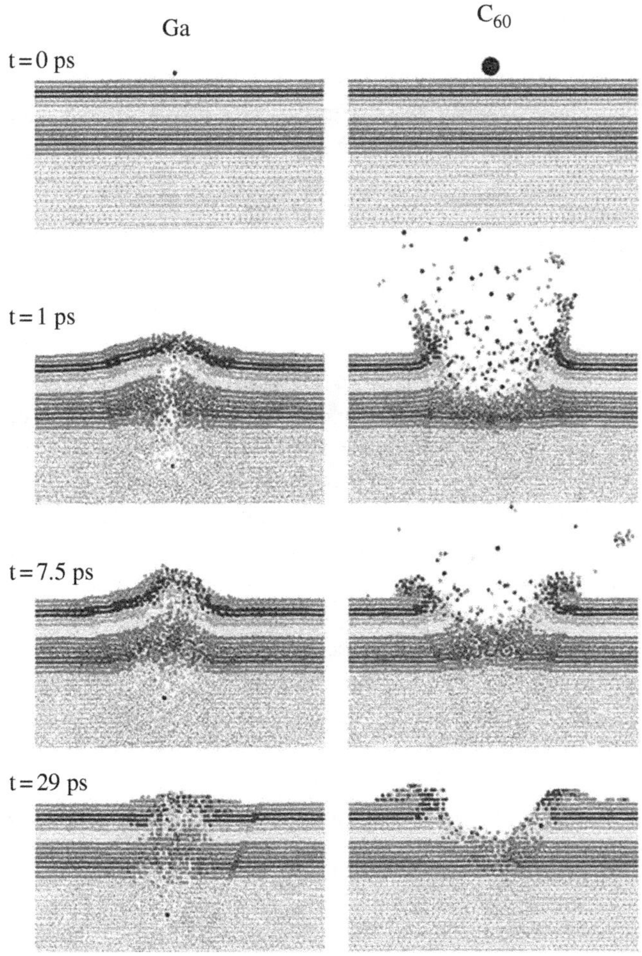

Figure 2.58 Cross-sectional view of the temporal evolution of a typical collision event leading to ejection of atoms due to 15 keV Ga and C_{60} bombardment of an Ag(111) surface at normal incidence. The atoms are colored by original layers in the substrate, and the projectile atoms are black. Postawa et al. [77]/with permission of American Chemical Society.

In addition to the atomic mixing effect, other factors can also affect the depth resolution. Recall the information depth of XPS associated with the IMFP. This is the intrinsic limitation of XPS. Other extrinsic factors such as the roughness of initial surface and ion-induced roughening, and sputtering interval will affect the depth resolution that can be achieved.

Then, what sputter depth should be reported? A convenient way is to simply multiply the sputter time with a known sputter rate obtained using a reference sample that is chemically and structurally similar to the sample under consideration. Another way is to measure the total sputter depth after the experiment and then convert the sputter time to the depth assuming a linear relationship between the sputter time and the depth (i.e., a constant sputtering rate). Once the sputter time scale is converted to the depth scale, then the concentration profile between the minimum and maximum intensity region can be fitted with the error function to get the mean value, which can be considered as the average depth of the layer analyzed.

Lastly, it should be noted that sputtering is an irreversible destructive process. The material under investigation can undergo not only physical changes (interlayer mixing and removal) but also chemical changes. Preferential sputtering of one or some of constituents leads to alteration in stoichiometry [75, 82]. It is difficult to predict what materials will undergo preferential sputtering and how readily it will occur. Empirical evidence shows that atoms with mass closer to bombarding ion mass will be more efficiently sputtered, leaving excess in other elements. Not only preferential "sputter removal" from the surface, but preferential "inward diffusion" is also possible. Figure 2.59 shows the sputter-depth profiles of the network modifier ions, sodium and calcium ions determined with XPS and normalized with the atomic-% of silicon, of soda lime silica glass. Since silicon is the network former (so immobile), it can be used as an internal reference. For the fractured and re-melt surfaces, the Na/Si ratio is around 0.3 initially. As soon as the surface is sputtered with Ar^+ ions for depth profiling, the Na/Si ratio drops to ~0.15. This is due to the inward diffusion of highly mobile Na^+ ions in the glass network upon collisions with Ar^+ ions (in part due to surface charging). This inward diffusion of Na^+ ions is much larger than the information depth at the KE of the Na KLL Auger electron. So, the Na concentration after sputtering is reduced in the XPS analysis. The degree of this inward diffusion appears to saturate at a steady-state value, which is Na/Si ≈ 0.15 in the sputtering condition of Figure 2.59. This kind of inward diffusion of Na^+ ions in the glass network can be reduced if a cluster ion gun (such as $C_{60}{}^+$) is employed [83].

The Ca/Si profiles of the fractured and re-melt surfaces show an opposite trend; the ratio increases from 0.13 to 0.23 over sputtering of ~25 nm. The exact mechanism for this increase is not well understood. Maybe the inward diffusion of Ca^{2+} is smaller than the information depth at the KE of the Ca 2*p* photoelectron, leading to enrichment of Ca^{2+} in the XPS analysis.

If someone attempts Na-depth profiling of the sodium-leached glass with an expectation that the Na/Si ratio would approach the bulk composition value at the boundary of the leached layer with the intact bulk glass, that person will never find the leached layer thickness and may reach a conclusion that the sodium is completely leached out throughout the entire glass. Obviously, this is a wrong conclusion. When the Na/Si ratio reaches the sputter-equilibrium value of the bulk glass (which is 0.15 in the sputtering condition of Figure 2.59), the sputtering process can be stopped [84]. Then, what would be the thickness of the sodium-leached layer? In Figure 2.59, we can see that the Na/Si ratio reaches ~0.15 at ~110 nm. But this

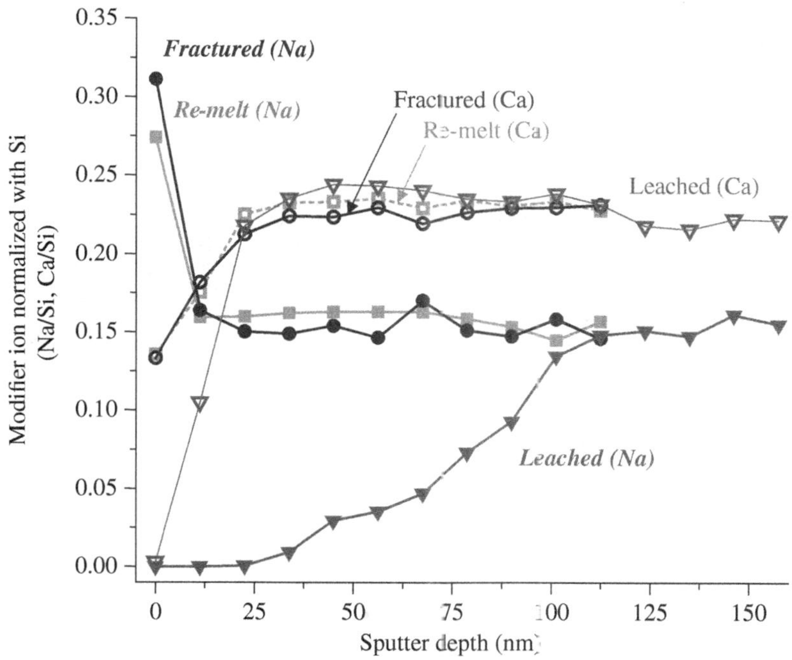

Figure 2.59 Ratio of network modifiers (Na and Ca) vs. network former (Si) concentrations of soda lime silicate glass surfaces: fractured in vacuum, pristine surface after re-melting, and sodium-leached surface by immersing the sample in pH1 solution. The atomic-% of each element was determined with XPS after Ar^+ ion sputtering.

should not be taken as the thickness of the leached layer. We should consider the interlayer mixing by sputtering and the information depth of the XPS (see Figure 2.56B). Thus, the Na/Si profile should be fitted with the error function to determine the leached layer thickness. Then, the thickness is found to be ~80 nm.

2.6 Chemical Analysis with X-ray Absorption Spectroscopy (XAS)

So far, we have studied the photoelectron emission process upon x-ray absorption. In commercial XPS systems, the energy of x-ray that can be used is fixed depending on the type of x-ray source (Section 2.2.1). This energy is high enough to induce emission of photoelectrons from core levels of many atoms into vacuum (Figure 2.2) and chemical analysis is done by measuring their KE and intensity (Figure 2.3). If the x-ray energy can be tuned across the binding energy of core electrons of specific elements, then we can measure the resonance absorption of x-ray as a function of the x-ray energy. The basic principle of x-ray absorption spectroscopy (XAS) is similar to that of UV–VIS absorption spectroscopy. The main difference is that while UV–VIS involves the electronic excitation from occupied molecular orbitals in the valence bands to unoccupied molecular orbitals (i.e., electronic transition within valence states), XAS involves the electronic transition from the core levels to the unoccupied orbitals (Figure 2.60A). XAS is done at a synchrotron radiation facility, since it requires using x-ray with tunable energy. Also, to get a good S/N ratio (see Section 2.2.5), it requires an extremely high brightness (high photon flux). XAS can be done for both bulk and surface analysis, depending on the detection method. Typically, it is done for chemical and structural analysis of a known atomic species in the sample, rather than to determine overall elemental compositions of an unknown sample. In addition, XAS can be done for *in situ* analysis of chemical processes and/or under extreme conditions of temperature and pressure for minor elements with trace abundance in various types of materials.

2.6.1 Resonant X-ray Absorption

Like UV–VIS spectroscopy, XAS is governed by the Beer–Lambert Law: $I_t = I_i e^{-\sigma d}$ where I_t and I_i are the intensity of the transmitted and incident beam, respectively, d is the sample thickness and σ is the absorptivity. The absorptivity here is essentially the same as the cross section discussed in eq. (2.23). The only difference from XPS is that the final state is not the free electron with high KE in the continuum states but the unfilled bound states near or just above the

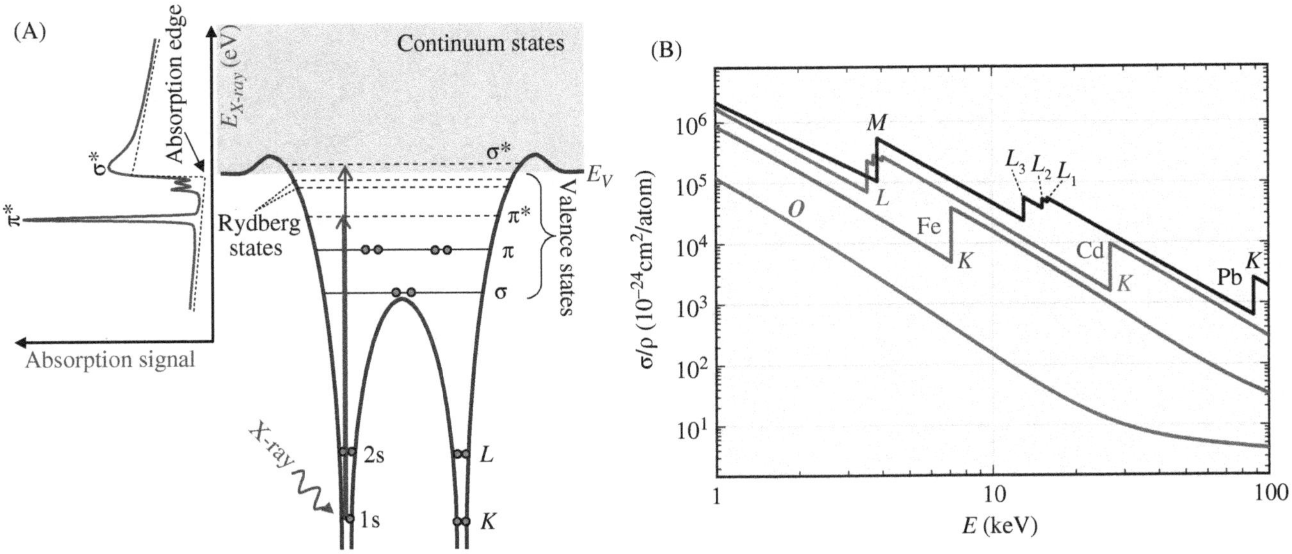

Figure 2.60 (A) Schematic electronic states and corresponding NEXAFS *K*-shell spectrum of a diatomic molecule. (B) Absorption cross-section (σ/ρ) for several elements over the x-ray energy range of 1–100 keV. Newville [85]/with permission of Mineralogical Society of America.

vacuum level (Figure 2.60A). The unfilled bound states are anti-bonding orbitals (such as σ* and π^*) of covalent compounds or conduction bands of metals and semiconductors.

In the x-ray energy regime, the absorptivity is a function of the atomic number (Z) and mass (m) of the element and the x-ray energy (E): $\sigma \approx \frac{\rho Z^4}{mE^3}$, where ρ is the sample density [85]. Figure 2.60B shows the energy dependence of σ/ρ for oxygen, iron, cadmium, and lead in the 1–100 keV range of x-ray energy [85]. The sharp rise in σ/ρ corresponds to the resonant absorption edge where the incident x-ray energy is equal to the binding energy difference between the core level electron and the final state to which the electron is excited. For a given atom, there could be a series of adsorption edges. If the electronic excitation is from $n = 1$, it is called the K-edge. If it is from $n = 2$, it is called L-edge. The L-edge can have three subsets: L_1, L_2, and L_3 involving 2*s*, $2p_{1/2}$, and $2p_{3/2}$ state, respectively. If the core level from which the electronic excitation occurs is $n = 3$, it is called the M-edge.

For molecules or solids, fine structures originating from the molecular orbitals of the atom to which the core-level excitation takes place, and interference patterns due to backscattered electrons from neighboring atoms are superimposed onto the atomic absorption edge. The technique probing the molecular orbital states is near-edge x-ray absorption fine structure (NEXAFS, also called x-ray absorption near-edge spectroscopy, XANES), which typically occurs in the x-ray energy range from ~10 eV below the absorption edge up to ~50 eV above the edge (Figure 2.60A) [86]. The technique studying the interference by backscattered electrons is extended x-ray absorption fine structure (EXAFS), which typically occurs over the extended energy range up to ~1000 eV above the edge [87].

The attenuation of x-ray due to absorption by the sample can work for bulk analysis but not for surface analysis (because *d* is so small). So, an alternative method should be used. Same as in the photoelectron emission process, the x-ray absorption process is accompanied by the relaxation processes discussed in Section 2.1.2 (Figure 2.6). Thus, the x-ray absorption yield can be quantified by measuring the Auger electron yield (AEY) or the x-ray fluorescence yield (FY). In the hard x-ray regime (>10 keV), x-ray fluorescence is more likely to occur than Auger emission, while Auger processes dominate in the soft x-ray regime (<2 keV). The Auger electron can undergo inelastic scattering processes and emerge from the sample with a KE less than its original energy; this is the same as inelastic scattering of photoelectrons (Figure 2.2). The inelastic Auger intensity will follow the elastic one. If the scattered electron emission yield is detected above a certain cutoff value, it is called partial electron yield (PEY). Because of the flexibility in choosing the cutoff electric field, PEY can avoid some of the interference problems encountered in AEY detection (such as detecting some photoemission peaks with KE falling into the Auger electron energy range). If the yield of electron emissions with all energies is measured, it is called total electron yield (TEY). The TEY signal is dominated by low-energy secondary electrons with KE below 20 eV. Except different signal-to-background and signal-to-noise ratios, AEY, PEY, TEY, ad FY give almost identical NEXAFS spectra.

One main advantage of AEY is the surface sensitivity since its intensity is governed by the Auger electrons generated within the information depth which is three times the IMFP, as discussed in Section 2.1.4 for the surface sensitivity of XPS (Figures 2.10 and 2.11). The main advantage of PEY and TEY is the simplicity of the experimental method and stronger signal intensity. The FY detection does not require vacuum; so, it is suitable for *in-situ* analysis in high pressure gases or liquid environments.

As indicted by the Beer–Lambert law, the measured signal depends on the incident x-ray intensity (I_i). Thus, the recorded spectra need to be corrected for variations of the incident intensity as a function of photon energy. The variation in x-ray flux with photon energy is common because of energy-dependent reflectivity change in x-ray optics (monochromator) in the synchrotron beamline. Also, the measured signal intensity of a specific atom may have a large background from other elements in the sample or substrate. The intensity normalization is typically done by dividing the experimental spectra with the detector signal measured with a reference grid. If the sample of interest is an adsorbate, then the clean substrate signal can be used for normalization.

2.6.2 Near-Edge X-ray Absorption Fine Structure (NEXAFS)

NEXAFS is used to determine the formal oxidation state and coordination chemistry of the absorbing atom and the orientation of adsorbed molecules [86, 88]. Figure 2.61 compares the XPS and NEAFS spectra of carbon monoxide (C≡O) on Ni(100) surface, from which we can see the main difference between these two techniques [89]. In the XPS spectrum taken with $h\nu = 550$ eV, the Auger signals of O KVV fall between the Ni valence band (VB) and Ni 3p photoelectron (PE) peaks (Figure 2.61A). The O KVV emission peak is centered at KE = 511 eV and has multiple components typical of CO molecules adsorbed on metal surfaces. Figure 2.61B displays the NEAXFS spectra recorded with the AEY mode (with the analyzer set to KE = 511 eV with a resolution of 8 eV) collected at two incident angles: $\theta = 4°$ and 90° with respect to the surface. As the x-ray photon energy is scanned from 500 eV to 600 eV, the Ni VB PE peak is first swept through the fixed KE window of the analyzer. Around $h\nu = 530$ eV, the threshold of the O 1s excitation is reached and the O KVV electron starts passing through

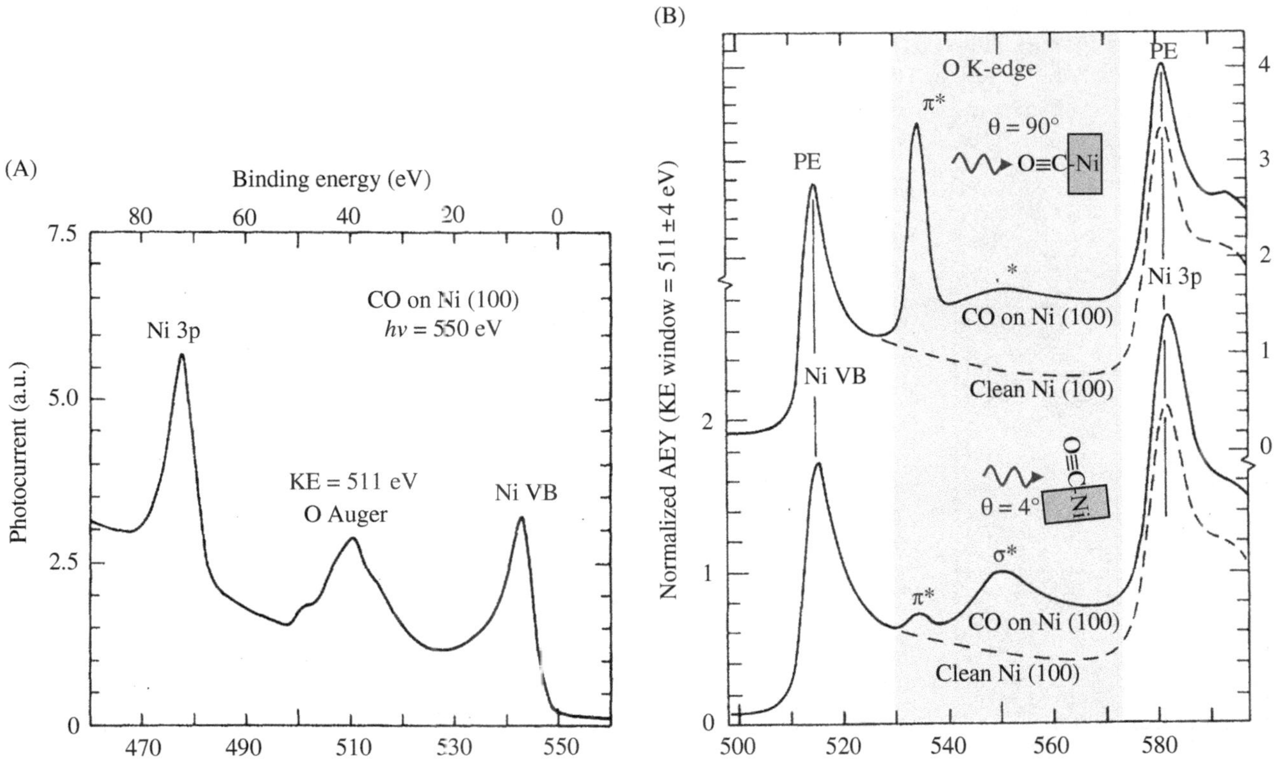

Figure 2.61 (A) XPS and (B) NEXAFS spectra of CO on Ni(100). The photon energy of the XPS spectrum is 550 eV, and the NEAFS spectrum is AEY at a kinetic energy of 511 eV with a resolution of 8 eV. Stöhr and Jaeger [89]/with permission of American Physical Society.

the analyzer. Between 530 and 570 eV, two pronounced absorption resonances are detected with different relative intensities at $\theta = 4°$ and 90°. As $h\nu$ increases further, the Ni 3p PE signal is swept through the analyzer and detected.

The x-ray beam from a synchrotron radiation source is polarized, which allows to determine the orientation of an adsorbate molecules on a surface. The dependence of the x-absorption cross section on the angle (φ) between the electric field of x-ray and the intra-molecular symmetry axis can be described as [89]:

$$\sigma(\varphi) \propto 1 + \frac{\beta}{2}\left(3\cos^2\varphi - 1\right) \tag{2.29}$$

where β is the asymmetry parameter of a given electronic excitation: 2 for $s \rightarrow \sigma^*$ and −1 for $s \rightarrow \pi^*$. In the case of C≡O on Ni(100), the molecule is adsorbed with its molecular axis normal to the surface. So, at the $\theta = 90°$ incident angle, the E-field of x-ray is aligned with the π orbital; so, the $1s \rightarrow \pi^*$ absorption signal is stronger. At the $\theta = 4°$ incident angle, the E-field of x-ray is aligned better with the σ bond; thus, the $1s \rightarrow \sigma^*$ absorption signal is stronger. If the adsorbate is ethylene or benzene, the result will be different. These molecules adsorb with the molecular axis or plane parallel to the surface. So, the $1s \rightarrow \sigma^*$ signal is larger at $\theta = 90°$ (because the x-ray E-field is parallel to the surface), and the $1s \rightarrow \pi^*$ signal becomes larger at $\theta = 4°$ (where the x-ray E-field is mostly normal to surface).

Similar to the valence band features in XPS (Section 2.3.4), the exact peak positions and shapes in NEXAFS spectra involving resonance with antibonding orbitals are difficult to predict. It requires first-principle theoretical calculations [86]. Still, there are some useful "rules of thumb" that can be used for peak interpretation for NEXAFS analyses of organic molecules. Figure 2.62 compares the experimental and theoretically calculated C–K edge and O–K edge spectra of carbon monoxide (C≡O), formaldehyde ($CH_2C{=}O$), and methanol (CH_3–OH) [86]. In the K-shell spectra of low-Z molecules, the $1s \rightarrow \pi^*$ peak, if present, is the lowest energy peak. During the excitation of the core electron by x-ray absorption, there is an upward relaxation shift of the 1s orbital and a downward shift of the outer orbitals due to an increased Coulombic interaction with the core hole. For that reason, the $1s \rightarrow \pi^*$ peak appears slightly below the ionization potential (IP) of the 1s electron. Obviously, the $1s \rightarrow \pi^*$ excitation is observed only for molecules with π bonding. For example, methanol does not show the π^* resonance peak. The natural width of the π^* resonance peak is governed by the lifetime of the excited state (see Section 2.4.2). Within a series of molecules containing the same atoms, the intensity of the $1s \rightarrow \pi^*$ peak increases with the bond order between the atoms. Similarly, the peak position of the $1s \rightarrow \sigma^*$ excitation also increase with the bond order. The σ* resonances have asymmetric shape which is a consequence of a combination of lifetime broadening and molecular vibration along the bond direction. The small features between the π^* resonance and the IP energy are excitations to Rydberg states (highly excited bound states of a Coulomb potential) or their hybridizations with hydrogen-derived antibonding orbitals.

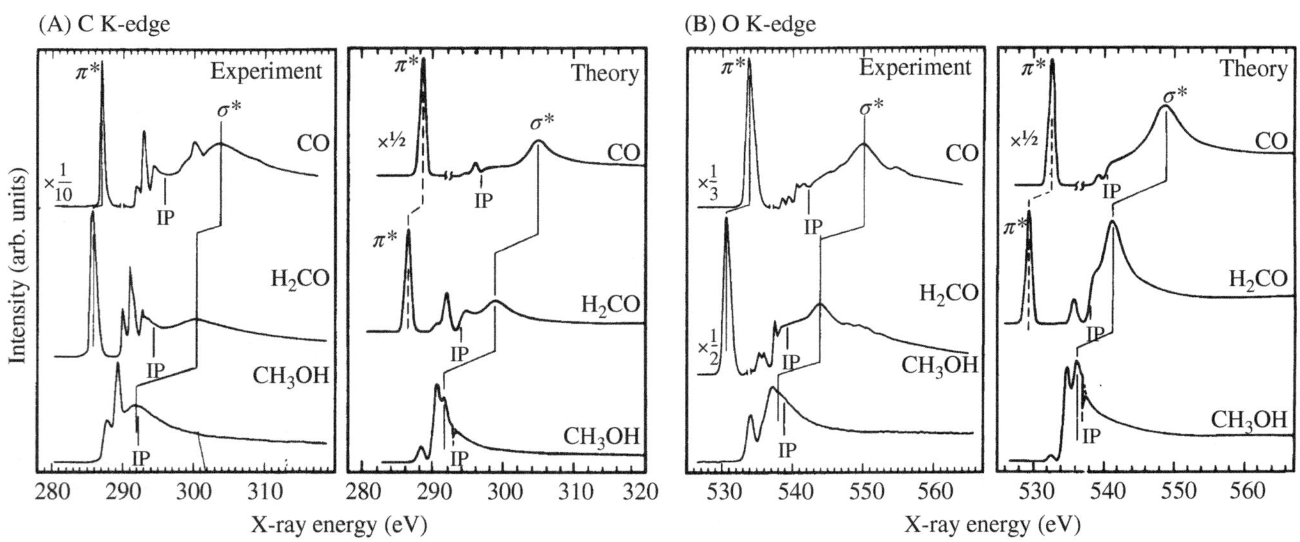

Figure 2.62 Comparison of experimental and theoretically calculated (A) C K-edge and (B) O-Kedge spectra of CO, H_2CO, and CH_3OH. Stöhr [86]/with permission of Springer Nature.

2.6.3 Extended X-ray Absorption Fine Structure (EXAFS)

EXAFS is used to determine the distances, coordination numbers, and species of the neighbors of the absorbing atom [85]. This can be understood in terms of the wave behavior of the photoelectron created through the x-ray absorption process. The *KE* of electrons excited to the continuum state above the vacuum level (E_V) in Figure 2.60A can be described as $KE = h\nu - BE = p^2/2m_e$, where p and m are the momentum and mass of electron, respectively. Then, we can calculate de Broglie wavelength: $\lambda_e = h/p = h/\sqrt{2m_e(h\nu - BE)}$. For a photoelectron with KE = 1 eV, the wavelength is 1.23 nm. For KE = 100 eV, $\lambda_e = 1.23$ Å; for KE = 1000 eV, $\lambda_e = 0.39$ Å. The wavefunction of this photoelectron propagates away from the atomic core where excitation occurs, and it is partially backscattered by the surrounding atoms in the medium and returns to the absorbing atom. The interference between the outgoing wave and the backscattered wave modulates the amplitude of the final state wavefunction, which in turn modulates the transition probability defined by eq. (2.23) and thus the absorption cross-section, $\sigma(h\nu)$. By analyzing the oscillation patterns in $\sigma(h\nu)$ above the absorption edge, the structural information of neighboring atoms can be obtained. This is the basis of EXAFS.

In a three-dimensional sample, the photoelectron wavefunction spreads radially out, and the scattering probability decreases as the distance from the photoelectron emitting atom increases. The oscillation patterns are produced mostly by atoms close to the x-ray absorbing atom, which makes EXAFS sensitive to the short-range order in local structures (i.e., distances and symmetry of the nearest and next nearest atoms). This is the main difference from XRD which is sensitive to the long-range order, so useful for analysis of crystalline lattices.

We can use an example shown in Figure 2.63 to grasp the EXAFS analysis process. Figure 2.63A shows the Fe K-edge EXFAS spectrum of FeO [85]. The first large peak near the absorption edge is the NEXAFS signal. After that, we can see oscillatory patterns in $\sigma(h\nu)$, up to ~3000 eV from the edge, originating from interferences with back-scattered electrons. The oscillatory pattern above the absorption edge can be obtained by subtraction the smooth (unmodulated) background function, $\sigma_o(h\nu)$, representing the absorption of an isolated atom (as shown in Figure 2.60B). This defines the fine-structure function, $\chi(k)$ [85]:

$$\chi(k) = \frac{\sigma(k) - \sigma_o(k)}{\Delta\sigma} \tag{2.30}$$

where $k = 2\pi/\lambda_e$ is the wavevector and $\Delta\sigma$ is the measured jump in the absorption at the threshold energy. In Figure 2.63A, we can see that the oscillation decays quickly as the energy increases. To emphasize the oscillation, $\chi(k)$ is often multiplied by k^2 or k^3, as shown in Figure 2.63B. The different frequencies apparent in the $\chi(k)$ oscillations correspond to different positions of atoms in the neighboring coordination shells. By applying the Fourier transform to $\chi(k)$ with proper scattering factors of neighboring atoms, the radial distribution of atoms around the x-ray absorbing atom is obtained (Figure 2.63C). The details of data processing algorithm are beyond the scope of this book and can be found at [87].

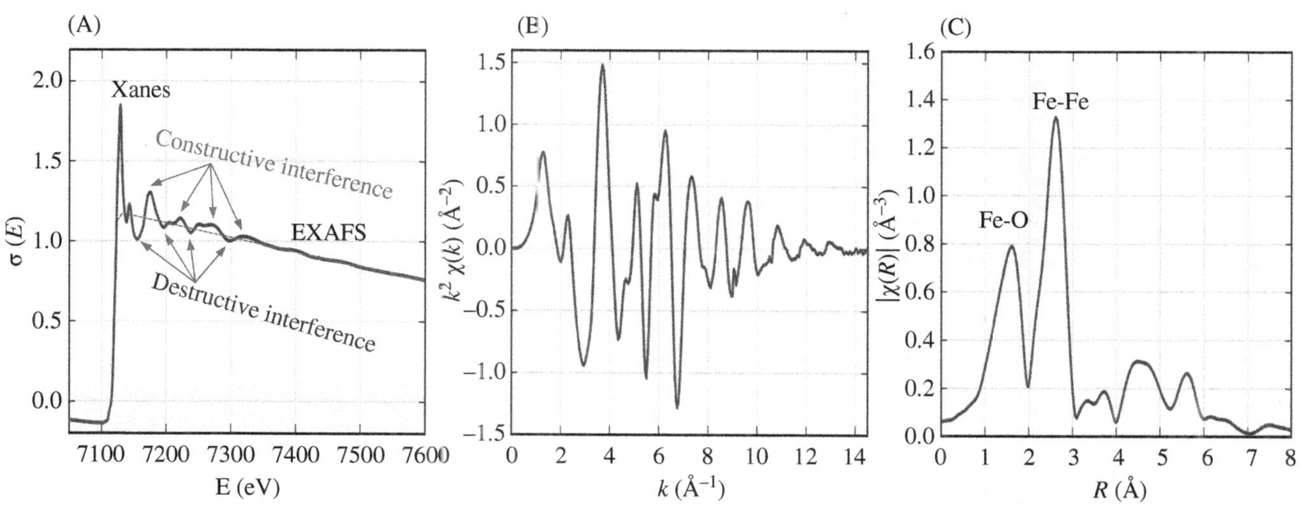

Figure 2.63 (A) Fe K-edge EXAFS spectrum of FeO, (B) isolated EXAFS weighted by k^2 to emphasize the oscillation in the high-k portion, and (C) Fourier transform of the k-weighted EXAFS showing the radial distribution of Fe–O and Fe–Fe neighbors. Newville [85]/with permission of Mineralogical Society of America.

2.7 Further Reading for XPS

In 2020, a group of XPS experts made significant efforts to organize a special topic collection featuring articles that address challenges in XPS analysis, with a hope to help researchers to recognize problematic and erroneous data and provide guides for best practices, thus assisting analysts in conducting and reporting reliable and useful information in the literature. The introductory paper is published in *Journal of Vacuum Science and Technology* in 2021 [1], and the full collection of articles related to surface analysis using XPS can also be found at https://pubs.aip.org/jva/collection/1440/Special-Topic-Collection-Reproducibility. Among them, the articles listed below will be helpful for students or new researchers who are about to start learning XPS or need to find proper ways of doing experiments and/or interpreting collected data. But it should be noted that these papers are just general guidance; without proper experiences, it would be difficult to fully understand and grasp the details described in these papers. So, it is very helpful to discuss with or at least check with experienced users at the beginning.

Stevie FA, Donley CL. Introduction to x-ray photoelectron spectroscopy. Journal of Vacuum Science & Technology A. 2020;38:063204. https://doi.org/10.1116/6.0000412

Stevie FA, Garcia R, Shallenberger J, Newman JG, Donley CL. Sample handling, preparation and mounting for XPS and other surface analytical techniques. Journal of Vacuum Science & Technology A. 2020;38:063202. https://doi.org/10.1116/6.0000421

Baer DR, Artyushkova K, Cohen H, Easton CD, Engelhard M, Gengenbach TR, et al. XPS guide: charge neutralization and binding energy referencing for insulating samples. Journal of Vacuum Science & Technology A. 2020;38:031204. https://doi.org/10.1116/6.0000057

Powell CJ. Practical guide for inelastic mean free paths, effective attenuation lengths, mean escape depths, and information depths in x-ray photoelectron spectroscopy. Journal of Vacuum Science & Technology A. 2020;38:023209. https://doi.org/10.1116/1.5141079

Engelhard MH, Baer DR, Herrera-Gomez A, Sherwood PMA. Introductory guide to backgrounds in XPS spectra and their impact on determining peak intensities. Journal of Vacuum Science & Technology A. 2020;38:063203. https://doi.org/10.1116/6.0000359

Major GH, Fairley N, Sherwood PMA, Linford MR, Terry J, Fernandez V, et al. Practical guide for curve fitting in x-ray photoelectron spectroscopy. Journal of Vacuum Science & Technology A. 2020;38:061203. https://doi.org/10.1116/6.0000377

Gengenbach TR, Major GH, Linford MR, Easton CD. Practical guides for x-ray photoelectron spectroscopy (XPS): interpreting the carbon 1s spectrum. Journal of Vacuum Science & Technology A. 2021;39:013204. https://doi.org/10.1116/6.0000682

Shard AG. Practical guides for x-ray photoelectron spectroscopy: Quantitative XPS. Journal of Vacuum Science & Technology A. 2020;38:041201. https://doi.org/10.1116/1.5141395

Brundle CR, Crist BV. X-ray photoelectron spectroscopy: a perspective on quantitation accuracy for composition analysis of homogeneous materials. Journal of Vacuum Science & Technology A. 2020;38:041001. https://doi.org/10.1116/1.5143897

Major GH, Avval TG, Moeini B, Pinto G, Shah D, Jain V, et al. Assessment of the frequency and nature of erroneous x-ray photoelectron spectroscopy analyses in the scientific literature. Journal of Vacuum Science & Technology A. 2020;38:061204. https://doi.org/10.1116/6.0000685

Major GH, Pinder JW, Austin DE, Baer DR, Castle SL, Čechal J, et al. Perspective on improving the quality of surface and material data analysis in the scientific literature with a focus on x-ray photoelectron spectroscopy (XPS). Journal of Vacuum Science & Technology A. 2023;41:038501. https://doi.org/10.1116/6.0002437 [45]

There are also many books specifically describing XPS and related techniques. A few examples are listed below:

van der Heide P. X-Ray Photoelectron Spectroscopy: An Introduction to Principles and Practices. John Wiley & Sons; 2011, https://doi.org/10.1002/9781118162897

Watts JF, Wolstenholme J. An Introduction to Surface Analysis by XPS and AES (Second Edition). John Wiley & Sons; 2019. 978-1-119-41758-3

Many XPS data are processed with the *CasaXPS* software [27]. Users can find many useful information and YouTube videos from their website (http://www.casaxps.com/). Not only XPS, but AES, SIMS, and LEIS data can also be processed with the *CasaXPS* program.

Moulder JF, Stickle, WF, Sobol E, Bomben KD. Handbook of X-ray Photoelectron Spectroscopy (Physical Electronics Division, Perkin-Elmer Corporation; 1992) can be found in the Internet. This handbook compiles standard spectra useful for identification and interpretation of XPS data. Element-specific references can be found at the "XPS Reference Pages" by the Surface Science Western lab (http://www.xpsfitting.com/) [47]. It would be extremely helpful to study the examples listed for each element of interest before doing actual peak fitting of real data. Also, it would be very helpful to study examples of bad peak fitting listed in "Bad Peak-fits Explained" in the xpslibrary.com website (https://xpslibrary.com/bad-peak-fits-explained/).

Other experiments that could be done with XPS systems, if a proper UV or electron source is installed, is ultraviolet photoelectron spectroscopy (UPS) and inverse photoelectron spectroscopy (IPES). A good tutorial article about UPS can be found at [90]. A good review article on using IPES for measuring unoccupied states of organic semiconductors can be found at [91].

Practice Problems

2.1 Why can't we see hydrogen in XPS?

2.2 Here is the Si 2*p* spectra of a series of SiO_2 layers thermally grown on Si wafers. Calculate the IMFP of the Si 2*p* photoelectron that has a kinetic energy of ~1387 eV.

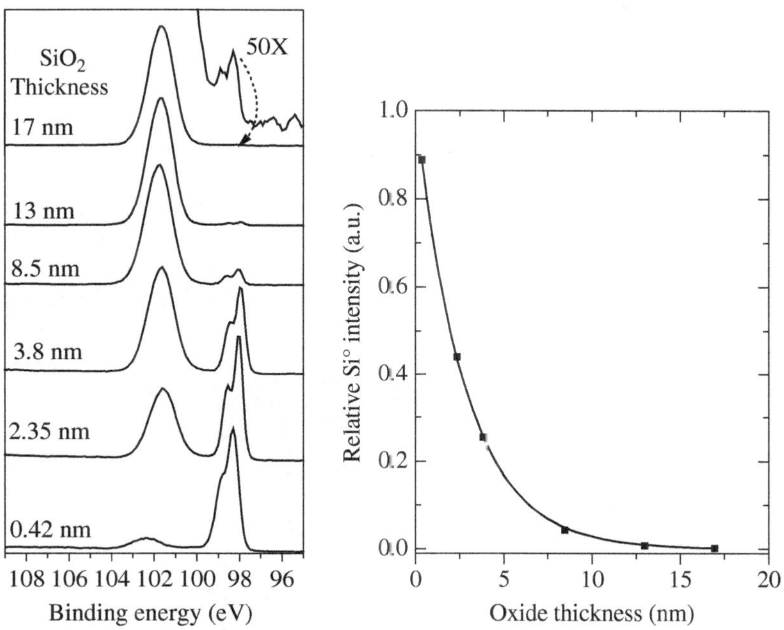

2.3 Use the Cu 2*p* XPS data shown in Figure 2.32A.

a) These compounds show four peaks. Assign each BE peak to electronic excitation process:
   i) 932–936 eV:
   ii) 940–945 eV:
   iii) 952–956 eV:
   iv) 962–964 eV:

b) Compare the peaks of CuF_2, $CuCl_2$, and $CuBr_2$ in the (i) range. You can see the peak position is systematically shifting. Explain the reason.

c) For the quantitative analysis, you need to determine the RSF of different elements using these reference samples. For that purpose, you need to compare the relative intensity of anion ligand and the copper peak, for example, the area Cl 2*p* and two peaks of the Cu signals out of these four peaks. Which two should be chosen and why? And, what RSF value should be used?

2.4 The Ce 3*d* XPS spectra of two cerium oxides are shown here. Based on the spin–orbit splitting principle, we expect a doublet peak with a 2:3 relative intensity ratio. But, we can see four peaks for Ce^{3+} and six peaks for Ce^{4+}. Mark peaks that belong to $3d_{3/2}$ and $3d_{5/2}$, respectively, for each oxidation state and explain the source of multiple peaks.

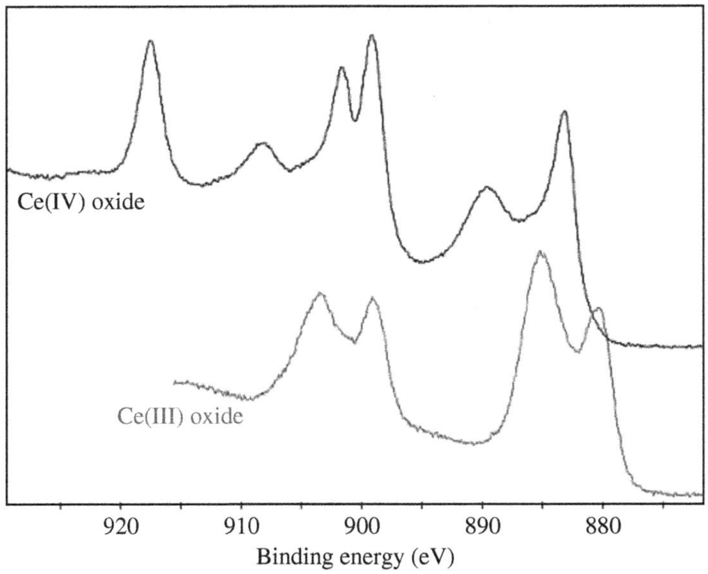

2.5 Shown below is the Ni 2*p* XPS spectra of a reference sample (oxide of nickel). The data are fitted following the examples that are commonly found in google search with a keyword "Ni 2*p* XPS spectra." But, they are all wrong. Discuss what mistakes are made in these peak fittings. Compare the spectra with some reference data found in Surf. Sci. Spectra 3, 231–238 (1994) https://doi.org/10.1116/1.1247751 or in XPS Reference Pages [47]. What is the reference material used in this XPS analysis. It is a single compound.

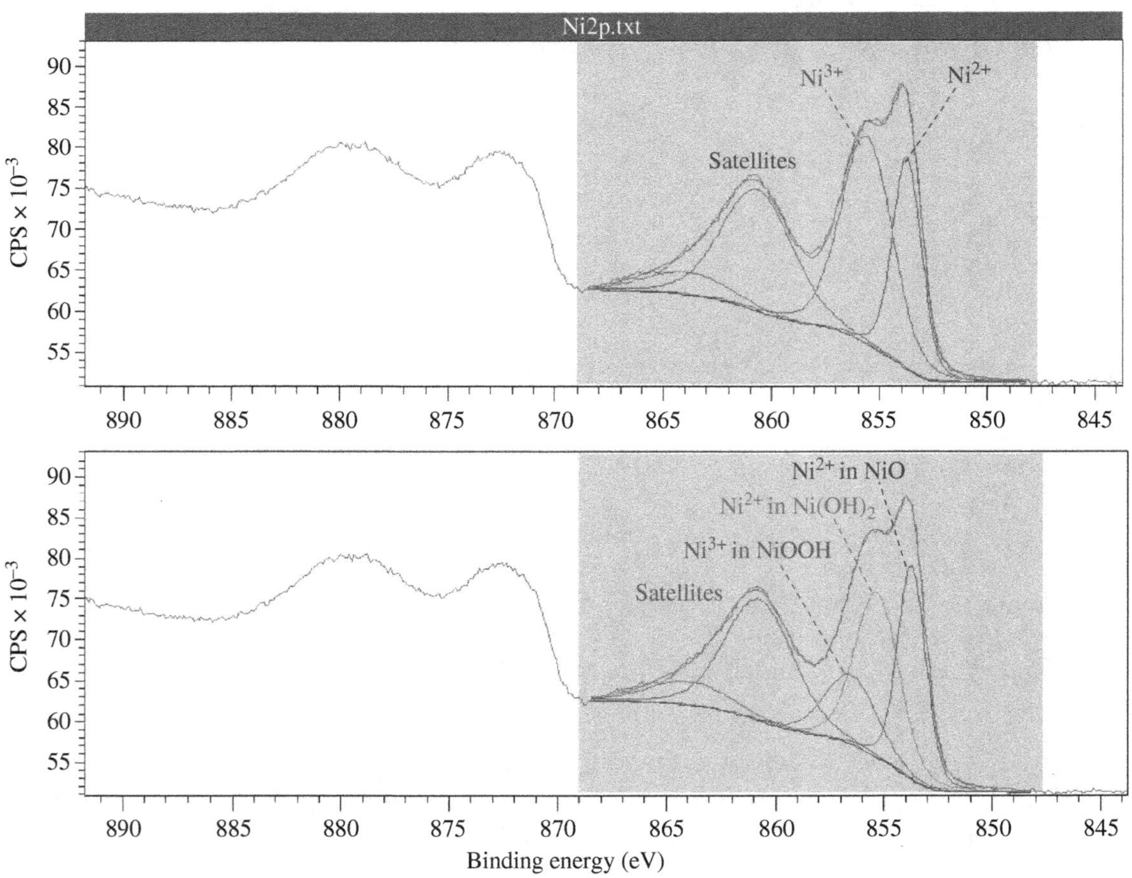

2.6 A WS_2 film (only a few layers) grown on an alumina substrate was analyzed in ultra-high vacuum (UHV) and NAP conditions (5 mbar of ambient air or water) using a *SPECS Enviro-ESCA*^{TM} system (Figure 2.20) equipped with a monochromatic Al-Kα excitation. The survey data from this analysis is given below. You can compare the spectra shown here with the reference spectra that can be found in Surface Science Spectra 25, 014002 (2018) https://doi.org/10.1116/1.5030093.

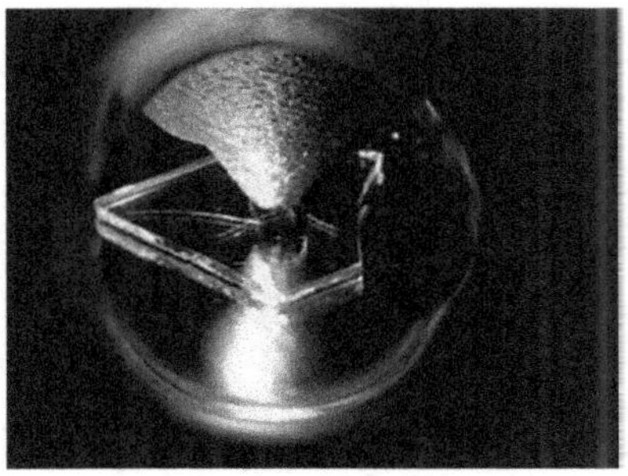

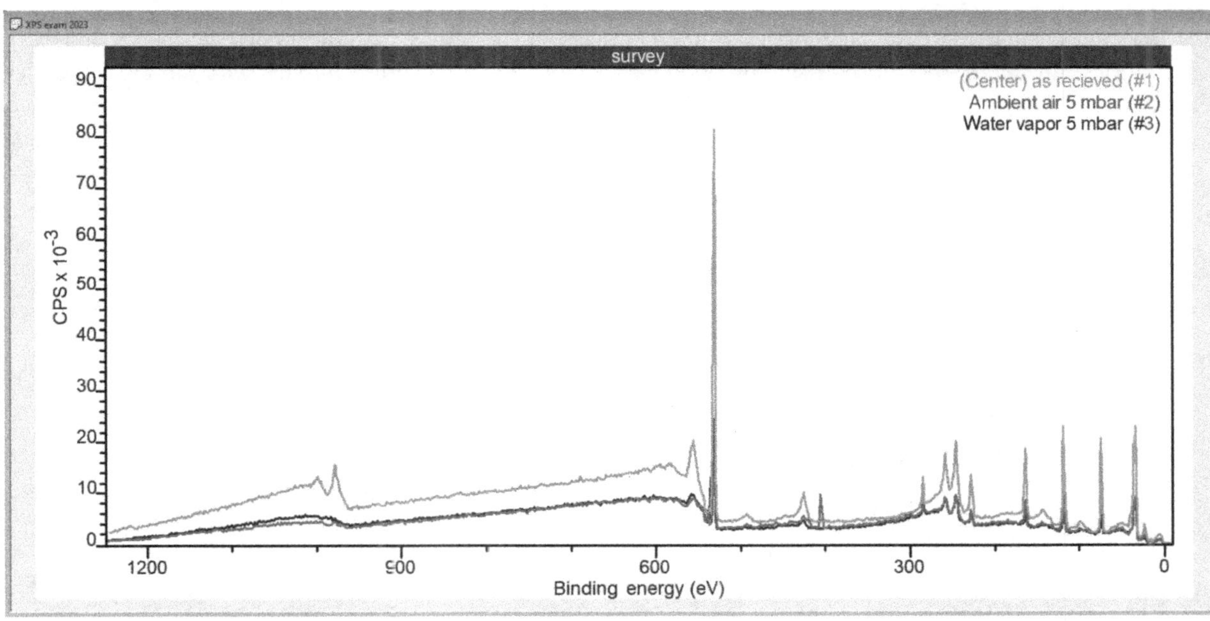

We determined the atomic percentage of all elements identified in each survey spectrum using the *CasaXPS*'s default function (downloaded from [27]). Here are the at-% values that *CasaXPS* suggested.

Peak	UHV survey (#1)	5 mbar air (#2)	5 mbar water (#3)
W 4d	1.82	2.20	1.74
S 2p	12.07	12.92	10.66
Al 2p	38.93	47.48	39.21
O 1s	47.18	37.40	48.39

Now we can see that the at-% ratio of W and S does not match with the stoichiometry of WS_2 at all. Upon *inspecting each region of the peak in the fitted survey spectrum*, it was found that *CasaXPS*'s "Create Regions" function did not define the spectral range. It picked the W $4d_{5/2}$ peak only, while it used the RSP value for both C $4d_{3/5}$ and $4d_{5/2}$. So, we tried two different methods to correct that error, and the updated results are shown below:

Method#1 – Adjust W 4*d* region to cover both $4d_{3/2}$ and $4d_{5/2}$.

Peak	UHV survey (#1)	5 mbar air (#2)	5 mbar water (#3)
W 4d	4.45	4.35	3.83
S 2p	11.75	12.64	10.43
Al 2p	37.88	46.44	38.38
O 1s	45.91	36.58	47.36

Method#2 – Keep the region same and use RSF of $4d_{5/2}$ only.

Peak	UHV survey (#1)	5 mbar air (#2)	5 mbar water (#3)
W $4d_{5/2}$	2.45	3.67	2.91
S 2p	11.99	12.73	10.53
Al 2p	38.67	46.77	38.75
O 1s	46.87	36.84	47.82

The results obtained with proper ranges of the W 4*d* peaks are still far from the stoichiometry ratio of W and S expected for tungsten disulfide. So, let us try using the W 4*f* peak in the survey spectrum collected with the 100 eV pass energy.

Peak	UHV survey (#1)	5 mbar air (#2)	5 mbar water (#3)
W f	3.07	3.99	3.30
S 2p	11.92	12.62	10.49
Al 2p	38.43	46.89	38.59
O 1s	46.58	36.51	47.62

The W:S stoichiometry ratio is still far from the expected 1:2 ratio. We can try the quantitative analysis using the high-resolution spectra collected with 50 eV pass energy. Here are the high-resolution spectra of W 4*f*, S 2*p*, Al 2*p*, and O1*s*:

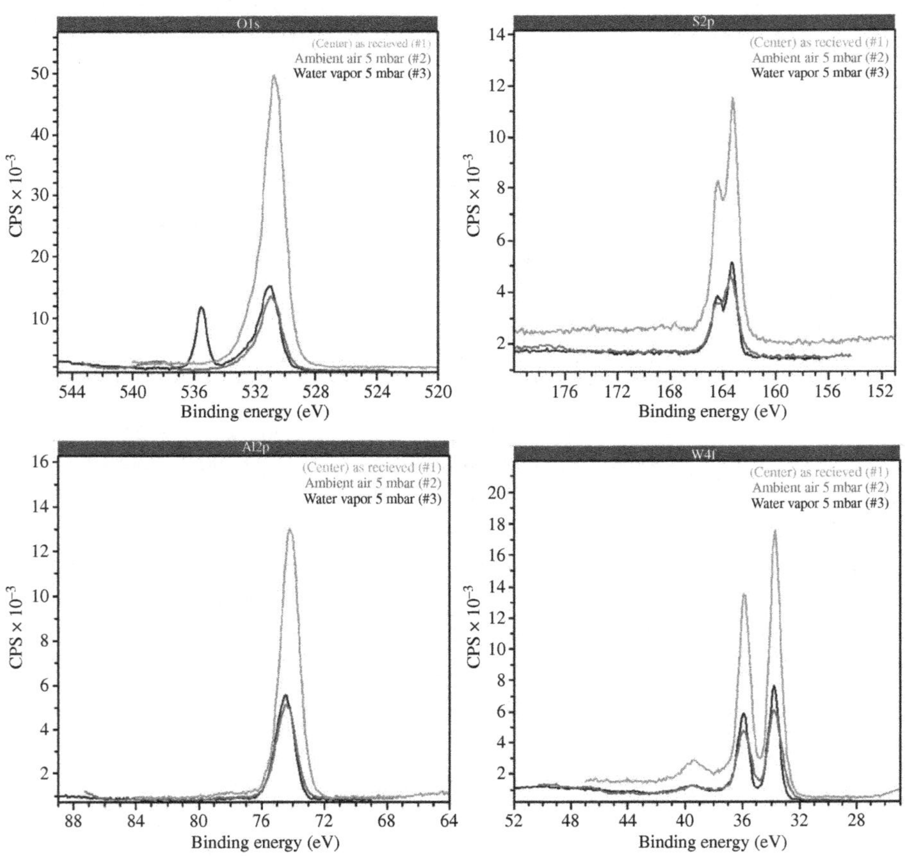

The table below shows the areas under the W 4*f* and S 2*p* peaks and the atomic ratio of these two elements calculated with their RSF values.

	UHV survey (#1)		5 mbar air (#2)		5 mbar water (3)	
Peak	**Peak area**	**Atomic ratio**	**Peak area**	**Atomic ratio**	**Peak area**	**Atomic ratio**
W 4f	30588	1	12404	1	11853	1
S 2p	15315	3.16	5624	2.86	5401	2.88

The atomic ratio of W:S is still not close to the stoichiometric ratio of WS_2. There are three possibilities that we can speculate – one is the deviation of the epitaxial thin film composition from the bulk stoichiometry, and another is the matrix effect because the tungsten atoms are buried under the sulfur atoms at the topmost layer. The third is due to an instrumental factor. If the latter is the major factor, what is the possible source that affects this quantitative analysis?

Here the survey spectra are replotted with an intensity normalization:

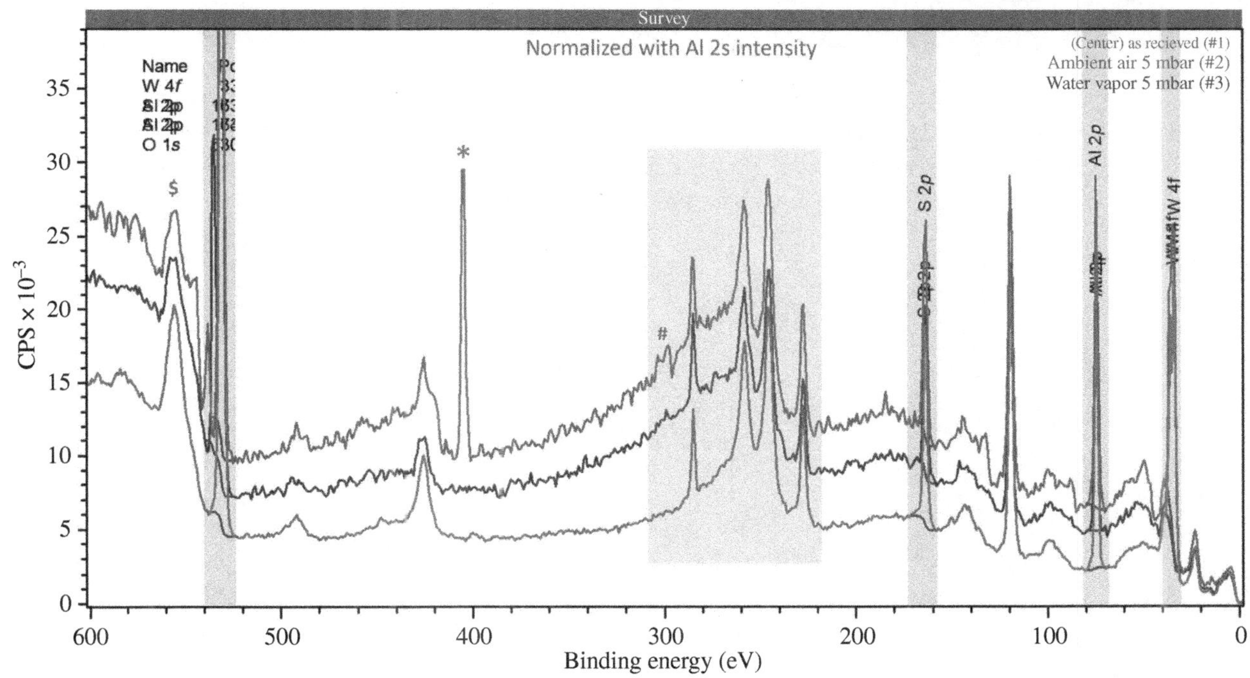

1) What is the origin of the peak marked with * in the survey spectrum collected in 5 mbar of ambient air?
2) What is the origin of the peak marked with $ at 555.5~556 eV?
3) Do you think the peak marked with # in the survey spectrum collected in the background pressure of 5 mbar of ambient air is real or noise?
4) Explain why the background in the shaded region is so large and even extended to the C 1*s* region. The source for this weird background shape can be seen in other places as well although they are manifested in a different way due to a certain reason.

Here the O1*s* spectra are replotted with an intensity normalization:

5) What is the origin of the extra peak at 538–539 eV in the spectrum collected in the background pressure of 5 mbar of ambient air?
6) What is the origin of the extra peak at 535.5 eV in the spectrum collected in the background pressure of 5 mbar of water vapor?
7) The main O 1*s* peak looks asymmetric. What is the origin of this asymmetric peak shape?

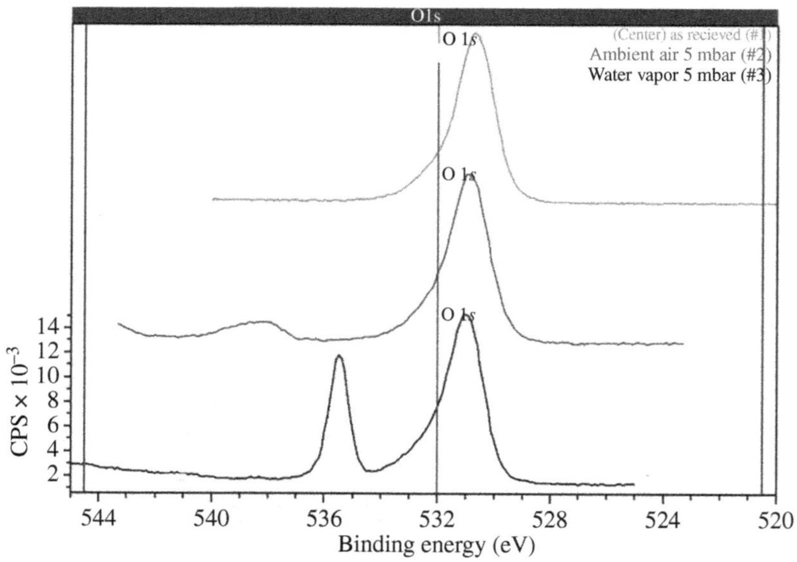

2.7 Sketch typical XPS and XAS spectra of a sample containing C (BE ≈ 285 eV), O (BE ≈ 532 eV), N (BE ≈ 400 eV), and F (BE ≈ 685 eV). The scale and unit of *x*-axis variables and the signal plotted in the *y*-axis should be stated. Also, the peak shape and baseline should be drawn with proper physical justification.

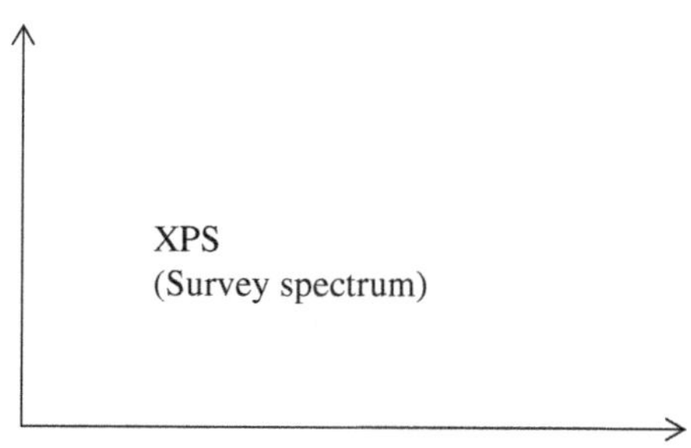

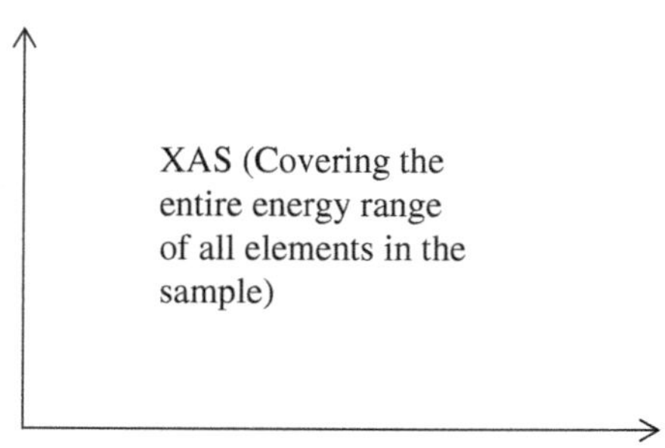

References

1 Baer DR, McGuire GE, Artyushkova K, Easton CD, Engelhard MH, Shard AG. Introduction to topical collection: reproducibility challenges and solutions with a focus on guides to XPS analysis. Journal of Vacuum Science & Technology A. 2021;39(2):021601.

2 Atomic Orbital. In Wikipedia; 2023, June 14. [Available from: https://en.wikipedia.org/wiki/Atomic_orbital].

3 Barat M, Lichten W. Extension of the electron-promotion model to asymmetric atomic collisions. Physical Review A. 1972;6(1):211–29.

4 Kotochigova S, Levine ZH, Shirley EL, Stiles MD, Clark CW. Atomic Reference Data for Electronic Structure Calculations 2009 [Available from: https://www.nist.gov/pml/atomic-reference-data-electronic-structure-calculations].

5 Bonzel HP, Kleint C. On the history of photoemission. Progress in Surface Science. 1995;49(2):107–53.

6 Watts JF, Wolstenholme J. Electron spectroscopy: some basic concepts. In: Watts JF and Wolstenholme J, editors. An Introduction to Surface Analysis by XPS and AES; 2019. https://doi.org/10.1002/9781119417651.ch1.

7 Russ JC. Fundamentals of Energy Dispersive X-ray Analysis. Cornwall, UK: Butterworth-Heinemann; 1984.

8 Scimeca M, Bischetti S, Lamsira HK, Bonfiglio R, Bonanno E. Energy dispersive X-ray (EDX) microanalysis: a powerful tool in biomedical research and diagnosis. European Journal of Histochemistry. 2018;62(1):2841.

9 Llovet X, Moy A, Pinard PT, Fournelle JH. Electron probe microanalysis: a review of recent developments and applications in materials science and engineering. Progress in Materials Science. 2021;116:100673.

10 Moulder JF, Chastain J. Handbook of X-ray Photoelectron Spectroscopy: A Reference Book of Standard Spectra for Identification and Interpretation of XPS. Eden Prairie, MN: Data Physical Electronics Division, Perkin-Elmer Corporation; 1992.

11 Seah MP, Dench WA. Quantitative electron spectroscopy of surfaces: a standard data base for electron inelastic mean free paths in solids. Surface and Interface Analysis. 1979;1(1):2–11.

12 Powell CJ. Practical guide for inelastic mean free paths, effective attenuation lengths, mean escape depths, and information depths in x-ray photoelectron spectroscopy. Journal of Vacuum Science & Technology A. 2020;38(2):023209.

13 Watts JF, Wolstenholme J. Electron spectrometer design. In: Watts JF and Wolstenholme J, editors. An Introduction to Surface Analysis by XPS and AES; 2019: .https://doi.org/10.1002/9781119417651.ch2.

14 Greczynski G, Hultman L. Reliable determination of chemical state in x-ray photoelectron spectroscopy based on sample-work-function referencing to adventitious carbon: resolving the myth of apparent constant binding energy of the C 1s peak. Applied Surface Science. 2018;451:99–103.

15 Baer DR, Artyushkova K, Cohen H, Easton CD, Engelhard M, Gengenbach TR, et al. XPS guide: charge neutralization and binding energy referencing for insulating samples. Journal of Vacuum Science & Technology A. 2020;38(3):031204.

16 Greczynski G, Hultman L. X-ray photoelectron spectroscopy: towards reliable binding energy referencing. Progress in Materials Science. 2020;107:100591.

17 Edwards L, Mack P, Morgan DJ. Recent advances in dual mode charge compensation for XPS analysis. Surface and Interface Analysis. 2019;51(9):925–33.

18 Kim SH, Kim J-H, Kang B-K, Uhm HS. Superhydrophobic CF_x coating via in-line atmospheric RF plasma of $He{-}CF_4{-}H_2$. Langmuir. 2005;21(26):12213–7.

19 Parvate S, Dixit P, Chattopadhyay S. Superhydrophobic surfaces: insights from theory and experiment. The Journal of Physical Chemistry B. 2020;124(8):1323–60.

20 Suzer S. Differential charging in X-ray photoelectron spectroscopy: a nuisance or a useful tool? Analytical Chemistry. 2003;75(24):7026–9.

21 Österlund L. Pressure gaps in heterogeneous catalysis. Heterogeneous Catalysts. 2021;1:225–51.

22 Salmeron M, Schlögl R. Ambient pressure photoelectron spectroscopy: a new tool for surface science and nanotechnology. Surface Science Reports. 2008;63(4):169–99.

23 Pielsticker L, Nicholls R, Beeg S, Hartwig C, Klihm G, Schlögl R, et al. Inelastic electron scattering by the gas phase in near ambient pressure XPS measurements. Surface and Interface Analysis. 2021;53(7):605–17.

24 Channeltron Electron Multiplier Handbook for Mass Spectrometry Applications. Galileo Electro-Optics Co; 1991 [Available from: https://www.triumf.ca/sites/default/files/ChannelBookBurle.pdf].

25 Unger WES, Stockmann JM, Senoner M, Weimann T, Bütefisch S, Passiu C, et al. Introduction to lateral resolution and analysis area measurements in XPS. Journal of Vacuum Science & Technology A. 2020;38(5):053206.

26 Naumkin AV, Kraut-Vass A, Gaarenstroom SW, Powell CJ. NIST X-ray Photoelectron Spectroscopy Database 2012 [Available from: https://srdata.nist.gov/xps/].

27 Fairley N. CasaXPS: Processing Software for XPS, AES, SIMS and More [Available from: http://www.casaxps.com/].

28 Gelius U, Basilier E, Svensson S, Bergmark T, Siegbahn K. A high resolution ESCA instrument with X-ray monochromator for gases and solids. Journal of Electron Spectroscopy and Related Phenomena. 1973;2(4):405–34.

29 Ratner BD, Castner DG. Electron Spectroscopy for Chemical Analysis. Surface Analysis – The Principal Techniques. p. 47–112.

30 Batich CD. Chemical derivatization and surface analysis. Applied Surface Science. 1988;32(1):57–73.

31 Laibinis PE, Graham RL, Biebuyck HA, Whitesides GM. X-ray Damage to CF_3CO_2-terminated organic monolayers on Si/Au: principal effect of electrons. Science. 1991;254(5034):981–3.

32 Beamson G, Briggs DR, editors. The XPS of Polymers Database. Wiley; 2000.

33 Deleuze M, Denis JP, Delhalle J, Pickup BT. Theoretical study of spectral differences in the XPS valence bands of polyethylene lamellae and films. The Journal of Physical Chemistry. 1993;97(19):5115–23.

34 Endo K, Kobayashi N, Aida M, Inoue C. Spectra analysis of XPS valence energy levels of polymers by the semi-empirical MO method–I. Polyvinylalcohol, polyethyleneoxide and polyvinylchloride. Journal of Physics and Chemistry of Solids. 1993;54(8):887–94.

35 Kim SH, Somorjai GA. Polymer growth mode on the heterogeneous Ziegler–Natta catalyst: active sites at the bottom of the growing polymer layer. Catalysis Letters. 2000;68(1):7–11.

36 Fuggle JC, Hillebrecht FU, Zeller R, Zołnierek Z, Bennett PA, Freiburg C. Electronic structure of Ni and Pd alloys. I. X-ray photoelectron spectroscopy of the valence bands. Physical Review B. 1983;27(4):2145–78.

37 Kowalczyk SP, Ley L, Martin RL, McFeely FR, Shirley DA. Relaxation and final-state structure in XPS of atoms, molecules, and metals. Faraday Discussions of the Chemical Society. 1975;60:7–17.
38 Biesinger MC. Advanced analysis of copper X-ray photoelectron spectra. Surface and Interface Analysis. 2017;49(13):1325–34.
39 Wagner CD. Chemical shifts of Auger lines, and the Auger parameter. Faraday Discussions of the Chemical Society. 1975;60:291–300.
40 Wagner CD, Zatko DA, Raymond RH. Use of the oxygen KLL Auger lines in identification of surface chemical states by electron spectroscopy for chemical analysis. Analytical Chemistry. 1980;52(9):1445–51.
41 Fadley CS, Shirley DA. Multiplet splitting of metal-atom electron binding energies. Physical Review A. 1970;2(4):1109–20.
42 Biesinger MC, Lau LWM, Gerson AR, Smart RSC. Resolving surface chemical states in XPS analysis of first row transition metals, oxides and hydroxides: Sc, Ti, V, Cu and Zn. Applied Surface Science. 2010;257(3):887–98.
43 Biesinger MC, Payne BP, Grosvenor AP, Lau LWM, Gerson AR, Smart RSC. Resolving surface chemical states in XPS analysis of first row transition metals, oxides and hydroxides: Cr, Mn, Fe, Co and Ni. Applied Surface Science. 2011;257(7):2717–30.
44 Yamashita T, Hayes P. Analysis of XPS spectra of Fe^{2+} and Fe^{3+} ions in oxide materials. Applied Surface Science. 2008;254(8):2441–9.
45 Major GH, Pinder JW, Austin DE, Baer DR, Castle SL, Čechal J, et al. Perspective on improving the quality of surface and material data analysis in the scientific literature with a focus on x-ray photoelectron spectroscopy (XPS). Journal of Vacuum Science & Technology A. 2023;41(3):038501.
46 Barlow AJ, Jones RT, McDonald AJ, Pigram PJ. XPSSurfA: an open collaborative XPS data repository using the CMSShub platform. Surface and Interface Analysis. 2018;50(5):527–40.
47 Biesinger M. X-ray Photoelectron Spectroscopy (XPS) Reference Pages: Surface Science Western lab [Available from: http://www.xpsfitting.com/].
48 Shirley DA. High-resolution X-Ray photoemission spectrum of the valence bands of gold. Physical Review B. 1972;5(12):4709–14.
49 Tougaard S. Practical guide to the use of backgrounds in quantitative XPS. Journal of Vacuum Science & Technology A. 2020;39(1):011201.
50 Herrera-Gomez A, Cabrera-German D, Dutoi AD, Vazquez-Lepe M, Aguirre-Tostado S, Pianetta P, et al. Intensity modulation of the Shirley background of the Cr 3p spectra with photon energies around the Cr 2p edge. Surface and Interface Analysis. 2018;50(2):246–52.
51 Engelhard MH, Baer DR, Herrera-Gomez A, Sherwood PMA. Introductory guide to backgrounds in XPS spectra and their impact on determining peak intensities. Journal of Vacuum Science & Technology A. 2020;38(6):063203.
52 Tougaard S, Greiner M. Method to correct ambient pressure XPS for the distortion caused by the gas. Applied Surface Science. 2020;530:147243.
53 Jain V, Biesinger MC, Linford MR. The Gaussian-Lorentzian sum, product, and convolution (Voigt) functions in the context of peak fitting X-ray photoelectron spectroscopy (XPS) narrow scans. Applied Surface Science. 2018;447:548–53.
54 Major GH, Fairley N, Sherwood PMA, Linford MR, Terry J, Fernandez V, et al. Practical guide for curve fitting in x-ray photoelectron spectroscopy. Journal of Vacuum Science & Technology A. 2020;38(6):061203.
55 Doniach S, Sunjic M. Many-electron singularity in X-ray photoemission and X-ray line spectra from metals. Journal of Physics C: Solid State Physics. 1970;3(2):285.
56 McCafferty E, Wightman JP. Determination of the concentration of surface hydroxyl groups on metal oxide films by a quantitative XPS method. Surface and Interface Analysis. 1998;26(8):549–64.
57 Banerjee J, Bojan V, Pantano CG, Kim SH. Effect of heat treatment on the surface chemical structure of glass: oxygen speciation from in situ XPS analysis. Journal of the American Ceramic Society. 2018;101(2):644–56.
58 Banerjee J, Kim SH, Pantano CG. Elemental areal density calculation and oxygen speciation for flat glass surfaces using x-ray photoelectron spectroscopy. Journal of Non-Crystalline Solids. 2016;450:185–93.
59 Gengenbach TR, Major GH, Linford MR, Easton CD. Practical guides for x-ray photoelectron spectroscopy (XPS): interpreting the carbon 1s spectrum. Journal of Vacuum Science & Technology A. 2021;39(1):013204.
60 Scofield JH. Hartree-Slater subshell photoionization cross-sections at 1254 and 1487 eV. Journal of Electron Spectroscopy and Related Phenomena. 1976;8(2):129–37.
61 Wagner CD, Davis LE, Zeller MV, Taylor JA, Raymond RH, Gale LH. Empirical atomic sensitivity factors for quantitative analysis by electron spectroscopy for chemical analysis. Surface and Interface Analysis. 1981;3(5):211–25.

62 Reilman RF, Msezane A, Manson ST. Relative intensities in photoelectron spectroscopy of atoms and molecules. Journal of Electron Spectroscopy and Related Phenomena. 1976;8(5):389–94.

63 Shard AG. Practical guides for x-ray photoelectron spectroscopy: quantitative XPS. Journal of Vacuum Science & Technology A. 2020;38(4):041201.

64 Reed BP, Cant DJH, Spencer SJ, Carmona-Carmona AJ, Bushell A, Herrera-Gómez A, et al. Versailles project on advanced materials and standards interlaboratory study on intensity calibration for x-ray photoelectron spectroscopy instruments using low-density polyethylene. Journal of Vacuum Science & Technology A. 2020;38(6):063208.

65 Seah MP. An accurate and simple universal curve for the energy-dependent electron inelastic mean free path. Surface and Interface Analysis. 2012;44(4):497–503.

66 Shinotsuka H, Tanuma S, Powell CJ, Penn DR. Calculations of electron inelastic mean free paths. XII. Data for 42 inorganic compounds over the 50 eV to 200 keV range with the full Penn algorithm. Surface and Interface Analysis. 2019;51(4):427–57.

67 Smith GC. Evaluation of a simple correction for the hydrocarbon contamination layer in quantitative surface analysis by XPS. Journal of Electron Spectroscopy and Related Phenomena. 2005;148(1):21–8.

68 Tougaard S. Improved XPS analysis by visual inspection of the survey spectrum. Surface and Interface Analysis. 2018;50(6):657–66.

69 Shallenberger JR, Smith NJ, Banerjee J. Derivation of empirical XPS relative sensitivity factors from silicate glasses. Surface and Interface Analysis. 2021;53(6):569–79.

70 Tougaard S. Energy loss in XPS: fundamental processes and applications for quantification, non-destructive depth profiling and 3D imaging. Journal of Electron Spectroscopy and Related Phenomena. 2010;178–179:128–53.

71 Ganta D, Sinha S, Haasch RT. 2-D material molybdenum disulfide analyzed by XPS. Surface Science Spectra. 2014;21(1):19–27.

72 Wheeler DR, Pepper SV. Effect of x-ray flux on polytetrafluoroethylene in x-ray photoelectron spectroscopy. Journal of Vacuum Science and Technology. 1982;20(2):226–32.

73 Akhter S, Cannon KC, White JM. Distribution of compositional defects in solvent cast polyvinyl alcohol films: angular dependent XPS and reactivity study. Applied Surface Science. 1990;44(1):49–57.

74 Livesey AK, Smith GC. The determination of depth profiles from angle-dependent XPS using maximum entropy data analysis. Journal of Electron Spectroscopy and Related Phenomena. 1994;67(3):439–61.

75 Hofmann S. Sputter depth profile analysis of interfaces. Reports on Progress in Physics. 1998;61(7):827.

76 Ziegler JF. SRIM - The Stopping and Range of Ions in Matter 2013 [Available from: http://www.srim.org/].

77 Postawa Z, Czerwinski B, Szewczyk M, Smiley EJ, Winograd N, Garrison BJ. Microscopic insights into the sputtering of Ag{111} induced by C60 and Ga bombardment. The Journal of Physical Chemistry B. 2004;108(23):7831–8.

78 Paruch RJ, Postawa Z, Garrison BJ. Seduction of finding universality in sputtering yields due to cluster bombardment of solids. Accounts of Chemical Research. 2015;48(9):2529–36.

79 Winograd N. Gas cluster ion beams for secondary ion mass spectrometry. Annual Review of Analytical Chemistry. 2018;11(1):29–48.

80 Weibel D, Wong S, Lockyer N, Blenkinsopp P, Hill R, Vickerman JC. A C60 primary ion beam system for time of flight secondary ion mass spectrometry: its development and secondary ion yield characteristics. Analytical Chemistry. 2003;75(7):1754–64.

81 Chang H-Y, Lin W-C, Chu P-C, Wang Y-K, Sogo M, Iida S-i, et al. X-ray photoelectron spectroscopy equipped with gas cluster ion beams for evaluation of the sputtering behavior of various nanomaterials. ACS Applied Nano Materials. 2022;5(3):4260–8.

82 Hryha E, Rutqvist E, Nyborg L. Stoichiometric vanadium oxides studied by XPS. Surface and Interface Analysis. 2012;44(8):1022–5.

83 Yamamoto Y, Yamamoto K. Precise XPS depth profile of soda-lime-silica glass using C60 ion beam. Journal of Non-Crystalline Solids. 2010;356(1):14–8.

84 Sheth N, Ngo D, Banerjee J, Zhou Y, Pantano CG, Kim SH. Probing hydrogen-bonding interactions of water molecules adsorbed on silica, sodium calcium silicate, and calcium aluminosilicate glasses. The Journal of Physical Chemistry C. 2018;122(31):17792–801.

85 Newville M. Fundamentals of XAFS. Reviews in Mineralogy and Geochemistry. 2014;78(1):33–74.

86 Stöhr J. NEXAFS Spectroscopy. Berlin, Heidelberg: Springer; 1992.

87 Teo BK. EXAFS: Basic Principles and Data Analysis. Berlin, Heidelberg: Springer; 1986.

88 Ade H, Hitchcock AP. NEXAFS microscopy and resonant scattering: composition and orientation probed in real and reciprocal space. Polymer. 2008;49(3):643–75.

89 Stöhr J, Jaeger R. Absorption-edge resonances, core-hole screening, and orientation of chemisorbed molecules: CO, NO, and N_2 on Ni(100). Physical Review B. 1982;26(8):4111–31.

90 Whitten JE. Ultraviolet photoelectron spectroscopy: practical aspects and best practices. Applied Surface Science Advances. 2023;13:100384.

91 Yoshida H. Principle and application of low energy inverse photoemission spectroscopy: a new method for measuring unoccupied states of organic semiconductors. Journal of Electron Spectroscopy and Related Phenomena. 2015;204:116–24.

3

Elemental Analysis via Electron Irradiation

X-ray photoelectron spectroscopy (XPS) is based on the photoemission of electrons from core orbitals of individual atoms upon absorption of x-ray. By measuring the kinetic energy of photoemitted electrons, we can determine their binding energy from which elemental analysis can be done. The electron emission from the core orbitals can also be induced by bombardment of high energy electrons, which will be accompanied by the same relaxation processes of the electronic structure that follow x-ray absorption which is described in Section 2.1.2 (Figure 2.6), i.e., Auger electron and x-ray emission. The technique measuring the x-ray emission is energy-dispersive x-ray spectroscopy (EDX or EDS) or electron probe microanalysis (EPMA). Elemental analysis based on EDX and EPMA will not be covered here since it is more like a bulk analysis (due to a deep probe depth). The detection principle and examples of EDX and EPMA can be found in various review articles and books [1–3]. Here, we will focus on Auger electron spectroscopy (AES) only. AES is useful for elemental analysis with a high lateral resolution (limited primarily by electron spot size and scattering) with relatively good surface sensitivity (~5 nm probe depth or even less). It can identify the chemical state of elements, be semi-quantitative without standards, and perform compositional depth profiling with ion sputtering (same as depth profiling in XPS, see Section 2.5.2). Since primary excitation and detection signals are electrons, AES is done in ultra-high vacuum (UHV) condition. Thus, the sample must be UHV compatible, and it works best with conducting materials. AES is rarely used for chemical analysis of organic materials since they are highly susceptible to electron beam damage.

3.1 Principle of Auger Electron Spectroscopy (AES)

3.1.1 Emission of Auger Electron

To induce the Auger electron emission, the core level of the element in the sample needs to be ionized, and this is done by irradiation of a high energy electron beam onto the sample (i.e., e-beam, instead of x-ray, in Figure 2.6A). Typical electron sources used in AES systems are tungsten (W) filament, lanthanum boride (LaB_6) crystal, or field emission gun (FEG). The W filament source can be operated at low electron energy (1–10 keV) with a beam diameter of several microns. The LaB_6 and FEG sources operates at much higher energy (20–30 keV) and can give much smaller beam diameters (≤20 nm). The electron beam current can vary from 100 nA down to ≤1 nA, depending on the sample to be analyzed and the spot size of the primary incident beam.

To ionize the core level of an element in the sample, the kinetic energy (E_I) of the incident electron beam should be larger than the threshold ionization energy (E_B) of the electron in that level. Figure 3.1A plots the ionization cross-section of a core electron (σ) multiplied with the square of its threshold energy (E_B^2) as a function of E_I/E_B calculated with two different quantum mechanics models [4]. The Gryzinski model predicts that the ionization cross-section is maximum at $E_I/E_B \approx 3$–5. The Casnati–Tatari–Baraldi (CTB) model predicts that the maximum is at $E_I/E_B \approx 2$–4, and the ionization cross-section is slightly higher at a higher E_I for a given E_B. The reason that there is a maximum in $\sigma \cdot E_B^2$ is because the governing mechanism for inelastic collision of electrons with atomic targets varies depending on the relation between the impacting electron velocity to the mean orbital velocity of the electron in the subshell.

When the core level electron is ejected through inelastic collision with the primary incident electron, the relaxation of electronic structures (Figure 2.6B) can cause emission of Auger electron or x-ray photon (Figure 2.6C). The emission

Surface and Interface Analysis: Principles and Applications, First Edition. Seong H. Kim.

Companion website: www.wiley.com/go/Kim

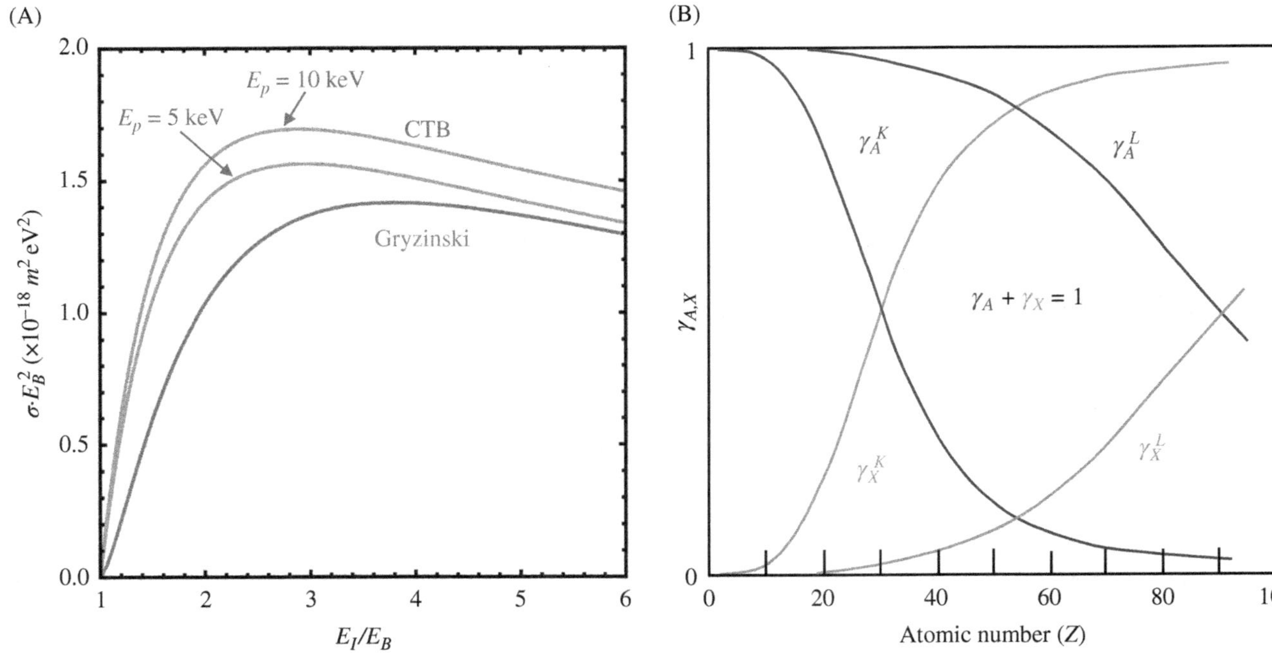

Figure 3.1 (A) Plot of $\sigma \cdot E_B^2$ vs. E_I/E_B for the cross sections of Gryzinski and CTB models. (B) Probabilities for Auger emission (γ_A) and x-ray emission (γ_X) as a function of atomic number.

probability of Auger electron (γ_A) and x-ray photon (γ_X) varies with the atomic number Z and the type of atomic level involved in initial ionization (K, L, etc.) (Figure 3.1B) [5]. The sum of these two probabilities is unity. In the K-level ionization of light elements, the Auger electron emission probability is larger than the x-ray emission probability. Thus, the detection sensitivity of light elements in the second row of the periodic table is high in AES and low in EDX. In contrast, for heavy elements, γ_X followed by the K-shell ionization is larger than γ_A. Thus, EDX is more sensitive to heavy elements in the sample. For elements in the fourth row and below in the periodic table, the Auger electrons initiated by the L-shell ionization are detected.

Figure 3.2A schematically shows the distribution of working ranges of primary electrons. Inelastically scattered primary electrons travel several microns in the sample; but if their energy falls below the threshold ionization energy (E_B), they cannot induce Auger electron or x-ray emission. The diameter of the analyzed zone is larger than the diameter of the primary beam due to scattering of incident electrons in the sample.

Typical energy analyzers used in AES are a cylindrical mirror analyzer (CMA) and a concentric hemispherical analyzer (CHA). The working principle of CHA is described in depth in Section 2.2.2. The working principle of CMA is basically the same (see Figure 2.14), although its geometry is different. CMA is more widely used for an instrument built specifically for AES because of its excellent transmission efficiency, compact size, and ease of usage; but its energy resolution is lower than CHA.

Figure 3.2B schematically illustrates the spectrum of all electrons detected as a function of their energy from the primary beam energy to the zero kinetic energy. The biggest peak at the highest energy is the elastically scattered electron. Just below that, several well-defined loss peaks can be seen. The inelastic scattering mechanisms for these loss peaks can vary depending on the specimen. Analyzing these loss peaks can provide the structural and chemical information; this is the electron energy loss spectroscopy (EELS) [6]. Then, the exponentially decaying featureless background signal follows, which is due to the backscattered electrons. In the low kinetic energy region, secondary electrons are detected, which is used for scanning electron microscopy (SEM). On top of the smoothly varying background signal, tiny Auger electron signals are superimposed.

As shown in Figure 3.2A, the primary electron beam ionizes elements in the top 1–2 μm region from the surface; but, the collected Auger electron signals are only the ones originating from topmost several nanometers. This surface sensitivity is due to the short inelastic mean free path (IMPF), same as the XPS discussed in Section 2.1.4. Typical kinetic energy (KE) of

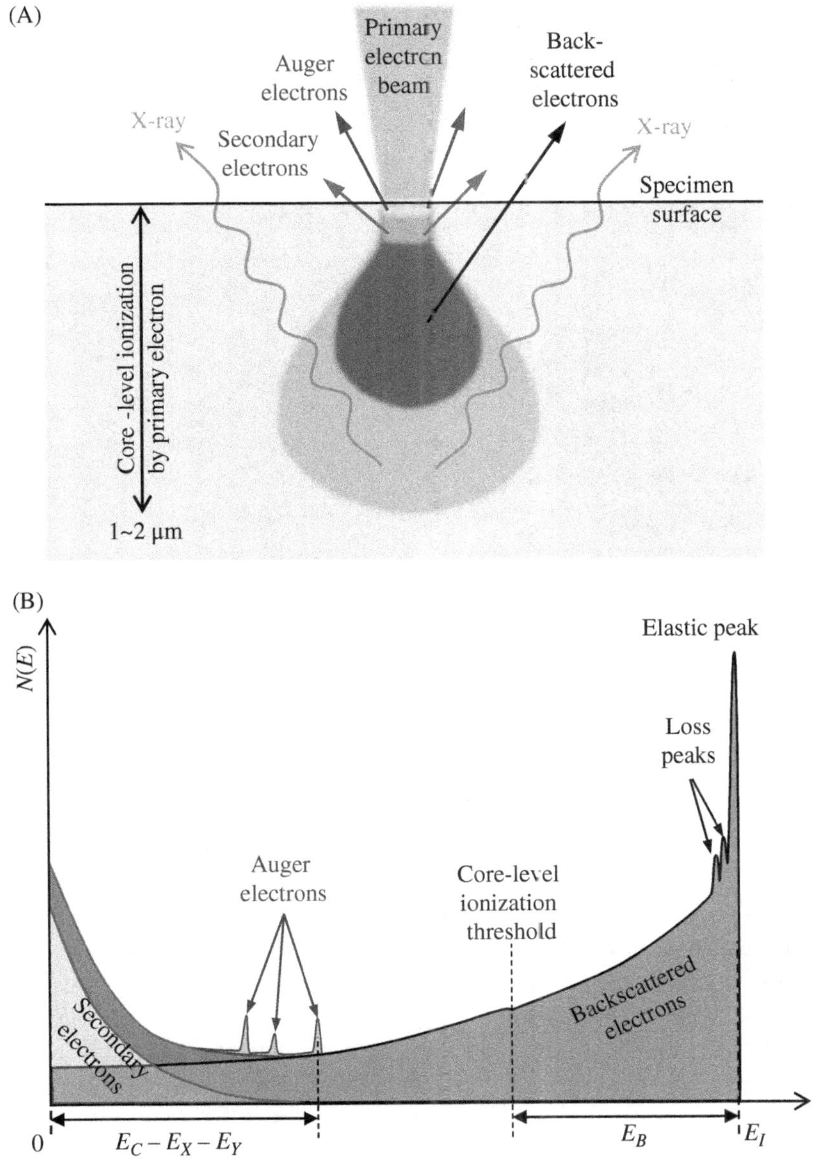

Figure 3.2 (A) Working ranges of primary, backscattered, secondary, and Auger electrons, along with x-ray emission. (B) Schematic representation of the detected electron spectrum.

Auger electrons is less than 1500 eV; thus, IMFP is on the order of 1 nm or less (Figure 2.11). Thus, the effective probe depth is typically 3–4 nm or less, depending on the kinetic energy of the detected Auger electron.

As shown schematically in Figure 3.2, Auger electron signals are quite small and often difficult to see if the collected spectrum is plotted in full scale along with the elastically scattered signal. Figure 3.3 plots an AES spectrum of a TiO_2 sample with some sulfur and carbon species on it, in the total electron scale as a function of the electron KE, $N(E)$. The Auger signals are barely visible in the raw spectrum. To clearly identify the Auger signal above the scattered electron background, the AES spectrum is often shown in the first derivative scale, $dN(E)/dE$. The inset in Figure 3.3 is obtained through digital data processing. In early instruments, a lock-in amplifier was used to generate the differential signal.

3.1.2 Kinetic Energy of Auger Electron

The Auger electrons are described with three letters, for example, KLL, LMM, and KVV. The first letter represents the core-level orbital where the hole is generated by the inelastic collision with the primary electron (Figure 2.6A). The second letter

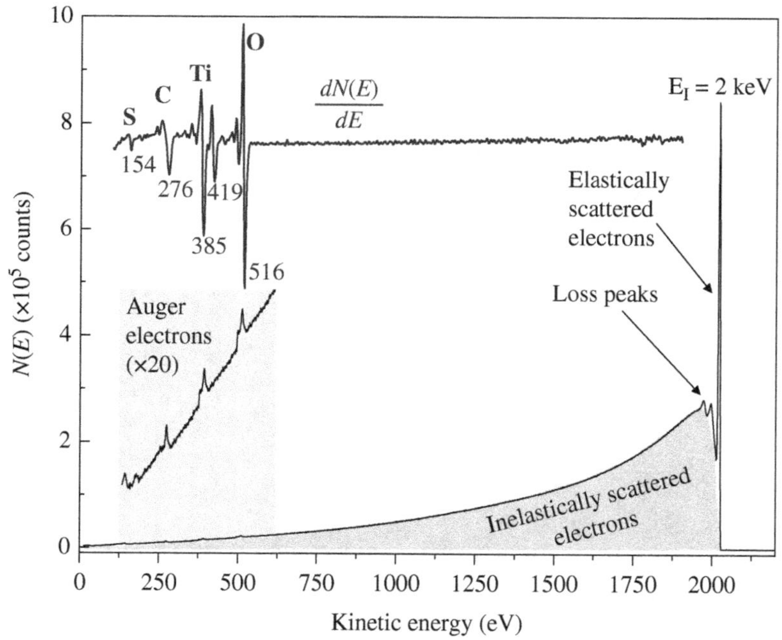

Figure 3.3 AES spectrum of TiO_2, with some C and S species on it, collected with $E_I = 2$ keV.

represents the orbital from which the secondary electronic transition takes place to fill the core hole (Figure 2.6B). The third letter represents the orbital from which the Auger electron is emitted (Figure 2.6C). Sometime, subscripts are used to denote the subshells; for example, L_1 for 2*s*, L_2 for $2p_{1/2}$, and L_3 for $2p_{3/2}$. When valence electrons are involved in the relaxation process, they are denoted with the symbol *V*.

In Section 2.1.2, eq. (2.3) depicts that the KE of Auger electron is equal to the difference between the energy of the core hole and the energy levels of the two outer electrons (which is marked in the abscissa in Figure 3.2B). This equation does not take into account the final state effects such as interaction energies between the core hole and the outer electrons or the interatomic relaxations discussed in Section 2.3.5 [7]. Thus, the calculation of the Auger electron KE is much more complex than eq. (2.3). There is an empirical equation that works reasonably well. It considers the displacement of electronic levels towards higher binding energy upon the core-level ionization of atom by the primary electron. The displacement effect is approximated by taking an average of the energies of the same orbitals in the atom from which Auger electron is emitted and the atom with one higher atomic number (i.e., $Z + 1$). Thus, eq. (2.3) can be modified to [7, 8]:

$$KE(CXY;Z) = E(C;Z) - \frac{E(X;Z) + E(X;Z+1)}{2} - \frac{E(Y;Z) + E(Y;Z+1)}{2} - \Phi_w^{spectr} \quad (3.1)$$

where Z is the atomic number of the atom from which Auger electron is emitted, C is the core level at which the initial ionization occurs, and X and Y are the energy levels involved in the relaxation that lead to the Auger electron emission. Since KE is measured with respect to the work function of the spectrometer (see Figure 2.7), Φ_w^{spectr} is subtracted from the energy differences. Typical work function of the metals used in spectrometer is ~4 eV.

We can check how well this empirical equation works. The binding energies of Ti ($Z = 22$) and V ($Z = 23$) are listed in Table 3.1 [9]. In Figure 3.3, the two main peaks of Ti are at 385 and 419 eV. First, we can use eq. (2.3) and calculate the measured KE for the following Auger transitions:

$$E(L_3) - E(M_1) - E(M_{4,5}) - \Phi_w^{spectr} = 455 - 59 - 3 - 4 = 389 \text{ eV}$$
$$E(L_2) - E(M_{2,3}) - E(M_{4,5}) - \Phi_w^{spectr} = 461 - 34 - 3 - 4 = 420 \text{ eV}$$

Now, let us use eq. (3.1) and calculate the KE for the same Auger transitions:

$$E(L_3) - E(M_1) - E(M_{4,5}) - \Phi_w^{spectr} = 455 - \frac{59 + 66}{2} - \frac{3 + 2}{2} - 4 = 386 \text{ eV}$$

Table 3.1 Binding energy (eV) of electrons in Ti and V.

	K	L_1	L_2	L_3	M_1	$M_{2,3}$	$M_{4,5}$
Ti (Z = 22)	4965	564	461	455	59	34	3
V (Z = 23)	5465	628	520	513	66	38	2

Source: Adapted from Mathieu [9].

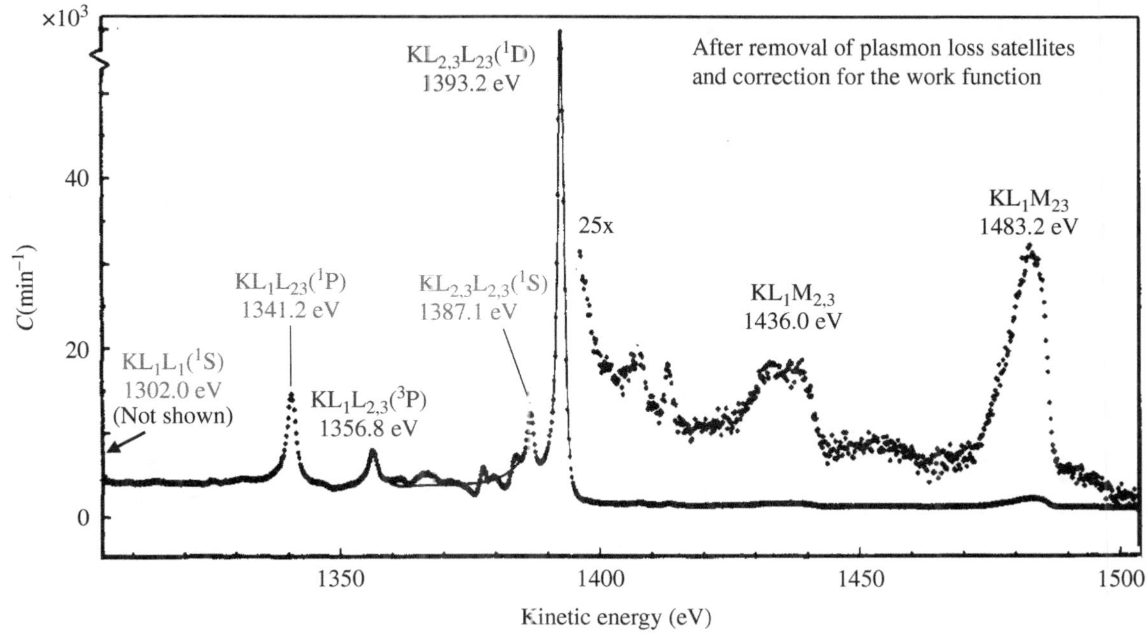

Figure 3.4 X-ray excited KLL and KLV Auger spectrum of Al metal. The plasmon loss satellites are removed from the experimental data and the KE scale is corrected for the work function (Φ_w^{spectr}) = 4.2 eV. Reproduced from van Attekum and Trooster [10]/with permission of IOP Publishing Ltd.

$$E(L_2) - E(M_{2,3}) - E(M_{4,5}) - \Phi_w^{spectr} = 461 - \frac{34 + 38}{2} - \frac{3 + 2}{2} - 4 = 418.5\,\text{eV}$$

We can see that eq. (3.1) predicts the KE of Auger electrons more closely than eq. (2.3).

Let us take a look at another example. The K, L_1, and $L_{2,3}$ binding energies of Al ($Z = 13$) are 1559.6, 117.7, and 73.1 eV, respectively [9]. When the KEs of KL_1L_1, KL_1L_2, and $KL_{23}L_{23}$ of Al are calculated using eq. (2.3), we get 1320, 1364.6, and 1409.2 eV, respectively. The L_1 and $L_{2,3}$ binding energies of Si ($Z = 14$) are 148.7 eV and 99.2 eV, respectively [9]. When eq. (3.1) is used, the KEs of Al Auger electrons are predicted to be 1289, 1336, and 1383 eV. Figure 3.4 displays the Auger spectrum of aluminum metal collected with x-ray irradiation [10]. Since the KE of the Auger electron is not a function of the core-level ionization source, we can still compare the calculated KEs with the experimental data. The KE values predicted using eq. (3.1) match well with the experimental data. Obviously, this simple calculation does not predict additional final state effects such as $KL_1L_{2,3}(^3P)$ and $KL_{2,3}L_{2,3}(^1D)$ in the experimental spectrum.

3.2 Qualitative Analysis with AES

Like XPS, AES can be used for the identification of elements present in the topmost surface region. Figure 3.5 shows the characteristic energy chart of principal Auger electrons of each element with atomic number (Z) of three and higher [11]. Since the Auger transition is a three-electron process, hydrogen and helium cannot produce any Auger electrons.

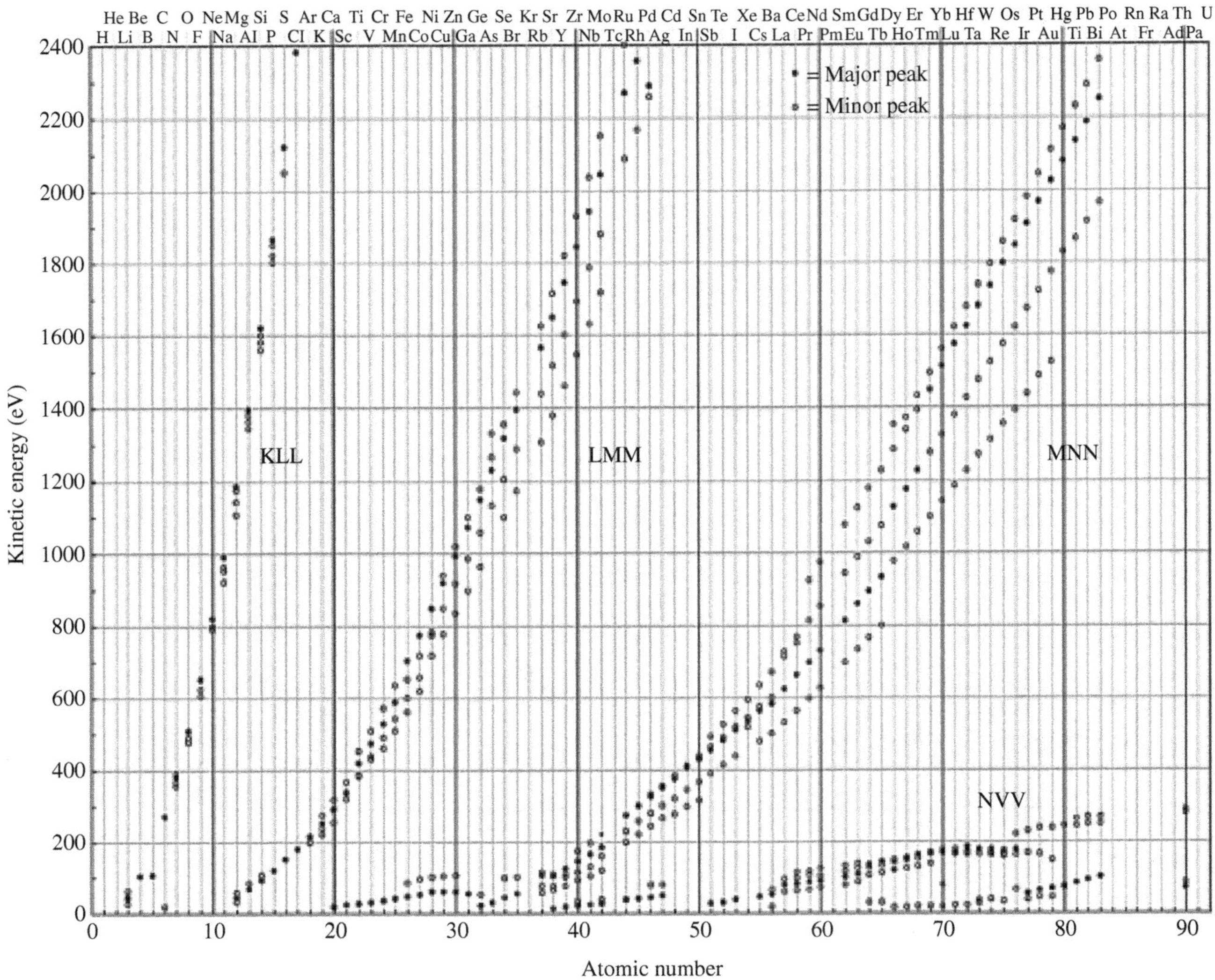

Figure 3.5 Chart of principal Auger electron energies of each element. Reproduced from Watts and Wolstenholme [11]/with permission of John Wiley & Sons.

Changes in the chemical environment of atoms in the topmost surface region of a solid can alter the KE and/or peak shape of Auger signals [8]. Like XPS, a shift in KE is expected when there is a charge transfer from one atom to another. Figure 3.6A compared the Al KLL Auger spectra of several aluminum compounds [12]. In AES, since three energy levels are involved in the Auger transition, often it is not straightforward to correlate the degree of chemical shift to the charge transfer or oxidation state [13]. So, it is used more like a fingerprint for comparison with reference data (like the valence spectra in XPS).

The C KLL Auger spectrum is particularly sensitive to the type of bond hybridization. Carbon has four valence electrons and exhibits a variety of bonding configurations. Figure 3.6B compares the carbon Auger spectra of diamond, graphite, tungsten carbide, and CO on W. These changes in peak position and shape are sufficiently distinctive to serve as a fingerprint for identification of the form of carbon at a specimen surface [14]. In the case of oxygen, the peak shape change is negligible and all O KLL spectra look almost identical (Figure 3.6C) [15].

Comparing the Auger spectra of an element (e.g., shown in Figure 3.6) with the corresponding core-level photoelectron lines of the same element in XPS, it can be noticed that chemical shifts in the Auger transition are different from those in the photoelectron spectrum. The final states of the doubly ionized atoms result in substantial differences in screening energy, which varies with the chemical environment. Thus, by analyzing both Auger transition peaks and the photoelectron peaks, more abundant chemical state information can be obtained [16]. This principle is already described in Section 2.3.5.

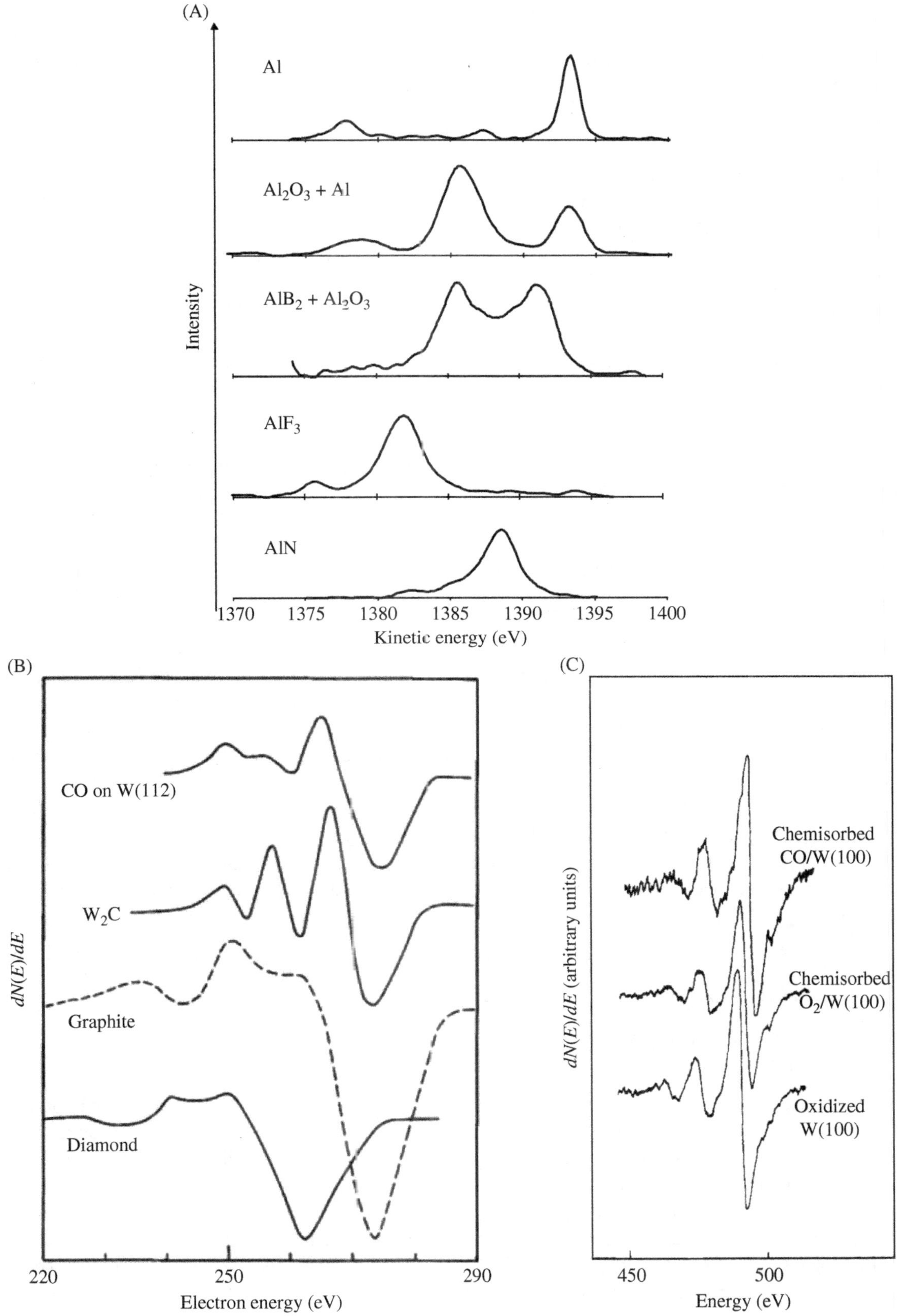

Figure 3.6 (A) Al KLL Auger spectra of Al metal and compounds. Reproduced from Timmermans et al. [12]/with permission of John Wiley & Sons. (B) C KLL Auger spectra from diamond, graphite, tungsten carbide, and CO on W(112). Reproduced from Haas et al. [14]/with permission of AIP Publishing LLC. (C) O KVV Auger spectra from chemisorbed CO, O_2, and oxide. Reproduced from Chesters et al. [15]/with permission of IOP Publishing Ltd.

An example of comparing the Auger KE and the core-level BE of copper compounds in a two-dimensional plot is shown in Figure 2.32. It can also be done by calculating the Auger parameter (α) which is the sum of the Auger KE and the core-level BE. The Auger parameters, as well as representative Auger KE, of almost all elements in various compound forms can be the NIST XPS Database [17].

3.3 Quantitative Analysis with AES

The quantitative analysis principle of AES is similar to that of XPS (eq. (2.21)), which is descried in Section 2.4.3. The equation correlating the Auger peak intensity of an element A (I_A) to its atomic concentration (N_A) can be written as:

$$I_A = I_p\,(1 + r_M)\,\sigma_A\,\gamma_A\,\frac{\Delta\Omega}{\pi}\,T\int_0^{\infty}\exp\left[-\frac{z}{\lambda\cos\theta}\right]dz\,D\,S\,y_S\,N_A \tag{3.2}$$

where I_p is the primary beam current, r_M is the backscattering factor, σ_A is the core-level ionization cross-section (described in Figure 3.1A), γ_A is the probability of Auger emission following the ionization (described in Figure 3.2B), $\frac{\Delta\Omega}{\pi}$ is the angular acceptance of the analyzer, T is the transmission function of the electron analyzer, z is the depth from the surface, λ is the inelastic mean free path in the analyte solid (see Figure 2.11), θ is the angle of emission from the surface normal direction (if the take-off angle from the surface plane is used, then cosθ is changed to sinθ; see Section 2.5.1), D is the electron multiplier amplification and counting efficiency, S is an effective surface area, y_S is the sample-specific parameter such as surface roughness, porosity, and matrix effects. In AES, primary electrons have a kinetic energy of 2–20 keV. Their penetration depth can be up to several microns (Figure 3.2A). During their trajectory, they can be backscattered (Figure 3.2B). Some of the backscattered electrons can in turn produce additional Auger electrons if they have sufficient energy. This effect is lumped into the r_M term, which depends on the binding energy of the core level (E_B) and the energy and incidence angle of the primary electron beam (E_I, ϑ_I).

In eq. (3.2), all terms are instrument-specific constant, except $(1 + r_M)\sigma_A\gamma_A$ and λ (IMFP). For a homogeneous sample, the product of these two is called atomic sensitivity factor. Figure 3.7 shows the relative sensitivity factor chart, normalized with the Cu LMM peak, for the Auger peak-to-peak height collected in a differentiated signal ($dN(E)/dE$) using a CMA with 10 keV primary beam [18]. Similar charts can also be found in the Handbook of Auger Electron Spectroscopy [19, 20]. The tabulated values for calculated Auger yields and sensitivity factors for KLL, NMM, MNN transitions with 1, 3, 5, and 10 kV primary beams can be found at [21]. If the background subtraction can be done properly, the raw signal ($N(E)$) is likely to be more accurate than the differentiated signal because the area under the peak is a better measure of the Auger intensity than the peak-to-peak height in the derivative spectra. For Auger signals with KE less than 200 eV or so, an internal calibration may be required since the electron multiplier gain (D in eq. (3.2)) may decrease and secondaries from the bulk can contribute to the Auger yield.

Applying eq. (3.2) to a binary compound AB, we can write the signal intensity ratio of two constituents, $\frac{I_{A,AB}}{I_{B,AB}}$, in the following form using the relative sensitivity factors (s_A and s_B):

$$\frac{I_{A,AB}}{I_{B,AB}} = \frac{N_A}{N_B}\left(\frac{1 + r_{A,AB}}{1 + r_{B,AB}}\,\frac{\sigma_A}{\sigma_B}\,\frac{\gamma_A}{\gamma_B}\,\frac{\lambda_{A,AB}}{\lambda_{B,AB}}\right) \approx \frac{N_A}{N_B}\,\frac{s_A}{s_B} \tag{3.3}$$

Note that the backscattering factors (r_M) and IMFP (λ) may vary depending on the sample. Altogether, it is called a matrix effect (y_S in eq. (3.2)). Assuming the matrix effect is negligible may introduce an error of 10~50% error on the absolute value of the calculated atomic concentration. For a better accuracy in compositional analysis, the composition-dependent backscattering and matrix effect should be corrected [22].

There are many other factors which can compromise quantification accuracy. The peak shape change with chemical state could affect the area calculation. Plasmon losses can introduce significant errors; an example is discussed in Section 2.4.4 for XPS analysis. Note that Figure 3.4 is the Al Auger spectrum after the plasmon signals are removed; the raw spectrum contains many plasmon peaks [9]. The influence of the transmission function of the analyzer could be significant especially when the KEs of Auger electrons are quite different.

The AES detection sensitivity is known to be approximately 0.1–1.0 at.%; but it should be noted that it varies depending on the spot size. If the primary beam is focused too tight, the analysis area can be too small to produce the spectrum with

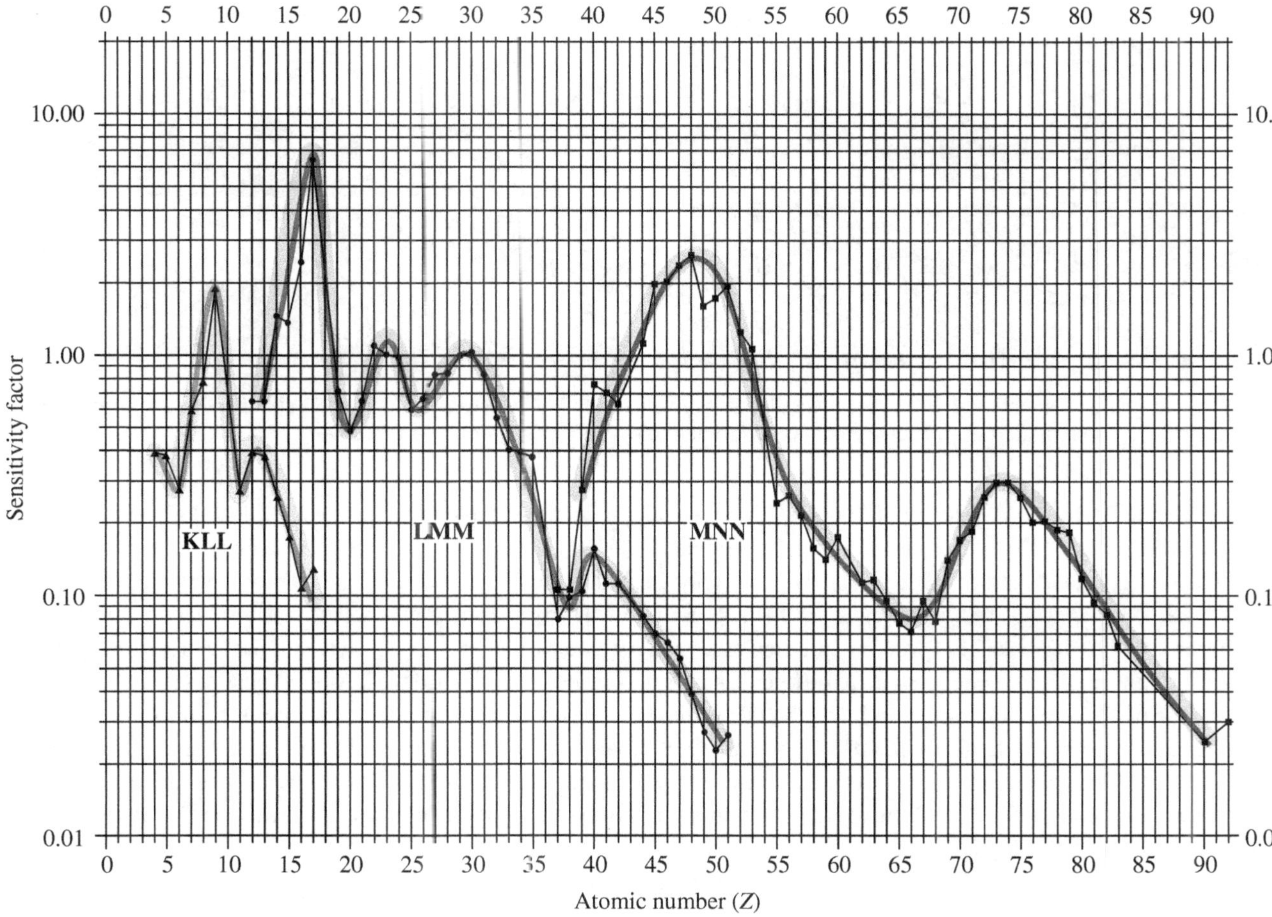

Figure 3.7 Relative sensitivity factors derived from Auger peak-to-peak height data using a CMA with a 10 keV primary beam. For relative sensitivities, the LMM peak of copper (Z = 29) is set to unity. Smooth lines are a guide to eyes. Joshi and Paul [18]/ASM International.

sufficiently high signal-to-noise (S/N) ratio. If the focused beam size of the primary incident electron is small and the relative sensitive of the element of interest is low, then even getting a sensitivity of 1 at.% could be challenging. For insulating specimens, sample charging could be so big that the peak positions may be shifted far from the position expected for a given element. If the sample is inhomogeneous or the surface is rough, differential charging may occur. The charge neutralization in AES is much more difficult than that in XPS. In XPS, insulating samples are positively charged, which can be compensated using a low-energy electron flood gun. In AES, insulating samples are negatively charged and a low-energy ion gun should be used, which may alter the sample composition during the analysis. Electron beam damage effects could be serious, inducing desorption of adsorbed species and/or decomposition of adventitiously adsorbed organic contaminants. Even graphite oxide can be damaged during the AES analysis if the irradiation time is prolonged [23]. It is well known that, during the SEM imaging which detects the total yield of secondary electrons emitted by the sample (Figure 3.2B), residual carbon species in the vacuum chamber can be adsorbed and dissociated in the electron beam irradiated area, leading to accumulation of carbon species [24]. The same can take place during the AES analysis.

3.4 Scanning Auger Mapping

One major advantage of AES over XPS is the high-resolution imaging, since the high-energy electron beam can be focused into a small spot. Using LaB_6 and FEG sources, it is possible to focus the primary beam down to ≤20 nm in diameter. It should be noted that tight focusing of electron beam is done with smaller beam current; thus, the detection sensitivity may be sacrificed when the image resolution is pushed to the limit.

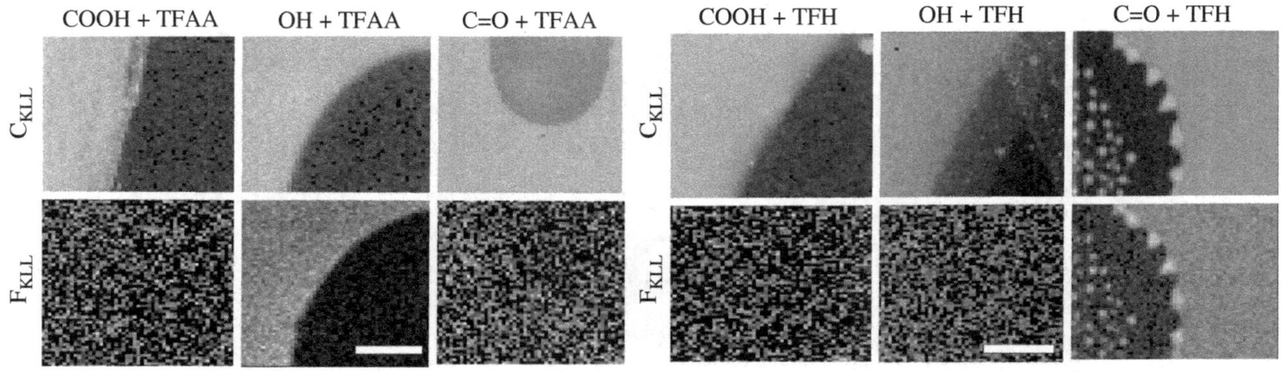

Figure 3.8 Scanning AES images of C-KLL and F-KLL signals from patterned PAA (COOH), PVA (OH), and PVMK (C=O) films after exposure to TFAA and TFH. The scale bar is 100 μm. Reproduced with permission from Yang et al. [25] / Elsevier.

Figure 3.8 shows the AES mapping of fluorine atoms introduced to the sample through the chemical derivatization reactions for XPS analysis [25]. The basic principle of chemical derivatization as well as a set of widely used reactions are described in Section 2.3.3. In a control experiment, polyvinyl alcohol (PVA with OH group), polyvinyl methyl ketone (PVMK with C=O group), and polyacrylic acid (PAA with COOH group) were exposed to trifluoroacetic acid (TFAA) and trifluoroethyl hydrazine (TFH) vapors and then analyzed with scanning AES. The edges of easily recognizable patterns were chosen to visualize the chemical state differences present on the polymer surfaces. After exposure to the TFAA derivatization agent, only the polymer containing the OH group (PVA) picks up the fluorine atoms and the fluorine uptake by other polymers containing C=O and COOH is negligible. Similarly, the TFH agent reacts only with the C=O containing polymer (PVMK) and does not react with the OH and COOH groups. These results clearly demonstrate the selective reactivity of the TFAA and TFH derivatizing agents. Such imaging can be done for all derivatization reactions listed in Table 2.2 before utilizing them for applications to specific specimens.

Figure 3.9 shows an example of high-resolution elemental mapping with scanning AES. The specimen is a stardust particulate from a nova [26]. From the SEM image, a polygrain structure is identified (marked with dotted line). Scanning Auger elemental maps show that the grain has inhomogeneous distributions of Mg, Si, and Fe and a very little amount of S in it. If this chemical mapping was attempted with EDX, the detected signal would be dominated by the chemical information of the underling bulk (matrix) specimen. If it was attempted with XPS, the particle would not have been resolved due to poor lateral resolution. Quantitative AES analysis showed that region 1 has a stoichiometry between olivine and pyroxene, and region 2 is a Si-rich silicate [26].

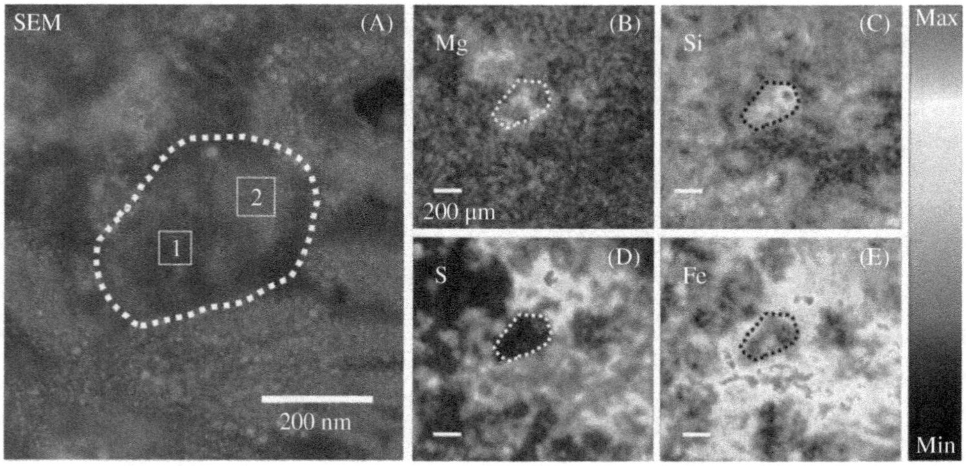

Figure 3.9 (A) SEM and scanning Auger elemental maps of (B) Mg, (C) Si, (D) S, and (E) Fe in a stardust grain from a nova (outlined with dotted line) on a fine-grained meteorite material of Solar System origin. All scale bars are 200 nm. Reproduced with permission from Leitner et al. [26] / American Astronomical Society.

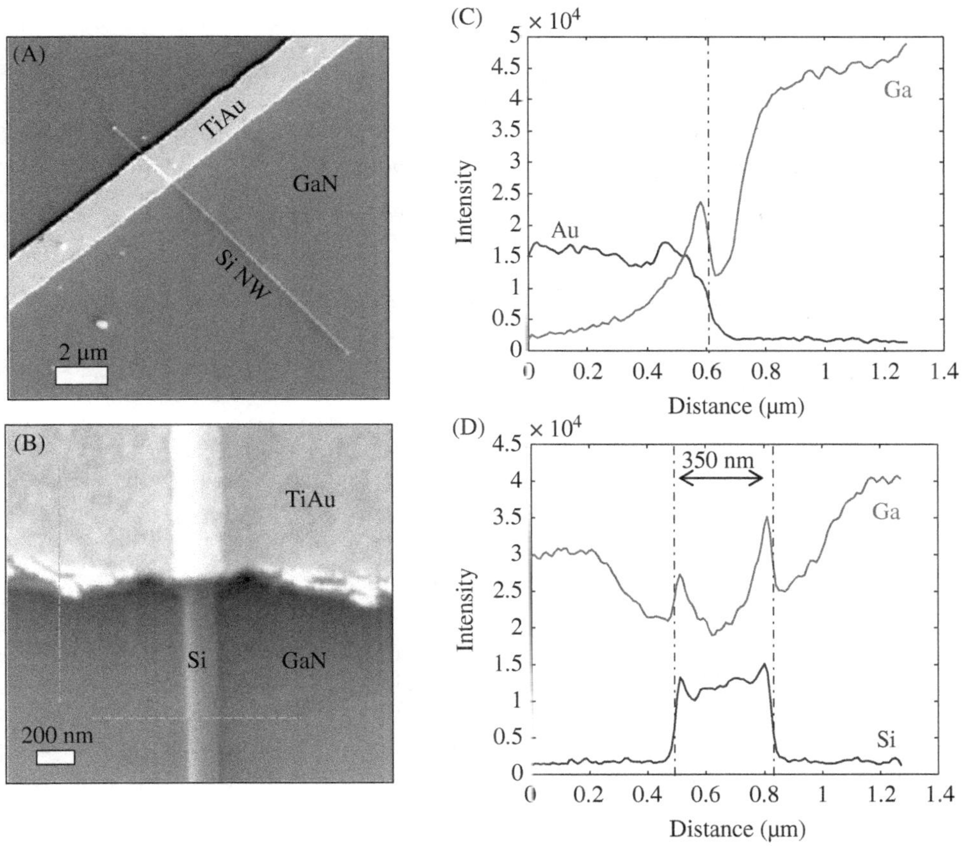

Figure 3.10 (A,B) SEM images of Si NW on a GaN substrate with a TiAu electrode. (C,D) Scanning Auger line profiles across (C) TiAu edge and (D) Si NM. Primary beam energy = 20 keV. By Courtesy of Dr. Bangzhi Liu.

The primary electron beam with a high kinetic energy can penetrate deeper and spread wider than the one with a low kinetic energy. Some of the backscattered electrons can have sufficient energy to induce Auger transitions in elements with low ionization energy. Figure 3.10 shows an example where such backscattered electrons cause an artifact. The sample is a silicon nanowire (NW) on a gallium nitride (GaN) substrate onto which a 120 nm thick TiAu electrical contact line is deposited. In the line scan across the TiAu edge on the GaN substrate, the Ga-LMM signal intensity is enhanced right at the edge although it is expected to decrease continuously as the primary beam scans across the edge from the substrate to the electrical contact. In the line scan across the Si NW, the Ga-LMM signal is enhanced at the edge of the Si NW. This enhancement was caused because the scattered electrons from the TiAu and Si edges induced the Auger transition of Ga. The Au-MNN and Si-KLL signals do not have such an artifact since their core-level ionization energy is larger.

3.5 Further Reading for AES

Nowadays, the usage of AES for surface analysis is much less than XPS. It is probably because most of elemental information that can be obtained with AES can be obtained with XPS. In terms of chemical state analysis, XPS is more versatile than AES. And the x-ray beam damage to the sample is much milder than the electron beam damage. Nonetheless, AES still has a niche where it outperforms XPS, such as elemental mapping with a higher spatial resolution. Also, AES can be done along with SEM; so, region-selective analysis can be done much more easily. Two good reference books with detailed instrumentation information and a large number of application examples are recommended for the people who want to study AES more: *Surface Analysis by Auger and x-Ray Photoelectron Spectroscopy* by D. Briggs and J. T. Grant [27] and *An Introduction to Surface Analysis by XPS and AES* by J. F. Watts and J. Wolstenholme [28].

Practice Problems

3.1 Why can AES be used to analyze hydrogen and helium?

3.2 Can an isolated Li atom produce the Auger electron signal? Can the Li metal produce Auger electron signal? How about Li_2O in which Li^{+1} has the $1s^2\ 2s^0$ electronic structure. Justify your answer.

3.3 Calculate the KE of the electron from the KL_1L_1, KL_1L_{23}, and $KL_{23}L_{23}$ Auger transitions of Mg. Compare the calculated KE with the experimental data in [9].

3.4 Why are the chemical shift and peak shape of Auger signals difficult to predict as compared to those of photoelectron signals in XPS?

3.5 Discuss the advantages and disadvantages of using $N(E)$ vs. $dN(E)/dE$ data in AES analysis.

3.6 What are possible experimental artifacts in AES analysis?

References

1 Russ JC. Fundamentals of Energy Dispersive X-ray Analysis. Russ JC, editor. Oxford, UK: Butterworth-Heinemann; 1984.

2 Scimeca M, Bischetti S, Lamsira HK, Bonfiglio R, Bonanno E. Energy dispersive X-ray (EDX) microanalysis: a powerful tool in biomedical research and diagnosis. European Journal of Histochemistry. 2018;62(1):2841.

3 Llovet X, Moy A, Pinard PT, Fournelle JH. Electron probe microanalysis: a review of recent developments and applications in materials science and engineering. Progress in Materials Science. 2021;116:100673.

4 Seah MP, Gilmore IS. Quantitative AES VII. The ionization cross-section in AES. Surface and Interface Analysis. 1998;26(11):815–24.

5 Krause MO. Atomic radiative and radiationless yields for K and L shells. Journal of Physical and Chemical Reference Data. 1979;8(2):307–27.

6 Egerton RF. Electron Energy-Loss Spectroscopy in the Electron Microscope. Boston, MA: Springer US; 2011.

7 Wagner CD. Chemical shifts of Auger lines, and the Auger parameter. Faraday Discussions of the Chemical Society. 1975;60:291–300.

8 Castle JE, Epler D. Chemical shifts in photo-excited Auger spectra. Proceedings of the Royal Society of London Series A, Mathematical and Physical Sciences. 1974;339(1616):49–72.

9 Mathieu HJ. Auger electron spectroscopy. Surface Analysis – The Principal Techniques. 2009. 9–45.

10 van Attekum PMTM, Trooster JM. A plasmon gain satellite in the KLL Auger spectrum of Mg and Al metal. Journal of Physics F: Metal Physics. 1978;8(7):L169.

11 Watts JF, Wolstenholme J. Appendix 1: Auger electron energies. In: An Introduction to Surface Analysis by XPS and AES. Hoboken, NJ, USA: John Wiley & Sons Ltd.; 2019, p. 247.

12 Timmermans B, Vaeck N, Hubin A, Reniers F. Chemical effects in Auger electron spectra of aluminium. Surface and Interface Analysis. 2002;34(1):356–9.

13 Davis GD, Natan M, Anderson KA. Study of titanium oxides using Auger line shapes. Applications of Surface Science. 1983;15(1):321–33.

14 Haas TW, Grant JT, Dooley GJ, III. Chemical effects in Auger electron spectroscopy. Journal of Applied Physics. 1972;43(4):1853–60.

15 Chesters MA, Hopkins BJ, Taylor PA. Chemical effects in oxygen Auger spectra. Journal of Physics C: Solid State Physics. 1976;9(12):L329.

16 Wagner CD, Joshi A. The Auger parameter, its utility and advantages: a review. Journal of Electron Spectroscopy and Related Phenomena. 1988;47:283–313.

17 Naumkin AV, Kraut-Vass A, Gaarenstroom SW, Powell CJ. NIST X-ray Photoelectron Spectroscopy Database 2012 [Available from: https://srdata.nist.gov/xps/].

18 Joshi A, Paul DF. Auger electron spectroscopy. In: Materials Characterization (Volume 10). Detroit, MI: ASM International; 2019. p. 675–98.

19 Davis LE. Handbook of Auger electron spectroscopy. In: Physical Electronics Division. Chanhassen, Minnesota: Physical Electronics Inc.; 1978.

20 Childs KD, Carlson BA, Lavanier LA, Moulder JF, Paul DF, Stickle WF, et al. Handbook of Auger Electron Spectroscopy: A Book of Reference Data for Identification and Interpretation in Auger Electron Spectroscopy. Chanhassen, Minnesota: Physical Electronics Inc.; 1995.

21 Mroczkowski S, Lichtman D. Calculated Auger yields and sensitivity factors for KLL–NOO transitions with 1–10 kV primary beams. Journal of Vacuum Science & Technology A. 1985;3(4):1860–5.

22 Streubel P, Berndt H. A practical method for determining backscattering factors and matrix correction in quantitative AES analysis of binary alloys. Surface and Interface Analysis. 1984;6(2):48–55.

23 Galuska AA, Madden HH, Allred RE. Electron spectroscopy of graphite, graphite oxide and amorphous carbon. Applied Surface Science. 1988;32(3):253–72.

24 Postek MT, Vladár AE, Purushotham KP. Does your SEM really tell the truth? How would you know? Part 2. Scanning. 2014;36(3):347–55.

25 Yang M, Marino MJ, Bojan VJ, Eryilmaz OL, Erdemir A, Kim SH. Quantification of oxygenated species on a diamond-like carbon (DLC) surface. Applied Surface Science. 2011;257(17):7633–8.

26 Leitner J, Kodolányi J, Hoppe P, Floss C. Laboratory analysis of presolar silicate stardust from a nova. The Astrophysical Journal Letters. 2012;754(2):L41.

27 Briggs D, Grant JT. Surface Analysis by Auger and X-ray Photoelectron Spectroscopy. Chichester, UK: IM Publications; 2003.

28 Watts JF, Wolstenholme J. An Introduction to Surface Analysis by XPS and AES. Chichester, UK: John Wiley & Sons, Ltd; 2003.

4

Elemental Analysis via Ion Irradiation

In Section 2.5.2, we have studied depth profiling with x-ray photoelectron spectroscopy (XPS). It is done by repeating the process of removing a certain amount of the surface layer via ion sputtering and then analyzing what is left in the surface with XPS. The same can be done with Auger electron spectroscopy (AES). Compositional analysis of the surface can also be done by analyzing the species being removed during the ion sputtering process. This is done by analyzing their masses. The mass analyzer sorts out ions based on their mass-to-charge ratios; thus, secondary ions sputtered off the surface during the bombardment with high-energy primary ions can be analyzed directly using a mass spectrometer. This is called secondary ion mass spectrometry (SIMS) [1]. A much larger fraction of sputtered species is neutral (in most cases, about two orders of magnitude more than ionic species), but they cannot be analyzed directly with SIMS. To analyze neutral species, a post-desorption ionization method should be employed (such as laser-induced photo-ionization); then, secondary neutral mass spectrometry (SNMS) can be done [2, 3]. Although SNMS can provide some advantages over SIMS, it is less popular than SIMS or not readily available in many analytical facilities, since it requires additional instrumentation for post-desorption ionization. This chapter will cover the principles of SIMS and a few selected application examples, which will be helpful to understand the literature where SIMS is used as a key analytical technique.

4.1 Principle of Secondary Ion Mass Spectrometry (SIMS)

4.1.1 Secondary Ion Emission During Ion Beam Sputtering

In SIMS, a wide range of primary ion beam source designs have been used. For noble gas ions (Ar^+ and Xe^+), an electron-bombardment source is used. For reactive ions such as O_2^+, a plasma source is used. A surface-ionization gun is used to produce a Cs^+ beam. This source can provide high sensitivity to electronegative species in the analyte. Widely used liquid metal guns such as gallium, indium, gold, or bismuth are field-ionization sources. These liquid metal sources can produce polyatomic cluster ions, for example, Au_3^+ and Bi_3^+, and specific ions are selected with a Wien filter and then accelerated to a desired energy.

When a beam of ions with high kinetic energy (KE) (typically from a few keV up to 30–40 keV) bombards a surface, their KE is transferred to the atoms in the solid by elastic and inelastic collisions. Through a cascade of collisions, an ensemble of atoms and their clusters are emitted from the surface, some of which are ionized during the emission process. The sputter yield (S) can be defined as the removal rate of surface atoms per impinging ions. The sputter yield will be influenced by many factors including the energy and incident angle of impinging ions as well as target material properties such as compositions and structures. Figure 4.1A schematically shows the dependence of sputter yield on the incident ion energy (E_I) [4, 5]. In a low-energy region, a threshold limit for a minimum detection limit (10^{-4}–10^{-5} atoms/ion) is in the order of 20–30 eV, which is about four times the heat of sublimation. Just above the threshold, S increases roughly in proportion to the square of E_I, then becomes linear to E_I, and finally levels off at $E_I \approx$ 10–100 keV. Beyond this plateau region, the sputter yield drops as the impinging ion penetrates deeper into the solid (so, ion implantation occurs), and less energy is recoiled back to the surface. The sputter yield also varies with the incident angle of the primary beam. For amorphous or polycrystalline materials, the sputter yield increases monotonically with increasing angle relative to the surface normal and reaches a maximum at angles between 60° and 80° [4, 5]. The angular distribution of sputtered ions tends to have a cosine distribution around the angle of specular reflection of the primary beam.

Surface and Interface Analysis: Principles and Applications, First Edition. Seong H. Kim.

Companion website: www.wiley.com/go/Kim

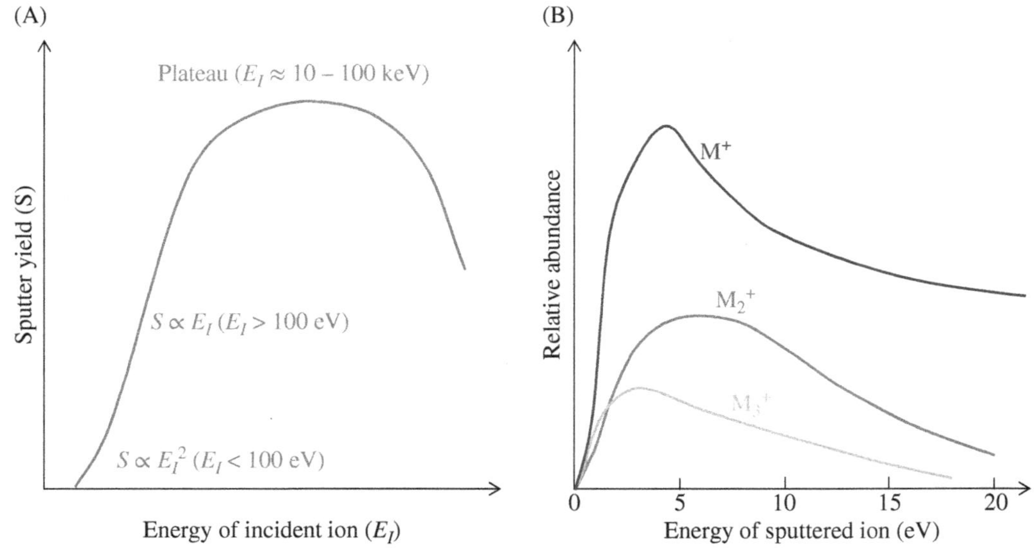

Figure 4.1 (A) Typical dependence of sputter yield on incident ion energy. (B) Typical energy distributions of M^+, M_2^+, and M_3^+ ions emitted during ion sputtering.

The sputtered ions will have a broad distribution of KE, which is again a function of the impinging ion species, energy, and incident angle [4, 5]. Figure 4.1B displays typical energy distribution of monoatomic and polyatomic ions [6]. The energy distributions of polyatomic ions (and molecular ions as well) are distinct from those of monoatomic ions, because some of the KE excites internal vibrational and rotational modes while mono atomic ions have all KE in translational modes.

4.1.2 Mass Spectrometric Detection of Secondary Ions

Three types of mass spectrometers are used in SIMS. Early instruments were mostly equipped with a quadrupole RF mass filter due to its small size. Although it is versatile, one limitation of this mass filter is that the RF frequency needs to be scanned; so it only allows the sequential transmission of ions of one mass while filtering out all other ions with different masses. Thus, it has a low transmission yield of ions through the quadrupole filter and takes time to collect the full spectrum over a wide range of mass. Furthermore, the mass range that it can filter is limited to several hundreds of amu.

Another type of mass analyzer is a magnetic filter. Ions are extracted from the sample using a high electrical bias and sent to a magnetic field zone. The charged particles moving in a magnetic field experience a force in the direction orthogonal to the magnetic flux line and its original traveling direction (recall Fleming's right-hand-rule from electromagnetics course). The force experienced by the particle is directly related to its velocity. Since all ions are accelerated with the same KE, their velocities in the magnetic field vary depending on their masses; thus, ions with different masses form different circular trajectories. One advantage of the magnetic sector device is that the entering trajectories of secondary ions are retained and projected to a position sensitive detector, allowing two-dimensional imaging in a short time. But the magnetic mass filter is bulky and cannot be baked out to achieve UHV conditions (due to limited thermal stability of magnets). Also, it still scans mass sequentially.

Nowadays, the most widely used mass filter is the time-of-flight (ToF) analyzer (Figure 4.2). In ToF-SIMS, pulses of secondary ions are extracted into a drift tube and accelerated to a given potential V_A (2–8 keV) so that all ions possess approximately the same KE. Then, they travel through the field-free space until they strike the detector. The heavier ions travel slower through the flight tube than the lighter ions. The ToF in the tube is measured and converted to the corresponding mass-to-charge (m/e) ratio using the KE equation shown in Figure 4.2. If the acceleration voltage is $V_A = 5$ keV and the drift tube length is $L = 2$ m, then ions with $m = 10$ amu, 100 amu, and 1000 amu arrive at the detector at ToF of 6.46 μs, 20.44 μs, and 64.63 μm, respectively. Thus, a mass spectrum of all ions can be generated from the ToF spectrum within a few hundred μs. Mass resolution is critically dependent upon the pulse length of the generated secondary ions (which is a function of the

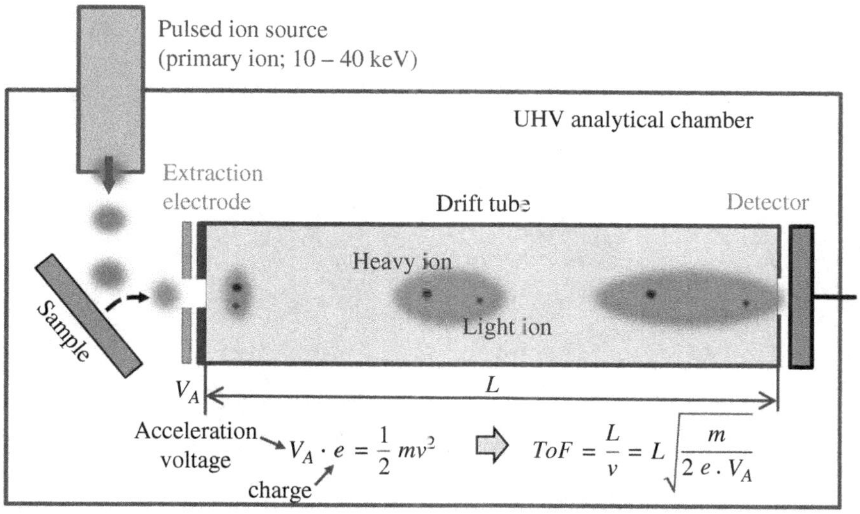

Figure 4.2 A schematic diagram of a ToF-SIMS instrument.

pulse width of the primary ion beam arriving at the surface; typically in the order of 1 ns) and their initial KE distribution (Figure 4.1B). The latter effect is compensated for using a reflection mirror in the flight tube which applies a gradually increasing retarding field [7]. The more energetic ions will penetrate deeper into the mirror before they are reflected, whilst the less energetic ions will take a shorter path. By tuning the retarding field in the reflection mirror, it is possible to force all ions of the same mass to arrive at the detector at the same time, despite their initial difference in KE. Thus, the drift tube with the reflection mirror can give a higher mass resolution.

The ToF mass analyzer has several advantages over other analyzers for the analysis of organic materials. Organic materials undergo substantial fragmentations under the bombardment of high-energy ions. Thus, it is important to record all fragment ions simultaneously. The ToF-SIMS can detect the full spectrum of a wide range of mass for every single pulse with a quite high mass resolution ($m/\Delta m \geq 1000$) up to around 10,000 amu. Since all ions entering the drift tube are detected, the transmission efficiency is significantly higher than the quadrupole and magnetic sector analyzers.

4.1.3 SIMS with Cluster Ion Source

There has been increasing interest in the use of polyatomic cluster ion sources, such as Au_n^+, Bi_n^+, SF_5^+, and C_{60}^+. The sputter yield is significantly enhanced as compared to the single-atom ion source [8]. The cluster ions break up as they impinge on the surface, and the projectile energy is distributed among all atoms. As a result, they penetrate the material much less and generate less chemical damage. Figure 4.3 depicts the time-lapse snapshots, from molecular dynamics simulations, of the sputtering processes for 15 keV Ga^+ and C_{60}^+ ions impinging three layers of benzene on silver (Ag) [9]. In the case of Ga^+, the primary ion penetrates about 4 nm deep into the sample (marked with an arrow in the 26 ps snapshot image), depositing the initial energy over a fairly long distance. Along the path of the Ga^+ projectile, a significant degree of atomic mixing occurs both in the lateral and vertical directions. The adsorbate molecules ejected from the surface are due to the energy recoiling back from the substrate. In the case of C_{60}^+, the dynamics associated with the cluster disintegration deposit most of the KE in the near surface region, thus restricting the damage to the substrate while producing a similar yield of adsorbate ejection yield. The sputter yield to the substate damage ratio is high enough that most of surface damage generated by the previous pulse is apparently removed by the subsequent pulse.

As a rule of thumb, one can expect that the cluster ion source enhances the secondary ion desorption yield with much less damage to the surface (such as sputter-induced atomic mixing and roughening). For these reasons, cluster ion sources are known to be superior to atomic sources for mass analysis of secondary ions and imaging of organic and polymeric materials [10]. However, there are some drawbacks with the cluster ion source such as C_{60}^+. Its beam current is low, which may require a longer data acquisition time [11].

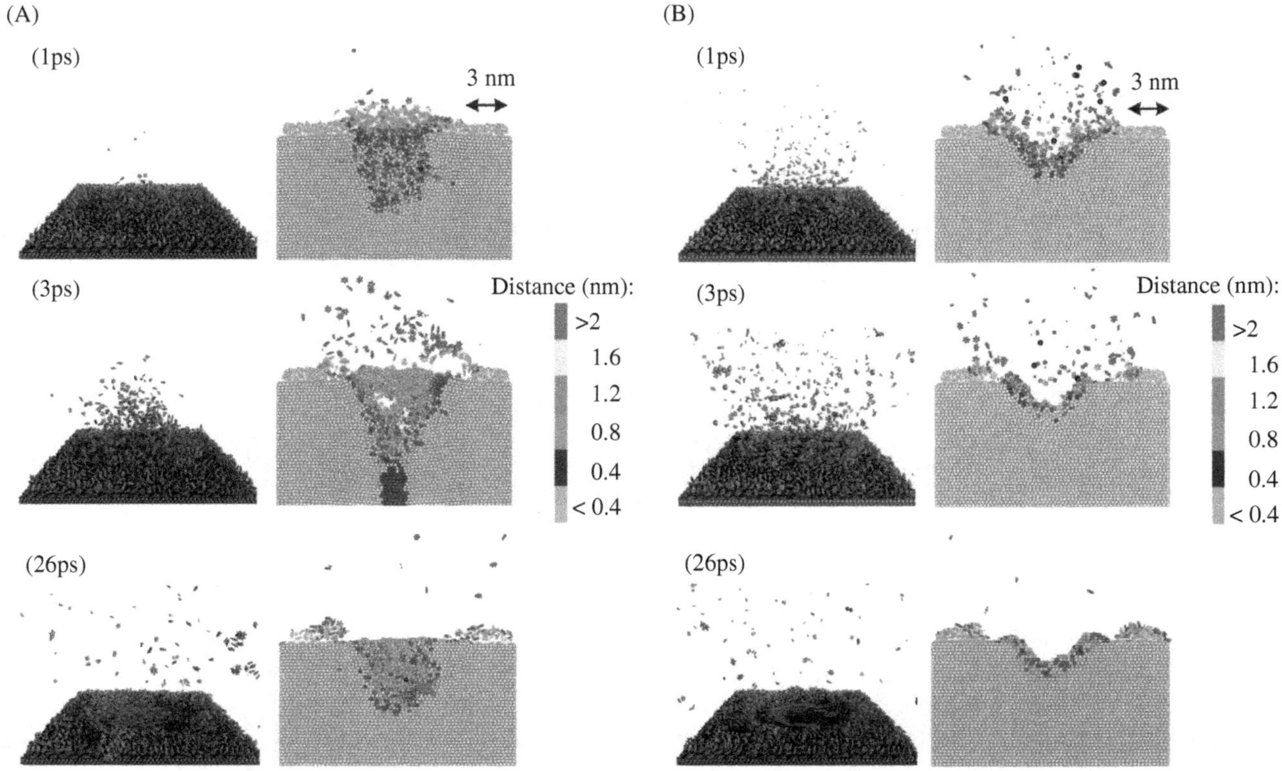

Figure 4.3 Snapshots for 15 keV (A) Ga^+ and (B) C_{60}^+ bombardment of three layers of benzene on silver (Ag) at three times. Left panel: perspective view of the target. Right panel: a slice through the center of the sample directly below the impact point. Reprinted with permission from [9]. Copyright 2008 John Wiley & Sons.

4.2 Qualitative Analysis with SIMS

4.2.1 Static SIMS

SIMS is essentially the destructive analysis method. But, if a very low dose of primary ion beam is used, the analysis can be done while maintaining the integrity of the surface layer within the time sale of the measurement. This operation mode is called *static* SIMS. If each impinging primary ion disturbs an area of 10 nm^2 (Figure 4.3), it will require 10^{13} impacts to affect an area of 1 cm^2 of the specimen surface. Typical damage cross sections for organic materials are on the order of 5×10^{-14} cm^{-2} [12]. Assuming the exponential decay of the topmost layer signal with the surface damage, we can estimate that the signal intensity would decrease by about 50% after a primary ion dose of 10^{13} ions/cm^2. The ion current of 1 nA/cm^2 corresponds to around 6.3×10^9 ions/s·cm^2. Thus, it would take 1600 s to reach the decay of the topmost layer signal by ~50%. If the SIMS measurement is done with a primary ion beam dose of 2×10^{12} ions/cm^2 (i.e., 320 s at 1 nA/cm^2), then it means the analysis is completed with approximately ~10% destruction of the topmost layer.

Using a liquid metal ion source, ToF-SIMS is widely used for elemental or chemical imaging with high spatial resolution. The liquid metal ion gun can be operated with beam diameters of typically 100–200 nm, in some cases down to ≤50 nm, at the sample surface, and digitally rastered across the area of interest with 256 × 256 or 512 × 512 pixels. At each pixel, it is possible to collect an intensity of a sing mass (m/e) using the quadrupole or magnetic mass analyzer, or a full mass spectrum with the ToF mass analyzer. Then, elemental or chemical state (fragment ions of molecular species) images can be generated.

When a specific ion is monitored, mass interferences can occur. This is when two ions have the same nominal mass (i.e., the same m/e ratio). For example, for the analysis of Fe in Si, $^{56}Fe^+$ signal can be interfered with $^{28}Si_2^+$ (diatomic Si ion). When nitrogen doping in Si is analyzed, the $^{28}Si^+$ ion can interfere with the $^{14}N_2^+$ signal. Similarly, the $^{16}O_2^+$ can interface with the detection of $^{32}S^+$. Although the analyte and interference ions may have the same nominal masses, their exact

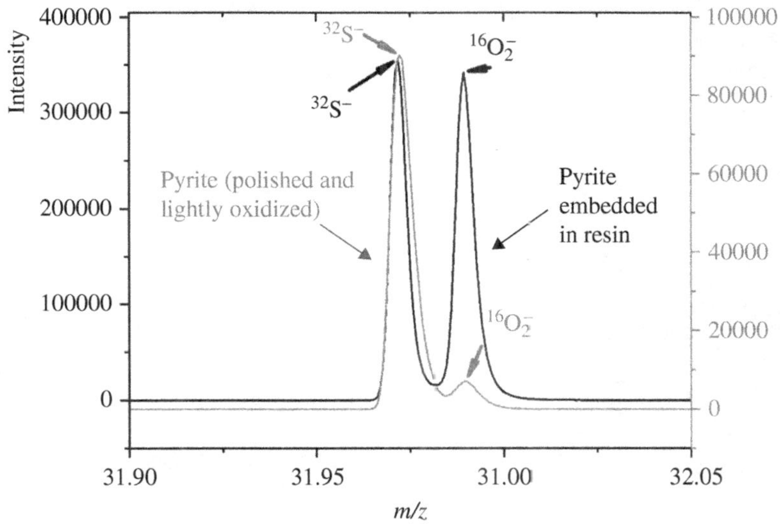

Figure 4.4 High-resolution ToF-SIMS spectra of two pyrite samples. Reproduced from Xu et al. [13]/with permission of Elsevier.

masses differ by a fraction of an atomic mass unit. So, they can be differentiated by increasing the mass resolution ($m/\Delta m$) of the spectrometer. Figure 4.4 shows an example where the $^{32}S^-$ ion of pyrite (FeS_2) is differentiated from the $^{16}O_2{}^-$ ion from the oxidized surface layer [13]. Alternatively, it is possible to use other isotope ions and subtract the contribution from interferent ions using the natural isotope abundance ratios, if they can be detected without any interference from other ions.

When the whole mass spectrum is obtained at each pixel, then hyperspectral image analysis is possible. The easiest way would be to look at signal intensities of a few selected m/e peaks in the mass spectrum. But, this requires *a priori* knowledge about the sample identify and may involve a user bias or miss other unexpected chemical information. For that reason, multivariate analysis is often used. By performing principal component analysis (PCA), one can find characteristic mass spectrum patterns that describe different variances in the data set. The principle of PCA is explained in Chapter 8 (for image processing, see Section 8.3.7).

An example of static ToF-SIMS imaging analysis is shown in Figure 4.5 [14]. The samples analyzed were silicon wafer surfaces on which friction tests were done in n-pentanol vapor (with $P/P_{sat} = 50\%$) with a 3 mm diameter silica sphere at two different loads (0.2 N and 0.7 N). In this environmental condition, the solid surfaces were covered with at least a monolayer of physiosorbed pentanol. When the applied load was 0.2 N (maximum Hertzian contact pressure $\approx$ 340 MPa), there was no visible wear mark in the wafer surface. When the load was increased to 0.7 N (which corresponds to a contact pressure of 510 MPa), visible wear marks were noticed. When these surfaces were analyzed with ToF-SIMS, the wear track exhibited a mass spectrum with multiple peaks with masses much larger than the pentanol molecule ($m/e = 88$ amu). The main principal component (PC1) found from PCA analysis of the entire spectra of the wear region showed bunches of multiple peaks separated by 13 ~ 14 amu, which is a characteristic feature of fragmentation of polymeric species (see Figure 4.5C). When the scores of PC1 were plotted in the image, it showed that those polymeric species were present only in the wear track (Figure 4.5B). When the wear track was analyzed with infrared spectroscopy, the C–H stretch peaks of organic species were detected, confirming the detection of polymeric species in ToF-SIMS. This implied that the wear of substrate is accompanied by the synthesis of polymeric species from adsorbed pentanol molecules. In the 0.2 N case, even though there was no clearly visible wear and IR spectroscopy did not detect any discernable signals of organic species, ToF-SIMS showed that there was a tiny amount of polymeric species in the sliding tract (Figure 4.5A), indicating that a small degree of substrate wear took place. This example shows how sensitive static ToF-SIMS analysis is to chemical changes in the surface layer.

Figure 4.6 displays ToF-SIMS analysis result of a different region tested in the P/P_{sat} 20% pentanol vapor with a 0.1 N load [14]. The PCA analysis of the entire mass spectra of all pixels in the collected ToF-SIMS image found three main PCs. One has the characteristic fragmentation pattern of polymeric species. The second one has mostly Si-containing ions. The third one is similar to the second one with somewhat different hydrocarbon-related signals. By combining the score plots of these three PCs, a montage image is constructed (Figure 4.6D) which shows the clear differences in chemical species present in different regions.

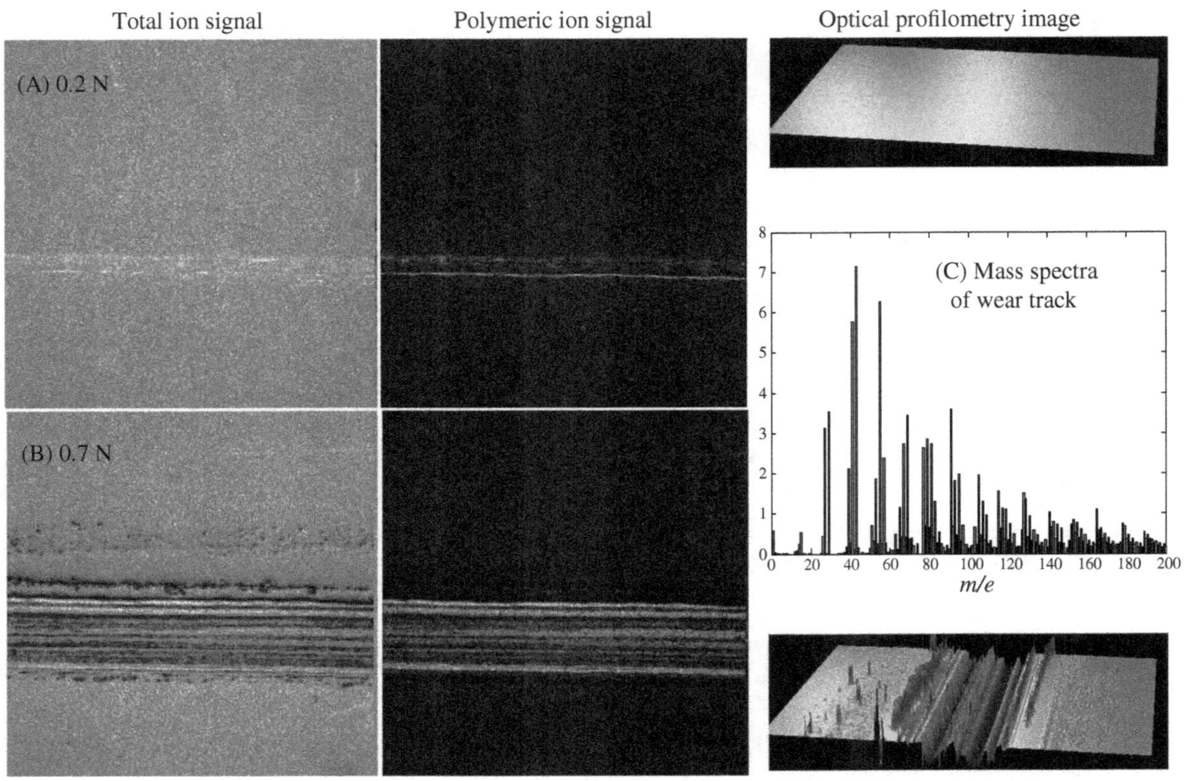

Figure 4.5 Total ion signal image, PC1 score image, optical profilometry image of the Si wafer surface after friction test in n-pentanol vapor environment with a silica sphere at (A) 0.2 N and (B) 0.7 N. (C) Characteristic mass spectrum found by PCA analysis representing the polymeric species in the wear track. ToF-SIMS analysis was done with a Bi_3^+ ion source. Reproduced with permission from Barnette et al. [14]/American Chemical Society.

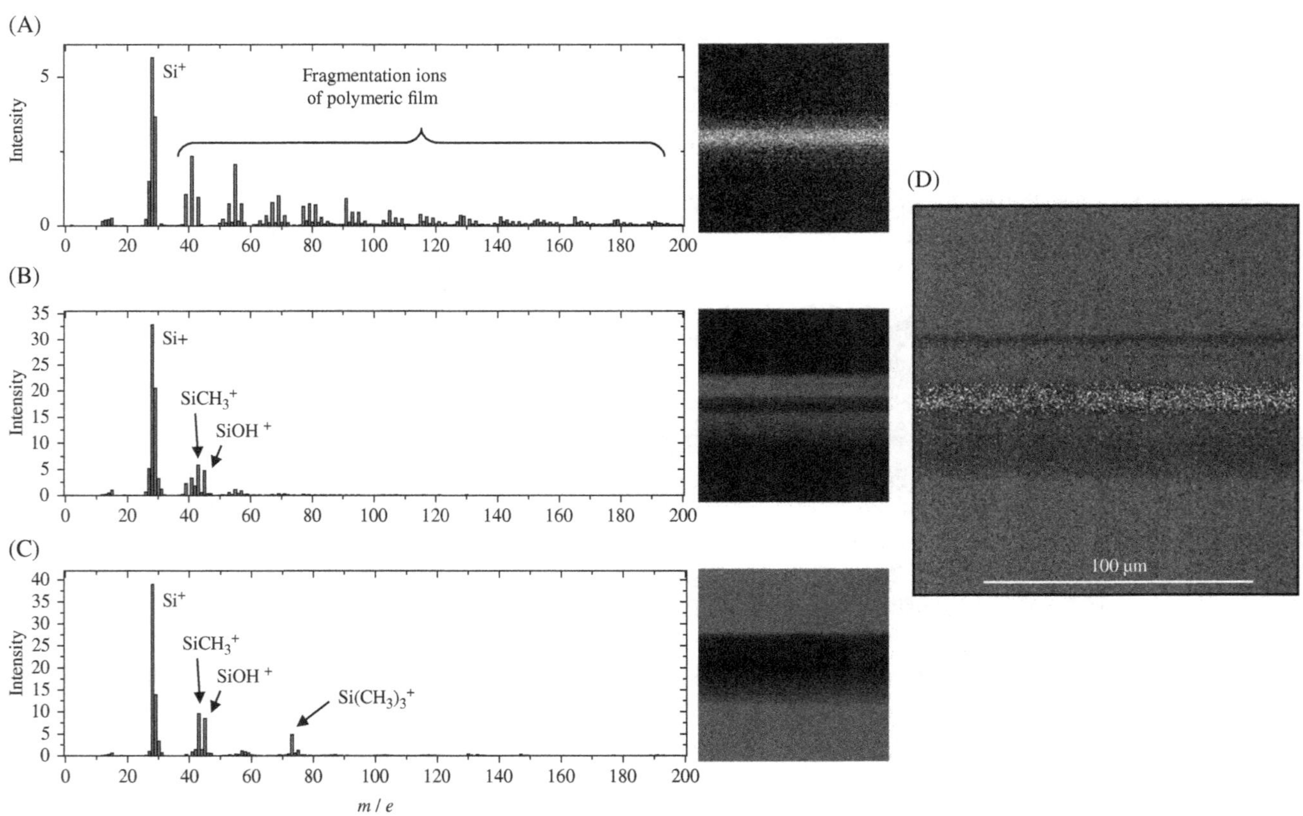

Figure 4.6 ToF-SIMS analysis of a wear track on a Si wafer made in pentanol vapor (P/P_{sat} = 20%) with a silica sphere with an applied load of 0.1 N. (A–C) Three main PCs and their score images: (A) polymeric species, (B) wear surface, and (C) intact surface. (D) Montage image displaying chemical differences within the imaged area. Reproduced from Barnette et al. [14]/with permission of American Chemical Society.

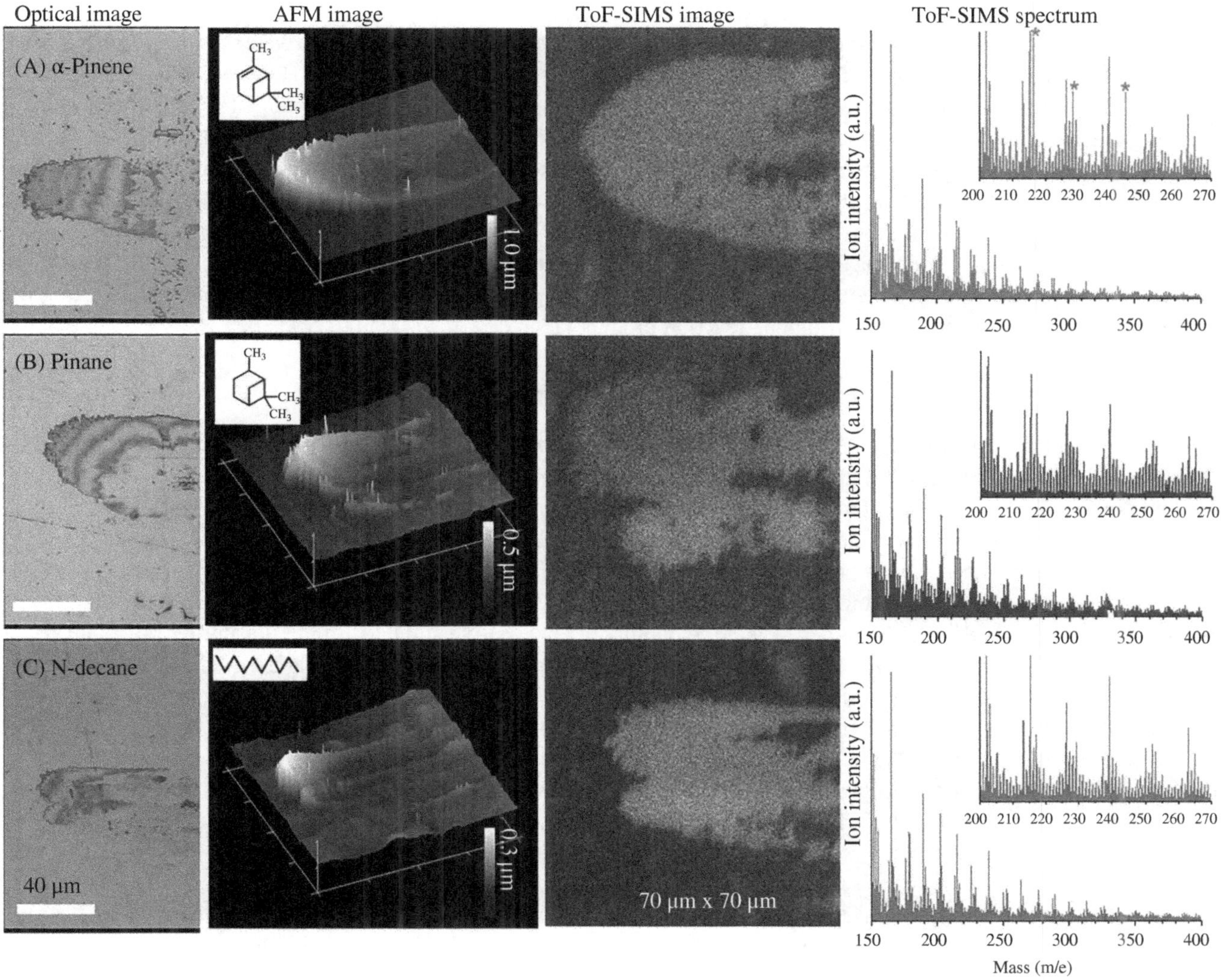

Figure 4.7 Optical, AFM, and ToF-SIMS images as well as characteristic mass spectra of polymeric products piled up at the end of the reciprocating sliding track friction-tested in (A) α-pinene, (B) pinane, and (C) n-decane vapor conditions (P/P_{sat} = 30%). The structures of precursor molecules adsorbed are shown in the inset of AFM image. In the ToF-SIMS image, the dark region is iron (m/e = 55.93 amu), slightly brighter spots in the iron region is chromium (m/e = 51.94 amu), and the brightest region (same shape as the AFM image) is polymeric species (m/e = 164–300 amu). ToF-SIMS analysis was done with Bi_3^{2+} ion (30 keV; total ion dose = 6.95×10^{12} ions/cm^2). Reproduced with permission from He et al. [15]/Elsevier.

In the previous examples, chemical reactions of adsorbate molecules are closely tied with the surface wear process. Figure 4.7 shows another example of ToF-SIMS analysis of polymeric products from adsorbed molecules upon frictional shear [15]. The substrate and ball were AISI 440C stainless steel. Friction tests were done in three different C_{10} hydrocarbon vapors ($P/P_{sat} = 30\%$, which was sufficient to form a full monolayer coverage on clean surface). The Hertzian contact pressure under the load was 600 MPa. When the tribo-tested tracts were imaged with optical microscope, there were some deposits with iridescent colors. The iridescent color means that the deposits have smooth surface with varying thicknesses (see Figure 10.9 in Chapter 10). Using atomic force microscopy (AFM), the topographic images were obtained, from which their volumes could be calculated. When the surface was rinsed with ethanol, the deposits were completely removed and there was no visible wear of the substrate at all. So, the deposits must be polymeric species produced by reactions of adsorbed molecules under frictional shear.

The polymeric nature of the reaction product was confirmed with ToF-SIMS. The images of the ToF-SIMS ion maps were very similar to the optical and AFM images. Outside the polymeric deposit region, the main ions detected were iron and chromium which are the main constituents of the stainless-steel surface. Within the deposit region, these inorganic ions of the substrate were not detected. Instead, organic ion peaks were detected with masses extending up to ~500 amu. The molecular weights of the adsorbed α-pinene, pinane, and n-decane molecules are 136 amu, 138 amu, and 144 amu, respectively. Also, the ions with masses higher than the molecular weight of the adsorbed molecules were separated roughly by 13 amu (CH) and 14 amu (CH_2), consistent with the fragmentation of polymeric hydrocarbons. However, due to the fragmentations of organic species by high-energy ion sputtering and a lower detection yield of the channeltron detector for higher mass organic ions, the maximum molecular weights of the polymeric species could not be determined. Since the stainless-steel surface cannot induce polymerization reactions of the adsorbed molecules at ambient conditions, they must be produced via tribochemical or mechanochemical mechanisms. It is noted that ions with $m/e = 216$ amu, 228 amu, and 244 amu were relatively more prominent in the mass spectrum of the product formed from α-pinene (marked with $*$ in Figure 4.6A), as compared to those of the products from pinane and n-decane. Since their mass differences are 12 amu and 16 amu, those peaks could be attributed to fragment ions containing carbonyl (C=O) groups, which was corroborated by detecting the 1720 cm^{-1} peak in IR analysis of the α-pinene reaction product [15].

4.2.2 Dynamic SIMS

In *dynamic* SIMS, a high current of primary ion beam is used, typically 1 $\mu A/cm^2$ or even larger, to sputter off the material from a specific location on the solid surface in the form of neutral and ionized atoms or atomic clusters. Then, by collecting the mass spectrum as a function of the sputter rate, depth profiling can be done along with lateral imaging. Alternatively, it can be done by stepwise sputter removal of a certain thickness and then performing static SIMS analysis at each sputter interval. The sputter rate can be approximated using an empirically tabulated correlation to the ion current density. Or the total depth sputtered can be measured after the depth profiling is completed using profilometry (Chapter 10) or AFM (Chapter 11), which will allow to calculate the depth at a given sputter time assuming the sputter rate is invariant during the sputtering process. The depth resolution is typically between 3–4 nm (see Figure 4.3) and 30–40 nm (see Section 2.5.2), which varies depending on the experimental condition and the sample type.

Figure 4.8 shows an example of dynamic SIMS imaging. The sample was a diamond-like carbon (DLC) coating on a steel ball which was used to test friction against another DLC coating in H_2-containing environment. The region that was in contact with the counter-surface showed a higher degree of hydrogen-containing species [16]. As the surface layer was sputtered off, the relative abundance of CH (13 amu) and C_2H (25 amu) species was changed. The oxygen signal was observed only in the topmost surface region suggesting that they might be oxygen-containing contaminants or an oxidized surface layer.

It should be noted that substantial sputtering is accompanied by various changes in surface composition and topography [17]. The primary ion can be implanted and induce beam damage. Enhanced diffusion of specific elements is also possible (see Figure 2.59). Preferential sputtering will change the stoichiometry of surface compositions. Redeposition of sputtered

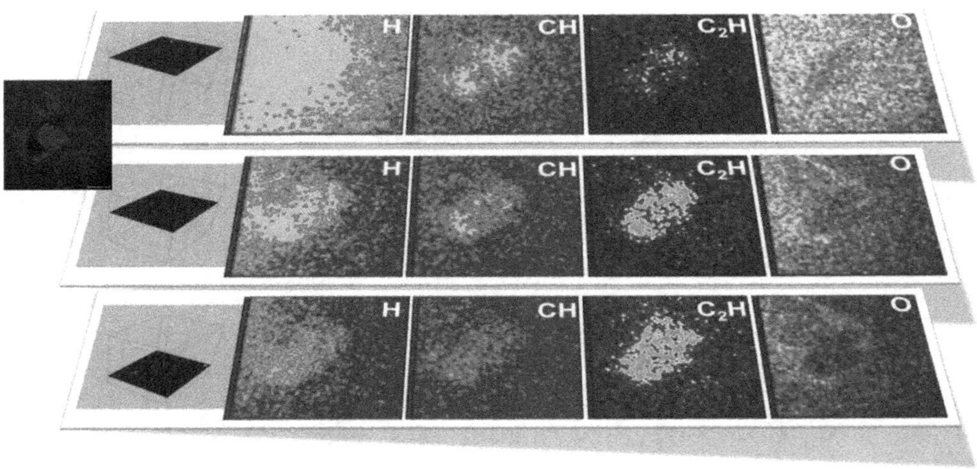

Figure 4.8 Dynamic SIMS images of the tribo-contact region of DLC tested in H_2 environment. The total ion signal image is shown in the left and the depth of the analyzed region is schematically shown. Courtesy of Dr. Ali Erdemir, Argonne National Laboratory.

material can also occur. Stochastic particle removal leads to roughening of the surface topography. It could further cause shadowing of certain surface features. For crystalline surfaces, the sputtering yield will vary depending on the azimuth angle of the impinging ion trajectory and the crystallographic orientation. So, when dynamic SIMS data are interpreted, these potential artifacts should be taken into account.

4.3 Quantitative Analysis with SIMS

In theory, the measured signal intensity of secondary ion can be converted to the concentration of analyte. But, in reality, this is often more complicated than what we have seen in XPS and AES. It is because the faction of ionic species desorbing from the surface is so low (on the order of 1% of all desorbing species); thus, a minute change in matrix chemistry can have a significant impact on the ionization yield of the desorbing species. This is called *matrix effect* [18]. The yield of secondary ions can be written as:

$$I_s(m) = I_p\, S(m)\, \alpha^{\pm}\ T\, D\, c(m) \tag{4.1}$$

where $I_s(m)$ is the secondary ion current of species with mass m, I_p is the primary ion flux, $S(m)$ is the sputter yield (Figure 4.1A), $\alpha^{\pm}$ is the probability to bear positive or negative charge, T is the transmission function of the analyzer, D is the detection sensitivity, and $c(m)$ is the fractional concentration of analyte with mass m. The $S(m)\alpha^{\pm}$ term is basically the atomic sensitivity. As in XPS and AES, we can define the relative sensitivity factor (RSF) of an ion A in a matrix M:

$$\frac{C_A}{C_M} = RSF\frac{I_A}{I_M} \tag{4.2}$$

where C and I are concentration and secondary ion intensity, respectively, and subscripts A and M are the analyte and the matrix, respectively.

Here, RSF can vary over many orders of magnitude depending on ionization potential for positive ions and electron affinity for negative ions. Figure 4.9A exhibits the charts plotting the relative secondary positive ion yield as a function of atomic

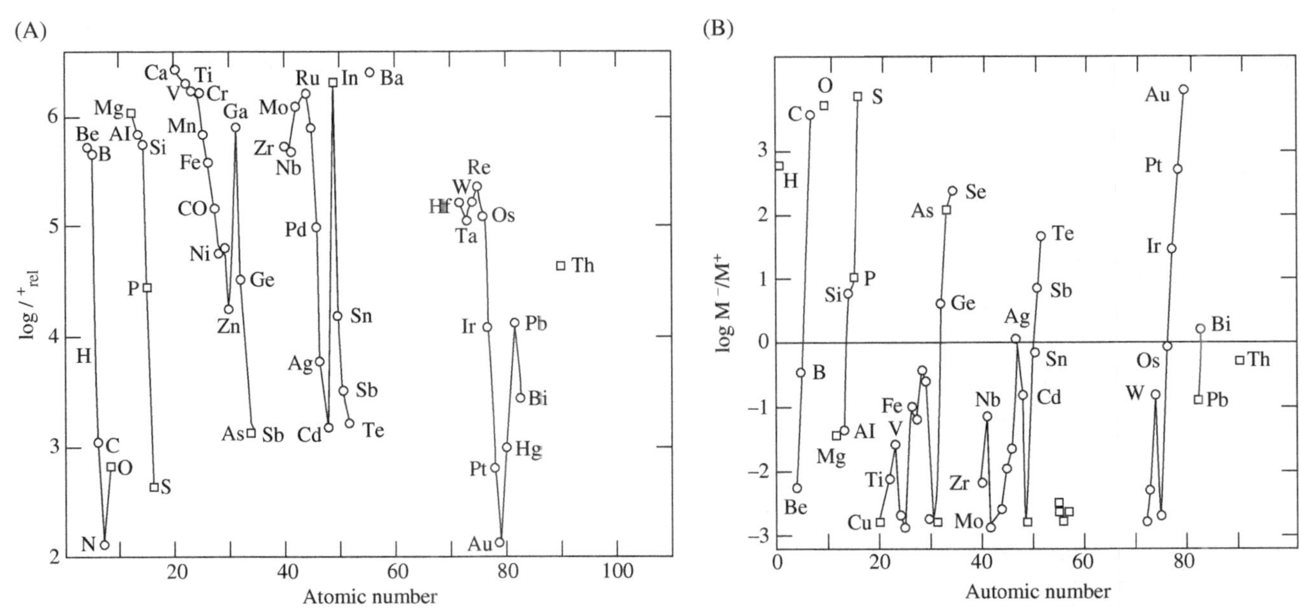

Figure 4.9 (A) Relative secondary positive ion yield as a function of atomic number (Z) for 1 nA of 13.5 keV O^- beam. (B) Ratio of negative ion yield (M^-) with 16.5 keV Cs^+ beam to positive ion yield (M^+) with 13.5 keV O^- beam. Reproduced from Storms et al. [19]/with permission of American Chemical Society.

number (Z) for 1 nA of 13.5 keV O^- bombardment [19]. It also varies with the primary beam type. The O^- bombardment increases the yield of positive ions, and the Cs^+ bombardment increases the yield of negative ions. The ratio of the negative ion emission yield vs. the positive ion emission yield can vary over several orders of magnitudes (Figure 4.9B) [19].

Even for the same element, depending on the matrix, RSF can vary over several orders of magnitude [20]. Oxidation alters the elemental ion yields to differing extends [11]. The relative capabilities of the desorbing species to capture or lose electrons or protons can vary drastically [21]. All these make quantification of absolute concentration complicated. If the maximum yield of secondary ions with the minimum consumption of sample material is required, primary beam energy dependent SIMS measurements are needed [22].

Matrix effects are even more complicated for organic materials. Different secondary ions from the same material may be enhanced or suppressed by different extents in mixtures, and different secondary ions appear to originate from different depths [23]. In an extreme case, it is possible that ejection of certain ions could be totally suppressed by the presence of other chemical species. In that case, detecting ions with opposite charge might be helpful.

When an insulating material is analyzed with a positive ion beam, the surface will become positively charged very quickly; the charging can increase the surface potential up to several hundred volts, which will affect the KE of the emitted ions beyond the acceptance window of the analyzer. Such charging problem can be compensated by irradiating the sample surface with a low-energy electron gun, the same as in the charge neutralization in XPS (Section 2.2.3). For negative ion detection, the surface potential needs to be negative for the negative ions to be released from the surface. In that case, a high flux of electrons can be used to keep the surface potential negative. Some other methods specific for samples could be devised to mitigate the charging problem.

4.4 Further Reading for SIMS

There are many good review papers and books on SIMS analysis. An easy-to-read one with lots of well-summarized examples is "Molecular Surface Mass Spectrometry by SIMS" by J. C. Vickerman in *Surface Analysis – The Principal Techniques* [11]. A more advanced version with full technical description is "Secondary Ion Mass Spectrometry: Basic Concepts, Instrumental Aspects, Applications and Trends" by A. Benninghoven, F. Rüdenauer, and H. Werner [1]. Even in qualitative analysis, the possible complication due to matrix effect should be considered. ToF-SIMS imaging and depth profiling can easily be overwhelmed by a large amount of data since every spectrum contains thousands of data points. Thus, processing data with multivariate analysis methods has become increasingly more common. When properly used, multivariate analysis can help the user identify the major sources of differences within a sample or between samples, determine whether and where certain compounds exist in the sample [24, 25]. The most frequently used multivariate analysis method is PCA; the basic principle of PCA is covered in Chapter 8.

Practice Problems

4.1 What are the main advantages of using a polyatomic cluster ion source over a monoatomic ion source?

4.2 In SIMS, the yield of secondary ions arising from different chemical species in a surface is not directly proportional to the chemical composition of the surface. Explain the reason.

4.3 Find a paper discussing the matrix effect in the literature and discuss how it could lead misinterpretation of the data if ignored.

4.4 Shown below is the ToF-SIMS depth profile data of ~22 nm thick SiO_2 on Si. The Si^+ signal appears higher in the SiO_2 phase (<220 Å) than the Si phase, although the atomic density of Si is higher in the Si phase (>220 Å). What is the reason for this counter-intuitive result?

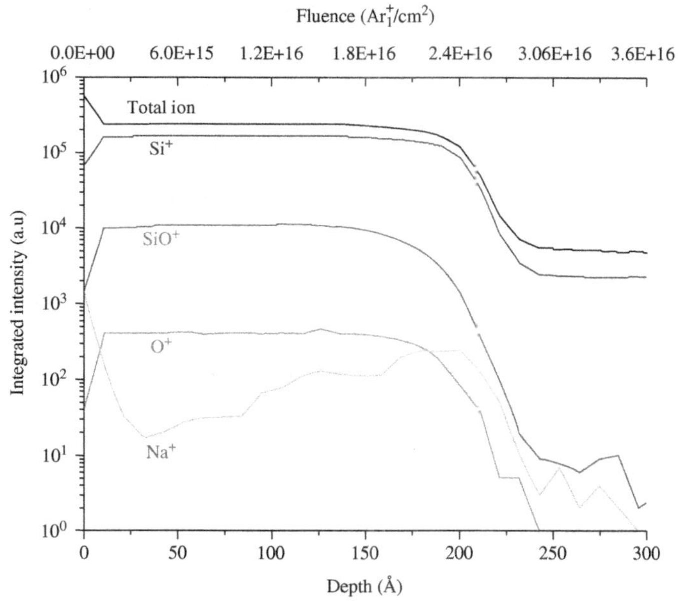

4.5 In the data shown above, determine the depth resolution. Just a rough estimate is sufficient.

References

1 Benninghoven A, Rüdenauer FG, Werner HW, editors. Secondary Ion Mass Spectrometry: Basic Concepts, Instrumental Aspects, Applications and Trends. New York, NY: Wliey; 1987.

2 Oechsner H. Surface analysis | secondary neutral mass spectrometry. In: Reference Module in Chemistry, Molecular Sciences and Chemical Engineering. Elsevier; 2013: 514–526.

3 Higashi Y. Quantitative depth profiling by laser-ionization sputtered neutral mass spectrometry. Spectrochimica Acta Part B: Atomic Spectroscopy. 1999;54(1):109–22.

4 Wasa K. 2 – Sputtering phenomena. In: Wasa K, Kanno I, Kotera H, editors. Handbook of Sputtering Technology (Second Edition). Oxford: William Andrew Publishing; 2012. p. 41–75.

5 Rauschenbach B. Sputtering. In: Low-Energy Ion Irradiation of Materials: Fundamentals and Application. Cham: Springer International Publishing; 2022. p. 123–74.

6 Slodzian G. Some problems encountered in secondary ion emission applied to elementary analysis. Surface Science. 1975;48(1):161–86.

7 Vickerman JC. Impact of mass spectrometry in surface analysis. Analyst. 1994;119(4):513–23.

8 Szakal C, Kozole J, Russo MF, Garrison BJ, Winograd N. Surface sensitivity in cluster-ion-induced sputtering. Physical Review Letters. 2006;96(21):216104.

9 Garrison BJ, Postawa Z. Computational view of surface based organic mass spectrometry. Mass Spectrometry Reviews. 2008;27(4):289–315.

10 Mahoney CM. Cluster secondary ion mass spectrometry of polymers and related materials. Mass Spectrometry Reviews. 2010;29(2):247–93.

11 Vickerman JC. Molecular surface mass spectrometry by SIMS. In: Vickerman JC, Gilmore IS, editors. Surface Analysis – The Principal Techniques. West Sussex, UK: John Wiley & Sons; 2009. p. 113–205.

12 Sichtermann W, Benninghoven A. Secondary ion formation from amino acids by proton and cation transfer. International Journal of Mass Spectrometry and Ion Physics. 1981;40(2):177–84.

13 Xu Z, Li S, Guo C, Li Z, Liang H. Change of surface oxide layer of pyrite samples prepared by resin embedding polishing method after long-term placement and its effect on TOF-SIMS analysis. Surfaces and Interfaces. 2024;46:104124.

14 Barnette AL, Asay DB, Ohlhausen JA, Dugger MT, Kim SH. Tribochemical polymerization of adsorbed *n*-pentanol on SiO_2 during rubbing: when does it occur and Is it responsible for effective vapor phase lubrication? Langmuir. 2010;26 (21):16299–304.

15 He X, Barthel AJ, Kim SH. Tribochemical synthesis of nano-lubricant films from adsorbed molecules at sliding solid interface: tribo-polymers from α-pinene, pinane, and n-decane. Surface Science. 2016;648:352–9.

16 Hofmann S. Sputter depth profile analysis of interfaces. Reports on Progress in Physics. 1998;61(7):827.

17 Eryilmaz OL, Erdemir A. On the hydrogen lubrication mechanism(s) of DLC films: an imaging TOF-SIMS study. Surface and Coatings Technology. 2008;203(5):750–5.

18 Deline VR, Katz W, Evans CA, Jr., Williams P. Mechanism of the SIMS matrix effect. Applied Physics Letters. 1978;33(9):832–5.

19 Storms HA, Brown KF, Stein JD. Evaluation of a cesium positive ion source for secondary ion mass spectrometry. Analytical Chemistry. 1977;49(13):2023–30.

20 Wilson RG. SIMS quantification in Si, GaAs, and diamond – an update. International Journal of Mass Spectrometry and Ion Processes. 1995;143:43–9.

21 Seah MP, Shard AG. The matrix effect in secondary ion mass spectrometry. Applied Surface Science. 2018;439:605–11.

22 Wittmaack K. Secondary ion emission from polymer layers by atomic and molecular ion bombardment: data evaluation based on linear-cascade sputtering theory. Applied Surface Science. 2006;252(19):6413–8.

23 Shard AG, Spencer SJ, Smith SA, Havelund R, Gilmore IS. The matrix effect in organic secondary ion mass spectrometry. International Journal of Mass Spectrometry. 2015;377:599–609.

24 Graham DJ, Castner DG. Multivariate analysis of ToF-SIMS data from multicomponent systems: the why, when, and how. Biointerphases. 2012;7(1):49.

25 Tyler BJ, Rayal G, Castner DG. Multivariate analysis strategies for processing ToF-SIMS images of biomaterials. Biomaterials. 2007;28(15):2412–23.

5

Light Propagation, Absorption, and Reflection

In previous chapters, we have studied elemental analysis of material surfaces. The surface chemistry of materials goes far beyond knowing the elemental composition. Very often, it is necessary to know molecular functionality of the surface. In XPS, XAS, AES, and SIMS, some degree of molecular-level speciation can be obtained, but quite limited. Perhaps the most widely used technique to get molecular information is vibrational spectroscopy. This is because chemically bonded atoms vibrate at their characteristic frequencies determined by the relative mass of atoms and the strength of chemical bonds connecting them. The energy associated with the excitation from the vibrationally ground state to an excited state falls in the infrared (IR) range of the photon energy scale (0.03–0.3 eV). Thus, we will study IR spectroscopy in the next chapter.

The vibrational excitation process is schematically illustrated in Figure 5.1 using the energy levels of carbon monoxide. As carbon and oxygen atoms bring together, their atomic orbitals (1*s*, 2*s*, 2*p*) are hybridized and form molecular orbitals, and all electrons of carbon and oxygen atoms are redistributed to these molecular orbitals (1σ, 2σ, 3σ, 4σ, 1π, and 5σ). At the equilibrium bond length (x_{eq}), the molecule has the lowest electronic energy. The molecule with the lowest-energy electronic configuration can have various vibrational states (n_v = 0, 1, 2, and so on). The occupancy of these vibrational states is governed by the Boltzmann distribution. At room temperature (T ≈ 300 K), thermal energy can be estimated with the gas constant: RT = 8.314 J/mol·K × 300 K = 2.5 kJ/mol = 26 meV/molecule. Since the thermal energy is much smaller than the vibrational state energy (for example, C-H stretch mode ≈ 2900 cm^{-1} = 360 meV), most molecules are in the vibrationally-ground state $n_v = 0$). Upon absorption of IR photon whose energy is in resonance with the vibrational energy difference, the molecule can be excited to the vibrationally excited state (n_v = 1).

Another technique that can give molecular information involves an excitation of electron from an occupied molecular orbital to an unoccupied orbital (Figure 5.1). The typical energy difference between the highest occupied molecular orbital (HOMO) and the lowest unoccupied orbital (LUMO) falls in the ultraviolet (UV) to visible (VIS) wavelength range (ca. 1.5–6 eV in the energy scale). The molecular orbitals are distinct based on the constituent atoms and their chemical bonding nature. Probing this electronic transition is UV–VIS absorption spectroscopy. Although the excitation processes probed are different, basic experimental principles of IR and UV–VIS spectroscopy share some common ground – optical spectroscopy. So, this chapter will review the optical theory to lay out the foundation for the physical principles in the following chapters. A specific focus is given to how light (which is electromagnetic wave) travels through matter and reflects/refracts at the interface of two bulk phases.

Some of the theoretical backgrounds discussed in this chapter are relevant to x-ray absorption, since x-ray is also electromagnetic wave. The electronic process associated with x-ray absorption is the excitation of electrons from core-level atomic orbitals to the vacuum level since the x-ray energy (ca. 100 eV–10's keV) is larger than the binding energies of core electrons. In Figure 5.1, note that the energy states of the C 1*s* and O 1*s* core orbitals in the molecular orbitals are slightly shifted from those in the atomic orbitals These small shifts are the basis for the initial state effect used to determine the oxidation state or chemical bonding nature in XPS (Section 2.3.2). The electron or photon emission during the relaxation process accompanying the electronic transition from the core level to the antibonding orbitals is what XAS (especially, NEXAFS) probes (Section 2.6.2).

In XPS (Chapter 2), the vacuum environment is critically needed because it involves detecting electrons (Chapter 1). In surface-sensitive XAS experiments, vacuum is needed since secondary electrons emitted during the relaxation of atomic orbitals are detected. AES (Chapter 3) and SIMS (Chapter 4) also require vacuum because they deal with electrons and ions,

Surface and Interface Analysis: Principles and Applications, First Edition. Seong H. Kim.

Companion website: www.wiley.com/go/Kim

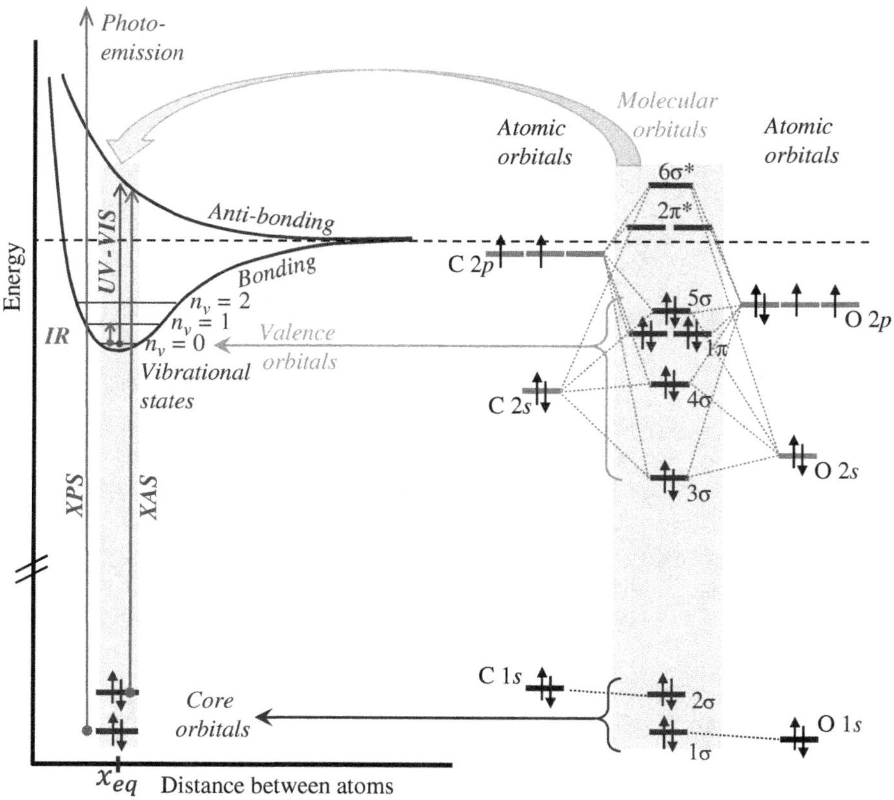

Figure 5.1 Schematic illustration of atomic and molecular orbitals for carbon monoxide and electronic transitions associated with IR, UV–Vis, and x-ray absorption.

respectively. The lifetime of these charged particles is very short in the gas phase and even shorter in the solid because their inelastic scattering probability upon collision with other atoms is very high. In the case of IR and UV–VIS spectroscopy, the "probe" going to the sample is photons and the "messenger" coming out of the sample is also photons. Since photons are chargeless, they can travel through the gas phase a long distance. Upon interactions with matter, the photon may be elastically scattered (e.g., reflected or diffracted), inelastically scattered (as in Raman spectroscopy), or absorbed. Typical penetration depths of photons can vary from a few hundred nanometers to tens of kilometers, depending on the frequency-dependent conductivity of the sample or the absorptivity. The use of polarized light presents another degree of freedom that may be exploited to attain sensitivity to molecular orientations. All these are advantageous, but at the same time, the surface sensitivity that originates from very short inelastic mean free path (IMFP) of charged particles is lost in optical spectroscopy. Nonetheless, surface sensitivity can still be achieved through a clever experimental design. Such principles will be discussed in the next chapter, which requires fundamental knowledge of how electromagnetic waves interact with and propagate through solid matters. For that, this chapter reviews the key concepts taught in sophomore-level undergraduate physics and partial differential equation courses and how they are linked to spectroscopy.

5.1 Propagation and Absorption of Electromagnetic Wave

5.1.1 Propagation of Light in a Non-absorbing Medium

Photon is an electromagnetic wave, and its energy is related to its frequency: $E = h\nu$ where $h = 6.626 \times 10^{-34}$ J·s is the Planck constant (see Section 2.1.1). The frequency (ν) is related to the wavelength (λ) through the speed of light (c) in a free space: $\nu = c/\lambda$, where $c = 2.998 \times 10^8$ m/s. The electric ($\boldsymbol{E}$) and magnetic ($\boldsymbol{H}$) fields of photon cause polarization of a medium through which it propagates. The displacement current (***D***) and magnetic induction (***B***) induced by ***E*** and ***H*** can be described as:

$$\boldsymbol{D} = \varepsilon \boldsymbol{E} \tag{5.1}$$

$$\boldsymbol{B} = \mu \boldsymbol{H} \tag{5.2}$$

where ε and μ are dielectric permittivity and magnetic permeability, respectively. For a free space (such as vacuum), the dielectric permittivity is $\varepsilon_o = 8.854 \times 10^{-12}$ C/V·m and the magnetic permeability is $\mu_o = 1.257 \times 10^{-6}$ N/A^2. Since electromagnetic wave is time dependent (note that it has a frequency), we need to consider Maxwell's equations to describe the electric and magnetic fields arising from varying distributions of electric charges and currents, and how those fields change in time. The following four are Maxwell's equations:

$$\int \vec{\boldsymbol{D}} \cdot d\vec{\boldsymbol{A}} = q \tag{5.3}$$

$$\int \vec{\boldsymbol{B}} \cdot d\vec{\boldsymbol{A}} = 0 \tag{5.4}$$

$$\oint \vec{\boldsymbol{E}} \cdot d\vec{\boldsymbol{l}} = -\frac{d}{dt}\left(\int \vec{\boldsymbol{B}} \cdot d\vec{\boldsymbol{A}}\right) \tag{5.5}$$

$$\oint \vec{\boldsymbol{H}} \cdot d\vec{\boldsymbol{l}} = \left(I + \frac{d}{dt}\left(\int \vec{\boldsymbol{D}} \cdot d\vec{\boldsymbol{A}}\right)\right) \tag{5.6}$$

Equation (5.3) is Gauss' law for electric fields, which depicts that the integral of the displacement current over the area enclosing a volume equals the total charge inside the volume. Equation (5.4) is Gauss' law for magnetic induction, which states that no magnetic monopole exists (recall that the simplest type of magnet is a dipole magnet). Equation (5.5) is Faraday's law of magnetic induction. The left-hand side is integrated around a closed line. It states that the total electric field induced around the circuit is proportional to the rate of change in magnetic flux going through the circuit. An example of using this law is electric generators. Equation (5 6) is Ampere's law describing the magnetic field around a circuit in terms of the current (I) in the circuit plus the rate of change in displacement current due to varying electric field through the circuit. An example of using this law is electric motors.

Let us apply these equations to a dielectric medium without any charge or real current. Then, eq. (5.3) states that the integration of displacement current over the closed surface is zero:

$$\oiint D_n dA = 0 \tag{5.7}$$

Here, the subscript n refers to the respective normal component, and the double integral with a circle means the integration over the surface enclosing a volume. Similarly, eq. (5.4) shows that the integration of magnetic induction over the closed surface is zero:

$$\oiint B_n dA = 0 \tag{5.8}$$

Consider an electromagnetic wave propagating along the x-direction in a Cartesian coordinate. The electric and magnetic fields of this wave are orthogonal to the propagation direction and vary only as a function of the position along the propagation direction (x) and time (t). Since these fields are orthogonal to each other, let us put ***E*** along the y-direction and ***B*** along the z-direction. It does not matter even if we use different directions as long as they are orthogonal to each other. This is schematically drawn in Figure 5.2. Calculate the change in electric field (E_s) over an infinitely small path (dl, marked with a dashed arrow) in the xy plane. Following the path, it can be written as:

$$E_s dl = \left[(E_y)_{x+dx} - (E_y)_x\right] dy = \left[\left(\frac{\partial E_y}{\partial x}\right) dx\right] dy = \left(\frac{\partial E_y}{\partial x}\right) dA \tag{5.9}$$

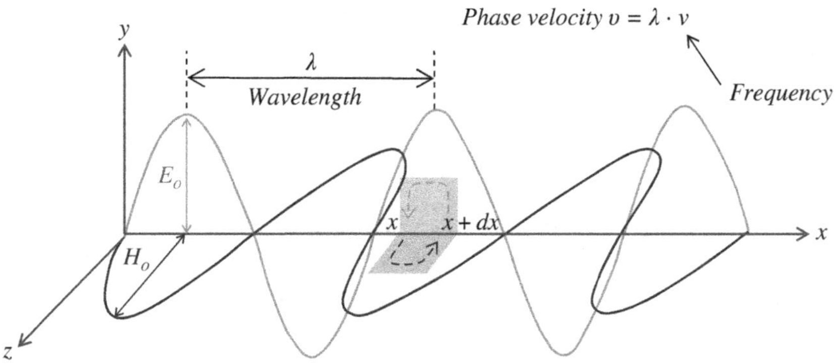

Figure 5.2 Electric and magnetic fields of an electromagnetic wave. The dashed lines are the infinitesimal paths considered to derive the relationship between transverse components (***E*** and ***H*** fields, respectively) propagating along the *x*-direction.

From eq. (5.5), we know that electric field along a closed path should be equal to the negative of the rate of magnetic flux change across the area defined by that path over time. Thus, we can write:

$$E_s dl = -\left(\frac{\partial B_n}{\partial t}\right) dA \rightarrow \left(\frac{\partial E_y}{\partial x}\right) = -\mu\left(\frac{\partial H_z}{\partial t}\right) \tag{5.10}$$

Next, calculate the change in magnetic field (H_s) over an infinitely small path (marked with a dashed line) in the *xz* plane. Following this path, it can be written as:

$$H_s dl = \left[(H_z)_x - (H_z)_{x+dx}\right] dz = -\left[\left(\frac{\partial H_z}{\partial x}\right) dx\right] dz = -\left(\frac{\partial H_z}{\partial x}\right) dA \tag{5.11}$$

From eq. (5.6), we know that magnetic field along a closed path should be equal to the rate of displacement current change across the area defined by that path over time; thus, we can write:

$$H_s dl = \left(\frac{\partial D_n}{\partial t}\right) dA \rightarrow \left(\frac{\partial H_z}{\partial x}\right) = -\varepsilon\left(\frac{\partial E_y}{\partial t}\right) \tag{5.12}$$

Putting eqs. (5.10) and (5.12) together, we can see a sort of positive feedback loop of electric fields creating magnetic fields, which in turn creates electric fields, which creates magnetic fields, and so on. Each is produced and influenced by the movement of the other.

Let us differentiate eq. (5.10) with respect to *x* (position) and differentiate eq. (5.12) with respect to *t* (time). Then, we can see there is a common term $\frac{\partial^2 H_z}{\partial x \partial t}$; by canceling this term from both equations, we get:

$$\frac{\partial^2 E_y}{\partial x^2} = \varepsilon\mu \frac{\partial^2 E_y}{\partial t^2} \tag{5.13}$$

With the same trick, we can see that H_z satisfy the equation of exactly the same form. This is the general *wave equation* for the propagation of an electromagnetic field with a periodic disturbance in a medium, and the proportionality constant in eq. (5.13) defines the phase velocity (υ) of this wave:

$$\upsilon = \sqrt{\frac{1}{\varepsilon\mu}} \tag{5.14}$$

The solution of eq. (5.13) can be written in a sine function:

$$E_y = E_o \sin \omega\left(t - \frac{x}{\upsilon}\right) \tag{5.15}$$

where E_o is the electric field strength and ω is the angular frequency, $\omega = 2\pi\nu$. Correspondingly, the solution of the magnetic field has the same form, with the H_o as the magnetic field strength:

$$H_z = H_o \sin \omega\left(t - \frac{x}{\upsilon}\right) \tag{5.16}$$

So, the electromagnetic wave is often described as a sine function (Figure 5.2).

If the medium under consideration is a free space (such as vacuum), the dielectric permittivity is $\varepsilon_o = 8.854 \times 10^{-12}$ C/V·m and the magnetic permeability is $\mu_o = 1.257 \times 10^{-6}$ N/A^2. Inserting these values into eq. (5.14), we get the speed of light in free space:

$$c = \sqrt{\frac{1}{\varepsilon_o \mu_o}} = \left[\left(8.854 \times 10^{-12} \frac{C}{V \cdot m}\right)\left(1.257 \times 10^{-6} \frac{N}{A^2}\right)\right]^{-1/2} = 2.998 \times 10^8 \text{ m/s} \tag{5.17}$$

The ratio of the speed of light (c) to the phase velocity in a dielectric medium (υ) is defined as the *refractive index* (n):

$$n = \frac{c}{\upsilon} \tag{5.18}$$

The higher n is, the slower υ will be. The relative dielectric permittivity (ε_r) of a dielectric material is the ratio of its magnitude to ε_o, that is, $\varepsilon_r = \varepsilon/\varepsilon_o$. In the frequency range relevant to IR and UV–VIS spectroscopy, μ is essentially the same as μ_o. Then, we can obtain the equation relating the refractive index of a dielectric material to its relative dielectric permittivity:

$$n = \frac{c}{\upsilon} = \sqrt{\frac{\varepsilon\mu}{\varepsilon_o\mu_o}} \doteq \sqrt{\varepsilon_r} \tag{5.19}$$

Note that the subscript r is often dropped for convenience. Remember this relationship because it shows the relationship between refractive index and dielectric permittivity. In the next section, we will study what determines the peak position of the absorption band in IR spectroscopy.

5.1.2 Absorption of IR via Vibrational Excitation

As a starting point, let us consider the simplest case – a diatomic molecule. The potential energy as a function of distance between two atoms has the shape of the bonding orbital curve in Figure 5.1. There are several forms of potential functions that can be used, which include Morse potential and Lennard–Jones potential. For typical IR spectroscopy, we can simply focus on the bottom part of the potential at the equilibrium bond length (x_{eq}), since most molecules are populated in the ground state ($n_\nu = 0$). In theory, all potential functions can be expanded with a Taylor series:

$$U(x) = U(x=0) + \left(\frac{dU}{dx}\right)_{x_{eq}} x + \frac{1}{2!}\left(\frac{d^2U}{dx^2}\right)_{x_{eq}} x^2 + \cdots \tag{5.20}$$

where x is the displacement from the equilibrium distance. In the right-hand side, the first term is a constant and is not important, since we can zero out the constant offset by re-defining zero potential. By definition, the second term $\left(\frac{dU}{dx}\right)$ is zero at the bottom of the potential well (at equilibrium). Assuming that the displacement of bond length from the equilibrium distance (x) is small, we can ignore higher-order terms with powers of three or greater. Then, we are left with the harmonic oscillator potential. If the diatomic molecule is represented with two masses connected with a spring with a spring constant of k (following the Hooke's law), the potential energy term can be written as $U(x) = \frac{1}{2}kx^2$.

Then, we can write the Schrödinger equation for a simple harmonic oscillator:

$$-\frac{\hbar^2}{2m_r}\frac{d^2\psi(x)}{dx^2} + \frac{1}{2}kx^2\psi(x) = E\psi(x) \tag{5.21}$$

where $\psi(x)$ is the wave function of electron (note that the time dependence is dropped here since we are interested in the standing waves allowed within the harmonic potential boundary), $\hbar$ is the reduced Planck constant ($\hbar = \frac{h}{2\pi} = 1.055 \times 10^{-34}$ J · s), m_r is the reduced mass ($m_r = \frac{m_1 m_2}{m_1 + m_2}$, where m_1 and m_2 are the mass of each atom), and E is the total energy. The first term in the left-hand side is the kinetic energy term and the second is the potential energy term. So, eq. (5.21) means that the sum of kinetic and potential energies should be the same as the total energy at all times, which is basically the energy conservation. Rearranging eq. (5.21), we get:

$$\frac{d^2\psi(x)}{dx^2} + \frac{2m_r}{\hbar^2}\left(E - \frac{1}{2}kx^2\right)\psi(x) = 0 \tag{5.22}$$

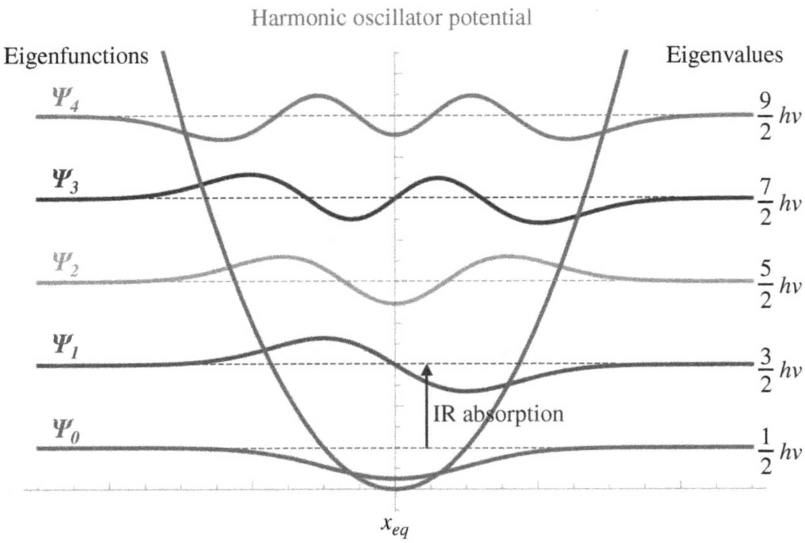

Figure 5.3 Eigen functions and eigen values of the Schrödinger equation of an electron in a diatomic molecule bound by the Harmonic potential.

In the differential equation class, we have solved $f''(x) + (a - x^2)f(x) = 0$; the Schrödinger equation has the same form. So, let us skip solving the Schrödinger equation here and go directly to the solution. The eigen values of the solution is given as:

$$E_{n_\nu} = \hbar\left(\frac{k}{m_r}\right)^{1/2}\left(n_\nu + \frac{1}{2}\right) = \hbar\omega\left(n_\nu + \frac{1}{2}\right) \tag{5.23}$$

where n_ν is the integer. When $n_\nu = 0$, it describes the ground state energy. From this, we can see that the resonance angular frequency is defined by the spring constant (k) of the chemical bond and the reduced mass (m_r) of atoms connected by the bond:

$$\omega_o = \sqrt{\frac{k}{m_r}}; \text{thus, } \nu_o = \frac{1}{2\pi}\sqrt{\frac{k}{m_r}} \tag{5.24}$$

The ψ_{n_ν} eigen functions (waveforms) are graphically shown in Figure 5.3.

In quantum mechanics, the allowed vibrational excitation is $\Delta n_\nu = \pm 1$. Since the most molecules are at the vibrational ground state ($n_\nu = 0$), the IR absorption will take place when the IR energy is equal to the energy difference between $n_\nu = 1$ and $n_\nu = 0$ states. In IR spectroscopy, the energy scale is expressed in the wavenumber (cm^{-1}) unit: $\overline{\omega} = 1/\lambda = \nu/c$. Thus, the peak position of the resonant IR absorption process is:

$$\overline{\omega}_{IR} = \frac{1}{2\pi c}\sqrt{\frac{k}{m_r}} \tag{5.25}$$

The IR absorption probability (a) is described in the following form:

$$a \propto |\langle\psi_1|\mu'|\psi_0\rangle|^2 \tag{5.26}$$

where ψ_0 and ψ_1 are the initial (ground state) and final (first-excited state) eigen functions and μ' is the operator describing the transient dipole moment change associated with the vibrational excitation upon IR absorption. This equation states that if the dipole moment change is zero (i.e., $\mu' = 0$) upon vibrational excitation, then the molecule cannot absorb IR; only the vibrational mode accompanied by dipole moment change can absorb IR. This tells us which vibrational modes are IR-active or inactive. For example, the symmetric stretch mode of carbon dioxide (O=C=O) is IR-inactive, while the asymmetric stretch mode of CO_2 is IR-active. Note that this equation has the same form as the equation describing the x-ray absorption cross section in eq. (2.23); the only difference is the operator.

The above result is for the diatomic molecule; but the same can be applied to polyatomic molecules using the spring constant of the normal mode (symmetry-allowed vibrational mode) and the reduced mass of atoms involved in the normal

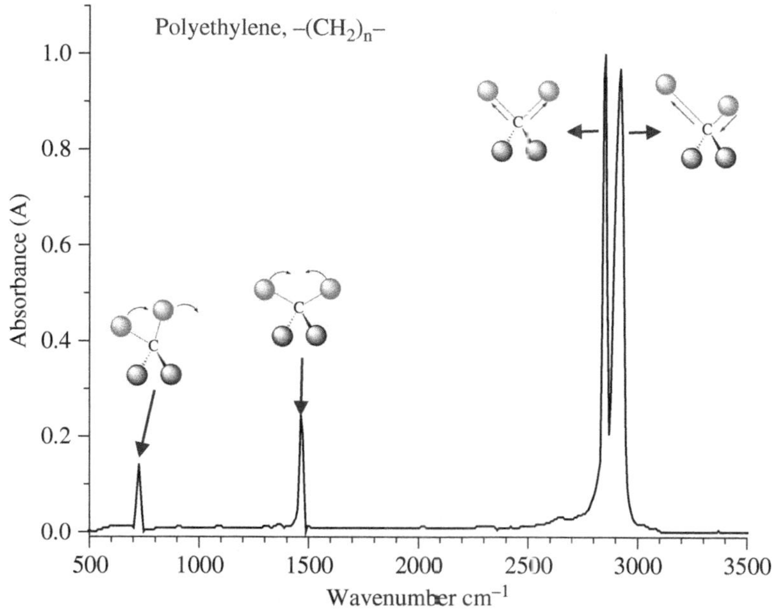

Figure 5.4 Transmission IR spectrum of polyethylene. The normal mode of each peak is illustrated schematically.

mode. Figure 5.4 shows the IR absorption spectrum of polyethylene, as an example. We can see four absorption bands in the 500–3500 cm^{-1} region. The 725, 1466, 2854, and 2924 cm^{-1} peaks correspond to the rocking, scissoring, symmetric stretch, and asymmetric stretch modes of CH_2 groups in the polymer. Their peak positions are determined by the effective spring constant and reduced mass of the CH_2 groups in individual normal modes, as depicted by eq. (5.25). In transmission spectrum, the signal intensity is often plotted in the absorbance (A) unit which is defined as $A = -\log(I_t/I_o)$, where I_o and I_t are the incident IR intensity and the transmitted IR intensity measured with the sample. If the sample thickness is b (i.e., path length of the beam through the sample), we can write the Beer–Lambert Law:

$$A = a\,b\,c \tag{5.27}$$

where a is the IR absorptivity defined by eq. (5.26) and c is the concentration. Thus, the intensity of each peak in Figure 5.4 is governed by the square of the integral of the function describing the overlap between the first-exited state wavefunction and the ground state wavefunction multiplied with the transient dipole moment change associated with each vibrational excitation.

5.1.3 Polarization of Medium by Electromagnetic Wave

We have learned how the light (electromagnetic wave) propagates through a medium (Section 5.1.1) and how it can be absorbed by a molecule when its energy is in resonance with the vibrational excitation of the molecule (Section 5.1.2). Now, let us put them together. When the light (which is the "probe" that we send to a material of interest) propagates through a medium containing molecules that can absorb light via a resonant excitation, how is the propagation of light (which is the "messenger" that we detect in spectroscopy) altered? The resonant process in IR spectroscopy is the vibrational excitation (Figure 5.3). The resonant process in UV–VIS spectroscopy is electronic excitation from an occupied valence orbital to an unoccupied orbital (Figure 5.1). The resonance process in XAS is electronic excitation from a core level to an unoccupied orbital (Figure 5.1). The resonance process in XPS is electronic excitation from a core level to the vacuum level (Figure 5.1).

Let us consider an atom placed in front of an electromagnetic wave propagating along the x-direction (Figure 5.5). Electrons in the atom will be polarized in response to the electric field of the wave. In Section 5.1.2, the magnetic permeability of a medium is essentially the same as that of a free space at optical frequencies. It means the response to the magnetic field of light is negligible and can be ignored in optical spectroscopy. The electron in the atom is bound to the nucleus by a spring which determines the natural oscillation frequency. So, upon a small displacement (x) from its equilibrium position, there will be a restoring force which can be described with Hooke's law: $F = -kx$, where k is the spring constant. Hooke's law

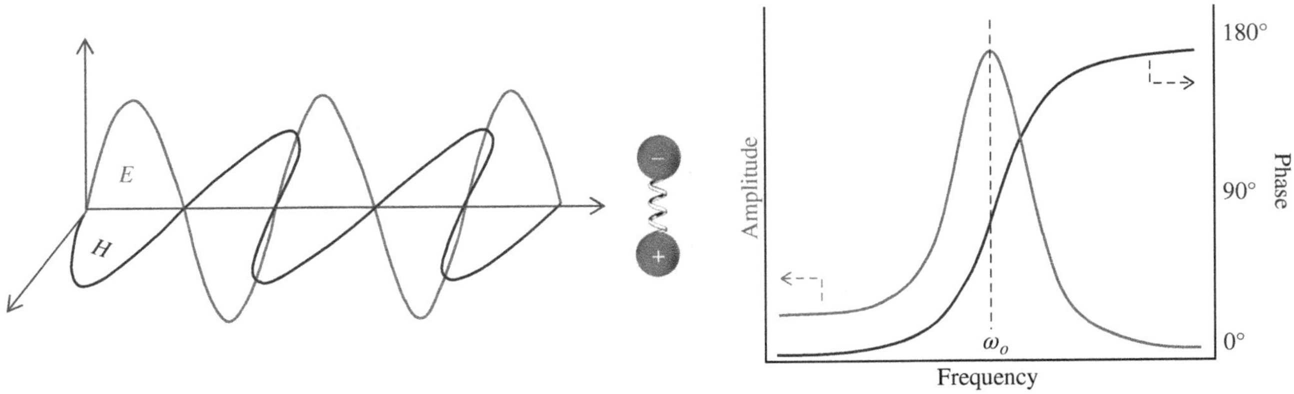

Figure 5.5 A model of a damped harmonic oscillator driven by a sinusoidal field and its behavior of electronic displacement as a function of frequency.

describes the restoring force due to the harmonic potential. Based on eq. (5.24), we know that the natural frequency is $\omega_o = \sqrt{k/m_r}$, where m_r is the reduced mass. Thus, the spring constant term can be replaced with the mass and natural frequency: $k = m_r\omega_o^2$. The dispersive behavior of a medium near an absorption region can be modeled as an electron undergoing damped periodic oscillations in the perturbing electric field. This "damped harmonic oscillator" is called the Lorentzian oscillator. Here, the damping rate is proportional to the displacement rate. The proportionality constant is called a damping factor (γ). This is the loss term. Then, we can set up the force balance ($F_{kinetic} + F_{loss} + F_{potential} = F_{electric}$) governing the electron displacement motion:

$$m_r\ddot{x}(t) + m_r\gamma\dot{x}(t) + m_r\omega_o^2 x(t) = -eE(t) \tag{5.28}$$

The left-hand side describes the forces involved in the damped oscillator and the right hand is the force driving the oscillation. Here, e is the charge of electron (1.602×10^{-19} C). These forces are time dependent since the electromagnetic wave (=light) varies at a certain frequency.

Again, this has the form of differential equation that we have solved in undergraduate math classes. The solution for $x(t)$ and $E(t)$ can be written in terms of Fourier components:

$$x(t) = \frac{1}{\sqrt{2\pi}}\int_{-\infty}^{\infty} x(\omega)e^{-i(\omega t-\varphi)}d\omega \tag{5.29}$$

$$E(t) = \frac{1}{\sqrt{2\pi}}\int_{-\infty}^{\infty} E(\omega)e^{-i\omega t}d\omega \tag{5.30}$$

where ω is the angular frequency of the light. Inserting $x(t)$ and $E(t)$ into eq. (5.28) gives the solution for the displacement:

$$x(t) = \frac{-eE(t)/m_r}{\omega_o^2 - \omega^2 - i\gamma\omega} \tag{5.31}$$

And the phase of displacement with respect to the electric field of the light is described as:

$$\varphi(\omega) = \tan^{-1}\frac{\gamma\omega}{\omega_o^2 - \omega^2} \tag{5.32}$$

The frequency dependence of the amplitude and phase of the damped oscillator driven by the electric field of light is schematically illustrated in Figure 5.5. The displacement given with eq. (5.31) is a complex number; its absolute magnitude is the amplitude in the schematic. The argument of this complex number is the phase. In the low-frequency limit, the denominator terms in eqs. (5.31) and (5.32) have a large positive value; thus, the amplitude is small, and the phase is nearly zero. As the frequency increases, these denominator terms become smaller, and both amplitude and phase increase. If the damping factor (γ) is zero, then the amplitude can increase to infinity as the driving frequency approaches the resonance frequency ($\omega \to \omega_o$). But, due to the damping (loss force), the amplitude remains finite. In other words, the magnitude and frequency dependence of displacement from the equilibrium position near the resonance frequency ($\omega = \omega_o$) is a function of the damping parameter. At the resonance, the term inside the arctangent function becomes infinity, so the phase between the displacement and the driving field becomes 90°. As ω becomes larger than ω_o, the magnitude of the denominators in both equations becomes larger again. This causes the amplitude to decrease as ω increases above ω_o. In the case of phase, the term inside the arctangent function is negative. So, the phase between the amplitude and the driving field is opposite. This means that the driving force is applied while the amplitude is decreasing, or vice versa. Thus, the kinetic form and the driving force term work against each other. In the high-frequency limit, the phase approaches 180°, and the amplitude decreases to the minimum value.

Now, we can convert the frequency-dependent displacement of an atom, $x(\omega)$, to derive the dielectric permittivity, $\varepsilon(\omega)$, that describes the polarization of a medium that we have studied in Section 5.1.1. If the time-dependent polarization of a single atom is given as $p(t) = -ex(t)$, and the number density of atoms in the medium is N, then the polarization per unit volume is $P(t) = Np(t)$. This term contributes the time-dependent displacement current: $D(t) = \varepsilon_o E(t) + P(t)$. Recall that ε_o is the dielectric permittivity of free space. The second term can have the contribution in resonance with the frequency of light, $P_{res}(\omega)$, and the nonresonant background contribution, P_{bg}. Using eq. (5.31), the resonant term can be written as:

$$P_{res}(\omega) = Np(\omega) = -Nex(\omega) = \frac{Ne^2}{m_r}\frac{1}{\omega_0^2 - \omega^2 - i\gamma\omega}E(\omega) \tag{5.33}$$

The nonresonant term, P_{bg}, can be described with a constant called electric susceptivity (χ). Then, the total electric displacement of a medium under the electric field of light can be expressed as:

$$\begin{aligned} D(\omega) &= \varepsilon_0 E(\omega) + \left[P_{bg}(\omega) + P_{res}(\omega)\right] \\ &= \varepsilon_0 E(\omega) + \left[\varepsilon_0 \chi E(\omega) + \frac{Ne^2}{m_r}\frac{1}{\omega_0^2 - \omega^2 - i\gamma\omega}E(\omega)\right] \\ &= \left[\varepsilon_0(1+\chi) + \frac{Ne^2}{m_r}\frac{1}{\omega_0^2 - \omega^2 - i\gamma\omega}\right]E(\omega) = \varepsilon(\omega)E(\omega) \end{aligned} \tag{5.34}$$

For dielectric materials, we often use relative permittivity which is defined as $\varepsilon_r = \frac{\varepsilon}{\varepsilon_o}$, as defined in eq. (5.19). Thus, we can write:

$$\varepsilon_r(\omega) = 1 + \chi + \frac{Ne^2}{\varepsilon_o m_r}\frac{1}{\omega_0^2 - \omega^2 - i\gamma\omega} \tag{5.35}$$

This complex number can be rearranged to $\varepsilon_r(\omega) = \varepsilon_1(\omega) + i\varepsilon_2(\omega)$. By manipulating eq. (5.35), we get the real and imaginary components of relative dielectric permittivity as:

$$\varepsilon_1(\omega) = 1 + \chi + \frac{Ne^2}{\varepsilon_o m_r}\frac{\omega_0^2 - \omega^2}{\left(\omega_0^2 - \omega^2\right)^2 + (\gamma\omega)^2} \tag{5.36}$$

$$\varepsilon_2(\omega) = \frac{Ne^2}{\varepsilon_o m_r}\frac{\gamma\omega}{\left(\omega_0^2 - \omega^2\right)^2 + (\gamma\omega)^2} \tag{5.37}$$

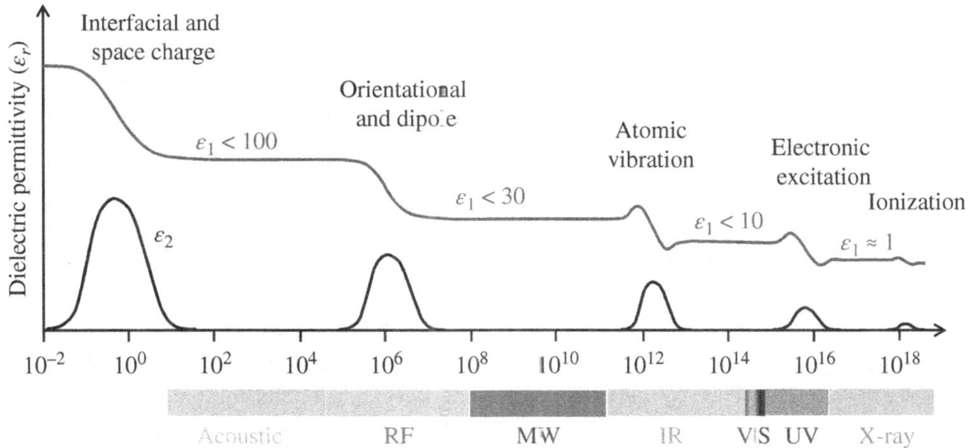

Figure 5.6 Graphical illustration relative dielectric permittivity change with frequency.

Figure 5.6 displays how real and imaginary components of relative dielectric permittivity of a medium change with the frequency of applied electric field. In the very low-frequency regime, ε_r is governed by interfacial and space charge. In this case, $\omega_o$will be the characteristic charge/discharge rate of the medium. In the radio frequency (RF) and microwave (MW) frequency regime, polarization due to ion and space charge movements become insignificant (because the field changes too fast compared to their mobilities), and reorientation of dipole becomes important in the overall polarization. In the IR frequency regime, the electric field change rate is too fast for dipole reorientation; instead, vibrational motions of atoms connected though chemical bonds play an important role. In the UV–VIS frequency regime, atomic vibrations are too slow to response to the oscillating electric field, and only electronic excitation from valence bands can resonate with the electric field. In the x-ray regime, photoionization takes place. Whenever the frequency of electric field passes over the frequency of one of these resonance processes, the real component of relative dielectric permittivity, $\varepsilon_1(\omega)$, decreases because the process contributing to the lower-frequency side of the resonance becomes inactive in the high frequency side. In the imaginary component of relative dielectric permittivity, $\varepsilon_2(\omega)$, there is a corresponding change (increase and then decrease) near the resonance frequency.

5.1.4 Refractive Index, Absorptivity, and Absorption Band Width

In spectroscopy, the frequency region of interest is near the resonant frequency. We can re-write the permittivity equations in terms of $\Delta\omega = \omega_o - \omega$ near the resonance. When $\omega \ll \omega_o$, the ω term in eq. (5.36) can be dropped, which leads to $\varepsilon_{st} = 1 + \chi + \frac{Ne^2}{\varepsilon_o m_r \omega_o^2}$. When $\omega \gg \omega_o$, the ω_o term drops out, and eq. (5.36) becomes $\varepsilon_\infty = 1 + \chi$. Using these terms, eqs. (5.36) and (5.37) can be re-written as:

$$\varepsilon_1(\Delta\omega) = \varepsilon_\infty + \frac{(\varepsilon_{st} - \varepsilon_\infty)}{2} \frac{\omega_o \Delta\omega}{\Delta\omega^2 + (\gamma/2)^2} \tag{5.38}$$

$$\varepsilon_2(\Delta\omega) = \frac{(\varepsilon_{st} - \varepsilon_\infty)}{2} \frac{\omega_o(\gamma/2)}{\Delta\omega^2 + (\gamma/2)^2} \tag{5.39}$$

We now know how the relative dielectric permittivity changes near the resonant frequency of a medium. Then, using eq. (5.19), we can calculate the refractive index of the medium. Since the relative dielectric permittivity is a complex number, the refractive index is also a complex number. By setting $\varepsilon_r = \varepsilon_1 + i\varepsilon_2 = (n + ik)^2$, we can derive:

$$n(\omega) = \sqrt{\frac{1}{2}\left(\sqrt{\varepsilon_1(\omega)^2 + \varepsilon_2(\omega)^2} + \varepsilon_1(\omega)\right)} \tag{5.40}$$

$$k(\omega) = \sqrt{\frac{1}{2}\left(\sqrt{\varepsilon_1(\omega)^2 + \varepsilon_2(\omega)^2} - \varepsilon_1(\omega)\right)} \tag{5.41}$$

Here, $n(\omega)$ and $k(\omega)$ are the real and imaginary components of complex refractive index, $\tilde{n}(\omega) = n(\omega) + ik(\omega)$. The latter part is related to the absorptivity, a, in the Beer–Lambert law in eq. (5.27):

$$a(\omega) = \frac{4\pi k(\omega)}{\lambda} \tag{5.42}$$

For a non-absorbing medium (i.e., absorptivity = 0), the imaginary component of refractive index is zero, that is, $k(\omega) = 0$. So, its refractive index is a simple real number, which gives $n(\omega) = \sqrt{\varepsilon_1(\omega)}$. We can also drive equations calculating the real and imaginary components of relative dielectric permittivity from those of refractive index:

$$\varepsilon_1(\omega) = n(\omega)^2 - k(\omega)^2 \tag{5.43}$$

$$\varepsilon_2(\omega) = 2n(\omega)\,k(\omega) \tag{5.44}$$

Equations (5.40)–(5.44) describe the relationship between relative dielectric permittivity and refractive index, and how they are related to the absorptivity that governs the intensity of absorption band.

Let us shift a gear and see how they are related to the width of absorption band using a simple example. A low-pressure sodium vapor lamp has been used as streetlights in cities. The number density (N) of sodium atom in the lamp is $10^{17}\ \text{m}^{-3}$, which corresponds to a partial pressure of 4.1×10^{-4} Pa at room temperature. The wavelength of its characteristic D2 line is 589.0 nm (so, it looks yellowish orange) and the full-width-half-max (FWHM) of this line is 100 MHz. Converting the wavelength to frequency, we get $\nu_o = 509.0$ THz and $\omega_o = 3198 \times 10^{12}$ rad/sec. Since the vapor density in the lamp is low, we can assume that, at frequencies far away from the resonance line (thus, at non-resonance frequencies), the real component of refractive index of the gas medium inside the lamp is close to one. For a simplicity, let us assume $n_{eff} = 1$.

Let us calculate the frequencies at which the real part of refractive index reaches max and min near the resonance peak. For that, we differentiate $\varepsilon_1(\omega)$ with respect to ω and set it to zero. Since we are interested in the changes near the resonance peak, we can use eq. (5.38):

$$\frac{d\varepsilon_1(\omega_o)}{d\omega} = \frac{d\varepsilon_1(\Delta\omega)}{d\Delta\omega} \propto \frac{\Delta\omega^2 - (\gamma/2)^2}{\left(\Delta\omega^2 + (\gamma/2)^2\right)^2} = 0 \tag{5.45}$$

The solution of eq. (5.45) is $\Delta\omega = \pm\gamma/2$. Thus, $\varepsilon_1(\Delta\omega)$ reaches its max and min at $\omega_o - \gamma/2$ and $\omega_o + \gamma/2$, respectively. This corresponds to the FWHM of the $\varepsilon_2(\Delta\omega)$ function. From this, we can see the damping factor (γ) is basically equal to 2π times the FWHM (in Hz) of the imaginary component of the relative dielectric permittivity. The deviation of the max and min values of $n(\Delta\omega)$ from n_{eff} can be calculated by inserting $\Delta\omega = \pm(\gamma/2)$ into eq. (5.38):

$$\Delta n = \sqrt{\varepsilon_1(\gamma/2)} - n_{eff} = \sqrt{\varepsilon_\infty + \frac{Ne^2}{2\varepsilon_o m_e \omega_o^2}\frac{\omega_o}{\gamma}} - n_{eff} = 3.96 \times 10^{-5} \tag{5.46}$$

Using eqs. (5.37) and (5.44), we can calculate the imaginary component of refractive index:

$$k(\omega_o) \approx \frac{\varepsilon_2(\omega_o)}{2\, n_{eff}} = \frac{1}{2\, n_{eff}}\frac{N e^2}{\varepsilon_o\, m_e}\frac{1}{\gamma\omega_o} = 7.92 \times 10^{-5} (\ll n_{\text{eff}}) \tag{5.47}$$

where m_e is the mass of electron (9.11×10^{-31} kg). Then, the absorptivity is calculated to be:

$$a(\nu_o) = \frac{4\pi k(\nu_o)}{\lambda} = 1.69 \times 10^3\ \text{m}^{-1} \tag{5.48}$$

From these calculations, we can plot the refractive index and absorptivity of the gas containing 10^{17} sodium atom/m^3 near the D2 resonance line, which is shown in Figure 5.7. These plots show the general peak shape of the refractive index and absorptivity of a weakly absorbing Lorentzian oscillator (note that we use $\Delta\omega = \omega - \omega_o$ as in Figure 5.7):

$$n(\Delta\omega) = n_{eff} - \Delta n \frac{\gamma\Delta\omega}{\Delta\omega^2 + (\gamma/2)^2} \tag{5.49}$$

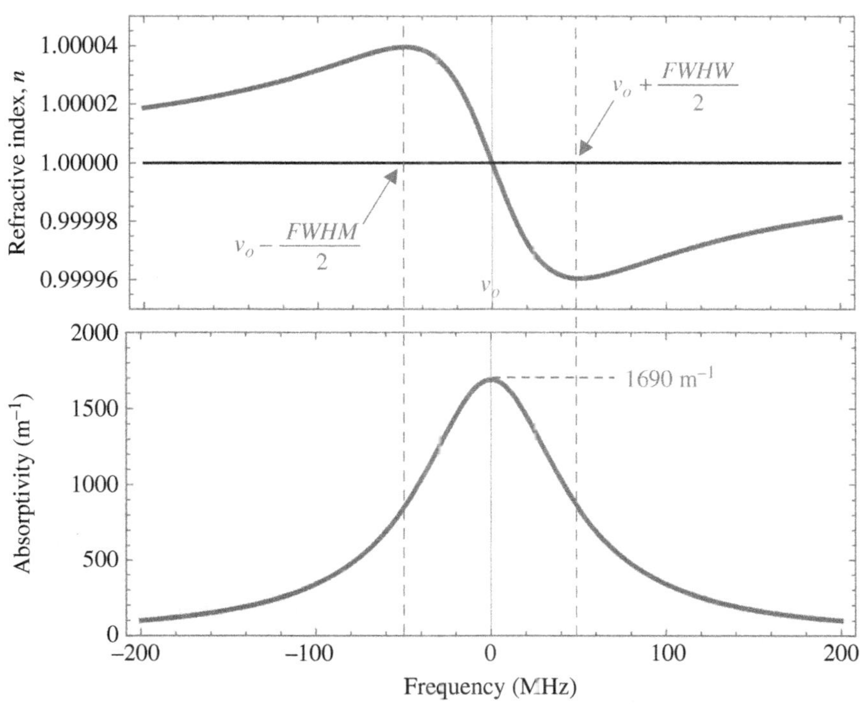

Figure 5.7 Absorptivity and refractive index, near the sodium-D2 line, of a gas-discharge lamp containing 4.1×10^{-4} Pa of sodium vapor. Note that the x-axis is shown in the $\nu - \nu_o$ scale, where ν_o = 509.0 THz (which corresponds to 589.0 nm).

$$a(\Delta\omega) = a(\omega_o)\frac{(\gamma/2)^2}{\Delta\omega^2 + (\gamma/2)^2} \tag{5.50}$$

Here, we can see that $n(\Delta\omega)$ and $a(\Delta\omega)$ are not independent to each other. Although the variance in magnitude is small in the example shown in Figure 5.7 (because here we consider a dilute system), $n(\Delta\omega)$ varies near the resonance peak such that, as frequency (ω) increases, its magnitude increases from n_{eff} to the maximum at $\omega_o - \gamma/2$, becomes equal to n_{eff} at ω_o, decreases to the minimum at $\omega_o + \gamma/2$, and then asymptotically approaches n_{eff}. Meantime, $a(\Delta\omega)$ increases from zero to the maximum value at ω_o and then decreases back to zero as ω passes ω_o and increases further. Note that these properties originate from the polarization (=output) of the medium under electric field (=input), which is described in eq. (5.28).

In the time domain, the impulse response is described by eq. (5.31). The output cannot respond to the input before the input has happened in the time domain. This is called *causality*. The causality can be expressed with the heavy-side function: $u(t) = 0$ when $t < 0$ and 1 when $t \geq 0$. Multiplying the heavy-side function to the impulse function in the time domain and then taking the Fourier transformation, we can see that *the real and imaginary parts of the response in the frequency domain are related to each other*. This is the Kramers–Kronig (KK) relation [1]. Another way of saying it is that the real and imaginary parts in the frequency domain are the projection of a physical (causal) response onto two different principal axes describing different aspects of the property. Thus, the real and imaginary parts of the polarization change upon light absorption are related to each other.

The KK relation for relative dielectric constant, $\varepsilon_r(\omega) = \varepsilon_1(\omega) + i\varepsilon_2(\omega)$, can be written as [1]:

$$\varepsilon_1(\omega) = 1 + \frac{2}{\pi}\mathcal{P}\int_0^{\infty} \frac{\Omega\,\varepsilon_2(\Omega)}{\Omega^2 - \omega^2}d\Omega \tag{5.51}$$

$$\varepsilon_2(\omega) = -\frac{2\omega}{\pi}\mathcal{P}\int_0^{\infty} \frac{\varepsilon_1(\Omega) - 1}{\Omega^2 - \omega^2}d\Omega \tag{5.52}$$

where $\mathcal{P}$ is the Cauchy principal value, which is required when calculating the integral numerically. Due to the KK relationship, the ε_1 curve goes up and down when there is a resonance peak in the ε_2 curve in Figure 5.6. This behavior is clearly seen in the optical frequency regime, i.e., the resonance absorption of IR and UV–VIS photons, but not obvious in the lower-frequency regime. That is because the damping factor (γ) of the slow response process involved in the low-frequency regime is too large.

When ε_1 and ε_2 in eqs. (5.51) and (5.52) are replaced with n and k, respectively, the equations become the KK relation for the complex refractive index, $\tilde{n}(\omega) = n(\omega) + ik(\omega)$:

$$n(\omega) = 1 + \frac{2}{\pi}\mathcal{P}\int_0^{\infty} \frac{\Omega\,k(\Omega)}{\Omega^2 - \omega^2}d\Omega \tag{5.53}$$

$$k(\omega) = -\frac{2\omega}{\pi}\mathcal{P}\int_0^{\infty} \frac{n(\Omega) - 1}{\Omega^2 - \omega^2}d\Omega \tag{5.54}$$

By combining these with eq. (5.42), we can get the KK relation for $n(\omega)$ and $a(\omega)$:

$$n(\omega) = 1 + \frac{c}{\pi}\mathcal{P}\int_0^{\infty} \frac{a(\Omega)}{\Omega^2 - \omega^2}d\Omega \tag{5.55}$$

$$a(\omega) = -\frac{4\omega^2}{\pi c}\mathcal{P}\int_0^{\infty} \frac{n(\Omega) - 1}{\Omega^2 - \omega^2}d\Omega \tag{5.56}$$

These KK relations are very important in optical characterizations. The KK relation provides a way of calculating the complex refractive index from the absorption measurement.

Now, let us get back to Figure 5.7. As shown in eq. (5.50), the width of the absorption band, $a(\omega)$, is directly governed by the damping parameter (γ). There are many factors that can contribute to γ. The first is the quantum mechanical uncertainty in the energy (ΔE) of the excited state with a finite lifetime (Δt) due to the spontaneous decay to the ground state. This determines the natural width of absorption line. It can be explained with Heisenberg's uncertainty principle, which states $\Delta E \cdot \Delta t \geq \hbar/2$. If the lifetime ($\Delta t$) of the excited state is very short, it means the damping rate is high. Since $\Delta\nu = \Delta E/h \geq (4\pi\Delta t)^{-1}$, it means that the energy uncertainty is large and so is the peak width. This is homogeneous broadening and produces the Lorentzian line shape. Equation (2.16) describes this line shape (see Section 2.4.2).

Frequent collisions with other molecules can also reduce the effective lifetime of a state, leading to broadening of peak width. So, the absorption band width becomes significantly larger for high-pressure gas or liquid phase, as compared to the low-pressure gas. Thermal motion of individual molecules in the gas becomes faster at higher temperature. Thus, the absorption band width becomes broader as temperature increases. These are inhomogeneous broadening. In the solid state, inhomogeneous broadening can also come from structural disorder. For example, in a crystal, there are lattice defects that cause statistically varying strain fields. These strain fields can result in statistically varying frequency shifts of the zero-phonon transitions of the solid. In organic crystals, the inhomogeneous broadening is on the order of a few wavenumbers. In inorganic glasses, it is on the order of several hundred wavenumbers. The inhomogeneous distribution is governed by the statistics of the distribution of such disorders; hence it gives the Gaussian line shape.

We can study how the damping factor (γ) affects the line shape using a hypothetical Lorentzian oscillator. Let us say this oscillator has a resonance frequency of $\nu_o = 100$ THz and relative dielectric permittivity of $\varepsilon_{st} = 11$ and $\varepsilon_\infty = 10$ at the low- and high-frequency limits and consider three cases with different damping factors, $\gamma = 1$, 5, and 100 THz. The relative dielectric permittivity around the resonance frequency can be calculated using eqs. (5.38) and (5.39). The refractive index in the same region can be calculated using eqs. (5.40) and (5.41). Equation (5.42) calculates the absorptivity as a function of frequency. For comparison, we calculate the reflectance upon normal incidence of the wave from air using the following equation (for derivation, see Section 5.2.3):

$$R_n(\omega) = \frac{(n(\omega) - 1)^2 + k(\omega)^2}{(n(\omega) + 1)^2 + k(\omega)^2} \tag{5.57}$$

Figure 5.8 plots the calculation results. In Figure 5.8A, we can see that as γ increases from 1 to 10 THz, the FWHM of $\varepsilon_2(\nu)$ increases from 1 to 10 THz; at the same time, the maximum intensity of ε_2 at 100 THz decreases from 99 to 10. As γ increases, the fluctuation in $\varepsilon_1(\nu)$ near the resonance frequency also decreases. In Figure 5.8B, the $n(\nu)$ curve has a shape similar to $\varepsilon_1(\nu)$, but the high-frequency side is distorted. While $\varepsilon_1(\nu)$ can go to the negative region, $n(\nu)$ cannot be smaller than 0. Concurrently, the high-frequency side of $k(\nu)$ is distorted from the $\varepsilon_2(\nu)$ curve. In Figure 5.8C, the $a(\nu)$ curve looks almost identical to the $k(\nu)$ shape; but they are not the same due to the $1/\lambda$ dependence of $a(\nu)$ in addition of the proportionality to $k(\nu)$. The absorptivity decreases significantly as the damping factor of the oscillator increases. In Figure 5.8C, we can see that reflectance is enhanced near and above the resonance band. In the case of $\gamma = 1$ THz, the reflectance enhancement is significant just above the resonance frequency. In the 100–105 THz region, $\varepsilon_1(\nu)$ is negative (Figure 5.8A). The negative ε_1 means that the electromagnetic wave cannot propagate in a given medium. As a result of this inability to propagate, the incident beam experiences strong reflection from the medium. This enhanced reflectance in the narrow frequency

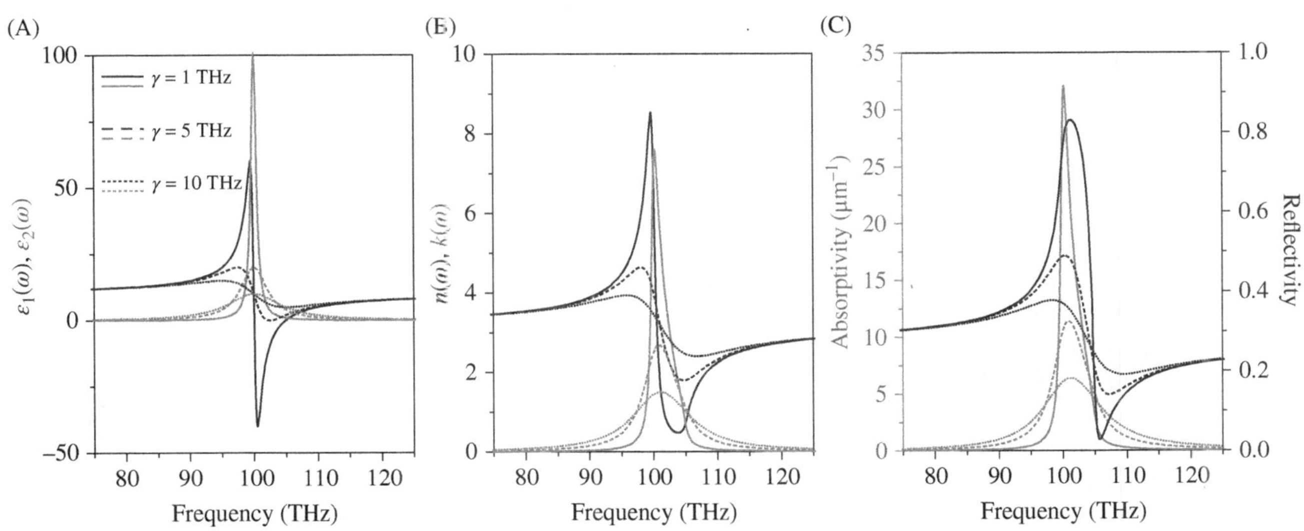

Figure 5.8 (A) Real and imaginary components of relative dielectric permittivity, (B) real and imaginary components of refractive index, (C) absorptivity and reflectance (at surface normal incidence) of a hypothetical oscillator with $\varepsilon_{st} = 11$, $\varepsilon_\infty = 10$, $\nu_o = 100$ THz and different damping factors: $\gamma = 1$, 5, and 10 THz.

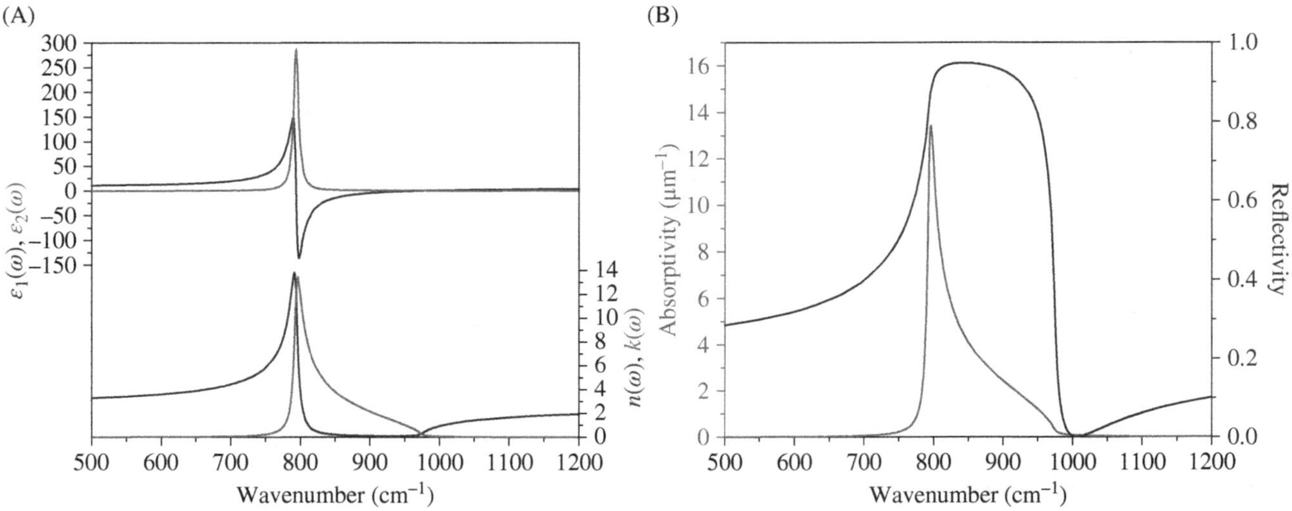

Figure 5.9 (A) Relative dielectric permittivity and refractive index and (B) absorptivity and normal-incidence reflectance of β-SiC with ε_{st} = 9.72, ε_∞ = 6.7, $\overline{\omega}_o$ = 793.6 cm^{-1}, and γ = 8.49 cm^{-1}.

region above the resonance peak is called the Reststrahlen effect (in German, "residual rays"). The Reststrahlen effect is larger for the oscillator with a smaller damping factor.

Figure 5.9 plots the relative dielectric permittivity, refractive index, absorptivity, and normal-incidence reflectance calculated for the β-SiC calculated with the physical constants extracted from the literature [2]. Since the inhomogeneous broadening is relatively small in the crystalline phase, damping factor is relatively small, compared to the resonance wavenumber. Also, the difference between ε_{st} and ε_∞ is quite large compared to the hypothetical example we simulated in Figure 5.8. So, ε_2 at the resonant wavenumber is very large, and ε_1 just above the resonance peak is extremely negative. This leads to highly asymmetric $\overline{\omega}$ dependence of n and k. The FWHM of the absorptivity is quite narrow, but it has a long tail in the high wavenumber side. The Reststrahlen effect is very significant and extended up to ~200 cm^{-1} above the resonance peak. Here we can see that the reflectance at ~1008 cm^{-1} is zero. What it means the IR with λ = 9.92 μm is not reflected from the β-SiC surface. If we check the refractive index at this wavenumber (Figure 5.9A), we can see that $n \approx 1$ and $k \approx 0$ at ~1008 cm^{-1}. Since the refractive index of the medium is the same as that of air, light does not see the interface and keeps propagating into the medium at that wavelength.

The examples discussed in this section show how the polarization of medium [$\varepsilon_r(\omega) = \varepsilon_1(\omega) + i\varepsilon_2(\omega)$] is related to the refractive index [$\tilde{n}(\omega) = n(\omega) + ik(\omega)$], how it changes near the resonant absorption band, how the shape and intensity of the absorption band [$a(\omega)$] are related to the damping factor (γ), and how all these factors influence the reflectance [$R_n(\omega)$] of the electromagnetic wave upon normal incidence into a medium.

5.2 Reflection/Refraction at Interface of Two Media

Spectroscopic IR analyses are carried out in various modes. The simplest way would be the transmission mode (Figure 5.10); so, it has been most extensively covered in undergraduate instrumental analysis courses. In the standard transmission-mode analysis with the normal incidence of light, the intensity decrease upon transmission through a sample can be described with the simple Beer–Lambert law. From eq. (5.27), we can see that as the thickness of a sample decreases, the absorbance also decreases proportionally. When the sample thickness is less than ~100 nm (thin films on a substrate), the intensity contrast achieved with the standard technique becomes extremely small. To achieve a sufficient sensitivity, one may stack multiple samples. However, simply stacking multiple samples along the beam path is not sufficient. As we see in Figures 5.8 and 5.9, a large fraction of light is reflected from the front surface of the sample. Therefore, the fraction of radiation arriving to the sample furthest from the first sample becomes quite small. This limitation can be avoided by doing the transmission experiment at an angle; this angle is the Brewster angle (BA). But this method does not work if the substrate absorbs or scatters the probe beam. Many materials reflect and absorb IR. In that case, we must rely on alternative methods. Those

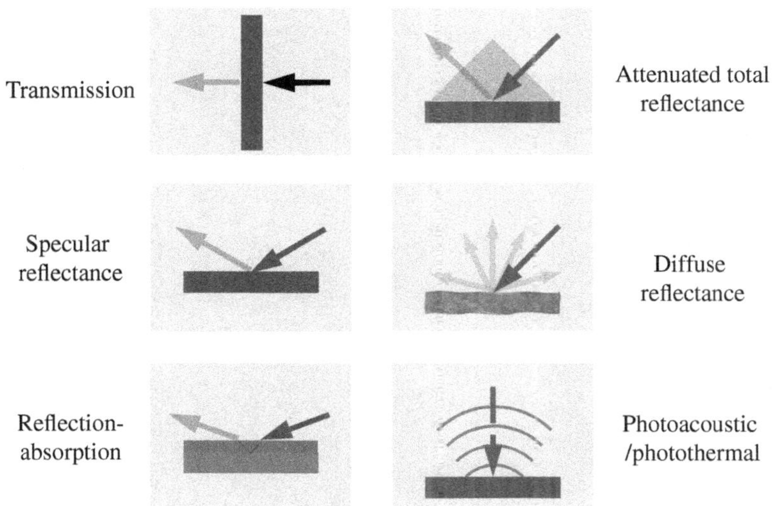

Figure 5.10 Schematic illustration of various modes of IR analysis.

include specular-reflection (SR), reflection–absorption (RA), attenuated total reflectance (ATR), diffuse reflectance (DR), and photoacoustic (PA) or photothermal (PT) method (Figure 5.10). In the next chapter, we will study the principles governing the BA-transmission, SR, RA, ATR, and DR modes. The PT method will be briefly touched in conjunction with IR-spectroscopic imaging at the end of next chapter. The PA method will not be covered, since the basic principle is somewhat similar to the PT method although the detection method is completely different (relying on detection of acoustic wave generated by the absorption of a modulated IR beam and propagating through the surrounding medium). The rest of this chapter derives important equations that are used in Chapter 6.

5.2.1 Snell's Law

In the previous sections, we have considered light (=electromagnetic wave) propagating through a medium. But, in spectroscopic experiment, light first must go through the boundary between the surrounding medium and the sample. An experimenter stays in the surrounding medium, sends the probe beam (=light) to the sample, and detects the light (=messenger signal) that is coming out of the sample. Even in the standard transmission-mode IR spectroscopy, the IR beam should pass through the interface between the surrounding medium and the sample. But this reflection at normal incidence of light, which is described by eq. (5.57), is not considered in typical transmission experiments because its effect is relatively small for organic material and can be factored out through the blank sample measurement (see the Introduction of Chapter 6). Unless the refractive index of the sample is the same as the surrounding, some fraction of light will be reflected. In the β-SiC examples shown in Figure 5.9, IR at all wavelengths except $\lambda = 9.92\ \mu m$ (which corresponds to $1008\ cm^{-1}$) is reflected from the sample surface.

Figure 5.11 describes what happens to the light arriving at the interface of two non-absorbing media with different refractive indices at an angle (θ_i) from the surface normal direction. Here we consider the case where the refractive index of medium 1 is smaller than that of medium 2, $n_1 < n_2$. In this example, the imaginary part (k) of complex refractive index drops out since we are considering the non-absorbing medium. At the interface, a portion of light is reflected back into medium 1 and the rest continues propagating into medium 2 but with an altered direction. This deflection upon passing the interface with a different refractive index medium is called refraction. The electric field waveform ($\boldsymbol{E}$) of light propagating in a medium can be described with the Euler formula, instead of sine function, eq. (5.15), derived in Section 5.1:

$$\boldsymbol{E} = E_o \cdot \exp\left[-i\left(\omega t - \boldsymbol{k} \cdot \vec{r}\right)\right] \tag{5.58}$$

where E_o is the amplitude of the field, ω is the angular frequency of the wave, $\vec{r}$ is the position along the propagation direction, and $\boldsymbol{k}$ is the wavevector which is defined as $\boldsymbol{k} = 2\pi n/\lambda = \omega/v$ and has the unit of wavenumber (not to be confused with k representing the imaginary part of refractive index). At the boundary of two media, the tangential components of $\boldsymbol{E}$ should be continuous. Thus, the exponent term of eq. (5.58) of the refracted beam should be equal to that of the incoming beam:

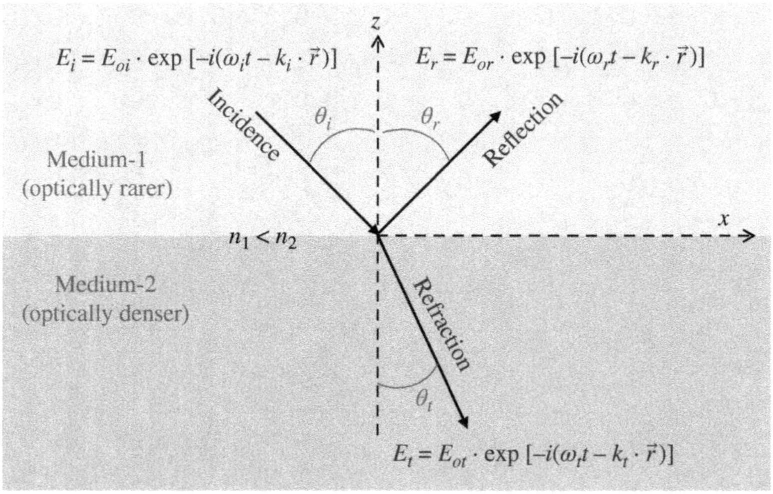

Figure 5.11 Incidence, reflection, and refraction of light at the interface of two media with different refractive indexed ($n_1 < n_2$).

$$\omega_i t - \boldsymbol{k}_i \cdot \vec{r} \sin\theta_i = \omega_t t - \boldsymbol{k}_t \cdot \vec{r} \sin\theta_t \tag{5.59}$$

To meet this condition at all values of t, the frequency should remain unaltered when a beam crosses the interface and enters a new medium; thus, $\omega_i = \omega_t$. Also, the wavevector component should be the same:

$$\boldsymbol{k}_i \sin\theta_i = \boldsymbol{k}_t \sin\theta_t \tag{5.60}$$

From the definition of refractive index, we know $\boldsymbol{k} = \omega/\upsilon = \omega\ n/c$. So, eq. (5.60) becomes:

$$n_1 \sin\theta_i = n_2 \sin\theta_t \tag{5.61}$$

This is Snell's law, which describes the degree of deflection of the transmitted beam as a function of the refractive indices of the two media. If we do the same for the reflected beam, we get $\theta_i = \theta_r$.

5.2.2 Fresnel Coefficients

Next, let us consider the photon flux (Figure 5.12). According to the law of energy conservation, the incident radiation flux per unit area of the beam (I_i, measured in W/cm^2) must be equal to the sum of the flux of the reflected light (I_r) and that of the refracted light (I_t).

$$I_i \cos\theta_i = I_r \cos\theta_r + I_t \cos\theta_t \tag{5.62}$$

Rearranging it, we get the relationship between the incoming and outgoing fluxes:

$$I_i = I_r + I_t \frac{\cos\theta_t}{\cos\theta_i} \tag{5.63}$$

Since the radiation flux per unit area (=intensity) is the time-averaged Poynting vector of the electromagnetic wave, it is proportional to the square of the electric field $\boldsymbol{E}$ and to the velocity $\upsilon = c/n$ of the radiation in the medium. Based on this, we can convert the flux to the field intensity:

$$I = \frac{\varepsilon}{2\pi} E^2 v = \frac{\varepsilon_0 n^2}{2\pi} E^2 \frac{c}{n} = \frac{\varepsilon_0 c}{2\pi} n E^2 \tag{5.64}$$

Combining eqs. (5.63) and (5.64), we get:

$$n_1 E_i^2 = n_1 E_r^2 + n_2 E_t^2 \frac{\cos\theta_t}{\cos\theta_i} \tag{5.65}$$

The electric field of the light can be divided into two components: parallel and perpendicular to the incidence plane that is defined by the propagation axes of the incident and reflected beams (see inset in Figure 5.12). The parallel and perpendicular

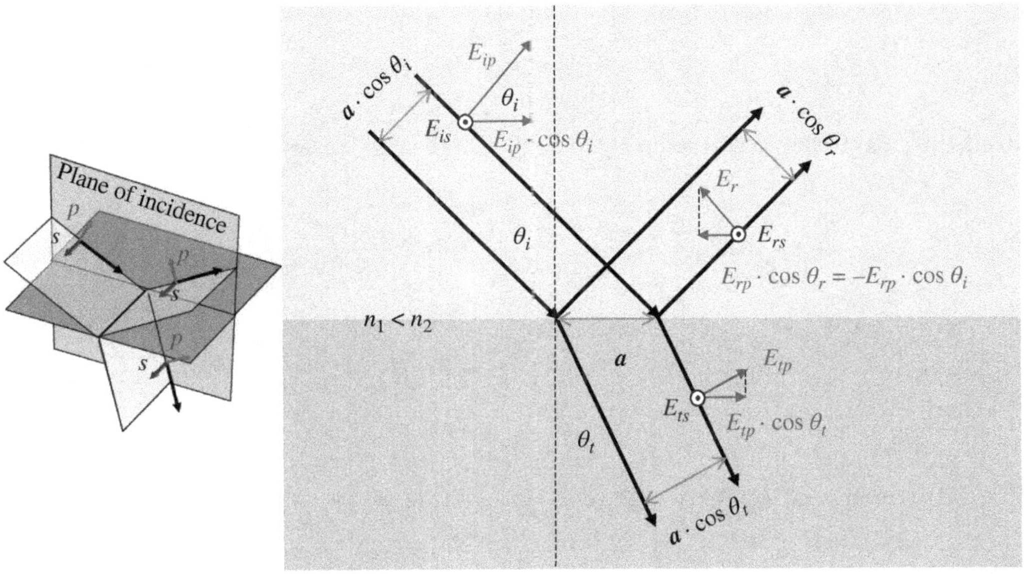

Figure 5.12 Mirror reflection of a parallel beam of rays on a plane phase boundary. The electric field can be divided into the *p* and *s* (from *senkrecht*, German for perpendicular) components that are parallel and perpendicular to the plane of incidence defined by the incident and reflected beam axes. In this schematic, the *s*-component is the field direction coming out of the page. The sign of polarization shown here is Verdet's convention. In the Fresnel's convention, the sign of the reflected beam is switched.

components are called *p*- and *s*-polarization, respectively. Since the incidence plane is perpendicular to the interface, the electric field of the *s*-polarization is parallel to the interface. The electric field of the *p*-polarization has both surface-parallel and perpendicular components. The surface-parallel component of the *p*-polarization is obtained by multiplying the cosine of the propagation axis angle with respect to the surface normal direction. Note that the sign of the *p*-polarization of the reflected light (E_{rp}) is opposite to that of the incident light (E_{ip}). We can set the continuity of the tangential components of E_s and E_p at the interface for the *s*- and *p*-polarizations separately:

$$E_{is} + E_{rs} = E_{ts} \tag{5.66}$$

$$E_{ip} - E_{rp} = E_{tp}\frac{\cos\theta_t}{\cos\theta_i} \tag{5.67}$$

Both *s*- and *p*-polarization components should follow the law of energy conservation, eq. (5.65). Let us write the energy conservation for the *s*-polarization component (eq. 5.63) and then simplify it using the continuity condition:

$$n_1E_{is}^2 - n_1E_{rs}^2 = n_1(E_{is} + E_{rs})(E_{is} - E_{rs}) = n_1(E_{ts})(E_{is} - E_{rs}) = n_2E_{ts}^2\frac{\cos\theta_t}{\cos\theta_i} \tag{5.68}$$

$$n_1(E_{is} - E_{rs}) = n_2(E_{is} + E_{rs})\frac{\cos\theta_t}{\cos\theta_i} \tag{5.69}$$

Rearranging this equation, we can find the relationship describing the ratio of E_{rs} to E_{is}:

$$r_s = \frac{E_{rs}}{E_{is}} = \frac{n_1\cos\theta_i - n_2\cos\theta_t}{n_1\cos\theta_i + n_2\cos\theta_t} \tag{5.70}$$

Using Snell's law, this can be further simplified to:

$$r_s = -\frac{\sin(\theta_i - \theta_t)}{\sin(\theta_i + \theta_t)} \tag{5.70a}$$

If we eliminate the E_{rs} term in eq. (5.69) using eq. (5.66), then we get the relationship describing the ratio of E_{ts} to E_{is}:

$$t_s = \frac{E_{ts}}{E_{is}} = \frac{2n_1\cos\theta_i}{n_1\cos\theta_i + n_2\cos\theta_t} \tag{5.71}$$

Again, using Snell's law, we can express this ratio in a different form:

$$t_s = \frac{2\cos\theta_i \sin\theta_t}{\sin(\theta_i + \theta_t)} \tag{5.71a}$$

We can repeat the same manipulations for the p-polarization. By combining the energy conservation and continuity equations, we get:

$$n_1 E_{ip}^2 - n_1 E_{rp}^2 = n_1 \left(E_{ip} + E_{rp}\right)\left(E_{ip} - E_{rp}\right) = n_1 \left(E_{ip} + E_{rp}\right)\left(E_{tp}\frac{\cos\theta_t}{\cos\theta_i}\right) = n_2 E_{tp}^2 \frac{\cos\theta_t}{\cos\theta_i} \tag{5.72}$$

$$n_1 \left(E_{ip} + E_{rp}\right) = n_2 E_{tp} \tag{5.73}$$

By eliminating the E_{tp} term in eq. (5.73) using eq. (5.67), we get the relationship describing the ration of E_{rp} to E_{ip}:

$$r_p = \frac{E_{rp}}{E_{ip}} = \frac{n_2\cos\theta_i - n_1\cos\theta_t}{n_2\cos\theta_i + n_1\cos\theta_t} = \frac{\tan(\theta_i - \theta_t)}{\tan(\theta_i + \theta_t)} \tag{5.74}$$

Similarly, we can eliminate the E_{rp} term in eq. (5.73) using eq. (5.67) to get the relationship describing the ration of E_{tp} to E_{ip}:

$$t_p = \frac{E_{tp}}{E_{ip}} = \frac{2n_1\cos\theta_i}{n_2\cos\theta_i + n_1\cos\theta_t} = \frac{2\cos\theta_i \sin\theta_t}{\sin(\theta_i + \theta_t)\cos(\theta_i - \theta_t)} \tag{5.75}$$

These four equations – r_s in eq. (5.70), t_s in eq. (5.71), r_p in eq. (5.74), and t_p in eq. (5.75) – are the Fresnel coefficients. We have derived these coefficients for non-absorbing media, that is, $k = 0$, so the refractive index terms used in the above equations are all real number. If the medium of interest absorbs light in a certain frequency region, we simple replace the refractive index terms in these equations with frequency-dependent complex refractive index, $\tilde{n}(\omega) = n(\omega) + ik(\omega)$. Then, almost all spectroscopy principles can be described with these Fresnel coefficients and the Snell's law, eq. (5.61).

Calculating or solving equations containing complex numbers by hand is complicated; but it can be done easily using math software such as *Mathematica* or *MATLAB*, since these programs can handle complex numbers as a whole. All examples covered in this book can be solved or computed easily with *Mathematica* or *MATLAB*, although they can be done with Microsoft Excel which is a spreadsheet software program that most students have on their computers.

5.2.3 Reflectance and Transmittance

Before we delve into spectroscopic phenomena, let us look into some properties of the Fresnel coefficients using a simple non-absorbing system (so, $k = 0$), such as the interface of air (n_1) and glass (n_2). Typical refractive index (n) of silicate glass materials (e.g., window glass) is around 1.5 in the visible wavelength region. Figure 5.13A plots the four Fresnel coefficients as a function of incident angle for an air/glass interface ($n_1 < n_2$). Near the normal incidence ($\theta_i \approx 0°$), r_s and r_p have the same absolute value, but their signs are opposite. The negative sign means the reversal of the phase relative to the incident wave upon reflection. The r_s coefficient starts from a negative value at the normal incidence and decreases toward -1 as θ_i increases to 90°. At $\theta_i = 90°$, the incident beam is essentially parallel to the surface, so the beam cannot cross the interface. The r_p coefficient starts from a positive value at the normal incidence and decreases toward -1 as θ_i increases to 90°. Thus, it will pass through zero. The angle where r_p becomes zero is called the Brewster angle (θ_p). At this angle, the reflected and refracted beams are perpendicular to each other ($\theta_p + \theta_t = \pi/2$), which leads to $\sin\theta_t = \cos\theta_p$. Thus, Snell's law at θ_p becomes:

$$\tan\theta_p = \frac{n_2}{n_1} \tag{5.76}$$

At the Brewster angle, the p-polarization component of the reflected beam vanishes; in other words, only the s-polarization component of incident light is reflected. For the air/glass interface shown in Figure 5.13A, θ_p is 56.3°. When an unpolarized light is irradiated to this glass at the 56.3° incident angle, then the reflected light is perfectly s-polarized, because the p-polarization component is not reflected at the Brewster angle. In other words, the p-polarization component transmits through the interface without any reflection at the Brewster angle.

The ratio of the total reflected beam intensity ($I_r A_r$) to the total incident beam intensity ($I_i A_i$) is called reflectance (R). Since $\theta_i = \theta_r$ in Figure 5.12, the footprint of the reflected beam (A_r) is the same as that of the incident beams (A_i). From eq. (5.64),

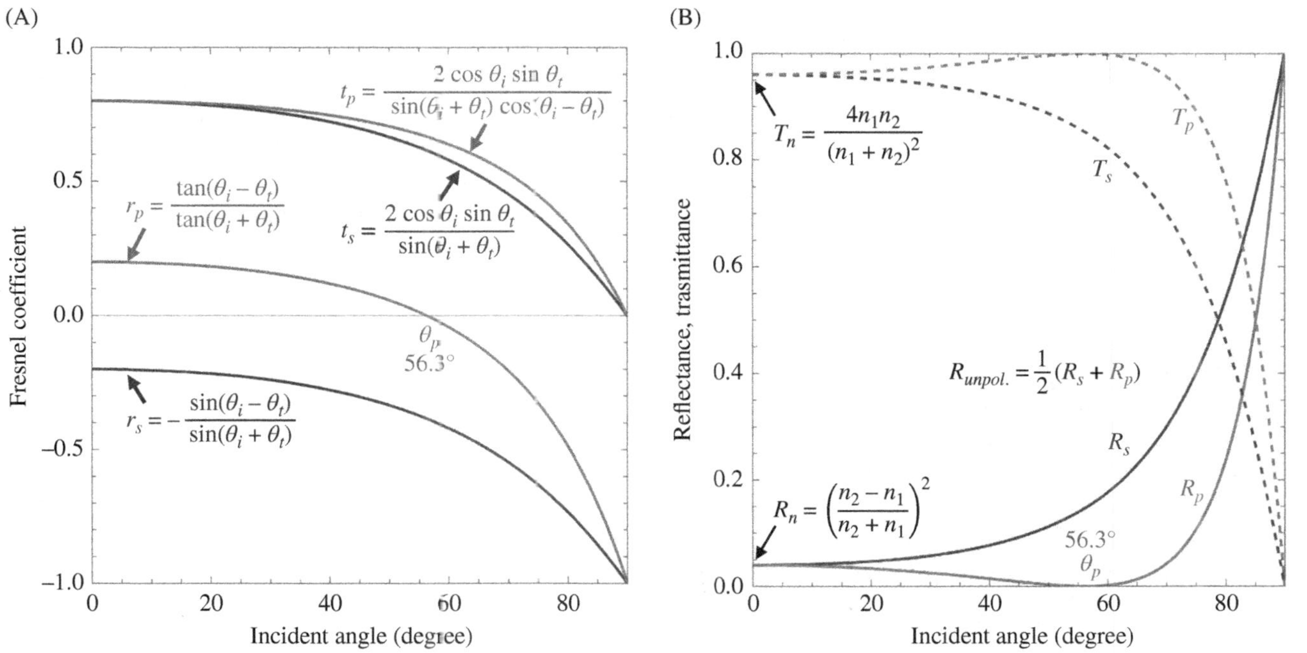

Figure 5.13 (A) Fresnel coefficients (r_s, t_s, r_p, and t_p) and (B) reflectance (R_s, R_p) and transmittance (T_s, T_p) as a function of incident angle (θ_i) for an air (n_1 = 1) / glass (n_2 = 1.5) interface.

we know that $I_r/I_i = |E_r|^2/|E_i|^2$. The right-hand side is simply the square of the Fresnel reflection coefficients (r_s and r_p for *s*- and *p*-polarizations, respectively). If the refractive index is complex number (as for absorbing media with nonzero *k*), r_s and r_p are complex numbers. Then we should multiply it with its conjugate. Thus, the reflectance for *s*- and *p*-polarized component becomes:

$$R_s(\theta_i) = r_s(\theta_i) \cdot r_s^*(\theta_i) \tag{5.77}$$

$$R_p(\theta_i) = r_p(\theta_i) \cdot r_p^*(\theta_i) \tag{5.78}$$

If the incident beam is completely unpolarized, then there are equal amounts of *s*- and *p*-components. So, the reflectance of unpolarized light is the average of these two:

$$R(\theta_i) = \frac{1}{2}\left(R_s(\theta_i) + R_p(\theta_i)\right) \tag{5.79}$$

At the normal incidence ($\theta_i = 0°$) into a non-absorbing medium, eq. (5.78) becomes $R_n = \left(\frac{n_2 - n_1}{n_2 + n_1}\right)^2$. For an absorbing medium ($k \neq 0$), R_n becomes eq. (5.57).

For the transmittance, we cannot simply take as $t_s(\theta_i) \cdot t_s^*(\theta_i)$ and $t_p(\theta_i) \cdot t_p^*(\theta_i)$. This is because the beam area is changed upon refraction ($\theta_i \neq \theta_t$), as shown in Figure 5.12. The transmittance of *s*-polarization (T_s) is calculated to be:

$$T_s(\theta_i) = \frac{I_{ts}(\theta_i)A_t(\theta_i)}{I_{is}(\theta_i)A_i(\theta_i)} = \frac{n_2|E_{ts}(\theta_i)|^2}{n_1|E_{is}(\theta_i)|^2}\frac{\cos\theta_t}{\cos\theta_i} = \frac{n_2\cos\theta_t}{n_1\cos\theta_i}\frac{|E_{ts}(\theta_i)|^2}{|E_{is}(\theta_i)|^2} = \left(\frac{n_2\cos\theta_t}{n_1\cos\theta_i}\right)t_s(\theta_i)\cdot t_s^*(\theta_i) \tag{5.80}$$

Similarly, the transmittance of *p*-polarization (T_p) is:

$$T_p(\theta_i) = \frac{I_{tp}(\theta_i)A_t(\theta_i)}{I_{ip}(\theta_i)A_i(\theta_i)} = \left(\frac{n_2\cos\theta_t}{n_1\cos\theta_i}\right)t_p(\theta_i)\cdot t_p^*(\theta_i) \tag{5.81}$$

According to the law of energy conservation, the sum of reflectance and transmittance should be one; thus, $R_s(\theta_i) + T_s(\theta_i) = 1$ and $R_p(\theta_i) + T_p(\theta_i) = 1$. For unpolarized light, $R(\theta_i) + T(\theta_i) = 1$. So, using this energy conservation principle would be an easier way to calculate the transmittance. Note that the transmittance calculated here is the flux of photon crossing the interface between the media 1 and 2 and should not confused with the flux of photon passing through the medium 2 and coming out to medium 1 in the opposite side (which is used in the calculation of the absorbance).

Figure 5.13B plots $R_s(\theta_i)$, $R_p(\theta_i)$, $T_s(\theta_i)$, and $T_p(\theta_i)$ for the air/glass interface. At $\theta_i = 0°$, the reflectance (R_n) is 0.04, which means only 4% of the incident beam will be back reflected in the visible wavelength (where $n_2 \approx 1.5$). It means 96% of the incident beam is transmitted. As θ_i increases from 0° to 90°, $R_s(\theta_i)$ increases from 0.04 to 1, while $R_p(\theta_i)$ decrease from 0.04 to 0 at $\theta_p = 56.3°$ and then increases to 1. Following the law of energy conservation, $T_s(\theta_i)$ and $T_p(\theta_i)$ are opposite to $R_s(\theta_i)$ and $R_p(\theta_i)$, respectively. We can see that $T_p(\theta_p) = 1$, which means that the p-polarized beam transmits through the interface without any reflection at $\theta_i = \theta_p$.

5.2.4 Total Internal Reflection and Evanescent Wave

In Figure 5.11, we have considered the case where the incident beam propagates from the optically rarer medium to the interface with the optically denser medium, that is, $n_1 < n_2$. What happens when the incident beam comes from the optically denser medium? This is illustrated in Figure 5.14. When $n_1 > n_2$, Snell's law dictates that the angle of refraction (θ_t) is larger than the angle of incidence (θ_i). Then, we can see that at a certain θ_i, θ_t becomes 90°. This incident angle is called the critical angle (θ_c); at this angle, the transmitted light becomes parallel to the surface. When θ_i is increased further, then the incidence beam is completely reflected back to medium 1 and there is no beam propagating into medium 2. This is called total internal reflection (TIR). From Snell's law, the critical angle is calculated to be:

$$\theta_c = \sin^{-1}\left(\frac{n_2}{n_1}\right) \tag{5.82}$$

An example of the TIR phenomenon is the beam steerers which are used to compress the beam path inside binoculars with large magnifications. This phenomenon is also used in fiber optics transmitting optical signals for a long distance without loss to the surrounding. If we submerge under the water in a swimming pool, we can see the ceiling directly above us but cannot see the people standing at the side of the pool. This is also due to the TIR phenomenon. The fingerprint sensor also works with this principle. The ridges on a finger act as locations where TIR is frustrated (because n_2 at that location is larger than the refractive index of air), so the sensor light is not reflected; but between the ridges, there is still TIR, so light is fully reflected. This allows imaging of the fingerprint when we touch the fingerprint sensor.

Now, let us look at the wave property associated with the TIR phenomenon. The wavevector ($\mathbf{k}_t$) in the second medium has x (surface parallel) and z (surface normal) components:

$$\mathbf{k}_{tx} = \mathbf{k}_t \sin\theta_t \tag{5.83}$$

$$\mathbf{k}_{tz} = \mathbf{k}_t \cos\theta_t \tag{5.84}$$

Recall that $\mathbf{k}_t = 2\pi n_2/\lambda$, so $\mathbf{k}_t$ is a positive real number. When $\theta_i > \theta_c$, Snell's law gives $\sin\theta_t = (n_1/n_2)\sin\theta_i$. So. $\mathbf{k}_{tx}$ is a real quantity for non-absorbing interface, although θ_t cannot be defined. In eq. (5.84), $\cos\theta_t = \sqrt{1-\sin^2\theta_t} = \sqrt{1-(n_1/n_2)^2\sin^2\theta_i}$ is a pure imaginary quantity. So, we can set $\mathbf{k}_{tz}$ to be $i\beta$ where β is:

$$\beta = \frac{2\pi}{\lambda}\sqrt{n_1^2\sin^2\theta_i - n_2^2} \tag{5.85}$$

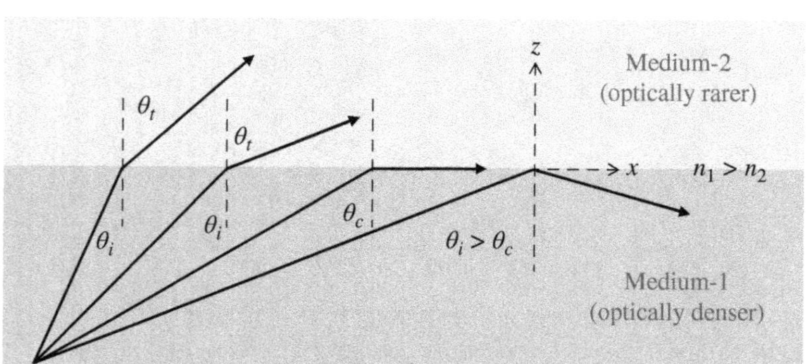

Figure 5.14 Propagation of light from optically denser medium to optically rarer medium.

Using this complex wavevector, $\mathbf{k}_t = k_{tx}\vec{x} + i\beta\vec{z}$, we can write the wave equation in the second medium as:

$$E_t(x, z, t) = E_0 e^{-i(\omega t - \mathbf{k}_t \cdot \vec{r})} = E_0 e^{-i(\omega t - (2\pi n_1/\lambda)\sin\theta_i\, x)} + E_0 e^{-i\omega t} e^{-\beta z} \quad (5.86)$$

Here, we can see that the wave is still propagating along the x (surface parallel) direction; but it decays exponentially along the z (surface normal) direction, although it still oscillates in the time domain. In other words, the z-component of the wave in the second medium does not propagate in space; it is localized. The exponentially decaying wave is called the *evanescent* wave, and the quantity β^{-1} is the characteristic length of this exponential decay.

Same as in the XPS case discussed in Section 2.1.4 (see Figure 2.10), this characteristic length is called the penetration depth (d_p) of evanescent wave:

$$d_p = \frac{\lambda}{2\pi\sqrt{n_1^2 \sin^2\theta_i - n_2^2}} \quad (5.87)$$

Figure 5.15A plots $E_t(z/d_p)/E_0 = e^{-z/d_p}$. For the interface of diamond ($n_1 = 2.42$) / water ($n_2 = 1.33$), the critical angle (θ_i) is 33.7°. At $\lambda = 532$ nm and $\theta_i = 45°$, d_p is calculated to be 78.6 nm. At $z = 0.693d_p$, the electric field decreases to 50% of the amplitude at the surface. It means 50% of the total electric field penetrating into the medium is attenuated at $0.693d_p$ from the interface. Total internal reflection fluorescence (TIR-F) microscopy uses this principle to detect fluorophores only close to the surface of TIR element without interferences from the ones inside the bulk phase of solution [3].

Because the intensity (I) is proportional to E^2 (eq. (5.64)), the evanescent wave intensity decreases two times faster than the electric field of the evanescent wave. Thus, the intensity penetration depth is:

$$d_{p,\text{int}} = \frac{\lambda}{4\pi\sqrt{n_1^2 \sin^2\theta_i - n_2^2}} \quad (5.88)$$

In ATR-IR spectroscopy (which will be discussed in the next section), the measured intensity is the sum of all signals from the entire penetration depth of the evanescent wave. When the area under the intensity curve from $z/d_{p,\text{int}} = 1$ to infinity is calculated, it becomes equal to the area of the $(1 - I_t(z/d_{p,\text{int}})/I_0)$ function integrated from $z = 0$ to $z/d_{p,\text{int}} = 1$ (see Figure 5.15A, although it is shown for the electric field amplitude decay, the general behavior is the same since both

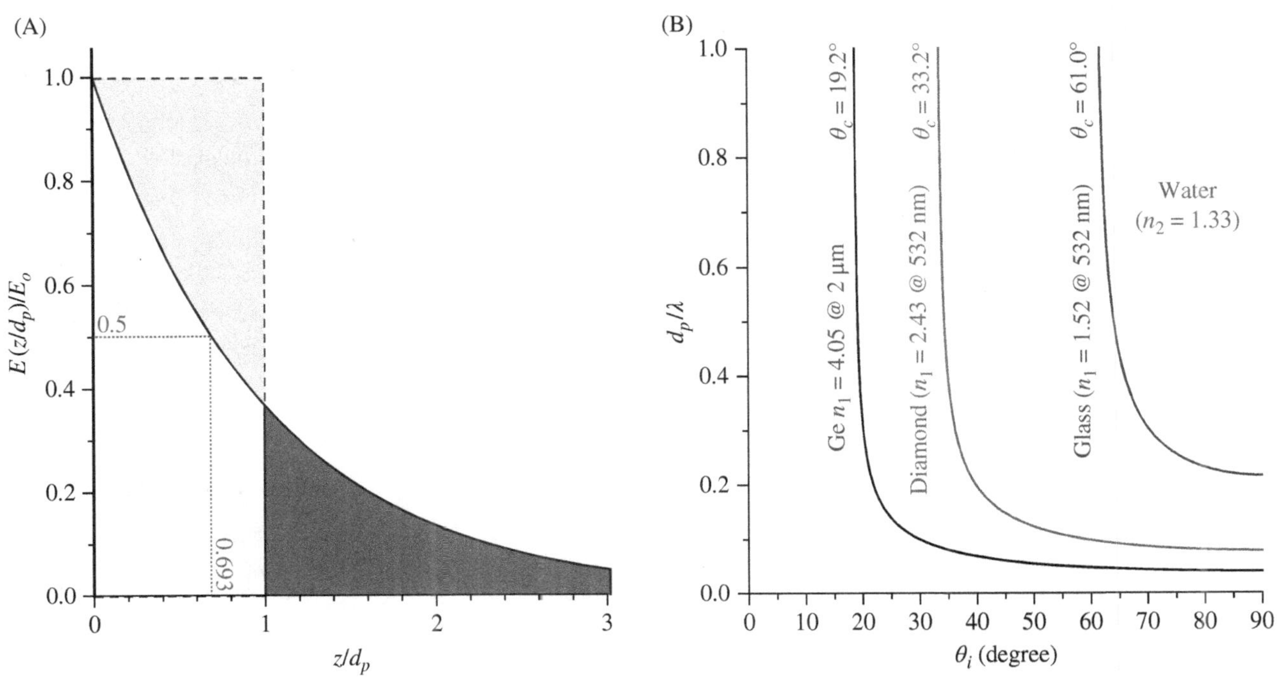

Figure 5.15 (A) Normalized electric field (E/E_o) as a function of normalized depth (z/d_p). (B) Penetration depth (d_p) normalized with wavelength (λ) as a function of incidence angle (θ_i). The refractive index of medium 1 (n_1) is shown for each curve, and medium 2 is water (n_2 = 1.33).

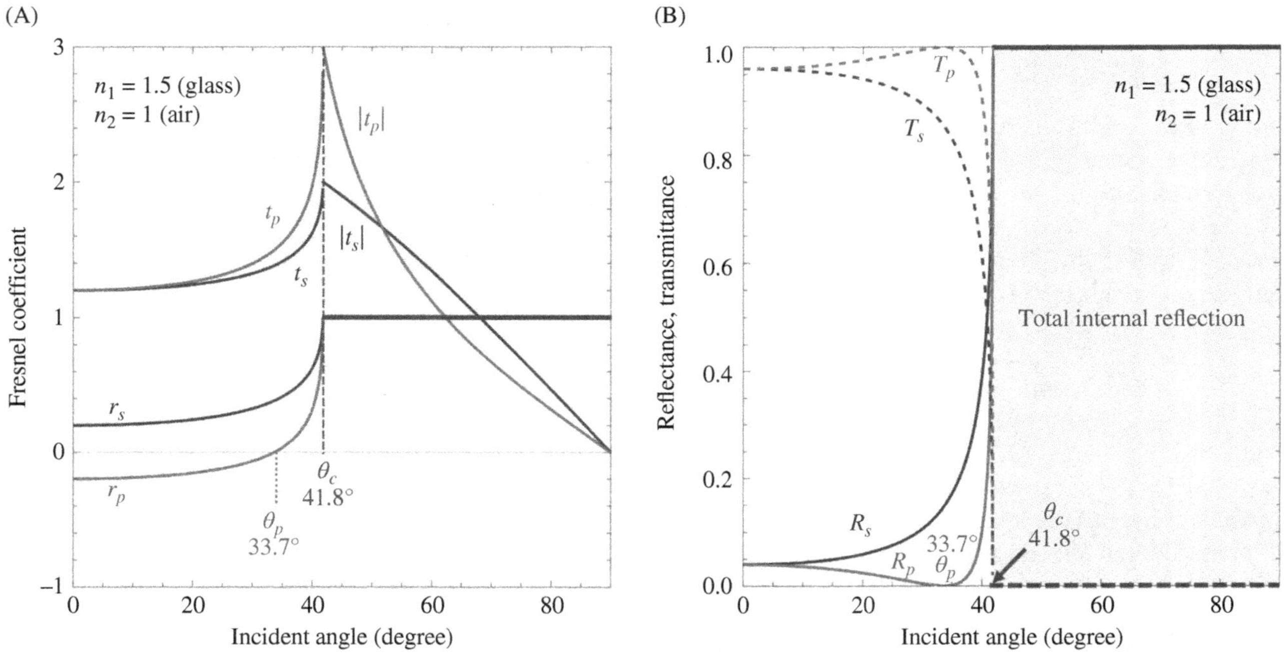

Figure 5.16 (A) Fresnel coefficients (r_s, t_s, r_p, and t_p) and (B) reflectance (R_s, R_p) and transmittance (T_s, T_p) as a function of incident angle (θ_i) for an glass (n_1 = 1.5) / air (n_1 = 1) interface. In (A), the t_s and t_p values plotted above the critical angle are their absolute value.

are the same exponential decay function). This means that the ATR-IR intensity of a film with a thickness of $d_{p,\text{int}}$ on a non-absorbing medium would be a half of the intensity measured for a bulk sample that is thicker than $5 \times d_{p,\text{int}}$. Also, it implies that the ATR-IR intensity of a bulk sample probed with the evanescent wave would be equal to the transmission intensity of the same sample with a thickness of $d_{p,\text{int}}$ probed with a beam propagating through the sample.

Figure 5.15B plots the penetration depth of the electric field normalized with wavelength (d_p/λ) as a function of incident angle (θ_i) for three different TIR elements (germanium, diamond, and glass) in contact with water. When θ_i is just above the critical angle (θ_c), the penetration depth is quite large. As θ_i increases further away from θ_c, the penetration depth decreases quickly and asymptotically approaches the limiting value.

When the incident angle (θ_i) is larger than the critical angle (θ_c), $\sin\theta_t = \frac{n_1}{n_2}\sin\theta_i > 1$ and $\cos\theta_t$ is an imaginary number. Thus, r_s in eq. (5.70) and r_p in eq. (5.74) are complex number. If we calculate $R_s = r_s \cdot r_s^*$ and $R_p = r_p \cdot r_p^*$ for $\theta_i > \theta_c$, we get $R_s = R_p = 1$; thus, transmittance above the critical angle becomes $T_s = T_p = 0$. This is shown for the glass ($n_1 = 1.5$) / air ($n_1 = 1$) interface in Figure 5.16.

5.2.5 Total Internal Reflection at the Absorbing Medium

So far, we have studied, for simplicity, how light is reflected and refracted at the interface of non-absorbing (transparent) media. Now, let us consider the reflection from the interface with an absorbing medium. The electric field of a plane-polarized wave traveling with a phase velocity υ in the positive x direction in a non-absorbing medium is given as:

$$E_x(x,t) = E_{xo}\mathrm{e}^{-i\omega(t-x/\upsilon)} = E_{xo}\mathrm{e}^{-i\omega(t-nx/c)} \tag{5.89}$$

This is the same as eq. (5.58). In an absorbing medium, the intensity experiences an attenuation in accordance with the Beer–Lambert law:

$$I(x) = I_o\mathrm{e}^{-(4\pi k/\lambda)x} \tag{5.90}$$

where k is the absorption index, which relates to the absorptivity, a, via eq. (5.42). In terms of electric field amplitude, this is written as:

$$E(x) = E_o\mathrm{e}^{-(2\pi k/\lambda)x} \tag{5.91}$$

Let E_{xo} be the electric field amplitude at $x = 0$. Then, at a point x, from eq. (5.89) and (5.90), it follows:

$$E_x(x,t) = E_{xo}\exp[-(2\pi k/\lambda)x]\exp[-i\omega(t-nx/c)] = E_{xo}\exp\left[-i\omega\left(t-\frac{n+ik}{c}x\right)\right] \quad (5.92)$$

So, we get the complex refractive index again, $\tilde{n} = n + ik$.

Snell's law and the Fresnel coefficients that we have derived for the interface of non-absorbing media are all valid for the absorbing media; we just have to use the appropriate complex refractive index, $\tilde{n} = n + ik$, in the n term of for eq. (5.61) for θ_t calculation, eq. (5.70) for r_s, eq. (5.71) for t_s, eq. (5.74) for r_p, and eq. (5.75) for t_p. As an example, let us consider a TIR case where the incident medium is diamond ($n_1 = 2.4$) and the second medium is water ($n_2 = 1.33$). Figure 5.17 plots the $r_s(\theta_i)$ and $r_p(\theta_i)$, and corresponding $R_s(\theta_i)$ and $R_p(\theta_i)$, for three different cases, calculated using Snell's law and Fresnel coefficient equations with $\tilde{n}_2 = n_2 + ik_2$. The first case is when the second medium is pure water (i.e., $k_2 = 0$ and $n_2 = 1.33$). Since the refractive index of the second medium is a real number, r_s and r_p have no imaginary part at $\theta_i < \theta_c$ (33.7°). Although they have imaginary components at $\theta_i > \theta_c$, the modulus (absolute value) is unity. So, the reflectance, R_s and R_p, above θ_c is one (i.e., total internal refection). The second case considered is when a small amount of chemical is added so that the aqueous phase can absorb the light that is reflected at the diamond/water interface. Let us assume that the concentration of the added chemical is such that the effective k_2 of the second medium is 0.01 (i.e., $\tilde{n}_2 = 1.33 + 0.01i$). Although the light cannot propagate into the aqueous phase when θ_i is larger than θ_c, the evanescent wave that decays exponentially as a function of distance from the interface can be absorbed in the second medium; this results in attenuation of total reflection (i.e., $R_s < 1$ and $R_p < 1$ at $\theta_i > 33.7°$). This is the main principle of the ATR-IR spectroscopy, which is the first subject to study in Chapter 6.

In Figure 5.17, when $\tilde{n}_2 = 1.33 + 0.01i$, the calculated critical angle is also a complex number ($\theta_c = 33.7 + 0.3i$) and the r_s and r_p coefficients are complex numbers as well even at $\theta_i < \theta_c$. When the absorption index (k_2) of the second medium is increased to further to 0.1 (i.e., $\tilde{n}_2 = 1.33 + 0.1i$), then the degree of attenuation of total reflection at $\theta_i > 33.7°$ becomes even larger. Again, the r_s and r_p coefficients are complex number at $\theta_i < \theta_c$. Another interesting trend to note here is that, when $k_2 = 0$, $R_p = 0$ at $\theta_i = 29°$ (θ_p); but when $k_2 = 0.1$, $R_p = 0.006$ at the same θ_i (Figure 5.17B). This is because r_p is a complex

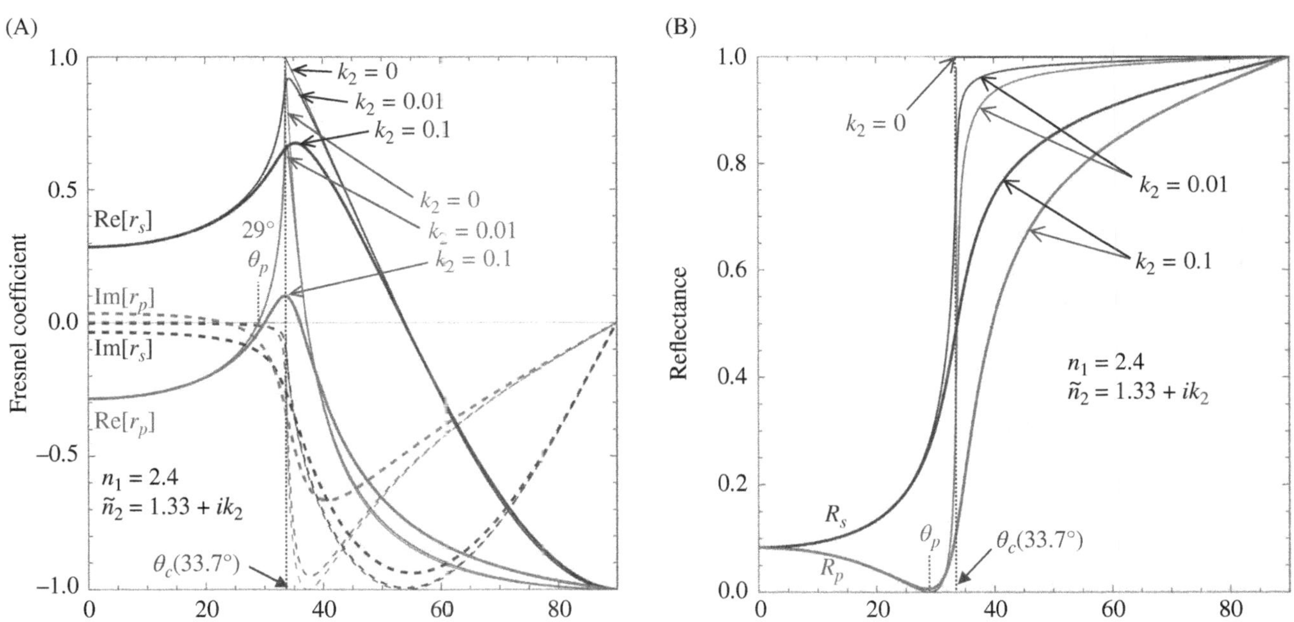

Figure 5.17 (A) Real and imaginary parts of r_s and r_p, and (B) reflectance, R_s and R_p, as a function of incident angle (θ_i) when the incident medium is diamond ($n_1 = 2.4$) and the reflecting medium is aqueous solution with different absorption index values ($\tilde{n}_2 = 1.33 + ik_2$ with $k_2 = 0$ (non-absorbing; thinnest lines), 0.01 (weakly absorbing; thicker lines), and 0.1 (strongly absorbing; thickest lines).

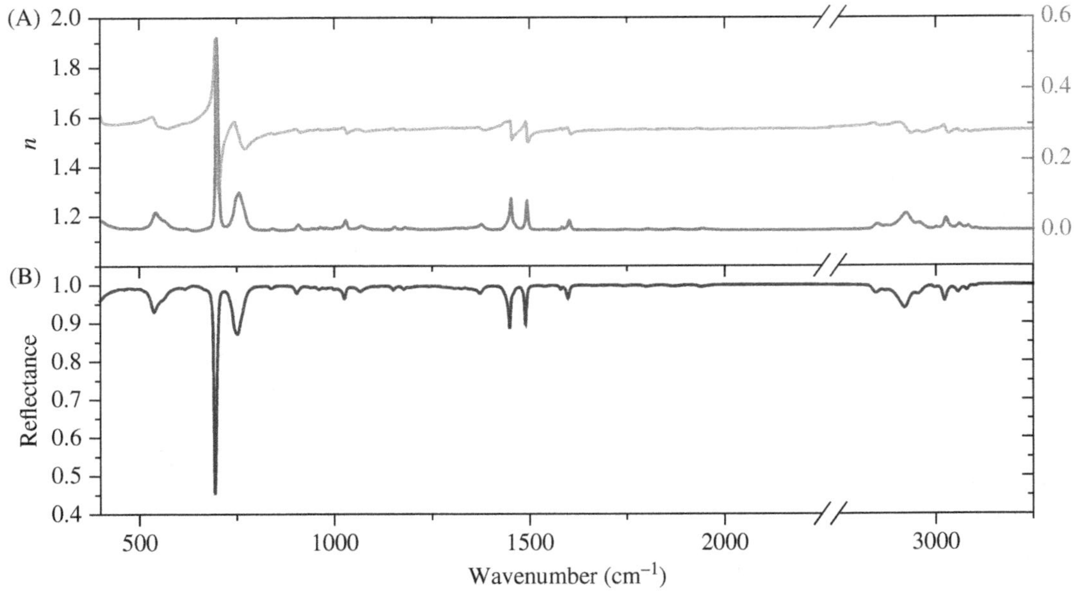

Figure 5.18 (A) Refractive index of polystyrene (PS), $\tilde{n}_2(\overline{\omega}) = n(\overline{\omega}) + ik(\overline{\omega})$, downloaded from Refractiveindex.INFO. Adapted from Polyanskiy [4]. (B) Theoretically calculated reflectance spectrum, $R(\overline{\omega})$, using eq. (5.79) for the case where the incident medium is silicon ($n_1 = 3.4$) and the incident angle (θ_i) is 45°.

number (Figure 5.17A); in fact, the calculated θ_p is also a complex number ($29.0 + 1.8i$ when $\tilde{n}_2 = 1.33 + 0.1i$). Overall, r_p and R_p are more markedly affected than r_s and R_s near and above θ_c when absorption occurs in the second medium ($k_2 \neq 0$).

Figure 5.17 is the simulation of R_s and R_p for non-absorbing, weakly absorbing, and strongly absorbing cases at one wavelength as a function of incident angle (θ_i). Using the same r_s and r_p Fresnel coefficient equations, we can calculate R_s and R_p for a fixed incident angle that is larger than the critical angle, $\theta_i > \theta_c$, as a function of wavenumber ($\overline{\omega}$) using the complex refractive index which varies with wavenumber, $\tilde{n}_2(\overline{\omega}) = n(\overline{\omega}) + ik(\overline{\omega})$. Then, we obtain the theoretical ATR-IR spectrum. The complex refractive index of many materials can be obtained from literatures or online database [4]. Figure 5.18 shows an example simulating the reflectance spectrum of the silicon/polystyrene (PS) interface using eq. (5.79), and appropriate terms defined there, with the complex refractive index downloaded from [4].

There is another interesting phenomenon that attenuates total reflectance even without absorption in the second medium. If a third medium with a high refractive index is placed within the evanescent wave penetration depth from the interface of the first and second media (i.e., creating another interface within the evanescent field), the evanescent wave can cross this new interface of the second and third media and continue propagating into the third medium. This is called the *frustrated* total internal reflection [5]. The frustrated total internal reflection is not widely used in spectroscopy and thus will not be discussed further.

5.2.6 Spectroscopy with Snell's Law, Fresnel Coefficients, and Refractive Index

As shown in Figure 5.18 for IR reflection at the silicon/PS interface, the calculation of theoretical spectrum can be done using Snell's law and the Fresnel coefficients: eqs. (5.61), (5.70), (5.71), (5.74), and (5.75). When a medium or layer absorbing light is involved, a complex refractive index is used. Then, for any given incident angle, reflectance and transmittance can be calculated as a function of wavenumber. The complex refractive index is related to the dielectric permittivity: eqs. (5.38)–(5.44). The real and imaginary parts of complex refractive index and dielectric permittivity are inter-dependent through the Kramers–Kronig (KK) relations: eqs. (5.51)–(5.56). If the absorption spectrum is obtained, the real part of the complex refractive index can be derived using the KK relationship. In IR spectroscopic analysis, it is important to remember that not only the resonance frequency and absorption probability described with eqs. (5.25) and (5.26) but the damping factor also plays an important role as we have seen in eqs. (5.36) and (5.37). *These are the basic equations used to simulate almost all reflection and transmission spectra for any given experimental geometry and wavelength.*

While this is an obvious statement, it cannot be over-emphasized how important it is to do actual calculations for some of the examples discussed in Chapter 6 where we will study how Snell's law and Fresnel coefficient equations are employed in ATR-IR, SR-IR, RAIRS, BAT-IR, DRFIT, and IR imaging coupled with AFM. The same equations are also involved in non-linear spectroscopy in Chapter 7 and reflectometry and ellipsometry in Chapter 9. *Once the calculations described in this chapter are done using Snell's law and the Fresnel coefficient equations, then almost all theoretical calculations discussed in the literature can be done with small incremental variations*; so, it is highly recommended to do the practice calculations using the parameters and equations shown in this chapter. If the calculations are done properly, then you will get the same results shown in the figures in this chapter. The calculations can be done with Microsoft Excel, if you know how to deal with complex numbers in Excel and derive equations for real and imaginary parts separately; but it is somewhat laborious. Instead, if *Mathematica* or *MATLAB* software is used, then calculations can be done easily using the equations in the format shown in this chapter without extra derivations.

5.3 Further Reading for IR and Raman Spectroscopy

The IR spectral interpretation will not be covered in this book. Readers interested in learning the basic principles and examples can find many good textbooks. A particularly useful one is *Infrared Spectral Interpretation: A Systematic Approach by* Brian Smith [6]. He also has an online workshop available at Spectroscopyonline.com [7], where readers can find general rules and interpretation examples for specific functional groups or materials. For basic principles and application examples of Raman spectroscopy, a good book would be *Modern Raman Spectroscopy – A Practical Approach* by E. Smith and G. Dent [8]. Many of the equations discussed in this chapter and the next chapter are also explained very well in [9]. Often, it is good to read two books on the same subject, because it helps build more comprehensive understanding of the subject matter.

Practice Problems

5.1 Calculate (a) dielectric constant $[\varepsilon_1(\omega) + i\varepsilon_2(\omega)]$, (b) phase of dielectric constant $[\varphi(\omega)]$, (c) refractive index $[n(\omega) + ik(\omega)]$, (d) absorptivity $[\alpha(\omega)]$, and (e) reflectance $[R(\omega)$ for $n_{air} = 1]$ for the following case: $\omega_0 = 120$ THz, $\varepsilon_{st} = 12.1$, $\varepsilon_\infty = 10$ for γ = (i) 0.1 THz, (ii) 1 THz, and (iii) 10 THz. Plot the results calculated with three different γ values in the same graph for comparison for ω from 80 to 150 THz.

5.2 Derive the following Fresnel coefficient terms:

$$r_p = \frac{\tan(\theta_i - \theta_t)}{\tan(\theta_i + \theta_t)} \text{ and } t_p = \frac{2\cos\theta_i \sin\theta_t}{\sin(\theta_i + \theta_t)\cos(\theta_i - \theta_t)}$$

When we know r_p, then we can calculate reflectance using $R = r_p \cdot r_p^*$. Can you calculate transmittance in the same way: $T = t_p \cdot t_p^*$?

5.3 What physical parameters govern the penetration depth (d_p) at which the amplitude of evanescence wave field decreases by 63.2%? Is d_p the same or different for *p*- and *s*-polarizations?

5.4 For the total internal reflection from an absorbing medium (Figure 5.17B), why is attenuation larger for the *p*-polarization than the *s*-polarization?

References

1 Lucarini V, Peiponen K-E, Saarinen JJ, Vartiainen EM. Kramers–Kronig relations and sum rules in linear optics. In: Kramers–Kronig Relations in Optical Materials Research. Berlin, Heidelberg: Springer; 2005. p. 27–48.
2 Pitman KM, Hofmeister AM, Corman AB, Speck AK. Optical properties of silicon carbide for astrophysical applications. Astronomy & Astrophysics. 2008;483(2):661–72.
3 Axelrod D. Total internal reflection fluorescence microscopy in cell biology. Traffic. 2001;2(11):764–74.
4 Polyanskiy MN. Refractive Index Database [Available from: https://refractiveindex.info].

5 Zhu S, Yu AW, Hawley D, Roy R. Frustrated total internal reflection: a demonstration and review. American Journal of Physics. 1986;54(7):601–7.

6 Smith BC. Infrared Spectral Interpretation. A Systematic Approach. Boca Raton, FL: CRC Press; 1998.

7 Smith BC. IR Spectral Interpretation Workshop [Available from: https://www.spectroscopyonline.com/authors/brian-c-smith].

8 Smith E, Dent G. Modern Raman Spectroscopy: A Practical Approach (Second Edition). John Wiley & Sons, Ltd; 2019.

9 Tolstoy VP, Chernyshova IV, Skryshevsky VA. Absorption and reflection of infrared radiation by ultrathin films. In: Handbook of Infrared Spectroscopy of Ultrathin Films. John Wiley & Sons, Inc; 2003. p. 1–78.

6

Spectroscopic Analysis via IR Reflection and Transmission

This chapter focuses primarily on vibrational spectroscopic analysis methods that can give a surface sensitivity with varying probe depths, which include attenuated total reflectance infrared (ATR-IR) spectroscopy (Figure 6.1A), specular reflection infrared (SR-IR) spectroscopy (Figure 6.1B), reflection–absorption infrared spectroscopy (RAIRS; Figure 6.1C), polarization-modulation RAIRS (PM-RAIRS), and Brewster-angle transmission (BAT) infrared spectroscopy (Figure 6.1D). These methods are discussed in Sections 6.1–6.4. They are employed for analysis of flat surfaces. For powders, diffuse-reflectance infrared Fourier transform (DRIFT) method is used, which is explained in Section 6.5. In Section 6.6, we will briefly touch on how the optical theories we have studied in the previous chapter are applied to IR-spectroscopic imaging by combining with scanning probe microscopy. Three representative techniques are photothermal (PT) infrared spectroscopy coupled with atomic force microscopy (AFM-IR), infrared nano-spectroscopy (nano-IR) based on scattering-type scanning near-field optical microscopy (s-SNOM), and polarization-induced force microscopy (PiFM). The same principles described for IR analysis can be employed for UV–VIS spectroscopy. Raman spectroscopy is another vibrational spectroscopic analysis. It is mostly based on inelastic scattering through the absorption–emission process induced by near-IR, visible, or UV light, rather than resonance absorption process described in Figure 5.1. So, it will not be covered in this chapter.

This chapter is about how to do surface-sensitive vibrational spectroscopy properly, and not how to interpret the measured spectral features. So, the peak assignment will not be discussed in depth. Nonetheless, it is important to mention here that the spectroscopic interpretation of peaks and bands detected in vibrational spectra *cannot be done by simply comparing the maximum intensity positions with tables summarizing "typical" absorption band positions or using the Beer–Lambert law, eq. (5.27).* This is because the absorption band position, intensity, and shape can vary depending on the experimental method. A few examples of such cases will be discussed in this chapter.

It is extremely important to do full theoretical calculations to understand and utilize surface-sensitive vibrational spectroscopy properly. In many books, equations with various forms are discussed in depth; those equations are usually derived with some assumptions or simplifications. Thus, one should check the applicability of those equations to the system under consideration. An alternative and even better way is *to use the genuine equations that can be employed universally.* The full set of those equations are Snell's law and the Fresnel coefficients, which we have derived in Sections 5.2.1 and 5.2.2, respectively. Since they will be used repeatedly throughout this chapter (and Chapter 9 as well), they are summarized again for an interface of two media (1 and 2) (see Figures 5.11 and 5.12):

$$\tilde{n}_1(\overline{\omega}) \sin\theta_i = \tilde{n}_2(\overline{\omega}) \sin\theta_t \tag{6.1}$$

$$r_s(\overline{\omega}, \theta_i) = \frac{\tilde{n}_1(\overline{\omega})\cos\theta_i - \tilde{n}_2(\overline{\omega})\cos\theta_t}{\tilde{n}_1(\overline{\omega})\cos\theta_i + \tilde{n}_2(\overline{\omega})\cos\theta_t} = -\frac{\sin(\theta_i - \theta_t)}{\sin(\theta_i + \theta_t)} \tag{6.2}$$

$$r_p(\overline{\omega}, \theta_i) = \frac{\tilde{n}_2(\overline{\omega})\cos\theta_i - \tilde{n}_1(\overline{\omega})\cos\theta_t}{\tilde{n}_2(\overline{\omega})\cos\theta_i + \tilde{n}_1(\overline{\omega})\cos\theta_t} = \frac{\tan(\theta_i - \theta_t)}{\tan(\theta_i + \theta_t)} \tag{6.3}$$

$$t_s(\overline{\omega}, \theta_i) = \frac{2\tilde{n}_1(\overline{\omega})\cos\theta_i}{\tilde{n}_1(\overline{\omega})\cos\theta_i + \tilde{n}_2(\overline{\omega})\cos\theta_t} = \frac{2\cos\theta_i \sin\theta_t}{\sin(\theta_i + \theta_t)} \tag{6.4}$$

$$t_p(\overline{\omega}, \theta_i) = \frac{2\tilde{n}_1(\overline{\omega})\cos\theta_i}{\tilde{n}_2(\overline{\omega})\cos\theta_i + \tilde{n}_1(\overline{\omega})\cos\theta_t} = \frac{2\cos\theta_i \sin\theta_t}{\sin(\theta_i + \theta_t)\cos(\theta_i - \theta_t)} \tag{6.5}$$

Surface and Interface Analysis: Principles and Applications, First Edition. Seong H. Kim.

Companion website: www.wiley.com/go/Kim

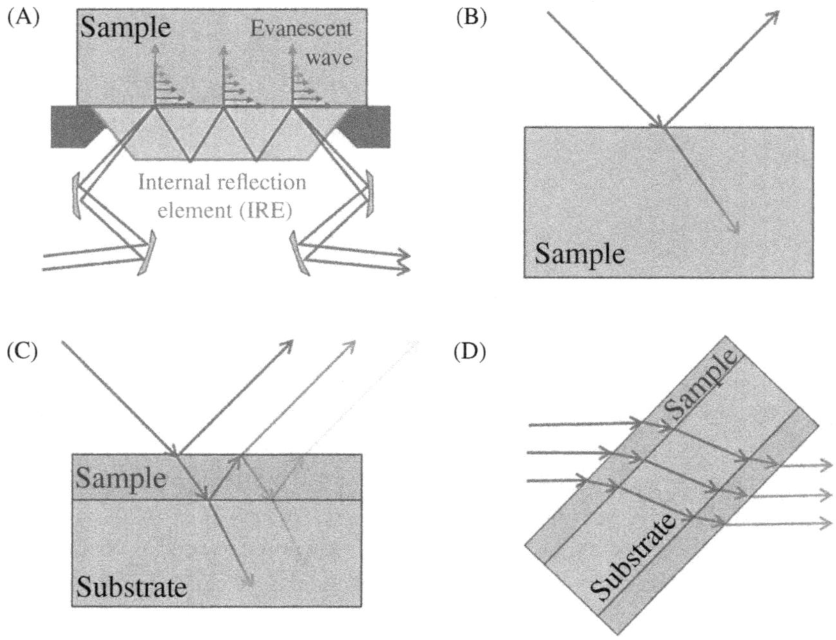

Figure 6.1 (A) Attenuated total reflectance (ATR)-IR spectroscopy, (B) specular reflection (SR)-IR spectroscopy, (C) reflection–absorption infrared spectroscopy (RAIRS), and (D) Brewster-angle transmission (BAT) IR spectroscopy.

The angles, θ_i and θ_t, are always defined from the normal direction of each reflecting/refracting interface. For an absorbing medium, the refractive index, $\tilde{n}_2(\overline{\omega}) = n_2(\overline{\omega}) + ik_2(\overline{\omega})$, is a complex number, so the refraction angle, θ_t, is also a complex number. If the incident medium is absorbing, then $\tilde{n}_1(\overline{\omega})$ is a complex number as well. The transmission coefficients are not used in the calculation for the reflection from a single interface; but they are needed in the calculation for multiple interfaces such as thin films on a substrate (Figure 6.1C,D), which will be discussed in Sections 6.3 and 6.4. All other forms of equations found in the literature are derived from these Snell's law and Fresnel coefficient equations with some assumptions for specific applications; but, there is no need to derive such equations (which are often complicated) as long as we can handle complex numbers using math software such as *Mathematica* or *MATLAB*. All theoretical calculations discussed in this chapter (and Chapter 9) are done using these five equations.

The reflectance of the s- and p-polarized light at the interface of two media with different refractive indices is calculated as:

$$R_s(\overline{\omega}, \theta_i) = r_s(\overline{\omega}, \theta_i) \cdot r_s^*(\overline{\omega}, \theta_i) \tag{6.6}$$

$$R_p(\overline{\omega}, \theta_i) = r_p(\overline{\omega}, \theta_i) \cdot r_p^*(\overline{\omega}, \theta_i) \tag{6.7}$$

For the unpolarized light, the total reflectance is the average of these two:

$$R(\overline{\omega}, \theta_i) = \left(R_s(\overline{\omega}, \theta_i) + R_p(\overline{\omega}, \theta_i)\right)/2 \tag{6.8}$$

The transmittance *at the interface between medium-1 and -2* is calculated easily from the flux conservation:

$$T(\overline{\omega}, \theta_i) = 1 - R(\overline{\omega}, \theta_i) \tag{6.9}$$

Note that eq. (6.9) calculates the fraction of photon flux *passing through the interface* between medium-1 and -2 (see Figures 5.11 and 5.12) and should not be confused with the transmittance measured in the typical absorbance measurement. The absorbance measurement involves the photon flux *passing through the medium-2* (=sample) and measured at the opposite side of the input beam. Figure 6.2 shows the case where light is passing through the sample. Here we should consider the following processes: (i) crossing the interface between medium-1 and 2 (which involves $T_{12} = 1 - R_{12}$), (ii) propagating through medium-2 in which attenuation by medium-2 (At_2) occurs, and (iii) crossing the interface between medium-2 and -1 (which involves $T_{21} = 1 - R_{21}$). Here, At_2 is described by eq. (5.90). Thus, the ratio of the input and output fluxes should be written as:

$$\frac{I_{out}}{I_{in}} = T_{12} \cdot At_2 \cdot T_{21} \tag{6.10}$$

In typical transmission spectroscopy, the intensity of light *passing through a blank sample* is measured first and taken as the intensity of the *input* beam (I_{in}). Thus, the transmittance measured through the sample is the attenuation by the sample (At_2) times the reflection loss of the beam at the front and back interfaces (which is $T_{12} \cdot T_{21}$). Even if the measurement uses a non-absorbing cuvette or container whose refractive index matches the real part of the refractive index of the sample, the reflection loss at the air/cuvette interface still exists. If the refractive index of cuvette or the real part of weakly-absorbing sample is 1.4, then $T_{1/2} \cdot T_{2/1}$ is 0.95. Thus, the measured transmittance can be differ by ~5% from the net attenuation by the sample (At_2). But, when we use the Beer–Lambert law, we normally consider the attenuation part only (which is given by eq. (5.90)):

$$A(\overline{\omega}) = -\log\left(\frac{I_{out}}{I_{in}}\right) = a(\overline{\omega}) \cdot b \cdot c \tag{6.11}$$

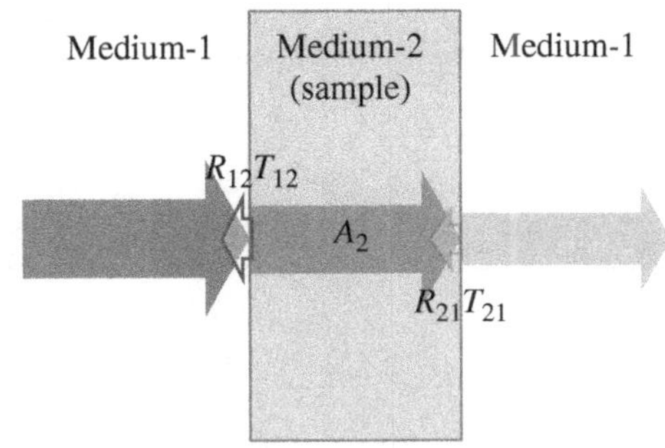

Figure 6.2 Reflection, transmission, and attenuation involved in the absorbance measurement experiment where the incident beam enters the sample in the surface-normal direction and the intensity of the beam passing through the sample is measured.

where $A(\overline{\omega})$ is the absorbance, the $\frac{I_{out}}{I_{in}}$ ratio is also called "transmittance" (which should not be confused with the transmittance in eq. (6.9)), $a(\overline{\omega})$ is the absorptivity defined as $\frac{4\pi k_2(\overline{\omega})}{\lambda}$ (see eq. (5.42)), b is the beam path length through the sample (=medium-2), and c is the analyte concentration (in volume fraction) within the unit length. Thus, strictly speaking, without considering the effect of passing through interfaces, using the Beer–Lambert law in the form of eq. (6.11) for the experimental data processing is bound to a small error.

6.1 Attenuated Total Reflectance Infrared (ATR-IR) Spectroscopy

6.1.1 Internal Reflection Element for ATR-IR

ATR-IR relies on the attenuation of total internal reflection of the light at the interface of the first and second media by the absorption in the second medium (Section 5.2.5). For that, an internal reflection element (IRE) with a refractive index higher than the sample refractive index is required (Figure 6.1A). Table 6.1 lists commercially available IREs along with their refractive indices in the mid-IR region. The IRE materials listed here are transparent in the mid-IR wavelength range and suitable for ATR-IR analyses of polymers and organic coatings, because the refractive indices of these samples are typically around 1.4–1.6 in the mid-IR region. Inorganic materials can also be analyzed as long as the refractive index of sample is lower than that of IRE, that is, $n_{sample} < n_{IRE}$, so that the incident angle of the instrument can be set larger than the critical angle (θ_c, eq. (5.82)). Since the penetration depth of evanescent wave into the sample is small, the sample should have intimate contact with the IRE surface. If there is a small gap between the IRE and sample surfaces, then frustrated total internal reflection can occur. Organic samples can be pressed against the IRE surface to have a good contact. If the sample surface is rough and cannot be flattened by compression against the IRE surface, the signal intensity is lower than what is expected. The ATR-IR signal increases as the number of reflections increases by making the length of the IRE larger along the beam path (as shown in Figure 6.1A), which intuitively makes sense.

6.1.2 Probe Depth in ATR-IR

As we discussed in Section 5.2.4, the penetration depth of the evanescent wave from the IRE interface into the sample is defined as:

$$d_p(\overline{\omega}, \theta_i) = \frac{\lambda(\overline{\omega})}{2\pi\sqrt{\tilde{n}_1(\overline{\omega})^2 \sin^2\theta_i - \tilde{n}_2(\overline{\omega})^2}} \tag{6.12}$$

The penetration depth shown in Table 6.1 is calculated assuming that the refractive index of the sample (n_2) is 1.5 and the incident angle at the IRE/sample interface is 45°. In the non-absorbing region of mid-IR wavelength, typical refractive indices of organic molecules are around 1.5. Same as in XPS (Figure 2.10A), the information depth or probe depth of ATR-IR is

Table 6.1 Commonly used IRE materials for ATR-IR experiment. The penetration depth of evanescent wave (d_p) is calculated using eq. (5.87) for the incident angle of 45°. The information depth is $3 \times d_p$.

IRE material	n_1 in mid-IR	d_p for $n_2 = 1.5$ @ 1000 cm^{-1}	@ 3000 cm^{-1}	Notes
ZnS	2.2	3.86 μm	1.29 μm	>850 cm^{-1}, not stable in low and high pH
Diamond	2.38	2.09 μm	695 nm	Inert, very hard; not transparent in the 1900–2400 cm^{-1} region due to a strong phonon band absorption
ZnSe	2.4	2.01 μm	668 nm	Not good for acid or base
AMTIR	2.5	1.70 μm	567 nm	Not compatible with high pH
Silicon	3.45	0.827 μm	276 nm	Not transparent at <1000 cm^{-1}
Germanium	4.0	0.663 μm	221 nm	Not transparent at <780 cm^{-1}

considered to be $3 \times d_p$. For the ATR-IR intensity, we should use $d_{p,int}(\overline{\omega}, \theta_i) = d_p(\overline{\omega}, \theta_i)/2$, as described in eq. (5.88). It should be noted that the probe depth of the ATR-IR analysis is not the same for all absorption bands at different $\overline{\omega}$'s. In the absorption band region, the refractive index is a complex number ($n_2 + ik_2$), and the penetration depth becomes shorter because k is not zero. Since $n_2(\overline{\omega})$ varies in a differential shape of $k_2(\overline{\omega})$ due to the Kramers–Kronig relation (see eqs. (5.53) and (5.54); Figures 5.7 and 5.18A), d_p varies in a complicated shape with $\overline{\omega}$ around the absorption band ($\overline{\omega}_o$).

Figure 6.3A replots the reflectance spectrum of polystyrene (PS) calculated with eq. (6.8) for the measurement with a silicon IRE at an incident angle of 45° (shown in Figure 5.18B) in the log(1/R) scale. Most commercial instruments display the ATR-IR spectrum in this scale with the y-axis marked as "absorbance unit" for easy comparison with the absorbance spectrum obtained in the transmission geometry (Figure 6.2). However, it should be remembered that this is just for convenience. The absorbance (A) is defined as log(1/T), where T is attenuation by the sample during the transmission through the sample without considering the reflection loss. So, log(1/R) and absorbance (A) are *dimensionally equivalent*, but they are *different* physical quantities. The reflectance measurement at a slanted incident angle ($\theta_i > 0$) is affected by both $n(\overline{\omega})$

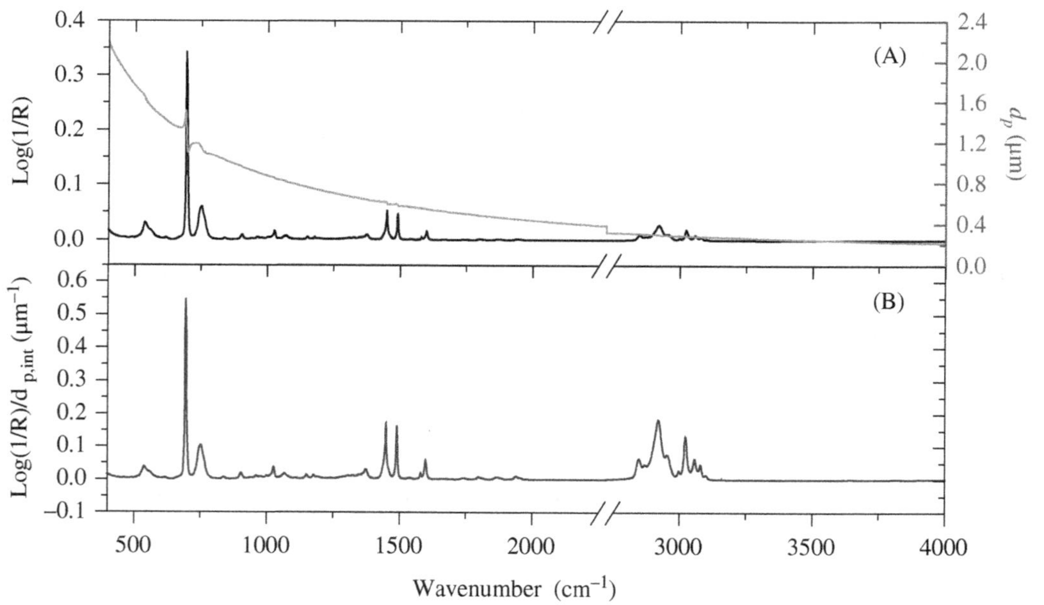

Figure 6.3 (A) Raw ATR-IR spectrum of polystyrene/Si IRE calculated using eq. (6.8) with its complex refractive index (see Figure 5.18A) and plotted in the log(1/R) scale. The penetration depth of the evanescent electric field (d_p) is also plotted for comparison. (B) ATR-IR spectrum of the same divided with the penetration depth of the evanescent wave intensity ($d_{p,int} = d_p/2$) calculated with the refractive index. The incident angle is 45°.

and $k(\varpi)$ of the sample, while the transmission measurement at the surface-normal incidence (i.e., $\theta_i = 0$) is mostly governed by the absorptivity, $a(\varpi) = 4\pi k(\varpi)/\lambda$.

In the ATR-IR spectrum shown in Figure 6.3A, the CH_2 asymmetric stretching band at 2930 cm^{-1} is about 16 times smaller than the aromatic ring bending peak at 698 cm^{-1}, although the absorptivity (a) ratio calculated with the data in Figure 5.18A is only a factor of 3.1. This is due to the difference in the penetration depth of evanescent wave at the wavelengths of these bands. The penetration depth of the evanescence field calculated with eq. (6.12) is also plotted in Figure 6.3A for comparison. If the main interest is only the relative intensities of various peaks in the ATR-IR spectrum, the penetration depth effect can be factored out by normalizing the ATR-IR spectrum with the penetration depth ($d_{p,\text{int}} = d_p/2$) of the evanescent wave intensity. This is shown in Figure 6.3B. In many commercial instruments, this is called "ATR correction.". Note that the commercial instrument does not know what sample is being analyzed; so, it uses an approximation value calculated assuming $n_2 = 1.5$ (as shown in Table 6.1), instead of using the real d_p of the sample.

When absorption is small, the exponential decay of the evanescent wave can be simplified into a linear equation. That is because e^{-x} can be expanded into a series, $1 - x + x^2/2 + O[x]^3$, and the terms higher than the first order become negligible when x is small. Using this simplification and to facilitate the quantitative interpretation of the ATR-IR signal intensity, Harrick defined the "effective path length (d_e)" of the evanescent wave as the "equivalent path length" that would give the same intensity in the transmission experiment [1]. By setting $R = e^{-a \cdot d_e} \cong 1 - a \cdot d_e$ for the low absorption case, d_e for the s- and p-polarizations can be derived as [1]:

$$d_{e,s} = \frac{n_1^2 n_2 \cos\theta_i}{\pi(n_1^2 - n_2^2)} \cdot \frac{\lambda/n_1}{\left(n_1^2 \sin^2\theta_i - n_2^2\right)^{1/2}} \tag{6.13}$$

$$d_{e,p} = \frac{n_1^2 n_2 \cos\theta_i}{\pi(n_1^2 - n_2^2)} \cdot \frac{2n_1^2 \sin^2\theta_i - n_2^2}{(n_1^2 + n_2^2)\sin^2\theta_i - n_2^2} \cdot \frac{\lambda/n_1}{\left(n_1^2 \sin^2\theta_i - n_2^2\right)^{1/2}} \tag{6.14}$$

The d_e for unpolarized beam is the average of these two, $(d_{e,s} + d_{e,p})/2$. Note that, unlike $d_{e,s}$ and $d_{e,p}$ derived here to normalize the signal intensity in polarized ATR-IR spectra (which will be discussed further in Figure 6.4), d_p has no polarization dependence. In other words, the penetration depth of the evanescent wave is the same for the electric fields of s- and p-polarizations. The difference in $d_{e,s}$ and $d_{e,p}$ comes from the difference in the magnitude of electric fields projects to the surface, which is converted to the "effective path length" in the transmission experiment.

Keep in mind that these equations are derived for the low absorption case. As the $a \cdot d_e$ value increases from 0.1 to 0.3, the error in R caused by neglecting the higher-order terms increase from 5% to 16%. Also, it assumes that n_2 is real and constant. These approximations would work for the absorption bands in the C–H stretch region in Figure 6.3A because log(1/R) is small and changes in $n(\varpi)$ is relatively small as shown in Figure 5.18A. But, it does not work well for the band with a strong absorptivity because $n(\varpi)$ varies with $k(\varpi)$ due to the Kramers–Kronig relation (eq. (5.53) and (5.54)). For example, we can see how drastically d_p fluctuates near the 698 cm^{-1} band in Figure 6.3A; although smaller, we can see the same fluctuations in d_p near the 1452 cm^{-1} and 1493 cm^{-1} peaks. In contrast, the d_e value changes smoothly in the same regions because n_2 is assumed to be constant.

Let us take a look at another example. Figure 6.4 plots the theoretical calculation results for the polarized ATR-IR measurement of water with three different IREs – silicon, diamond, and sapphire – using refractive index values downloaded from online database [2]. The incident angle for silicon and diamond is set to 45°, but that for sapphire, it is to 65° (since the critical angle for sapphire/water interface is 51°). The R_s and R_p can be calculated with eqs. (6.6) and (6.7), respectively, and plotted in the $-log$ scale. The band peaked at ~1640 cm^{-1} is the H–O–H bending mode and the broad band spanning from 3000 to 3700 cm^{-1} is the O–H stretch mode. If the OH group is not hydrogen bonded to other H_2O molecules, its stretch mode appears at 3700 cm^{-1}. This is called free-OH. But, when the OH group is hydrogen bonded, its vibrational energy is red-shifted to lower wavenumbers. The degree of this shift depends on the strength of hydrogen bonding interaction [3, 4]. For liquid water, the OH stretch band is centered around 3400 cm^{-1}, as shown in the $k(\varpi)$ spectrum in Figure 6.4A. For ice water, it is centered around 3250 cm^{-1} and relatively narrower and taller than the OH stretch band of liquid water [5, 6].

Figure 6.4B shows the polarized ATR-IR spectra of water probed with silicon, diamond, and alumina IREs. There are two aspects that we should discuss here. First, the ATR-IR intensities of the absorption bands of water are larger in the log (1/R_p) spectrum than the log(1/R_s) spectrum in all the three cases. Recall that the electric field of the p-polarization has both surface-normal and surface-parallel components, while that of the s-polarization has the surface-parallel component

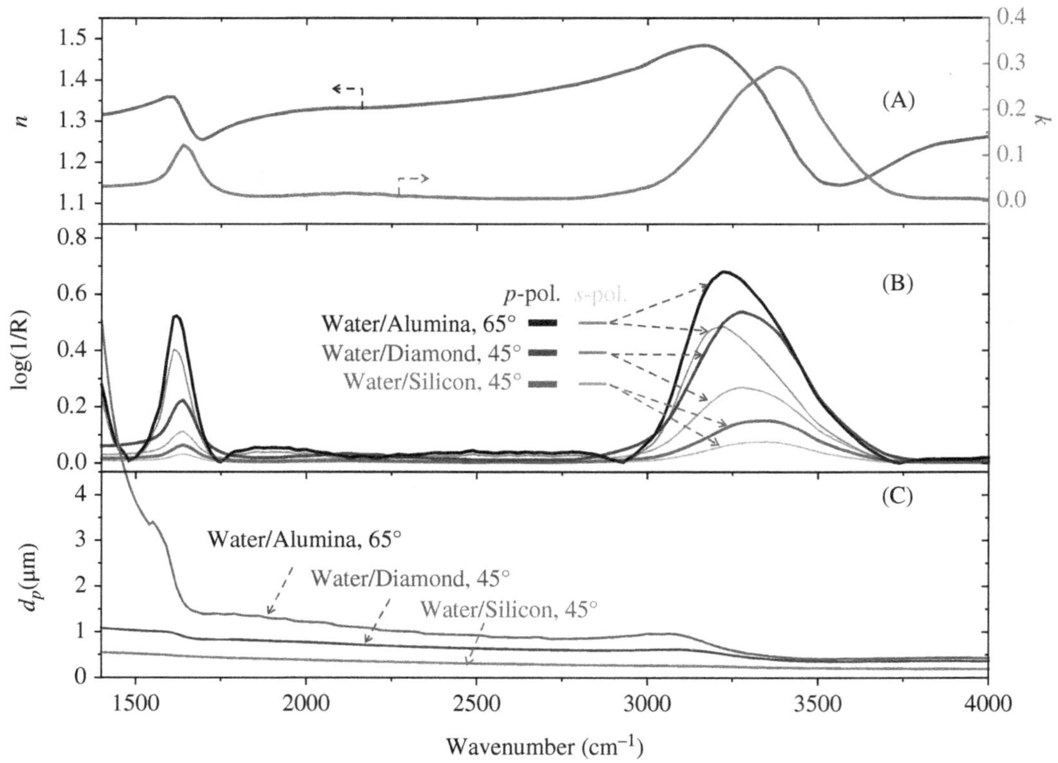

Figure 6.4 (A) Refractive index, $n(\varpi) + ik(\varpi)$, of liquid water. (B) Theoretically calculated ATR-IR spectra of water probed with three different IRE materials at *p*- and *s*-polarizations. The incident angle is 45° for silicon and diamond and 65° for alumina. (C) Absolute value of complex penetration depth. The refractive indices used in the calculation are from online database [2].

only (see Figure 5.12). The relative ratio of the electric field amplitudes projected to the surface plane by the *p*- and *s*-polarizations will vary depending on the refractive index difference of two media and the incident angle [7, 8].

The differences in the electric fields of *p*- and *s*-polarizations projected to the surface-normal and -parallel directions are reflected in the derivation of the effective path lengths, which leads to eq. (6.13) for $d_{e,s}$ and eq. (6.14) for $d_{e,p}$. When $n_1 = 3.45$ (Si) and $n_2 = 1.33$, $d_{e,p}/d_{e,s} = 2.0$ at $\theta_i = 45°$; so, the intensity in the *p*-polarized ATR-IR spectrum of water/Si is approximately two times larger than that in the *s*-polarized spectrum in Figure 6.4B. When $n_1 = 1.7$ (alumina) and $n_2 = 1.33$, $d_{e,p}/d_{e,s} = 1.45$ at $\theta_i = 65°$; so, the intensity in the *p*-polarized ATR-IR spectrum of water/alumina is expected to be 1.45 times larger than that in the *s*-polarized spectrum. However, the actual difference in the theoretically calculated spectra is 1.39 in the OH stretch region and 1.28 in the H_2O bending region in Figure 6.4B, smaller than the prediction made with the weak absorption approximation. This comparison reiterates the limitation of the $d_{e,s}$ and $d_{e,p}$ concept due to the assumption made in the derivation.

Second, the shape of the OH stretch band of water is different when the ATR-IR spectrum is collected with different IRE materials. In the case of ATR-IR spectrum calculated for the alumina IRE, the peak maximum is shifted to ~3250 cm^{-1} although the absorption maximum in the $k(\varpi)$ spectrum of bulk water is at ~3400 cm^{-1}. If the principle of ATR-IR is not considered, one may interpret that the hydrogen bonding interactions of water near the alumina interface is very strong. But, it is simply an optical effect due to the change in $n(\varpi)$ around the absorption band (Figure 6.4A), and not due to a chemical effect of the alumina surface. Figure 6.4C plots the penetration depth as a function of wavenumber, $d_p(\varpi)$. The lower wavenumber side of the absorption band has a larger d_p than the higher wavenumber side. This is because $n(\varpi)$ is larger in the lower wavenumber side of the absorption band and lower in the higher wavenumber side compared to the value in the non-absorbing region. If one thinks that the ATR-IR spectrum would be similar to the transmission IR spectrum since both are plotted in the absorbance unit, this difference could be called as the "anomalous" dispersion effect. But it is not an anomaly, it is exactly what is expected based on the wavenumber-dependent dispersion of refractive index at the absorption band, that is, the Kramers–Kronig relation (see Section 5.1.4). In the ATR-IR spectrum of water/alumina, the H_2O bending band appears to be significantly enhanced as compared to the OH stretch band. Again, this is the "expected" behavior since the real part of the refractive index of alumina decreases and the imaginary part start increasing as the IR wavenumber decreases below 1700 cm^{-1} [2]. For that reason, the d_p of the water/alumina interface drastically increases below 1700 cm^{-1}.

The examples shown in Figures 6.3 and 6.4 epitomize potential artifacts that may lead to inaccurate or inadequate interpretations if one attempts to construe the ATR-IR spectrum using the Beer–Lambert law concept without knowing that log (1/R) is not the same as log(1/T). The effective path depth, d_e, of the evanescent wave probing the sample in contact with the IRE, which can be estimated with eqs. (6.13) and (6.14), is just an approximation that works only when the absorption by the sample is weak. So, normalizing log(1/R) with d_e might be convenient and sufficient in certain cases, especially, when the signal intensity is small (i.e., log(1/R) ≤ 0.1); but the error becomes significant for large absorption bands.

ATR-IR has sufficient sensitivity to detect a monolayer of organic molecules chemisorbed on the IRE surface. Figure 6.5 displays an example of ATR-IR study for the formation of organic monolayer on a silicon IRE [9]. The silicon surface is typically covered with a native oxide layer, unless the oxidized layer is intentionally removed by chemical treatment. When the IRE surface is exposed to a CCl_4 solution containing methyl 23-(trichlorosilyl)tridecanoate (MTC), the characteristic peaks of MTC are observed in Figure 6.5A. The O=C—OC ester band at 1742 cm^{-1} is quite strong and comparable to the CH_2 symmetric stretch band at 2855 cm^{-1}. This is because the penetration depth (d_p) of evanescent wave is larger at the lower wavenumber (i.e., longer wavelength) region. When the solution phase is removed, a monolayer of MTC chemically bonded to the native oxide surface is detected in the ATR-IR spectrum (Figure 6.5B). The signal intensity is now significantly reduced since there is only one molecular layer of MTC within the information depth ($3d_p$) from the IRE surface. Now, the 1742 cm^{-1} ester band intensity is quite small, because the penetration depth difference does not matter anymore. When the sample is treated with a solution of $LiAlH_4$ in THF, then the ester group is reduced to the OH group; so, the 1742 cm^{-1} peak disappears and the broad OH band appears in the 3300–3600 cm^{-1} region in Figure 6.5C.

6.1.3 In Situ ATR-IR Analysis of Adsorption Isotherm

The exponential decay of the evanescent wave provides an opportunity to detect, *in situ*, the molecular species interacting with the IRE surface which is in contact with a medium containing the analyte of interest. A thin layer of other functional materials can be deposited on the IRE surface to study specific chemistries. When the analyte is adsorbed onto the surface from the liquid phase, its concentration can be adjusted such that the signal from the species unbound to the surface is smaller than the signal from the ones bound to the surface. The density of gas is about three orders of magnitude lower than that of liquid. Thus, when the analyte is adsorbed from the gas phase, the signal from the molecules present within the penetration depth of evanescent wave is so weak and negligible as long as the partial pressure of analyte is low enough. This makes ATR-IR quite powerful to detect molecular species that are in equilibrium with the gas phase without suffering from interferences from the bulk (background) signal.

An example of *in situ* ATR-IR study measuring adsorption isotherm using this "surface-region sensitivity" to detect the adsorbate species only is shown in Figure 6.6 [10]. Figure 6.6A displays the ATR-IR spectra of water adsorbed on a clean native oxide surface of the Si IRE (with $\theta_i = 45°$) from ambient air with various relative humidity (RH) conditions. At a very low RH, the OH stretch band of the adsorbed water layer is centered at ~3230 cm^{-1}. As RH increases, the ATR-IR signal intensity increases, and at the same time, another band grows at ~3400 cm^{-1} and eventually becomes dominant. The thickness of the adsorbed layer can be estimated by comparing the area of the adsorbed water signal with the penetration depth into the bulk water. As discussed in Figure 5.15A, the total signal intensity of a bulk material probed by the evanescent wave is equivalent to the signal of the same material with a thickness of d_p probed by the beam propagating without exponential decay. Thus, by multiplying d_p to the ratio of the adsorbate signal area (A_{ads}) to the bulk water signal area (A_{bulk}), one can estimate the "effective" thickness (t_{eff}) of the adsorbate layer [11]:

$$t_{eff} \cong \frac{A_{ads}}{A_{bulk}} \times d_p \tag{6.15}$$

Here, the reason to call it effective thickness is because the refractive index of the adsorbed layer can be different from that of the bulk, but the actual difference is not known. Another way of estimating the effective thickness is to compare the peak area in the experimental ATR-IR spectra with the area of the theoretical spectrum calculated as a function of layer thickness [12]. The theoretical calculation of the adsorbate spectrum can be done with the equations that will be derived in Section 6.3 (associated with the experimental geometry shown in Figure 6.1C). If the dimension of the IRE used is large so there are multiple internal reflections, then the measured ATR-IR intensity in the log(1/R) scale increases roughly in proportion to

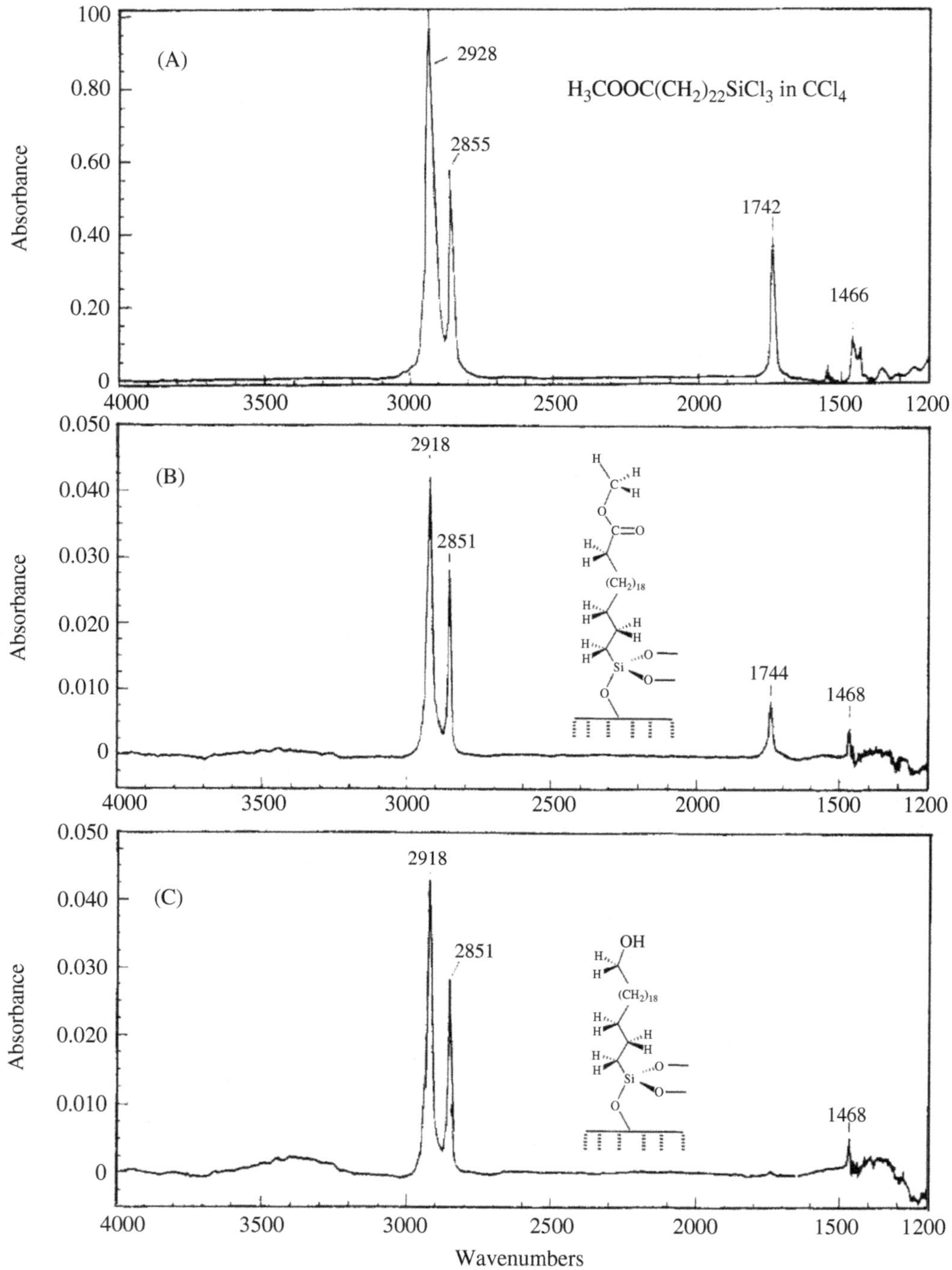

Figure 6.5 ATR-IR spectra of (A) MTC in CCl_4, (B) MTC monolayer, and (C) MTC monolayer reduced with $LiAlH_4$. The IRE was Si and the incident angle was 45°. Based on the dimension of the IRE, the total number of internal reflections is calculated to be 8; thus, the single reflection intensity is about eight times smaller than the measured intensity. Reproduced from [9]/with permission of American Chemical Society.

the number of multiple reflections [8]. Using these methods and considering multiple reflections, the measured ATR-IR intensity in Figure 6.6A can be converted to the effective thickness of the adsorbed layer, which is plotted in Figure 6.6B [10]. As we can see here, the thickness of the adsorbed water layer is 2–3 orders of magnitude smaller than the characteristic penetration depth (d_p) into the bulk water. So, the optical effect originating from the ϖ dependence of n (that we have discussed in Figure 6.4B) is insignificant here.

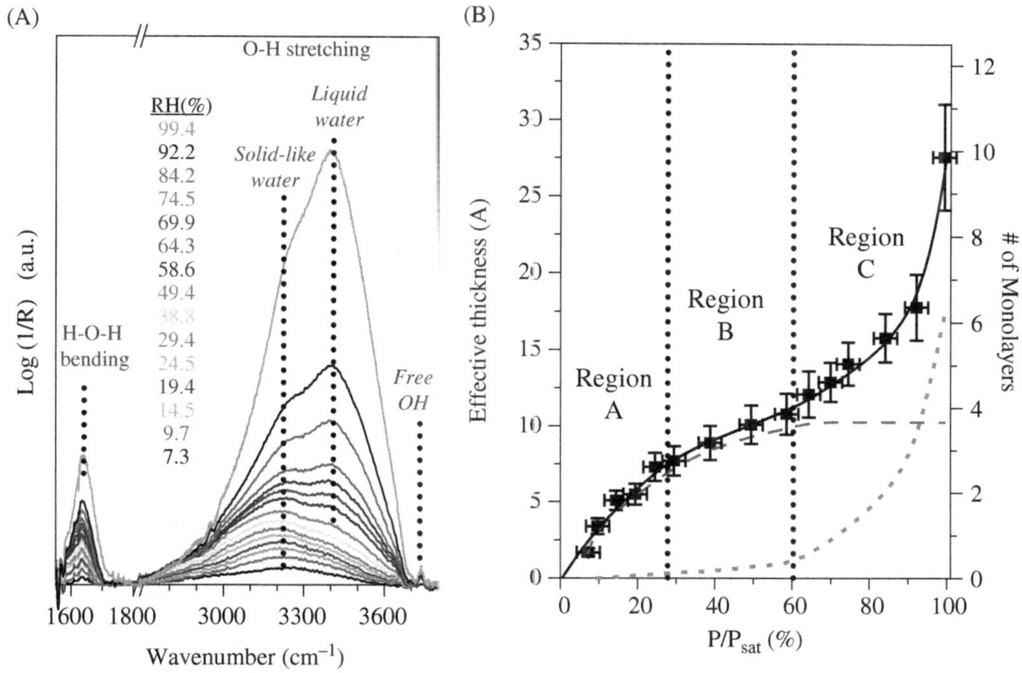

Figure 6.6 (A) ATR-IR spectra of water adsorbed on silicon oxide at various relative humidity (RH) conditions at room temperature. (B) Isotherm of water adsorption on silicon oxide. Square symbols are the total thickness of the adsorbed water layer calculated from the intensity of the H–O–H bending vibration peak. The solid line is drawn to guide eyes. The dashed and dotted lines are the thickness of the solid-like water and liquid water layers, respectively. The thickness of each component is calculated by deconvoluting the observed O–H stretching peaks into two components centered at 3230 and 3400 cm^{-1}. Replotted with the data in [10]/American Chemical Society.

Ruling out the optical artifact, the spectral shape change observed as a function of RH in Figure 6.6A can be attributed to the chemical effect. The clean silicon oxide surface is terminated with silanol groups [13]. When the water molecules adsorb on this surface, they will have hydrogen bonding interactions with those surface silanol groups. Since the silanol groups are not mobile, the hydrogen bonding interactions at the silica surface would be similar to those in bulk ice at low temperature. In fact, the OH stretch mode of water in bulk ice appears at $\sim$3230 cm^{-1} [5, 6]. So, the component centered at $\sim$3230 cm^{-1} is often assigned to the contribution from the "ice-like" or "solid-like" structure. The ATR-IR spectra of thin water layers formed in low RH conditions marked as Region-A in Figure 6.6B are dominated by this component (Figure 6.7A). As the thickness of adsorbed water layer increases, the thermal energy of the system disrupts the solid-like structure, driving to more disordered "liquid-like" structure that is equilibrated at room temperature. So, the OH stretch band of the liquid water appears in the intermediate RH condition which is marked as Region-B in Figure 6.6B. The transition from the region dominated by the solid-like structure to the region with growing contribution from the liquid-like structure appears to occur at 2–3 layers of adsorbed water (Figure 6.7A). As RH increases further above 60%, the thickness of the adsorbed water layer grows exponentially with RH (Region-C in Figure 6.6B) and the OH stretch band shape is dominated by the liquid-like feature (Figure 6.6A). This is the RH region where the driving force for condensation of liquid water becomes significant (Figure 6.7A).

The ATR-IR measurement of water adsorption isotherm can be done at different substrate temperatures and the calculated adsorbed layer thickness can be processed with the Clausius–Clapeyron equation to estimate the isosteric heat of adsorption. The ice-like structure of the adsorbed water layer in the low RH condition is found to have the isosteric heat of adsorption much higher than the heat of vaporization of liquid water ($\sim$44 kJ/mol) at room temperature, while the thicker layers dominated with the liquid-like structure has the isosteric heat similar to the latent heat of bulk water evaporation (Figure 6.7B) [14]. The corresponding isosteric entropy of adsorption for the ice-like water is lower than the entropy associated with the water condensation [14]. This finding from the ATR-IR study is in agreement with the result from near-ambient pressure x-ray photoelectron spectroscopy (NAP-XPS) measurements of water adsorption on $TiO_2(110)$ surfaces [15]. The physical property of this ice-like water layer was also found to be different from that of liquid water and induce an interfacial adhesion much stronger than what is expected from the surface tension of liquid water [16]. Also, the friction

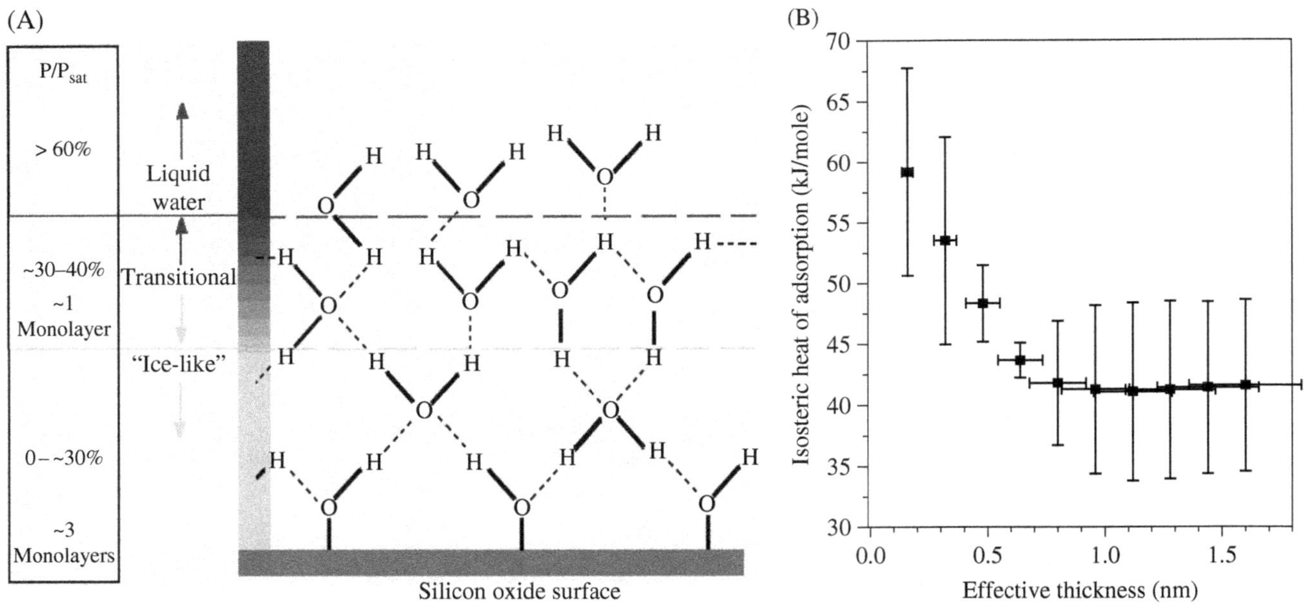

Figure 6.7 (A) Schematic illustration of RH-dependence of the adsorbed water layer structure on a clean silicon oxide surface. The regions divided by horizontal lines correspond to the regions marked in Figure 6.5B. The dotted and solid lines represent hydrogen bonds and covalent bonds, respectively (note that illustration is not to scale) [10]/American Chemical Society. (B) Isosteric heat of adsorption of water on silicon oxide calculated with the Clausius–Clapeyron relation using the isotherm data measured at different temperatures. Redrawn with the data in [14].

behavior of the adsorbed water layer with the ice-like structure was found to be different from that of the water layer with the liquid-like structure [17].

As shown in Figure 6.4, the intensity of the absorption peak in the *s*- and *p*-polarized ATR-IR spectra are different. This property can be used to find the tilt angle of the dipole moment responsible for IR absorption [18]. The ratio of the signal intensity in the *s*- to *p*-polarized spectra (A_s/A_p) is called dichroic ratio (DR):

$$DR = \frac{A_s}{A_p} = \frac{\log(1/R_s)}{\log(1/R_p)} = \frac{E_y^2 \sin^2\alpha}{E_x^2 \sin^2\alpha + 2E_z^2 \cos^2\alpha} \tag{6.16}$$

where E_x, E_y, and E_z are the amplitude of the electric fields and α is the average tilt angle of the IR transition dipole moment (see eq. (5.26)) relative to the surface-normal direction. Here, the IR incident plane is set in the xz plane and the surface is in the xy plane. So, the *s*-polarization has the electric field along the y-direction only, and the *p*-polarization has the field along the x- and z-components (see Figure 5.12). Based on the ATR-IR thin film theory, the electric components can be expressed as [7]:

$$E_x = \frac{2\cos\theta_i}{\left(1-n_{31}^2\right)^{1/2}} \frac{\left(\sin^2\theta_i - n_{31}^2\right)^{1/2}}{\left[\left(1+n_{31}^2\right)\sin^2\theta_i - n_{31}^2\right]^{1/2}} \tag{6.17}$$

$$E_y = \frac{2\cos\theta_i}{\left(1-n_{31}^2\right)^{1/2}} \tag{6.18}$$

$$E_z = \frac{2\cos\theta_i}{\left(1-n_{31}^2\right)^{1/2}} \frac{n_{32}^2 \sin\theta_i}{\left[\left(1+n_{31}^2\right)\sin^2\theta_i - n_{31}^2\right]^{1/2}} \tag{6.19}$$

where n_{31} and n_{32} are the ratios of the refractive indices study of air (n_3) to silicon (n_1), and air to water (n_2), respectively, and θ_i is the incident angle.

Using eq. (6.16), the theoretical DR can be calculated and plotted as a function of average tilt angle (α) of the IR-absorbing dipole. For the water layer probed with the Si IRE at $\theta_i = 45°$, the calculated DR is zero when $\alpha = 0°$ (perpendicular to the surface), increases in a sigmoidal shape as α increases (tilting toward the surface), and reaches 1.1 when $\alpha = 90°$ (parallel to

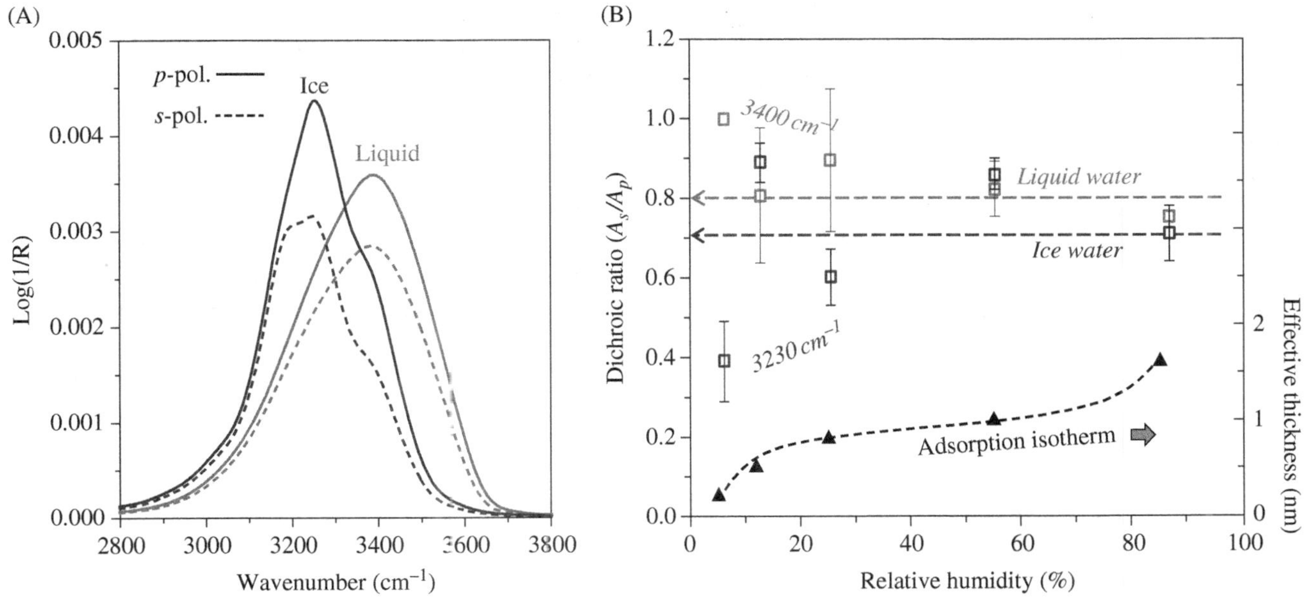

Figure 6.8 (A) Simulated *s*- and *p*-polarized ATR-IR spectra for 1-nm thick film of ice and liquid water on a silicon surface. (B) Dichroic ratio of the ice-like component at 3230 cm^{-1} and the liquid-like component at 3400 cm^{-1} in the ATR-IR spectra of water layer adsorbed on a Si IRE surface covered with native oxide. The theoretical dichroic ratios for randomly oriented molecules are shown with horizontal arrows. Replotted with the data in [12].

the surface) [12]. Figure 6.8A plots the theoretically calculated *s*- and *p*-polarized ATR-IR spectra of a 1-nm thick film of ice and liquid water on a silicon IRE [12]. While the ratio of the *s*-polarized ATR-IR intensity to the *p*-polarized one is ~0.5 for bulk water probed with the Si IRE (Figure 6.4B), it is larger (that is, the difference is smaller) for the calculated spectra of the adsorbed water layer: DR = 0.73 for the ice layer and 0.80 for the water layer. Since the refractive indices of bulk ice and liquid are used in this calculation and water molecules in these bulk phases are randomly oriented, these DR values can be assumed to be the value expected for the random orientation of water molecules in the 1-nm thick film of adsorbed water with the ice and liquid structures, respectively (which are marked with the horizontal arrows in Figure 6.8B).

In Figure 6.8B, the ice-like component in the adsorbed water layer formed at 5% RH shows a DR of ~0.4, which suggests the tilt angle of the OH stretch dipole to be ~35° from the surface-normal direction. Meantime, the liquid-like component has a DR of ~1.0, corresponding to $\alpha \approx 70°$. This high degree of anisotropy must be related to the low isosteric entropy associated with the high isosteric enthalpy of adsorption (Figure 6.7B). As the thickness of water layer increases above 1 nm in RH > 50%, the DR values of both ice-like and liquid-like components approach the value expected for the random orientation [12].

The example discussed here (Figures 6 6–6.8) shows what kind of information can be obtained through *in situ* ATR-IR study. Under the condition where the molecular species within the penetration depth (d_p) of the probe wavelength dominates over the contribution from the bulk phase, ATR-IR can be used to obtain the effective thickness and structure of the adsorbed layer, isosteric heat of adsorption, and tilt angle of molecular species in the layer. There are numerous examples employing such capabilities of ATR-IR in various systems including biological, electrochemical, geochemical, and catalytic systems [19–22]. The theoretical aspects discussed in this section will be sufficient to read and understand the analytical principles described in the literature.

6.2 Specular Reflection Infrared (SR-IR) Spectroscopy

6.2.1 External Reflection from an Absorbing Medium

The reflectance spectrum measured in the incident side of the interface (medium-1) is sensitive to the absorption of the medium-2 in the transmission side of the interface (Figure 6.1B). One may think that reflectance would be attenuated when light is reflected off an absorbing medium, as compared to the reflection from a non-absorbing medium. But this is a misconception originating from the mistake of mixing the principle associated with absorption of light *passing through the entire sample* (Figure 6.2) with the reflection of light at *the surface of the sample* (Figure 6.1B). Probably, the reason for making such a mistake is because the term "transmittance" is used in discussion of both the cases; but the transmittances defined in these two cases are different. In the external reflection (Figure 6.1B), reflectance is enhanced near the absorption band due to the resonance response of the sample. We have studied this effect in Figures 5.8 and 5.9 (see Section 5.1.4).

To see this from a different perspective, we can calculate reflectance of a hypothetical glass with and without chromophores that can absorb the irradiated light. Let us assume that the real part of the refractive index of the hypothetical glass is not affected by the chromophores in the bulk. For simplicity, the refractive index of the absorbing glass can be set to be $1.5 + i1.0$, while that of the non-absorbing glass is 1.5. Figure 6.9 compares the reflectance of *s*- and *p*-polarized light (R_s and R_p) as a function of the incidence angle (θ_i). Note that the non-absorbing case in Figure 6.9 is the same as the one in Figure 5.13B. Here we can see that both R_s and R_p are enhanced when the glass medium absorbs the light, as compared to the non-absorbing glass. In the non-absorbing case, R_p is zero at the Brewster angle (θ_p). In the absorbing case, the Brewster angle is now a complex number because the refractive index of the sample is a complex number and thus $r_p(\theta_p)$ is not zero anymore. Since R_s and R_p are enhanced at the interface of the absorbing sample, T_s and T_p (i.e., the fraction of light crossing the interface and entering the sample phase (Figure 6.1B) are decreased; the transmitted beam is further attenuated as it propagates through the sample due to absorption in the sample (Figure 6.2). Again, remember that *the Beer–Lambert law does not include the reflection and transmission at the interface of the sample (eqs. (6.8)* and *(6.9)); it considers the absorption inside the sample only (eq. (6.11)). The reflectance spectrum is a function of n,k, and θ_i, while the absorbance spectrum is a function of k*. Mathematically, it can be expressed as $R(\overline{\omega}) = f(n(\overline{\omega}), k(\overline{\omega}), \theta_i)$ and $A(\overline{\omega}) \cong f(k(\overline{\omega}), b)$, where b is the beam path length inside the sample (which is the sample thickness if the incidence angle is 0°).

One can think that the enhanced reflection of the absorbing medium is due to the "absorption and re-emission" of light by the sample. The absorption of IR induces the polarization of vibrational dipole inside the sample at the resonance frequency (ω_o) and this oscillating dipole can emit electromagnetic wave back to the external medium. Since the energy should be conserved, the frequency of the incident and reflected lights should be the same, that is, $\omega_i = \omega_r$ (see Section 5.2.1). Since the momentum (wave vector) should be conserved, the incident and reflected angles should be the same, that is, $\theta_i = \theta_r$. This is called the specular reflection. Thus, by measuring the reflectance (R) as a function of IR wavenumber ($\overline{\omega}$) at the specular reflection condition, we can obtain the vibrational spectroscopic information of the sample. Same as the ATR-IR case, the specular reflection IR (SR-IR) spectrum, $R(\overline{\omega})$, can be calculated, using eqs. (6.6)–(6.8), for any given incident angle of the experimental geometry and the refractive index of the material under consideration.

Since the IR beam intensity is attenuated as it propagates inside the sample following the Beer–Lambert law (eqs. (5.42) and (6.11)), the contribution to the enhancement of reflectance due to the polarization of the sample decreases exponentially as the distance from the surface increases. From eq. (5.90), the characteristic decay or attenuation length along the

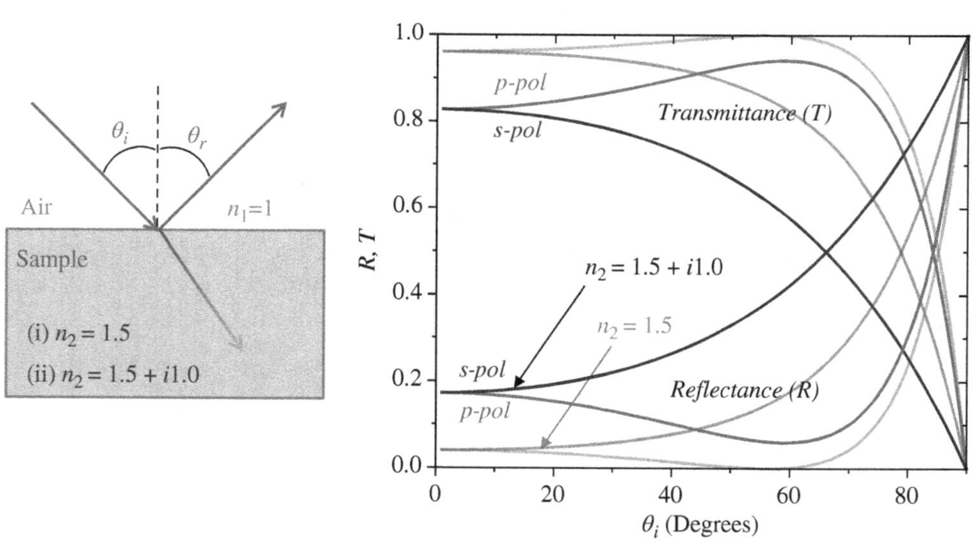

Figure 6.9 Reflectance (R_{12}) and transmittance (T_{12}), as a function of incident angle (θ_i), at the surfaces of two hypothetical glasses (medium-2): one does not absorb light (k_2 = 0; light lines) and the other strongly absorbs the irradiated light (k_2 = 1; dark lines).

propagation direction is the inverse of the absorptivity, $1/a(\varpi)$. Since SR-IR is usually done at a slanted angle, the effective probe depth from the surface can be defined as:

$$d_p = \cos\theta_i/a(\varpi) = \cos\theta_i\lambda(\varpi)/4\pi k(\varpi) \tag{6.20}$$

Then, using the same argument used in XPS (Figure 2.10) and ATR-IR (Figure 5.15), we can define the information depth of SR-IR as $3 \times d_p$. In other words, the region deeper than this distance from the surface will not be polarized much by the incoming beam and thus will not contribute to the enhanced intensity in the $R(\varpi)$ spectrum.

6.2.2 ATR-IR vs. SR-IR of Organic Materials

Although both ATR-IR and SR-IR probe the same vibrational modes of the sample, their raw spectra look quite different in the reflectance (R) unit. But, if they are normalized with the penetration depth in the $-\log(1/R)$ unit, they look similar and quite comparable. We can see this by doing theoretical calculation of the SR-IR spectra of PS and comparing the SR-IR results with the ATR-IR spectra shown in Figures 5.18 and 6.3. Figure 6.10A plots the polarized SR-IR spectra of PS (=medium-2) probed in air (=medium-1) at three incident angles (θ_i = 15°, 45°, and 75°). As we see in Figure 6.10A, the overall R_s(15°) and R_p(15°) are small and similar to each other. This is because the incident angle (θ_i = 15°) is close to the surface-normal direction (Figure 5.13B). At θ_i = 45°, R_p(45°) is much smaller than R_s(45°) and even smaller than R_s(15°). This is because the incident angle is close to the Brewster angle (θ_p = 57°). At a glancing incident angle (θ_i = 75°), R_s(75°) is very large, and R_p(75°) is also enhanced. Intensity-wide, R_p(75°) is similar to R_s(45°), but its peak shape is opposite to R_p(15°) and R_p(45°). For example, as ϖ passes through 698 cm^{-1}, R_p(75°) decreases and then increases, while R_p(15°) and R_p(45°) increase first and then decrease. This is because the phase of r_p changes as it phases through the Brewster angle (Figure 5.13A).

In the case of ATR-IR, the reflectance spectrum looks similar to the negative peaks of the $k(\varpi)$ spectrum (Figure 5.18A); so, when the ATR-IR spectrum is plotted in the log(1/R) scale, it looks similar to the $k(\varpi)$ spectrum (Figure 6.3A), and even the relative intensities among the peaks are almost identical to the $k(\varpi)$ spectrum when it is normalized with $d_{p,\text{int}}$ (Figure 6.3B). In the case of SR-IR, the reflectance spectrum looks closer to the real part, $n(\varpi)$, of the complex refractive index (Figure 5.18A). This makes sense because the magnitude of $n(\varpi)$ is larger than that of $k(\varpi)$. Since the effect of $k(\varpi)$ is smaller than that of $n(\varpi)$ in $R(\varpi)$, the absorption-like band is difficult to see in the SR-IR spectrum plotted in the log(1/R) scale (Figure 6.10B). But, if the log(1/$R(\theta_i)$) spectrum is normalized with the effective probe depth (eq. (6.20)), we get the spectrum that looks almost identical to the absorbance spectrum (Figure 6.10C). Although difficult to see due to the overlap of three spectra, the log(1/R(15°))/d_p(15°) spectrum looks almost identical to the ATR-IR spectrum normalized with $d_{p,\text{int}}$

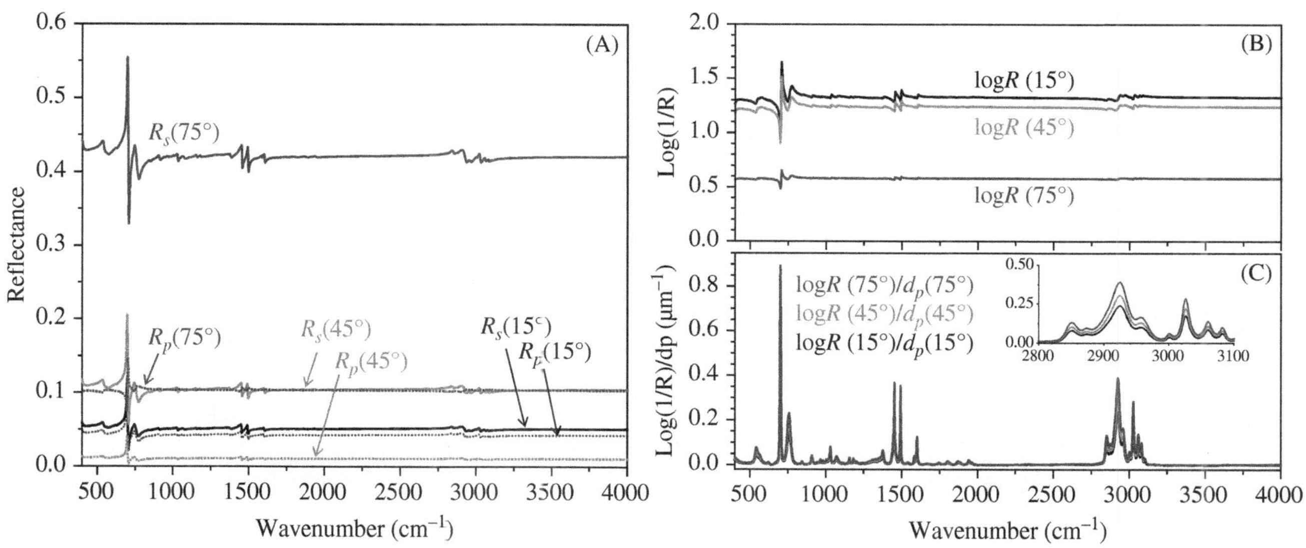

Figure 6.10 (A) Polarized SR-IR spectra of polystyrene calculated using eqs. (6.6) and (6.7) with its complex refractive index (see Figure 5.18A) at incident angles of 15°, 45°, and 75°. The incident medium is air (n_1 = 1). (B) Unpolarized SR-IR spectra plotted in the log(1/R) scale. (C) Unpolarized SR-IR spectra normalized with the effective probe depth at each incident angle.

(Figure 6.3B). For example, in Figure 6.10C, the 698 cm^{-1} intensity in the log(1/R(15°))/d_p(15°) spectrum is ~0.43 μm^{-1}, which is almost the same as the 698 cm^{-1} peak in the spectrum shown in Figure 6.3B.

6.2.3 ATR-IR vs. SR-IR of Dielectric Inorganic Materials

For organic materials, the magnitude of $k(\overline{\omega})$ of most absorption bands is smaller than that of $n(\overline{\omega})$, so their SR-IR spectra are dominated by the $n(\overline{\omega})$ contribution. But, in the case of inorganic dielectric materials, $k(\overline{\omega})$ of the absorption band is comparable to $n(\overline{\omega})$. So, the contribution from $k(\overline{\omega})$ can be seen even in the raw $R(\overline{\omega})$ spectrum plotted in the reflectance scale. Here, we are going to study how $n(\overline{\omega})$ and $k(\overline{\omega})$ affect the spectral shape of inorganic materials in SR-IR. Since they affect the ATR-IR analysis as well (as we have seen in Figure 6.4), it would be didactive to compare the ATR-IR and SR-IR spectra together. Figure 6.11A displays the ATR-IR spectra of a soda lime silica (SLS) window glass measured with diamond and germanium IREs, and Figure 6.11B exhibits the SR-IR spectra of the same glass measured at two different incident angles, 20° and 45° [24].

Although the same material (SLS window glass) is analyzed, the spectral shapes are drastically different in Figure 6.11. To an inexperienced eye, they may look like different materials. In the IR literature of silicate glass, the main peak at ~1060 cm^{-1} is assigned to the stretch mode of the Si–O–Si (bridging oxygen, BO) network, and the shoulder peak at 900~950 cm^{-1} is attributed to the stretch mode of the Si–O$^-$ (nonbridging oxygen, NBO) group disrupting the BO network [25]. In the SR-IR spectra, we can see the main band at ~1060 cm^{-1} (Figure 6.11B), but this band is missing or appears to be significantly red-shifted to ~910 cm^{-1} in the ATR-IR spectrum collected with the diamond IRE (Figure 6.11A). Based on this, one may think the difference between ATR-IR and SR-IR spectra means that the surface region of the glass probed with ATR-IR is different from the bulk region probed with SR-IR [26]. However, this is an erroneous interpretation, and such mistakes are commonly found in literature. Note that interpreting the reflectance IR without considering the convolution due to contributions from both $n(\overline{\omega})$ and $k(\overline{\omega})$ can lead to an incorrect conclusion. The rest of this section explains the reason for that using theoretical calculations.

Using eqs. (6.6)–(6.8) and the refractive index of an SLS found in the literature (Figure 6.12A) [23], we can calculate theoretical ATR-IR and SR-IR spectra. The calculation results are shown in Figure 6.12B,C [25]. Note that the refractive index of glass can vary substantially depending on the glass composition and thermal history [27]. The $k(\overline{\omega})$ spectrum shows very strong absorption bands of SLS: Si–O–Si bending mode at 460 cm^{-1} and BO stretch mode at 1030 cm^{-1} (Figure 6.12A). In these absorption band regions, the $k(\overline{\omega})$ values are comparable with $n(\overline{\omega})$. Note that the sum of these values is close to the refractive index of diamond (n_1 = 2.4). For that reason, total internal reflection condition does not meet in the strong

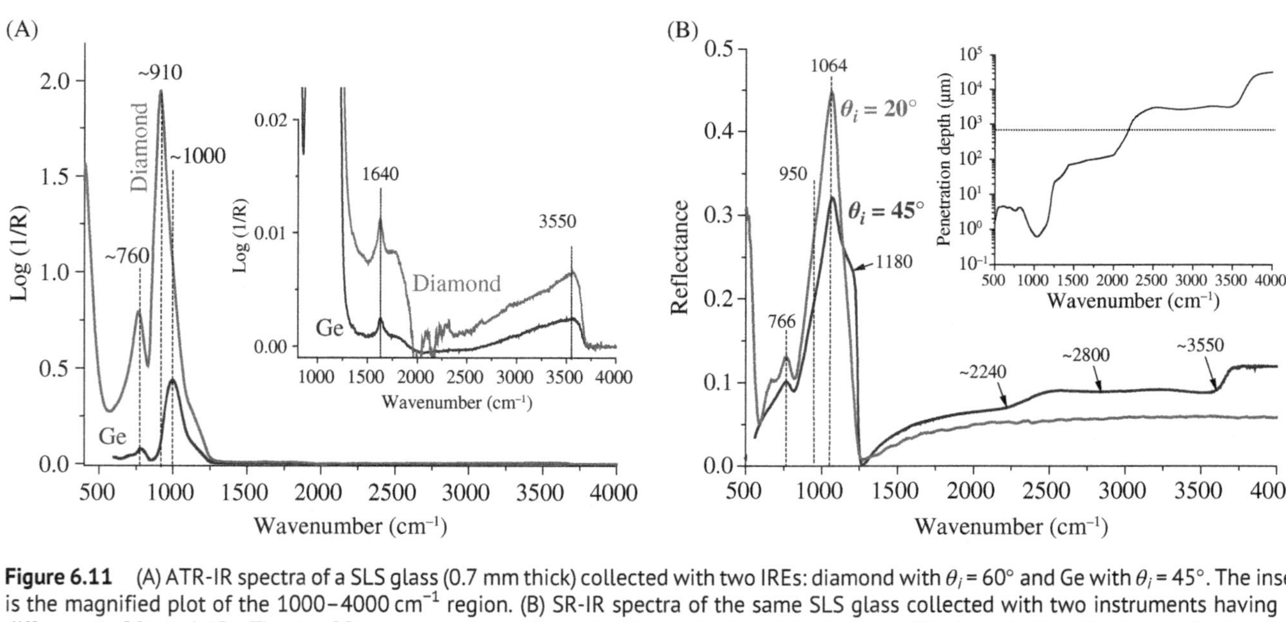

Figure 6.11 (A) ATR-IR spectra of a SLS glass (0.7 mm thick) collected with two IREs: diamond with θ_i = 60° and Ge with θ_i = 45°. The inset is the magnified plot of the 1000–4000 cm^{-1} region. (B) SR-IR spectra of the same SLS glass collected with two instruments having different θ_i: 20° and 45°. The θ_i = 20° spectrum was collected with a reflective objective lens. The inset is the effective probe depth (=characteristic attenuation depth) of IR inside the sample due to absorption, $k(\overline{\omega})$, calculated with the refractive index of an SLS found in the literature [23]. Replotted with the data in [24].

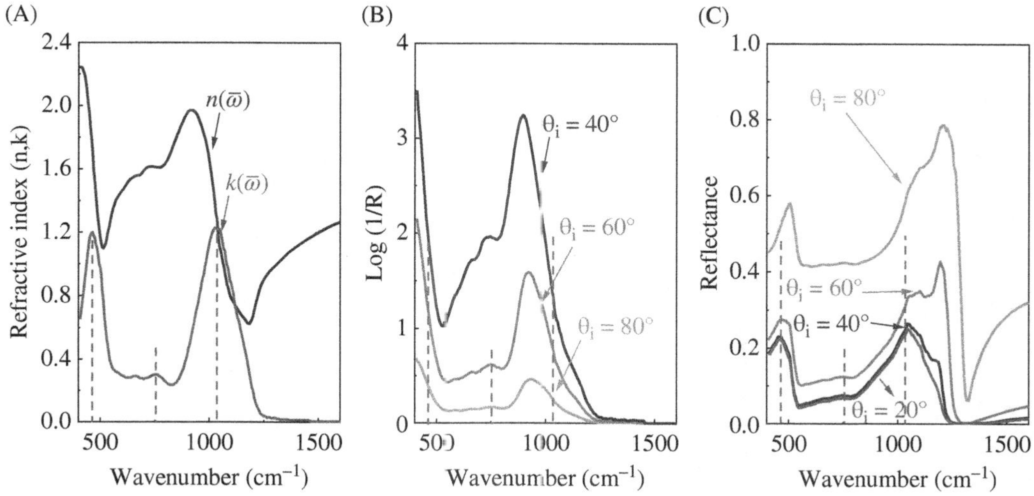

Figure 6.12 (A) Refractive index, $n(\overline{\omega}) + ik(\overline{\omega})$, of an SLS glass from [23]/with permission of ELSEVIER. (B) Calculated ATR-IR spectra of SLS/diamond at three different incident angles above the critical angle. (C) Calculated SR-IR spectra of SLS probed in air at four different incident angles. Reproduced from [25]/with permission of John Wiley & Sons.

absorption bands regions for the ATR-IR measurement of SLS with the diamond IRE. Due to the Kramers–Kronig relation, n is the maximum at the lower wavenumber side of the full-width-half-maximum (FWHM) of the absorption band (in the case of the BO stretch mode, it is ~910 cm^{-1}). At that wavenumber region, most of the IR beam passes through the IRE/sample interface and propagates into the sample, and a very little is reflected (so, $R \ll 1$). That is the reason that the log($1/R$) is the largest at ~910 cm^{-1} in the ATR-IR spectra (Figure 6.12B), not because the sample has a larger absorption in that region. Thus, the absence of the BO stretch band in Figure 6.11A should not be interpreted as the dissociation of Si–O–Si bonds, and the enhancement of the ~910 cm^{-1} band should not be interpreted as the abundance of the NBO groups.

In the calculated SR-IR spectra (Figure 6.12C), the R(20°) spectrum resembles quite closely the $k(\overline{\omega})$ spectrum. Some minor differences are the enhancement of the 950 cm^{-1} shoulder and a slight blue-shift of the maximum position of the BO stretch band to ~1060 cm^{-1}. This is due to the convolution of $n(\overline{\omega})$ and $k(\overline{\omega})$ in the reflectance spectrum. Once again, the strong positive peak at 460 and 1060 cm^{-1} in the reflectance spectrum should not be interpreted with the Beer–Lambert law concept; remember that this law is applicable only to the absorbance spectrum where the intensity is attenuated inside the sample (eq. (6.11)), and not to the reflectance spectrum where the intensity is enhanced by the resonance with the absorption band of the sample (eq. (6.8); see Figure 6.9). The Beer–Lambert law works in the negative log scale; but the SR-IR spectrum is plotted in the linear scale.

As the incident angle (θ_i) increases, the overall SR-IR intensity increases and the shoulder band at 1180–1200 cm^{-1} increases and becomes dominant at high incident angles (Figure 6.12C). In the vibrational spectroscopy literature of silica and silicate glass, this shoulder band becoming prominent at higher θ_i is often called the longitudinal optical (LO) mode. This assignment has a problem and let us take a look at why. In an ionic crystal, the lattice vibration (i.e., phonon) can be split into transverse optical (TO) and LO modes due to the long-range periodic order, and the LO mode can be observed only in the glancing incident angle with the p-polarization only. In contrast, glass is amorphous and does not have any long-range order. Thus, glass cannot have "phonon" and the TO–LO splitting concept of the crystalline lattice cannot be applied to the vibration mode of amorphous glass [28].

Probably, the reason that the side band in the high wavenumber region is called the LO mode comes from a coincidence of peak position. In a dielectric material, the reflectance of the high wavenumber side of an absorption band is enhanced due to the Reststrahlen effect (see Figures 5.8 and 5.9). The high-frequency end of this effect is often called the frequency of LO wave (ω_{LO}). When an amorphous silica is deposited on a silicon wafer and the transmission spectrum is measured at the Brewster angle with the p-polarized IR, then additional band is observed near the wavenumber *coincident with* the ω_{LO} position of silica, but the same is not observed in the s-polarized transmission spectrum; so, the additional band at ~1200 cm^{-1} was thought to be the LO mode (Figure 6.13A). Then, by comparing the SR-IR spectra collected at oblique incident angles with unpolarized IR (Figures 6.11B and 6.12C) with the p-polarized transmission-IR spectra of thin films,

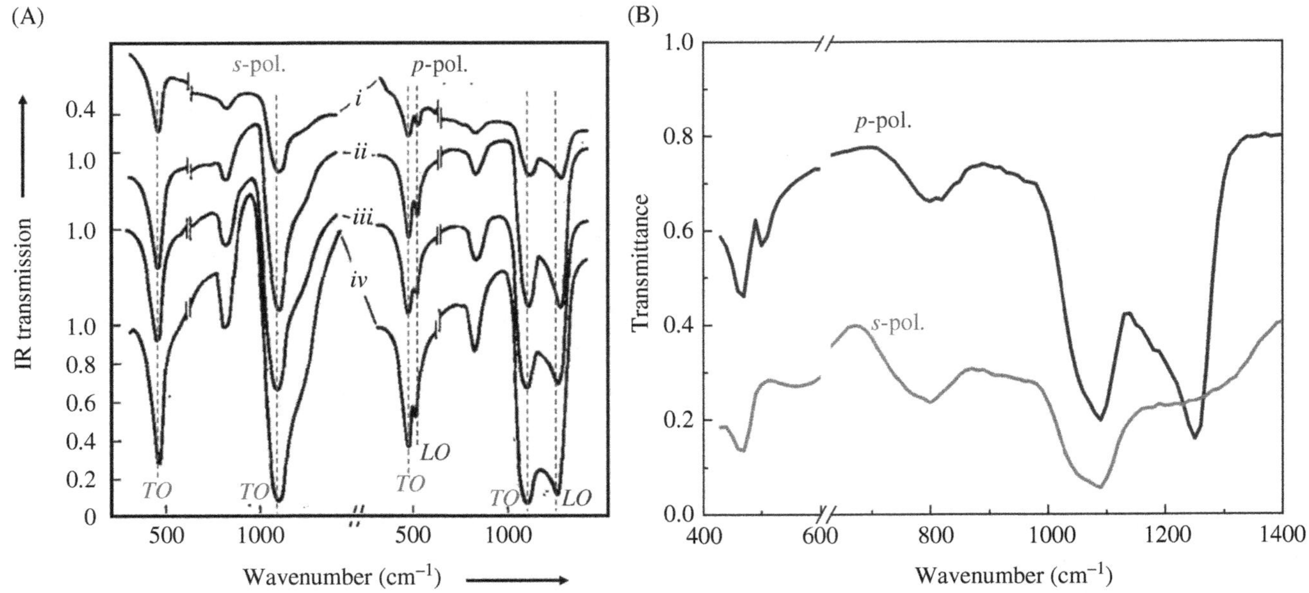

Figure 6.13 (A) Transmission IR spectra of amorphous silica films produced via (i) chemical vapor deposition, (ii) thermal oxidation, (iii) sputter deposition, and (iv) steam oxidation on a silicon substrate collected at an incidence angle of 55° with *s*- and *p*-polarizations. (B) Theoretical *s*- and *p*-polarization transmission IR spectra of a 500-nm thick silica film on a 500-μm silicon wafer calculated using the Fresnel coefficient equations and the refractive indices of silica and silicon. The calculation result is the average of calculations done for a range of incidence angles of around 55°, mimicking the real experimental setup. Reproduced from [24]/with permission of John Wiley & Sons.

it was proposed that the 1180–1200 cm^{-1} band observed in the *p*-polarized SR-IR spectrum of silica and silicate glasses could be assigned to the LO mode.

However, the enhanced peak in the *p*-polarization transmission in Figure 6.13A is solely due to the macroscopic polarization across the thin film caused by its strong absorption band, which is called the Berreman effect [29]. This effect can be seen in the theoretical transmission spectrum calculated using the refractive indices of silica and silicon, which is shown in Figure 6.13B (details of calculation method will be discussed in Section 6.4). The same cannot be observed in the SR-IR spectrum of bulk glass. In fact, if we measure the polarized SR-IR spectra of the SLS glass or theoretically calculate the $R_s(\varpi)$ and $R_p(\varpi)$ using the refractive index shown in Figure 6.12A, the enhancement of 1180–1200 cm^{-1} band is observed in both polarizations, and not in the *p*-polarization only (Figure 6.14) [25]. Thus, in SR-IR analysis of glass, this side band in the high wavenumber region of the absorption band cannot be assigned to the LO mode using the analogy of the phonon mode splitting in crystalline materials.

The origin for the reflectance enhancement in the 1180–1200 cm^{-1} region in both $R_s(\varpi)$ and $R_p(\varpi)$ can be found from the Fresnel coefficient calculations. This is the region where the $n(\varpi)$ is the smallest (Figure 6.12A) and the imaginary part of $r_s(\varpi)$ and $r_p(\varpi)$ has the largest negative value. Due to the large magnitude in $\mathrm{Im}[r_s(\varpi)]$ and $\mathrm{Im}\left[r_p(\varpi)\right]$, $R_s(\varpi)$ and $R_p(\varpi)$ become larger in this region, as θ_i increases. This explains the growth of a shoulder peak at 1180 cm^{-1} in the SR-IR spectrum of $\theta_i = 45°$ compared to that of $\theta_i = 20°$ in Figure 6.11B.

In Figure 6.11B, the reflectance is nearly zero in the 1250–1270 cm^{-1} region. This is the region where both real and imaginary parts of $r_s(\varpi)$ and $r_p(\varpi)$ of SLS are close to zero. Note that this is the region where $k(\varpi)$ is quite small and $n(\varpi)$ is close to 1 (Figure 6.12A). Because the refractive index difference from air is so small, the IR beam in the 1250–1270 cm^{-1} region passes through the interface without much reflection (Figure 6.12C).

Let us look at the spectral region above the glass network vibrations in the ATR-IR and SR-IR spectra. In Figure 6.11A, the ATR-IR spectra show a relatively sharp peak at H_2O bending region (1640 cm^{-1}) and an extremely broad peak in the OH region with the maximum intensity at 3550 cm^{-1}. These bands are due to the molecular water adsorbed on the glass surface and the hydroxyl groups in the subsurface region of the glass [30]. The OH stretch band is extremely broad because of a broad distribution of hydrogen bonding interactions of OH groups inside the glass [3, 4]. Note that the noisy signals in the 1900–2400 cm^{-1} region of the glass/diamond ATR-IR spectrum is due to the attenuation of IR beam by the diamond lattice vibration modes (so, no transmission of IR in that region).

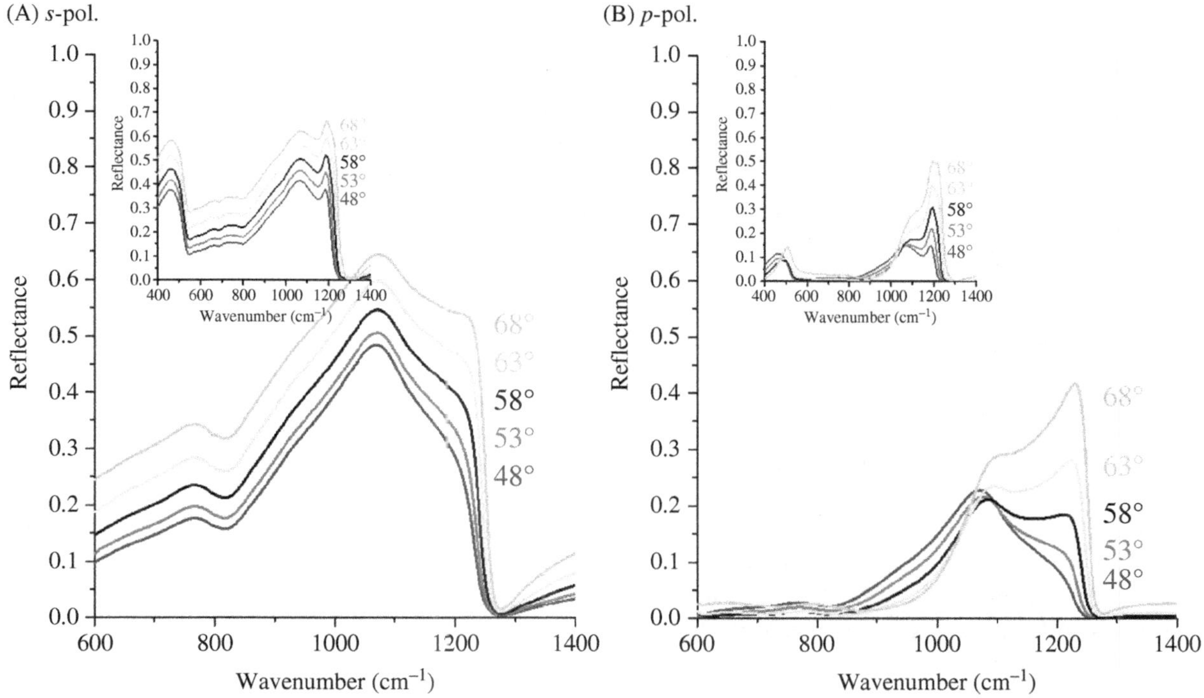

Figure 6.14 Comparison of (A) *s*-polarized and (B) *p*-polarized SR-IR spectra of a polished SLS glass at incidence angles around the Brewster angle (58°). The main panels are the SR-IR spectra measured with an SLS glass, and the insets are the $R_s(\overline{\omega})$ and $R_p(\overline{\omega})$ spectra simulated using the refractive index shown in Figure 6.12A. Reproduced from [25]/with permission of John Wiley & Sons.

In the SR-IR spectrum collected at $\theta_i = 45°$ in Figure 6.11B, we can see that there is a broad OH stretch band peaked at ~3550 cm^{-1}, but this peak is in the negative side of the baseline. Then, one may ask why the resonance band appears as a negative peak in the SR-IR spectrum, which is opposite to what we have discussed earlier in this section. The inset to Figure 6.11B plots the characteristic attenuation depth of IR (d_p) as a function of $\overline{\omega}$ (eq. (6.20)). When the wavenumber is higher than ~2240 cm^{-1}, d_p becomes larger than the thickness of the sample used (marked with a dotted line in the figure). That means, the reflection from the backside of the sample also reaches the detector in the incident medium. Thus, the spectral region higher than 2240 cm^{-1} is dominated by the absorption of IR traveling through the sample and then reflected by the backside of the sample. If the backside is roughened or in contact with a liquid with the same refractive index or the sample is very thick (thicker than the IR attenuation depth), the backside reflection contribution disappears, and the SR-IR spectrum does not show any intensity change at ~2240 cm^{-1} [24]. This explains the reason that there is a sudden step near 2240 cm^{-1} in the SR-IR spectrum of the 0.7-μm thick SLS sample and the absorption band of the internal OH groups appears in the negative shape. In Figure 6.11B, the SR-IR spectrum collected at $\theta_i = 20°$ does not have a step change near 2240 cm^{-1} region. Note that, as mentioned in the caption, this spectrum was collected with an IR microscope equipped with a reflective objective lens [24]. Since the focal depth of this objective lens was much smaller than the thickness of the sample, the back reflection signal was naturally discriminated.

All these details that we have reviewed in Figures 6.11 and 6.14 show the red-flag against interpreting the spectral features in reflectance spectrum using the simple Beer–Lambert law concept without considering the convolution effects of $n(\overline{\omega})$ and $k(\overline{\omega})$ as well as the effective probe depth. If the SR-IR spectra shown in Figure 6.12C is plotted in the log(1/*R*) scale normalized with d_p (eq. (6.20)), then the spectral features look similar to the $k(\overline{\omega})$ spectrum (same as the example shown in Figure 6.10C). Such normalization process is possible only when the complex refractive index of sample is known. For an unknown sample, the refractive index should be determined independently. In theory, the $k(\overline{\omega})$ spectrum can be obtained from the $R(\overline{\omega})$ spectrum using the Kramers–Kronig (KK) transformation. Many commercial instruments have this function in the operation software, but they assume that the data are collected at the normal incident angle ($\theta_i = 0°$), although the actual measurement is done at an oblique angle ($\theta_i > 10°$). It is because there are two unknowns to determine ($n(\overline{\omega})$ and $k(\overline{\omega})$) from one spectrum. This limitation can be circumvented if $R(\overline{\omega})$ is measured at least two incident angles and these two spectra are numerically solved with the Fresnel coefficient equations [27]. Alternatively, one can use IR variable-angle spectroscopic ellipsometry, which is explained in Chapter 9.

Even if such calculations are not possible, spectral interpretation is possible with keeping the following rule of thumb in mind: due to the convolution of $n(\varpi)$ in the SR-IR measurement, the maximum peak position in the $R(\varpi)$ spectrum is always blue-shifted from that in the k-spectrum (as we have seen in Figure 6.12C), and the degree of this shift varies depending on the sharpness of the absorption band as well as the incidence angle in the SR-IR measurement. And the intensity in R of SR-IR (and log(1/R) of ATR-IR as well) should not be interpreted using the Beer–Lambert law concept.

6.3 Reflection Absorption Infrared Spectroscopy (RAIRS)

6.3.1 RAIRS of a Thin Layer on a Flat Surface

So far, we have used the Fresnel coefficient equations to describe the reflection at a single interface of two media with different refractive indices (Figures 6.1A,B). If an adsorbate layer is formed or a thin film is applied on a substrate (Figure 6.1C), then we must deal with two parallel interfaces – one is the air/overlayer interface (between medium-1 and -2) and the other is the overlayer/substrate interface (between medium-2 and -3). These two interfaces are separated by the thickness of the medium-2 (d_2). This is schematically illustrated in Figure 6.15.

The electromagnetic wave propagating in medium-1 toward these interfaces can be described as (eq. (5.58)):

$$E_i(r,t) = E_o \exp\left[-i\left(\omega t - \mathbf{k}\cdot\vec{r}\right)\right] \tag{6.21}$$

where E_o is the amplitude of the field, ω is the angular frequency of the wave, r is the position along the propagation direction, $\mathbf{k}$ is the wavevector which is defined as $\mathbf{k} = 2\pi\tilde{n}/\lambda$. The phase of this beam is denoted as $\phi = -\mathbf{k}\cdot r$. The beams reflected from the top and bottom of the medium-2, which can be denoted as E_{r1} and E_{r2}, respectively, will have a phase difference due to the difference in travel distance and phase velocity. The phase difference (β) between E_{r1} traveling from point-A to point-B and E_{r2} traveling from point-A to point-D is:

$$\beta = \mathbf{k}_2\cdot r_2 - \mathbf{k}_1\cdot r_1 = \frac{2\pi}{\lambda}(\tilde{n}_2\cdot r_2 - \tilde{n}_1\cdot r_1) \tag{6.22}$$

Using the trigonometric relations, we can find:

$$r_2 = \frac{d_2}{\cos\theta_2} \tag{6.23}$$

$$r_1 = \sin\theta_1\cdot y = \sin\theta_1\cdot(\sin\theta_2\cdot r_2) = \frac{\sin\theta_1\sin\theta_2\, d_2}{\cos\theta_2} \tag{6.24}$$

By inserting eqs. (6.23) and (6.24) into eq. (6.22) and simplifying the equation using Snell's law, we get:

$$\beta = \frac{2\pi d_2}{\lambda}\tilde{n}_2\cos\theta_2 = \frac{2\pi d_2}{\lambda}\sqrt{\tilde{n}_2^2 - \tilde{n}_1^2\sin^2(\theta_1)} \tag{6.25}$$

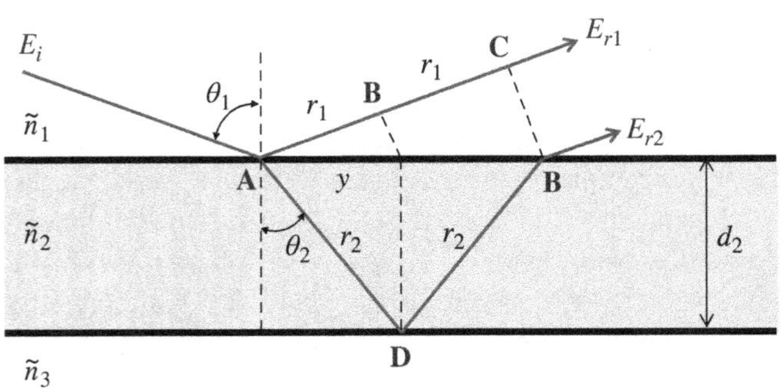

Figure 6.15 Reflection of an incident electromagnetic wave E_i. It is reflected and transmitted at the $\tilde{n}_1/\tilde{n}_2$ interface; the transmitted beam is reflected at the $\tilde{n}_2/\tilde{n}_3$ interface and then transmitted at the $\tilde{n}_2/\tilde{n}_1$ interface.

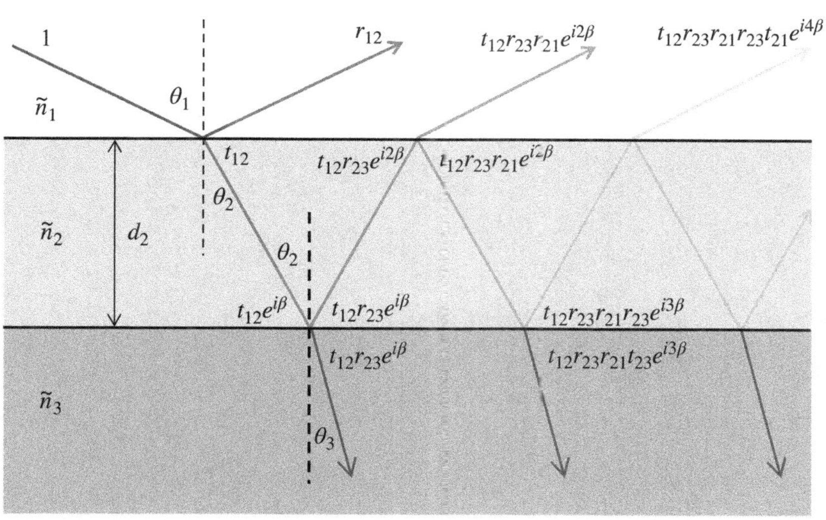

Figure 6.16 Illustration of a three-medium optical system ($\tilde{n}_1/\tilde{n}_2/\tilde{n}_3$), where an incident wave (whose intensity is normalized to 1) is being reflected and transmitted due to the boundaries separated by d_2.

The phase difference between E_{r1} traveling from point-B to point-C and E_{r2} traveling from point-C to point-B is the same as β.

When the reflectance is measured, it has the contribution from all beams reflected and refracted multiple times between two interfaces separated by distance d_2, which is illustrated in Figure 6.16. At each interface, we can set up Snell's law (eq. (6.1)) and Fresnel coefficients (eqs. (6.2)–(6.5)). The phase shift due to propagation through a medium with a different refractive index (thus at a different phase velocity; eq. (5.18)) can be described by the $e^{i\beta}$. The fraction of the incident E_o-field reflected and transmitted at the $\tilde{n}_1/\tilde{n}_2$ interface is r_{12} and t_{12}, respectively. The transmitted beam arriving at the $\tilde{n}_2/\tilde{n}_3$ interface is $t_{12}e^{i\beta}$. The fraction of this beam reflected and transmitted at that interface, with respect to the incident field, is $t_{12}r_{23}e^{i\beta}$ and $t_{12}t_{23}e^{i\beta}$, respectively. The reflected beam arrives at the $\tilde{n}_2/\tilde{n}_1$ interface with another phase shift, so it becomes $t_{12}r_{23}e^{i2\beta}$. Then, the fraction of this beam transmitted to the $\tilde{n}_1$ medium and reflected back to the $\tilde{n}_2$ medium is $t_{12}r_{23}t_{21}e^{i2\beta}$ and $t_{12}r_{23}r_{21}e^{i2\beta}$, respectively. This process continues until the electric field is completely attenuated. Thus, the total reflection coefficient (r_{123}) of two interfaces of the $\tilde{n}_1/\tilde{n}_2/\tilde{n}_3$ system can be written in an infinite series:

$$r_{123} = r_{12} + t_{12}r_{23}t_{21}e^{2i\beta} + t_{12}r_{23}r_{21}r_{23}t_{21}e^{4i\beta} + \cdots = r_{12} + t_{12}r_{23}t_{21}e^{2i\beta}\left[\sum_{j=0}^{\infty}\left(r_{21}r_{23}e^{2i\beta}\right)^j\right] \tag{6.26}$$

Since the magnitude of $r_{21}r_{23}e^{2i\beta}$ is <1, we can use the following relationship for the sum of the geometric progression:

$$\sum_{j=0}^{\infty} q^j = \frac{1}{1-q} \text{ when } 0 < q < 1 \tag{6.27}$$

Then, eq. (6.26) becomes:

$$r_{123} = \frac{r_{12} + r_{23}(t_{12}t_{21} - r_{12}r_{21})e^{2i\beta}}{1 - r_{21}r_{23}e^{2i\beta}} \tag{6.28}$$

We can simplify this equation using the Stokes relations [31], which originate from the time-reversal symmetry of Maxwell's equations. Figure 6.17 can be used to understand the Stokes relations. The first panel shows the beam (E) propagating forward in time. Let us envision the beam running backwards in time, so that there are two incident beams of amplitudes, $r_{12}E$ and $t_{12}E$. The $r_{12}E$ beam will be reflected and transmitted at the interface, which can be described as r_{12}^2E and $t_{12}r_{12}E$, respectively. Similarly, the $t_{12}E$ beam will be reflected and transmitted at the interface, which can be described as $r_{21}t_{12}E$ and $t_{21}t_{12}E$, respectively. Then, the time-reversal symmetry depicts the following relationships:

$$r_{12}^2E + t_{21}t_{12}E = E \tag{6.29}$$

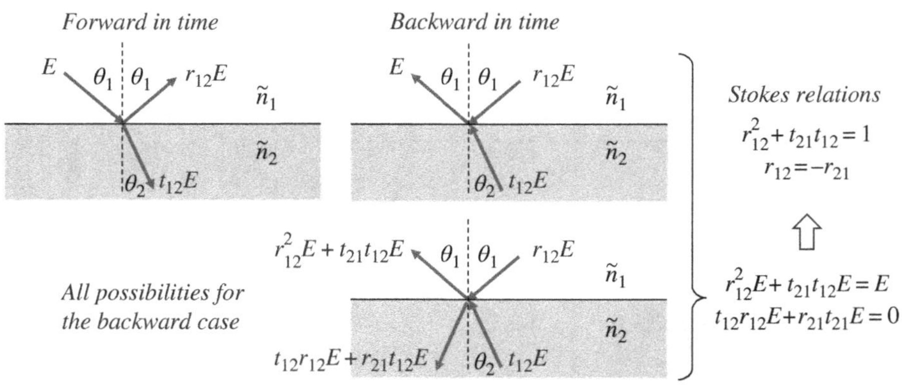

Figure 6.17 Time-reversal argument in Stokes relations.

$$t_{12}r_{12}E + r_{21}t_{12}E = 0 \tag{6.30}$$

Canceling the common terms in the left- and right-hand sides, we find the Stokes relations:

$$r_{12}^2 + t_{21}t_{12} = 1 \tag{6.31}$$

$$r_{12} = -r_{21} \tag{6.32}$$

These relations connect the r_{12} and t_{12} coefficients for the beam originating in the $\tilde{n}_1$ medium and the r_{21} and t_{21} coefficients for the beam originating in the $\tilde{n}_2$ medium. Using eqs. (6.31) and (6.32), we can simplify eq. (6.28) to the following form:

$$r_{123} = \frac{r_{12} + r_{23}e^{2i\beta}}{1 + r_{12}r_{23}e^{2i\beta}} \tag{6.33}$$

Note that if there is only one interface, then the r_{23} term drops out and eq. (6.33) becomes eq. (6.2) for *s*-polarization and eq. (6.3) for *p*-polarization. This equation retains the same form for widely diverse conditions, including the case of strongly absorbing medium-2 and -3, or total internal reflection at either boundary [32].

The r_{123} calculation should be done separately for *s*- and *p*-polarizations (i.e., $r_{s,123}$ and $r_{p,123}$) using the corresponding Fresnel coefficients (eqs. (6.2)–(6.3)) and proper incident and transmission angles at each interface (θ_1 and θ_2 for the $\tilde{n}_1/\tilde{n}_2$ interface and θ_2 and θ_3 for the $\tilde{n}_2/\tilde{n}_3$ interface, which are calculated with eq. (6.1)). Same as in the single-interface system, the reflectance of *s*- and *p*-polarizations at the $\tilde{n}_1/\tilde{n}_2/\tilde{n}_3$ two-interface system can be obtained by multiplying r_{123} with its conjugate (same as eqs. (6.6) and (6.7)):

$$R_{s,123} = r_{s,123} \cdot r_{s,123}^* \tag{6.34}$$

$$R_{p,123} = r_{p,123} \cdot r_{p,123}^* \tag{6.35}$$

Equations (6.34) and (6.35) calculate the theoretical spectra of the *s*- and *p*-polarized RAIRS measurement for the sample geometry shown in Figure 6.1C. The total reflectance of unpolarized beam is the average of these *s*- and *p*-polarizations (same as eq. (6.9)). Note that eq. (6.33) is also used to calculate the phase difference of the *s*- and *p*-polarization reflections, which is the core principle of ellipsometry that is discussed in Chapter 9.

In ATR-IR, the reflectance spectrum is plotted in the $\log(1/R)$ or $-\log R$ scale, because it gives the positive peaks in the spectrum which can be compared with the absorbance intensity if it is normalized with the proper penetration depth (see Figure 6.3). In SR-IR, the reflectance spectrum is typically plotted in the linear scale (as shown in Figures 6.11B, 6.12C, and 6.14). Again, if the SR-IR spectrum is plotted in the $-\log R$ scale and then normalized with the characteristic attenuation depth, it gives a spectrum that can be compared with the absorbance spectrum (see Figure 6.10). For the same reason, the RAIRS spectrum is typically plotted in the log scale. In the ATR-IR and SR-IR, the blank spectrum is measured first and then used to normalize the beam profile (emission spectrum of the IR source and the gas-phase water and CO_2 absorption in the beam path). So, even if that process is not mentioned in the equation, it is already included in the measurement step. In the case of RAIRS, the blank spectrum is the reflectance of the clean substrate (without medium-2 in Figures 6.15 and 6.16), which can be denoted as R_o. So, the RAIRS spectrum is plotted in the $-\log(R/R_o)$. If the spectral change is very small (which is true when the overlayer is thin and much smaller than the wavelength of the IR probe beam), this expression can be simplified in the following form:

$$-\log\left(\frac{R}{R_0}\right) = -\log\left(\frac{R_{123}}{R_{13}}\right) \cong 1 - \frac{R_{123}}{R_{13}} = \frac{R_{13} - R_{123}}{R_{13}} = \frac{\Delta R}{R_o} \tag{6.36}$$

Recall that, for the $\tilde{n}_1/\tilde{n}_2/\tilde{n}_3$ two-interface system, the sample spectrum (R) is $R_{s,123}$ and $R_{p,123}$, and the blank spectrum (R_o) is $R_{s,13}$ and $R_{p,13}$ (because the clean substrate means the absence of the $\tilde{n}_2$ layer).

6.3.2 RAIRS of a Thin Organic Layer on Dielectric Materials

Now, let us see how this new equation that we just derived is used in RAIRS analysis of thin films on dielectric substrates. For comparison with ATR-IR and SR-IR, we will consider a 1-nm thick PS film on a silicon (Si) wafer and a SLS glass. Then, we can reuse the refractive indices of these materials that we have used in Figures 6.3, 6.10, 6.12, and 6.14. The top panel of Figure 6.18A shows the difference signal ΔR ($= R_o - R$) of s- and p-polarizations calculated for the 2925 cm^{-1} peak (CH_2 asymmetric stretch mode) of a 1-nm thick PS film on Si. The s-polarization signal ($\Delta R_s = R_{s,13} - R_{s,123}$) is negative at all incident angles, which means that the s-reflectance of the air/1-nm PS/Si interface is larger than that of the air/Si interface. Its magnitude is the largest at the near-normal incident angle ($\theta_1 \approx 0$), and remains constant up to $\theta_1 \approx 45°$, and then monotonically decreases to zero as the incident angle increases to 90°. In contrast, the p-reflectance ($\Delta R_p = R_{p,13} - R_{p,123}$) exhibits more complicated dependence on the incident angle. It starts from the same negative value as the s-reflectance at the near-normal incidence and then its magnitude decreases fast as θ_1 increases from zero and becomes zero at $\theta_1 \approx 43°$. As θ_1 increases further, $R_{p,13} - R_{p,123}$ becomes positive and its magnitude reaches a positive maximum value at $\theta_1 \approx 61°$ and then decreases to zero again near the Brewster angle of the air/Si interface. Then, $R_{p,13} - R_{p,123}$ reaches a negative maximum value at $\theta_1 \approx 86°$ and decreases back to zero as θ_1 approaches 90°. This complicated pattern is due to the different incident

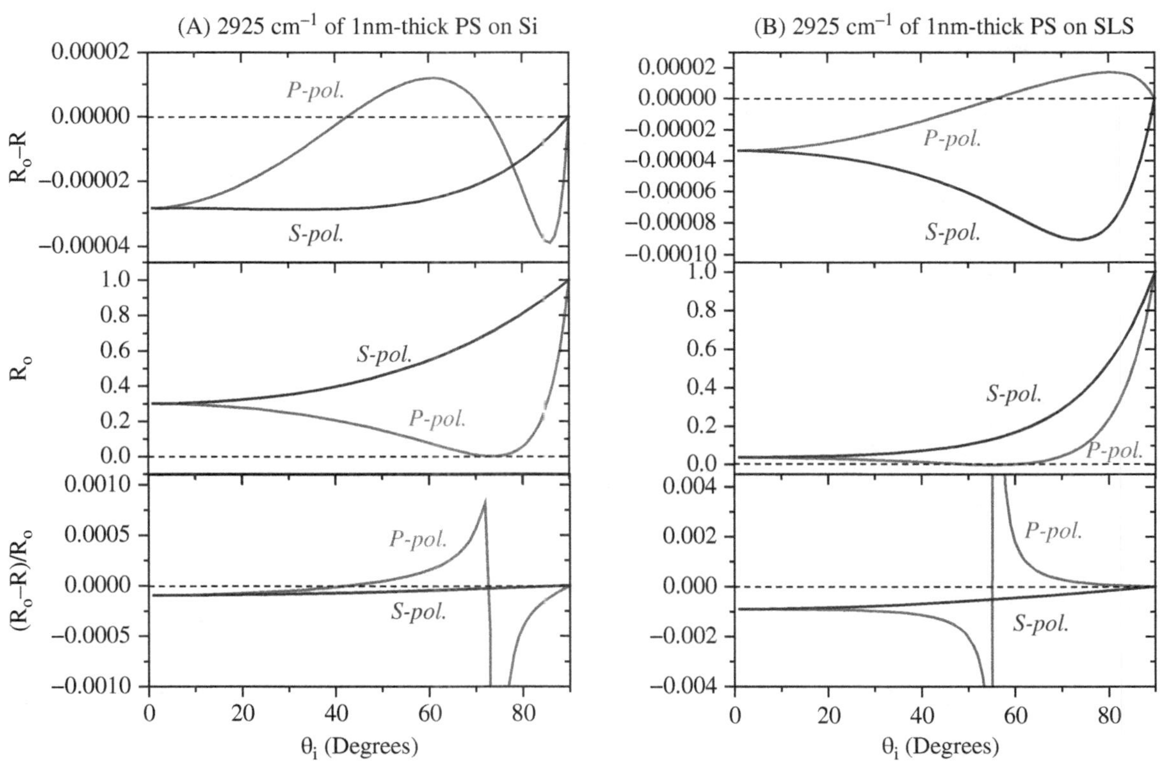

Figure 6.18 Reflectance difference between clean and thin film samples (ΔR), reflectance of clean sample (R_o), and normalized signal ($\Delta R/R_o$=− log(R/R_o)) of s- and p-polarized IR calculated for the CH_2 asymmetric stretch peak (2925 cm^{-1}) of a 1-nm thick PS film on (A) Si and (B) SLS substrates. The refractive indices used are the same as the ones used in refractive indices of these materials in previous calculations shown in Figures 6.3, 6.10, 6.12, and 6.14.

angle dependence of the two components probed by the p-polarized light within the incident (XZ) plane – one is the surface-parallel component probed by the E_x field and the surface-normal component probed by the E_z field [33, 34].

The middle panel of Figure 6.18A shows the $R_o(\theta_i)$ of the air/Si interface, which is the same as the SR-IR spectra of the air/Si interface. The bottom panel plots the incident angle dependence in the logarithmic scale, $-\log\left(\frac{R}{R_o}\right) = \Delta R/R_o$. Note that although this is dimensionally equivalent to absorbance, it is not the same as the absorbance in the transmission experiment described in eq. (6.11). In this scale, the $\Delta R_s/R_{s,o}$ signal appears to be extremely small, but this is mainly due to the large magnitude of $R_{s,o}$. Near the Brewster angle of air/Si interface, the $\Delta R_p/R_{p,o}$ magnitude is extremely large because $R_{p,o}$ approaches zero, and its sign changes from the positive to the negative as θ_1 passes the Brewster angle. From this theoretical calculation, it is obvious that the $\Delta R_p/R_{p,o}$ measurement is difficult (since the signal is nearly zero) if the incident angle is about 45° when the PS thin thickness on a Si substrate is 1 nm.

In the case of the 1-nm thick PS film on an SLS glass substrate (Figure 6.18B), ΔR_s is negative and shows a maximum at $\theta_1 \approx 74°$, while ΔR_p starts as a negative value at near-zero incident angle, becomes zero at the Brewster angle ($\theta_1 \approx 56°$) and then stays positive above this angle. Again, the sign change is due to the different incident angle dependence of the surface-parallel and -normal components probed by the E_x and E_z fields, respectively, within the incident plane. When the RAIRS data is plotted in the logarithmic scale after normalization with the reflectance of the clean surface (bottom panel of Figure 6.18B), the s-reflectance is negative, and its magnitude decreases monotonically as the incidence angle increases. The p-reflectance shows the inversion of sign at the Brewster angle and its value is largest just below and above the Brewster angle because $R_{p,o}$ is the smallest near the Brewster angle.

To produce the full RAIRS spectrum, the calculation shown in Figure 6.18 can be done for the entire spectral range (i.e., 400–4000 cm^{-1}) at a given incident angle. This is shown in Figure 6.19 for the 1-nm thick PS film on the Si and SLS surfaces. When the incident angle is small ($\theta_1 = 15°$), the $\Delta R_s/R_{s,o}$ and $\Delta R_p/R_{p,o}$ are almost identical (top panels of Figure 6.19). In the case of the 1-nm PS/SLS, the large variance in the baseline between 400 and 1400 cm^{-1} results from the change in the reflectance signal in the SLS absorption band region due to the presence of the 1-nm thick dielectric layer. At the intermediate

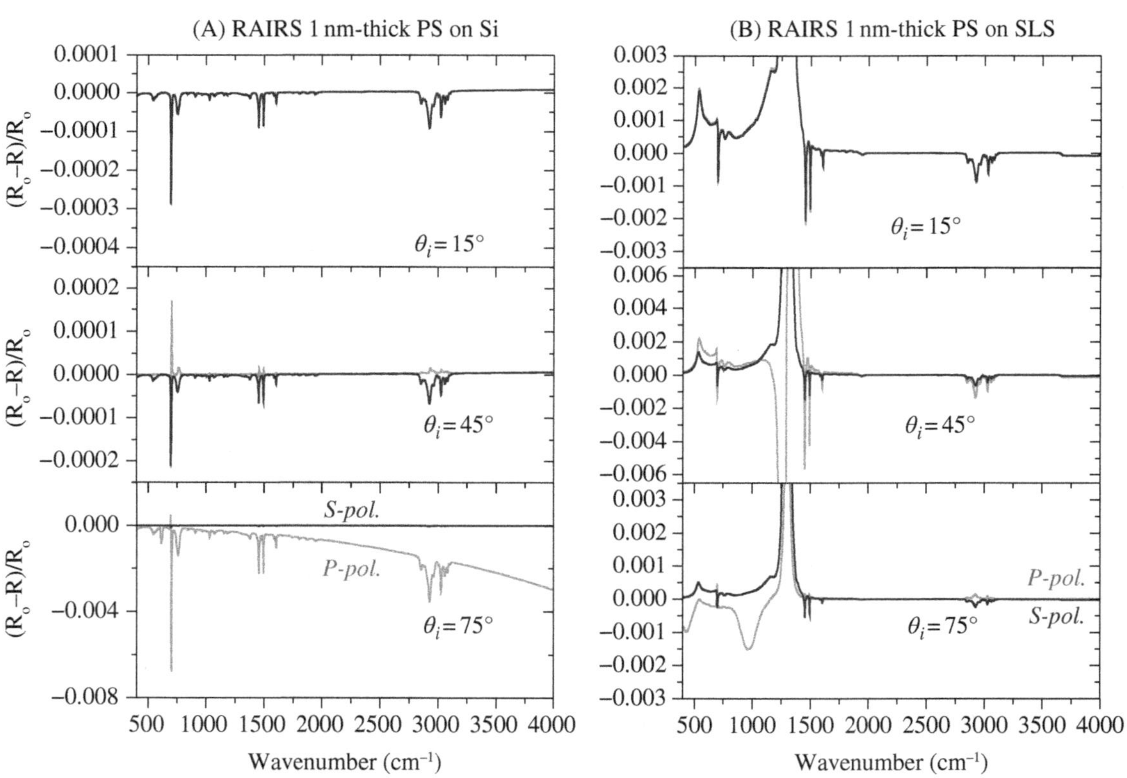

Figure 6.19 RAIRS spectra of a 1-nm thick PS film on (A) Si and (B) SLS substrates calculated for the incident angle of 15°, 45°, and 75°. The refractive indices used are the same as the ones used in refractive indices of these materials in previous calculations shown in Figures 6.3, 6.10, 6.12, and 6.14.

incident angle ($\theta_1 = 45°$; middle panel of Figure 6.19A), the Si substrate shows a positive $\Delta R_p/R_{p,o}$ spectrum (sign inversion as predicted in Figure 6.18A), while the $\Delta R_s/R_{s,o}$ still shows a negative spectrum. At this incident angle, the $\Delta R_p/R_{p,o}$ spectrum of the 1-nm PS/SLS looks larger than the $\Delta R_s/R_{s,o}$ spectrum (middle panel in Figure 6.19B), although ΔR_p is smaller than ΔR_s (as shown in the top panel of Figure 6.18B). This is simply because the $R_{p,o}$ (45°) signal is smaller because the incident angle is close to the Brewster angle. When the incident angle is large ($\theta_1 = 75°$), the $\Delta R_p/R_{p,o}$ of the 1-nm PS/Si shows large negative intensities (bottom panel of Figure 6.19); again, this is simply because the $R_{p,o}(75°)$ signal is so small near the Brewster angle. The actual difference in the reflectance is larger for the *s*-polarization, that is, $\Delta R_s(75°) > \Delta R_p(75°)$ (which can be seen in the top panel of Figure 6.18A).

Often, the signal intensity in RAIRS plotted in the $-\log\left(\frac{R}{R_o}\right)$ or $\Delta R/R_o$ scale is interpreted with the Beer–Lambert law concept (because it is dimensionally equivalent to "absorbance"); again, this practice is likely to lead to an incorrect interpretation. The RAIRS signal intensity of an organic thin film on a dielectric substrate material does not necessarily increase in proportion to the thickness of the film. Figures 6.20 plots the theoretical RAIRS intensity of the 2925 cm^{-1} peak of PS on the Si substrate, as a function of IR incident angle (ϑ_i), for the film thickness of 1, 10, 20, and 40 nm. The inversion pattern of the ΔR_p signal from negative to positive and then back to negative sign as (θ_i) increases from 0° to 90° can be seen for 1- and 10-nm thick films; but that pattern is changed for the films thicker than 20 nm. When the PS film thickness is 40 nm, then both ΔR_p and ΔR_s signals are positive at small incident angles.

Figure 6.21 plots the RAIRS intensity of the 2925 cm^{-1} peak of PS on the Si substrate, calculated for three incident angles, as a function of the PS film thickness. Note that when θ_i is 15°, the 2925 cm^{-1} signal is nearly zero when the PS film thickness is ~18 nm; this is a consequence due to the inversion of RAIRS signal phase as the film thickness increases from 0 to >20 nm. Similarly, at $\theta_i = 45°$, the *s*-polarized RAIRS signal becomes zero when the film thickness is ~20 nm; in contrast, the *p*-polarized signal is always positive at $\theta_i = 45°$, but its increase is not linear to the film thickness. At $\theta_i = 75°$, the normalized *s*-polarized RAIRS signal appears to be nearly zero, while the normalized *p*-polarized signal is very large. The absolute difference is indeed larger for ΔR_p than ΔR_s; this difference is amplified in the normalized signal (i.e., $\Delta R_p/R_{p,0}$) because $R_{p,0}$ is nearly zero at $\theta_i = 75°$.

These examples clearly show that the RAIRS spectra of organic layers on dielectric materials should not be interpreted with the Beer–Lambert law. In other words, *without comparing the experimental spectra with the theoretical spectra calculated with eq. (6.36), simply using the Beer–Lambert law concept is not adequate for quantitative analysis of RAIRS spectra for dielectric materials.*

6.3.3 RAIRS (or SR-IR) of a Dielectric Material After Environmental Exposure

Being able to theoretically calculate or predict the RAIRS spectrum of the system under consideration is extremely important for proper interpretation of spectral features; otherwise, it is likely to make a mistake by interpreting experimental artifacts as meaningful signals or peaks. Note that the experimental geometry of RAIRS is technically the same as that of SR-IR discussed in Section 6.2; both measure the reflectance (intensity) change at the specular reflection angle. SR-IR is described with eqs. (6.6) and (6.7) which deal with only one interface. If a surface layer is formed at the dielectric material interface due to exposure to reactive environmental conditions, its chemical composition or structure will be distinct from the bulk phase, which leads to a change in refractive index. Then, the SR-IR spectrum cannot be described with eqs. (6.6) and (6.7) anymore; it should be described with eqs. (6.34) and (6.35).

An example of such a case is shown in Figure 6.22. When an SLS glass is exposed to aqueous solution, sodium ions can leach out (see Figure 6.22A) [35]. Since sodium ions are the network modifier in the SLS glass network, its loss is inevitably accompanied by changes in chemical and physical properties [36, 37]. SR-IR has been widely used to analyze structural changes in the glass network by detecting changes in vibrational spectral features. Quite often, the leaching of sodium ions results in a small increase in the OH stretch band region (2500–3600 cm^{-1}) and a notable increase in the shoulder peak (~900 cm^{-1}) of the Si–O–Si (BO) stretch band (see Figure 6.22B) [35]. As mentioned in Section 6.2.3, the 900–950 cm^{-1} shoulder band of SLS is often attributed to the Si-NBO stretch [25]. This assignment originated from the observation that upon increasing the alkali ion concentrations in alkali silicate glass, the 900–950 cm^{-1} intensity increases in SR-IR analysis [38]. Then, using the Beer–Lambert law concept, this 900–950 cm^{-1} band was thought to be the stretch mode of the $Si–O^-$ (NBO) group. But it should be noted that the experimental data were the reflectance spectra (not even in the – *log* (*R*) scale); thus, the signal intensity will not be linearly proportional to the $Si–O^-$ (NBO) concentration. Nonetheless, this interpretation has been prevailing in literature. And, since the reduced mass of Si–OH would be similar to that of $Si–O^-$ (NBO),

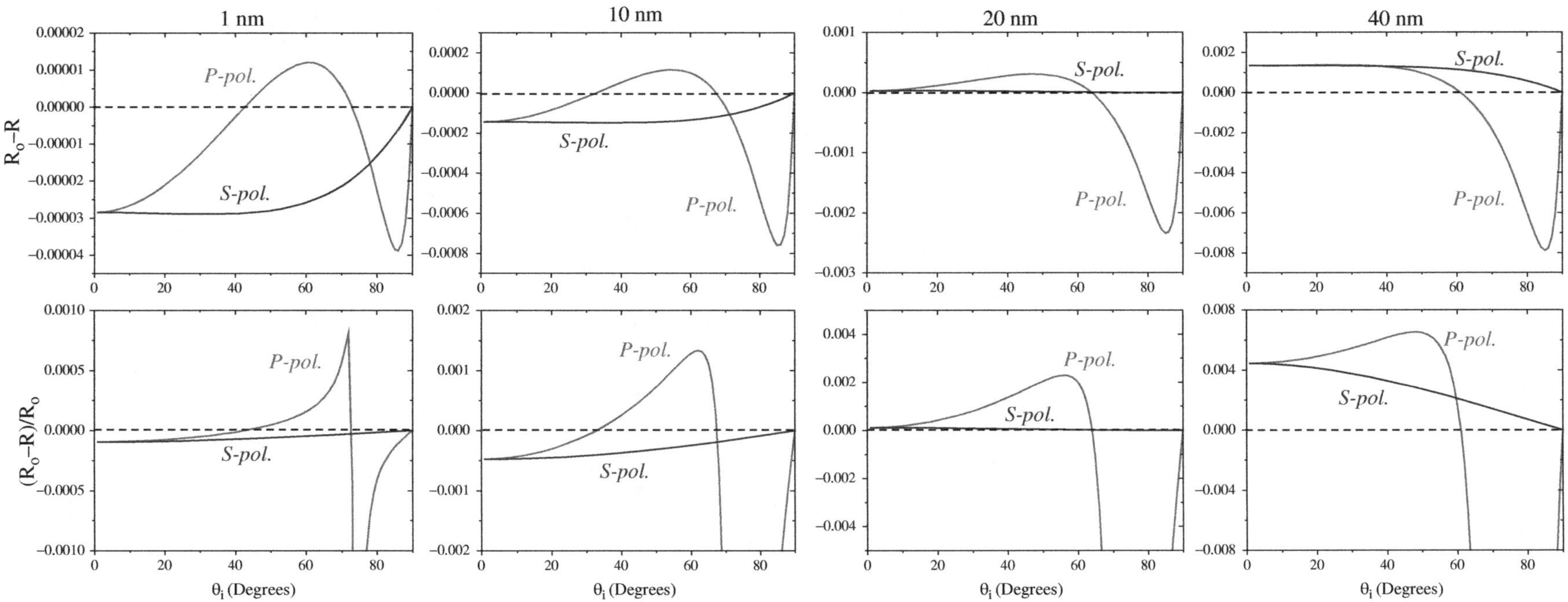

Figure 6.20 RAIRS signal intensity as a function of incident angle (θ_i) for the 2925 cm^{-1} peak of a PS film with a thickness of 1, 10, 20, and 40 nm on a Si wafer in *s*- and *p*-polarizations.

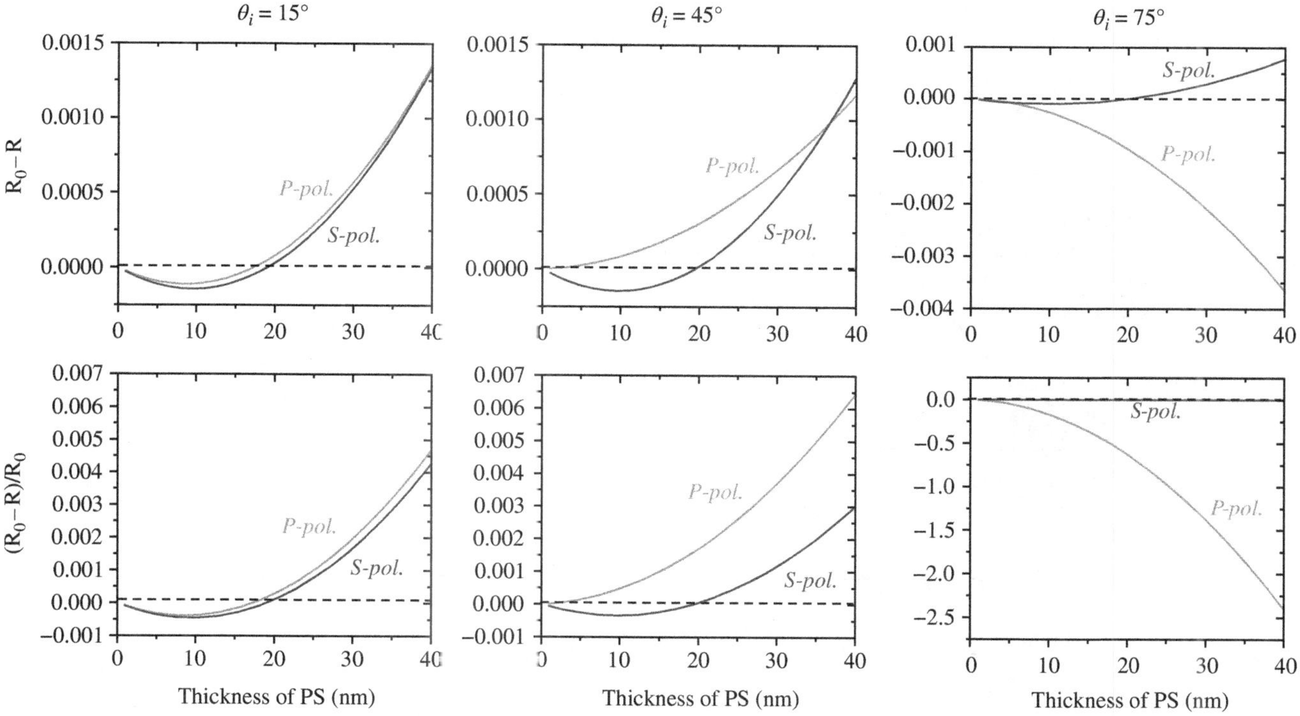

Figure 6.21 RAIRS signal intensity calculated for the 2925 cm^{-1} peak of a PS film on a Si wafer as a function of PS film thickness at θ_i = 15°, 45°, and 75° in *s*- and *p*-polarizations.

the Si–OH stretch band position was expected to be similar to that of the Si–O$^-$ (NBO) stretch band. Then, one may argue the newly growing peak in the SR-IR spectra of the Na-leached SLS glass could be also attributed to the stretch mode of Si–OH group. At a quick glance, this argument might make sense. Since the charge neutrality needs to be met, the leaching of Na$^+$ from the glass network should be accompanied by the ingress of H$^+$ from the aqueous solution, which will react with Si–O$^-$ forming Si–OH. The increase of the SR-IR intensity in the spectral region close to the OH stretch band may be viewed as evidence for the formation of the Si–OH group. This interpretation comes with one question that cannot be answered. If all Si–O$^-$ groups losing Na$^+$ ions are converted to Si–OH groups, then the Si–OH concentration should be the same as the Si–O$^-$ concentration. At least, the Si–OH concentration cannot be larger than the original Si–O$^-$ concentration. Then, why does the shoulder peak at ~900 cm^{-1} increase in SR-IR upon conversion of Si–O$^-$ to Si–OH?

For a long time, this question has been ignored or has not even been asked, but we now can answer this question unambiguously; all it takes is to calculate the theoretical SR-IR spectrum using eqs. (6.34) and (6.35) with the refractive index of the Na-leached layer. The challenge here is that the exact refractive index of the Na-leached layer was not known in the past. Recently, it has been estimated from the s-SNOM coupled with Fourier transform infrared (FT-IR) spectroscopy (see Section 6.6.3) and molecular dynamics (MD) simulations [35]. Figure 6.23A compares the reflective index of the protonated silica glass from MD simulations [35] with that of the pristine SLS glass [23]. Note that the protonated silica glass has a quite significant amount of Si–OH group (the simulated composition is $H_{0.4}SiO_{2.2}$), but it does not show any significant peak or shoulder in the ~900 cm^{-1} region. However, when the SR-IR spectra are calculated with the refractive indices of pristine SLS and protonated silica, it can be seen that the ~900 cm^{-1} shoulder region is enhanced when the surface layer (protonated silica) thickness is between 50 and 200 nm (Figure 6.23B), which is consistent with the trend observed experimentally (Figure 6.22B). When the refractive index of the Na-leached layer determined from the s-SNOM FT-IR analysis was used, the same trend was observed [35]. These results mean that the enhancement of the ~900 cm^{-1} shoulder in the SR-IR spectrum of the Na-leached SLS glass should not be attributed to the Si–OH stretch band or interpreted as an increase in the Si–OH concentration; it is simply the consequence of optical interference of the IR beams reflected from the air/leached layer interface and the leached layer/bulk interface (as described in Figure 6.16). These results invite a new interpretation for the origin of the 900–950 cm^{-1} shoulder peak of the SLS and alkali silicate glasses. Further details into the peak assignment can be found in Ref. [35]. This example will be re-visited in Section 9.4.3 (see Figure 9.23) when we study spectroscopic ellipsometry.

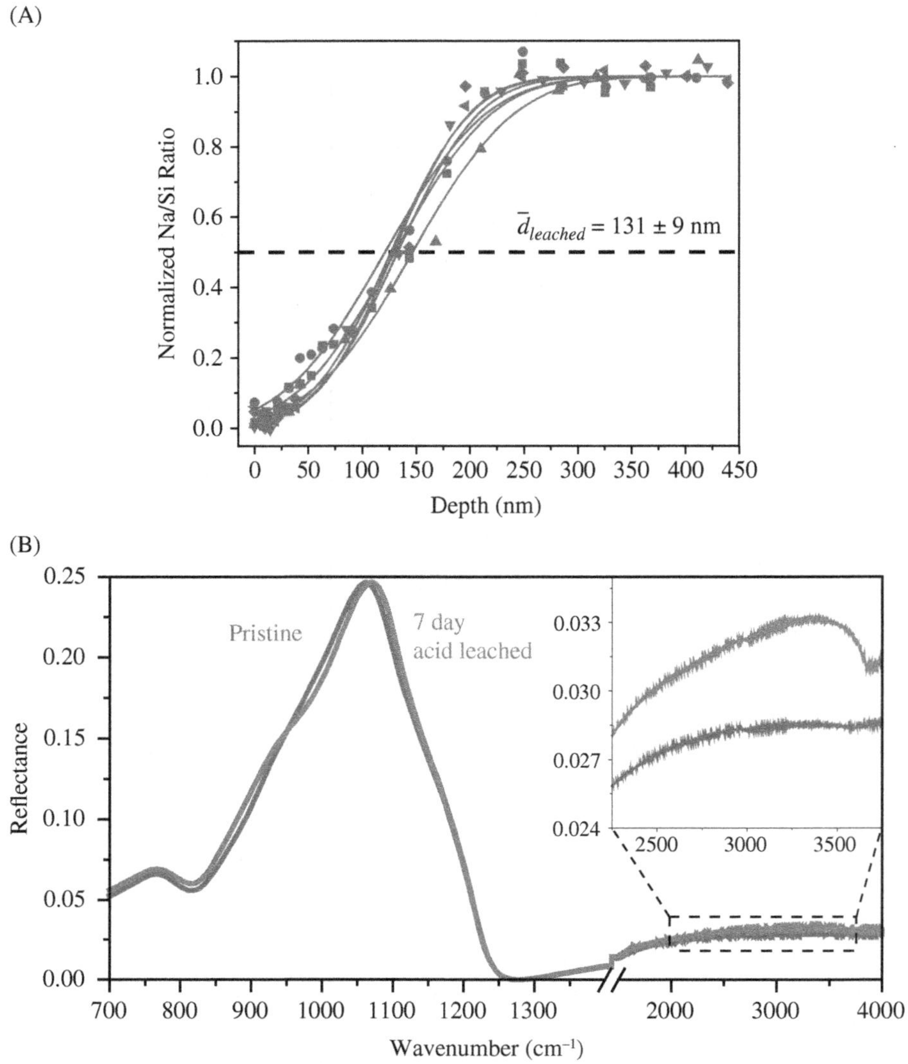

Figure 6.22 (A) XPS depth profile of Na, normalized with Si, of an SLS glass treated with pH 1 aqueous solution at 90 °C for 1 week. The Na-leach layer thickness is estimated from fitting of the Na profile with the error function (see Section 2.5.2). (B) Unpolarized SR-IR spectra of pristine and Na-leached SLS glass surfaces. Incident angle = 30°. Reproduced from [35]/with permission of ELSEVIER.

Even if the surface layer does not have any strong absorption band in the lower wavenumber side of the main Si–O–Si stretch band, the optical interference effect can produce very rich features in the SR-IR spectra. Figure 6.24A displays SR-IR spectra of sodium calcium boroaluminosilicate glass, called International Simple Glass (ISG), corroded in an aqueous solution for varying periods of time [39]. As corrosion time increases, a great deal of spectral changes occur in the lower wavenumber side of the main Si–O–Si stretch band. The corrosion of boroaluminosilicate glass studied here proceeds with not only the leach of alkali ions but also the dissolution of boron species, which leads to the formation of the so-called alteration layer (AL). In the SR-IR analysis of boroaluminosilicate glasses, the band at ~1400 cm^{-1} is assigned to the B-O stretch mode [39]. Then, it would be odd to see that this band initially decreases at the initial stage of corrosion (≲3 days) and then grows even bigger than the original intensity as the AL thickness increases further in 14 and 28 days (Figure 6.24A). In order to check if these spectral changes observed in SR-IR are real or not, it is necessary to know the thickness and refractive index of the AL. The former can be obtained from spectroscopic ellipsometry (SE), and the latter can be obtained by infrared variable-angle spectroscopic ellipsometry (IR-VASE). The principles of these methods are discussed in Chapter 9; here, let us just focus on the SR-IR simulation.

Figure 6.24B compares the refractive indices of the pristine ISG and the AL layer determined with IR-VASE [39]. Quite contrary to the spectral patterns observed in SR-IR (Figure 6.24A), the $k(\overline{\omega})$ spectrum of AL shows no strong absorption in the lower wavenumber side of the main Si–O–Si absorption band and the peak position of this band is blue-shifted

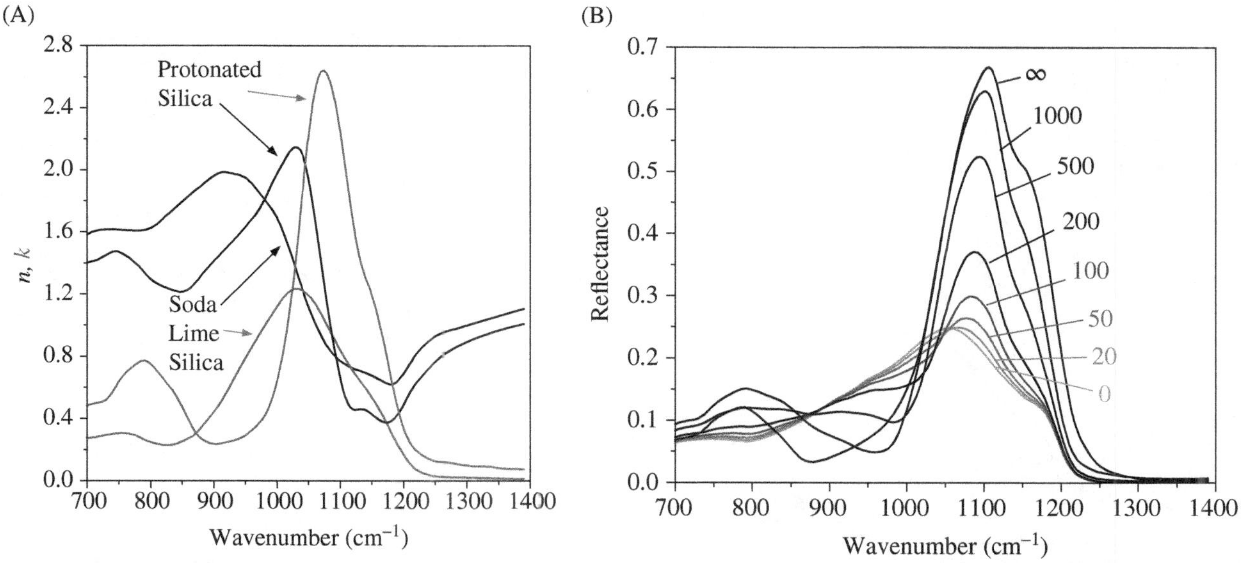

Figure 6.23 (A) Comparison of experimentally determined refractive index of a pristine SLS glass (from [23]) and computationally obtained refractive index of a protonated silica (from [35]). (B) Unpolarized SR-IR spectra, calculated using eqs. (6.34) and (6.35) and refractive indices shown in (A), for Na-leached SLS glass as a function of the leached lay.

by 31 cm^{-1}. Also, there is no B-O absorption band in the $k(\overline{\omega})$ spectrum of AL. Then, how could rich spectral features show up in the SR-IR spectra shown in Figure 6.24A? That is simply due to the optical interferences of IR beams reflected from the air/AL interface and the AL/bulk interface. Figure 6.24C displays the theoretical SR-IR spectra calculated using eqs. (6.34) and (6.35) with the refractive indices shown in Figure 6.24B and the AL layer thickness predetermined with SE [39]. All fine details observed in the experimental spectra can be seen in the theoretically calculated spectra.

Once again, these examples clearly demonstrate that the SR-IR spectra should not be interpreted using the Beer–Lambert law concept. The reflectance spectrum is essentially different from the absorbance spectrum. In order to interpret SR-IR and RAIRS spectra properly and accurately, comparison with theoretical calculations is essential. This calculation can be done using eqs. (6.34) and (6.35).

6.3.4 RAIRS of a Thin Organic Layer on Metals

The RAIRS analysis of analytes on a metallic surface is also described by eq. (6.36), but we need to consider additional interactions that are absent in the dielectric material. The light–matter interactions considered in Chapter 5 and earlier in Chapter 6 deal with the polarization of "bound" electrons (Figure 6.25A). Since they are bound to specific atoms, their polarizations in the matter are closely related to the motions of atoms. That is the main principle that by measuring the polarization of electrons in a matter using the electromagnetic wave in the infrared wavelength range, we can obtain spectroscopic information associated with vibrational motions of atoms that are chemically bonded. On metallic surfaces, we need to consider the response of "free" electrons in the conduction band of the metal as well (Figure 6.25B). Since the conduction band is delocalized and continuous, electrons in this band can move freely and collectively in the material. The response of free electrons to the applied electric field can be described with the following equation:

$$m_e \cdot \frac{d^2x(t)}{dt^2} + m_e\gamma \cdot \frac{dx(t)}{dt} = e \cdot Ee^{-i\omega t} \tag{6.37}$$

where m_e is the effective mass of electron and γ is the damping factor. Note that, if we drop the potential term of the damped harmonic oscillator equation in eq. (5.28), it becomes eq. (6.37). Since the free electrons are not bound locally, the restoring force term of the Lorentzian oscillator is dropped. If eq. (6.37) is solved in the same manner used to solve eq. (5.28), we get the following equation:

$$\varepsilon_{r,free}(\omega) = 1 - \frac{n_e e^2}{\varepsilon_0 m_e} \cdot \frac{1}{\omega^2 + i\omega\gamma} \tag{6.38}$$

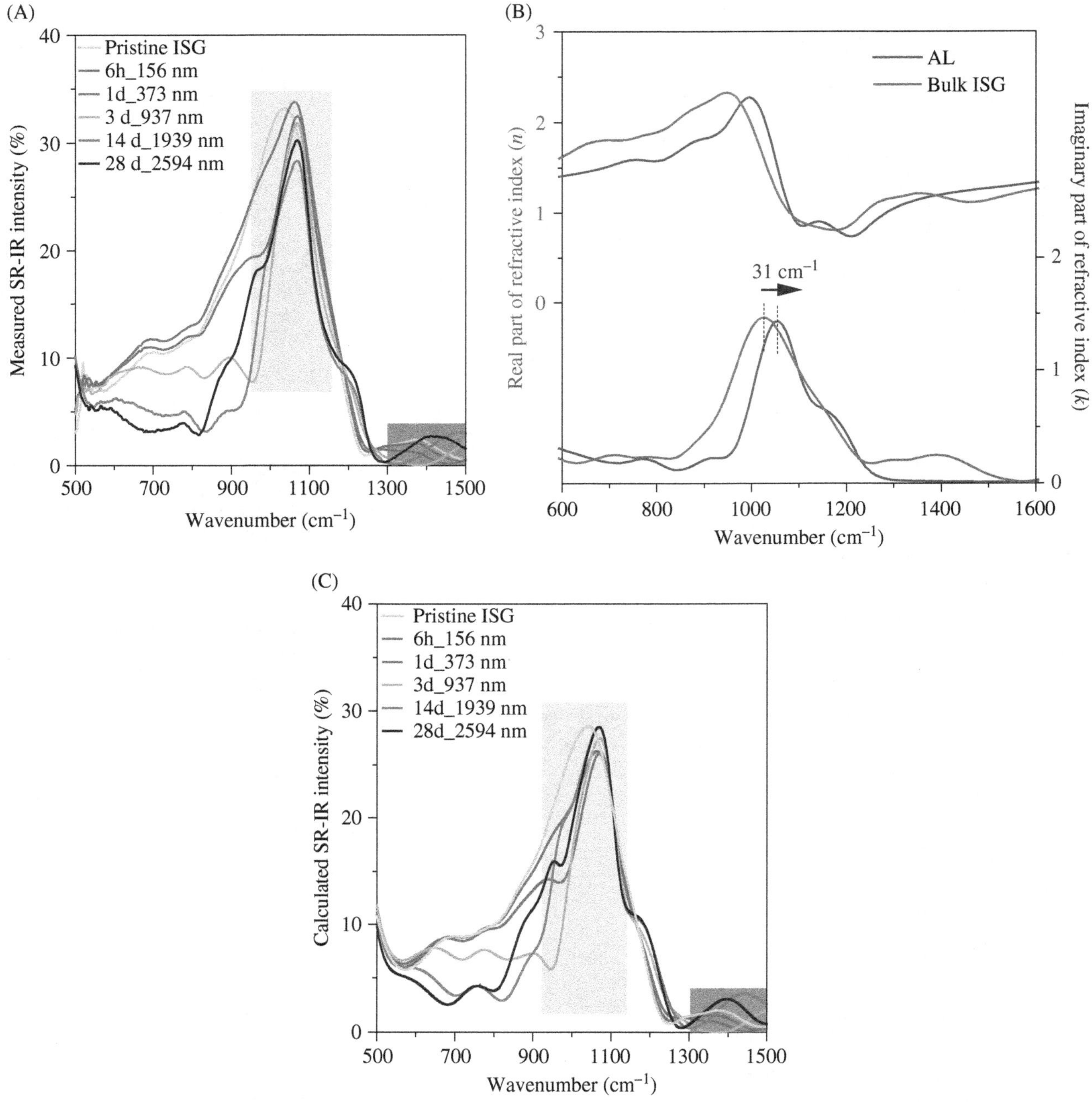

Figure 6.24 (A) Experimental SR-IR spectra of the ISG samples corroded in an aqueous solution. The thickness next to the time is determined from SE measurements. (B) Complex refractive index of the pristine bulk ISG (gray lines) and the alteration layer (AL, black lines) obtained from IR-VASE analysis of ISG sample corroded in the same aqueous solution for 34 days. (C) Theoretical SR-IR spectra calculated using an optical model of air/AL/pristine ISG and the predetermined AL thickness. Reproduced from [39]/with permission of ELSEVIER.

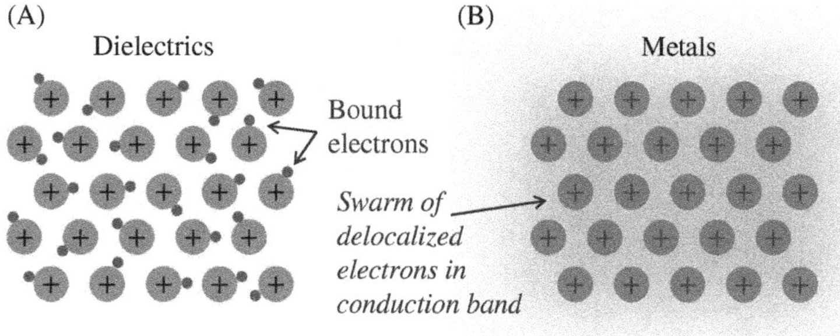

Figure 6.25 Schematic illustration of (A) bound electrons governing dielectric responses and (B) free electrons in conduction band of metal.

where n_e is the number density of free electrons in the conduction band. This is known as the Drude model. Here, we can define the plasma frequency, ω_p, as:

$$\omega_p = \sqrt{\frac{n_e e^2}{\varepsilon_o m_e}} \tag{6.39}$$

Then, eq. (6.38) is simplified to:

$$\varepsilon_{r,free}(\omega) = 1 - \frac{\omega_p^2}{\omega^2 + i\omega\gamma_p} \tag{6.40}$$

where the damping term of free electron is denoted as γ_p, which is a function of the Fermi velocity (v_F) and the mean free path (l) of an electron between successive collisions ($\gamma_p = v_F/l$). Equation (6.40) describes the polarization of free electrons in the conduction band in response to the electric field of the incident electromagnetic wave. The ω_p and γ_p values of various metals can be found in the literature [40, 41].

Typically, γ_p is much smaller than ω_p. So, when ω is smaller than ω_p ($\omega < \omega_p$), $\varepsilon_{r,free}(\omega)$ becomes negative (Figure 6.26A). Then, the refractive index becomes imaginary, that is, $n(\omega) \approx 0$ and $k(\omega) > 0$. As is the evanescent wave case discussed in Section 5.24 (especially, see the discussion for the propagation along the z direction, eq. (5.86)), the electromagnetic wave

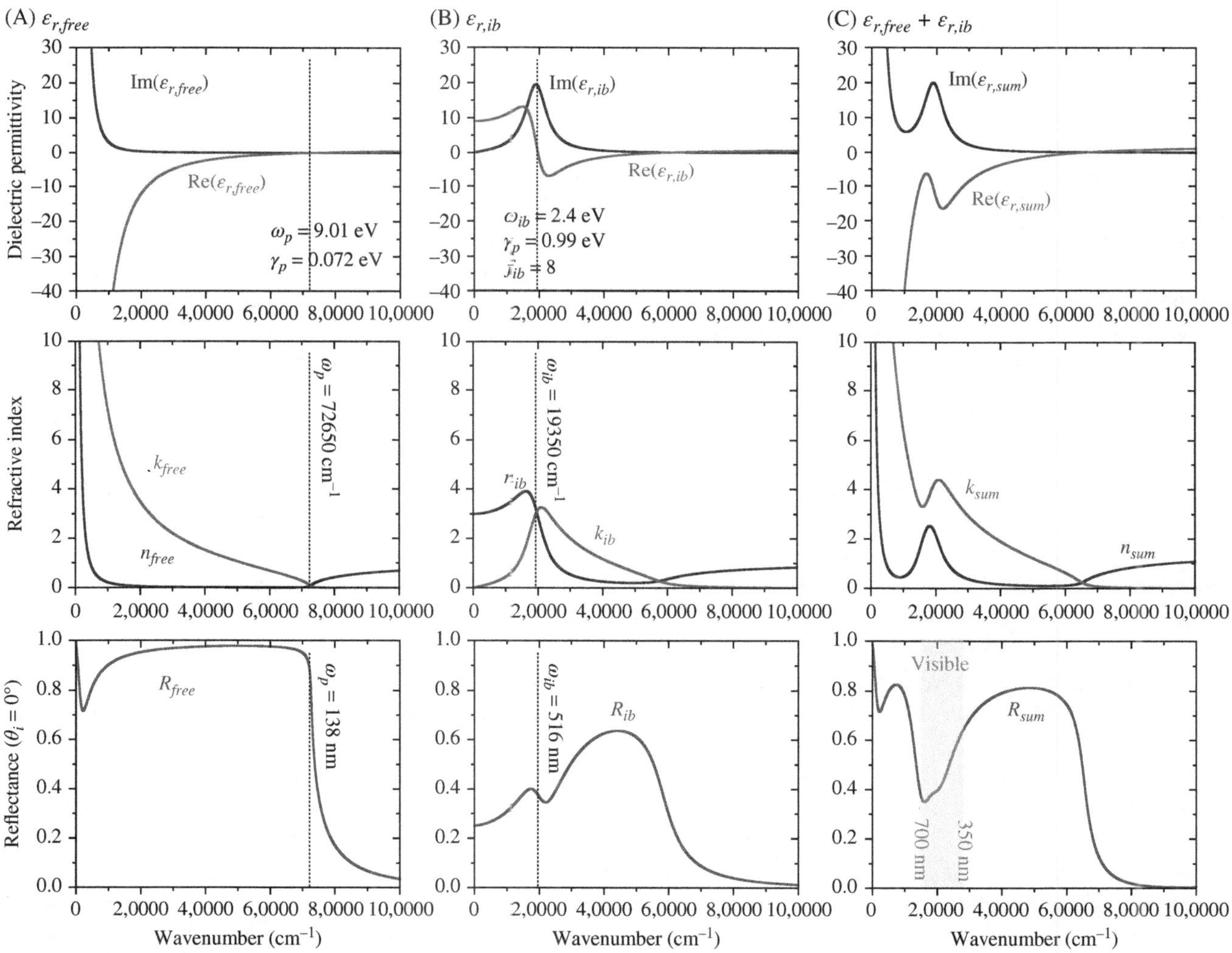

Figure 6.26 Theoretically calculated complex dielectric permittivity, complex refractive index, and reflectance at normal incidence for (A) free electrons calculated with the Drude model, (B) interband transitions calculated with the Lorentzian model, and (C) sum of both contributions. The parameter values used in calculations are taken from [38, 40].

with frequency less than ω_p cannot propagate into the metallic medium, and it will be reflected. In other words, electrons in the metal can move fast enough to effectively screen the electric field of the light whose frequency (oscillating rate) is smaller than the plasma frequency of the metal. The light of frequencies higher than ω_p is transmitted into the metal, because the electrons in the conduction band cannot respond fast enough to screen it. This is illustrated in Figure 6.26A, which plots the real and imaginary parts of $\varepsilon_{r,free}(\omega)$ calculated for gold using eq. (6.40) and the corresponding refractive index (calculated with eqs. (5.40) and (5.41)) and the reflectance at normal incident (calculated with eq. (5.57)). The ω_p and γ_p values for Au are taken from [40], which are 9.01 and 0.72 eV, respectively.

When the applied electromagnetic wave energy is high enough, then electronic transition from an occupied orbital to the unoccupied orbital in the conduction band can occur. This is called an interband transition. In the case of gold, one of the prominent interband transitions is the electronic excitation from the 5*d* orbital to the 6*s* orbital, which occurs with the threshold at 2.4 eV (=516 nm = 19,350 cm^{-1}). Due to this interband transition, the color of gold is yellowish. In the case of silver, the most significant one is the 4*d* ➔ 5*s* transition, which occurs at 4.02 eV (=302 nm). Since this energy is in the UV range, all visible wavelength lights are reflected. So, the color of silver is bright white-gray. Since the interband electronic transition involves a bound-state electron, its response to the applied electromagnetic wave can be described by the Lorentzian model (same as eq. (5.28)):

$$m_e \cdot \frac{d^2x(t)}{dt^2} + m_e\gamma_{ib} \cdot \frac{dx(t)}{dt} + m_e{\omega_{ib}}^2 x(t) = e \cdot Ee^{-i\omega t} \tag{6.41}$$

where ω_{ib} and γ_{ib} are the resonant frequency and damping factor of the interband transition process. Solving this equation, we can get the interband transition contribution to the dielectric permittivity:

$$\varepsilon_{r,ib}(\omega) = 1 + \frac{f_i \cdot \omega_{ib}^2}{\omega_{ib}^2 - \omega^2 - i\omega\gamma_{ib}} \tag{6.42}$$

Here, the f_i term is added to reflect the relative intensity of a given (*i*th) transition. Figure 6.26B plots the real and imaginary parts of $\varepsilon_{r,ib}(\omega)$ calculated for gold using eq. (6.42) and the corresponding refractive index and reflectance at normal incident. The f_i, ω_{ib}, and γ_{ib} values for Au are set to be 8, 2.4, and 0.98 eV, respectively, based on [42]. As discussed in Section 5.1.4, the n_{ib} above ω_{ib} is nearly zero and the reflectance is high above ω_{ib}, which is the Reststrahlen effect. Unlike the free electron response, the bound electron response gives a nonzero positive value in the real part of complex dielectric permittivity below ω_{ib}, and thus n_{ib} has a nonzero positive value in this region.

The overall response of all electrons in metal is the sum of all contributions from the Drude and Lorentzian models:

$$\varepsilon_r(\omega) = \varepsilon_{r,free}(\omega) + \sum \varepsilon_{r,ib}(\omega) \tag{6.43}$$

Figure 6.26C plots the sum of the free-electron contribution and the 2.4 eV interband contribution of gold. Note that there are multiple interband transitions in gold; thus, the real part of refractive index does not decrease fast to zero above the 20,000 cm^{-1} [42]. Anyhow, we can see that the reflectance of gold is low at 516 nm, which is the reason that gold appears yellowish. In the case of silver, this interband peak is in the UV region, so it does not assume any color.

Obviously, the energy of mid-IR used in the RAIRS analysis (400–4000 cm^{-1}) is far below the plasma frequency (ω_p) and interband transition frequency (ω_{ib}) of most metals [40–42]. So, the electric field of the IR beam is effectively screened by free elections in the conduction band; but, the effect of screening is different for the *s*- and *p*-polarized beams. This can be seen clearly by plotting the Fresnel reflection coefficients which is the ratio of the reflected electric field to the incident electric field (eqs. (5.70) and (5.74)). Figure 6.27A plots the real and imaginary parts of r_s and r_p calculated as a function of incident angle (θ_i) of IR at the wavelength of typical C-H stretch mode (2925 cm^{-1} = 3.419 μm) for gold. The refractive index of Au at this wavelength is 0.68 + 23.2 *i* [2]. Figure 6.27B plots the phase of the calculated r_s and r_p. From this calculation, we can see that r_s is nearly −1 at all incident angles and its phase is very close to −180°. It means that the *s*-polarized electric field of the reflected beam has the same amplitude as the incident beam and they are out of phase. Thus, these two cancel each other and the net field at the gold surface is nearly zero [43]. In contrast, the r_p is close to one at low incident angles and its phase is close to zero. This means the *p*-polarized electric fields of the incident and reflected beams will effectively add up, enhancing the net field along the surface-normal direction [43]. The real part of r_p decreases fast at $\theta_i > 80°$, passes zero at $\theta_i \approx 87.5°$ (at which the imaginary part reaches the maximum, that is, the phase is 90°), and then approaches −1 as θ_i increases further toward 90°.

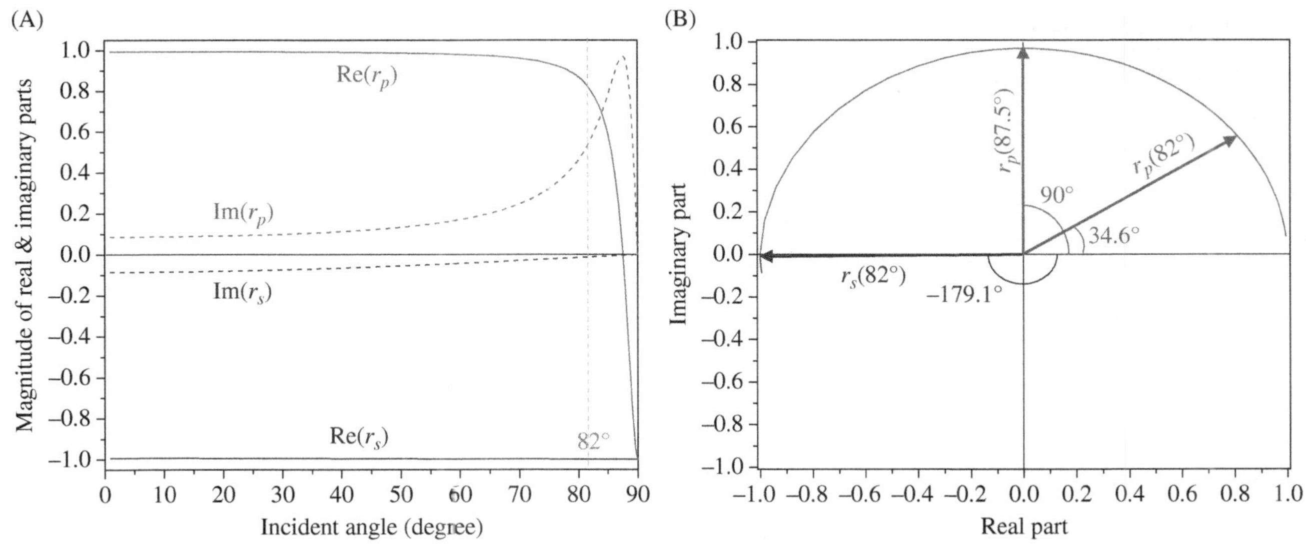

Figure 6.27 (A) Real and imaginary parts of Fresnel reflection coefficients (r_s, r_p), as a function of incident angle, for Au at 2925 cm^{-1} (=3.419 μm). (B) Phase of Fresnel reflection coefficients shown in (A).

This has a very significant consequence in RAIRS analysis of adsorbates on metallic surfaces. The transient dipole moment (μ') of a molecular vibration can be divided into surface-parallel and surface-normal components. Since their dipoles oscillate in the IR frequency range (far below ω_p), the surface-parallel component will be fully screened by the image charge formed by free electrons in the conduction band of the metal substrate (Figure 6.28A). Thus, its effective dipole moment will be zero (i.e., $\mu'_{\parallel,eff}(\omega_{IR}) = 0$). In contrast, the surface-normal component of dipole moments vibrating in the IR frequency will be enhanced by its image charge in the metal (Figure 6.28B). This leads to an enhancement of effective dipole, $\mu'_{\perp,eff}(\omega_{IR}) \approx 2\mu'_{\perp}(\omega_o)$. Thus, RAIRS on metal can detect only the surface-normal component of adsorbate vibrations (in fact, its signal is enhanced), while it cannot detect the surface-parallel component. This is so-called the metal selection rule [44].

We can calculate the RAIRS intensity theoretically using eqs. (6.34)–(6.36). These equations are much simpler than the ones originally derived by R. G. Greener [43]. Figure 6.29 plots $\Delta R = R_o - R$ $(= R_{13} - R_{123})$, R_o $(= R_{13})$, and $\Delta R/R_o$ $(=-\log(R/R_o))$ calculated for the CH_2 asymmetric stretch peak at 2925 cm^{-1} for a 1-nm thick PS on gold (Au) and aluminum (Al). Here, media denoted as 1, 2, and 3 are air, 1-nm thick PS, and metal substrates, respectively. As the metal section rule has predicted (Figure 6.28A), the difference signal in the s-polarization is zero. This is because the surface-parallel component of the E-field oscillating at the IR frequency is fully cancelled out by the image charge induced in the metal surface that oscillates at the same frequency with the opposite phase (Figure 6.27). In the p-polarization case, the difference signal intensity increases as the incident angle increases, reaches the maximum at $\theta_i = 82°$ and then decreases to zero as θ_i approaches 86° (top panels in Figure 6.29). This means that the RAIRS experiment for a thin layer on a metallic substrate can get the highest

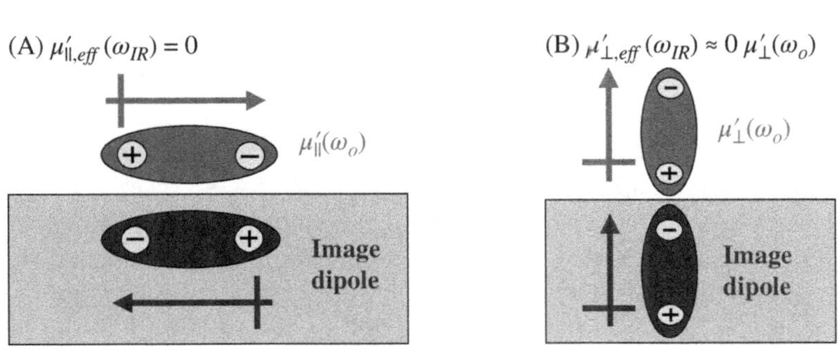

Figure 6.28 Metal selection rule in RAIRS analysis of an adsorbate on a metallic substrate. (A) Cancellation of the surface-parallel component and (B) enhancement of the surface-normal component of a transient dipole moment oscillating in the IR frequency region.

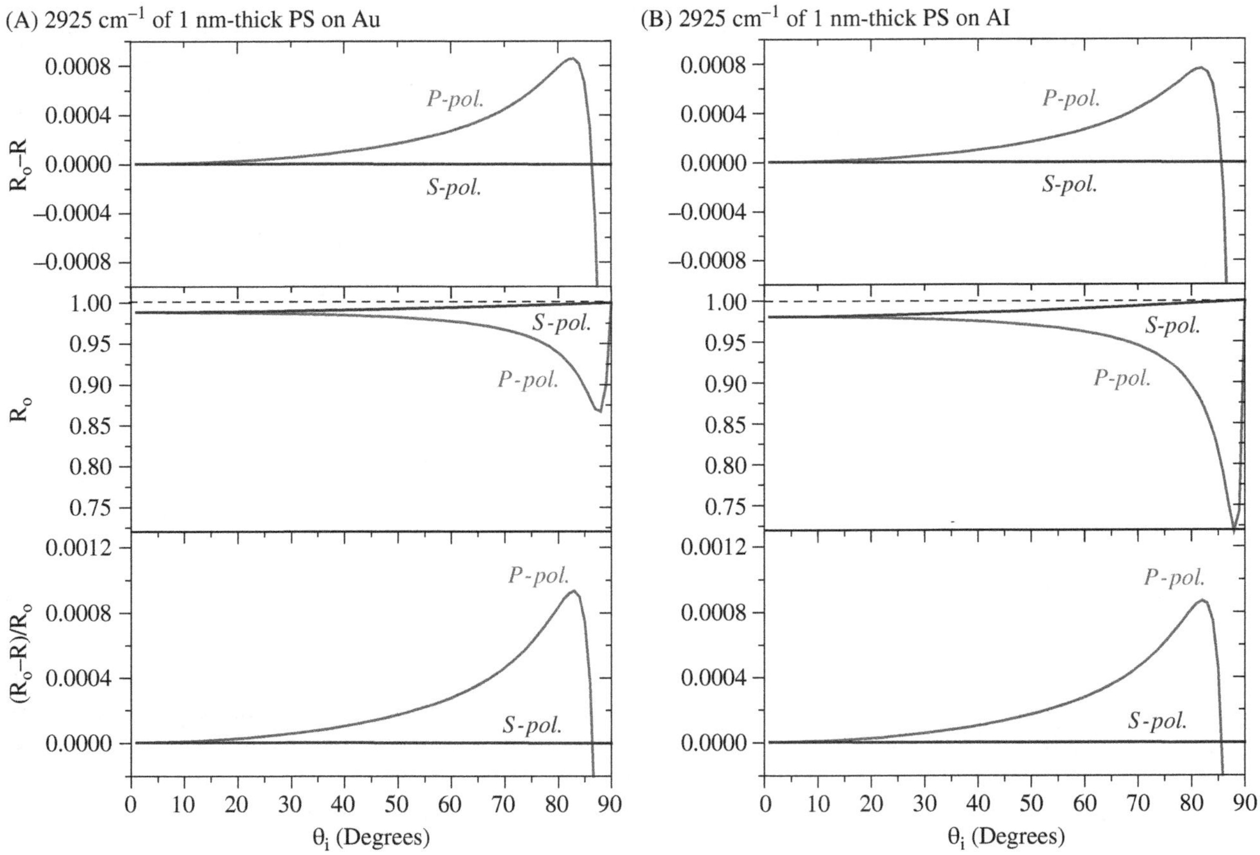

Figure 6.29 Reflectance difference between clean and thin film samples (ΔR), reflectance of clean sample (R_o), and normalized signal ($\Delta R/R_o = -\log(R/R_o)$) of *s*- and *p*-polarized IR calculated for the CH_2 asymmetric stretch peak (2925 cm^{-1}) of a 1-nm thick PS film on (A) Si and (B) SLS substrates. The refractive index of PS used is the same as the one in Figure 6.18. The refractive indices of Au and Al are taken from [2].

sensitivity when $\theta_i = 82°$. This optimum incident angle is invariant for most metals. The difference signal at $\theta_i = 82°$ is slightly larger for the Au surface than the Al surface. Note that the magnitude of the difference signal is even larger at $\theta_i \approx 88°$ (with the negative sign); but, practically it is difficult to do RAIRS experiment at this extreme angle.

For quantitative analysis, the difference signal needs to be normalized with the reflectance of the clean surface, which is shown in the middle panel in Figure 6.29. Since Al is more lossy than Au, the reflectance of the *s*- and *p*-polarized light is smaller. Note that the *s*-reflectance is nearly one in all incident angles; in contrast, the *p*-reflectance decreases to the minimum value at $\theta_i \approx 87.5°$ (where the phase of reflected beam is 90° with respect to the incident beam; Figure 6.27B) and then increases to one at $\theta_i = 90°$.

The bottom panels in Figure 6.29 display the theoretically calculated $\Delta R/R_o$, which is the same as $-\log(R/R_o)$, of the 2925 cm^{-1} peak of the 1-nm thick PS film on Au and Al surfaces. At all incident angles below 80°, the normalized intensity is essentially identical between these two metal substrates and a minor difference can be seen near the optimum incident angle (82°) and higher.

Figure 6.30 compares the full RAIRS spectra of 1-nm thick PS film on Au and Al calculated for the incident angles of 50°, 65°, and 80°. Unlike the RAIRS spectra of the same film on dielectric materials (Figure 6.19), the RAIRS spectrum of an adsorbate on a metallic surface always shows positive peaks. If the RAIRS experiment is carried out at the incident angle of 87–88°, the spectrum would have larger peaks in the negative direction; but, doing measurements at this extremely high incident angle is practically difficult. As expected from the metal selection rule, the *s*-polarized RAIRS spectra do not show any signal. The signal intensity in the *p*-polarized RAIRS spectrum increases as the incident angle increases. In most IR instrument, the incident beam is focused with an elliptical mirror with a cone angle of $\pm 7\sim 9°$. For this reason, the mean incident angle of the *p*-polarized beam is chosen to be 80–82°. The small decrease in the baseline in the nonabsorption

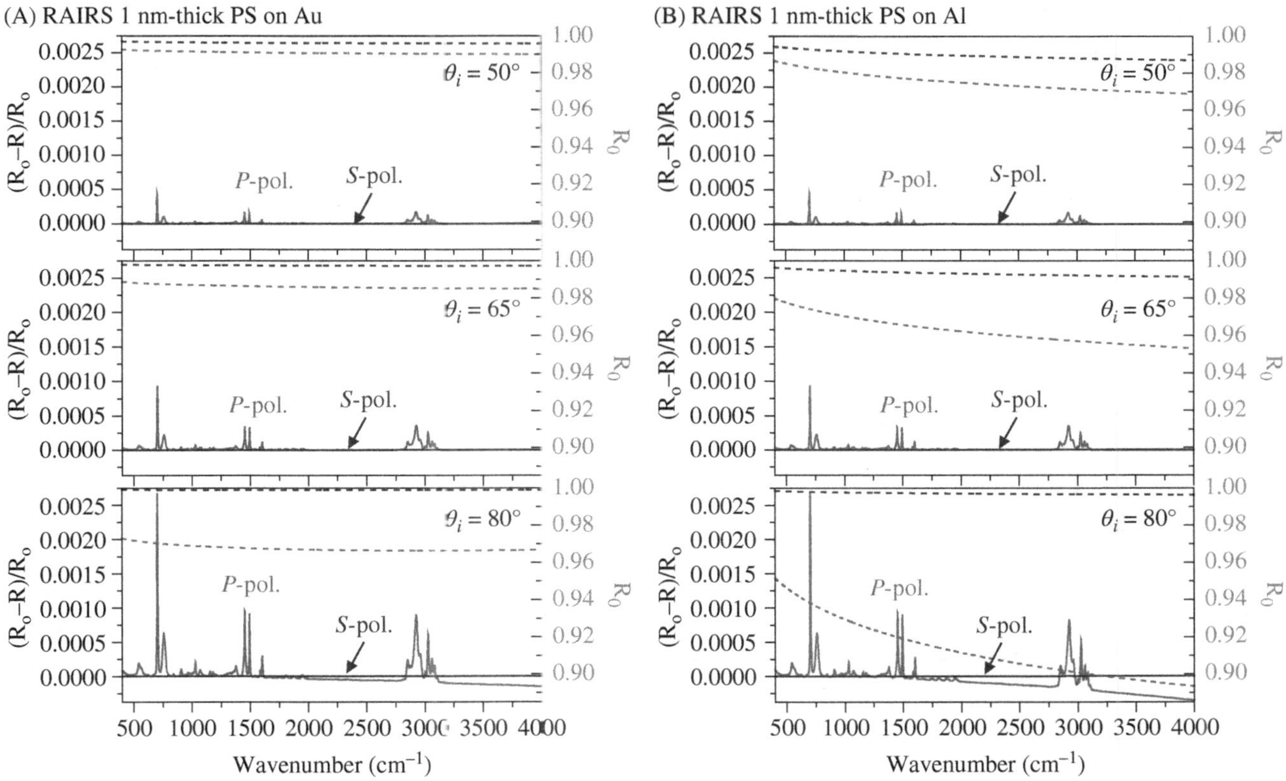

Figure 6.30 RAIRS spectra of a 1-nm thick PS film on (A) Au and (B) Al substrates calculated for the incident angle of 50°, 65°, and 80°. The refractive index of PS used is the same as the one in Figure 6.18. The refractive indices of Au and Al are taken from [2]. The dotted lines are $R_{s,o}$ and $R_{p,o}$, plotted with the secondary axis in the right side; the one close to $R = 1$ is $R_{s,o}$.

region at high incident angles (e.g., see $\theta_i = 80°$ in Figure 6.30) is due to the reflectance change in the presence of dielectric layer on the metal surface.

In all calculations that we have discussed so far, we assume that the refractive index of the adsorbate layer is the same as that of the homogeneous bulk phase. In reality, the refractive index of the adsorbate (molecularly thin layer) may deviate from that of the corresponding bulk phase especially if strong interactions between the adsorbate and the substrate surface alter polarization behavior of the adsorbate molecule. Chemisorbed species falls in this case; for example, the CO molecules adsorbed on metal surfaces show different absorption peaks depending on the adsorption geometry [45]. In the case of physisorbed species, such perturbation is likely to be negligible or at least much smaller.

It is also possible that adsorbate molecules may be aligned anisotropically. Such anisotropy is often observed for self-assembled monolayers (SAMs) of long-chain organic molecules. By comparing the theoretically calculated RAIRS intensity [32, 46] with the experimentally determined intensity, one can determine the molecular tilt angle of SAMs on metals with respect to the surface-normal direction. The intensity (I) of a given vibrational mode is proportional to the square of the scalar product of the electric field and the dipole moment derivative (or transient dipole) with respect to the normal coordinate, that is, $I \propto \left|\frac{\partial \mu}{\partial q} \cdot E\right|^2$. If the molecules at the 2D surface are azimuthally isotropic or indistinguishable, its RAIRS intensity will vary according to $\cos^2\vartheta$, where ϑ is the average tile angle of molecules from the surface normal. Note that when the molecule is tilted by the magic angle (54.74°), then it is indistinguishable from the random orientation. The magic angle is geometrically the same as the angle between the space diagonal of a cube (i.e., $a = b = c = 1$) and any of its three connecting edges (a or b or c), which is $\cos^{-1}\frac{1}{\sqrt{3}} = 54.74°$. Thus, at the magic angle, the contributions from all x, y, and z components are equal, and the net effect will be the same as the sum of all contributions in the system with randomly oriented molecules. In other words, the theoretical intensity (I_{calcd}) calculated with the refractive index of the isotropic bulk phase is the same as the observed intensity (I_{obsd}) of the molecules at the magic angle. When the adsorbate structure is anisotropic but otherwise identical with the bulk structure, the alignment of the transient dipole moment along the

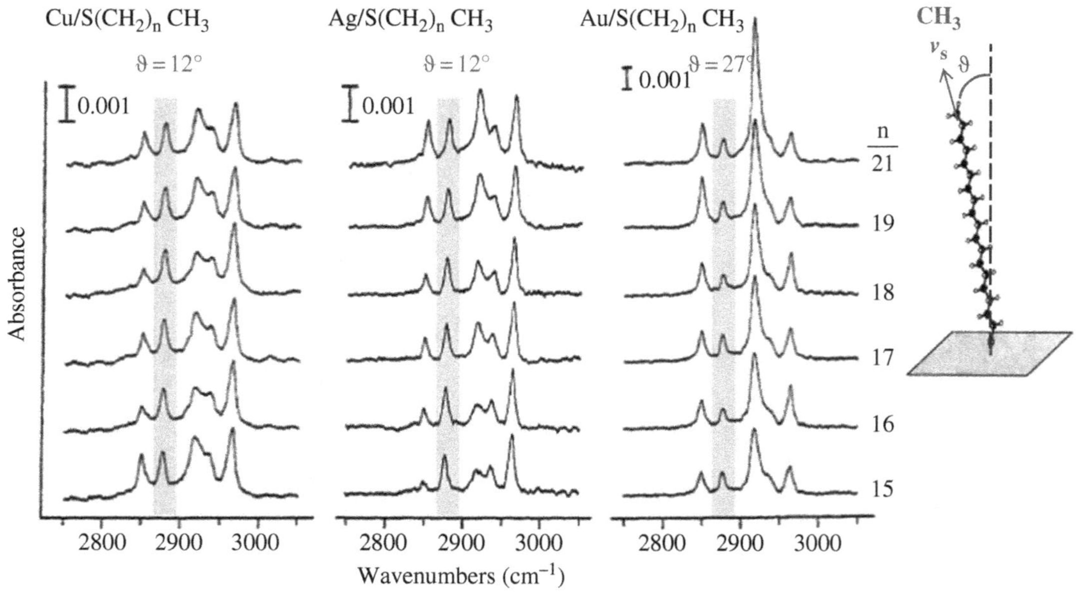

Figure 6.31 RAIRS spectra of monolayers derived from exposure of surfaces of copper, silver, and gold to *n*-alkanethiols, $CH_3(CH_2)_nSH$. Reproduced from [47]/with permission of American Chemical Society. The CH_3 symmetric stretch peak (ν_s) is highlighted.

surface-normal direction ($\vartheta = 0°$) will give the I_{obsd} value that is three times the I_{calcd} for the intensity of the random orientation case. All this can be described with the following equation:

$$\frac{I_{obsd}}{3I_{calcd}} = \cos^2\vartheta \tag{6.44}$$

Figure 6.31 shows an example where this principle is used to determine the tilt angle of *n*-alkanethiol SAMs on Cu, Ag, and Au surfaces [47]. Assuming the all-*trans* conformation of the CH_2 groups in the alkyl chains, the transient dipole moment of the symmetric stretch (ν_s) of the terminal CH_3 group would be fairly well aligned with the molecular axis. The experimentally observed CH_3 ν_s intensity can be compared with the theoretical intensity calculated for the isotropic orientation using the refractive index of the bulk. Then, using eq. (6.44), the molecular tilt angle can be estimated.

6.3.5 Polarization Modulation (PM)-RAIRS of a Thin Layer on Metals

RAIRS is a static measurement where the *s*- and *p*-polarized spectra are measured separately. If one intends to measure the spectrum of an adsorbate film on a metal surface in the presence of the same analyte or other potential interferent molecules in the beam path above the sample, the *s*- and *p*-polarization spectra should be measured in sequence and then the *s*-polarization spectrum could be subtracted from the *p*-polarization spectrum to obtain the adsorbate spectrum. In theory, this is plausible since the *p*-polarization spectrum contains both gas and adsorbate signals, while the *s*-polarization spectrum contains the gas-phase signal only (based on the metal selection rule, Figure 6.28). However, in reality, this approach is subject to instrumental and background fluctuations during the sequential measurements. Quite often, the adsorbate signal is even smaller than the fluctuation of the gas-phase signal. To overcome this problem, the "polarization modulation (PM)" technique is used; this method is called PM-RAIRS. The surface sensitivity of PM-RAIRS comes from the fact that difference of the *p*- and *s*-polarization spectra is measured simultaneously by modulating the polarization of the incident IR beam at high frequency with a photoelastic modulator (PEM). The system configuration is shown in Figure 6.32. Basically, the same configuration can also be used (obviously with a different sample geometry) to measure the vibrational linear dichroism (VLD) and the vibrational circular dichroism (VCD) [48].

The PEM unit consists of a ZnSe crystal which is periodically stressed by compression and stretch at 37 or 50 kHz. Then, the linear polarization is switched between vertical and horizontal directions at a frequency of 74 or 100 kHz. This high-frequency modulation is added on top of the low-frequency Fourier modulation produced by the moving interferometer. Thus, the raw detector signal contains time-variant contributions from both polarizations. These two are separated using dual-channel synchronous sampling demodulator (SSD). The low-frequency signal measured with a low-pass filter

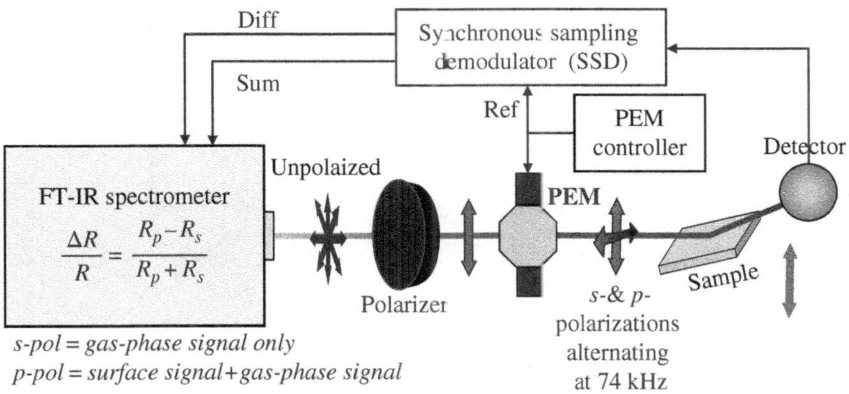

Figure 6.32 System configuration for PM-RAIRS experiment.

gives the normal IR spectrum which is the sum of both *p*- and *s*-polarizations, and the high-frequency signal measured with a high-pass filter gives the difference between the *p*- and *s*-polarizations. Then, the experimental PM-RAIRS spectrum is calculated by dividing the difference signal ($R_p - R_s$) with the sum signal ($R_p + R_s$) as follows [49, 50]:

$$\frac{(R_p - R_s)}{(R_p + R_s)_{flim}} = \frac{J_2(\varphi_o)g\{rI_p - I_s\}}{rI_p + I_s} \tag{6.45}$$

where $J_2(\varphi_o)$ is the second-order Bessel function with the maximum dephasing φ_o introduced by the PEM, I_p and I_s are the intensities of reflected *p*- and *s*-polarized light, respectively, *g* is the ratio of overall gain for the two channels, and *r* is the ratio of overall responses of the electronic setup for *p*- and *s*- polarizations. In order to eliminate the factors $J_2(\varphi_o)$ and g, eq. (6.45) is normalized with the equivalent PM-RAIRS signal from the bare surface collected with the same experimental setup. Then, the normalized signal can be compared with the RAIRS signal of the film ($\log(1/R_p)$) [51]:

$$\frac{(R_p - R_s)}{(R_p + R_s)_{norm}} = \frac{(R_p - R_s)}{(R_p + R_s)_{film}} \Big/ \frac{(R_p - R_s)}{(R_p + R_s)_{bare}} = 1 + \frac{2r\rho\ \ln(10)}{1 - r^2\rho^2}\log(1/R_p) \tag{6.46}$$

where ρ is the reflectivity ratio of *p*- and *s*-polarizations on the bare metal surface. For a typical gold surface, $\rho \approx 0.95$ in the mid-IR range and $r \approx 1$; then, the normalized PM-RAIRS signal intensity can be converted to the equivalent RAIRS signal using eq. (6.46): $\log(1/R_p) \approx 0.0223\left(\frac{(R_p - R_s)}{(R_p - R_s)_{norm}} - 1\right)$.

Figure 6.33 displays the sum spectra and normalized difference spectra obtained from PM-RAIRS measurements of n-propanol adsorbed on a gold film as a function of n-propanol partial pressure with respect to its saturation pressure (P/P_{sat}). The sum signal decreases as the P/P_{sat} increases due to the gas-phase absorption. Since the sum signal is dominated by the absorption by the gas-phase molecule, it shows a sharp free OH signal of the alcohol group without any hydrogen bonding interactions at 3670 cm^{-1}. The difference spectra, normalized with the bare gold spectrum obtained in the absence of n-propanol vapor, show positive signals of the n-propanol molecules adsorbed on the gold surface. When the normalized difference signal intensity is plotted against P/P_{sat}, we can see a convex point near $P/P_{sat} \approx 0.2$. Based on Type-II adsorption isotherm (also called BET isotherm) [52], this knee point can be attributed to the formation of a full monolayer of n-propanol physisorbed on the gold surface. The normalized difference spectra do not show the OH stretch band. This could mean that the adsorbed alcohol molecules have the OH group parallel to the gold surface (Figure 6.28A). As the P/P_{sat} approaches the saturation point, the normalized signal intensity increases rapidly. This is due to the condensation of n-propanol molecules. At this point, the broad OH signals centered at 3350–3400 cm^{-1} appear because multilayers of n-propanol are formed through hydrogen bonding interactions among the molecules. At the near saturation pressure, the normalized difference signal intensity is ~1.065, which can be converted to 0.0014 in the $\log(1/R_p)$ scale using eq. (6.46). When this value is compared with the alkyl signal intensity of the PS film on gold collected at the IR incident angle of 80° (Figure 6.30A), the thickness of the condensed n-propanol layer is estimated to be slightly larger than 1 nm. More accurate estimation can be done by calculating the theoretical RAIRS intensity of n-propanol on gold using the complex refractive index of n-propanol using eq. (6.36).

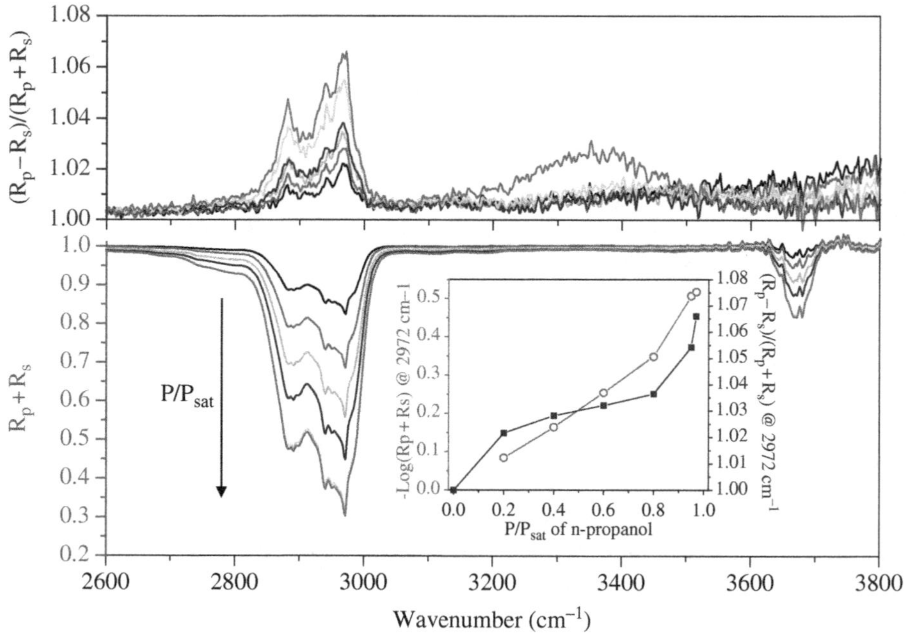

Figure 6.33 Sum and normalized difference spectra of PM-RAIRS of n-propanol on gold as a function of partial pressure of n-propanol with respect to its saturation vapor pressure (P/P_{sat}). The inset plots the sum signal intensity in the log(1/R) scale (open circle) and the normalized difference signal intensity of the 2972 cm^{-1} peak (filled square) vs. P/P_{sat} of n-propanol. The incident angle of IR is 81°.

6.4 Brewster-Angle Transmission (BAT) Infrared Spectroscopy

So far, we have considered the IR analysis in the reflection geometry. This can be applied to analyze adsorbate layers or thin films on flat substrates, regardless of whether the substrate is transparent to IR or not and at any incident angles (although a certain incident angle giving a higher S/N ratio is preferred). If the substrate is transparent to the probe beam (in our case, IR), then the experiment can be done in the transmission mode, but it has to be done at a specific angle. This is schematically illustrated in Figure 6.1D. Let us derive the theoretical equation involved in the transmission-mode measurement. Figure 6.34 depicts a three-layer system in which medium-1 and -5 are the gas phase, medium-2 and -4 are the adsorbate layers or thin films on the front and back sides of a substrate (medium-3). Then, r_{ij} and t_{ij} are the reflection and transmission Fresnel coefficients, respectively, as defined by eqs. (6.2)–(6.5) at the interface of i-th and j-th media. The incident and transmission angles at each ij interface can be calculated using eq. (6.1) from the initial incident angle from medium-1 (θ_i) and the refractive index of each medium: 1(=5), 2(=4), and 3.

As we discussed in Section 6.3.1, we need to consider the multiple reflections between two consecutive interfaces, which will propagate through all interfaces. For the interfaces of media-1|2 and -2|3 in Figure 6.34, which are separated by a thickness d_2, we already have derived the reflection coefficient (r_{123}) in eq. 6.33 (see Figure 6.16). In the same way, the transmission coefficient (t_{123}) can be derived. For the sake of comparison, r_{123} is shown again along with t_{123}:

$$r_{123} = \frac{r_{12} + r_{23}e^{2i\beta_2}}{1 + r_{12}r_{23}e^{2i\beta_2}} \tag{6.33}$$

$$t_{123} = \frac{t_{12}t_{23}e^{i\beta_2}}{1 + r_{12}r_{23}e^{2i\beta_2}} \tag{6.47}$$

where $\beta_2 = \frac{2\pi d_2}{\lambda}\tilde{n}_2 \cos\theta_2 = \frac{2\pi d_2}{\lambda}\sqrt{\tilde{n}_2^2 - \tilde{n}_1^2 \sin^2(\theta_1)}$, as defined in eq. (6.25). Note that subscripts p and s are omitted for the simplicity of expression; the calculations for p- and s-polarizations should be done separately using the proper Fresnel coefficients for each polarization.

In the case shown in Figure 6.34, there are four interfaces, that is, 1(gas)|2(absorbing layer)|3(transparent substrate)|4 (=2)|5(=1). So, we have to consider multiple reflections at each interface along the IR beam path. The overall reflection

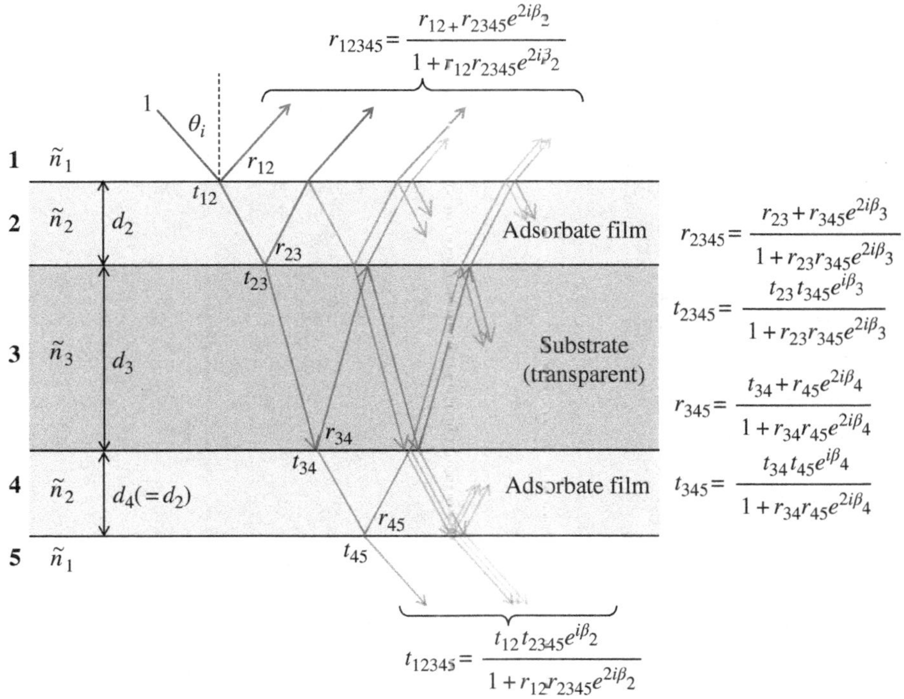

Figure 6.34 Fresnel coefficients in transmission infrared spectroscopy probing thin films adsorbed on a transparent substrate. At the Brewster angle of the substrate ($\theta_i = \theta_p$), r_{12345} becomes zero (see Figure 5.13), and the IR beam is fully transmitted.

(r_{12345}) and transmission (t_{12345}) coefficients can be calculated by expanding the two-interface model from the bottom pair, that is, by calculating r_{345} and t_{345} first, and then r_{2345} and t_{2345}, and then finally r_{12345} and t_{12345} [53]:

$$r_{12345} = \frac{r_{12} + r_{2345}e^{2i\beta_2}}{1 + r_{12}r_{2345}e^{2i\beta_2}} \quad \text{where } r_{2345} = \frac{r_{23} + r_{345}e^{2i\beta_3}}{1 + r_{23}r_{345}e^{2i\beta_3}} \quad \text{where } r_{345} = \frac{r_{34} + r_{45}e^{2i\beta_4}}{1 + r_{34}r_{45}e^{2i\beta_4}} \tag{6.48}$$

$$t_{12345} = \frac{t_{12}t_{2345}e^{i\beta_2}}{1 + r_{12}r_{2345}e^{2i\beta_2}} \quad \text{where } t_{2345} = \frac{t_{23}t_{345}e^{i\beta_3}}{1 + r_{23}r_{345}e^{2i\beta_3}} \quad \text{where } t_{345} = \frac{t_{34}t_{45}e^{i\beta_4}}{1 + r_{34}r_{45}e^{2i\beta_4}} \tag{6.49}$$

Here, β_3 and β_4 are calculated in the same way as β_2 with the proper refractive indices and the thickness following the notations shown in Figure 6.34. Then, the total reflectance and transmittance are simply the product of the overall coefficients and their conjugates, $R = r_{12345} \cdot r_{12345}^*$ and $T = t_{12345} \cdot t_{12345}^*$, respectively. As we compare the calculated R with the experimental reflectance spectrum in Sections 6.1–6.3, we can compare the calculated T with the experimental transmission spectrum. Using this model, the theoretical spectrum of adsorbate layers or thin films on a transparent substrate can be simulated for a given incident angle (θ_i), polarization (p or s), and effective thickness ($d_2 = d_4$), if the refractive indices of the substrate ($\tilde{n}_3$) and the adsorbate layer or thin film ($\tilde{n}_2 = \tilde{n}_4$) are known.

Figure 6.35 shows an example where the adsorbate layer is liquid water with a thickness (d_2) of 0.5 nm on a calcium boroaluminosilicate glass (commercial glass used for a display panel) [53]. The same calculation can be done for other substrate materials (such as SiO_2, Al_2O_3, TiO_2, AlN, and Si_3N_4) as long as their refractive index is known [2, 54]. The refractive index of liquid water is taken from Ref. [55]. If the adsorbed water layer assumes the ice-like structure, then the refractive index of ice water can be used [56]. If the adsorbate spectrum overlaps with the gas-phase spectrum, distinguishing the adsorbate signals from the measured spectrum is extremely difficult. This is the main difference between the PM-RAIRS experiment probing adsorbate species on metal which are in equilibrium with the gas phase (see Section 6.3.5). In the case of water, the gas-phase molecules do not have any hydrogen bonding interactions because they are all isolated (so their peaks are very sharp with ro-vibrational excitations [57]), while the adsorbate species are influenced by hydrogen bonding interactions. So, the adsorbate spectrum has a broad and continuous feature with its center red-shifted from the gas-phase signals, making it possible to differentiate the adsorbate features from the gas-phase peaks. In the case of organic thin films with known refractive index [2] (such as PS films that we have considered so far), eqs. (6.48) and (6.49) work well to produce the theoretical absorbance spectrum.

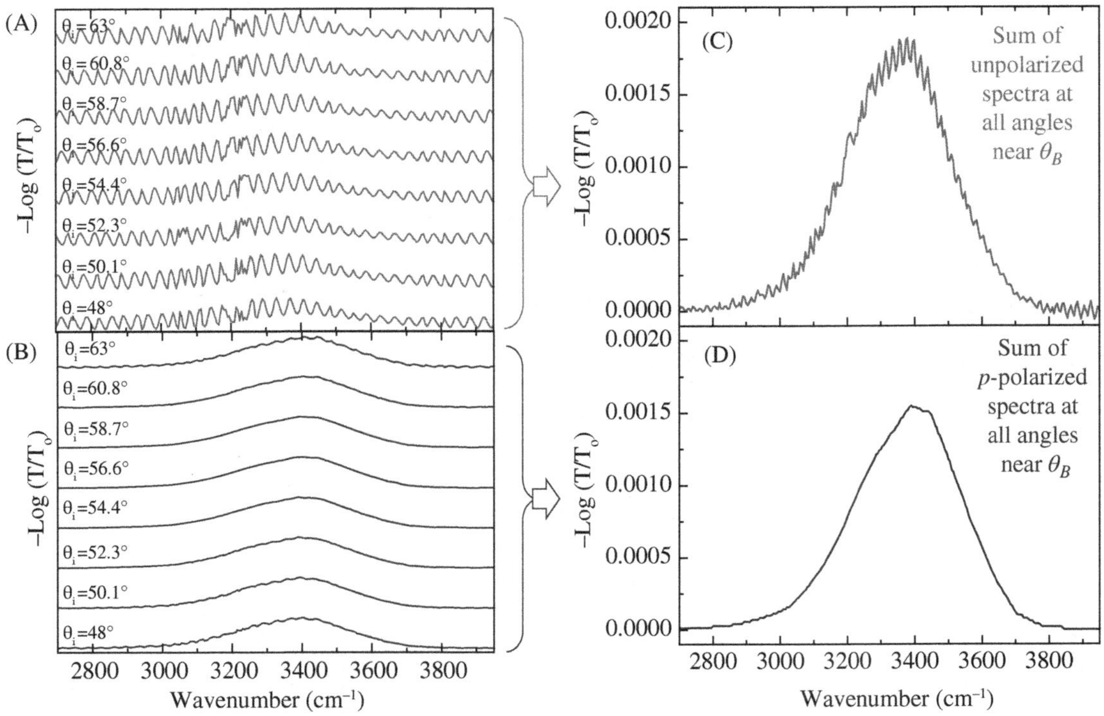

Figure 6.35 Theoretical transmission spectra of the 0.5-nm thick (=d_2) liquid water layer on a calcium boroaluminosilicate glass panel calculated for (A,C) unpolarized and (B,D) *p*-polarized light at (A,B) various incident angles near the Brewster angle (θ_i = 55°) and (C,D) weighted sum of all incident angles. The spectra are normalized with the transmission spectrum calculated for the bare glass ($T_o = t_{135} \cdot t_{135}^*$). The weighting factor can be estimated from the cross-section length of the elliptical reflection lens focusing the IR beam onto the sample. Replotted graphs of the data in Ref. [53].

In Figure 6.35A, it is noted that the $-Log(T/T_o)$ spectra calculated at various incident angles of the unpolarized IR have undulating features on top of the liquid-phase water spectrum. The magnitudes of the undulating features are even larger than the liquid water signal intensity at a given incident angle. These features are due to the interferences of multiple reflected *s*-polarized light ($r_{s,12} \neq r_{s,23} \neq r_{s,34} \neq r_{s,45} \neq 0$). Recall that for the unpolarized light, $T = (t_{s,12345} \cdot t_{s,12345}^* + t_{p,12345} \cdot t_{p,12345}^*)/2$. When these spectra are averaged with the weighting factor estimated for the beam with a broad range of incident angles (because the incident light is focused with a parabolic mirror in real experiment), the undulating features are significantly reduced. This is because the phases of the signal at different incident angles are continuously shifting, but they are not completely suppressed (Figure 6.35C). In contrast, the *p*-polarized $-Log(T/T_o)$ spectra calculated for incident angles near the Brewster angle ($\theta_{B,13} = 55^{o°}$ for the gas/substrate interface considered here) have no undulating features (Figure 6.35B). This is because the reflectivity of *p*-polarized light is zero ($r_{p,12} \approx r_{p,23} \approx r_{p,34} \approx r_{p,45} \approx 0$ at $\theta_i = \theta_{B,13}$), which eliminates the interference effects of multiple reflection; thus, the transmission components of each interface become dominant (Figure 6.1D). The theoretical spectra calculated for the incident angle far from $\theta_{B,13}$ have tiny undulations, but their phases are opposite for the ones above and below $\theta_{B,13}$. So, when all contributing signals are averaged, the calculated adsorbate spectrum has no undulating feature (Figure 6.35D). For this reason, the transmission experiment measuring the spectra of adsorbates or thin films on transparent substrates are done at the Brewster angle of the substrate with the *p*-polarized beam. This is called the BAT experiment.

Figure 6.36 displays the experimental data obtained with the unpolarized and *p*-polarized IR beams. The undulating features in the unpolarized BAT spectra (Figure 6.36A) are quite similar to the theoretically calculated one (Figure 6.35C). The undulation feature makes quantitative analysis of the experimental data difficult. When the BAT experiment is done with the *p*-polarized light, the undulation is completely gone (Figure 6.36B). Note that the spectral shapes are slightly different between the two experiments. The theoretically calculated spectra in Figure 6.35B,D are slightly different as well. This is because the glass spectra are different in the *s*- and *p*-polarizations (see Figure 6.14) and interference effects vary with the wavelength. The experimental data show somewhat larger differences, which could be due to small differences in the glass surface condition.

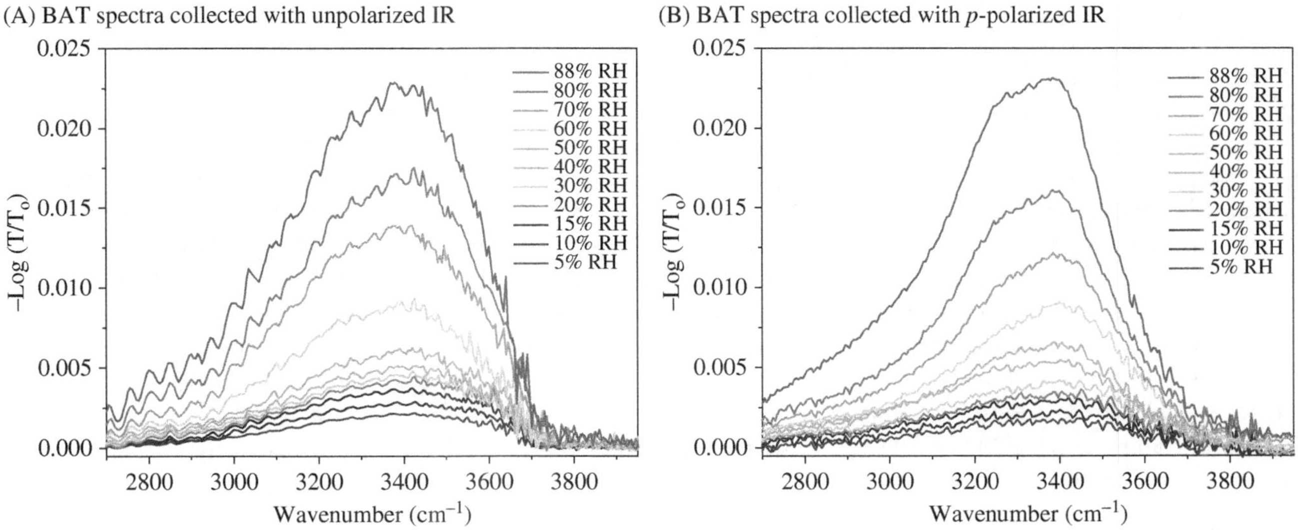

Figure 6.36 Transmission IR spectra, collected with (A) unpolarized and (B) *p*-polarized IR at the Brewster incident angle (θ_B = 55°), of water layers adsorbed on an assembly of five parallel plates of calcium boroaluminosilicate glass at various relative humidity (RH) conditions. In the experiment, the gas-phase spectra were collected at each RH condition without the glass samples in the beam path and then subtracted from the spectra measured with the glass samples to remove the gas-phase water signals. The T_o spectrum was with the bare glass samples in the dry condition (0% RH). Reproduced from [53]/John Wiley & Sons.

Figure 6.37 shows how the *p*-polarized BAT-IR spectra can be analyzed quantitatively. The peak position and shape of the OH stretch band of the condensed water is highly sensitive to the degree or strength of the hydrogen bonding interaction of the OH group with neighboring molecules [58]. One can assume that the water molecules with hydrogen bonding interactions similar to the ones in the liquid phase (which have the average number of hydrogen bonds between 2.2 and 3.4 [59]) will show the OH stretch band similar to the liquid water (with a broad band peaked at ~3400 cm^{-1}). Similarly, the water

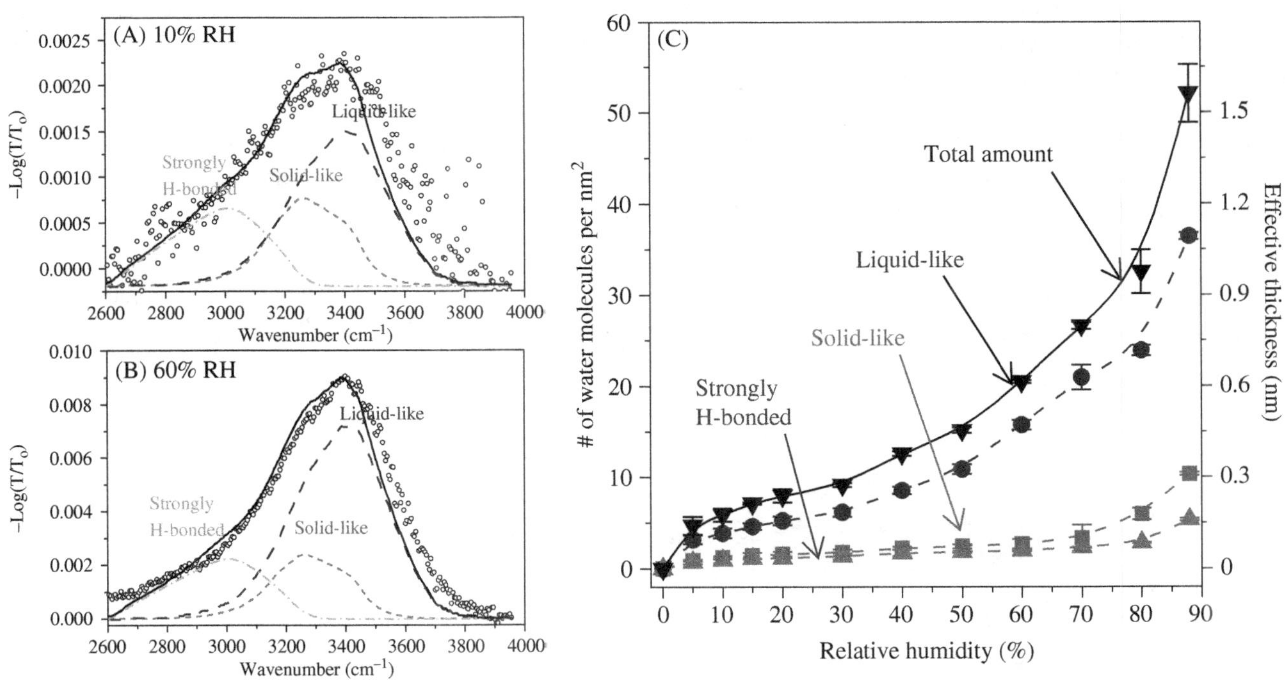

Figure 6.37 (A,B) BAT-IR spectra of water layers on a calcium boroaluminosilicate glass at (A) 10% and (B) 60% RH conditions. The dotted data are from Figure 6.36B, and the solid and dashed lines are the fit results. (C) Adsorption isotherm of water on the calcium boroaluminosilicate glass at room temperature calculated from the BAT-IR spectra in Figure 6.36B. Replotted using the data in Ref. [53].

molecules in the ice-like structure (with four hydrogen bonds per molecule) will show the OH stretch band similar to the ice water (with a band peaked at ~3250 cm^{-1}). If there are water molecules having even stronger hydrogen bonding interactions (due to the presence of surface-bound OH groups), there will be significant intensities in the wavenumber region below the typical ice water spectrum [3, 4]. Figure 6.37A,B shows the deconvolution of the experimental data assuming the presence of these three components in the adsorbed water layer: liquid-like water, ice-like water, and strongly H-bonded water. The hydrogen bonding interaction changes not only the absorption band position but also the absorptivity because the damping factor changes [60, 61]. When the peak area of each component is converted to the "effective" thickness of each component using appropriate absorptivity, one can get the adsorption isotherm of water, which is shown in Figure 6.37C [53]. The adsorbed water layer amounts or thicknesses shown here are calculated with the absorptivity estimated from the bulk phase water (liquid and ice); it is likely that the interactions with the dielectric substrate may change the polarization behavior of water (as in the case of strongly H-bonded ones in the first monolayer), which would alter the refractive index of the adsorbed water. Also, the calculation is done assuming the adsorbed layer is homogenous. If the adsorbate layer thickness is not uniform (e.g., due to the formation of islands), this will introduce an error in the thickness estimation. These are some fine details that one should keep in mind when analyzing or interpreting the adsorption isotherm data obtained from infrared spectroscopy.

6.5 Diffuse-Reflectance Infrared Fourier Transform (DRIFT) Spectroscopy

So far, we have considered the "optically flat" interface, where the reflected and transmitted lights propagate along the specific directions which are governed by the incident angle of light. When IR analysis is to be done for "rough" surfaces or powders, SR-IR and RAIRS methods will not work. ATR-IR can still work as long as the sample can be compressed tightly against the IRE surface. But, if the interaction of the analyte surface with molecules in the surrounding gas or liquid phase is to be studied *in situ*, the ATR-IR method would not work well. In that case, DRIFT method can be used [62, 63]. Unlike ATR-IR (Section 6.1), SR-IR (Section 6.2), RAIRS (Section 6.3), and BAT-IR (Section 6.4) that we have studied, diffuse reflection lacks an explicit theoretical description due to the complexity originating from the inhomogeneity of the reflecting surface. When the surface roughness or powder size is comparable to the wavelength of light, the incident light will be scattered by the sample in addition to absorption by the sample. Thus, the diffuse reflectance spectrum is a function of not only the complex refractive index of the sample but also the scattering characteristics of the sample. Although the exact solution cannot be derived, an approximate solution can still be derived, which provides an insight into the DRIFT analysis. Note that the theoretical derivation discussed in this section cannot be employed to describe the scattering from roughened metallic surfaces. Each facet of the rough metal surface will have specular reflection at its front surface; the only reason that the rough metal surface looks "scattering" is because different facets reflect the incident beam in different directions.

Let us consider densely packed powders of a dielectric material (such as a packed bed of fine oxide particles). A fraction of the incident light will be reflected from the front surface of each particle and the rest will penetrate the particle. The light

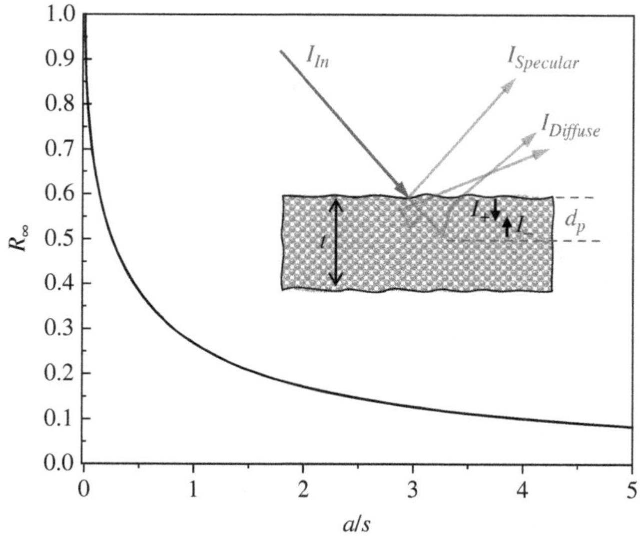

Figure 6.38 Diffuse reflectance (R_∞) vs. the absorption to scattering coefficient ratio (a/s). The inset depicts the diffuse reflectance process.

propagating through the powder sample will get scattered due to the inhomogeneity of the powder (see inset of Figure 6.38). If the sample is randomly inhomogeneous, the phases of the scattered waves are all random. This means the phase relationship (Figures 6.16 and 6.34) drops out in averaging of the radiation amplitude over time. Instead, diffuse reflectance is described by the intensity of the light propagating through the randomly inhomogeneous medium. We can describe the amount of light absorbed by the sample (I_{abs}) during the propagation along the infinitesimal path (dx) as:

$$\frac{dI_{abs}(x)}{dx} = -a \cdot I(x) \tag{6.50}$$

where the proportionality constant is the absorption coefficient, $a = 4\pi k(\overline{\omega})/\lambda$. Recall that $k(\omega)$ is the imaginary part of the complex refractive index of the sample. Similarly, the amount of scattered light along the infinitesimal path can be described as:

$$\frac{dI_{sc}(x)}{dx} = -s \cdot I(x) \tag{6.51}$$

where s is the scattering coefficient. If an internally inhomogeneous sample is illuminated, there will be forward (I_+) and backward (I_-) propagations of the light within the sample. By extending the concept of eqs. (6.50) and (6.51), the changes in the forward and backward components at a given position x can be expressed as [64]:

$$\frac{dI_+(x)}{dx} = -(a+s) \cdot I_+(x) + sI_-(x) \tag{6.52}$$

$$\frac{dI_-(x)}{dx} = -(a+s) \cdot I_-(x) - sI_+(x) \tag{6.53}$$

These first-order differential equations can be solved to obtain the equation governing the diffuse reflectance [64, 65]. As the sample thickness approaches infinity ($t \to \infty$), the solution of these equations gives the reflectance in the following form [64, 65]:

$$R_\infty = 1 + \frac{a}{s} - \sqrt{\frac{a}{s}\left(\frac{a}{s} + 2\right)} \tag{6.54}$$

Figure 6.38 plots eq. (6.54). Note that as $a/s \to 0$, $R_\infty \to 1$. Conversely, as a/s increases from zero, R_∞ decreases fast, implying a high sensitivity of diffuse reflectance to weak absorption systems.

The experimental reflectance spectrum, which is described in eq. (6.54), is processed with the Kubelka–Munk (KM) transform to give an absorption-like spectrum:

$$\frac{a}{s} = \frac{(1-R_\infty)^2}{2R_\infty} \tag{6.55}$$

This formula describes diffuse reflectance of a semi-infinite sample with negligible front surface reflectance.

For the randomly shaped powders, the grain size and the spacing between the grains can be assumed to be on the same order of magnitude. Then, both parameters can be represented by a single parameter, d. If d is small (i.e., $ad \ll 1$), one can find $s = \frac{R}{d}$, where $R \propto \sin^2(nad)$ [65]. When the particle size is much smaller than the wavelength ($d \ll \lambda$), then $\sin^2(nad) \approx (nad)^2$; thus, s is proportional to n^2a^2d. If d is larger than λ, the $\sin^2(nad)$ term oscillates as d changes. The average of this oscillating value over the spread of d converges to ½ [65]. Then, s becomes proportional to $1/d$. From this, one can see how the DRFIT intensity will vary with the particle size. The scattering coefficient is the largest when the particle size is comparable to the wavelength; in that condition, the detection sensitivity of diffusion reflectance is the largest (Figure 6.38).

In DRIFT, the penetration depth at which the IR intensity decays to e^{-1} can be expressed as:

$$d_p = 1/\sqrt{a(a+2s)} \tag{6.56}$$

In the absence of scattering ($s = 0$), eq. (6.56) becomes the same as the definition of IR attenuation inside a homogenous sample. In experiment with powders, it is difficult to determine the absolute value of s. So, it is difficult to convert the measured DRIFT intensity to effective thickness or amount of surface species using theoretical calculations. Thus, one must construct the calibration curve for quantitative analysis. But, if the particle size fluctuates among the calibration samples, then the calibration may not be reliable. Even if the particle size is invariant, a variation in packing density can affect the DRFIT sensitivity [66]. Note that all derivations discussed so far are for weakly absorbing materials. Inorganic powders with high absorptivity can have very complicated peak shapes near their own absorption bands due to the Reststrahlen effect (see Section 5.1.4), which makes the quantitative interpretation of the Kubelka–Munk transform intensity difficult [67].

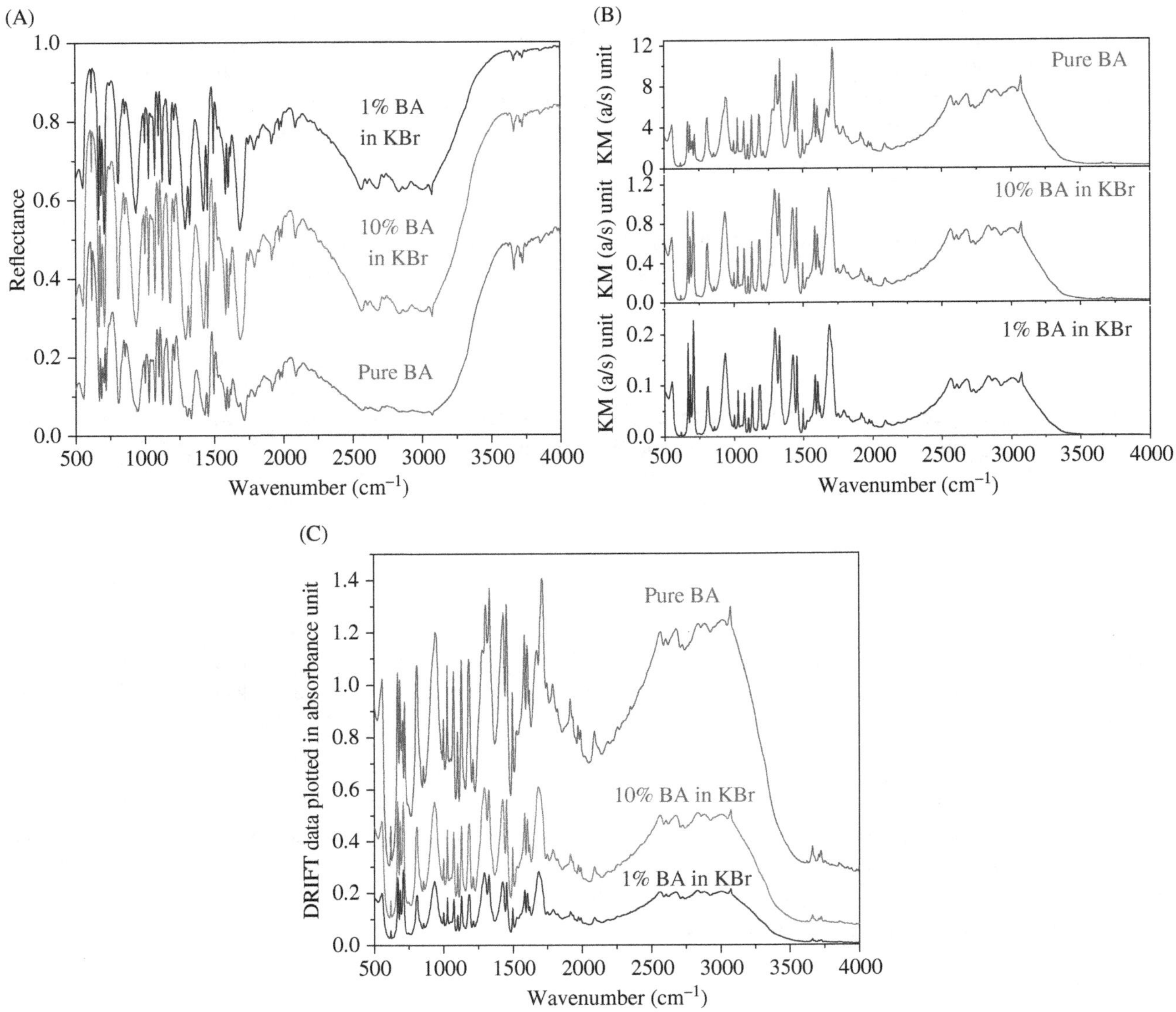

Figure 6.39 DRIFT spectra of benzoic acid in (A) reflectance unit and (B) KM transform unit. In (C), the same DRIFT spectra are plotted in the absorbance unit as found in many papers in literature, which is physically meaningless. By courtesy of Jordan Meyet, Penn State Materials Characterization Lab.

For the demonstration purpose, the DRIFT data of benzoic acid (BA) sample prepared in three different concentrations were collected and plotted in three different units in Figure 6.39. The BA powders were ground using a mortar and pestle and packed in a commercial DRIFT cell (Harrick Scientific's Praying Mantis™ accessary). The powder bed was leveled with a spatula. Figure 6.39A displays the raw DRIFT spectra of BA powders without and with dilution in a spectrograde KBr matrix. In order to do quantitative analysis, the spectra are converted to the KM unit (in a/s) using eq. (6.55), which are plotted in Figure 6.39B. We can see that the intensity in the KM unit is roughly, but not exactly, proportional to the BA concentration. The deviation from linearity is due to difference in particle size, which is difficult to avoid. The relative intensities among the low and high wavenumber peaks also change. This suggests that the average particle size (obtained by grinding) is smaller than or comparable to the wavelength range of IR; so, the s term in the KM unit varies with the IR wavelength (i.e., $\propto a^2d \propto d/\lambda^2$). If the particle size was much larger than 20 μm (= 500 cm^{-1}), then the wavelength dependence of s would be small ($s \propto 1/d$), and the relative intensities among different modes would be relatively constant for the three samples analyzed here. Also, the shape of the bands at ~670 cm^{-1}, ~1280 cm^{-1}, and 1690 cm^{-1} are different between the sample ground without KBr matrix (pure) and the ones ground with KBr matrix (10% and 1%). This suggests that these modes are very sensitive to the sample preparation (probably fracture or crushing of particles or ionic interactions between K^+ and $-COO^-$ during grinding). This shows how sensitive the DRIFT analysis is to the sample preparation method, which complicates the data interpretation or introduces artifacts in the data.

It is not uncommon to see papers in the literature reporting the DRIFT spectra in the absorbance unit. The instrument software can display the data in the absorbance unit, or sometimes that is the default setting. Note that this does not mean the DRFIT spectra can be plotted or interpreted in the absorbance unit. It is fundamentally incorrect. Figure 6.39C shows the reason. If the DRIFT spectra are plotted in the absorbance unit, the signal intensity does not match with the concentration of analyte in the sample at all. In Figure 6.39C, the intensity of 1% BA in KBr is only six times smaller than that of pure BA; it should be ~100 times smaller.

6.6 IR Spectroscopic Imaging

6.6.1 Microscopic IR Imaging

Infrared spectroscopy can be combined with an optical microscope to conduct spatially resolved chemical analysis. This allows nondestructive and label-free chemical imaging. The electromagnetic wave in the mid-IR wavelength does not penetrate through typical lens materials (which are made of oxide glasses); most refractive objective lens cannot be used to focus IR to the sample surface. It is possible to make a refractive objective lens with IR-transparent materials such as BaF_2 or CaF_2; but these have a cutoff wavelength below which IR is attenuated by the absorption. The more commonly used lens for microscopic analysis in the mid-IR region is the reflective objective lens with the reverse-Cassegrain design (also called Schwarzschild design). In this design, collimated light passes through the aperture hole in the center of a concave parabolic primary mirror to a convex hyperbolic secondary mirror. The secondary mirror then reflects and diverges the beam to fill the primary mirror, and the primary mirror focuses the beam to a small focal spot in the object plane. The light reflected from or transmitted through the sample can be processed through the Fourier transform or dispersed with a monochromator if a broad band light source is used. Alternatively, a monochromatic source can be used for intensity measurement at a specific absorption band position. Then, the IR intensity or spectrum is recorded as a function of the sample location to conduct chemical mapping. Many excellent review papers can be found covering the fundamentals of the technique, instrumentation, and applications [68, 69].

When the imaging is done by focusing the probing light, the spatial resolution is limited by the diffraction limit ($d_{Abbe} = \lambda/2NA$, where NA is the numerical aperture of the lens) and the pixel size of the detector [68]. So, it is difficult to obtain a spatial resolution smaller than a few microns in the mid-IR imaging using the objective lens alone. To overcome this limitation, several techniques combining IR spectroscopy with AFM were developed and are now commercially available. They are summarized in Table 6.2. The details of AFM are discussed in Chapter 11. The basic principle of each

Table 6.2 Comparison of commercially available spectroscopic imaging techniques in the mid-IR spectral region.

Acronym of technique	(Photothermal) AFM-IR	Nano-IR (s-SNOM)	PiFM
Detection mechanism	Height change of the sample due to thermal expansion of the subsurface region upon IR absorption	Alteration of near-field scattering from the apex of the probe tip, which is a function of the dielectric constant of the sample underneath the tip	Coulombic force between the induced dipole of the tip and the sample and/or photothermal expansion of the sample
AFM operation mode*	Contact, tapping, or force-modulation (FM)	Tapping	Noncontact
Lateral resolution**	>20 nm	~10 nm	~10 nm
Effective probe depth**	Contact: up to micron Tapping & FM: >30 nm	~50 nm (on the order of tip curvature)	~20 nm
Correlation with typical absorbance spectrum, $a(\omega) = 4\pi k(\omega)/\lambda$	Good when $n(\omega)$ is relatively invariant and much larger than $k(\omega)$ works well for organic samples	Raw scattering signals (amplitude and phase) can be converted to $\varepsilon_1 + i\varepsilon_2$, then to $n(\omega) + ik(\omega)$	If polarization force is dominant, the data correlates with $n(\omega)$; if photothermal effect is dominant, it correlates with $k(\omega)$

* See Chapter 11.
** Best reported in literature or an estimate; note that this is obtained for control samples with fine tuning of operation modes and cannot be guaranteed for all types of samples.

technique will be briefly explained in the context of what we have studied so far in Chapters 5 and 6. Full derivations of the spectroscopic detection principles will be skipped here because they are beyond the scope of this book, and they can be found in the literature.

6.6.2 Photothermal Atomic Force Microscopy (AFM)-IR

AFM-IR is basically the photothermal resonance infrared (PT-IR) spectroscopy. Upon the resonance absorption of photon, the damped oscillator dissipates a small fraction of the absorbed energy through heat, which increases the sample temperature. The change in sample temperature can alter various physical properties including refractive index or volume of the sample, which can be detected optically [70, 71]. AFM-IR is just another method that measures a tiny deflection of the AFM cantilever in contact with the sample that undergoes the photoinduced thermal expansion process. Since the IR absorption is enhanced when the IR energy is in resonance with the absorption band, the height change detected with AFM is also enhanced at the absorption band. Since the size of the probe tip underneath the AFM cantilever is on the order of 10's nm, the lateral resolution can be, in theory, as small as the tip size [72, 73]. Under the electric dipole approximation which is valid when $n(\omega) \gg k(\omega)$ and the sample size is much smaller than λ, the absorbed power of the electromagnetic wave can be expressed as [74]:

$$P_{abs} = \frac{2\pi}{\lambda} c\varepsilon_o \frac{9n(\varpi)k(\varpi)}{\left(n(\varpi)^2 + 2\right)^2} |E_{in}(\varpi)|^2 V \tag{6.57}$$

where $E_{in}(\varpi)$ is the incident electric field, λ is the wavelength of the incident light, and V is the sample volume. For typical organic materials, the change in $n(\varpi)$ near the absorption band is very small compared to that of $k(\varpi)$. In other words, the dispersive effect in $n(\varpi)$ is weak around the absorption band. In that case, eq. (6.57) implies that P_{abs} is practically proportional to $k(\varpi)/\lambda$, as in the typical absorbance measured in the transmission IR spectroscopy. Note that this assumption is not valid for inorganic dielectric materials, which will be discussed later.

The temperature change in the sample can be described with the Fourier heat equation [74]:

$$\rho C_p \frac{\partial T}{\partial t} = \frac{P_{abs}\Pi\left(t_p\right)}{V} + \kappa_{th}\nabla^2 T \tag{6.58}$$

where ρ, C_p, and κ_{th} are the density, heat capacity, and thermal conductivity of the sample, respectively, Π is a rectangle function of IR pulse length t_p, and ∇^2 is the second derivative with respect to the lab coordinate. The solution of eq. (6.58) depicts that when the IR laser pulse duration (t_p) is shorter than the thermal diffusion time of the sample ($t_{td} = R^2/\alpha$, where R is the radius of the IR irradiated area and $\alpha = \kappa_{th}/\rho C_p$ is the thermal diffusivity), the temperature inside the sample increases linearly to the maximum value ($\Delta T_{max} = P_{abs}/\rho C_p V$) during t_p and then decays exponentially with the characteristic decay time of t_{td} (Figure 6.40A). The temperature increase in the sample ($\Delta T(t)$) induces an internal strain, resulting in expansion of the heated zone which can be detected as a height change ($\Delta h(t)$) with AFM [72, 73]. The degree of thermal expansion will be governed by the coefficient of thermal expansion (CTE) of the material. Thus, one can expect that a material with higher CTE is likely to give a stronger AFM-IR signal. An example of AFM-IR analysis is shown in Figure 6.40B.

In order to detect $\Delta h(t)$ without interferences from natural vibrations of the experimental system or thermal drift, the AFM-IR experiment is done with a pulsed IR beam with a certain repetition frequency (ν_r) and detecting the $\Delta h(\varpi)$ signal oscillating at the same frequency as ν_r. By tuning ν_r to one of the AFM cantilever resonance frequencies ($\nu_{cl,1}$) while keeping the force applied to the sample by the cantilever constant, the amplitude of the detected $\Delta h(\varpi)$ signal is enhanced. This is called resonance-enhanced contact mode. AFM-IR can be done in the tapping mode in which the AFM cantilever is oscillated at a fixed frequency and amplitude and makes intermittent contacts with the sample surface. This mode uses the heterodyne detection of the cantilever resonant excitation ($\nu_{cl,1}$) by non-linear mixing of the cantilever oscillation at another resonance frequency ($\nu_{cl,2}$) with the sample photothermal expansion at ν_r. By using a high frequency for the photothermal modulation ($\nu_r \gg \nu_{cl,1}$), background signal can be suppressed, and spatial resolution can be improved. The same heterodyne detection principle can be used in the force modulation (FM) mode in which non-linear mixing is done with modulation of the force applied to the sample by the cantilever in the contact mode. The FM mode is also called surface-sensitive mode since it gives much better surface sensitivity than the resonance-enhanced contact mode [75].

The ultimate resolution of AFM-IR can be as small as the tip diameter, which can be on the order of 10s nm; this is the reason that AFM-IR can achieve the lateral resolution much smaller than the diffraction limit (d_{Abbe}). However, it should be noted that the practical limit of the achievable resolution in AFM-IR is highly dependent on the AFM operation mode and

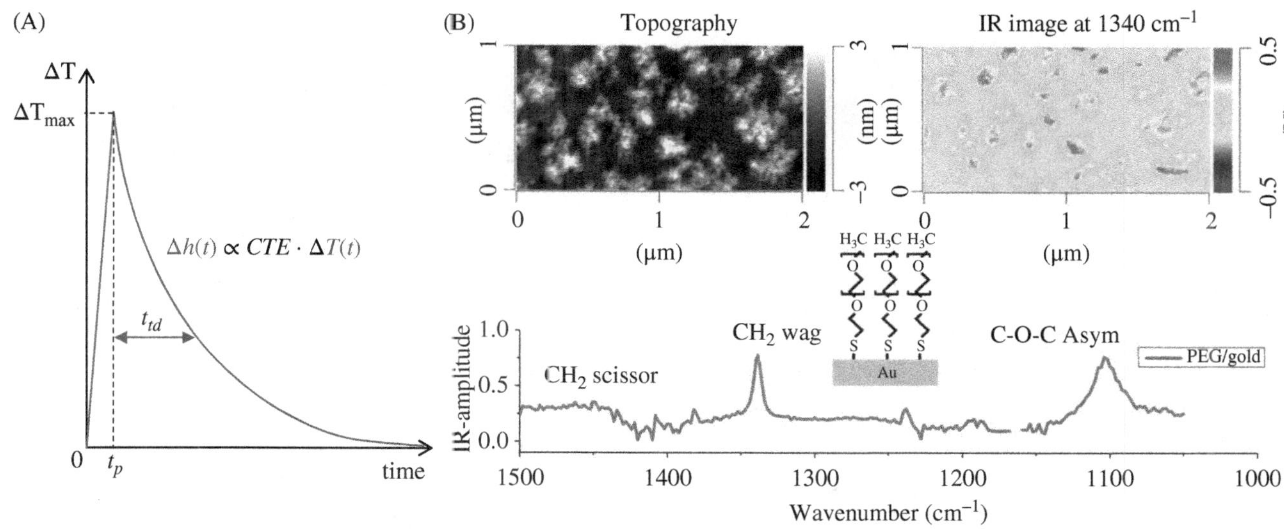

Figure 6.40 (A) A schematic illustration of the sample temperature illuminated by an IR laser pulse with a pulse duration (t_p) shorter than the characteristic thermal diffusion time (t_{td}). (B) AFM-IR of a self-assembled monolayer of polyethylene glycol (PEG) on gold. Shown here are AFM topographic image, the intensity map of AFM-IR signal at 1340 cm^{-1}, and AFM-IR spectrum of the monolayer. Reproduced from [73]/with permission of American Chemical Society.

the sample properties. Since the IR absorption by the sample is detected through the local height change of the sample, the magnitude of the AFM-IR signal is highly dependent on the competition between the sample heating rate by IR pulses vs. the thermal relaxation rate through heat conduction to the surrounding as well as the mechanical coupling of the thermal expansion dynamics of the sample with the cantilever vibration dynamics. The elastic modulus and mechanical damping behavior of the sample will have strong effects on AFM-IR detection sensitivity; note that these properties have no impact of typical IR experiments where changes in light intensity are detected. For inhomogeneous materials, expansion of non-absorbing regions due to thermal diffusion from neighboring absorbing regions will also affect the lateral resolution. The effective probe depth varies significantly depending on the AFM operation mode. In the contact mode which typically done at the contact resonance of the tip–sample interface, the probe depth is a strong function of the IR beam attenuation inside the sample; thus, the probe depth would be similar to (or on the same order of magnitude as) the optical probe depth discussed in Section 6.2.1. In the tapping and FM mode, the IR beam with a very high repetition frequency is used and the AFM-IR signal is measured in a heterodyne detection mode; in this case, the effective probe depth can be reduced to ~25–50 nm, depending on the IR repetition frequency, the cantilever driving frequency, and signal mixing between these two. Further details of AFM-IR operation mode are beyond the scope of this book; readers interested in this subject can find many review papers in the literature for self-study [75].

Based on the linear correlation between P_{abs} described in eq. (6.57) and $\Delta h(\varpi)$, the AFM-IR spectrum is expected to resemble the typical absorbance spectrum of the sample, $A(\varpi) \propto a(\varpi) = 4\pi k(\varpi)/\lambda$. This is valid only for organic materials because eq. (6.57) is derived with the assumption that $\sqrt[3]{V}$ (sample size which is undefined in eq. (6.57)) $\ll \lambda$ and $k(\varpi) \ll n(\varpi) \approx n$ (negligible dispersion). But, these assumptions are not valid for inorganic materials with large dielectric constants. For example, in Figure 6.41A, we can see that the theoretical spectrum calculated with eq. (6.57) and the $n(\varpi)$ and $k(\varpi)$ values of silica do not match with the AFM-IR spectrum of silica glass collected in the contact mode [76]. Obviously, we know that the assumption of $k(\varpi) \ll n(\varpi) \approx n$ is not valid here. And, the irradiated volume is not smaller than the wavelength of IR. The radius of the heated area by the IR beam ($R = 3r_o$, where r_o is the Gaussian beam raius) irradiated at a glancing angle is much larger than λ (Figure 6.41B). Thus, the P_{abs} cannot be estimated using eq. (6.57). In order to simulate the IR absorption more accurately, one has to consider the transmittance of the beam across the air/sample interface, which can be calculated from the reflectance ($R_i(\varpi)$) using eq. (6.9). Considering the transmittance at the interface and the thermal properties of silica, the Fourier heat equation for the geometry shown in Figure 6.41B gives the following solution [76]:

$$\theta(r,z,t) = e^{t/z_d^2}\left[e^{\frac{z}{z_d}}\operatorname{erfc}\left(\frac{\sqrt{t}}{z_d} + \frac{z}{2\sqrt{t}}\right) + e^{\frac{-z}{z_d}}\operatorname{erfc}\left(\frac{\sqrt{t}}{z_d} - \frac{z}{2\sqrt{t}}\right)\right]\sum_{n=0}^{\infty}\frac{f(\beta_n)}{J_1^2(\beta_n)}J_0(\beta_n r)e^{-\beta_n^2 t} \tag{6.59}$$

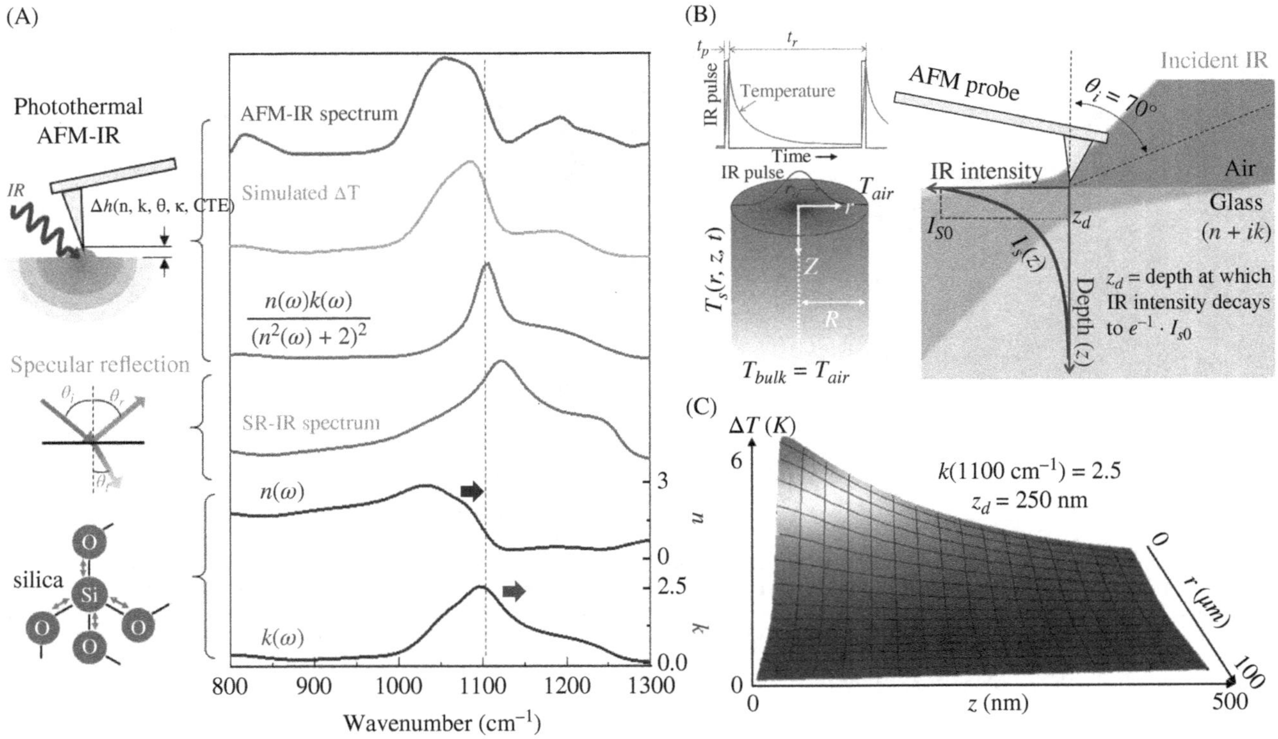

Figure 6.41 (A) Comparison of complex refractive index, SR-IR, and AFM-IR spectra of a fused quartz. The spectrum predicted using eq. (6.57) and the temperature change (ΔT) calculated using eq. (6.59) are shown for comparison. (B) Schematic diagram of experimental geometry and dimensions of AFM-IR. The narrow-band IR pulse with a temporal pulse width t_p and a repetition time t_r is focused from the side of the AFM probe with an elliptical reflection mirror. (C) Radial and depth distribution of temperature inside the fused quartz irradiated with a 1100 cm^{-1} pulse (2 mW, t_p = 100 ns) and a beam waist (r_o) of 50 μm. Reproduced from [76]/with permission of American Chemical Society.

where $\theta(r,z,t) = \kappa_{th}(T_S - T_{air})/(R/r)^2 s_{\max}$ is the dimensionless temperature with $s_{\max} = \frac{I_{in}(1-R_i)}{\pi r_0^2 z_d}$ (where T_S and T_{air} are the sample surface and air temperatures, respectively, and z_d is the characteristic attenuation depth of IR due to absorption; all independent variables are dimensionless), $J_0(\beta_n)$ and $J_1(\beta_n)$ are the zeroth- and first-order Bessel functions, and $f(\beta_n)$ is a convolution of radial and zeroth-order Bessel functions. Figure 6.41C plots the radial and depth distribution of temperature inside the fused quartz irradiated with a single pulse of a 1100 cm^{-1} laser beam (2 mW) at t_p = 100 ns at an incident angle $\theta_i = 70°$. The temperature increase in the center region of the irradiated area is estimated to be ~6 K and it decays over ~$1.5r_o$ radially and ~$2z_d$ in depth. When the average temperature change over z_d by a single IR pulse is calculated as a function of wavenumber using $n(\overline{\omega})$ and $k(\overline{\omega})$, the $\Delta T(\overline{\omega})$ spectrum matches relatively well with the experimental AFM-IR spectrum (Figure 6.41A). The calculated spectrum also explains the difference from the SR-IR spectrum; AFM-IR is governed by the absorption of the transmitted beam followed by thermal expansion, while SR-IR is governed by the reflected beam which has zero contribution to the sample temperature change. This example demonstrates the importance of full consideration of optical properties of inorganic dielectric materials in AFM-IR analysis.

6.6.3 Nano-IR Spectroscopy Based on Scattering-Type Scanning Near-field Optical Microscopy (s-SNOM)

The nano-IR analysis is based on the s-SNOM technique. It uses an AFM probe coated with a metallic thin film. The illuminated apex of the tip can exhibit an enhanced optical field in its vicinity, and this near-field is modified by the presence of the sample. Thus, the scattered light measured in the far field carries the near-field information on the tip–sample local interactions. This is the basis of the s-SNOM principle. The highly concentrated near-field at the metallic tip apex also enhances Raman scattering, which is the basis of tip-enhanced Raman spectroscopy (TERS) [77, 78]. The sample illumination geometry is the same as the one shown in Figure 6.41B. Instead of detecting the thermal expansion of the sample surface, nano-IR detects the backscattered IR beam in the far field. The near-field interaction is confined in the gap between the tip apex and the sample which is sinusoidally modulated under the tapping mode AFM operation. The typical amplitude of the cantilever modulation (tapping) is about 2–3 times of the tip radius (a_t). The backscattering from the illuminated tip surface and the sample outside the probing region is barely modulated by the tip apex; thus, their contribution in the back-scattering signal can be filtered out electronically from the detected signal.

In theoretical derivation, the AFM probe tip ($a_t \ll \lambda$) is modeled by a sphere with a polarizability described by [79]:

$$\alpha(\varpi) = 4\pi a_t^3 \frac{\varepsilon_t(\varpi) - 1}{\varepsilon_t(\varpi) + 2} \tag{6.60}$$

where $\varepsilon_t(\varpi)$ is the dielectric function of the metal coated on the tip. In the mid-IR range, the dielectric constant of metal has a relatively large negative value (see Section 6.3.4) and its frequency dependence is relatively weak. Based on the antenna theory, the dipole induced in this sphere by the incident electric field (E_i) is along the surface-normal direction. This field can induce the polarization (mirror dipole, as illustrated in Figure 6.28B) of the sample. In the electrostatic limit and negligible retardation, the polarizability of the sample can be described by [79]:

$$\beta_s(\varpi) = \frac{\varepsilon_s(\varpi) - 1}{\varepsilon_s(\varpi) + 1} \tag{6.61}$$

where $\varepsilon_s(\varpi)$ is the dielectric function of the sample. The electric field of the mirror dipole decreases with the third power of the tip–sample distance (z). The interaction of this mirror dipole with the tip dipole leads to an effective polarizability of the coupled tip–sample [79]:

$$\alpha_{eff}(\varpi) = \frac{\alpha(\varpi)(1 + \beta_s(\varpi))}{1 - \alpha(\varpi)\beta_s(\varpi)/\left(16\pi(a_t + z)^3\right)} \tag{6.62}$$

The scattered near-field (E_s) is proportional to the product of $\alpha_{eff}(\varpi)$ and the incident electric field (E_{in}). Since the dielectric function is a complex function, $\alpha_{eff}(\varpi)$ is also a complex function: $s(\varpi)e^{i\varphi(\varpi)}$, where $s(\varpi)$ and $\varphi(\varpi)$ are the amplitude and phase of the scattered near-field. In the mid-IR region, $\varepsilon_t(\varpi)$ is relatively frequency independent and can be replaced with a constant. In eq. (6.62), when z is two or three times larger than a_t (i.e., $a_t^3/(a_t + z)^3 \ll 1$), the second term in the denominator is much smaller than 1. This allows us to simplify the entire equation using the Taylor series, $\frac{1}{1-x} \cong 1 + x$. Thus, eq. (6.62) can be simplified to [80]:

$$\alpha_{eff}(\varpi) = s(\varpi)e^{i\varphi(\varpi)} \cong B\frac{\varepsilon_s(\varpi) - 1}{\varepsilon_s(\varpi) + 1} \tag{6.63}$$

where B is a constant governed by experimental conditions (i.e., the dielectric constant and radius of the tip and the oscillation amplitude and frequency of the tip). From the amplitude and phase of the scattered near-field, $s(\varpi)e^{i\varphi(\varpi)}$, one can calculate the complex dielectric function of the sample, $\varepsilon_1(\varpi) + i\varepsilon_2(\varpi)$, from which the complex refractive index of the sample, $n(\varpi) + ik(\varpi)$, can be calculated using eqs. (5.40) and (5.41). In general, the amplitude signal $s(\varpi)$ resembles the dispersive part of the refractive index and the phase signal $\varphi(\varpi)$ resembles the absorptivity spectrum [81]. The separation of the scattered near-field signal from the spurious far-field signal (measured with regular optics far away from the sample) can be done by demodulation of the detector signal at a higher harmonics of the tip oscillation frequency. For quantification of the signal, the s-SNOM signal of the sample can be normalized with that of a reference sample (which can be a clean substrate). Since the $\alpha_{eff}(\varpi)$ depends on the third power $(a + z)^{-3}$, a better surface sensitivity can be achieved when a smaller modulation amplitude is used.

Figure 6.42 shows an example of converting the amplitude and phase of the scattering signal from s-SNOM to the complex dielectric constant which then can be converted to the complex refractive index [80]. Figure 6.42A shows the topographic analysis result of a nanoscratch line made on a silica glass by a cono-spherical nanoindent tip sliding with an applied load of 7 mN. The raw scattering signals (Figure 6.42B) were baseline corrected and shifted so that, in the Argand plot (see Section 7.3.9), the imaginary component stayed positive, and the real component was relatively symmetric. To obtain the dielectric function (Figure 6.42C) and the refractive index (Figure 6.42D), the reconstructed amplitude and phase signals were processed with the following equation:

$$\varepsilon_s = Q\frac{1 + (S \cdot e^{i \cdot \varphi}/B^2)}{1 - (S \cdot e^{i \cdot \varphi}/B^2)} = \varepsilon_1 + i\varepsilon_2 = (n + ik)^2 \tag{6.64}$$

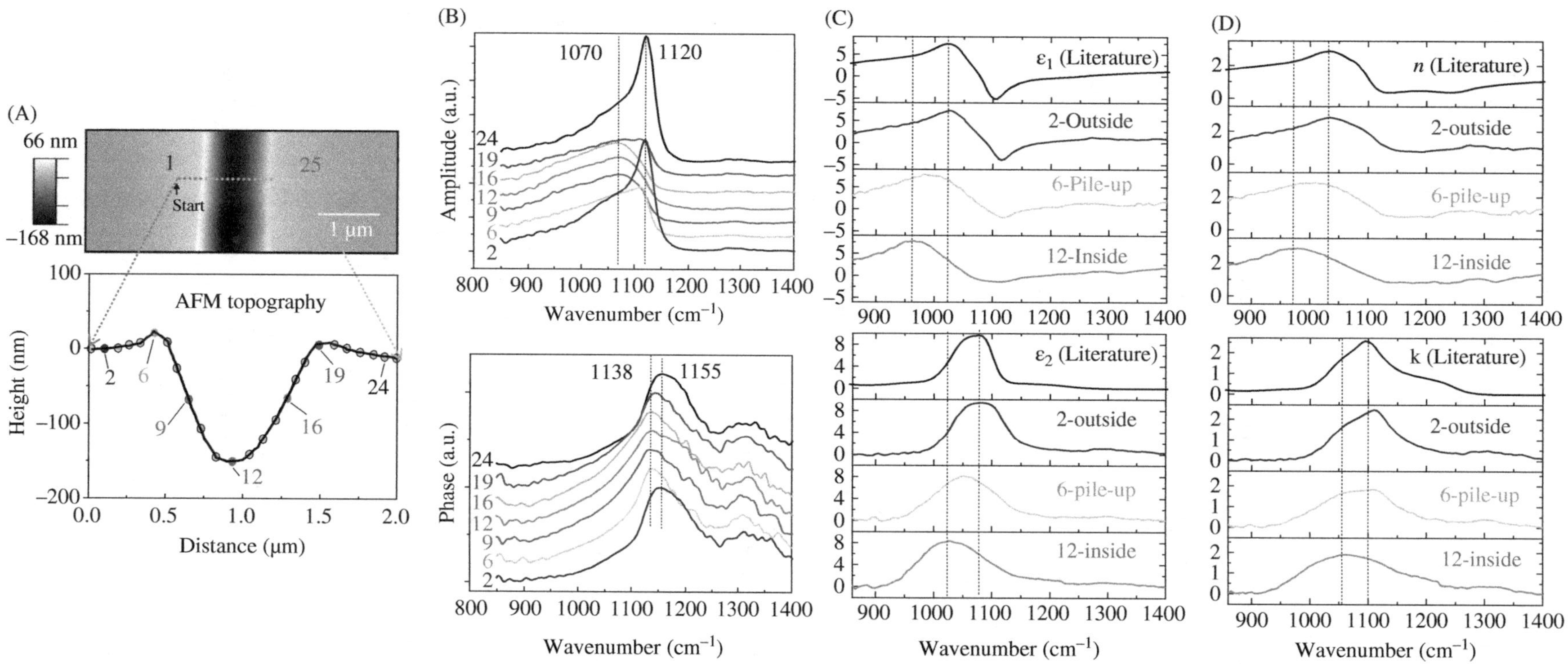

Figure 6.42 (A) AFM topography images and cross-sectional profile of a nanoscratch line on a silica glass surface. The nano-FTIR signals were measured at 25 locations (separated by 80 nm) across the scratch line marked in the image. (B) Amplitude (S_2) and phase (φ_2) spectra of the second-harmonic scattering signals at selected locations. Comparison of (C) ε_1 and ε_2 spectra and (D) n and k spectra calculated from the measured S_2 and φ_2 using eq. (6.64). Reproduced from [80]/with permission of ELSEVIER.

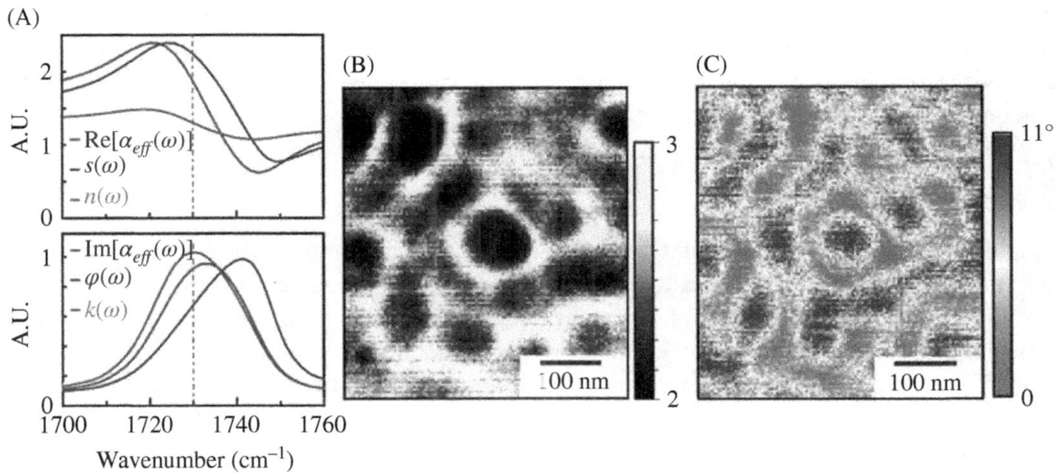

Figure 6.43 Example of s-SNOM analysis of polymer surfaces. (A) Comparison of the amplitude $s(\varpi)$ and phase $\varphi(\varpi)$ spectra of PMMA with $\text{Re}\left[\alpha_{eff}(\varpi)\right]$ and $\text{Im}\left[\alpha_{eff}(\varpi)\right]$ as well as $n(\varpi)$ and $k(\varpi)$. (B) $s(1730\ \text{cm}^{-1})$ and (C) $\varphi(1730\ \text{cm}^{-1})$ maps of the carbonyl resonance in a block copolymer. Reprinted with permiss on from [31]. Copyright 2004 American Chemical Society.

where Q is a scaling factor to match with the reference spectrum of a known sample. The peak shape and position in the scattering signal spectra (Figure 6.42B) are quite different from the known absorbance spectra of silica glass, but the processed ε_1 and ε_2 spectra (Figure 6.42C) and the n and k spectra (Figure 6.42D) are quite comparable to the corresponding spectra of silica glass. This large change in the spectral feature in the scattering signal is because the absorptivity of the sample is quite large. Due to the Kramers–Kronig relation, both real and imaginary parts of the dielectric function vary significantly around the absorption band. Thus, the accurate interpretation of the scattering spectrum should be done after the scattering signals are converted to the optical constant. In Figure 6.42D, the red-shift of the Si–O–Si stretch mode in the nanoscratch line suggests an elongation of the Si–O bond length distribution [80].

Figure 6.43 shows an example of s-SNOM-based nano-IR analysis of polymer samples [82]. In Figure 6.43A, the amplitude $s(\varpi)$ and phase $\varphi(\varpi)$ spectra of polymethylmethacrylate (PMMA) are compared with the calculated $\text{Re}\left[\alpha_{eff}(\varpi)\right]$ and $\text{Im}\left[\alpha_{eff}(\varpi)\right]$ spectra as well as the $n(\varpi)$ and $k(\varpi)$ spectra measured with ellipsometry. The amplitude $s(\varpi)$ spectrum is dominated by the dispersive characteristic of $n(\varpi)$. The phase $\varphi(\varpi)$ spectrum is quite close to the $k(\varpi)$ spectrum. This similarity is because the absorptivity of the sample is quite small and thus the variation of the real part of the dielectric function around the absorption band is relatively small (which is the main difference from Figure 6.42). In this case, the phase spectrum can be used for chemical identification although the apparent shift in the peak position can complicate the data interpretation. When the nano-IR analysis is performed for the inorganic dielectric materials, the spectral shift between $\varphi(\varpi)$ and $k(\varpi)$ is even larger. In such cases, the $s(\varpi)$ and $\varphi(\varpi)$ spectra can be processed with eq. (6.63) to obtain the $\varepsilon_1(\varpi)$ and $\varepsilon_2(\varpi)$ spectra of the sample, which then can be converted to the $n(\varpi)$ and $k(\varpi)$ spectra [80].

6.6.4 Polarization-Induced Force Microscopy (PiFM)

PiFM measures a tiny change in the attractive force exerted by the dipole in the sample which is induced by the near-field of the dipole of the metal-coated AFM probe oscillating by the electric field of the incident light. This near-field dipole interaction is a complex function of dielectric properties of the tip and the sample and the dynamic coupling with the oscillating probe above the sample. Since it measures the attractive force gradient between the tip and the sample, PiFM is operated in noncontact mode. The description of the photoinduced force is a lot more complicated than photothermal AFM-IR and s-SONM nano-IR, but it is still worthy to review an approximate solution. To simplify the problem, several assumptions can be taken. First, both the tip and the sample are assumed to be point dipoles. In a simple mutual interaction picture, the induced dipole moments in each of these two can be depicted as:

$$\mu_t(\varpi) = \alpha_t(\varpi)(E_{in}(\varpi) + E_s(\varpi)) = \alpha_t'(\varpi) + i\alpha_t''(\varpi) \quad (6.65)$$

$$\mu_s(\varpi) = \alpha_s(\varpi)(E_{in}(\varpi) + E_t(\varpi)) = \alpha_s'(\varpi) + i\alpha_s''(\varpi) \quad (6.66)$$

where $\mu_i(\varpi)$, $\alpha_i(\varpi)$, and $E_i(\varpi)$ are the dipole moment, polarizability, and induced electric field of the medium noted by subscript i (t = tip and s = sample), respectively. The force acting on the tip by the induced dipole of the sample can be assumed to be dominated by the coupling along the surface-normal (z) direction only. In the limit of near field, the time-averaged force acting on the tip can be expressed as [83]:

$$F_{dip}(z) \approx -\frac{3\,\mathrm{Re}\left[\alpha_t^*(\varpi)\alpha_s(\varpi)\right]}{2\pi\varepsilon_0 z^4}|E_{in,z}(\varpi)|^2 + \frac{\boldsymbol{k}}{2}\mathrm{Im}[\alpha_t(\varpi)]|E_{in,z}(\varpi)|^2 \tag{6.67}$$

where $\boldsymbol{k}$ is the wavevector as defined in eq. (5.58). In the right-hand side, the first term is the attractive *local* force due to field gradient, and the second term is the constant repulsive scattering force due to radiation pressure (as in the optical tweezer force). The repulsive force extends over the dimension of the incident electric field, which is on the order of μm and thus nonlocal. The attractive force bears a steep distance dependence (z^{-4}), rendering a high lateral resolution and depth sensitivity. It is proportional to the real part of complex polarizability product of the tip and the sample, given as $\mathrm{Re}\left[\alpha_t^*(\varpi)\alpha_s(\varpi)\right] \approx \alpha_t'(\varpi)\alpha_s'(\varpi) + \alpha_t''(\varpi)\alpha_s''(\varpi)$, and carried the spectroscopic information of the sample. The $\alpha_t''(\omega)$ contribution becomes large when the incident light resonates with the plasmonic frequency of the metallic tip (see Figure 6.26). Far away from the plasmon resonance (which is typically in the visible region for most metals), the $\alpha_t''(\varpi)$ term is negligible and thus the contribution from the imaginary part of the polarizability is not detected. In the IR range, the $\alpha_t'(\varpi)$ part is still substantial; thus, the dominant term in the attractive force is $\alpha_t'(\varpi)\alpha_s'(\varpi)$ and the PiFM spectrum near the absorption band of the sample follows a dispersive line shape. An example is shown in Figure 6.44A which compares the PiFM signal of silicon 2,3-naphthalocyanine (SiNc) dye with the real and imaginary parts of its optical response [84].

Although the PiFM is originally designed to detect a small change in the polarization-induced attractive force upon the IR absorption by the sample in the noncontact mode, the IR-absorbed sample spontaneously undergoes thermal expansion (see Section 6.6.2), changing the tip–sample gap distance and eventually modulating the tip–sample interaction force. In the limit of small oscillation of the tip, this thermal expansion contribution can be expressed as [85]:

$$F_{th}(z) \propto \frac{\partial F_{ts}}{\partial z}\left(\Delta L_t(z) + \Delta L_g(z)\right) \tag{6.68}$$

where F_{ts} is the tip–sample interaction force (which could be estimated with the van der Waals interaction model), and $\Delta L_t(z)$ and $\Delta L_g(z)$ are the tip-enhanced thermal expansion portion by the near-field and the global expansion portion due to the IR absorption by the entire sample, respectively. The latter term monotonically increases with respect to the sample thickness and is not a function of the tip–sample distance. While the homodyne PiFM detection is directly proportional to the total thermal expansion (thus, volume sensitive), the heterodyne detection is sensitive to the tip-enhanced term

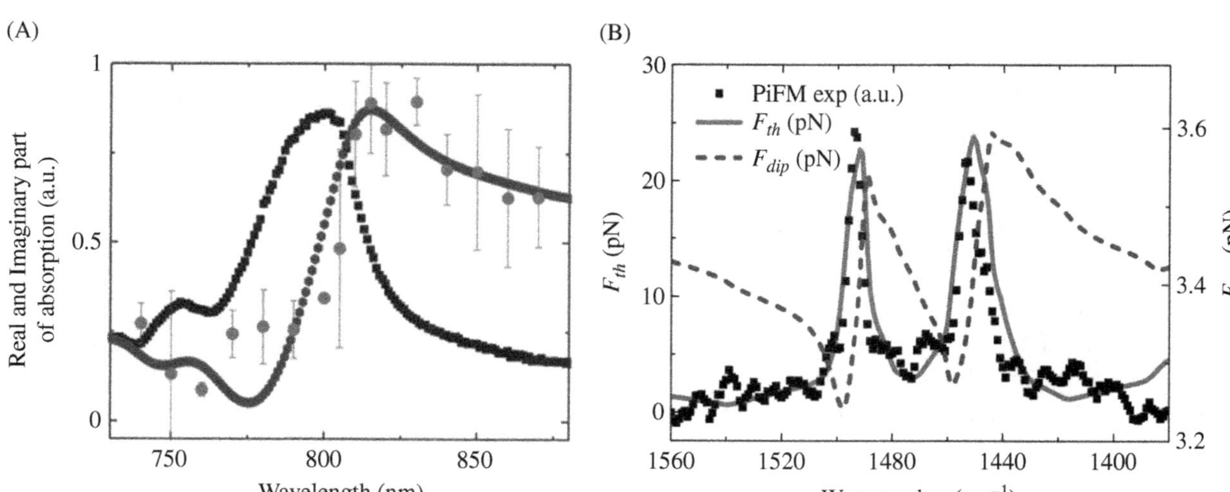

Figure 6.44 (A) Measured PiFM spectrum (dots with error bars) for a SiNc dye and the real (circles) and imaginary (squares) parts of the optical response near the absorption band at 810 nm. Reproduced from [84]/with permission of American Chemical Society.
(B) Calculation of the induced dipole force (dashed line) and the thermal expansion force (solid line) along with the PiFM data (square dots) for a 60-nm thick PS on a Si substrate. Reproduced from [85]/with permission of PNAS.

in the small oscillation condition (making it surface sensitive) [86]. The $F_{th}(z)$ force can be one or two orders of magnitude larger than the $F_{dip}(z)$ force for most organic materials because they have relatively large CTE values. An example is shown in Figure 6.44B where the PiFM signal of PS is compared with the calculated $F_{th}(z)$ and $F_{dip}(z)$ forces [85]. Readers interested in further details can find many review papers and study them independently [86].

Figure 6.45 demonstrates the interplay of these two contributions in PiFM. The region analyzed here is an edge of a PS coating with a thickness gradient partially covering an Au nanowire placed on a glass substrate (Figure 6.45A,B). Commercially available metal-coated AFM tips are often contaminated with a thin layer of polymethyldisiloxane (PDMS) which originates from the adhesive used in the tip case. When PiFM imaging is done at 1730 cm^{-1} where there are no absorption resonances of PS, PDMS, glass, and Au, the signal contrast in the image is dominated by the $F_{dip}(z)$ contribution which probes the dielectric constant differences (Figure 6.45C). The PiFM signal of Au underneath the PS layer is detected until the PS overlayer thickness is ~150 nm. Since there is no substantial difference in the dielectric contact of glass and PS at this wavenumber, the PS film and the glass substrate do not show any contrast. When PiFM imaging is done at 1268 cm^{-1} where only PDMS absorbs the irradiated beam, the image contrast is enhanced due to the $F_{th}(z)$ contribution from the PDMS contaminant at the tip surface, and the detection of the PiFM signal from Au underneath the PS surface is extended to ~250 nm (Figure 6.45D). When PiFM imaging is done at 1020 cm^{-1}, the $F_{th}(z)$ contribution of the glass substrate dominates the image contrast (Figure 6.45E). Significantly richer contrasts can be seen in the PiFM intensity map collected at 1492 cm^{-1} which is one of the absorption bands of PS (Figure 6.45F). On the Au nanowire region exposed outside the PS film, the contrast against the glass region is mainly due to the $F_{dip}(z)$ contribution. On the PS film region above the glass substrate, the signal intensity increases gradually with the PS thickness up to ~100 nm and then levels off. This is because the $\frac{\partial F_{ts}}{\partial z}\Delta L_t(z)$

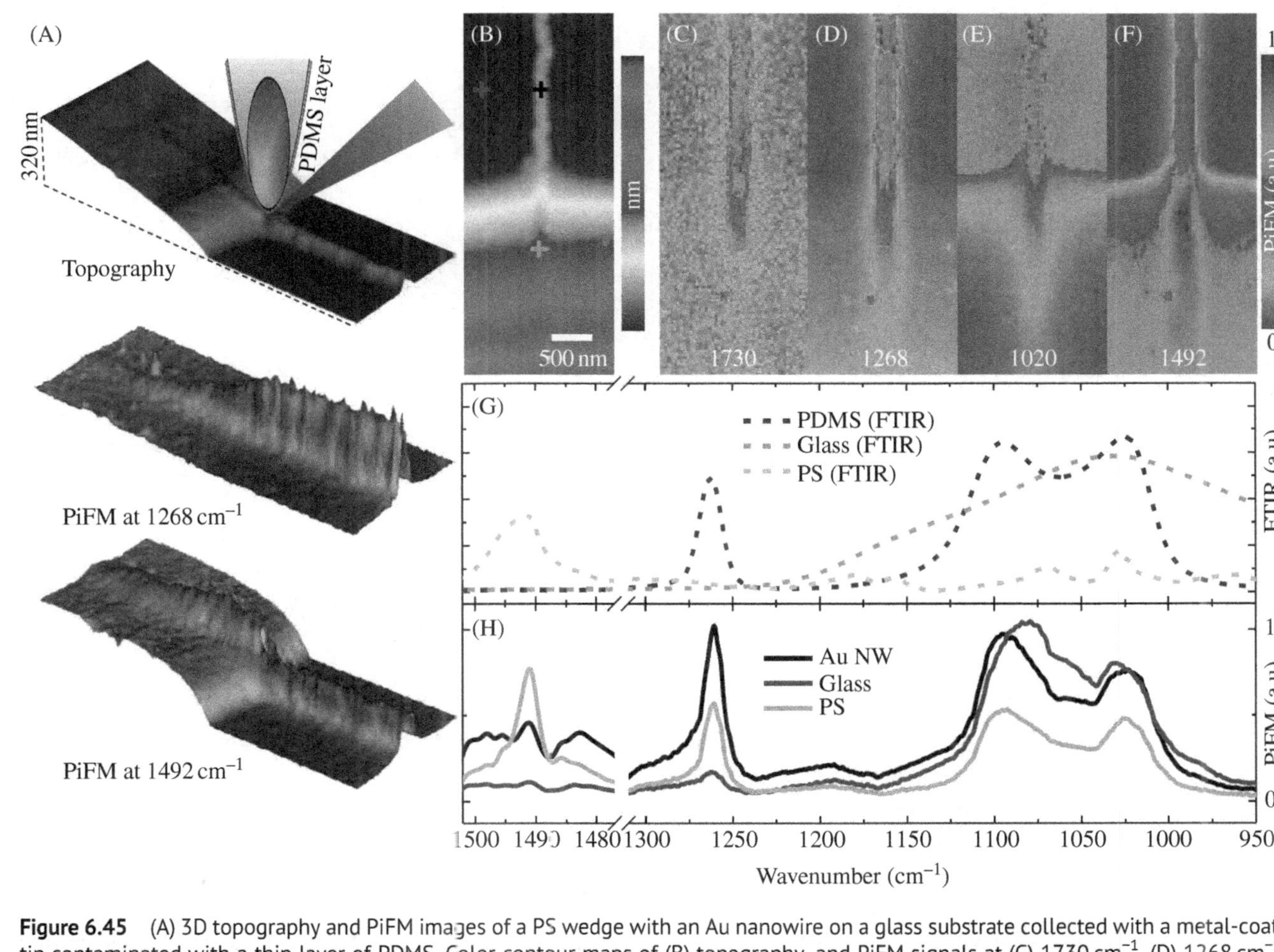

Figure 6.45 (A) 3D topography and PiFM images of a PS wedge with an Au nanowire on a glass substrate collected with a metal-coated tip contaminated with a thin layer of PDMS. Color contour maps of (B) topography, and PiFM signals at (C) 1730 cm^{-1}, (D) 1268 cm^{-1}, (E) 1020 cm^{-1}, and (F) 1492 cm^{-1}. (G) FTIR spectra of PDMS, glass, and PS. (H) PiFM spectra collected at the positions on the Au nanowire, the glass surface, and the PS film (marked with + in (B)). Reproduced from [85]/with permission of PNAS.

contribution saturates at that thickness in the specific detection mode used for imaging. The PiFM signal of the PS film region above the Au nanowire is much larger than that above the glass region. It saturates at ~70 nm of the PS film and then gradually decreases to the level of the thick PS region. This shows different tip-induced thermal expansion effects of the PiFM signal measured at the surface depending on the material beneath the surface. Thus, *the PiFM signal may deviate drastically from what is expected from the absorption spectrum of the topmost region of the sample.*

Vibrational spectroscopic imaging has become popular and powerful because it is a label-free chemical imaging method; unlike other methods, it detects and monitors sample-specific vibrational spectral features. In XPS analysis, chemical derivatization is needed to resolve functional groups whose binding energies overlap to each other (Section 2.3.3). In UV–VIS spectroscopy, fluorescent labeling to specific moieties in the sample is often needed to differentiate them from other components [87, 88]. Although, in principle, label-free chemical identification is possible in vibrational spectroscopic imaging, it highly depends on the sample and molecular groups that we need to identify. If characteristic bands of analytes cannot be clearly resolved, chemical identification could be difficult to do. In spectroscopic imaging, the signal-to-noise (S/N) ratio decreases as the spatial resolution is increased (because the overall signal intensity per pixel decreases as the pixel size decreases). So, there is a practical limit in the achievable resolution, which varies depending on the sample and may be significantly larger than what is advertised by the instrument manufacturer. In the case of IR imaging coupled with AFM, the detected signal is not the photon intensity change due to absorption by the sample. They are the local changes in sample height in the case of photothermal AFM-IR, near-field scattering intensity in nano-IR, and polarization force (which can be enhanced due to thermal expansion) in PiFM. So, the correlation between the detected signal and the absorption characteristics of the sample should be checked for spectral interpretation; that correlation may not be straightforward. The example in Figure 6.45 demonstrates the complexity of the intensity interpretation in chemical mapping with PiFM. The same caution or warning pertains to photothermal AFM-IR and s-SNOM nano-IR.

6.7 Further Reading for Surface-Sensitive IR Spectroscopy

We have studied how Snell's law and the Fresnel coefficient equations are used for surface-sensitive vibrational spectroscopy. From this, we have learned that the IR spectra obtained with the reflection geometries cannot be interpreted with Beer–Lambert's law. Further details can be found at [89]. Readers interested in learning more examples of interpreting IR spectra of ultrathin films can study [90].

Practice Problems

6.1 Let us consider the following hypothetical oscillators:

	ε_{st}	ε	γ	ω_o
Paraffin (isolated)	2.3	2.2	1 cm^{-1}	800 cm^{-1}
Paraffin (condensed)	2.3	2.2	4 cm^{-1}	800 cm^{-1}
Crystalline oxide	11	10	4 cm^{-1}	800 cm^{-1}
Amorphous oxide	11	10	40 cm^{-1}	800 cm^{-1}

a) Plot the SR-IR spectrum of each case at an incidence angle of 45° in the reflectance (R) scale from 600 to 1000 cm^{-1}. Assume that the thickness of the sample is infinity.
b) Plot the transmission-IR spectrum of each case at the surface-normal incidence in the absorbance ($A = -\log[I_T/I_{in}]$) scale from 600 to 1000 cm^{-1}. Assume that the thickness of the sample is 10 μm.
c) Compare the SR-IR and transmission-IR spectra. Of course, the peak intensity will be different since one is for the bulk material and the other is for a thin sample. Adjust the scales of the y-axes of your plots to show the spectral features clearly. Check the other aspects of the spectral features; can you support the idea that the SR-IR spectrum can be fully interpreted by simply comparing with the peak position in the transmission spectrum, at least for these sets of hypothetical samples?

6.2 a) In spectroscopic analysis with a reflection geometry, spectra are often plotted in the $-\log(I_R/I_{in})$ format. Can that quantity be assumed to be the same as the absorbance, $A = -\log(I_T/I_{in})$ in the transmission geometry? The answer should be properly explained.

b) In RAIRS analysis of molecules adsorbed on a metal surface, why cannot you detect vibrational modes parallel to the metal surface? Provide two reasons based on electromagnetism principles.

6.3 Simulate the RAIRS measurement of a 1-nm thick adsorbate layer (assume $n_2 = 1.4$) on a gold surface being probed with the s-polarized IR beam (assume $\lambda = 5.5\ \mu m$) and prove that RAIRS with the s-polarization is an inefficient way to detect surface species on a metal. This also proves that PM-RAIRS can detect the adsorbate layer on a metallic surface *in-situ*, that is, in the presence of the same molecule in the gas phase.

6.4 An alternative way to estimate the molecular tilt angle of the CH_3 group in RAIRS is to compare the relative intensity ratio of the symmetric (ν_s) and asymmetric (ν_{cs}) stretch modes of the CH_3 chain end group. The IR transition dipoles of these two modes are orthogonal to each other. So, if we know the relative oscillator strengths of these modes, we can calculate the tilt angle (ϑ) of the CH_3 molecular axis from the surface-normal direction from the experimentally measured $I(\nu_s)/I(\nu_{as})$ ratio using the following relationship:

$$\frac{a_s^2}{a_{as}^2}\frac{\cos^2(\vartheta)}{\cos^2(\vartheta + 90^o)} = \frac{I_{obsd}(\nu_s)}{I_{obsd}(\nu_{as})}$$

where a_s^2/a_{as}^2 is the ratio of their relative oscillator strengths. If we know the refractive index of the bulk, we can calculate theoretical RAIRS intensity ratio, $I_{calcd}(\nu_s)/I_{calcd}(\nu_{as})$, using eq. (6.36). Then, by setting $\vartheta = 54.7°$ (since the bulk phase is isotropic) in eq. (6.45), we can calculate a_s^2/a_{as}^2.

a) Download the refractive index of n-octane from [2] in the mid-IR region and calculate the RAIRS spectrum of a 2-nm thick n-octane liquid film on an Au surface. Then, fit the ν_s and ν_{as} of the CH_3 group.

b) From (a), we can calculate the a_s^2/a_{as}^2 ratio.

c) Assume that we are now having 1-octanethiol SAM on an Au surface. Calculate the $I(\nu_s)/I(\nu_{as})$ ratio as a function of the octyl group tilt angle from the surface-normal direction. For simplicity, we assume that the methylene groups in the SAM molecule are all *trans* conformation.

References

1 Harrick NJ, du Pré FK. Effective thickness of bulk materials and of thin films for internal reflection spectroscopy. Applied Optics. 1966;5(11):1739–43.

2 Polyanskiy MN. Refractive Index Database [Available from: https://refractiveindex.info].

3 Novak A, editor. Hydrogen Bonding in Solids Correlation of Spectroscopic and Crystallographic Data. Berlin, Heidelberg: Springer; 1974.

4 Libowitzky E. Correlation of O-H stretching frequencies and O-H...O hydrogen bond lengths in minerals. Monatshefte für Chemie / Chemical Monthly. 1999;130(8):1047–59.

5 Irvine WM, Pollack JB. Infrared optical properties of water and ice spheres. Icarus. 1968;8(1):324–60.

6 Ewing GE. Thin film water. The Journal of Physical Chemistry B. 2004;108(41):15953–61.

7 Harrick NJ. Electric field strengths at totally reflecting interfaces. Journal of the Optical Society of America. 1965;55(7):851–7.

8 Suëtaka W, Yates JT. Internal Reflection Spectroscopy. Surface Infrared and Raman Spectroscopy: Methods and Applications. Boston, MA: Springer US; 1995. p. 117–61.

9 Tillman N, Ulman A, Penner TL. Formation of multilayers by self-assembly. Langmuir. 1989;5(1):101–11.

10 Asay DB, Kim SH. Evolution of the adsorbed water layer structure on silicon oxide at room temperature. The Journal of Physical Chemistry B. 2005;109(35):16760–3.

11 Chen L, He X, Liu H, Qian L, Kim SH. Water adsorption on hydrophilic and hydrophobic surfaces of silicon. The Journal of Physical Chemistry C. 2018;122(21):11385–91.

12 Barnette AL, Asay DB, Kim SH. Average molecular orientations in the adsorbed water layers on silicon oxide in ambient conditions. Physical Chemistry Chemical Physics. 2008;10(32):4981–6.
13 Banerjee J, Kim SH, Pantano CG. Elemental areal density calculation and oxygen speciation for flat glass surfaces using x-ray photoelectron spectroscopy. Journal of Non-Crystalline Solids. 2016;450:185–93.
14 Asay DB, Barnette AL, Kim SH. Effects of surface chemistry on structure and thermodynamics of water layers at solid–vapor interfaces. The Journal of Physical Chemistry C. 2009;113(6):2128–33.
15 Ketteler G, Yamamoto S, Bluhm H, Andersson K, Starr DE, Ogletree DF, et al. The nature of water nucleation sites on $TiO_2(110)$ surfaces revealed by ambient pressure x-ray photoelectron spectroscopy. The Journal of Physical Chemistry C. 2007;111(23):8278–82.
16 Asay DB, Kim SH. Effects of adsorbed water layer structure on adhesion force of silicon oxide nanoasperity contact in humid ambient. The Journal of Chemical Physics. 2006;124(17):174712.
17 Chen L, Xiao C, Yu B, Kim SH, Qian L. What governs friction of silicon oxide in humid environment: contact area between solids, water meniscus around the contact, or water layer structure? Langmuir. 2017;33(38):9673–9.
18 Ras RHA, Schoonheydt RA, Johnston CT. Relation between s-polarized and p-polarized internal reflection spectra: application for the spectral resolution of perpendicular vibrational modes. The Journal of Physical Chemistry A. 2007;111(36):8787–91.
19 Goormaghtigh E, Raussens V, Ruysschaert J-M. Attenuated total reflection infrared spectroscopy of proteins and lipids in biological membranes. Biochimica et Biophysica Acta (BBA) – Reviews on Biomembranes. 1999;1422(2):105–85.
20 Lorenz-Fonfria VA. Infrared difference spectroscopy of proteins: from bands to bonds. Chemical Reviews. 2020;120 (7):3466–576.
21 Wang H, Zhou Y-W, Cai W-B. Recent applications of in situ ATR-IR spectroscopy in interfacial electrochemistry. Current Opinion in Electrochemistry. 2017;1(1):73–9.
22 Lefèvre G. In situ Fourier-transform infrared spectroscopy studies of inorganic ions adsorption on metal oxides and hydroxides. Advances in Colloid and Interface Science. 2004;107(2):109–23.
23 Rubin M. Optical properties of soda lime silica glasses. Solar Energy Materials. 1985;12(4):275–88.
24 Amma S-i, Luo J, Pantano CG, Kim SH. Specular reflectance (SR) and attenuated total reflectance (ATR) infrared (IR) spectroscopy of transparent flat glass surfaces: a case study for soda lime float glass. Journal of Non-Crystalline Solids. 2015;428:189–96.
25 Liu H, Kaya H, Lin Y-T, Ogrinc A, Kim SH. Vibrational spectroscopy analysis of silica and silicate glass networks. Journal of the American Ceramic Society. 2022;105(4):2355–84.
26 Uchino T, Sakka T, Hotta K, Iwasaki M. Attenuated total reflectance Fourier-transform infrared spectra of a hydrated sodium silicate glass. Journal of the American Ceramic Society. 1989;72(11):2173–5.
27 Luo J, Smith NJ, Pantano CG, Kim SH. Complex refractive index of silica, silicate, borosilicate, and boroaluminosilicate glasses – analysis of glass network vibration modes with specular-reflection IR spectroscopy. Journal of Non-Crystalline Solids. 2018;494:94–103.
28 Tuschel DD. Why are the Raman spectra of crystalline and amorphous solids different?Spectroscopy. 2017;32:26–33.
29 Harbecke B, Heinz B, Grosse P. Optical properties of thin films and the Berreman effect. Applied Physics A. 1985;38(4):263–7.
30 Sheth N, Ngo D, Banerjee J, Zhou Y, Pantano CG, Kim SH. Probing hydrogen-bonding interactions of water molecules adsorbed on silica, sodium calcium silicate, and calcium aluminosilicate glasses. The Journal of Physical Chemistry C. 2018;122(31):17792–801.
31 Shepelevich VV. Stokes relations in optically active media. Journal of Applied Spectroscopy. 1979;30(2):240–2.
32 Hansen WN. Electric fields produced by the propagation of plane coherent electromagnetic radiation in a stratified medium. Journal of the Optical Society of America. 1968;58(3):380–90.
33 Brunner H, Mayer U, Hoffmann H. External reflection infrared spectroscopy of anisotropic adsorbate layers on dielectric substrates. Applied Spectroscopy. 1997;51(2):209–17.
34 Mielczarski JA, Yoon RH. Fourier transform infrared external reflection study of molecular orientation in spontaneously adsorbed layers on low-absorption substrates. The Journal of Physical Chemistry. 1989;93(5):2034–8.
35 Ogrinc AL, Zhou Y, Hahn SH, Lin Y-T, Kim SH. Revealing the structure of the sodium-leached layer of soda lime silica glass: a comprehensive spectroscopic analysis. Journal of Non-Crystalline Solids. 2023;600:121989.
36 Hahn SH, van Duin ACT. Surface reactivity and leaching of a sodium silicate glass under an aqueous environment: a ReaxFF molecular dynamics study. The Journal of Physical Chemistry C. 2019;123(25):15606–17.
37 Sheth N, Hahn SH, Ngo D, Howzen A, Bermejo R, van Duin ACT, et al. Influence of acid leaching surface treatment on indentation cracking of soda lime silicate glass. Journal of Non-Crystalline Solids. 2020;543:120144.

38 Sanders DM, Person WB, Hench LL. Quantitative analysis of glass structure with the use of infrared reflection spectra. Applied Spectroscopy. 1974;28(3):247–55.

39 Kaya H, Ngo D, Smith NJ, Gin S, Kim SH. Network structure in alteration layer of boroaluminosilicate glass formed by aqueous corrosion. Journal of Non-Crystalline Solids. 2021;553:120494.

40 Ordal MA, Bell RJ, Alexander RW, Long LL, Querry MR. Optical properties of fourteen metals in the infrared and far infrared: Al, Co, Cu, Au, Fe, Pb, Mo, Ni, Pd, Pt, Ag, Ti, V, and W. Applied Optics. 1985;24(24):4493–9.

41 Ordal MA, Long LL, Bell RJ, Bell SE, Bell RR, Alexander RW, et al. Optical properties of the metals Al, Co, Cu, Au, Fe, Pb, Ni, Pd, Pt, Ag, Ti, and W in the infrared and far infrared. Applied Optics. 1983;22(7):1099–119.

42 Kolwas K, Derkachova A. Impact of the interband transitions in gold and silver on the dynamics of propagating and localized surface plasmons. Nanomaterials. 2020;10(7):1411.

43 Greenler RG. Infrared study of adsorbed molecules on metal surfaces by reflection techniques. The Journal of Chemical Physics. 1966;44(1):310–5.

44 Zaera F. New advances in the use of infrared absorption spectroscopy for the characterization of heterogeneous catalytic reactions. Chemical Society Reviews. 2014;43(22):7624–63.

45 de Jong AM, Niemantsverdriet JW. The adsorption of CO on Rh(100): reflection absorption infrared spectroscopy, low energy electron diffraction, and thermal desorption spectroscopy. The Journal of Chemical Physics. 1994;101(11):10126–33.

46 Yeh P. Optics of anisotropic layered media: a new 4 × 4 matrix algebra. Surface Science. 1980;96(1):41–53.

47 Laibinis PE, Whitesides GM, Allara DL, Tao YT, Parikh AN, Nuzzo RG. Comparison of the structures and wetting properties of self-assembled monolayers of n-alkanethiols on the coinage metal surfaces, copper, silver, and gold. Journal of the American Chemical Society. 1991;113(19):7152–67.

48 Nafie LA. Vibrational optical activity: from discovery and development to future challenges. Chirality. 2020;32(5):667–92.

49 Buffeteau T, Desbat B, Blaudez D, Turlet JM. Calibration procedure to derive IRRAS spectra from PM-IRRAS spectra. Applied Spectroscopy. 2000;54(11):1646–50.

50 Buffeteau T, Desbat B, Turlet JM. Polarization modulation FT-IR spectroscopy of surfaces and ultra-thin films: experimental procedure and quantitative analysis. Applied Spectroscopy. 1991;45(3):380–9.

51 Jiang EY. Advanced FT-IR Spectroscopy: Principles, Experiments and Applications, Based on Research-Grade Nicholet™ FT-IR Spectrometers. Madison, WI: Thermo Fisher Scientific Inc.; 2007.

52 Adamson AW, Gast AP. Physical Chemistry of Surfaces (Sixth Edition). New York: Wiley; 1997.

53 Lin Y-T, Smith NJ, Banerjee J, Agnello G, Manley RG, Walczak WJ, et al. Water adsorption on silica and calcium-boroaluminosilicate glass surfaces – thickness and hydrogen bonding of water layer. Journal of the American Ceramic Society. 2021;104(3):1568–80.

54 Kischkat J, Peters S, Gruska B, Semtsiv M, Chashnikova M, Klinkmüller M, et al. Mid-infrared optical properties of thin films of aluminum oxide, titanium dioxide, silicon dioxide, aluminum nitride, and silicon nitride. Applied Optics. 2012;51(28):6789–98.

55 Hale GM, Querry MR. Optical constants of water in the 200-nm to 200-μm wavelength region. Applied Optics. 1973;12(3):555–63.

56 Warren SG, Brandt RE. Optical constants of ice from the ultraviolet to the microwave: a revised compilation. Journal of Geophysical Research: Atmospheres. 2008;113, D14220.

57 Atkins P, Paula Jd. Atkins' Physical Chemistry (Eight Edition). W. H. Freeman, New York, NY; 2006.

58 Ford TA, Falk M. Hydrogen bonding in water and ice. Canadian Journal of Chemistry. 1968;46(22):3579–86.

59 Muthachikavil AV, Peng B, Kontogeorgis GM, Liang X. Distinguishing weak and strong hydrogen bonds in liquid water – a potential of mean force-based approach. The Journal of Physical Chemistry B. 2021;125(26):7187–98.

60 Libowitzky E, Rossman GR. An IR absorption calibration for water in minerals. American Mineralogist. 1997;82 (11–12):1111–5.

61 Balan E, Refson K, Blanchard M, Delattre S, Lazzeri M, Ingrin J, et al. Theoretical infrared absorption coefficient of OH groups in minerals. American Mineralogist. 2008;93(5–6):950–3.

62 Wang D, Zhang L, Kamasamudram K, Epling WS. In situ-DRIFTS study of selective catalytic reduction of NOx by NH_3 over Cu-exchanged SAPO-34. ACS Catalysis. 2013;3(5):871–81.

63 Zhang Q, Mo S, Li J, Sun Y, Zhang M, Chen P, et al. In situ DRIFT spectroscopy insights into the reaction mechanism of CO and toluene co-oxidation over Pt-based catalysts. Catalysis Science & Technology. 2019;9(17):4538–51.

64 Kubelka P, Munk F. An article on optics of paint layers. Zeitschrift für Physik 1931;12(593–601):259–74.

65 Milosevic M, Berets SL. A review of FT-IR diffuse reflection sampling considerations. Applied Spectroscopy Reviews. 2002;37(4):347–64.

66 Yeboah SA, Wang S-H, Griffiths PR. Effect of pressure on diffuse reflectance infrared spectra of compressed powders. Applied Spectroscopy. 1984;38(2):259–64.

67 Eickhoff T, Grosse P, Theiss W. Diffuse reflectance spectroscopy of powders. Vibrational Spectroscopy. 1990;1(2):229–33.

68 Wrobel TP, Bhargava R. Infrared spectroscopic imaging advances as an analytical technology for biomedical sciences. Analytical Chemistry. 2018;90(3):1444–63.

69 Gierlinger N. New insights into plant cell walls by vibrational microspectroscopy. Applied Spectroscopy Reviews. 2017;53:1–35.

70 Jackson WB, Amer NM, Boccara AC, Fournier D. Photothermal deflection spectroscopy and detection. Applied Optics. 1981;20(8):1333–44.

71 Olmstead MA, Amer NM, Kohn S, Fournier D, Boccara AC. Photothermal displacement spectroscopy: an optical probe for solids and surfaces. Applied Physics A. 1983;32(3):141–54.

72 Kurouski D, Dazzi A, Zenobi R, Centrone A. Infrared and Raman chemical imaging and spectroscopy at the nanoscale. Chemical Society Reviews. 2020;49(11):3315–47.

73 Dazzi A, Prater CB. AFM-IR: technology and applications in nanoscale infrared spectroscopy and chemical imaging. Chemical Reviews. 2017;117(7):5146–73.

74 Dazzi A, Glotin F, Carminati R. Theory of infrared nanospectroscopy by photothermal induced resonance. Journal of Applied Physics. 2010;107(12):124519.

75 Schwartz JJ, Jakob DS, Centrone A. A guide to nanoscale IR spectroscopy: resonance enhanced transduction in contact and tapping mode AFM-IR. Chemical Society Reviews. 2022;51(13):5248–67.

76 Lin Y-T, He H, Kaya H, Liu H, Ngo D, Smith NJ, et al. Photothermal atomic force microscopy coupled with infrared spectroscopy (AFM-IR) analysis of high extinction coefficient materials: a case study with silica and silicate glasses. Analytical Chemistry. 2022;94(13):5231–9.

77 Pettinger B, Schambach P, Villagómez CJ, Scott N. Tip-enhanced Raman spectroscopy: near-fields acting on a few molecules. Annual Review of Physical Chemistry. 2012;63(1):379–99.

78 Verma P. Tip-enhanced Raman spectroscopy: technique and recent advances. Chemical Reviews. 2017;117(9):6447–66.

79 Fritz K, Hillenbrand R. Near-field microscopy by elastic light scattering from a tip. Philosophical Transactions: Mathematical, Physical and Engineering Sciences. 2004;362(1817):787–805.

80 He H, Chen Z, Lin Y-T, Hahn SH, Yu J, van Duin ACT, et al. Subsurface structural change of silica upon nanoscale physical contact: chemical plasticity beyond topographic elasticity. Acta Materialia. 2021;208:116694.

81 Taubner T, Hillenbrand R, Keilmann F. Nanoscale polymer recognition by spectral signature in scattering infrared near-field microscopy. Applied Physics Letters. 2004;85(21):5064–6.

82 Muller EA, Pollard B, Raschke MB. Infrared chemical nano-imaging: accessing structure, coupling, and dynamics on molecular length scales. The Journal of Physical Chemistry Letters. 2015;6(7):1275–84.

83 Jahng J, Tork Ladani F, Khan R, Potma E. Photo-induced force for spectroscopic imaging at the nanoscale. Proc. SPIE, Complex Light and Optical Forces X; 2016; 9764: 9764J.

84 Jahng J, Fishman DA, Park S, Nowak DB, Morrison WA, Wickramasinghe HK, et al. Linear and nonlinear optical spectroscopy at the nanoscale with photoinduced force microscopy. Accounts of Chemical Research. 2015;48(10):2671–9.

85 Jahng J, Potma EO, Lee ES. Nanoscale spectroscopic origins of photoinduced tip-sample force in the midinfrared. Proceedings of the National Academy of Sciences. 2019;116(52):26359–66.

86 Sifat AA, Jahng J, Potma EO. Photo-induced force microscopy (PiFM) – principles and implementations. Chemical Society Reviews. 2022;51(11):4208–22.

87 Toseland CP. Fluorescent labeling and modification of proteins. Journal of Chemical Biology. 2013;6(3):85–95.

88 Stone MRL, Butler MS, Phetsang W, Cooper MA, Blaskovich MAT. Fluorescent antibiotics: new research tools to fight antibiotic resistance. Trends in Biotechnology. 2018;36(5):523–36.

89 Mayerhöfer TG, Pahlow S, Popp J. The Bouguer–Beer–Lambert law: shining light on the obscure. ChemPhysChem. 2020;21(18):2029–46.

90 Tolstoy VP, Chernyshova IV, Skryshevsky VA. Interpretation of IR spectra of ultrathin films. In: Handbook of Infrared Spectroscopy of Ultrathin Films. John Wiley & Sons, Inc; 2003. p. 140–306.

7

Buried Interface Analysis via Nonlinear Spectroscopy

In Chapters 5 and 6, we have studied how the complex refractive index of materials and the Fresnel coefficient equations are derived for reflection and transmission at interfaces and used in vibrational spectroscopic analysis in the mid-IR wavelength region. The techniques covered in Chapter 6 include ATR-IR, SR-IR, RAIRS, PM-RAIRS, BAT-IR, and DRIFT. The extension of IR vibrational spectroscopy to chemical imaging is briefly described at the end of Chapter 6, which touches the use of the reflective objective lens to couple with optical microscopy and the combination with scanning probe microscopy techniques such as photothermal AFM-IR, s-SNOM nano-IR, and PiFM. All these techniques can give molecular information on chemical species or their states in the surface region. The probe depth of "subsurface region" (=distance from the topmost surface) varies depending on the method used and the sample. The techniques described in Chapter 6 work well when there is no interference signal from the surrounding medium (so, no analyte in the surrounding medium), or if the spectral features of surface species are shifted enough to easily differentiate from the same species in the subsurface region or the surrounding phase. Otherwise, the surface signal will be swamped in the noise level of the bulk signal. PM-RAIRS might be an exception since the metal selection rule, coupled with the synchronous sampling demodulator technique, allows the subtraction of the gas-phase signal from the total signal containing both surface and gas species; but it would not work for nonmetallic surfaces. ATR-IR can also probe the interface region in the presence of the same analyte in the surrounding medium since the IR beam propagates inside the solid substrate, and the evanescent wave is confined in a short distance from the substrate surface; but it does not work for the solid that is not transparent to IR.

In this chapter, we will study nonlinear spectroscopy which is *intrinsically sensitive to surface species or phase boundaries as long as the bulk phases are centrosymmetric or amorphous*, thus, suitable for analyses of gas/solid, liquid/solid, solid/solid, gas/liquid, and (immiscible) liquid/liquid interfaces [1–3]. These are often called "buried" interfaces. Of course, the probing light should be able to reach the interface of the two phases; otherwise, the technique will not work. Since we've been discussing vibrational spectroscopy in the previous chapter, we will focus on vibrational sum frequency generation (SFG) spectroscopy in this chapter. But a good portion of the general principle discussed here is applicable to second-harmonic generation (SHG) technique as well. Both SFG and SHG analyses can be applied to detect *crystalline domains without centrosymmetry interspersed in amorphous matrices* and that principle will also be touched at the end of this chapter. The main difference between SFG and SHG is that SFG is molecule-specific (since one of the input beams resonates with the vibrational excitation of the analyte), while SHG does not requires resonance with specific excitations (it just detects noncentrosymmetric dielectric responses of the system).

7.1 Nonlinear vs. Linear Optical Responses

In derivation of Snell's law in Section 5.2.1, we have learned that the frequency of the electromagnetic beam should remain unaltered when the beam crosses the interface and propagates into a new medium, that is, $\omega_{in} = \omega_{out}$. Only its intensity can vary upon reflection at the interface. Invariance in frequency is the same for absorption during the propagation through an absorbing medium (see eqs. (5.42 and 5.92)). This is the key feature of *linear spectroscopy* described in Chapters 5 and 6. In a bulk medium without permanent dipole, the polarization, $P(t)$, induced by the electric field of light, $E(t)$, can be described as:

$$P(t) = \varepsilon_o \chi^{(1)} E(t) \tag{7.1}$$

Surface and Interface Analysis: Principles and Applications, First Edition. Seong H. Kim.

Companion website: www.wiley.com/go/Kim

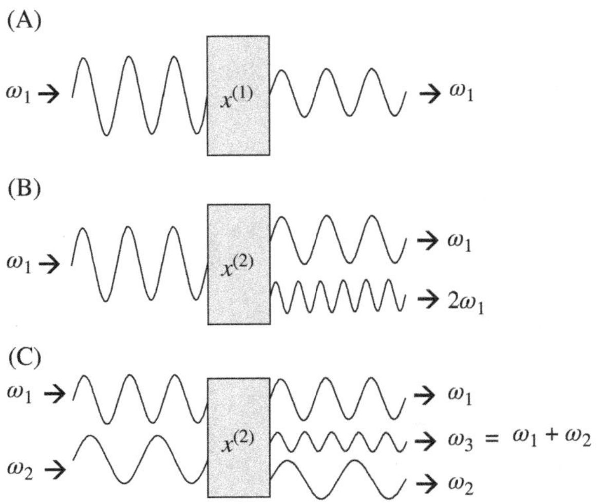

Figure 7.1 (A) Linear and (B,C) nonlinear optical responses of a medium and corresponding absorption and emission processes. The SHG and SFG processes are illustrated in (B) and (C), respectively.

which is essentially the same as eq. (5.33). Here, $\chi^{(1)}$ is the linear susceptibility, which corresponds to the resonant term in eq. (5.35). Thus, the induced dipole oscillates at the same frequency as the incident electric field. Then, the reflected and transmitted beams have the same frequency as the incident beam, which is schematically illustrated in Figure 7.1A. This is always observed regardless of the intensity of the electric field.

One of the key differences of the nonlinear optical response from the linear response is that it can generate physical responses at frequencies that are different from the input frequency (Figures 7.1B,C). In principle, the polarization of material can have nonlinear responses to the electric field. This can be described by generalizing eq. (7.1) through a power series:

$$P(t) = \varepsilon_o\left(\chi^{(1)}E(t) + \chi^{(2)}E^2(t) + \chi^{(3)}E^3(t) + \cdots\right) \tag{7.2}$$

where $\chi^{(2)}$ and $\chi^{(3)}$ are known as the second- and third-order nonlinear susceptibilities, respectively. These terms are quite small compared to $\chi^{(1)}$. Let us make a rough estimate of the magnitudes of these higher-order terms to see how small they are as compared to the linear susceptibility term. One way is to estimate the electric field at which the magnitude of $\chi^{(2)}E^2(t)$ becomes comparable to $\chi^{(1)}E(t)$, so that the second-order response can be measured with a detector used to measure the first-order response. From eq. (7.2), we can see that it will be when $\chi^{(2)}E(t) \approx \chi^{(1)}$. That would be when the amplitude of the applied electric field is of the order of the characteristic electric field strength of atom (E_{at}).

For simplicity, we can consider the hydrogen atom in which one electron is bound by one proton. If the atomic radius is a_o, the Coulombic force between these two will be $F = \frac{1}{4\pi\varepsilon_o}\frac{e^2}{a_o^2}$. Since $F = qE$ (q = charge), the electric field produced by the nucleus to the electron is $E_{at} = \frac{1}{4\pi\varepsilon_o}\frac{e}{a_o^2}$. We can estimate a_o from the Hamiltonian (H, which is the sum of the kinetic and potential energies) of the atomic system [4]:

$$H = \frac{p^2}{2m_e} - \frac{e^2}{r} \tag{7.3}$$

where p and m_e are the momentum and mass of electron, respectively, and r is the radius. Using the Heisenberg uncertainty principle $\Delta x \Delta p \geq \hbar/2$, we could replace p with $\hbar/r$. Then, the radius value (a_o) that minimizes the Hamiltonian energy can be found by setting $\frac{dH}{dr} = 0$. This leads to the following:

$$a_o = \frac{\hbar^2}{m_e e^2} \tag{7.4}$$

Numerically, we find that $a_o = 0.5$ Å $= 0.5 \times 10^{-10}$ m. Then, we get $E_{at} \approx 5.7 \times 10^{11}$ V/m.

For organics or simple oxides (such as silica and silicate), the refractive index in the visible wavelength is $n \approx 1.5$, then the relative dielectric constant is $\varepsilon_r = n^2 \approx 2.3$. Since $\varepsilon_r = 1 + 4\pi\chi^{(1)}$, we can estimate $\chi^{(1)} \approx 0.1$. From this, we get $\chi^{(2)} \approx \chi^{(1)}/E_{at} \approx 0.2 \times 10^{-12}$ m/V (i.e., 0.2 pm/V). In fact, this estimate is within the right order of magnitude of the second-order nonlinear susceptibility of α-quartz ($\chi^{(2)}_{111} = 2d_{11} = 0.6$ pm/V) [5]. In this calculation, we have assumed the nonlinear susceptibility to be a scalar, but it is a tensor. So, the subscript in $\chi^{(2)}_{ijk}$ indicates a specific tensor component, which will be discussed in Section 7.2.3.

From this estimation, we learn how small the nonlinear contributions are to the susceptibility. Unless the applied electric field is excessively large, the nonlinear terms are all essentially zero in comparison to the linear response. This is schematically illustrated in Figure 7.2. Figure 7.2A,B shows the normalized waveforms of the first- and second-harmonic terms, respectively. Figure 7.2C shows the mixing of these two with a very little contribution from the second-harmonic term.

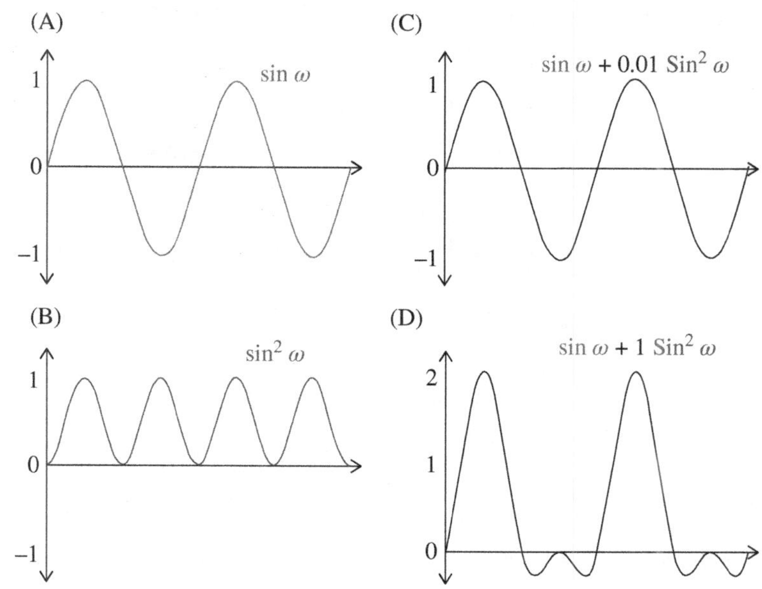

Figure 7.2 Waveforms of (A) sinω, (B) sin$^2\omega$, (C) sinω + 0.01 sin$^2\omega$, and (D) sinω + sin$^2\omega$.

To the naked eyes, this waveform is basically indistinguishable from the one in Figure 7.2A. This is the reason that eq. (7.2) is reduced to eq. (7.1) under ordinary illumination conditions (such as lamp or continuous-wave (CW) laser). Figure 7.2D displays the wave form when the second-harmonic contribution is as big as the first-harmonic contribution, which could be viewed as an extreme case. This illustration shows that, to make the nonlinear effect significant, we must apply an extremely high electric field (so, the magnitude of $\chi^{(2)}E^2(t)$ becomes measurable). Such a field can be obtained with a pulsed laser. If a femto-second laser output is 100 μJ/pulse with a pulse duration of 100 fs and this is focused into a 0.1 mm^2 area (i.e., radius ~200 μm), the power per pulse is 10^{10} W/mm^2. The conversion of intensity (=power per unit area) to the electric field amplitude can be done using eq. (5.64), which can be written as $I = \frac{c\varepsilon_o}{2}E^2$ for the free space. Then, the electric field in the irradiated area is calculated to be ~3×10^9 V/m. If a pico-second laser output is 100 mJ/pulse with a pulse duration of 30 ps and this is focused into the same 0.1 mm^2 area, then the electric field in the irradiated area is ~5×10^9 V/m. Then, the $\chi^{(2)}E^2(t)$ response becomes large enough to be measured with detectors used to measure the $\chi^{(1)}E(t)$ response.

If such a large power is delivered to the sample continuously, the sample would be damaged. In fact, tightly focusing a few mW CW laser beam into a small area can damage most of the organic samples. In Raman spectroscopy with the CW laser excitation, the sample damage by the focused beam is often a main obstacle [6]. But, in the case of pulsed laser irradiation, the beam damage can be avoided since the duration of the high-power beam is extremely short (10^{-14}–10^{-11} s). Of course, if the peak power of the incoming pulse is too high (i.e., via increasing the pulse energy or focusing the beam too tightly into a small area), the beam damage of the sample can still occur.

Next, let us study how the nonlinear responses of the medium are manifested. In eq. (7.2), the incident electric field oscillating at a frequency ω_1 can be expressed as:

$$E(t) = E\cos\omega_1 t \tag{7.5}$$

Insert eq. (7.5) into eq. (7.2):

$$P(t) = \varepsilon_o\left(\chi^{(1)}E\cos\omega_1 t + \chi^{(2)}E^2\cos^2\omega_1 t + \chi^{(3)}E^3\cos^3\omega_1 t + \cdots\right) \tag{7.6}$$

Here, the first term on the right-hand side is the first-order polarization, $P^{(1)}(t)$, that we have studied in Chapters 5 and 6. The second term in eq. (7.6) can be rewritten using the trigonometric identify rule, $\cos^2 x = (\cos 2x + 1)/2$, as:

$$P^{(2)}(t) = \varepsilon_o\chi^{(2)}E^2\cos^2\omega_1 t = \varepsilon_o\chi^{(2)}E^2\left(\frac{1}{2}\cos 2\omega_1 t + \frac{1}{2}\right) \tag{7.7}$$

We can see that the second-order polarization consists of a contribution at frequency 2ω and a contribution independent on frequency. The former leads to the generation of radiation at the second-harmonic frequency, which is called SHG (Figure 7.1B). The latter leads to a static electric field across the medium, which is called optical rectification (OR).

Now, let us consider the case where the medium is irradiated by two incident beams with different frequencies, which can be described as:

$$E(t) = E_1 \cos \omega_1 t + E_2 \cos \omega_2 t \tag{7.8}$$

Insert eq. (7.8) into eq. (7.2) and rewrite the second-order term using the trigonometric identity rule:

$$\begin{aligned} P^{(2)}(t) &= \varepsilon_o \chi^{(2)} (E_1 \cos \omega_1 t + E_2 \cos \omega_2 t)^2 \\ &= \varepsilon_o \chi^{(2)} (E_1^2 \cos^2 \omega_1 t + 2E_1 E_2 \cos \omega_1 t \cos \omega_2 t + E_2^2 \cos^2 \omega_2 t) \\ &= \varepsilon_o \chi^{(2)} \left(E_1^2 \frac{(\cos 2\omega_1 t + 1)}{2} + 2E_1 E_2 \frac{(\cos(\omega_1 + \omega_2)t + \cos(\omega_1 - \omega_2)t)}{2} + E_2^2 \frac{(\cos 2\omega_2 t + 1)}{2} \right) \\ &= \varepsilon_o \chi^{(2)} \left(\frac{E_1^2}{2} \cos 2\omega_1 t + \frac{E_2^2}{2} \cos 2\omega_2 t + E_1 E_2 \cos(\omega_1 + \omega_2)t + E_1 E_2 \cos(\omega_1 - \omega_2)t + \frac{E_1^2 + E_2^2}{2} \right) \end{aligned} \tag{7.9}$$

Here, on the last line, the first two terms are the SHG contributions from two incident beams, the third term is the SFG contribution (Figure 7.1C), the fourth term is the difference frequency generation (DFG) contribution, and the fifth term is the OR contribution.

We can insert eq. (7.5) into eq. (7.2) and rewrite the third-order polarization term using the trigonometric identity rule as:

$$P^{(3)}(t) = \varepsilon_o \chi^{(3)} E^3 \cos^3 \omega_1 t = \varepsilon_o \chi^{(3)} \left(\frac{E^3}{4} \cos 3\omega_1 t + \frac{3E^3}{4} \cos \omega_1 t \right) \tag{7.10}$$

Here, the first term on the right-hand side is the third harmonic generation (THG). Note that the second term has the same frequency as the first-order polarization. This is the intensity-dependent nonlinear contribution to the refractive index, $n = n_o + n_2 I$, where $n_2 = \frac{3}{2n_o^2 \varepsilon_o c} \chi^{(3)}$ [5].

We shall see in Section 7.3.1 that in centrosymmetric systems (i.e., having inversion symmetry), the second-order susceptibility is zero, $\chi^{(2)} = 0$; so, SHG and SFG can occur only in noncentrosymmetric systems. The interface of the two bulk phases naturally produces the noncentrosymmetric environment; thus, SHG and SFG are intrinsically sensitive to the interface. In this book, we will study vibrational SFG spectroscopy only since it is interface-sensitive and provides molecular information (Section 7.3). Note that SHG and SFG are sensitive to not only the interface but also nanoscale domains in the bulk phase if such domains have noncentrosymmetric components (Section 7.4). The DFG process is used in the optical parametric oscillation to generate high-intensity laser pulses in the mid-IR range where other laser sources of tunable radiation are not readily available. The third-order polarization involving $\chi^{(3)}$ can occur in both centrosymmetric and noncentrosymmetric media; since the bulk signal would dominate in THG, it will not be covered in this book. Readers who are interested in this subject can further study other nonlinear optics books.

7.2 Sum Frequency Generation (SFG) Process

7.2.1 Wave Equation Describing the SFG Process

Now let us take a deeper look into the frequency dependence of the polarization of the medium irradiated by the oscillatory electric field. It can be derived from the Maxwell equation [7, 8]. But, since we have already derived the propagation of electromagnetic wave in an optical medium in Section 5.1.1 and the polarization of the medium by the electromagnetic wave propagating through it in Section 5.1.3, we can use that knowledge here. We can combine eqs. (5.13) and (5.34) to write the wave equation of a polarizable medium:

$$\nabla^2 E - \frac{1}{\varepsilon_o c^2} \frac{\partial^2}{\partial t^2} D = 0 \tag{7.11}$$

where $D = \varepsilon_o E + P$ is the total electric displacement of the medium. For simplicity in writing, the frequency argument is suppressed. The polarization term can be split into the linear part (the first order in eq. (7.2)) and the nonlinear part (all higher-order terms in eq. (7.2)):

$$P = P^{(1)} + P^{NL} \tag{7.12}$$

Insert eq. (7.12) into eq. (7.11) and use eq. (5.34) to replace the linear component of the displacement current; then, we get the following expression:

$$\nabla^2 E - \frac{\varepsilon_r}{c^2}\frac{\partial^2 E}{\partial t^2} = \frac{1}{\varepsilon_o c^2}\frac{\partial^2 P^{NL}}{\partial t^2} \tag{7.13}$$

We next predict the generic form of the solution to this equation. In particular, we consider the SFG case, $\omega_3 = \omega_1 + \omega_2$, in a lossless medium since we are interested in interface-sensitive spectroscopy; as we shall see in Sections 7.2.2 and 7.3.1, the interface sensitivity can come from two factors of the SFG process – coherence requirement and noncentrosymmetry requirement. Again, for the simplicity in writing, we will replace the P^{NL} symbol in eq. (7.13) with P_3 to focus on the nonlinear response at ω_3 frequency. The plane wave oscillating with a frequency ω_3 and propagating in the z-direction can be written as:

$$E_3(z,t) = A_3 e^{i(k_3 z - \omega_3 t)} \tag{7.14}$$

where $k_3 = \frac{n_3\omega_3}{c}$ and $n_3 = \sqrt{\varepsilon_r(\omega_3)}$. Note that ω_3 is in the angular frequency unit. When the nonlinear source term is small, the solution to eq. (7.13) is in the form of eq. (7.14) with a slowly varying amplitude term as a function of z. We can express the P^{NL} term in eq. (7.13) as [7]:

$$P_3(z,t) = P_3(z)e^{-i\omega_3 t} \tag{7.15}$$

where $P_3(z) = 2\varepsilon_o\chi^{(2)}E_1E_2$ as described in eq. (7.9). The factor of 2 is added with the assumption that the nonlinear susceptibility possesses intrinsic permutation symmetry, that is, it does not matter which is the first field and which is the second in the product. Since $E_i(z,t) = A_i e^{ik_i z}e^{-i\omega_3 t}$ (where $i = 1,2$; incident beams), the $P_3(z)$ term in eq. (7.15) can be written as:

$$P_3(z) = 2\varepsilon_o\chi^{(2)}A_1A_2e^{i(k_1+k_2)z} \tag{7.16}$$

Since the conversion of the ω_1 and ω_2 incident fields to the ω_3 sum frequency field is small, we can assume that the A_1 and A_2 terms of the incident fields are constant.

Substituting eqs. (7.14)–(7.16) into eq. (7.13) and then rearranging the resulting equation will give the following expression [7]:

$$\frac{d^2A_3}{\partial z^2} + 2ik_3\frac{dA_3}{dz} = -\frac{2\chi^{(2)}\omega_3^2}{c^2}A_1A_2e^{i(k_1+k_2-k_3)z} \tag{7.17}$$

The first term on the left-hand side of eq. (7.17) can be neglected since it is smaller than the second term under the slowly varying amplitude approximation. Then, we get the following relationship:

$$\frac{dA_3}{dz} = \frac{i\chi^{(2)}\omega_3^2}{k_3c^2}A_1A_2e^{i\Delta k\cdot z} \tag{7.18}$$

where $\Delta k = k_1 + k_2 - k_3$, which describes the wavevector (momentum) mismatch among three waves (ω_1, ω_2, and ω_3). By integrating eq. (7.18) from $z = 0$ (incident plane) to $z = L$ (exit plane of the medium), we can calculate the amplitude of the ω_3 radiation produced by the medium:

$$A_3(L) = \frac{i\chi^{(2)}\omega_3^2A_1A_2}{k_3c^2}\int_0^L e^{i\Delta k\cdot z}dz = \frac{i\chi^{(2)}\omega_3^2A_1A_2}{k_3c^2}\left(\frac{e^{i\Delta kL}-1}{i\Delta k}\right) \tag{7.19}$$

Recall that this is for the normal incident angle case (Figure 7.1C). When the incoming ω_1 and ω_2 beams are off the normal direction, then the angular dependence term is added [8].

7.2.2 Phase Matching Condition and SFG Intensity

The intensity of the electromagnetic wave is given by the magnitude of the time-averaged Poynting vector (eq. 5.64), which can be written as:

$$I_i = \frac{1}{2\pi}\varepsilon_o n_i c|A_i|^2, \quad \text{where } i = 1,2,3 \tag{7.20}$$

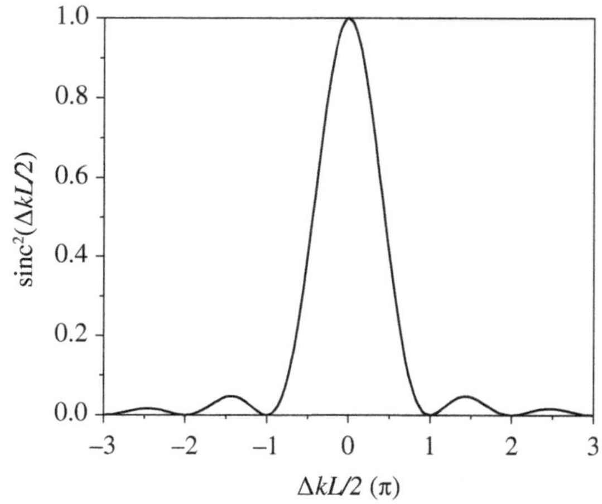

Figure 7.3 Effect of wavevector mismatch on the SFG efficiency.

Thus, from eq. (7.19), we obtain the following equation:

$$I_3 = \frac{2\pi {\chi^{(2)}}^2 \omega_3^2 I_1 I_2}{n_1 n_2 n_3 \varepsilon_0 c^3} \left| \frac{e^{i\Delta kL} - 1}{\Delta k} \right|^2 \tag{7.21}$$

The squared modulus term on the right-hand side of eq. (7.21) can be expressed as [7]:

$$\begin{aligned} \left| \frac{e^{i\Delta kL} - 1}{\Delta k} \right|^2 &= L^2 \left(\frac{e^{i\Delta kL} - 1}{\Delta kL} \right) \left(\frac{e^{-i\Delta kL} - 1}{\Delta kL} \right) \\ &= L^2 \frac{2(1 - \cos \Delta kL)}{\Delta kL^2} \\ &= L^2 \frac{\sin^2(\Delta kL/2)}{(\Delta kL/2)^2} = L^2 \mathrm{sinc}^2(\Delta kL/2) \end{aligned} \tag{7.22}$$

Thus, eq. (7.21) can be rewritten as:

$$I_3 = \frac{2\pi {\chi^{(2)}}^2 \omega_3^2 I_1 I_2}{n_1 n_2 n_3 \varepsilon_0 c^3} L^2 \mathrm{sinc}^2(\Delta kL/2) \tag{7.23}$$

The sinc-squared function in eq. (7.23) plays an important role in determining the efficiency of the SFG process. Figure 7.3 plots the $\mathrm{sinc}^2(\Delta kL/2)$ term, which is known as the phase mismatch factor. This plot shows that the efficiency of the SFG process ($\omega_3 = \omega_1 + \omega_2$, which is a three-wave mixing process) is maximum when $\Delta k = 0$, and it decays with some oscillatory fluctuations as Δk increases. When $\Delta kL/2 = \pi/2$, $\mathrm{sinc}^2(\Delta kL/2)$ is only 40% of the maximum value (which is close to e^{-1}). From this, one can define the characteristic length, which is called coherence length ($L_c = \pi/\Delta k$), for the efficient SFG process as:

$$L_c = \pi/\Delta k \tag{7.24}$$

Alternatively, the coherence length can be defined as $L_c \approx 2/\Delta k$ (which comes from setting $\Delta kL/2 = 1$). In any case, this means that if L is larger than L_c, the outgoing ω_3 wave can be out of phase with its driving polarization, and its energy can flow back into the ω_1 and ω_2 waves. For this reason, Δk $(= k_1 + k_2 - k_3)$ is called the phase mismatch. The sinc-squared function in eq. (7.23) depicts that the efficiency of the SFG process decreases dramatically when the phase-matching condition, $\Delta k = 0$, is not satisfied.

We now employ this phase-matching requirement and find the experimental condition in which the SFG signal can be generated effectively at two-dimensional (2D) interface. Continuing the vibrational spectroscopy theme that we have studied in Chapter 6, a vibrationally-resonance SFG process is considered here. To probe vibrational modes of analyte molecules, one of the two input beams is in the mid-IR range. The IR wavelength needs to be tunable to be in resonance with the vibrational absorption bands. Such tunable-wavelength IR beams with a pulse width in the femto- to pico-second range are generated through the optical parametric generation/amplification (OPG/OPA) process [9, 10], which involves the DFG process briefly discussed in Section 7.1. The other input beam is called the up-conversion beam. If a Ti:sapphire laser system is used for the SFG experiment, the up-conversion beam is 800 nm. If a Nd-YAG laser system is used, it can be 1064 nm or 532 nm. Although its wavelength can vary from near-IR to visible ranges, the up-conversion beam is conventionally called "visible" to differentiate it from the IR beam being tuned to find the resonance with molecular vibrations. Using this convention, the sum frequency process is described as:

$$\omega_{SF} = \omega_{VIS} + \omega_{IR} \tag{7.25}$$

Then, the phase-matching condition is written as:

$$\Delta \vec{k} = \vec{k}_{VIS} + \vec{k}_{IR} - \vec{k}_{SF} = 0 \tag{7.26}$$

Here, we now use the vector form, instead of the scalar form since it is the momentum vector of the electromagnetic wave. One can say that eqs. (7.25) and (7.26) are the energy conservation and momentum conservation, respectively, of the SFG process.

Figure 7.4 schematically depicts the reflection and refraction of the IR and VIS beams at the 2D interface as well as the emission of the SFG signal from the interface [11]. In the reflection mode experiment, the phase-matching condition along the surface-parallel direction (x-axis in Figure 7.4) is:

$$\vec{k}_{x,SF} = \vec{k}_{x,VIS} + \vec{k}_{x,IR} \tag{7.27}$$

which can be rewritten, using $k = \omega n/c$, as:

$$n_{1,SF}\omega_{SF} \sin\theta_{1,SF} = n_{1,VIS}\omega_{VIS} \sin\theta_{1,VIS} + n_{1,IR}\omega_{IR} \sin\theta_{1,IR} \tag{7.28}$$

Since $\left|\vec{k}_{VIS}\right| > \left|\vec{k}_{IR}\right|$, the SFG beam is emitted in the direction close to the VIS reflection beam. The signs of the IR incidence angle and the SFG emission angle are defined with respect to the direction of the VIS incidence and reflection, respectively. Thus, the $\vartheta_{1,IR}$ sign is positive if IR co-propagates along with VIS and negative if IR counter-propagates with VIS. For example, say that the $\theta_{1,VIS}$ of a 532-nm VIS beam is 60° and the $\theta_{1,IR}$ of a 2900-cm^{-1} IR beam is 54°; then, the $\theta_{1,SF}$ of the 460.89-nm SFG emission is 59.14° if IR co-propagates with VIS, and 39.95° if IR counter-propagates with VIS. The counter-propagation geometry is shown in the n_1 phase of Figure 7.4.

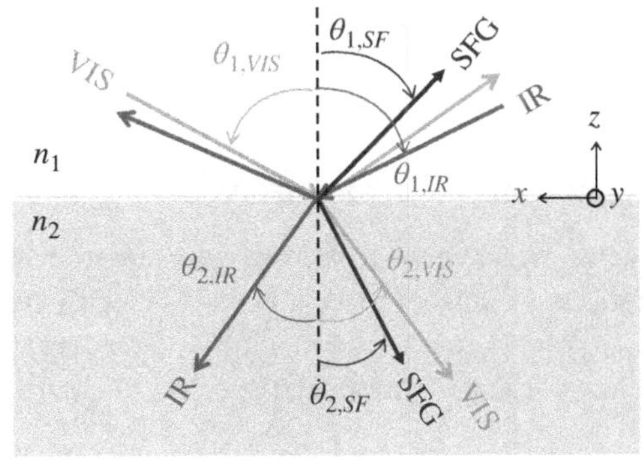

Figure 7.4 Incident and refraction angles of IR and visible input beams propagating in opposite directions and emission angles of SFG beam. The θ_{IR} sign is plus if IR arrives from the same x direction of VIS, and minus if IR arrives from the opposite direction of VIS. The latter case is shown in this figure. The refractive index of each medium is shown as n_1 and n_2. Reproduced from Choi et al. [11]/with permission of American Chemical Society.

The phase-matching requirement of the SFG process (Figure 7.3) depicts that the SFG signal emission is the largest (most efficient) at this angle and negligible at all other angles. In other words, if the SFG signal is detected at the phase-matching angle, the entire irradiated area by the IR and VIS pulses is in coherence ($L_c \to \infty$ when $\Delta\vec{k}_x = 0$; but obviously no signal from the region outside the irradiated area). Since the SFG requires both IR and VIS, the two incident pulses should be spatially and temporally overlapped; otherwise, the SFG signal cannot be generated.

In the transmission geometry, eq. (7.27) becomes:

$$n_{2,SF}\omega_{SF} \sin\theta_{2,SF} = n_{2,VIS}\omega_{VIS} \sin\theta_{2,VIS} + n_{2,IR}\omega_{IR} \sin\theta_{2,IR} \tag{7.29}$$

Here, $\theta_{2,\ VIS}$ and $\theta_{2,\ IR}$ are calculated from Snell's law (eq. 5.61). Using the same example used for the reflection geometry ($\theta_{532\ \mathrm{nm}} = 60°$ and $\theta_{2,900\ \mathrm{cm}^{-1}} = 54°$) and assuming $n_1 = 1$ (air) and $n_2 = 1.5$ (constant), we calculate $\theta_{460.89\ \mathrm{nm}}$ to be 32.79° for the co-propagation of the incident IR and VIS beams in the n_1 phase, and 23.84° for the counter-propagation of the incident beams. The counter-propagation geometry is shown in the n_2 phase of Figure 7.4. This is the angle where the SFG emission signal should be measured because of the phase-matching requirement (Figure 7.3), if the experiment is done in the transmission mode. In reality, n_2 is not constant; due to dispersion of refractive index, it varies with ω. Thus, we should use $n_{2,SF}$ at ω_{SF}, $n_{2,VIS}$ at ω_{VIS}, and $n_{2,IR}$ at ω_{IR} for the accurate calculation of $\theta_{2,SF}$.

Note that at the phase-matching angle along the x-direction (i.e., $\Delta\vec{k}_x = 0$), the momentum vector along the surface-normal direction (z-axis in Figure 7.4) cannot be conserved. The $\vec{k}_z$ magnitude is calculated using cosines of the propagation angles. For the reflection mode, the phase mismatch can be written as:

$$\Delta k_{z,\mathrm{refl}} = (n_{1,VIS}\omega_{VIS} \cos\theta_{1,VIS} + n_{1,IR}\omega_{IR} \cos\theta_{1,IR} + n_{1,SF}\omega_{SF} \cos\theta_{1,SF})/c \tag{7.30}$$

Note that the angular frequency should be used for ω here. The phase mismatch for the transmission mode is expressed as:

$$\Delta k_{z,trans} = |n_{2,VIS}\omega_{VIS} \cos\theta_{2,VIS} + n_{2,IR}\omega_{IR} \cos\theta_{2,IR} - n_{2,SF}\omega_{SF} \cos\theta_{2,SF}|/c \tag{7.31}$$

Then, the coherence length along the surface-normal direction can be defined as $L_{c,\ z} = \pi/\Delta k_z$. Using the values used in the calculation for $\Delta\vec{k}_x = 0$, $L_{c,z}$ is estimated to be 152 nm in the reflection mode and 1.08 μm in the transmission mode. The contribution from the subsurface region beyond this coherence length ($L_{c,z}$) to the measured SFG intensity is negligible since the coherence condition (Figure 7.3) cannot be met. This is one reason that SFG is sensitive to the interface region. If the subsurface region has an inversion symmetry or is completely random, SFG can be sensitive to interfacial molecules only, that is, interface specific (see Section 7.3.1) [8]. But, if there is a noncentrosymmetric electric field or structure in the subsurface region, then the SFG signal will have contributions from both the surface layer and the $L_{c,z}$-deep subsurface region [8].

7.2.3 Second-Order Susceptibility

So far, we have discussed the wave equation and phase-matching condition for SFG in the scalar form. To describe the SFG process, it is necessary to define the electric fields of incident IR and VIS beams (ω_{IR} and ω_{VIS}) projected to the medium, the ω_{SF} polarization of the medium induced by these fields, and emission of SF photons from the induced polarization in the Cartesian lab coordinate. Thus, the SFG term of eq. (7.9) should be written in the following form:

$$\boldsymbol{P}^{(2)}_{i,SF} = \varepsilon_o \chi^{(2)}_{ijk} \boldsymbol{E}_{j,VIS} \boldsymbol{E}_{k,IR} \tag{7.32}$$

where each subscript i, j, and k can be x, y, or z of the lab coordinate system. Thus, the second-order nonlinear susceptibility, $\chi^{(2)}_{ijk}$, is a third-order tensor having 27 (=3^3) components in the Cartesian space, each corresponding to a different combination of applied and induced electric field vectors. By tradition, the first, second, and third subscripts describe the directions of the electric fields of SF, VIS, and IR beams, respectively. For example, if the $\boldsymbol{E}$-fields of the incident ω_{IR} and ω_{VIS} beams are along the z and y-directions, respectively, and the $\boldsymbol{E}$-field of the induced polarization at ω_{SF} is in the x-direction, then the relevant nonlinear susceptibility term is $\chi^{(2)}_{xyz}$. Using this notation, the full description of the induced polarization is the sum of all possible 27 combinations of individual contributions:

$$\boldsymbol{P}^{(2)}_{SF} = \sum_{i}^{x,y,z} \boldsymbol{P}^{(2)}_{i,SF} = \varepsilon_o \sum_{i}^{x,y,z} \sum_{j}^{x,y,z} \sum_{k}^{x,y,z} \chi^{(2)}_{ijk} \boldsymbol{E}_{j,VIS} \boldsymbol{E}_{k,IR} \tag{7.33}$$

Each i, j, and k terms in eq. (7.33) describe the formation of the induced polarization by the incident $\boldsymbol{E}$-field projected to the surface. The projection of the $\boldsymbol{E}$-field from the incident medium (n_1 in Figure 7.4) to the surface can be described using the Fresnel coefficient equations derived in Section 5.2.2. For the s- and p-polarizations of the incoming beam, the projected $\boldsymbol{E}$-field in the x, y, and z directions can be calculated to be [12]:

$$\boldsymbol{E}_x = (E^I_x + E^R_x)\hat{\boldsymbol{x}} = \mp \cos\theta_i (1 - r_p) E^I_p \hat{\boldsymbol{x}} \tag{7.34}$$

$$\boldsymbol{E}_y = \left(E^I_y + E^R_y\right)\hat{\boldsymbol{y}} = (1 + r_s) E^I_s \hat{\boldsymbol{y}} \tag{7.35}$$

$$\boldsymbol{E}_z = (E^I_z + E^R_z)\hat{\boldsymbol{z}} = \sin\theta_i (1 + r_p) E^I_p \hat{\boldsymbol{z}} \tag{7.36}$$

where E^I and $E^R$are the incident and reflected beams, r_s and r_p are the Fresnel reflection coefficients, and $\hat{\boldsymbol{x}}$, $\hat{\boldsymbol{y}}$, and $\hat{\boldsymbol{z}}$ are the unit vectors of the Cartesian coordinate. The induced polarization oscillating at ω_{SF} frequency in the second medium (n_2 in Figure 7.4) can emit the SF photon to the direction defined by the phase-matching condition, that is, eq. (7.28) in the reflection direction or eq. (7.29) in the transmission direction, which can be described as:

$$\boldsymbol{E}_{i,SF} = L_i \boldsymbol{P}^{(2)}_{i,SF} \tag{7.37}$$

Here, the proportionality constant, L_i, can be derived in the same way used in the derivation of the Fresnel coefficient, that is, by imposing the continuity of $\boldsymbol{E}$-fields across an interface with the phase-matching restrictions on the SF emission angle ($\theta_{1,SF}$). We will skip the full derivation [13], and then focus on the use of the derived formula.

In the SFG experiment (Figure 7.5), the intensity of the emitted photon, not the $\boldsymbol{E}_{SF}$ field, is measured. Combining all terms described in eq. (7.37) and following the steps described in Sections 7.2.1–7.2.2, we can derive the equation describing the SF intensity. For the interface study, the sinc-squared function in eq. (7.23) can be replaced with $\sec^2\theta_{SF}$ because the SF emission is detected at the phase-matching direction [8] and drop the L^2 term since the SF intensity is determined by the symmetry argument within $L_{c,z}$ (= $\pi/\Delta k_z$: see eqs. (7.30) and (7.31) for Δk_z). Then, the resulting equation can be expressed as:

$$I_{SF} = \frac{2\pi\omega^2_{SF} \sec^2\theta_{SF}}{n_2(\omega_{SF}) n_2(\omega_{VIS}) n_2(\omega_{IR}) \varepsilon_o c^3} \left|\chi^{(2)}_{eff}\right|^2 I_{VIS} I_{IR} \tag{7.38}$$

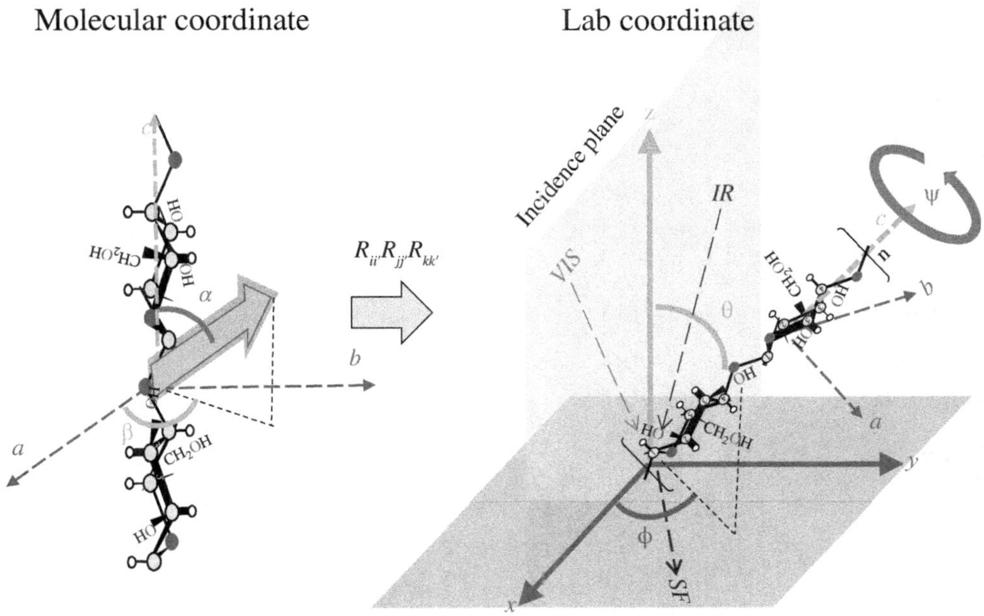

Figure 7.5 Schematic illustration of molecular axes (*a b*, and *c*) placed in the (*x*, *y*, and *z*) lab coordinate. Reproduced from Choi et al. [11]/with permission of American Chemical Society.

where I_{SF}, I_{VIS}, and I_{IR} are the intensities of SF, VIS, and IR, respectively, $n_2(\omega_{SF})$, $n_2(\omega_{VIS})$, and $n_2(\omega_{IR})$ are the refractive index of the medium at each frequency, and $\chi^{(2)}_{eff}$ is the effective second-order nonlinear susceptibility that can be written as:

$$\chi^{(2)}_{eff} = \sum_{i,j,k}^{x,y,z} L_i^{SF} L_j^{VIS} L_k^{IR} \hat{e}_{SF}\hat{e}_{VIS}\hat{e}_{IR}\chi^{(2)}_{ijk} \tag{7.39}$$

where $L_i^{SF} L_j^{VIS} L_k^{IR}$ and $\hat{e}_{SF}\hat{e}_{VIS}\hat{e}_{IR}$ are the terms describing the tensorial Fresnel factors and the electric field vector projections, respectively, of SFG, VIS, and IR beams. The products of these two terms describe all proportionality constants involved in eqs. (7.33) and (7.37). Each term in the tensorial Fresnel factor product in eq. (7.39) can be expressed as [11]:

In the reflection geometry:

For the SF term:

$$L_x^{SF,R} = \frac{2n_1(\omega_{SF})\cos\theta_{2,SF}}{n_1(\omega_{SF})\cos\theta_{2,SF} + n_2(\omega_{SF})\cos\theta_{1,SF}} \tag{7.40}$$

$$L_y^{SF,R} = \frac{2n_1(\omega_{SF})\cos\theta_{1,SF}}{n_1(\omega_{SF})\cos\theta_{1,SF} + n_2(\omega_{SF})\cos\theta_{2,SF}} \tag{7.41}$$

$$L_z^{SF,R} = \frac{2n_2(\omega_{SF})\cos\theta_{1,SF}}{n_1(\omega_{SF})\cos\theta_{2,SF} + n_2(\omega_{SF})\cos\theta_{1,SF}} \cdot \left(\frac{n_1(\omega_{SF})}{n_2(\omega_{SF})}\right)^2 \tag{7.42}$$

For the VIS and IR terms (ξ = VIS or IR):

$$L_x^{\xi,R} = \frac{2n_1(\omega_\xi)\cos\theta_{2,\xi}}{n_1(\omega_\xi)\cos\theta_{2,\xi} + n_2(\omega_\xi)\cos\theta_{1,\xi}} \tag{7.43}$$

$$L_y^{\xi,R} = \frac{2n_1(\omega_\xi)\cos\theta_{1,\xi}}{n_1(\omega_\xi)\cos\theta_{1,\xi} + n_2(\omega_\xi)\cos\theta_{2,\xi}} \tag{7.44}$$

$$L_z^{\xi,R} = \frac{2n_2(\omega_\xi)\cos\theta_{1,\xi}}{n_1(\omega_\xi)\cos\theta_{2,\xi} + n_2(\omega_\xi)\cos\theta_{1,\xi}} \cdot \left(\frac{n_1(\omega_\xi)}{n_2(\omega_\xi)}\right)^2 \tag{7.45}$$

In the transmission geometry:

For the SF term:

$$L_x^{SF,T} = \frac{2n_2(\omega_{SF})\cos\theta_{1,SF}}{n_1(\omega_{SF})\cos\theta_{2,SF} + n_2(\omega_{SF})\cos\theta_{1,SF}} \tag{7.46}$$

$$L_y^{SF,T} = \frac{2n_2(\omega_{SF})\cos\theta_{2,SF}}{n_1(\omega_{SF})\cos\theta_{1,SF} + n_2(\omega_{SF})\cos\theta_{2,SF}} \tag{7.47}$$

$$L_z^{SF,T} = \frac{2n_1(\omega_{SF})\cos\theta_{2,SF}}{n_1(\omega_{SF})\cos\theta_{2,SF} + n_2(\omega_{SF})\cos\theta_{1,SF}} \cdot \left(\frac{n_1(\omega_{SF})}{n_2(\omega_{SF})}\right)^2 \tag{7.48}$$

For the VIS and IR terms (ξ = VIS or IR):

$$L_x^{\xi,T} = \frac{2n_1(\omega_\xi)\cos\theta_{2,\xi}}{n_1(\omega_\xi)\cos\theta_{2,\xi} + n_2(\omega_\xi)\cos\theta_{1,\xi}} \tag{7.49}$$

$$L_y^{\xi,T} = \frac{2n_1(\omega_\xi)\cos\theta_{1,\xi}}{n_1(\omega_\xi)\cos\theta_{1,\xi} + n_2(\omega_\xi)\cos\theta_{2,\xi}} \tag{7.50}$$

$$L_z^{\xi,T} = \frac{2n_2(\omega_\xi)\cos\theta_{1,\xi}}{n_1(\omega_\xi)\cos\theta_{2,\xi} + n_2(\omega_\xi)\cos\theta_{1,\xi}} \cdot \left(\frac{n_1(\omega_\xi)}{n_2(\omega_\xi)}\right)^2 \tag{7.51}$$

Each term of the electric vector projections in eq. (7.39) can be expressed as:

$$\hat{e}_{SF,x} = -\cos(\theta_{1,SF}); \hat{e}_{SF,y} = 1; \hat{e}_{SF,z} = -\sin(\theta_{1,SF}) \text{ for reflection} \tag{7.52}$$

$$\hat{e}_{SF,x} = \cos(\theta_{2,SF}); \hat{e}_{SF,y} = 1; \hat{e}_{SF,z} = \sin(\theta_{2,SF}) \text{ for transmission} \tag{7.53}$$

$$\hat{e}_{VIS,x} = \cos(\theta_{1,VIS}); \hat{e}_{VIS,y} = 1; \hat{e}_{VIS,z} = \sin(\theta_{1,VIS}) \text{ for incidence} \tag{7.54}$$

$$\hat{e}_{IR,x} = \pm\cos(\theta_{1,IR}); \hat{e}_{IR,y} = 1; \hat{e}_{IR,z} = \sin(\theta_{1,IR}) \text{ for incidence} \tag{7.55}$$

In eq. (7.55), the + sign is used for the co-propagation geometry and the – sign is used for the counter-propagation geometry (which is the case shown in Figure 7.4). An example showing how eqs. (7.40)–(7.55) are used is given in Table 7.1 for a transmission-mode SFG experiment of an interface of water (medium 1) and organic matter (medium 2) using the counter propagating beam geometry.

In the experiment, the $\boldsymbol{E}$-field of the incoming and outgoing light is controlled in the *s*- and *p*-polarizations, and not in the $\boldsymbol{E}_x$, $\boldsymbol{E}_y$, and $\boldsymbol{E}_z$ directions independently. So, the experiment is done in a combination of *s*- and *p*-polarizations of two incoming beams and the emitted beam; so, there are eight possible combinations, which are *sss*, *ssp*, *sps*, *pps*, *spp*, *psp*, *pss*, and *ppp*. The order of three consecutive polarization symbols follows the polarization of SF, VIS, and IR beams, respectively. In the lab coordinate described in Figure 7.5 where the incident plane is in the *xz* plane, the *s*-polarization can probe the $\boldsymbol{E}_y$ component only, and the *p*-polarization can probe the contributions from both $\boldsymbol{E}_x$ and $\boldsymbol{E}_z$. From this, we can sort out what $\chi_{ijk}^{(2)}$ components can be probed in each polarization combination in SFG experiment, which is presented in Table 7.2. The sign in front of each $\chi_{ijk}^{(2)}$ is determined by the sign of the $\hat{e}_{SF}\hat{e}_{VIS}\hat{e}_{IR}$ product.

Table 7.1 Values calculated for the tensorial Fresnel factor and electric vector projection of the SFG, VIS, and IR beams for the transmission mode in the counter-propagation geometry with $\theta_{1,VIS}$ = +18° and $\theta_{1,IR}$ = −18°. The refractive index of water (n_1) and organic matter (n_2) used here are $n_{1,SF}$ = 1.3310 and $n_{2,SF}$ = 1.4670 at 650 nm, $n_{1,VIS}$ = 1.3290 and $n_{2,VIS}$ = 1.4640 at 800 nm, and $n_{1,IR}$ = 1.4200 and $n_{2,IR}$ = 1.4578 at 2944 cm^{-1} (= 3.4 μm).

	SFG		VIS		IR	
Component	L_i^{SF}	$\hat{e}_{SF}$	L_j^{VIS}	$\hat{e}_{VIS}$	L_k^{IR}	$\hat{e}_{IR}$
x	1.047	0.986	0.956	0.951	0.988	−0.951
y	1.050	1	0.947	1	0.986	1
z	0.953	0.170	0.860	0.309	0.960	0.309

Table 7.2 All $\chi_{ijk}^{(2)}$ terms possibly contributing each polarization combination for the co-propagation and counter-propagation geometry of experiment [11]. The order of three consecutive polarization symbols follows the polarization of SF, VIS, and IR beams, respectively. The laser incidence plane is the xz plane. The sign in front of each $\chi_{ijk}^{(2)}$ is determined by the sign of the $\hat{e}_{SF}\hat{e}_{VIS}\hat{e}_{IR}$ product; note that magnitude of $L_i^{SF}L_j^{VIS}L_k^{IR}\hat{e}_{SF}\hat{e}_{VIS}\hat{e}_{IR}$ is dropped here since it varies with the incident angles, wavelengths, and refractive indices.

		Co-propagation	Counter-propagation
Reflection	*sss*	$\chi_{yyy}^{(2)}$	$\chi_{yyy}^{(2)}$
	ssp	$\chi_{yyx}^{(2)} + \chi_{yyz}^{(2)}$	$-\chi_{yyx}^{(2)} + \chi_{yyz}^{(2)}$
	sps	$\chi_{yxy}^{(2)} + \chi_{yzy}^{(2)}$	$\chi_{yxy}^{(2)} + \chi_{yzy}^{(2)}$
	pss	$-\chi_{xyy}^{(2)} + \chi_{zyy}^{(2)}$	$-\chi_{xyy}^{(2)} + \chi_{zyy}^{(2)}$
	spp	$\chi_{yxx}^{(2)} + \chi_{yxz}^{(2)} + \chi_{yzx}^{(2)} + \chi_{yzz}^{(2)}$	$-\chi_{yxx}^{(2)} + \chi_{yxz}^{(2)} - \chi_{yzx}^{(2)} + \chi_{yzz}^{(2)}$
	psp	$-\chi_{xyx}^{(2)} - \chi_{xyz}^{(2)} + \chi_{zyx}^{(2)} + \chi_{zyz}^{(2)}$	$\chi_{xyx}^{(2)} - \chi_{xyz}^{(2)} - \chi_{zyx}^{(2)} + \chi_{zyz}^{(2)}$
	pps	$-\chi_{xxy}^{(2)} - \chi_{xzy}^{(2)} + \chi_{zxy}^{(2)} + \chi_{zzy}^{(2)}$	$-\chi_{xxy}^{(2)} - \chi_{xzy}^{(2)} + \chi_{zxy}^{(2)} + \chi_{zzy}^{(2)}$
	ppp	$-\chi_{xxx}^{(2)} - \chi_{xxz}^{(2)} - \chi_{xzx}^{(2)} - \chi_{xzz}^{(2)} + \chi_{zxx}^{(2)} + \chi_{zxz}^{(2)} + \chi_{zzx}^{(2)} + \chi_{zzz}^{(2)}$	$\chi_{xxx}^{(2)} - \chi_{xxz}^{(2)} + \chi_{xzx}^{(2)} - \chi_{xzz}^{(2)} - \chi_{zxx}^{(2)} + \chi_{zxz}^{(2)} - \chi_{zzx}^{(2)} + \chi_{zzz}^{(2)}$
Transmission	*sss*	$\chi_{yyy}^{(2)}$	$\chi_{yyy}^{(2)}$
	ssp	$\chi_{yyx}^{(2)} + \chi_{yyz}^{(2)}$	$-\chi_{yyx}^{(2)} + \chi_{yyz}^{(2)}$
	sps	$\chi_{yxy}^{(2)} + \chi_{yzy}^{(2)}$	$\chi_{yxy}^{(2)} + \chi_{yzy}^{(2)}$
	pss	$\chi_{xyy}^{(2)} + \chi_{zyy}^{(2)}$	$\chi_{xyy}^{(2)} + \chi_{zyy}^{(2)}$
	spp	$\chi_{yxx}^{(2)} + \chi_{yxz}^{(2)} + \chi_{yzx}^{(2)} + \chi_{yzz}^{(2)}$	$-\chi_{yxx}^{(2)} + \chi_{yxz}^{(2)} - \chi_{yzx}^{(2)} + \chi_{yzz}^{(2)}$
	psp	$\chi_{xyx}^{(2)} + \chi_{xyz}^{(2)} + \chi_{zyx}^{(2)} + \chi_{zyz}^{(2)}$	$-\chi_{xyx}^{(2)} + \chi_{xyz}^{(2)} - \chi_{zyx}^{(2)} + \chi_{zyz}^{(2)}$
	pps	$\chi_{xxy}^{(2)} + \chi_{xzy}^{(2)} + \chi_{zxy}^{(2)} + \chi_{zzy}^{(2)}$	$\chi_{xxy}^{(2)} + \chi_{xzy}^{(2)} + \chi_{zxy}^{(2)} + \chi_{zzy}^{(2)}$
	ppp	$\chi_{xxx}^{(2)} + \chi_{xxz}^{(2)} + \chi_{xzx}^{(2)} + \chi_{xzz}^{(2)} + \chi_{zxx}^{(2)} + \chi_{zxz}^{(2)} + \chi_{zzx}^{(2)} + \chi_{zzz}^{(2)}$	$-\chi_{xxx}^{(2)} + \chi_{xxz}^{(2)} - \chi_{xzx}^{(2)} + \chi_{xzz}^{(2)} - \chi_{zxx}^{(2)} + \chi_{zxz}^{(2)} - \chi_{zzx}^{(2)} + \chi_{zzz}^{(2)}$

Source: Choi et al. [11]/American Chemical Society.

7.2.4 Molecular Hyperpolarizability

The $L_i^{SF}L_j^{VIS}L_k^{IR}\hat{e}_{SF}\hat{e}_{VIS}\hat{e}_{IR}$ term in eq. (7.39) describes the experimental geometry of the SFG measurement; the $\chi_{ijk}^{(2)}$ term in eq. (7.39) describes how the property of molecule can be probed in SFG experiment at a given geometry. This third-rank tensor, $\chi_{ijk}^{(2)}$, can be related to the molecular hyperpolarizability tensor ($\beta_{i',j',k'}$) through:

$$\chi_{ijk}^{(2)} \propto N \sum\nolimits_{i',j',k'} \left\langle R_{ii'} R_{jj'} R_{kk'} \beta_{i',j',k'} \right\rangle \tag{7.56}$$

where N is the number density of the SFG-active modes within the probe volume defined by the coherence requirement (Section 7.2.2), and $R_{ii'}R_{jj'}R_{kk'}$ is the product of Euler matrices transforming the i', j', k' molecular coordinate (a, b, c) to the i, j, k lab coordinate (x, y, z). The Euler matrix term is expressed with the polar coordinate angles (θ, ϕ, and ψ) describing the relative orientation of the principal axes of (a, b, c) with respect to those of (x, y, z) [14], where θ, ϕ, and ψ are called tilt, azimuth, and twist angles, respectively (Figure 7.5). The most widely used rotational transformation in SFG experiment is the ZXZ sequence [14]. Then, the Euler transform matrix has the following form:

$$R_{ii'}R_{jj'}R_{kk'} = \begin{pmatrix} \cos\phi\cos\psi - \cos\theta\sin\phi\sin\psi & -\cos\theta\sin\phi\cos\psi - \cos\phi\sin\psi & \sin\theta\sin\phi \\ \sin\phi\cos\psi + \cos\theta\cos\phi\sin\psi & \cos\theta\cos\phi\cos\psi - \sin\phi\sin\psi & -\sin\theta\cos\phi \\ \sin\theta\sin\psi & \sin\theta\cos\psi & \cos\theta \end{pmatrix} \tag{7.57}$$

In eq. (7.56), $\beta_{i',j',k'}$ is proportional to the product of the polarizability derivative $\left(\frac{\partial\alpha_{i'j'}}{\partial Q_q}\right)$ and dipole derivative $\left(\frac{\partial\mu_{k'}}{\partial Q_q}\right)$ tensors along the normal mode coordinate (Q_q) of the analyte molecule [15], which can be expressed in the following matrix form:

$$\left(\frac{\partial\alpha_{i'j'}}{\partial Q_q}\right) = \begin{pmatrix} \alpha'_{aa} & \alpha'_{ab} & \alpha'_{ac} \\ \alpha'_{ba} & \alpha'_{bb} & \alpha'_{bc} \\ \alpha'_{ca} & \alpha'_{cb} & \alpha'_{cc} \end{pmatrix} \text{ and} \left(\frac{\partial\mu_{k'}}{\partial Q_q}\right) = \begin{pmatrix} \mu'_a \\ \mu'_b \\ \mu'_c \end{pmatrix} \tag{7.58}$$

The $\left(\frac{\partial\alpha_{i'j'}}{\partial Q_q}\right)$ and $\left(\frac{\partial\mu_{k'}}{\partial Q_q}\right)$ tensors can be obtained from density function theory (DFT) calculations [16] or polarized Raman and IR data [17]. Then, the $\beta_{i',j',k'}$ tensor is $\left(\frac{\partial\alpha_{i'j'}}{\partial Q_q}\right) \otimes \left(\frac{\partial\mu_{k'}}{\partial Q_q}\right)$, which gives the following expression:

$$\beta_{i',j',k'} = \begin{pmatrix} \begin{pmatrix}\beta_{aaa}\\ \beta_{aab}\\ \beta_{aac}\end{pmatrix} & \begin{pmatrix}\beta_{aba}\\ \beta_{abb}\\ \beta_{abc}\end{pmatrix} & \begin{pmatrix}\beta_{aca}\\ \beta_{acb}\\ \beta_{acc}\end{pmatrix} \\ \begin{pmatrix}\beta_{baa}\\ \beta_{bab}\\ \beta_{bac}\end{pmatrix} & \begin{pmatrix}\beta_{bba}\\ \beta_{bbb}\\ \beta_{bbc}\end{pmatrix} & \begin{pmatrix}\beta_{bca}\\ \beta_{bcb}\\ \beta_{bcc}\end{pmatrix} \\ \begin{pmatrix}\beta_{caa}\\ \beta_{cab}\\ \beta_{cac}\end{pmatrix} & \begin{pmatrix}\beta_{cba}\\ \beta_{cbb}\\ \beta_{cbc}\end{pmatrix} & \begin{pmatrix}\beta_{cca}\\ \beta_{ccb}\\ \beta_{ccc}\end{pmatrix} \end{pmatrix} = \begin{pmatrix} \begin{pmatrix}\alpha'_{aa}\mu'_a\\ \alpha'_{aa}\mu'_b\\ \alpha'_{aa}\mu'_c\end{pmatrix} & \begin{pmatrix}\alpha'_{ab}\mu'_a\\ \alpha'_{ab}\mu'_b\\ \alpha'_{ab}\mu'_c\end{pmatrix} & \begin{pmatrix}\alpha'_{ac}\mu'_a\\ \alpha'_{ac}\mu'_b\\ \alpha'_{ac}\mu'_c\end{pmatrix} \\ \begin{pmatrix}\alpha'_{ba}\mu'_a\\ \alpha'_{ba}\mu'_b\\ \alpha'_{ba}\mu'_c\end{pmatrix} & \begin{pmatrix}\alpha'_{bb}\mu'_a\\ \alpha'_{bb}\mu'_b\\ \alpha'_{bb}\mu'_c\end{pmatrix} & \begin{pmatrix}\alpha'_{bc}\mu'_a\\ \alpha'_{bc}\mu'_b\\ \alpha'_{bc}\mu'_c\end{pmatrix} \\ \begin{pmatrix}\alpha'_{ca}\mu'_a\\ \alpha'_{ca}\mu'_b\\ \alpha'_{ca}\mu'_c\end{pmatrix} & \begin{pmatrix}\alpha'_{cb}\mu'_a\\ \alpha'_{cb}\mu'_b\\ \alpha'_{cb}\mu'_c\end{pmatrix} & \begin{pmatrix}\alpha'_{cc}\mu'_a\\ \alpha'_{cc}\mu'_b\\ \alpha'_{cc}\mu'_c\end{pmatrix} \end{pmatrix} \tag{7.59}$$

The statistically weighted average value for each component in the $\beta_{i',j',k'}$ tensor can be obtained with the consideration of the Boltzmann distribution of various conformers within the molecular coordinate [16]. Another way to estimate $\beta_{i',j',k'}$ is based on the single-bond polarizability derivative (SBPD) model and the experimentally measured Raman depolarization ratio [14]. This approach requires an assumption that the vibrational modes of individual functional groups are not coupled with those of neighboring groups; then, specific symmetry rules can be imposed to each vibrational mode to find the non-zero $\beta_{i',j',k'}$ terms.

Once the full set of $\beta_{i',j',k'}$ is properly described, then $\chi^{(2)}_{ijk}$ can be calculated for any given tilt, azimuth, and twist angles using eq. (7.56). It is difficult to fully list all 27 terms involved in each $\chi^{(2)}_{ijk}(\omega_q)$ term of eq. (7.56), because there are total $27 \times 27 = 729$ components [13]. So, here, we show only a few terms in, for example, $\chi^{(2)}_{xxx}$ to help readers to gain conceptual understanding of the mathematical process. Equation (7.60) shows three terms of $\chi^{(2)}_{xxx}$ for a rod-shaped SFG-active domain with the a, b, c molecular coordinate placed in the x, y, z lab coordinate with the azimuthal (ϕ; angle between a and x axes), tilt (θ; angle between c and z axes), and rotational (ψ; rotation about c-axis) angles as defined in Figure 7.5:

$$\chi^{(2)}_{xxx} = R_{xa}R_{xa}R_{xa} \cdot \beta_{aaa} + R_{xa}R_{xb}R_{xb} \cdot \beta_{abb} + R_{xc}R_{xc}R_{xc} \cdot \beta_{ccc} + 24 \text{ terms} \tag{7.60}$$

where $R_{xa}R_{xa}R_{xa} = (\cos[\theta]\cos[\phi]\cos[\psi] - \sin[\phi]\sin[\psi])^3$, $R_{xa}R_{xb}R_{xb} = (-\cos[\psi]\sin[\phi] - \cos[\theta]\cos[\phi]\sin[\psi])^2(\cos[\theta]\cos[\phi]\cos[\psi] - \sin[\phi]\sin[\psi])$, and $R_{xc}R_{xc}R_{xc} = \cos[\phi]^3\sin[\theta]^3$ and so on [13]. This kind of calculation would be cumbersome if one must do it by hand. But it can be done easily and accurately if one uses mathematical programs such as *Mathematica* or *MATLAB* [11, 16].

The polarizability and dipole derivatives of the vibrational mode, whose product determines the molecular hyperpolarizability, $\beta_{i',j',k'}$, in eq. (7.56), are basically associated with the optical processes occurring in Raman and IR spectroscopy, respectively, which are schematically illustrated in Figure 7.6. They can be expressed using the quantum mechanical terms as:

$$\left(\frac{\partial\alpha_{i'j'}}{\partial Q_q}\right) \propto \sum_s \left[\frac{\langle g|\mu'_\alpha|s\rangle\langle s|\mu'_\beta|v\rangle}{(\omega_{SF} - \omega_{gs} + i\Gamma_s)(\omega_{VIS} - \omega_{sv} + i\Gamma_s)} + \text{permuted terms}\right] \tag{7.61}$$

$$\left(\frac{\partial \mu_{k'}}{\partial Q_q}\right) \propto \frac{\langle v|\mu'_{\gamma}|g\rangle}{\omega_{IR} - \omega_q + i\Gamma_q} \tag{7.62}$$

where the first and last letters in the angle brackets designate the electronic functions of the final and initial states of the transition, respectively, and the letter between two vertical lines represents the electric dipole operator. Here, ω_q and Γ_q are the resonant wavenumber and damping factor of the q-th normal mode; similarly, ω_{sv} and ω_{gs} are the wavenumbers of the $v \rightarrow s$ and $s \rightarrow v$ electronic transitions and Γ_s is the damping factor of the virtual state.

The damping factor represents the population decay rate of the excited state. Since the virtual state is not the eigenfunction of the electronic wave function, Γ_s is significantly larger than Γ_q. If both ω_{IR} and ω_{VIS} are not in resonance with any allowed excited state (as in Figure 7.6A), then the damping factors in eqs. (7.61) and (7.62) are large, so $\beta_{i'j'k'}$ in eq. (7.56) is small, and thus I_{SF} in eq. (7.38) is negligibly small. But, if one of the input beams is in resonance with the allowed state (as in Figure 7.6B), then $\beta_{i'j',k'}$ can be large and so is $\chi^{(2)}_{ijk}$; thus, I_{SF} can be significant.

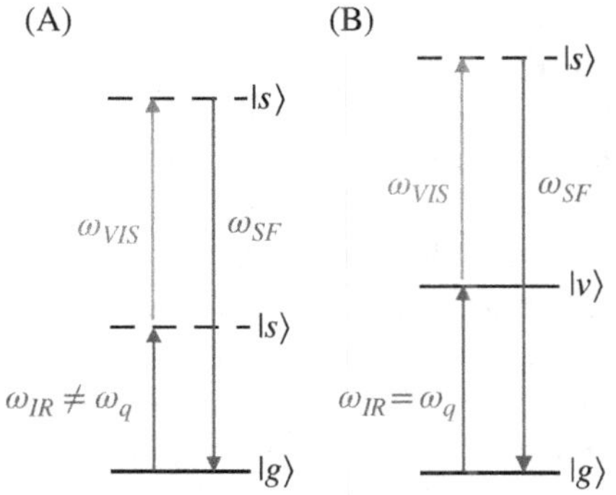

Figure 7.6 Description of SFG process with ω_{IR} in (A) non-resonance and (B) resonance with the qth vibrational mode of the sample. The symbols $|g\rangle$, $|v\rangle$, and $|s\rangle$ represent the ground state, a vibrationally excited state, and a virtually excited state, respectively. The solid and dashed horizonal lines represent the allowed and forbidden electronic state, respectively. The vertical upward and downward arrows correspond to the transitions associated with photon absorption and emission processes, respectively.

In Figure 7.6, the transition caused by ω_{IR} is the vibrational excitation from the ground state ($n_v = 0$) to the excited state ($n_v = 1$) of the analyte. The transitions associated with ω_{VIS} absorption and ω_{SF} emission are essentially the same as anti-Stoke Raman process. Thus, if the SF signal is recorded as a function of ω_{IR}, its intensity is essentially zero or at the noise level except in the IR absorption band region (unless there is a non-resonance background signal; see Section 7.3.6). Since I_{SF} is proportional to $\left|\chi^{(2)}_{ijk}\right|^2$ in eq. (7.38) and $\chi^{(2)}_{ijk}$ is proportional to N in eq. (7.56), the SF signal intensity is proportional to N^2, that is, the square of the concentration of the SFG-active molecule within the SFG probe volume (= beam area $\times L_{c,z}$).

7.2.5 Noncentrosymmetry Requirement at the Molecule Level

The polarizability and dipole derivative tensors described in eqs. (7.61) and (7.62) correspond to the Raman transition and IR absorption tensors, respectively. The latter term is already shown in eq. (5.26). Since $\chi^{(2)}_{ijk}$ is determined by the product of these two tensors, the vibrational mode to be probed with SFG should be Raman-active and IR-active. Otherwise, it cannot produce SF signal. Then, what kind of molecules can have vibrational modes that are active in both Raman and IR? For this, we need to use the group theory [18]. We can skip details of the group theory and review only a few examples to understand the essential points relevant to the SFG process.

Let us look at *trans*-1,2-dichloroethene to begin with. This molecule has an inversion symmetry. In other words, the inversion of all atomic positions through the central point gives the identical shape. In the group theory [18], this molecule falls into the C_{2h} point group. We can look up the character table for C_{2h} point group, which is shown in Table 7.3 [18]. The character table provides all the information we need to know. In the first row of the table, E, C_2, i, and σ_h represent symmetry

Table 7.3 Character table for point group C_{2h}.

	E	C_2	i	σ_h	Basis functions: Raman	Basis functions: IR	$\beta_{i'j'k'}$
A_g	1	1	1	1	x^2, y^2, z^2, xy		All $\beta_{i'j',k'}$ are zero since either $\left(\frac{\partial \alpha_{i'j'}}{\partial Q_q}\right)$ or $\left(\frac{\partial \mu_{k'}}{\partial Q_q}\right)$ is zero.
B_g	1	−1	1	−1	yz, xz		
A_u	1	1	−1	−1		z	
B_u	1	−1	−1	1		x, y	

operations of identity, twofold rotation, inversion, and horizontal reflection (mirror plane), respectively. The numbers 1 and −1 represent symmetric and antisymmetric, respectively. The symbols in the first column represent the irreducible representations. All normal modes of molecular vibrations can be described by one of these irreducible representations. The symbols in the basis function column represent specific physical properties. For example, x, y, z are the first-order functions, and $x^2, y^2, z^2, xy, yz, xz$ are the second-order functions. In other textbooks, the character table often has the rotational axes (R_x, R_y, R_z) as well; but they are omitted here since they are not relevant to the vibrational processes detected by SFG. Further details can be learned by going through the books on group theory [18].

The polarizability derivatives described by eq. (7.61) have the properties of the second-order function, and the dipole derivatives described by eq. (7.62) have those of the first-order function. When we look at the character table of point group C_{2h} (Table 7.3), we see no irreversible representation has both first- and second-order functions in the basis set. That means no normal mode can be both IR-active and Raman-active in the molecule with the C_{2h} point group; they are either IR-active only or Raman-active only. Because one of the derivative terms will be always zero for the normal mode of this molecule, their product (i.e., the numerator term of $\beta_{i',j',k'}$) is always zero; so, the C_{2h} molecule is *SFG-inactive*. This is the characteristic of all molecules that have inversion symmetry (i.e., having the i operation in the point group table) [18].

7.2.6 Nonzero Hyperpolarizability Components from Molecular Symmetry

Now, let us consider water molecule (H_2O) or methylene group ($-CH_2-$) which falls into the point group of C_{2v}. The character table of C_{2v} point group has two vertical reflections in the xz and yz planes and does not have the inversion symmetry (Table 7.4) [18]. In this point group, three irreducible representations have both first- and second-order functions. It means the normal modes of vibrations with A_1, B_1, or B_2 symmetry are both IR- and Raman-active; so, they can have nonzero $\beta_{i',j',k'}$ components and be SFG-active. The normal mode with A_2 symmetry is Raman-active but IR-inactive; thus, it is SFG-inactive.

From this symmetry argument, we can find, out of 27 elements in $\beta_{i',j',k'}$, which ones can have nonzero values. If the molecular coordinate (a, b, c) is aligned along the principal axes (x, y, z) of the point group, we can replace the symbols in the basis function with the molecular axes. The i' and j' subscripts are from the Raman tensor; note that these two are interchangeable. The k' subscript corresponds to the IR tensor. In the molecule with C_{2v} point group, the symmetric stretch (Figure 7.7A) and in-plane bending modes have the A_1 symmetry; these normal modes will have nonzero β_{aac}, β_{bbc}, and β_{ccc}, and zero for all other $\beta_{i',j',k'}$ elements. The asymmetric stretch mode (Figure 7.7B) has the B_1 symmetry; thus, it will have $\beta_{aca} = \beta_{caa} \neq 0$. The out-of-plane bending mode has the B_2 symmetry and $\beta_{bcb} = \beta_{cbb} \neq 0$ [14, 15].

Table 7.4 Character table for point group C_{2v} and corresponding $\beta_{i',j',k'}$ elements.

	E	C_2	$\sigma_v(xz)$	$\sigma_v(yz)$	Basis functions: Raman	Basis functions: IR	$\beta_{i',j',k'}$
A_1	1	1	1	1	x^2, y^2, z^2	z	$\beta_{aac}, \beta_{bbc}, \beta_{ccc}$
A_2	1	1	−1	−1	xy		
B_1	1	−1	1	−1	xz	x	$\beta_{aca} = \beta_{caa}$
B_2	1	−1	−1	1	yz	y	$\beta_{bcb} = \beta_{cbb}$

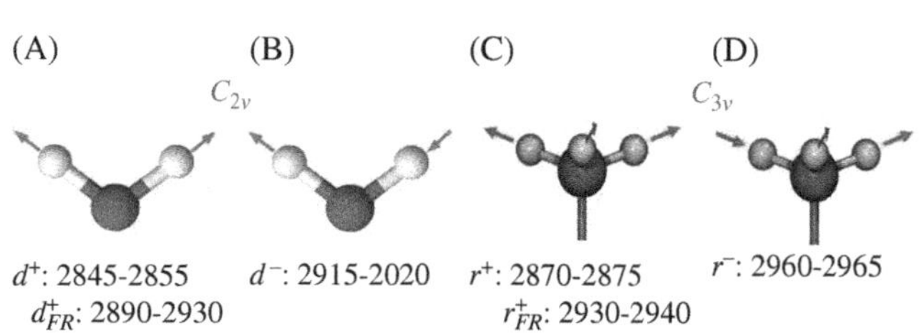

Figure 7.7 (A) Symmetric and (B) asymmetric stretch modes of functional group with C_{2v} point group and (C) symmetric and (D) asymmetric stretch modes of function group with C_{3v} point group.

Table 7.5 Character table for point group C_{3v} and corresponding $\beta_{i'j'k'}$ elements.

	E	$2C_3$	$3\sigma_v$	Basis functions: Raman	Basis functions: IR	$\beta_{i'j'k'}$
A_1	1	1	1	x^2+y^2, z^2	z	$\beta_{aac}=\beta_{bbc}, \beta_{ccc}$
A_2	1	1	−1			
E	1	−1	1	(yz, xz) (x^2-y^2, xy)	(x, y)	$\beta_{aca}=\beta_{caa}=\beta_{bcb}=\beta_{cbb}, \beta_{aaa}=-\beta_{abb}=-\beta_{bab}=-\beta_{bba}$ (or $\beta_{aab}=\beta_{aba}=\beta_{baa}=-\beta_{bbb}$, if a and b directions are switched)

Let us take a look at another example: ammonia (NH_3) or methyl ($-CH_3$) group with C_{3v} point group (Table 7.5) [18]. The symmetric stretch mode (Figure 7.7C) has A_1 symmetry, which can have $\beta_{aac}=\beta_{bbc}$ and β_{ccc} only. The asymmetric stretch mode has E symmetry. This means it is doubly degenerated, that is, two asymmetric stretch modes with their dipoles along the x- and y-directions are indistinguishable and energetically identical (Figure 7.7D). The asymmetric stretch modes can have $\beta_{aca}=\beta_{caa}=\beta_{bcb}=\beta_{cbb}$, and $\beta_{aaa}=-\beta_{abb}=-\beta_{bab}=-\beta_{bba}$ if the zero degree of C_3 rotation is aligned with the a-axis, or $\beta_{aab}=\beta_{aba}=\beta_{baa}=-\beta_{bbb}$ if it is aligned with the b-axis [14–16].

From these examples, we can see that, out of 27 $\beta_{i'j'k'}$ elements, only few of them are actually important due to the molecular symmetry. We will revisit this symmetry-based reduction of degrees of freedom in Section 7.3.3.

7.3 SFG Spectroscopy Probing 2D Interface

7.3.1 Interface Specificity

Although the $\chi^{(2)}_{ijk}$ tensor can have maximum 27 components, many of them are practically zero or negligible due to symmetry constraints. We have looked at the example of a centrosymmetric molecule (i.e., having inversion symmetry) in Section 7.2.5. Similarly, in a centrosymmetric environment, all x, y, z directions are equivalent; thus, the value of $\chi^{(2)}_{ijk}$ for two opposing directions must be identical, that is:

$$\chi^{(2)}_{ijk}=\chi^{(2)}_{-i-j-k} \tag{7.63}$$

However, since $\chi^{(2)}_{ijk}$ is a third-rank tensor, a change in the sign of the three axes (i.e., $x\to -x$, $y\to -y$, and $z\to -z$) must give a value with the reverse sign:

$$\chi^{(2)}_{ijk}=-\chi^{(2)}_{-i-j-k} \tag{7.64}$$

To satisfy both eqs. (7.63) and (7.64), $\chi^{(2)}_{ijk}$ should be zero. This means that, in a centrosymmetric medium, SFG is forbidden. In a completely random structure, the positive and negative directions are indistinguishable; thus, SFG is forbidden in the completely random medium as well.

The interface of the two phases is inherently noncentrosymmetric. Even if the bulk phase is completely random (such as liquid or gas) or centrosymmetric (such as NaCl crystal), the axis along the surface-normal direction breaks the randomness or centrosymmetry because two sides of the interface are different. So, SFG is allowed at the interface, which is the reason that SFG is intrinsically sensitive to the interfacial species. In this case, even if the coherence length along the surface-normal direction ($L_{c,z}=\pi/\Delta k_z$) is on the order of 100 nm (see Section 7.2.2), the contribution from the subsurface is essentially zero. Exceptions are when the subsurface region has a noncentrosymmetric structure [19] or an electric field gradient varying as a function of depth from the surface [20, 21]. In such cases, if the absolute phase of the subsurface signal is known, the subsurface signal can be used to determine the phase of surface signals [19–21].

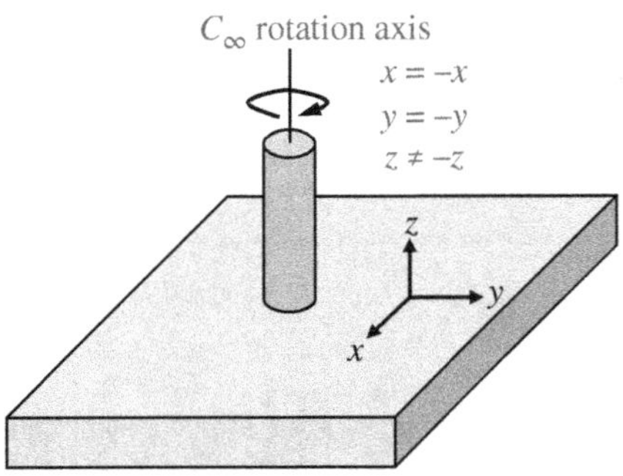

Figure 7.8 A planar surface which is isotropic and symmetric about the surface-normal axis.

7.3.2 Nonzero Second-Order Susceptibility Contributing SFG at 2D Interface

Let us consider a planar surface that is isotropic and symmetric about the surface-normal axis. This system contains a C_∞ rotation axis along the surface normal (z) direction, as depicted in Figure 7.8. In this system, there is an inversion symmetry within the xy plane (i.e., $x = -x$ and $y = -y$); but there is no inversion symmetry along the z-axis (i.e., $z \neq -z$). To find a pattern that gives nonzero $\chi^{(2)}_{ijk}$, let us do some simple symmetry operation. In the case of $\chi^{(2)}_{xxx}$, reverting the sign of x-axis gives $-\chi^{(2)}_{-x-x-x}$ since it is a third-order tensor (same as $(-1)\times(-1)\times(-1) = -1$), but, this should be the same as $\chi^{(2)}_{xxx}$ since there is an inversion symmetry along the x-axis. Thus, $\chi^{(2)}_{xxx}$ should be zero. Similarly, $\chi^{(2)}_{yyy}$ should be zero. In the case of $\chi^{(2)}_{xzz}$, reverting the sign of x-axis gives $-\chi^{(2)}_{-xzz}$, but it should be the same as $\chi^{(2)}_{xzz}$ due to the inversion symmetry along the x-axis. Thus, $\chi^{(2)}_{xzz}$ should be zero. Similarly, $\chi^{(2)}_{yzz}$ should be zero. In the case of $\chi^{(2)}_{xxz}$, reverting the sign of x-axis gives $\chi^{(2)}_{-x-xz}$ (same sign since the sign in x is changed twice), which is identical to $\chi^{(2)}_{xxz}$ due to the inversion symmetry within the inversion symmetry along the x-axis. So, $\chi^{(2)}_{xxz}$ can have a nonzero value. Similarly, $\chi^{(2)}_{yyz}$ can have a nonzero value. In the case of $\chi^{(2)}_{zzz}$, changing the sign of x or y does not matter since there is no x or y component involved. So, $\chi^{(2)}_{zzz}$ can have a nonzero value.

From these examples, we see that, in a planar surface that is isotropic within the surface plane (Figure 7.8), the $\chi^{(2)}_{ijk}$ terms that can have nonzero values are the ones having quadratic terms in either x or y contribution and the one that has no x or y contribution. Thus, out of 27 possible components of $\chi^{(2)}_{ijk}$, the nonzero components are only $\chi^{(2)}_{xxz}$, $\chi^{(2)}_{yyz}$, $\chi^{(2)}_{xzx}$, $\chi^{(2)}_{yzy}$, $\chi^{(2)}_{zxx}$, $\chi^{(2)}_{zyy}$, and $\chi^{(2)}_{zzz}$. Since x and y are indistinguishable in the 2D interface with C_∞ symmetry, we can see that $\chi^{(2)}_{xxz} = \chi^{(2)}_{yyz}$, $\chi^{(2)}_{xzx} = \chi^{(2)}_{yzy}$, and $\chi^{(2)}_{zxx} = \chi^{(2)}_{zyy}$. In Section 7.2.6, we see that the first two subscripts in the Raman component in $\beta_{i'j',k'}$ are exchangeable; that leads to $\chi^{(2)}_{yzy} = \chi^{(2)}_{zyy}$ in the susceptibility tensor. Thus, there will be total three independent values in the nonzero $\chi^{(2)}_{ijk}$ set.

Then, the long list of 27 $\chi^{(2)}_{ijk}$ terms shown in Table 7.2 that can contribute to eight possible polarization combinations can be reduced to 7 $\chi^{(2)}_{ijk}$ terms for a molecule present at 2D interfaces of dielectric materials. The nonzero terms for the 2D interface with C_∞ symmetry are shown in Table 7.6. The $\chi^{(2)}_{yyz}$, $\chi^{(2)}_{yzy}$, and $\chi^{(2)}_{zyy}$ terms are probed with the *ssp*, *sps*, and *pss* polarization

Table 7.6 Nonzero $\chi^{(2)}_{ijk}$ terms of an achiral molecule at a planar surface (*xy* plane) symmetric about the surface normal (*z*)-axis, contributing four possible polarization combinations in various experimental geometries used for SFG measurement. The order of three consecutive polarization symbols follows the polarization of SF, VIS, and IR beams, respectively, and the laser incidence plane is the *xz* plane. On metallic surfaces, only *ssp* and *ppp* polarizations can be used since the E-filed of the *s*-polarized IR (as in *sps* and *pss*) is effectively zero due to the metal selection rule.

		Co-propagation	Counter-propagation
Reflection	*ssp*	$\chi^{(2)}_{yyz}$	$\chi^{(2)}_{yyz}$
	sps	$\chi^{(2)}_{yzy}$	$\chi^{(2)}_{yzy}$
	pss	$\chi^{(2)}_{zyy}$	$\chi^{(2)}_{zyy}$
	ppp	$-\chi^{(2)}_{xxz} - \chi^{(2)}_{xzx} + \chi^{(2)}_{zxx} + \chi^{(2)}_{zzz}$	$-\chi^{(2)}_{xxz} + \chi^{(2)}_{xzx} - \chi^{(2)}_{zxx} + \chi^{(2)}_{zzz}$
Transmission	*ssp*	$\chi^{(2)}_{yyz}$	$\chi^{(2)}_{yyz}$
	sps	$\chi^{(2)}_{yzy}$	$\chi^{(2)}_{yzy}$
	pss	$\chi^{(2)}_{zyy}$	$\chi^{(2)}_{zyy}$
	ppp	$\chi^{(2)}_{xxz} + \chi^{(2)}_{xzx} + \chi^{(2)}_{zxx} + \chi^{(2)}_{zzz}$	$\chi^{(2)}_{xxz} - \chi^{(2)}_{xzx} - \chi^{(2)}_{zxx} + \chi^{(2)}_{zzz}$

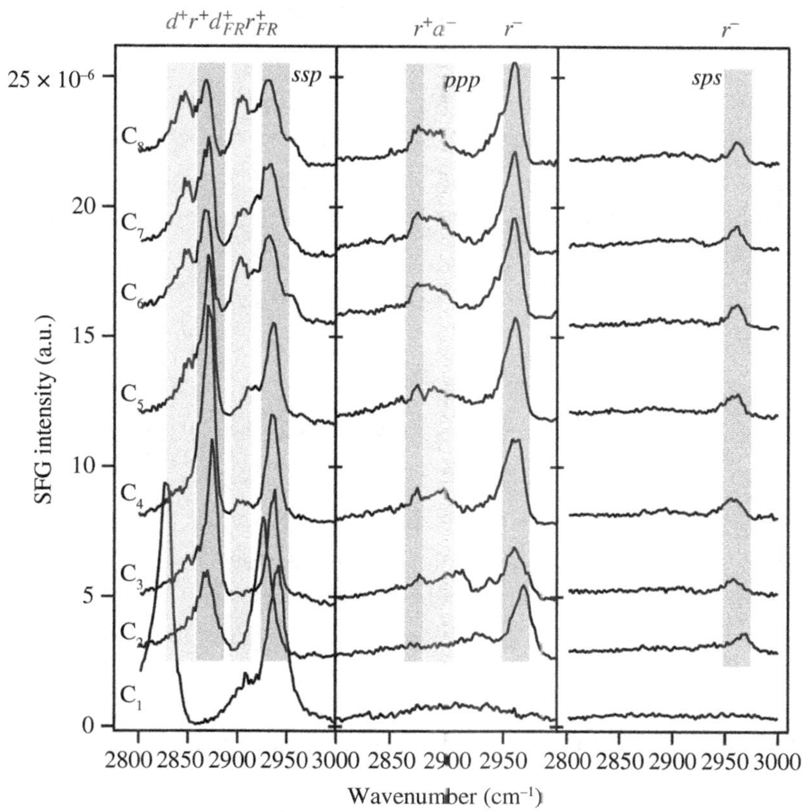

Figure 7.9 SFG spectra of *ssp*, *ppp*, and *sps* for vapor/primary alcohol interfaces with varying alkyl chain length from C1 (methanol) to C8 (n-octanol). The *pss* spectra are not listed because they are almost identical with the *sps* spectra. The visible and IR incident angles. Typical peak position ranges are marked with shaded regions for d^+, r^+, d^-, and r^- (see Figure 7.7 for the vibrational mode notation); d^+_{FR} and r^+_{FR} are the Fermi resonance peaks of d^+ and r^+ mixing with overtones of their symmetric bending vibration modes, respectively. Reproduced from Lu et al. [22]/with permission of American Chemical Society.

combinations, respectively. The *ppp* polarization signal has contributions from the $\chi^{(2)}_{xxz}$, $\chi^{(2)}_{xzx}$, $\chi^{(2)}_{zxx}$, and $\chi^{(2)}_{zzz}$ components with varying signs depending on the propagation directions of the incident VIS and IR beams and the detection direction of the generated SF signal.

Figure 7.9 displays an exemplary set of SFG spectra of vapor/liquid interfaces of a series of n-alcohols from methanol (C1) to n-octanol (C8) collected with *ssp*, *ppp*, and *sps* polarizations in a reflection geometry with $\theta_{VIS(532nm)} = 60°$, $\theta_{IR} = 55°$, and $\theta_{SF} = 59°$ [22]. Since $\chi^{(2)}_{yzy} = \chi^{(2)}_{zyy}$, the *sps* spectra are almost identical to the *pss* spectra (which are not shown). Even if the same molecules are probed, the SFG spectral features vary drastically depending on the polarization combination because different $\chi^{(2)}_{ijk}$ components are probed. Typical peak position ranges for the symmetric and asymmetric stretches of CH_2 (d^+ and d^-) and the symmetric and asymmetric stretches of CH_3 (r^+ and r^-) are marked with shaded regions. The regions marked with d^+_{FR} and r^+_{FR} are typical ranges for Fermi resonance peaks originating from quantum mechanical wavefunction mixing of the d^+ band with the overtone of the symmetric CH_2 bending mode and mixing of the r^+ band with the overtone of the symmetric CH_3 bending mode vibration modes, respectively [22]. Such mixing processes occur because the stretch mode and the overtone of bending mode have the same symmetry and similar energy.

If the substrate is a metal, then the metal selection rule (see Figure 6.28 in Section 6.2.4) depicts that the electric field of the *s*-polarized IR beam projected to the surface is negligible. Since the IR frequency is far below the plasma frequency of metallic substrates, the surface-parallel electric field component is fully cancelled by the image dipole formed by polarization of free electrons in the conduction band. For that reason, the experiment with the *sps* and *pss* polarizations cannot give any SFG signal on metallic surfaces. Thus, only *ssp* (probing $\chi^{(2)}_{yyz}$) and *ppp* (probing $\chi^{(2)}_{xxz}$ and $\chi^{(2)}_{zzz}$) measurements are possible on the metallic surface.

7.3.3 Predicting $\chi_{ijk}^{(2)}$ for Functional Groups with C_{2v} and C_{3v} Point Groups Present in the 2D Interface with C_∞ Symmetry

Although the full set of equations that relate $\chi_{ijk}^{(2)}$ to $\beta_{i',j',k'}$ have 729 terms in total [13], we have learned that only few of them can actually make major contributions due to the molecular symmetry (Section 7.2.5) and the noncentrosymmetry requirement at the 2D interface with C_∞ symmetry (Section 7.3.2). Let us take a deeper look at how the $\chi_{ijk}^{(2)}$ tensor terms can be obtained from the symmetry argument. Putting together eqs. (7.38), (7.39), and (7.56), one can rewrite the intensity equation for any given experimental geometry in the following expression:

$$\frac{I_{SF}}{I_{VIS}I_{IR}} \propto \left|\chi_{eff}^{(2)}\right|^2 \propto \left|\sum_{i,j,k}^{x,y,z}\chi_{ijk}^{(2)}\right|^2 \propto N^2\left|\sum_{i',j',k'}\left\langle R_{ii'}R_{jj'}R_{kk'}\beta_{i',j',k'}\right\rangle\right|^2 \tag{7.65}$$

Here, all tensorial Fresnel coefficient and electric vector terms (described in eqs. (7.40)–(7.55)) are lumped into a constant and dropped out because they do not change in a given experimental setup. Out of 27 $\chi_{ijk}^{(2)}$ elements, only 7, $\chi_{xxz}^{(2)}$, $\chi_{yyz}^{(2)}$, $\chi_{xzx}^{(2)}$, $\chi_{yzy}^{(2)}$, $\chi_{zxx}^{(2)}$, $\chi_{zyy}^{(2)}$, and $\chi_{zzz}^{(2)}$, can have nonzero values (Table 7.6), and some of them are identical ($\chi_{xxz}^{(2)} = \chi_{yyz}^{(2)}$ and $\chi_{xzx}^{(2)} = \chi_{yzy}^{(2)} = \chi_{zxx}^{(2)} = \chi_{zyy}^{(2)}$) at the 2D interface with C_∞ symmetry. Also, the molecules at the C_∞ interface can rotate freely about their twisting angle (ψ) and azimuth angle (ϕ). So, the summation over all possible ψ and ϕ angles can be replaced with integration over 0 to 2π of $R_{ii'}R_{jj'}R_{kk'}\beta_{i',j',k'}$ over ψ and ϕ, normalized with $\int_0^{2\pi}\int_0^{2\pi}d\psi d\phi$. Then, $\chi_{ijk}^{(2)}$ becomes a function of θ only, that is, varying as a function of molecular tilt angle with respect to the surface-normal axis. This is the basic principle that SFG can be used to find the tilt angle of a specific functional group present at the 2D interface, which will be discussed later in Section 7.3.6. The functional group specificity comes from the resonance of ω_{IR} with the vibrational excitation of that group (ω_q; see eq. (7.62)).

For a molecule with C_{2v} point group (such as CH_2), the symmetric stretching mode (Figure 7.7A) has A_1 symmetry. From Table 7.4, we can see that, out of 27 $\beta_{i',j',k'}$ elements, only β_{aac}, β_{bbc}, and β_{ccc} can have nonzero values. Doing the integration over ψ and ϕ using the nonzero $\beta_{i',j',k'}$ elements only and dropping the N^2 term for the sake of discussion, we get the following relationships:

$$\chi_{xxz}^{(2)} = \chi_{yyz}^{(2)} = \frac{1}{8}\cos\theta(3\beta_{aac} + 3\beta_{bbc} + 2\beta_{ccc} + (\beta_{aac} + \beta_{bbc} - 2\beta_{ccc})\cos 2\theta) \tag{7.66}$$

$$\chi_{xzx}^{(2)} = \chi_{yzy}^{(2)} = \chi_{zxx}^{(2)} = \chi_{zyy}^{(2)} = -\frac{1}{4}(\beta_{aac} + \beta_{bbc} - 2\beta_{ccc})\cos\theta\sin^2\theta \tag{7.67}$$

$$\chi_{zzz}^{(2)} = \frac{1}{4}\cos\theta(\beta_{aac} + \beta_{bbc} + 2\beta_{ccc} - (\beta_{aac} + \beta_{bbc} - 2\beta_{ccc})\cos 2\theta) \tag{7.68}$$

The asymmetric stretching mode of the C_{2v} molecule (Figure 7.7B) has B_1 symmetry if the stretching bonds are in the *ac* plane. Then, we can set all $\beta_{i',j',k'}$ elements are zero except $\beta_{aca} = \beta_{caa}$ (Table 7.4). Note that if the vibrating bonds are placed in the *bc* plane, then we use $\beta_{bcb} = \beta_{cbb}$ since the asymmetric stretch assumes B_2 symmetry. By doing the same integration over ψ and ϕ with $\beta_{aca} = \beta_{caa} \neq 0$ and setting all other $\beta_{i',j',k'} = 0$, we get the following relationships for the asymmetric stretch mode of the C_{2v} molecule:

$$\chi_{xxz}^{(2)} = \chi_{yyz}^{(2)} = -\frac{1}{2}\beta_{aca}\cos\theta\sin^2\theta \tag{7.69}$$

$$\chi_{xzx}^{(2)} = \chi_{yzy}^{(2)} = \chi_{zxx}^{(2)} = \chi_{zyy}^{(2)} = \frac{1}{2}\beta_{aca}\cos^3\theta \tag{7.70}$$

$$\chi_{zzz}^{(2)} = \beta_{aca}\cos\theta\sin^2\theta \tag{7.71}$$

For a molecule with C_{3v} point group (such as CH_3), the symmetric stretching mode (Figure 7.7C) has A_1 symmetry in which only $\beta_{aac} = \beta_{bbc}$ and β_{ccc} can have nonzero values (Table 7.5). Then, we find the following relationships:

$$\chi_{xxz}^{(2)} = \chi_{yyz}^{(2)} = \frac{1}{4}\cos\theta(3\beta_{aac} + \beta_{ccc} + (\beta_{aac} - \beta_{ccc})\cos 2\theta) \tag{7.72}$$

$$\chi^{(2)}_{xzx} = \chi^{(2)}_{yzy} = \chi^{(2)}_{zxx} = \chi^{(2)}_{zyy} = -\frac{1}{2}(\beta_{acc} - \beta_{ccc})\cos\theta\sin^2\theta \qquad (7.73)$$

$$\chi^{(2)}_{zzz} = \frac{1}{4}((\beta_{aac} + 3\beta_{ccc})\cos\theta + (-\beta_{aac} + \beta_{ccc})\cos 3\theta) \qquad (7.74)$$

Similarly, we can obtain the relationships between the nonlinear susceptibility and the hyperpolarizability for the asymmetric stretching mode of the C_{3v} molecule (Figure 7.7D) which has E symmetry. Using $\beta_{aca} = \beta_{caa} = \beta_{bcb} = \beta_{cbb}$ and $\beta_{aaa} = -\beta_{abb} = -\beta_{bab} = -\beta_{bba}$ (Table 7.5), the following relationships are derived:

$$\chi^{(2)}_{xxz} = \chi^{(2)}_{yyz} = -\beta_{aca}\cos\theta\sin^2\theta \qquad (7.75)$$

$$\chi^{(2)}_{xzx} = \chi^{(2)}_{yzy} = \chi^{(2)}_{zxx} = \chi^{(2)}_{zyy} = \beta_{aca}\cos^3\theta \qquad (7.76)$$

$$\chi^{(2)}_{zzz} = \beta_{aca}\sin\theta\sin 2\theta \qquad (7.77)$$

Note that switching the zero degree of the C_3 rotation to the b-axis does not make any difference since all terms having $\beta_{aaa} = -\beta_{abb} = -\beta_{bab} = -\beta_{bba}$ cancel out each other within the symmetry that we are considering (i.e., C_{3v} point group in C_∞ environment).

This method can be employed for molecules with different point groups. Using the character table of group theory [18], we can find the nonzero elements in the $\beta_{i',j',k'}$ hyperpolarizability tensor, which can be used to calculate the nonzero elements in the $\chi^{(2)}_{ijk}$ second-order nonlinear susceptibility that are listed in Table 7.6 [15].

7.3.4 Calculating SFG Signal Intensity of C_{3v} Functional Group in C_∞ Interface

Let us do actual calculation, to see exactly how it works, for the SFG signal intensity of the symmetric and asymmetric stretch modes of a CH_3 group that perfectly follows the symmetry of C_{3v} point group. For that, we choose iodomethane (CH_3I) and use DFT calculations to obtain the Raman and IR tensors [16]. This molecule is chosen because the mass difference between CH_3 and I is large; so, vibrational coupling between these two is negligible. The calculation results are shown in Table 7.7. Note that in this calculation, all possible tensors are calculated since it is done with *Mathematica* using the DFT calculation results obtained using *Gaussian-16*, a commercial quantum chemistry package [16]. All $\beta_{i',j',k'}$ values are normalized with β_{ccc} of CH_3^s.

Table 7.7 Polarizability and dipole derivative tensors and hyperpolarizability tensor of CH_3 symmetric (CH_3^s) and asymmetric (CH_3^{as}) modes of CH_3I obtained from DFT calculations. The threefold rotation axis of the CH_3 group is aligned with the (vertical) z-axis of the lab coordinate and one of the C—H bonds is in the yz plane. For the polarizability and dipole derivative tensors of the asymmetric mode, only one of the energetically degenerate E modes of CH_3I is shown; the other one is orthogonal to this one. The bold components in $\beta_{i',j',k'}$ are the ones predicted to have nonzero values for the C_{3v} molecule at the C_∞ interface [16].

CH_3I	CH_3^s	CH_3^{as}
$\left(\frac{\partial \alpha_{i'j'}}{\partial Q_q}\right) \otimes \left(\frac{\partial \mu_{k'}}{\partial Q_q}\right)$	$\begin{pmatrix} -1.68 & 0.00 & 0.00 \\ 0.00 & -1.68 & 0.00 \\ 0.00 & 0.00 & -1.75 \end{pmatrix} \otimes \begin{pmatrix} 0.00 \\ 0.00 \\ 3.71 \end{pmatrix}$	$\begin{pmatrix} 0.46 & 0.87 & 1.18 \\ 0.87 & -0.47 & 0.63 \\ 1.18 & 0.63 & 0.00 \end{pmatrix} \otimes \begin{pmatrix} -1.14 \\ -0.61 \\ 0.00 \end{pmatrix}$
$\beta_{i',j',k'}$ (normalized with β_{ccc})	$\begin{pmatrix} 0.00 \\ 0.00 \\ \mathbf{0.96} \end{pmatrix} \begin{pmatrix} 0.00 \\ 0.00 \\ 0.00 \end{pmatrix} \begin{pmatrix} 0.00 \\ 0.00 \\ 0.00 \end{pmatrix}$	$\begin{pmatrix} 0.00 \\ \mathbf{0.20} \\ 0.00 \end{pmatrix} \begin{pmatrix} \mathbf{0.20} \\ 0.00 \\ 0.00 \end{pmatrix} \begin{pmatrix} \mathbf{0.27} \\ 0.00 \\ 0.00 \end{pmatrix}$
	$\begin{pmatrix} 0.00 \\ 0.00 \\ 0.00 \end{pmatrix} \begin{pmatrix} 0.00 \\ 0.00 \\ \mathbf{0.96} \end{pmatrix} \begin{pmatrix} 0.00 \\ 0.00 \\ 0.00 \end{pmatrix}$	$\begin{pmatrix} \mathbf{0.20} \\ 0.00 \\ 0.00 \end{pmatrix} \begin{pmatrix} 0.00 \\ \mathbf{-0.20} \\ 0.00 \end{pmatrix} \begin{pmatrix} 0.00 \\ \mathbf{0.27} \\ 0.00 \end{pmatrix}$
	$\begin{pmatrix} 0.00 \\ 0.00 \\ 0.00 \end{pmatrix} \begin{pmatrix} 0.00 \\ 0.00 \\ 0.00 \end{pmatrix} \begin{pmatrix} 0.00 \\ 0.00 \\ \mathbf{1.00} \end{pmatrix}$	$\begin{pmatrix} \mathbf{0.27} \\ 0.00 \\ 0.00 \end{pmatrix} \begin{pmatrix} 0.00 \\ \mathbf{0.27} \\ 0.00 \end{pmatrix} \begin{pmatrix} 0.00 \\ 0.00 \\ 0.00 \end{pmatrix}$

Here, the denominator terms of eqs. (7.61) and (7.62), (i.e., wavenumber dependence) are dropped for simplicity, and only the numerator terms are considered. For the symmetric stretch mode, only $\beta_{aac} = \beta_{bbc}$ and β_{ccc} are found to have nonzero values. For the asymmetric stretch mode, we see the nonzero elements for $\beta_{aca} = \beta_{bcb} = \beta_{caa} = \beta_{cbb}$ (= 0.27) and $\beta_{aab} = \beta_{aba} = \beta_{baa} = -\beta_{bbb}$ (= 0.20, which is equivalent to $\beta_{aaa} = -\beta_{aab} = -\beta_{bba} = -\beta_{bab}$ with switching of the a and b directionality). All other terms are zero due to the cancellation of two orthogonal asymmetric modes. This is exactly what we have predicted in Table 7.5. Then, using eqs. (7.72)–(7.77), we can calculate $\chi^{(2)}_{xxz} = \chi^{(2)}_{yyz}, \chi^{(2)}_{xzx} = \chi^{(2)}_{yzy} = \chi^{(2)}_{zxx} = \chi^{(2)}_{zyy}$, and $\chi^{(2)}_{zzz}$. Finally, we can calculate $\frac{I_{SF}}{I_{VIS}I_{IR}} \propto \left|\chi^{(2)}_{eff}\right|^2$ using eq. (7.39) to compare the relative SFG intensities of selected vibrational modes collected with different polarizations for a given experimental geometry. The SFG peak intensities of the CH_3 stretch modes calculated as a function of the tilt angle of CH_3 group with respect to the surface normal axis are shown in Figure 7.10A [16].

In the *ssp* experiment (Figure 7.10A), the symmetric stretch mode of CH_3 (A_1 mode of C_{3v}) is peaked at $\theta = 0°$ and decays with the $\cos^2(\theta)$ function as θ increases. This can be easily understood if we look at the $\chi^{(2)}_{yyz}$ tensor in the C_∞ symmetry system. From Table 7.7, we see that the $(\beta_{aac} - \beta_{ccc})$ term is negligible compared to $3\beta_{aac} + \beta_{ccc}$ in eq. (7.72). This makes the tilt angle dependence of the CH_3^s *ssp* signal mode to be proportional to $\cos^2\theta$.

In contrast, the *ssp* signal of the asymmetric stretch mode (E mode of C_{3v}) is zero at $\theta = 0°$ and peaked at $\theta = 54.74°$ (Figure 7.10A). This is because the maximum of the $\cos(\theta)\sin^2(\theta)$ function in eq. (7.75) is at $\theta = 54.74°$. Note that this angle equals the magic angle at which the tilt angle is indistinguishable from the random case. Also, the maximum intensity of the CH_3^{as} *ssp* signal at $\theta = 54.74°$ is significantly smaller than that of CH_3^s *ssp* at $\theta = 0°$. That is because the maximum value of $\chi^{(2)}_{yyz}$ is proportional to $\beta_{aac}{}^2$ for CH_3^{as} and $(3\beta_{aac} + \beta_{ccc})^2$ for CH_3^s; note that here we have $\beta_{aac} \approx \beta_{ccc}$ from DFT calculations for CH_3I (Table 7.7).

The CH_3^{as} *ppp* signal is also peaked at $\theta = 54.74°$; this is the consequence of the convolution of contributions from $\chi^{(2)}_{xxz}, \chi^{(2)}_{xzx}$, $\chi^{(2)}_{zxx}$, and $\chi^{(2)}_{zzz}$ (Table 7.6). Note that although the magnitudes of $\chi^{(2)}_{xzx}$ and $\chi^{(2)}_{zxx}$ are the same (as shown in eq. (7.76)), they do not cancel each other perfectly since the magnitudes of $L_x^{SF}L_z^{VIS}L_x^{IR}\hat{e}_{SF,x}\hat{e}_{VIS,z}\hat{e}_{IR,x}$ and $L_z^{SF}L_x^{VIS}L_x^{IR}\hat{e}_{SF,z}\hat{e}_{VIS,x}\hat{e}_{IR,x}$ are not the same.

The CH_3^s *sps* and *pss* intensities are essentially zero because $\beta_{aac} - \beta_{ccc}$ is nearly zero in eq. (7.73) for the perfect C_{3v} point group. The *sps* and *pss* signals are almost identical for the CH_3^{as} mode, which is because the magnitudes of $\chi^{(2)}_{yzy}$ and $\chi^{(2)}_{zyy}$ are the same. The slight difference comes from the Fresnel factors and electric vector projections as described in eqs. (7.40)–(7.51). The CH_3^{as} *sps* intensity is peaked at $\theta = 0°$ and decays following the $\cos^2(\theta)$ function as predicted by eq. (7.76).

Alternatively, the hyperpolarizability tensor components can be estimated from the SBPD model and the experimentally measured Raman depolarization ratio [14]. Then, the relative magnitudes of nonzero $\beta_{i'j'k'}$ terms are different from the values from DFT calculations, but the general trends in the calculated SFG intensity are almost identical [16].

From Figure 7.10A, one can see that the symmetric stretch mode of CH_3 is strong in the *ssp* spectrum, while the asymmetric stretch mode of CH_3 is stronger in the *ppp* spectrum than in the *ssp* spectrum. The same trend is observed for the symmetric and asymmetric stretch modes of CH_2 group with C_{2v} point group in the C_∞ interface [22]. This polarization dependence pattern of symmetric and asymmetric modes can be quite useful in vibrational mode assignment. For example, the d^- peak range of CH_2 overlaps with the d^+_{FR} peak range in Figure 7.9. Although both *ssp* and *ppp* spectra show peaks in the same range, the ones in the *ssp* spectrum are assigned to d^+_{FR} because the asymmetric mode is negligible in the *ssp* spectrum, and the ones in the *ppp* spectrum are assigned to d^- because the symmetric mode is negligible in the *ppp* spectrum [22].

7.3.5 Intramolecular Vibrational Coupling Among Neighboring Groups

It should be noted that vibrations of functional groups in a molecule are not isolated from each other, and the normal modes of adjacent groups are coupled. So, the symmetry of the normal mode of vibration deviates from the ideal case considered in the previous section. These intramolecular coupling of vibrations of neighboring functional groups makes the relative intensity of *ssp*, *psp*, *pss*, and *ppp* signals of SFG vary from the prediction made with the ideal symmetry argument.

Let us take a look at the CH_3 group vibration in ethyl iodide (CH_3CH_2I), as an example. Its Raman and IR tensors as well as molecular hyperpolarizability tensor obtained from DFT calculations are shown in Table 7.8 [16]. We can see that the elements that are zero in Table 7.7 are not zero in Table 7.8. The values of the nonzero $\beta_{i'j'k'}$ terms of CH_3I in Table 7.7 are also slightly changed in Table 7.8 of CH_3CH_2I. This is because the normal modes involving CH_3 group vibrations cannot be fully described with the A_1 and E symmetry of C_{3v} point group in CH_3CH_2I due to coupling with the CH_2 vibrations.

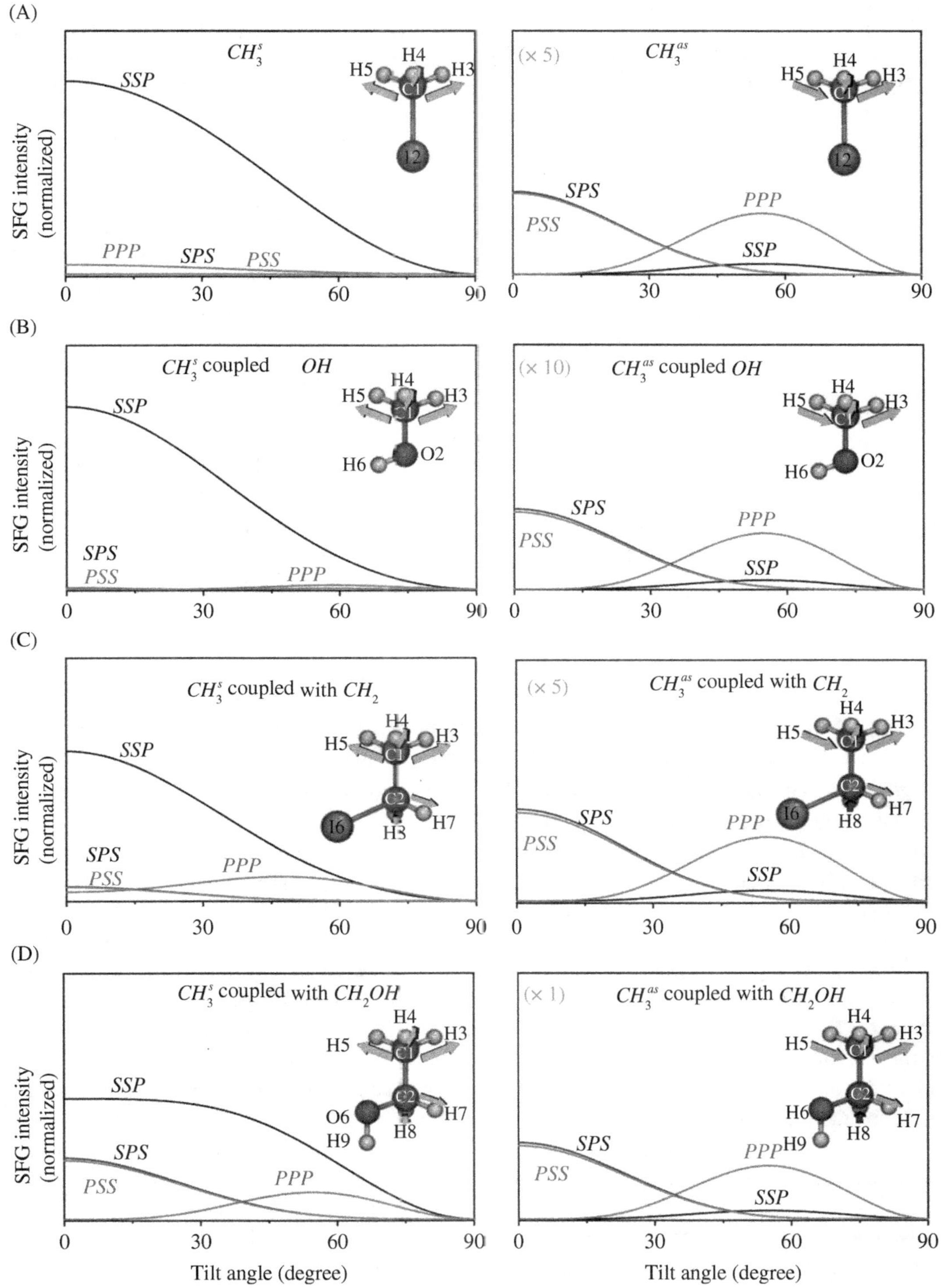

Figure 7.10 Comparison of the SFG intensities at *ssp*, *ppp*, *sps*, and *pss* polarizations for the symmetric and asymmetric stretch modes of the CH_3 group in (A) CH_3I, (B) CH_3OH, (C) CH_3CH_2I, and (D) CH_3CH_2OH as a function of the tilt angle (θ) of the CH_3 threefold rotation axis with respect to the surface normal direction, calculated using the molecular hyperpolarizability tensors obtained from DFT calculations. The calculation is for the case where the molecule is placed in the 2D surface with C_∞ symmetry. The incidence angles of the co-propagating I_{IR} and I_{VIS} beams are 56° and 60°, respectively, with respect to the surface-normal direction. The asymmetric mode signals are multiplied with the factor shown at the top of *y*-axis for comparison with the symmetric mode of the same molecule. Reproduced from Choi et al. [16]/with permission of American Chemical Society.

Table 7.8 Polarizability and dipole derivative tensors and hyperpolarizability tensor of CH_3 symmetric (CH_3^s) and asymmetric (CH_3^{as}) modes coupled with CH_2 vibrations in CH_3CH_2I. The threefold rotation axis of the CH_3 group is aligned with the (vertical) z-axis of the lab coordinate and one of the C—H bonds is in the yz plane. The polarizability and dipole derivative tensors' values shown here are for a 60° H-C-C-I dihedral angle (δ). The $\beta_{i'j',k'}$ tensor is the average of three staggered conformers (δ = 60°, 180°, and 300°). The bold components in $\beta_{i'j',k'}$ are the ones predicted to have nonzero values for the C_{3v} molecule at the C_∞ interface [16].

CH_3CH_2I	CH_3^s	CH_3^{as}
$\left(\frac{\partial \alpha_{i'j'}}{\partial Q_q}\right) \otimes \left(\frac{\partial \mu_{k'}}{\partial Q_q}\right)$	$\begin{pmatrix} 2.58 & 0.74 & 1.29 \\ 0.74 & 1.67 & 0.72 \\ 1.29 & 0.72 & 2.17 \end{pmatrix} \otimes \begin{pmatrix} -2.55 \\ -1.43 \\ -3.73 \end{pmatrix}$	$\begin{pmatrix} 0.90 & -0.55 & -0.74 \\ -0.55 & -0.90 & 1.31 \\ -0.74 & 1.31 & 0.00 \end{pmatrix} \otimes \begin{pmatrix} 1.60 \\ -2.84 \\ -0.01 \end{pmatrix}$
$\beta_{i'j',k'}$ (normalized with β_{ccc})	$\begin{pmatrix} 0.01 \\ 0.16 \\ \mathbf{0.98} \end{pmatrix} \begin{pmatrix} 0.16 \\ -0.01 \\ 0.00 \end{pmatrix} \begin{pmatrix} 0.27 \\ 0.00 \\ 0.00 \end{pmatrix}$ $\begin{pmatrix} 0.16 \\ -0.01 \\ 0.00 \end{pmatrix} \begin{pmatrix} -0.01 \\ -0.16 \\ \mathbf{0.98} \end{pmatrix} \begin{pmatrix} 0.00 \\ 0.27 \\ 0.00 \end{pmatrix}$ $\begin{pmatrix} 0.27 \\ 0.00 \\ 0.00 \end{pmatrix} \begin{pmatrix} 0.00 \\ 0.27 \\ 0.00 \end{pmatrix} \begin{pmatrix} 0.00 \\ 0.00 \\ \mathbf{1.00} \end{pmatrix}$	$\begin{pmatrix} 0.01 \\ \mathbf{0.21} \\ 0.00 \end{pmatrix} \begin{pmatrix} \mathbf{0.21} \\ -0.01 \\ 0.00 \end{pmatrix} \begin{pmatrix} \mathbf{0.30} \\ 0.00 \\ 0.00 \end{pmatrix}$ $\begin{pmatrix} \mathbf{0.21} \\ -0.01 \\ 0.00 \end{pmatrix} \begin{pmatrix} -0.01 \\ \mathbf{-0.21} \\ 0.00 \end{pmatrix} \begin{pmatrix} 0.00 \\ \mathbf{0.30} \\ 0.00 \end{pmatrix}$ $\begin{pmatrix} \mathbf{0.30} \\ 0.00 \\ 0.00 \end{pmatrix} \begin{pmatrix} 0.00 \\ \mathbf{0.30} \\ 0.00 \end{pmatrix} \begin{pmatrix} 0.00 \\ 0.00 \\ 0.00 \end{pmatrix}$

Source: Choi et al. [16]/American Chemical Society.

In the case of CH_3OH where the iodine atom of CH_3I is replaced with the hydroxyl group (OH), the C–H vibrations of CH_3 are coupled with the O–H stretch vibration. Due to the high polarity of the OH group, the values of the nonzero $\beta_{i'j',k'}$ terms of CH_3I in Table 7.7 are significantly changed in the CH_3OH system [16]. In the case of CH_3CH_2OH, the changes in $\beta_{i'j',k'}$ are even larger due to vibrational coupling between CH_3– and –CH_2OH [16].

The theoretically calculated SFG signal intensities of the normal modes involving symmetric and asymmetric stretch of the CH_3 group in CH_3OH, CH_3CH_2I, and CH_3CH_2OH are plotted in Figures 7.10B–D [16]. There are some significant deviations or differences from the perfect C_{3v} vibrations shown in Figure 7.10A. In the case of CH_3OH, the asymmetric vibration peak intensities are even smaller than those of the CH_3I. This is the consequence of the coupling of the asymmetric stretch of CH_3 with the OH stretch whose dipole is off the C_3 rotation axis. The theoretically calculated ratio of the CH_3^{as}-*ppp* intensity to the CH_3^s-*ssp* intensity is two times smaller for CH_3OH (Figure 7.10B) than CH_3I (Figure 7.10A). Since the OH group of methanol is pointing toward the liquid phase at the vapor/liquid interface, it will have hydrogen bonding interactions with other OH groups. Then, the damping factor would increase significantly for the OH group, which will also damp the adjacent CH_3 group vibration; this will make the CH_3^{as}-*ppp* signal intensity even smaller. This is the reason that, in the *ppp* spectrum of CH_3OH in Figure 7.9, the r^- peak of CH_3 is below the detection limit. The reason that the r^+ peak position of CH_3 is much different for CH_3OH than longer alcohols is also due to the coupling with the polar OH group. For alcohol molecules with longer chains, this coupling is negligible because there are CH_2 groups between the terminal CH_3 and OH groups.

In contrast to the *ppp* spectrum of CH_3OH, the *ppp* spectrum of CH_3CH_2OH shows a relatively strong r^- peak at 2970 cm^{-1}; in fact, its intensity is even comparable to the intensity of the r^+ peak at 2870 cm^{-1} in the *ssp* spectrum of CH_3CH_2OH (Figure 7.9). Although weak, the r^+peak can be seen above the noise level in the *ppp* spectrum. These trends can be explained if the experimental data are compared with the theoretical SFG signal intensities of CH_2CH_3OH calculated using the hyperpolarizability tensors obtained from DFT (Figure 7.10D); they could not be explained if the data are compared with those of CH_3I which has the perfect C_{3v} symmetry (Figure 7.10A).

Unlike CH_3OH, n-alcohols with alkyl chain longer than two show the r^+ signal in the *ppp* spectra (Figure 7.9), which is deviating the prediction made with the perfect C_{3v} symmetry (Figure 7.10A). The reason for that can be seen in the theoretically calculated SFG intensities for CH_3^s for CH_3CH_2I and CH_3CH_2OH shown in Figures 7.10C,D. It is because of the coupling of the CH_3 vibration with the adjacent CH_2 vibration. The theoretical calculation predicts that even with this coupling, the r^+ peak intensity will be negligible in the *ppp* spectrum if the tilt angle of the CH_3 molecular axis is close to the surface-normal direction (i.e., $\theta < 30°$). The sizable intensity of the r^+ mode in the *ppp* spectrum implies that the CH_3 groups are fairly tilted away from the surface-normal direction. The fact that the r^- mode intensity is quite smaller in the *sps* spectra than in the *ppp* spectra of n-alcohols in Figure 7.9 also suggests that the CH_3 chain ends of these alcohol molecules are significantly tilted away from the surface-normal direction (see Figure 7.10C). If the CH_3 group is aligned quite close to the surface-normal direction, then the r^- mode intensity would be larger in the *sps* spectrum than in the *ppp* spectrum.

7.3.6 Determining Molecular Tilt Angle from SFG Intensity

As briefly mentioned in the last example in the previous section, the relative SFG intensities of selected vibrational modes can be used to determine the molecular tilt angle with respect to the surface-normal direction [14]. In the literature, the ratio of the CH_3 asymmetric peak in the *ppp* and *sps* spectra ($I^{as}_{ppp}/I^{as}_{sps}$) or the ratio of the CH_3 symmetric peak in the *ssp* spectrum vs. the CH_3 asymmetric peak in the *ppp* spectrum (I^{s}_{ssp}/I^{as}_{ppp}) are most extensively used to determine the CH_3 tilt angle with respect to the surface-normal direction. Figure 7.11 plots the theoretically calculated ratios of $I^{as}_{ppp}/I^{as}_{sps}$ and I^{s}_{ssp}/I^{as}_{ppp} as a function of CH_3 group tilt angle from the surface-normal direction (θ). The $I^{as}_{ppp}/I^{as}_{sps}$ ratio is useful for the case when θ is less than $\sim 50°$; beyond this, it is difficult to use because the denominator term is too small (see Figure 7.10) and becomes difficult to measure with a good S/N ratio In contrast, the I^{s}_{ssp}/I^{as}_{ppp} ratio works well only for high tilt angles; when θ is small, then the I^{as}_{ppp} intensity is too small (see Figure 7.10) and becomes difficult to measure reliably. Figure 7.11 shows that the $I^{as}_{ppp}/I^{as}_{sps}$ ratio is fairly independent of model used to calculate the theoretical value; but the I^{s}_{ssp}/I^{as}_{ppp} ratio is sensitive to which model is used in the theoretical calculation [16]. The reason for this discrepancy can be seen clearly by comparing the theoretical plots in Figure 7.10.

When the CH_3CH_2OH data shown in Figures 7.9 and 7.11 are used, the tilt angle of the CH_3 chain end from the normal direction of the vapor/liquid interface is found to be 43–48° [16]. But there is another degree of freedom to be considered. All calculations demonstrated so far are done with the assumption that the molecular orientation has a delta function, that is, there is no spread in the orientation distribution function. In reality, there will be a statistical distribution around the average value [23]. The effect of such orientational distributions can be calculated by multiplying the Gaussian distribution function to the tilt angle when the Euler matrix term is integrated in eq. (7.65) (see Section 7.3.3). Figure 7.12 plots the $I^{as}_{ppp}/I^{as}_{sps}$ and I^{s}_{ssp}/I^{as}_{ppp} ratios as a function of average tilt angle ($\overline{\theta}$) with different Gaussian standard deviations ($\overline{\theta} \pm \sigma_\theta$ with $\sigma_\theta = 0°, 10°, 20°, 30°$, and $40°$). This calculation shows that as the tilt angle distribution becomes broader (i.e., σ_θ increases), the $\overline{\theta}$-dependence of $I^{as}_{ppp}/I^{as}_{sps}$ and I^{s}_{ssp}/I^{as}_{ppp} becomes weaker [16]. Depending on the spread of the distribution, the average value of the tilt angle changes.

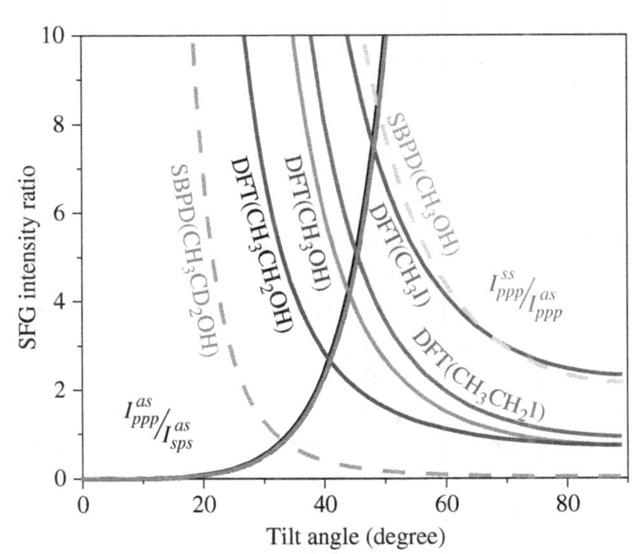

Figure 7.11 Relative intensities of the *ssp* signal of the symmetric CH_3 stretch vs. the *ppp* signal of the asymmetric CH_3 stretch (I^{s}_{ssp}/I^{as}_{ppp}) and the *ppp* vs. *sps* signals of the asymmetric CH_3 stretch ($I^{as}_{ppp}/I^{as}_{sps}$) predicted for CH_3I, CH_3OH, CH_3CH_2I, CH_3CH_2OH, and CH_3CD_2OH placed in the 2D interface with C_∞ symmetry. The solid lines are from DFT calculations, and the dashed lines are from the SBPD model. The $I^{as}_{ppp}/I^{as}_{sps}$ ratios of two models are essentially identical. Reproduced from Choi et al. [16]/with permission of American Chemical Society.

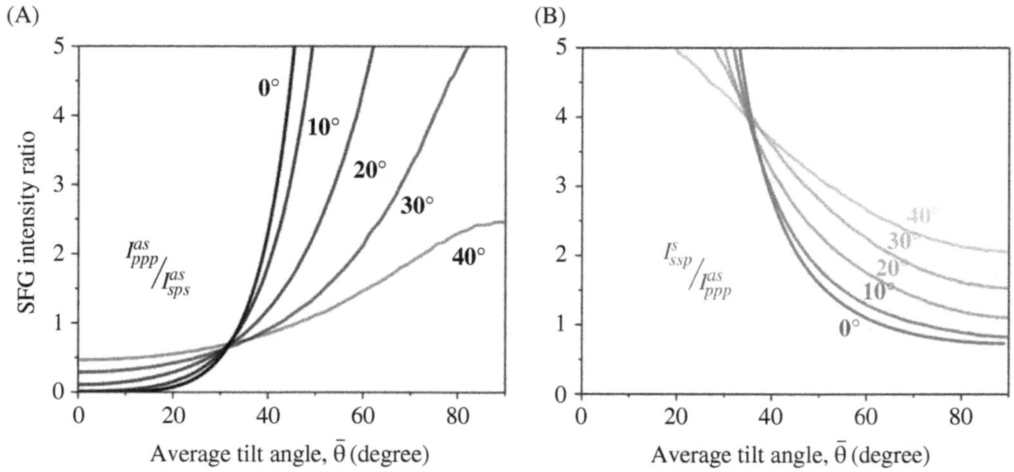

Figure 7.12 SFG intensity ratio for (A) $I^{as}_{ppp}/I^{as}_{sps}$ and (B) I^{s}_{ssp}/I^{as}_{ppp} predicted for the air/liquid interface of CH_3CH_2OH having the Gaussian distribution of tilt angle with the standard deviation of σ_θ = 0°, 10°, 20°, 30°, and 40°. Reproduced from Choi et al. [16]/with permission of American Chemical Society.

7.3.7 Molecular Conformation and Compositional Excess at Interfaces

Due to the interface specificity (Section 7.3.1), vibrational SFG spectroscopy has been widely used to study molecular conformation, orientation, and compositional excess at various interfaces [1–3]. In-depth review with advanced theoretical discussions can be found in the book by Y. R. Shen [24]. Here, we will discuss only two selected examples to get conceptual comprehension of SFG principles that we have studied so far.

In self-assembled monolayers (SAMs) of long-chain organic molecules, the degree of packing among alkyl chains will govern the conformation of alkyl chains. In order to achieve the maximum packing density of SAM molecules chemically bonded to the surface atoms at a flat 2D surface, all alkyl chains should assume the fully extended conformation, as shown with cartoon in Figure 7.13A [12]. In this case, all CH_2 groups are in the *trans* conformation and the CH_3 chain end groups point fairly close to the surface normal direction (see Figure 6.31). As the packing density decreases, some C—C bonds can rotate by 120°, assuming the *gauche* conformation (Figure 7.13B). The energy difference between *trans* and *gauche* conformations is only ~3.3 kJ/mol, which is comparable to thermal energy at room temperature (RT ~ 2.5 kJ/mol). Thus, in a low-density SAM, most alkyl chains will twist and flex, adopting a broad distribution of *trans* and *gauche* conformations (Figure 7.13C). The introduction of gauche defects allows more CH_3 end groups to tilt toward the surface. When the molecular density is really low so that most of them are isolated from each other (so, they are not self-assembled anymore), molecules will assume fairly random conformations (Figure 7.13D).

The degree of chain packing is manifested as spectral shape change in SFG [12]. In the all-*trans* conformation, the majority of CH_2 groups in the alkyl chains lie in a locally centrosymmetric environment, which makes them SFG-inactive. In contrast, most CH_3 terminal groups will point away from the surface, creating a highly noncentrosymmetric order. Thus, the *ppp*-SFG spectrum will have the CH_3 spectral features predominantly, and the CH_2 features are insignificant even though the *ppp* polarization can probe the surface-parallel components (because it can have contributions from $\chi^{(2)}_{xzx}$ and $\chi^{(2)}_{zxx}$; see Table 7.6). A hypothetical *ppp*-SFG spectrum of the all-*trans* SAM is schematically drawn in Figure 7.13A. As the degree of disorder increases, the fraction of *gauche* defects increases and such defects break the local symmetry of CH_2 groups, making them SFG-active. At the same time, the degree of noncentrosymmetric order of CH_3 groups at the chain end decreases. Thus, the CH_2 peak intensities increase and the CH_3 intensities decrease as the molecular density decreases in SAM (Figure 7.13B,C). When the surface coverage is too low that molecules are fully isolated, then the overall intensity becomes extremely low (because of low surface coverage) and both CH_2 and CH_3 features appear in the *ppp*-SFG spectrum (Figure 7.13D). If the molecules are oriented parallel to the surface (which would maximize the van der Waals interactions with the surface), then the CH_2 features may become stronger than the CH_3 features in the *ppp*-SFG spectrum (because there are more CH_2 groups than CH_3 groups in long-chain alkyl molecules).

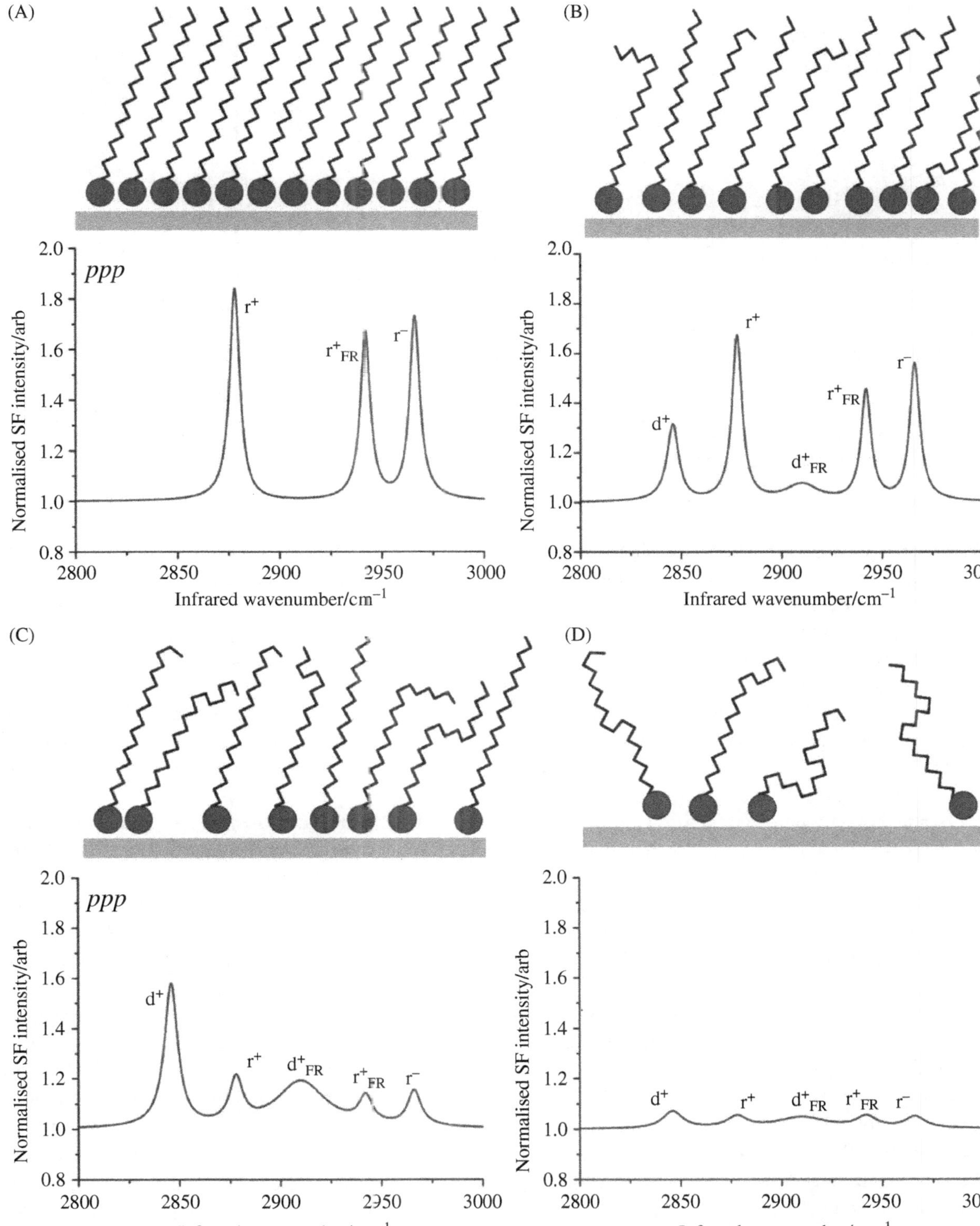

Figure 7.13 An illustration representation of the effect of chain packing in SAM on a SFG spectrum with *ppp* polarization spectrum: (A) perfect packing, (B) with small defects, (C) poorly packed, and (D) low coverage with negligible packing. Reproduced from Lambert et al. [12]/with permission of Taylor and Francis Group.

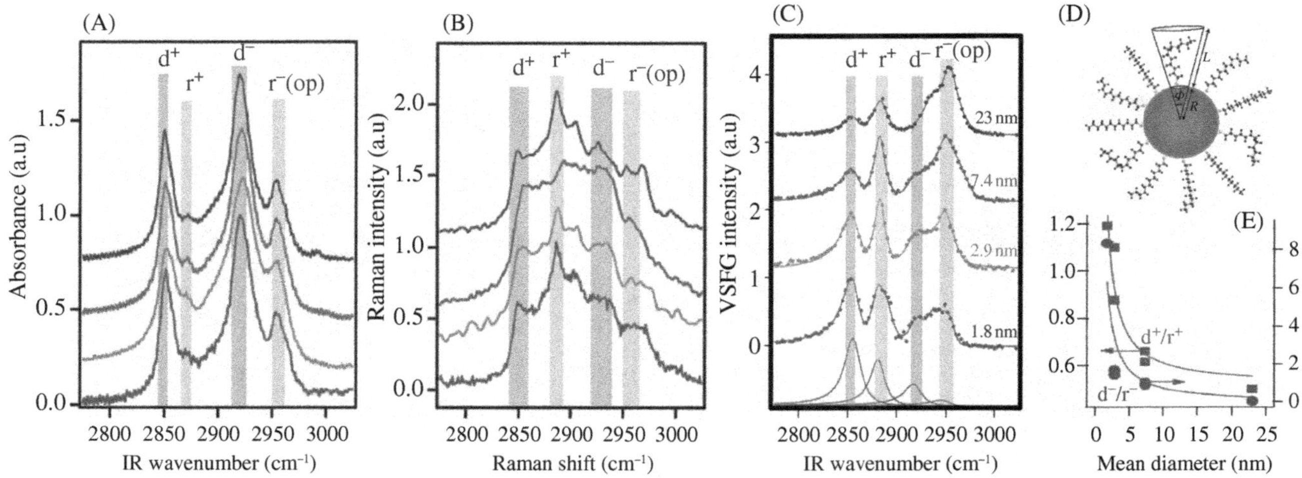

Figure 7.14 (A) IR, (B) Raman, and (C) *ssp*-SFG spectra of dodecanethiol SAM on gold nanoparticles of four different sizes. (D) Schematic illustration of SAM on nanoparticle. (E) Ratio of CH_2/CH_3 (d/r) peak intensities in the SFG spectra. Reproduced from Weeraman et al. [25]/ with permission of American Chemical Society.

Figure 7.14 displays an experimental dataset showing the chain packing effects [25]. In this experiment, dodecane thiol SAMs were prepared on gold nanoparticles of different diameters and nanoparticles were deposited on a CaF_2 window by drop-casting from a dilute suspension solution. The FT-IR spectra are mostly dominated by the CH_2 peaks and there is no dependence on the particle size (Figure 7.14A). The Raman spectra also show both CH_2 and CH_3 peaks with no dependence of their relative intensities on the particle size (Figure 7.14B). But the SFG spectra show changes in the CH_2/CH_3 peak ratio as the particle size changes (Figure 7.14C). For the smallest particle, the head space for alkyl chain of each surface-bound thiol molecule is quite large (Figure 7.14D). As the particle size increases (so, the surface becomes flatter), the head space becomes smaller and the degree of alkyl chain packing increases. This is equivalent to the trend seen in the direction from Figure 7.13C to Figure 7.13A. That explains the decrease in the CH_2/CH_3 peak intensity ratio as the mean diameter of nanoparticle increases (Figure 7.14D).

The next example is vapor–liquid equilibrium (VLE) and vapor–adsorbate equilibrium (VAE). Figure 7.15A is the VLE diagram of n-propanol (PrOH)–water mixture as a function of liquid and vapor mole fractions (x and y, respectively) of n-propanol at room temperature [26]. Since this is a binary mixture, the mole fraction of water becomes $1-x$ for the liquid phase, and $1-y$ for the vapor phase. This system shows the azeotrope point where the vapor pressure of the mixture is highest and the mole fraction of propanol in liquid is the same as that in vapor (i.e., $x_{PrOH} = y_{PrOH} = 0.36$). The VAE behavior can be measured using the *in situ* ATR-IR as described in Section 6.1.3. The ATR-IR spectra of the adsorbate layers formed on a silica surface exposed to the saturation vapor pressure of propanol/water mixtures are shown in Figure 7.15B. By deconvoluting the alkyl peaks of n-propanol and the OH peak of water and dividing their peak areas to the proper normalization factors (see Section 6.1.3), the optically equivalent thicknesses of n-propanol and water can be obtained at given mole fractions of n-propanol vapor. The result is shown in Figure 7.15C. Below the azeotrope concentration ($y_{PrOH} < 0.36$), the total thickness of n-propanol is less than one monolayer (ML). Above the azeotropic point, the total thickness of n-propanol is about 1–1.5 ML. The total thickness of water decreases from ~6 nm at $y_{PrOH} = 0$ (i.e., $y_{water} = 1$) to ~0.5 nm at $y_{PrOH} = 0$. Since the diameter of H_2O molecule is about 0.3 nm, this change corresponds to a drop from ~20 ML of water to ~2 ML of water as the vapor mole fraction changes across the azeotropic point. The total thickness of the binary adsorbate layer decreases as the PrOH concentration increases, although the total vapor pressure of the mixture is kept at the saturation point of each concentration. The relative concentration of PrOH vs. water in the binary adsorbate layer in VAE qualitatively follows the trend in the *xy* diagram of VLE (Figure 7.15D).

Unlike ATR-IR which measures the total amount of analytes within the probe depth (Section 6.1.2), SFG measures the molecules at the interface (Section 7.3.1) – thus, SFG probes the molecular species at the vapor/liquid interface in VLE and the sum of contributions from the vapor/adsorbate interface and the adsorbate/solid interface in VAE. Figure 7.16 displays the *ssp*-SFG spectra of the vapor/liquid interface and the vapor/adsorbate/solid double-interface of PrOH–water mixtures [26]. At the vapor/liquid interface (Figure 7.16A), it is noted that as soon as a small amount of PrOH is added to water, the

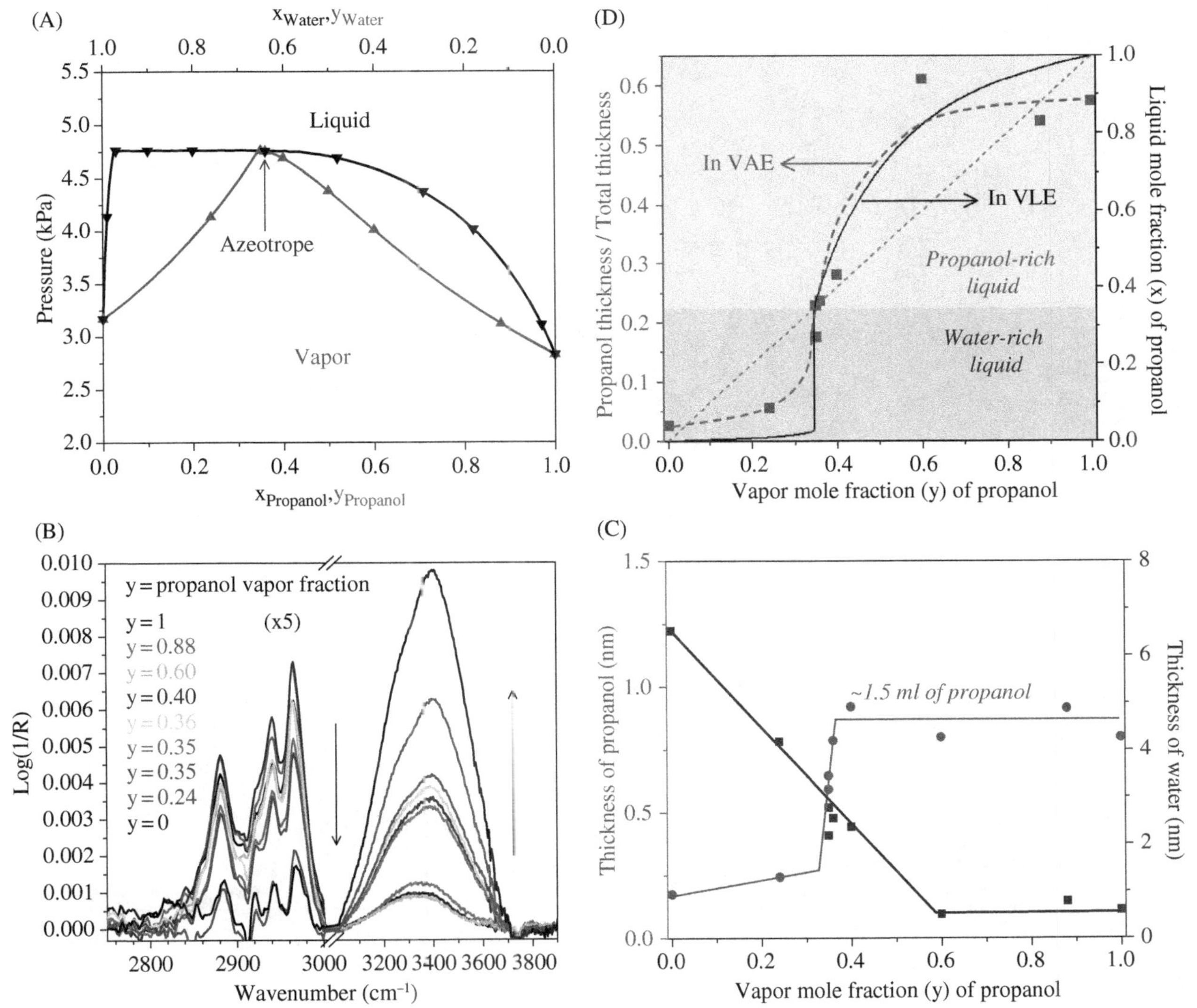

Figure 7.15 (A) Vapor/liquid equilibrium (VLE) curves for n-propanol and water mixtures at 25 °C. (B) ATR-IR spectra of the binary adsorbate layers of n-propanol and water on silica while the vapor mole fraction of y_{PrOH} is decreased from 1 to 0. (C) Effective thickness of n-propanol and water in the binary adsorbate layer on silica. (D) Relative amount of n-propanol in VLE and comparison with *xy* diagram of VLE. Reproduced from Barnette and Kim [26]/with permission of American Chemical Society.

propyl SFG signal increases drastically toward its maximum intensity (Figure 7.17A) and the OH stretch vibration of H_2O drops to a very small intensity. Although propanol also has the OH group, its contribution to the SFG spectral feature is negligible since it is pointing toward the bulk liquid and having hydrogen bonding with other OH groups. The data suggest that although propanol is completely miscible with water, they are preferentially segregated to the surface at this low concentration. This is because the water surface tension is much higher than the propanol surface tension; so, the mixture solution minimizes its free energy by covering its surface with propanol molecules (see Section 1.3). For that reason, the partial pressure of propanol in the vapor phase becomes even higher than the saturation vapor pressure of pure propanol (Figure 7.15A), and the bulk liquid phase becomes water-rich in this low concentration regime (Figure 7.15D).

For the binary co-adsorbate layer in equilibrium with the vapor mixture at saturation condition, the propyl SFG signal increases rather monotonically with the propanol concentration in the vapor phase (Figure 7.16B). But it is not linearly proportional to the vapor concentration. Near the azeotropic point, a noticeable change is observed even with a small change in y_{PrOH} (Figure 7.17A). Below the azeotropic point, the n-propanol thickness is less than one ML; thus, all adsorbed propanol is expected to be at the vapor/adsorbate interface, not inside the adsorbate layer which is about 4–6-nm thick (Figure 7.15C). Above the azeotropic point, the total coverage of n-propanol remains fairly constant at 1–1.5 ML, while

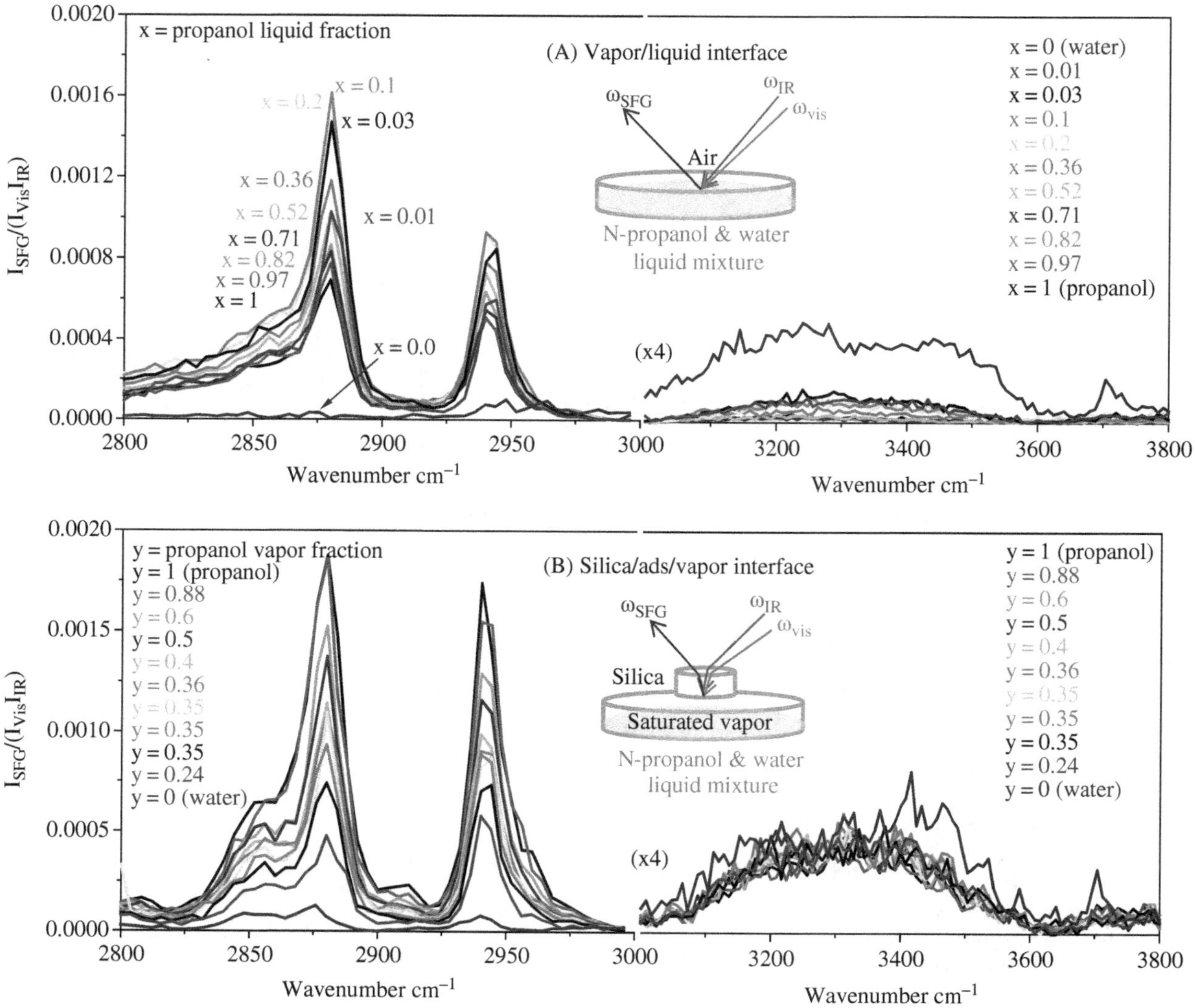

Figure 7.16 *ssp*-SFG spectra of the (A) n-propanol–water liquid/air interface as a function of liquid mole fraction (*x*) of n-propanol and (B) n-propanol–water binary adsorbate layer on silica as a function of vapor mole fraction (*y*) of n-propanol. Reproduced from Barnette and Kim [26]/with permission of American Chemical Society.

the adsorbed water amount decreases to about 2 ML (Figure 7.15C). So, the majority of n-propanol still remains at the vapor/adsorbate interface, and the water molecules stay beneath the n-propanol layer (Figure 7.17B). Based on this picture, the OH signal observed in the *ssp*-SFG spectra at $y_{PrOH} \geq 0.24$ (Figure 7.16B) must be due to the water molecules at the adsorbate/silica interface. At $y_{PrOH} = 0$, we can see a sharp peak at 3700 cm^{-1}; this is the free OH group of H_2O at the vapor/adsorbate interface, which disappears as soon as n-propanol is co-adsorbed (Figure 7.16B).

7.3.8 Probing Chiral Molecules at Interface

Certain molecules have a chirality. It means there is no mirror plane (σ in Tables 7.3–7.5) in the molecular symmetry; thus, the chiral molecule is not superimposable with its mirror image. A good example is amino acids consisting of proteins. All amino acids, except glycine (HOOC–CH_2–NH_2), have the chiral center at the C_α position (between carbonyl carbon and amine). In glycine, the central carbon connecting COOH and NH_2 groups is CH_2, so it has a mirror plane along the C–C–N plane (perpendicular to the C_2 rotational axis of the CH_2 group). All other amino acids have one side group attached to the central carbon, HOOC–C_αH(R)–NH_2. Due to the lack of mirror plane symmetry, chiral molecules can have additional nonzero components in the second-order nonlinear susceptibility that would be zero for achiral molecules.

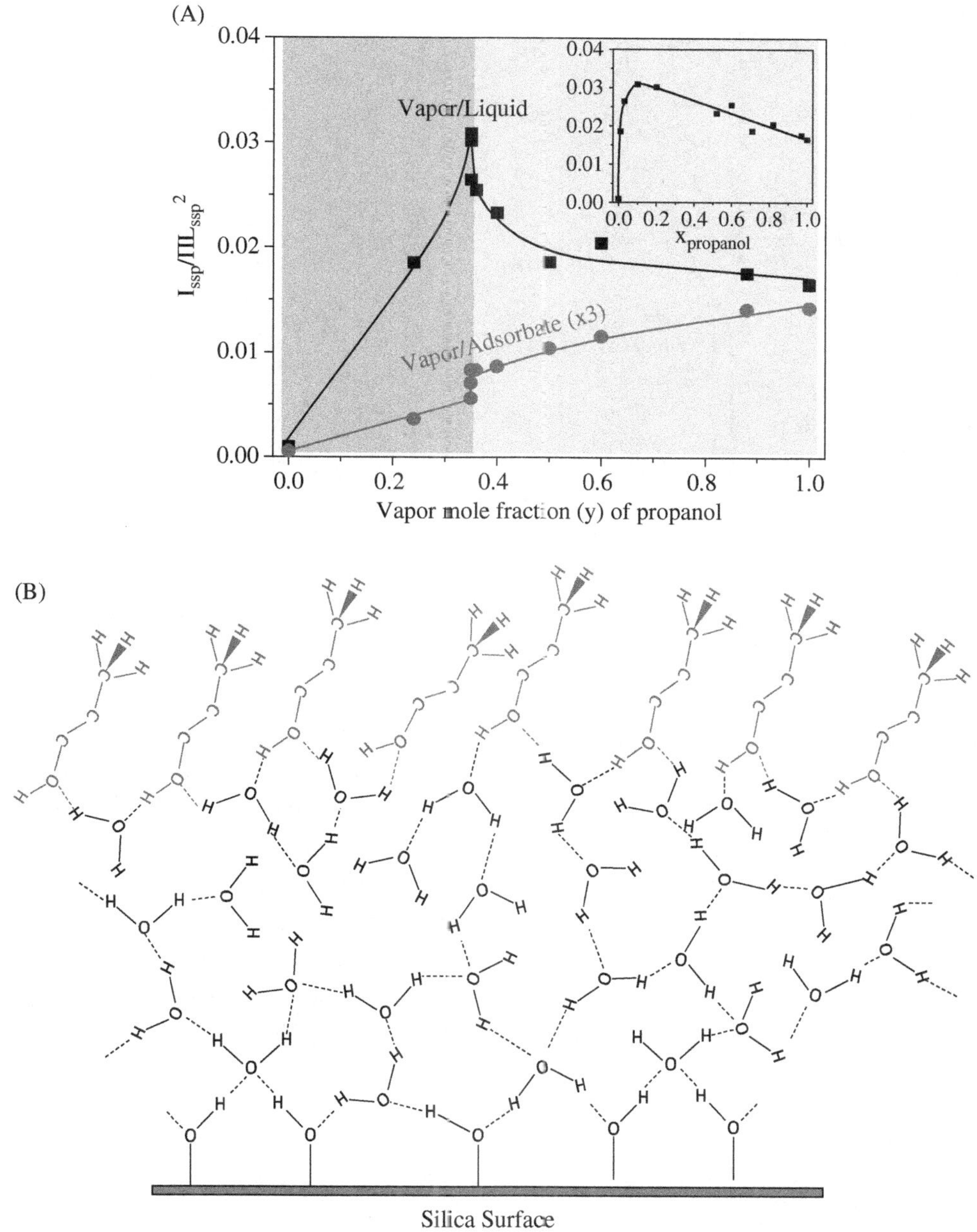

Figure 7.17 (A) SFG peak intensity of the CH_3 symmetric stretch of propanol, normalized by the Fresnel coefficients (eqs. (7.41), (7.44), and (7.45)), as a function of vapor mole fraction (y) of n-propanol for VLE and VAE. The inset shows the plot as a function of liquid mole fraction (x) of n-propanol for the VLE. (B) Schematic illustration of the structure for the binary adsorbate layer of n-propanol and water on silica. Reproduced from Barnette and Kim [26]/with permission of American Chemical Society.

To find the nonzero chiral components in $\beta_{i',j',k'}$, let us consider a simple hypothetical molecule (A) and its mirror image ($\overline{A}$) produced by reflection about the molecular ac plane shown in Figure 7.18. One is called right-handed (R) enantiomer, and the other is left-handed (L) enantiomer. Since the molecule is chiral, after the reflection through a mirror, the sign of the axis perpendicular to the mirror plane should be changed. The transformation matrix (M_b) for the σ_{ac} reflection of a chiral molecule as well as the operation of that matrix to the polarizability and dipole derivatives [27] are shown in Figure 7.18. Here, we can see that the components with only one or three b in the subscript (β_{aab}, β_{aba}, β_{abc}, β_{acb}, β_{baa}, β_{bac}, β_{bbb}, β_{bca}, β_{bcc}, β_{cab}, β_{cba}, β_{cbc}, and β_{ccb}) change the sign upon the σ_{ac} reflection. Thus, they can have nonzero values if we consider the σ_{ac} reflection, and all other components should be zero. Similarly, in the reflection through the σ_{bc} plane, all components other than the ones with only one a or three a's in the subscript should be zero. In the reflection through the σ_{ab} plane, all components

Transformation Matrix (M_b) over σ_{ac} $M_b = \begin{pmatrix} 1 & 0 & 0 \\ 0 & -1 & 0 \\ 0 & 0 & 1 \end{pmatrix}$

$$M_b^{-1}\left(\frac{\partial \alpha_{i'j'}}{\partial Q_q}\right)_A M_b = M_b^{-1}\begin{pmatrix} \alpha_{aa} & \alpha_{ab} & \alpha_{ac} \\ \alpha_{ab} & \alpha_{bb} & \alpha_{bc} \\ \alpha_{ac} & \alpha_{bc} & \alpha_{cc} \end{pmatrix}_A M_b = \begin{pmatrix} \alpha_{aa} & -\alpha_{ab} & \alpha_{ac} \\ -\alpha_{ab} & \alpha_{bb} & -\alpha_{bc} \\ \alpha_{ac} & -\alpha_{bc} & \alpha_{cc} \end{pmatrix}_{\bar{A}} = \left(\frac{\partial \alpha_{i'j'}}{\partial Q_q}\right)_{\bar{A}}$$

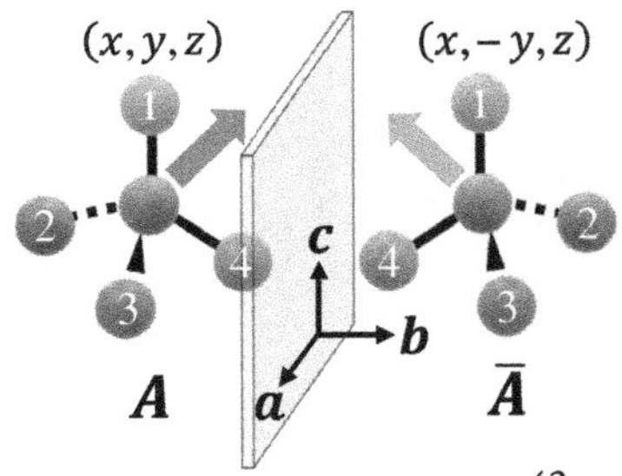

$$M_b\left(\frac{\partial \mu_{k'}}{\partial Q_q}\right)_A = M_b\begin{pmatrix} \mu_a \\ \mu_b \\ \mu_c \end{pmatrix}_A = \begin{pmatrix} \mu_a \\ -\mu_b \\ \mu_c \end{pmatrix}_{\bar{A}} = \left(\frac{\partial \mu_{k'}}{\partial Q_q}\right)_{\bar{A}}$$

$$\left(\frac{\partial \alpha_{i'j'}}{\partial Q_q}\right)_{\bar{A}} \otimes \left(\frac{\partial \mu_{k'}}{\partial Q_q}\right)_{\bar{A}} = \begin{pmatrix} \begin{pmatrix} \alpha'_{aa}\mu'_a \\ -\alpha'_{aa}\mu'_b \\ \alpha'_{aa}\mu'_c \end{pmatrix} & \begin{pmatrix} -\alpha'_{ab}\mu'_a \\ \alpha'_{ab}\mu'_b \\ -\alpha'_{ab}\mu'_c \end{pmatrix} & \begin{pmatrix} \alpha'_{ac}\mu'_a \\ -\alpha'_{ac}\mu'_b \\ \alpha'_{ac}\mu'_c \end{pmatrix} \\ \begin{pmatrix} -\alpha'_{ba}\mu'_a \\ \alpha'_{ba}\mu'_b \\ -\alpha'_{ba}\mu'_c \end{pmatrix} & \begin{pmatrix} \alpha'_{bb}\mu'_a \\ -\alpha'_{bb}\mu'_b \\ \alpha'_{bb}\mu'_c \end{pmatrix} & \begin{pmatrix} -\alpha'_{bc}\mu'_a \\ \alpha'_{bc}\mu'_b \\ -\alpha'_{bc}\mu'_c \end{pmatrix} \\ \begin{pmatrix} \alpha'_{ca}\mu'_a \\ -\alpha'_{ca}\mu'_b \\ \alpha'_{ca}\mu'_c \end{pmatrix} & \begin{pmatrix} -\alpha'_{cb}\mu'_a \\ \alpha'_{cb}\mu'_b \\ -\alpha'_{cb}\mu'_c \end{pmatrix} & \begin{pmatrix} \alpha'_{cc}\mu'_a \\ -\alpha'_{cc}\mu'_b \\ \alpha'_{cc}\mu'_c \end{pmatrix} \end{pmatrix}_{\bar{A}}$$

Figure 7.18 Polarizability and dipole derivative tensors of a chiral molecule and its mirror image. The arrows represent the same hypothetical dipole vector before and after mirror reflection.

other than the ones with only one c or three c's in the subscript can have nonzero values. Then, we find that the only components that change their signs in all three σ_{ac}, σ_{bc}, and σ_{ab} operations are the ones with one a, one b, and one c in the subscript (i.e., $\beta_{abc}, \beta_{acb}, \beta_{bac}, \beta_{bca}, \beta_{cab}$, and β_{cba}), which can be written as $\beta_{i',j',k'}$ with $i' \neq j' \neq k'$. These are the chiral elements of molecular hyperpolarizability [28].

Using the same argument, we can find that $\chi^{(2)}_{ijk}$ with $i \neq j \neq k$ are the chiral elements of second-order nonlinear susceptibility [28]. Following the method described in Section 7.3.3, we can find the relationships between the chiral second-order nonlinear susceptibility elements ($\chi^{(2)}_{ijk}$ with $i \neq j \neq k$) and the chiral hyperpolarizability elements ($\beta_{i',j',k'}$ with $i' \neq j' \neq k'$) for a chiral molecule present at the 2D surface with C_∞ symmetry. Note that since the Raman tensor is symmetric (e.g., see Table 7.8), i' and j' are interchangeable. This makes $\beta_{abc} = \beta_{bac}$, $\beta_{cab} = \beta_{acb}$, and $\beta_{cba} = \beta_{bca}$. Then, we get the following relationships:

$$\chi^{(2)}_{xyz} = \chi^{(2)}_{yxz} = 0 \tag{7.78}$$

$$\chi^{(2)}_{xzy} = \chi^{(2)}_{zxy} = \frac{1}{8}(\beta_{acb} - \beta_{bca})\,(1 + 3\cos 2\theta)) \tag{7.79}$$

$$\chi^{(2)}_{yzx} = \chi^{(2)}_{zyx} = -\frac{1}{8}(\beta_{acb} - \beta_{bca})\,(1 + 3\cos 2\theta)) \tag{7.80}$$

These chiral $\chi^{(2)}_{ijk}$ elements with $i \neq j \neq k$ can be probed in the *spp*, *psp*, and *pps* polarization combinations, as shown in Table 7.9. Note that the *spp* and *psp* polarizations probe essentially the same tensor element; the only difference between these two would be the Fresnel coefficient terms ($L_i^{SF} L_j^{VIS} L_k^{IR}$). Thus, the *spp* and *psp* spectra would have similar spectral shapes, but different intensities. In the reflection experiment, the *pps* signal is expected to be small since $\chi^{(2)}_{xzy} = \chi^{(2)}_{zxy}$; note that their contributions would not be completely cancelled out because their Fresnel coefficient terms are not the same (see Table 7.1). If the molecular tilt angle is 54.74° (which is the magic angle of the random distribution), eqs. (7.79) and (7.80) also becomes zero. Thus, the SFG signal will be zero. In other words, even chiral molecules at the 2D interface would not generate SFG signal if the tilt angle distribution is completely random.

Note that eqs. (7.78)–(7.80) are obtained with the assumption there is no preferential ordering in azimuth and twisting angles. If there is a preferential orientation in the azimuth angle, we should multiply a proper distribution function to eq. (7.56) and then integrate the resulting equation, followed by proper normalization. Similarly, if there is a preferential twist angle due to lateral packing of molecules or SFG-active domains with respect to the 2D surface, then a proper function

Table 7.9 Three polarization combinations probing nonzero $\chi^{(2)}_{ijk}$ terms of a chiral molecule. The order of three consecutive polarization symbols follows the polarization of SF, VIS, and IR beams, respectively, and the laser incidence plane is the *xz* plane. The sign of each $\chi^{(2)}_{ijk}$ with $i \neq j \neq k$ is from the sign of the $\hat{e}_{SF}\hat{e}_{VIS}\hat{e}_{IR}$ product. On metallic surfaces, only *spp* and *psp* polarization can be used due to the metal selection rule. Note that $\chi^{(2)}_{xyz} = \chi^{(2)}_{yxz} = 0$ if the chiral molecule is present at the 2D surface with C_∞ symmetry (eq. 7.78).

		Co-propagation	Counter-propagation
Reflection	*spp*	$\chi^{(2)}_{yxz} + \chi^{(2)}_{yzx}$	$\chi^{(2)}_{yxz} - \chi^{(2)}_{yzx}$
	psp	$-\chi^{(2)}_{xyz} + \chi^{(2)}_{zyx}$	$-\chi^{(2)}_{xyz} - \chi^{(2)}_{zyx}$
	pps	$-\chi^{(2)}_{xzy} + \chi^{(2)}_{zxy}$	$-\chi^{(2)}_{xzy} + \chi^{(2)}_{zxy}$
Transmission	*spp*	$\chi^{(2)}_{yxz} + \chi^{(2)}_{yzx}$	$\chi^{(2)}_{yxz} - \chi^{(2)}_{yzx}$
	psp	$\chi^{(2)}_{xyz} + \chi^{(2)}_{zyx}$	$\chi^{(2)}_{xyz} - \chi^{(2)}_{zyx}$
	pps	$\chi^{(2)}_{xzy} + \chi^{(2)}_{zxy}$	$\chi^{(2)}_{xzy} + \chi^{(2)}_{zxy}$

describing such a structural order should be multiplied to eq. (7.56) and then integrated. The same principle has already been explained in Section 7.3.6 when we discuss the effect of tilt angle distribution on the SFG signal intensity.

The chiral elements of the two enantiomers (*R* and *L*) differ only in their signs, that is, $\chi^{(2)}_{xzy,R} = -\chi^{(2)}_{xzy,L}$. For that reason, a racemic mixture of two enantiomers cannot generate chiral-sensitive SFG signal. Also, the experimentally measured intensity of SFG is proportional to the square of these tensors; so, the intensity measurement alone cannot distinguish two enantiomers. To distinguish them, we need to conduct the phase-sensitive SFG experiment (see Section 7.3.9).

Figure 7.19 demonstrates the distinction of achiral and chiral components in the system using SFG [29]. The system shown here is cellulose nanocrystals (CNCs) attracted to a monolayer of dioctadecyldimethylammonium (DODA) bromide at the vapor/water interface prepared using a Langmuir trough. Cellulose is a crystalline aggregate of long linear

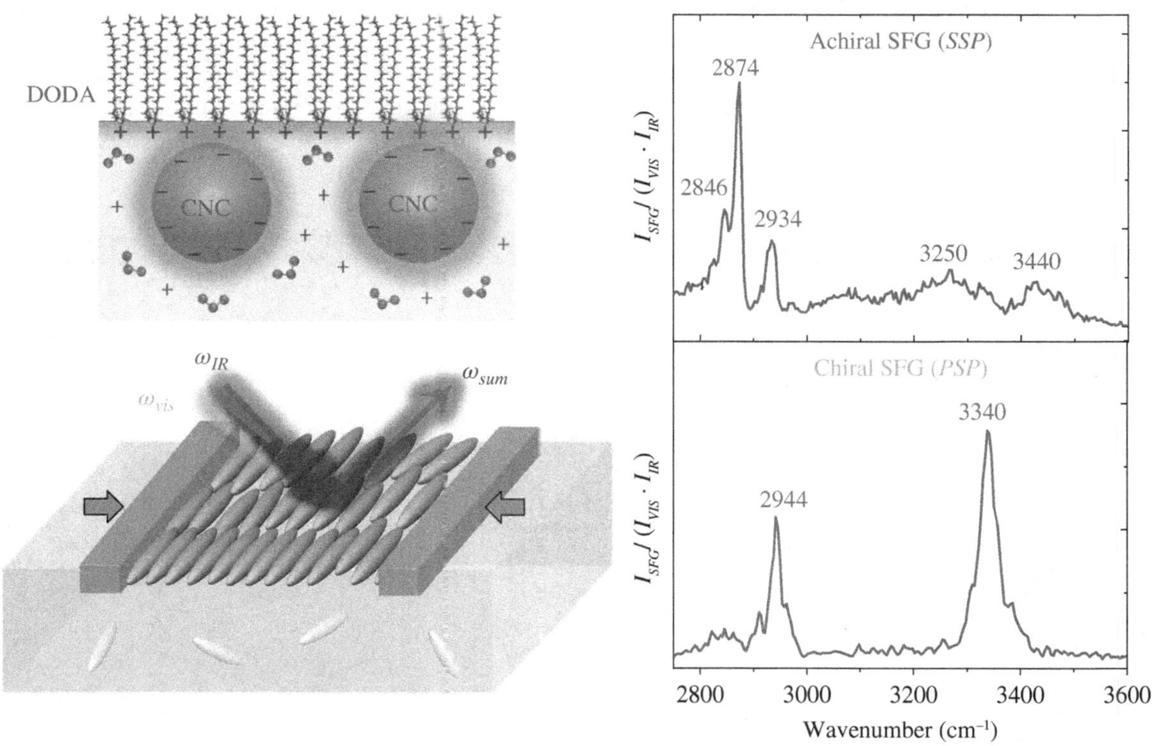

Figure 7.19 Schematic illustration of the *in situ* SFG measurement of CNC and DODA at the air/water interface with a SP = 60 mN/m in a Langmuir trough, and SFG spectra collected using the *ssp* and *psp* polarizations. Reproduced from Chae et al. [29]/with permission of American Chemical Society.

polysaccharide chains which consist of β-1,4-linked glucose units. Since glucose is chiral and the cellulose chains in the crystalline domain have the same directionality, crystalline cellulose is also chiral. CNCs prepared by hydrolysis of cellulosic materials using sulfuric acid contain sulfate groups at the surface; thus, they are negatively charged. DODA is an amphiphilic cationic surfactant, and it covers the air/water interface when it is dissolved in water. When CNCs are added into the aqueous phase covered with the DODA layer, they are attracted to the air/water interface which is positively charged due to the ammonium ion of DODA. Then, by bringing the barriers close together in the Langmuir trough, the DODA/CNC layer can be compressed to a certain surface pressure (SP). The SFG spectra in Figure 7.19 was taken at SP = 60 mN/m. In the *ssp* spectrum, the SFG signals from achiral species are dominant. So, the alkyl stretch region in the *ssp* spectrum shows mostly the vibrational peaks of the self-assembled long-chain alkyl groups of DODA (2846 cm^{-1}, 2874 cm^{-1}, and 2934 cm^{-1}) and the OH stretch region shows the hydrogen-bonded OH modes of water beneath the DODA layer (3250 cm^{-1} and 3440 cm^{-1}). In contrast, the *psp* spectrum shows only SFG signals of cellulose (2944 cm^{-1} and 3340 cm^{-1}); there is no signal from DODA and water [29].

Figure 7.20 compares the *psp*-SFG spectra of model proteins and peptides with different secondary structures at the air/water interface [30]. These data demonstrate that SFG can readily distinguish the secondary structures of proteins, namely, α-helix, 3_{10}-helix, parallel β-sheet, antiparallel β-sheet, and random coils. The same can be done with IR and Raman spectroscopy through peak deconvolution [31]; but, often it is quite challenging due to overlap of bands from different structures. In chiral SFG, the polarization selection rule makes it easier to distinguish different structures. The helical secondary structures of proteins show amide-A band (N–H stretch) only, but no amide-I band (C=O stretch), in the *psp*-SFG spectrum. The parallel β-sheet shows the amide-I band in the *psp* spectrum; but, its amide-A band is below the detection limit because its tilt angle happened to be close to 54.7° in the analyzed sample (see eq. (7.80)). The antiparallel β-sheet shows both amide-I and amide-A peaks in the *psp* spectrum. The random coil shows no SFG signal in the *psp* polarization, even though individual amino acids in the coil are chiral (as predicted by inserting the magic angle to eq. (7.80)).

7.3.9 Spectral Intensity and Shape of SFG Signal

Until this point, we have derived the SFG equations considering three processes — (i) projection of the electric fields of the incident beams to a medium (Section 7.2.3), (ii) polarization of the medium at the sum frequency of two incident beams with one of them in resonance with the vibrational excitation process of a molecule in the medium (Sections 7.2.4–7.2.6 and 7.3.4–7.3.5), and (iii) emission of sum-frequency photon from the polarized medium (Sections 7.2.1–7.2.2) — for 2D interfaces (Sections 7.3.1–7.3.3 and 7.3.6–7.3.7). In doing so, we have considered the relationship between the hyperpolarizability of molecule (Section 7.2.6) and the second-order nonlinear susceptibility in the lab coordinate as a function of tilt, azimuth, and twist angle distributions (Sections 7.3.3 and 7.3.8). For the 2D interface with C_{∞} symmetry, the azimuth and twist angle

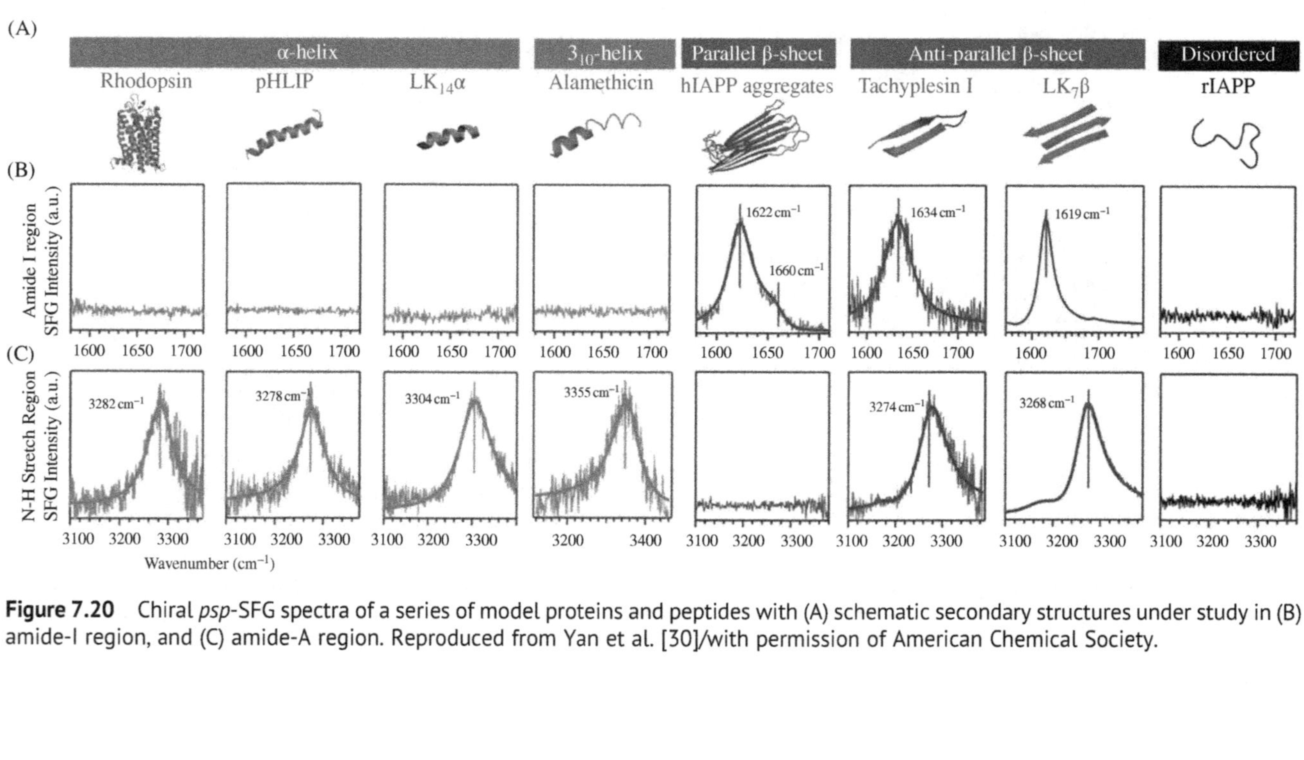

Figure 7.20 Chiral *psp*-SFG spectra of a series of model proteins and peptides with (A) schematic secondary structures under study in (B) amide-I region, and (C) amide-A region. Reproduced from Yan et al. [30]/with permission of American Chemical Society.

dependences drop out and only the tilt angle dependence remains (Sections 7.3.3–7.3.6). Now, let us study how the peak intensity and shape of SFG are related to the concentration of analyte and the surrounding medium in which the analyte is present.

The first factor affecting the SFG peak intensity is the oscillator strength, which is the term that goes to the numerators of eqs. (7.61) and (7.62). This is equivalent to the absorptivity in linear spectroscopy (i.e., the a term in eqs. (5.27) and (5.42)). Since the SFG spectroscopy probes vibrational excitations of analyte (i.e., the resonant IR absorption process; Figure 7.6), the denominator term of eq. (7.62) would have a larger effect to the SFG signal than the denominator terms of eq. (7.61). From this, we can see that, not only the oscillator strength (which is discussed in Sections 7.3.3 and 7.3.8), but the damping parameter (Γ) of the vibrational mode will also have a significant effect governing the SFG signal intensity and shape. Another term to consider is the number density of the oscillator (i.e., the concentration of analyte; the N term in eq. (7.56)).

Let us see how these terms affect the SFG signal intensity as the IR beam wavenumber (ω_{IR}) is scanned across a resonant absorption band. First, let us consider one oscillator (q^{th} normal mode) and set the oscillator strength times the number density to be unity, for simplicity. Then, the nonlinear susceptibility near the resonance peak, $\chi^{(2)}_{R,q}(\omega_{IR})$, can be rewritten as:

$$\chi^{(2)}_{R,q}(\omega_{IR}) \propto \frac{1}{(\omega_q - \omega_{IR}) - i\Gamma_q} = \frac{\omega_q - \omega_{IR}}{(\omega_q - \omega_{IR})^2 + \Gamma_q^2} + i\frac{\Gamma_q}{(\omega_q - \omega_{IR})^2 + \Gamma_q^2} = A_R(\omega_{IR})e^{i\delta(\omega_{IR})} \tag{7.81}$$

Figure 7.21A plots the real and imaginary parts of nonlinear susceptibility for a hypothetical alkyl stretch mode with $\omega_q = 2920\ \text{cm}^{-1}$ and $\Gamma_q = 5\ \text{cm}^{-1}$. As in the case of dielectric permittivity studied in Section 5.1.4, the imaginary part of nonlinear susceptibility at the peak absorption position ($\omega_{IR} = \omega_q$) varies in proportion to Γ_q^{-1}. That is because the

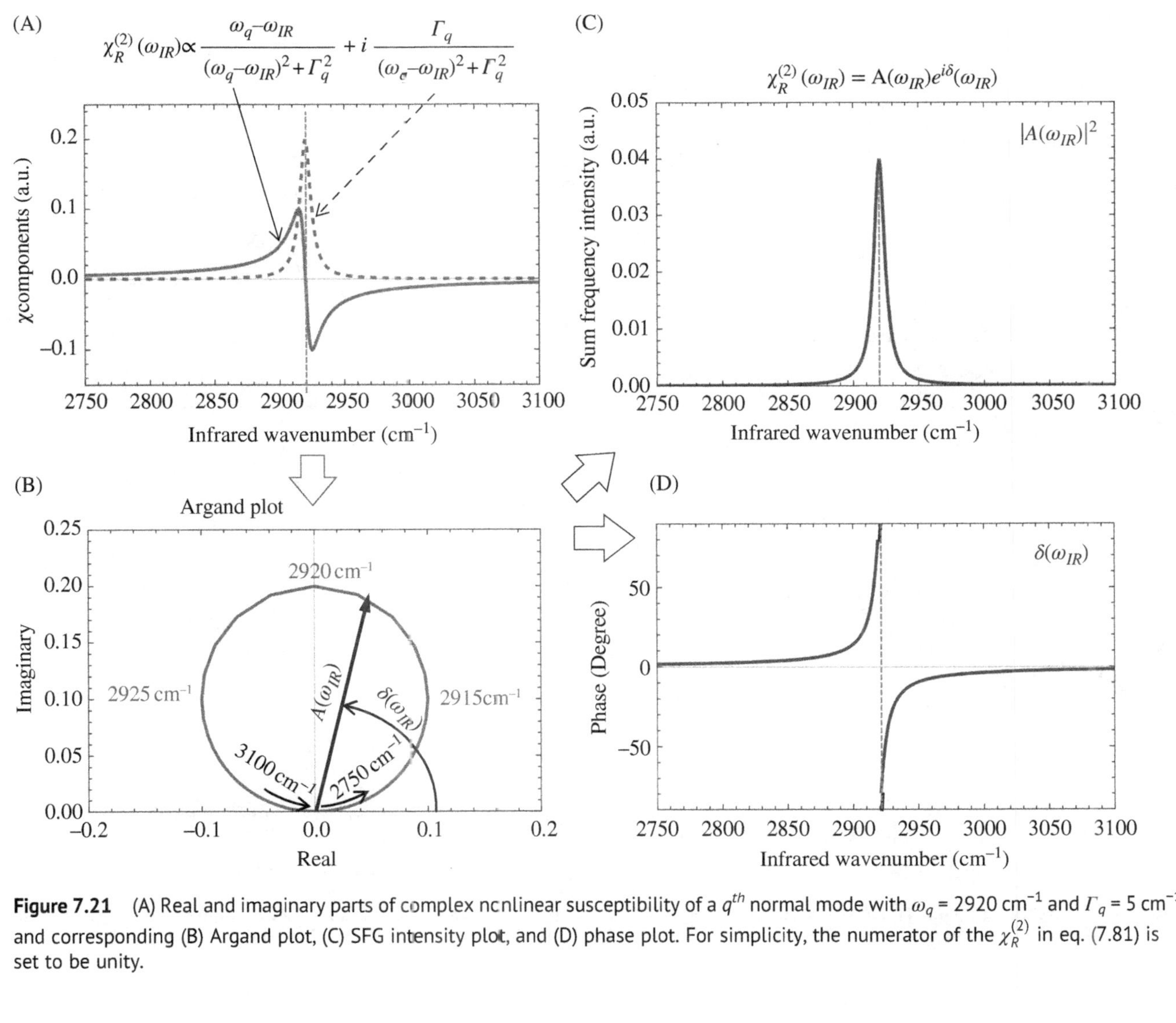

Figure 7.21 (A) Real and imaginary parts of complex nonlinear susceptibility of a qth normal mode with $\omega_q = 2920\ \text{cm}^{-1}$ and $\Gamma_q = 5\ \text{cm}^{-1}$ and corresponding (B) Argand plot, (C) SFG intensity plot, and (D) phase plot. For simplicity, the numerator of the $\chi_R^{(2)}$ in eq. (7.81) is set to be unity.

denominator term in eq. (7.81) has the same formula as that in $\varepsilon_2(\omega)$ in eqs. (5.37) and (5.39). Thus, as Γ_q increases, the SFG signal becomes not only broader but also smaller.

We can plot the imaginary part of this complex susceptibility against its real part at each wavenumber (ω_{IR}); this is called Argand plot (Figure 7.21B). In the Argand plot, an arrow can be drawn from the origin (0, 0) to the data point ($\mathrm{Re}\left[\chi^{(2)}_{R,q}(\omega_{IR})\right], \mathrm{Im}\left[\chi^{(2)}_{R,q}(\omega_{IR})\right]$) at each wavenumber. The length of this arrow and its angle from the real axis (=x-axis) are the magnitude ($A_R(\omega_{IR})$) and phase ($\delta(\omega_{IR})$), respectively, of the complex susceptibility expressed in the Euler form (the last part in eq. (7.81)). As shown in eq. (7.38), the SFG intensity is proportional to $\left|\chi^{(2)}_{eff}\right|^2$. Thus, the SFG signal ($I_{SFG}(\omega_{IR})$) will be proportional to $|A_R(\omega_{IR})|^2$, which is plotted in Figure 7.21C. The SFG spectral feature has the Lorentzian shape. Since $A_R(\omega_{IR} = \omega_q) \propto \Gamma_q^{-1}$, the maximum intensity at the resonance peak position is proportional to Γ_q^{-2}. The phase of the resonant SFG signal changes its sign as ω_{IR} passes across ω_q, as shown in Figure 7.21D.

Figure 7.22A shows the effect of the number density (N; concentration) of the SFG-active species. As N increases, $\mathrm{Im}\left[\chi^{(2)}_{R,q}(\omega_{IR})\right]$ and $A_R(\omega_{IR})$ increase linearly, but $I_{SFG}(\omega_{IR})$ increases in proportion to the square of the number density

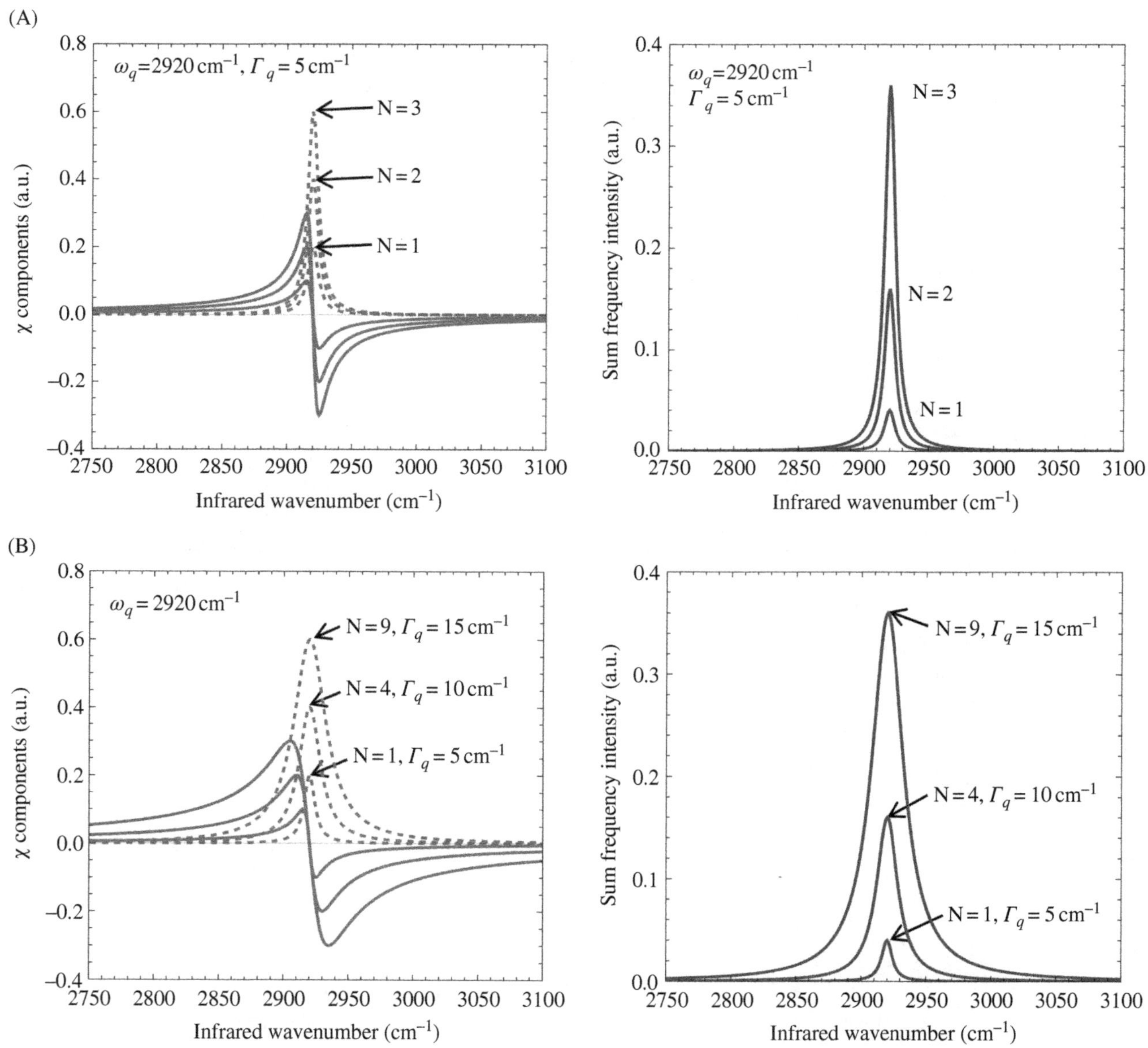

Figure 7.22 Real and imaginary parts of nonlinear susceptibility and SFG intensity plot as a function of wavenumber near a hypothetical absorption band for (A) different number densities of the same oscillator and (B) different number densities and damping factors for a given absorption peak.

(i.e., $\propto N^2$). Figure 7.22B shows the convolution effect due to changes in number density (N) and damping parameter (Γ_q). When both parameters change, the SFG signal intensity varies in proportion to N^2/Γ_q^2.

There is another effect to consider, which is the nonresonant background. Liquids and amorphous insulators are isotropic, and their electronic polarization is negligible at frequencies away from the resonance absorption band. Thus, they are SFG-inactive and do not produce any background signal. However, metals can generate strong nonresonant background signal due to the interband transition (see Section 6.3.4). If the up-conversion beam energy (ω_{IR}) or SFG beam energy (ω_{SF}) is within the interband transition range, the highest energy levels ($|s\rangle$ in Figure 7.6) is in resonance with the electronic state of the substrate, making the contribution from eq. (7.61) significant. This makes the nonresonant substrate signal large. If the substrate medium is noncentrosymmetric [19] or has a field gradient within the coherence length [20], it can also generate a measurable nonresonant SFG signal. So, the interference between the nonresonant and resonant signals alters the SFG peak intensity and shape.

In such cases, the effective nonlinear susceptibility can be written as the sum of the nonresonant background contribution, $\chi_{NR}^{(2)}$, and all resonant absorption processes, $\chi_{R,q}^{(2)}(\omega_{IR})$, which can be written as:

$$\chi_{\text{eff}}^{(2)}(\omega_{IR}) = \chi_{NR}^{(2)} + \sum_q \chi_{R,q}^{(2)}(\omega_{IR}) \tag{7.82}$$

Here, the nonresonant background term, which is not a function of ω_{IR} since it is not in resonance with the vibration excitation of analyte of our interest, can be written in the following form:

$$\chi_{NR}^{(2)} = A_{NR}e^{i\delta_{NR}} \tag{7.83}$$

where A_{NR} and δ_{NR} are the magnitude and phase of the nonresonant background signal from the substrate or surrounding medium, respectively.

Figure 7.23 displays the Argand plots and SFG spectra of a hypothetical system with three vibrational absorption bands at 2850 cm^{-1}, 2920 cm^{-1}, and 2965 cm^{-1} with the same damping factor ($\Gamma_1 = \Gamma_2 = \Gamma_3 = 5$ cm^{-1}). In this example, the ratio of oscillator strength times concentration is set to be $A_1 : A_2 : A_3 = 1{:}2{:}1$, and five different nonresonant background cases are considered: without any background (i.e., $A_{NR} = 0$) and with a fixed magnitude background (say, $A_{NR} = 0.3$) having different

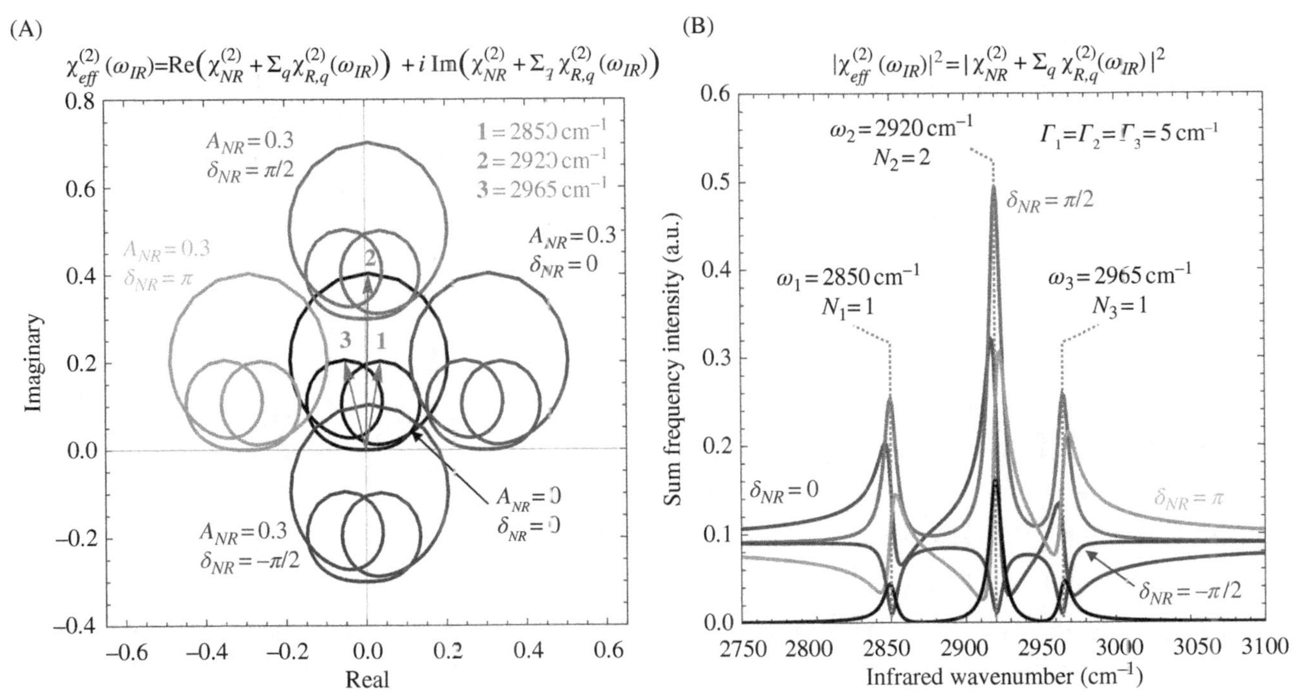

Figure 7.23 (A) Argand plots and (B) SFG spectra of a hypothetical system with three resonant peaks at 2850 cm^{-1}, 2920 cm^{-1}, and 2965 cm^{-1} with the (oscillator strength × abundance) ratio of 1:2:1 and the same damping factor (5 cm^{-1}) in the absence of nonresonant background and in the presence of nonresonant backgrounds of A_{NR} = 0.3 and δ_{NR} = 0, $\pi/2$, π, and $3\pi/2$ (which is equivalent to $-\pi/2$).

phases ($\delta_{NR} = 0$, $\pi/2$, π, and $3\pi/2$). In the absence of any background, the three resonance peaks have typical Lorentzian shapes, and their relative intensity ratio is $A_1^2 : A_2^2 : A_3^2$ in the Argand plot (which is the same as $N_1^2 : N_2^2 : N_3^2$ since the damping factors are same for three peaks in this example). When the nonresonant background contribution exists, the peak shapes and/or relative intensities change drastically. First of all, we see that the baseline of the SFG spectrum is shifted to $|A_{NR}|^2$. When the phase of nonresonant background is zero, the peak shape is distorted and becomes highly asymmetric, and the maximum intensity position is slightly red-shifted from the center of the absorption band. When δ_{NR} is π, the distortion direction is the opposite of the $\delta_{NR} = 0°$ case. When $\delta_{NR} = \pi/2$, the nonresonant signal is in-phase with the resonant signals; thus, the SFG signal intensities are enhanced. The ratio of SFG intensities at three peak positions is now changed to $\left(A_{NR} + \frac{N_1}{\Gamma_q}\right)^2 : \left(A_{NR} + \frac{N_2}{\Gamma_q}\right)^2 : \left(A_{NR} + \frac{N_3}{\Gamma_q}\right)^2$. When $\delta_{NR} = 3\pi/2$ (which is equivalent to $-\pi/2$), the nonresonant signal is

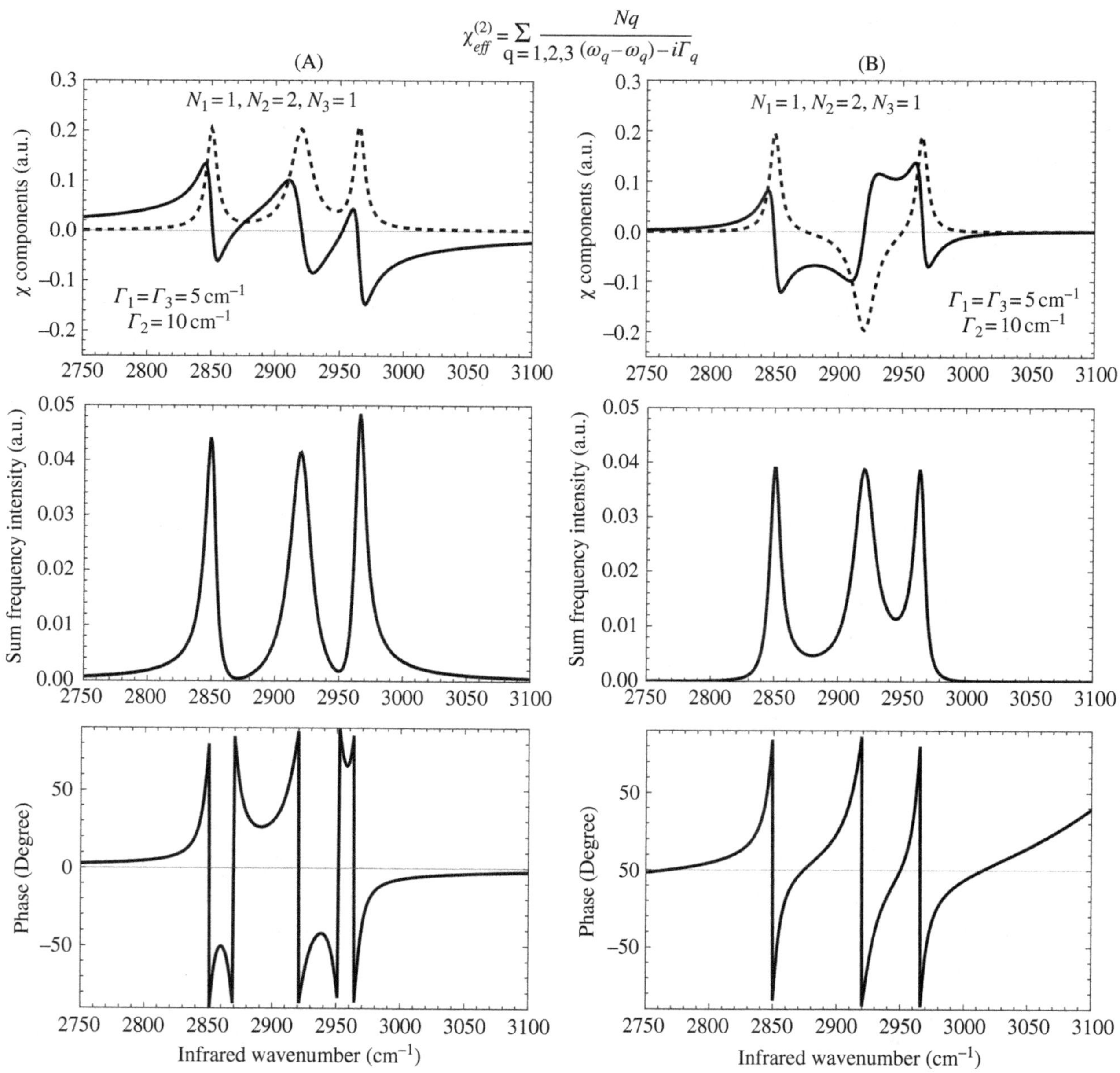

Figure 7.24 Nonlinear susceptibility as well as intensity and phase spectra of SFG signal for a hypothetical system with three resonant peaks at 2850 cm^{-1}, 2920 cm^{-1}, and 2965 cm^{-1} with the same damping factor (5 cm^{-1}) and the same oscillator strength magnitude: (A) all three peaks have the same phase and (B) the center peak has the phase opposite to the other two peaks.

out-of-phase with the resonant signals. In this case, the SFG peaks become negative from the baseline and their intensity ratio becomes $\left(\frac{N_1}{\Gamma_q} - A_{NR}\right)^2 : \left(\frac{N_2}{\Gamma_q} - A_{NR}\right)^2 : \left(\frac{N_3}{\Gamma_q} - A_{NR}\right)^2$.

The magnitude and the phase of the nonresonant signal depend on the material that generates the background signal. If the phase of the background signal is known for the bulk material, the phase of the analyte signal can be determined based on the peak shape of the detected SFG spectrum. Such background signal can be generated *in-situ* by the substrate onto which the analyte molecules adsorbed [19], or by *ex-situ* using a nonlinear crystal with a known phase. There are several ways of conducting phase-sensitive SFG analysis [32, 33], which will not be covered here since it is beyond the scope of this textbook.

The interference effect can also occur among the resonant peaks that are close to each other. Figure 7.24 compares the case with three peaks with all positive phases and the case with the same three peaks but with the center peak having the phase opposite to the other two peaks. In the former case, the imaginary parts of the nonlinear susceptibility of all three peaks are on the positive side and the sign of the real part changes between two adjacent peaks. So, the phase of the SFG signal changes by 180° in the region between adjacent peaks where the imaginary part is close to zero. Then, the SFG intensity becomes nearly zero at that crossover point (Figure 7.24A). In the latter case, the imaginary part of nonlinear susceptibility changes the sign as ω_{IR} moves from the first peak to the second peak; at the crossover point of the imaginary part sign change, the real part of nonlinear susceptibility has a nonzero value, making the SFG signal between the adjacent peaks float from zero (Figure 7.24B). Thus, if the SFG spectrum is collected with high enough signal-to-noise ratio, the relative phases of individual peaks could be determined even if there is no nonresonant background with a known phase. Deeper discussion on this subject can be found in the review paper by H.F. Wang [34].

7.4 SFG Spectroscopy Probing Noncentrosymmetric Domains in 3D Bulk

Not only molecules at 2D interfaces which naturally produce noncentrosymmetric environment along the surface normal direction, but noncentrosymmetric nanoscale domains interspersed in amorphous matrices can also be studied with SFG. This capability has been demonstrated in studies of various crystalline or semicrystalline biopolymers such as cellulose [35, 36], chitin [37, 38], amylose [39, 40], collagen [41, 42], silk [43, 44], and amyloid [45, 46]. These biopolymers have noncentrosymmetric crystal structures, so they are all SFG-active. Although they can be studied with linear spectroscopy such as IR and Raman [31, 47], SFG can provide many advantages that come from the noncentrosymmetry requirement, especially discriminating amorphous matrix components and selectively detecting noncentrosymmetric domains, and more importantly their assembly and organization at nano- to mesoscale dimensions [47]. In this section, we will review how SFG can be used to unveil the nano- and mesoscale structural information that cannot be obtained with other experimental techniques. It uses cellulose and other biopolymers as an example, but the same can be done with other systems.

7.4.1 Random Quasi-phase Matching Condition

When SFG is used to study noncentrosymmetric domains that are dispersed in an amorphous 3D matrix, we need to take into account additional terms or parameters that we could ignore in the 2D interface case based on the C_∞ symmetry argument. In the second-order susceptibility ($\chi^{(2)}_{ijk}$), many terms that are zero in the C_∞ symmetry are not zero anymore in 3D. For example, for achiral molecules in the 2D system with C_∞ symmetry, only *ssp*, *psp*, *spp*, and *ppp* polarizations are used for SFG detection (Table 7.6). But, in the 3D system with noncentrosymmetric domains, SFG signals can be detected in all eight possible polarizations (Table 7.2). Moreover, depending on the point group of crystal structure, the nonzero components in the hyperpolarizability ($\beta_{i'j'k'}$) vary [5, 27]. In the 2D interface analysis, the synchronization factor, that is, the sync-squared function in eq. (7.23), is used to determine the phase-matching angle at which the SFG signal will be emitted. If the bulk phase is isotropic and there is no field gradient from the 2D interface, the coherence length along the depth direction becomes unimportant and the SFG signal is generated mostly at the interface only (see Section 7.2.2). In the 3D case, the coherence length ($L_c = \frac{\pi}{\Delta k}$, where Δk is the wavevector mismatch) effectively defines the probe volume which in turn determines the number of SFG active-domains (assuming they are smaller than L_c) that can have constructive or destructive interferences.

There is an important variation in the phase matching condition when we consider SFG analyses of noncentrosymmetric domains interspersed in a 3D matrix. The IR and up-conversion beams propagate much deeper into the bulk than L_c from

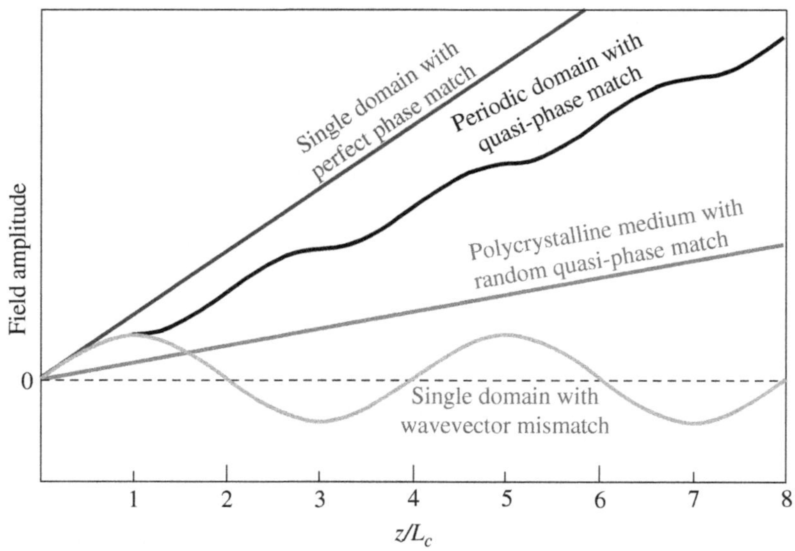

Figure 7.25 Schematic illustration of field amplitude variation as a function of medium thickness in a nonlinear optical response for four different phase matching conditions.

the external surface of the sample. Recall that, in Section 7.2.2, we learn that the SFG process is effective only when the phases of all wavevectors match (Figure 7.3). It means the phases of two input waves and the generated wave need to match at the exit plane of the medium. For a single crystalline medium, the amplitude of the SF beam will change in an oscillatory manner as the medium thickness (z) increases over L_c due to the cyclic nature of the phase as a function of thickness; this case is shown for a single domain with varying size (z/L_c) in Figure 7.25. If one can make a hypothetical material in which the refractive index can be altered or adjusted to always meet the phase matching condition at the exit plane as the medium thickness changes, it might be possible to have the field strength of the generated wave to grow linearly with the medium thickness; see the case of single domain with perfect phase match in Figure 7.25. But such a material does not exist. Instead, engineers can make a periodic material in which the thickness of the crystallographic domains with alternating polarity matches the coherent buildup length (L_c) of the nonlinear interaction. Then, the crystal polarity at the exit plane reverses every time the field amplitude of the generated wave is about to decrease as a consequence of the wavevector mismatch; this allows the field amplitude to grow monotonically as the waves propagate through the material. This is the quasi-phase matching; see the case of periodic domain with quasi-phase match in Figure 7.25 [7].

Now, let us consider a polycrystalline material containing many small crystalline domains interspersed with random orientations and varying distances in a host matrix. In such a medium, the nonlinearly generated waves from different domains will have random phases and interfere neither constructively nor destructively (as in DRIFT, see Section 6.5). Then, the intensities arising from individual domains will be summed up without interference; this leads to an increase of total intensity in proportion to the number of domains within the medium or across the thickness of the sample. This is the principle of random quasi-phase matching [48], which is shown schematically in Figure 7.25. Obviously, the efficiency of the random quasi-phase matching process is less than that of the true quasi-phase matching process; but its effect is not insignificant.

7.4.2 Selectivity to Noncentrosymmetric Crystalline Domine in 3D Matrix

Let us take a look at an example. Figure 7.26A compares the IR, Raman, and SFG spectra of microcrystalline cellulose [35]. Cellulose is a homopolymer of glucose units joined by β-1,4-glycosidic linkages; cellulose chains are synthesized by cellulose synthase complexes (CSCs) in plant cell walls which extrude multiple chains simultaneously in close proximity [49]. Through interchain hydrogen bonding and hydrophobic interactions, the coextruded chains aggregate into a fibril with a long-range order. Although the diameters of these fibrils are in the nanometer scale, they are historically called microfibrils in the plant biology community. The glucose units are chiral, and the crystalline unit cell of naturally produced cellulose has no inversion symmetry. Thus, crystalline cellulose is SFG-active [35].

In Figure 7.26A, the SFG spectral features of commercially available microcrystalline cellulose (called Avicel) are quite different from the IR and Raman spectra. There could be many reasons, but the most prominent one would be the noncentrosymmetry requirement at the molecular level. Equation (7.58) depicts that only the vibrational modes that are both IR-active and Raman-active can be SFG-active. For this reason, the strong peaks in IR may not be observed in the SFG

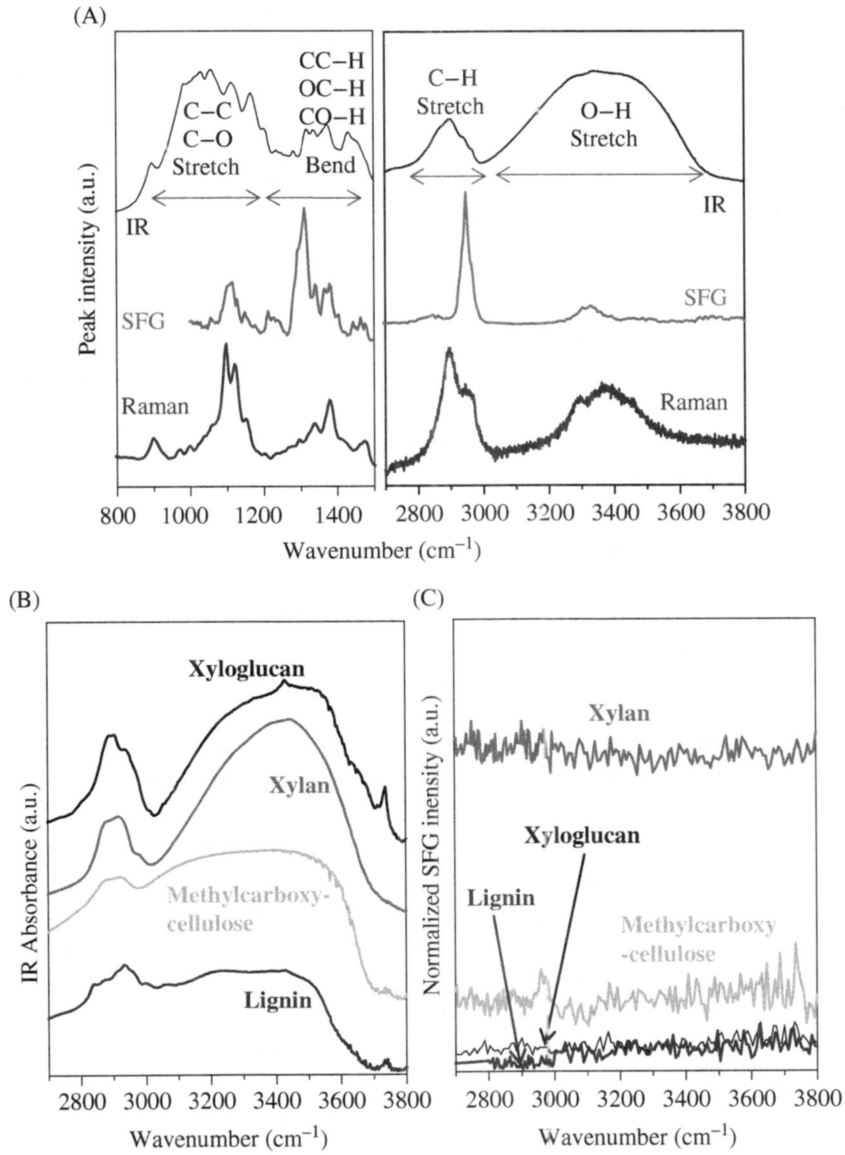

Figure 7.26 (A) Comparison of IR, Raman, and *ssp*-SFG spectra of microcrystalline cellulose (Avicel). Comparison of (B) IR spectra and (C) SFG spectra of hemicelluloses, lignin, and amorphous cellulose-derivative. Reproduced from Barnette et al. [35]/with permission of American Chemical Society.

spectrum if they are not Raman-active, and vice versa. In both IR and Raman spectra, the C–H stretch modes are centered at ~2900 cm^{-1}; but they are weak in the SFG spectrum. The absence of the ~2900 cm^{-1} peak in the cellulose SFG spectrum is not fully understood, but it is speculated to be symmetry cancellation of the axial C–H groups in the opposite side of the glucose ring. In the SFG spectrum of cellulose, the peak at ~2944 cm^{-1} is strong, although it is very weak or minor in both IR and Raman spectra. Similarly, the peak at ~1310 cm^{-1} is strong in SFG, while it is very weak or minor in both IR and Raman spectra. The OH stretch region is extremely broad in IR and Raman spectra due to amorphous phase and absorbed water signals; but the OH peak in SFG is rather sharp and well resolved, although its signal intensity is lower for this specific sample (note that the CH and OH peak intensities and shapes vary drastically depending on the nanoscale structure and mesoscale assembly). From these comparisons, we can see that not all vibrational modes of the noncentrosymmetric crystal consisting with chiral monomeric units are SFG-active; only certain modes that strictly meet the specific requirements for SFG can be detected [47].

Figure 7.26B,C exhibits the IR and SFG spectra, respectively, of two hemicellulose (xylan and xyloglucan) and lignin, which are the other major components in plant cell walls, and a cellulose derivative (methylcarboxy cellulose) [35]. All these samples are amorphous; thus, their SFG signals are barely seen or at the noise level, while they show very strong peaks in IR spectra. This demonstrates how drastically and effectively SFG can discriminate the amorphous phase, which allows selective detection of SFG-active domains dispersed in amorphous matrix without interference from the matrix. Since the cellulose IR signal (Figure 7.26A) and the hemicellulose IR features (Figure 7.26B) are similar to each other, it would be practically very challenging (if not impossible) to deconvolute the contributions from each species if they are all mixed together (as in plant cell walls).

From Figure 7.26, we can see that, since SFG selectively detects crystalline cellulose component without spectral interferences from amorphous components in the sample, the raw intensity can be converted to the cellulose content without peak deconvolution. This is an advantage of SFG over IR and Raman analyses. However, the conversion of the SFG intensity to the concentration may not be straight forward. First, the SFG intensity from a volume corresponding to a single coherence length will be proportional to the square of the number density of the SFG-active components in it (see eq. (7.56)). The probe beam penetration depth (attenuation depth) is larger than the coherence length; thus, the random-quasi phase matching condition predicts that the contributions from multiple coherence domains would be added linearly (Figure 7.26). Although amorphous components in the sample do not generate SFG signals (as demonstrated with xylan and lignin spectra in Figure 7.26C), they can absorb and attenuate the IR beam needed to generate SFG signal from the SFG-active domains. If the amorphous domain absorbs the SFG signal (which is in the UV region if the up-conversion beam is in the visible region; for example, 460 nm = 532 nm + 2945 cm^{-1}), then the SFG signal generated inside the sample may be attenuated before it is emitted to the detector. In addition, scattering of incident and generated beams is also possible if the sample is not homogeneous [9]. For these reasons, the quantitative measure of the SFG-active domain concentration is rather done through preparing a set of calibration samples with known concentrations using appropriate reference samples and comparing the measured SFG intensity with the calibration curve [50–52].

Figure 7.27A,B displays the SFG spectra of cellulose detected at different locations (bottom, bottom-middle, middle, middle-top, and top) along the inflorescence stems of six-week and eight-week old *Arabidopsis* [53]. The six-week old stem is still developing, so the cellulose content varies along the stem from the bottom (fully developed) region to the top (newer) region. The cellulose SFG signal intensity also varies accordingly, but nonlinearly. In the case of eight-week old stem, the cellulose content is relatively constant along the stem and so is the overall intensity of the SFG signal. As mentioned above, it is difficult to directly convert the SFG signal of 3D samples to an absolute concentration of the SFG-active domain inside the sample. When the SFG intensity data of Figure 7.27A,B were compared with the calibration dataset prepared by mixing commercially available microcrystalline cellulose, xylan, and lignin (Figure 7.26), the SFG intensity can be converted to a concentration in the unit of "Avicel equivalent %". Figure 7.27C shows that the SFG-determined concentration correlates well with the cellulose content determined through chemical analysis.

Figure 7.27D plots the OH/CH SFG intensity ratio of cellulose detected at different locations along the stems. It is noted that this ratio is not constant even though the SFG signal is generated by cellulose only in the sample. This shows that a simple Beer–Lambert law type concept does not work in SFG, which will be discussed further in the following sections.

7.4.3 Sensitivity to Nanoscale Order in Crystalline Domain

Since SFG is sensitive to the noncentrosymmetry in the crystalline domains, a small change in the crystal structure can be manifested as a large change in SFG spectrum. Figure 7.28 compares the IR, Raman, and SFG spectra of celluloses with different polymorphic structures [54]. Cellulose Iβ is naturally produced by plants. Cellulose II is produced by treating cellulose Iβ with a high-concentration NaOH solution, which is called mercerization. Cellulose III_I and III_{II} are produced by liquid ammonia treatment of cellulose Iβ and II, respectively. In the CH stretch region, the IR and Raman spectra do not show any discernable features among four types of crystalline celluloses which have different conformations of the exocyclic $^6CH_2$ groups. In the SFG spectra, microcrystalline cellulose Iβ shows a main peak at 2944 cm^{-1}, while all other polymorphs show the main peak at 2960 cm^{-1} with a shoulder at 2980 cm^{-1}. From the x-ray diffraction (XRD) and nuclear magnetic resonance (NMR) studies, it is known that the exocyclic $^6CH_2$ group conformation of cellulose Iβ is different from the other three, namely *trans-gauche* (*tg*) conformation for Iβ and *gauche-trans* (*gt*) conformation for the other three phases. Thus, the SFG spectral difference in the CH stretch region could be attributed to this conformation difference.

In the OH stretch region of IR spectra, cellulose II and III_I have small additional peaks at 3440 cm^{-1} and/or 3480 cm^{-1}, respectively. The SFG spectrum of cellulose Iβ does not have these peaks at all. In contrast, cellulose II and III_I do not have the 3320-cm^{-1} peak that cellulose Iβ has; instead, the SFG spectrum of cellulose II shows a weak peak at 3450 cm^{-1} and that of cellulose III_I shows a very strong peak at 3480 cm^{-1}. In cellulose Iβ, the hydrogen bonds of all OH groups are within the *ab* plane of the unit cell; but in cellulose II and III_I, they are tilted away from the *ab* plane and thus the hydrogen bonding

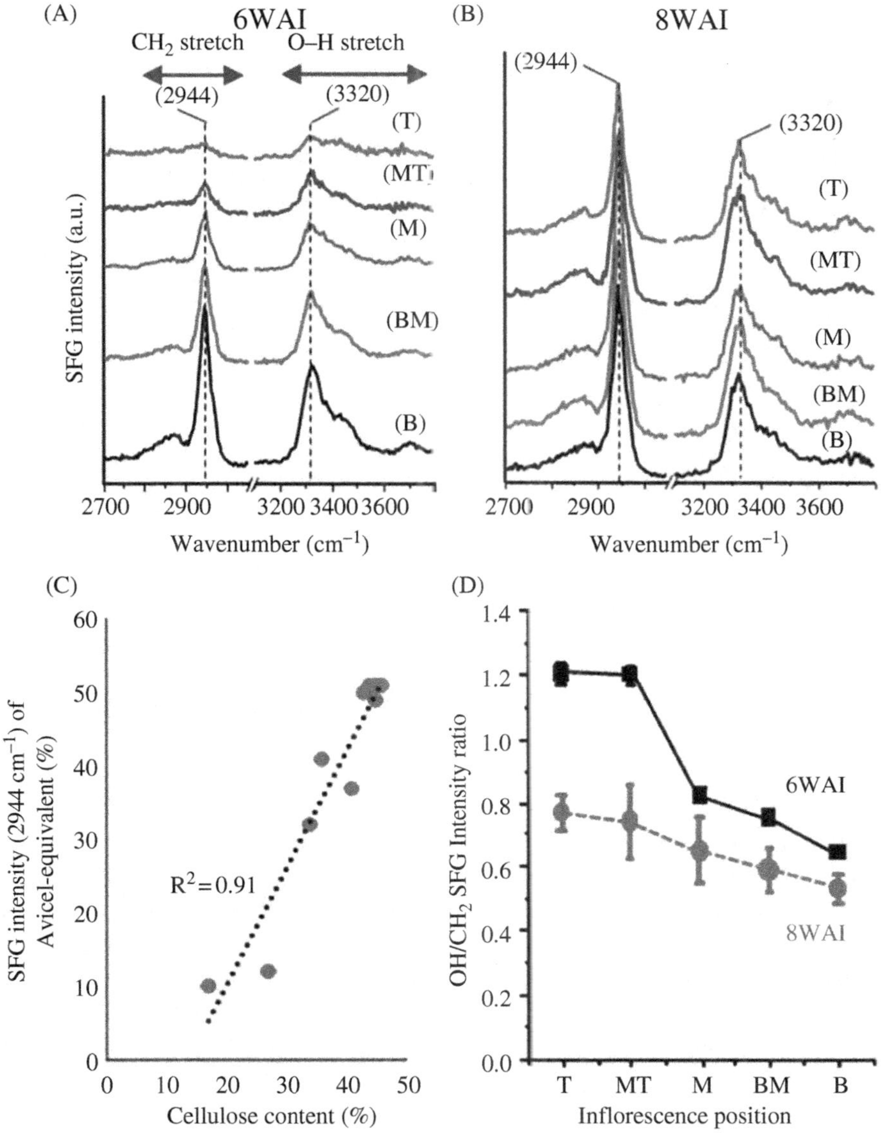

Figure 7.27 *ssp*-SFG spectra of cellulose in five different regions, along the height, in inflorescence stems of (A) six-week old and (B) eight-week old *Arabidopsis*. (C) Correlations of SFG intensity of the 2944 cm^{-1} peak (normalized with a calibration curve) with the crystalline cellulose content determined through chemical analysis. (D) Intensity ratio of OH/CH peaks (3320 cm^{-1}/2944 cm^{-1}) of cellulose in different locations along the stem. Reproduced from Park et al. [53]/with permission of Oxford University Press.

distances are longer. So, the OH stretch bands involving the interchain hydrogen bonding interactions are blue-shifted [55]. In cellulose II, the adjacent cellulose chains in the unit cell have the opposite polarity, while cellulose III_I has the same polarity for all chains as in the cellulose Iβ structure. For that reason, the OH SFG intensity of cellulose is very weak for cellulose II, while it is very strong for cellulose III_I. Cellulose III_{II} is produced from cellulose II; thus, it has a lower crystallinity and assumes the antiparallel chain directionality in the unit cell. For that reason, the OH signal in the SFG spectrum of cellulose III_{II} is negligible. From these comparisons, we can see that SFG can be more sensitive than IR or Raman in terms of distinguishing polymorphic structures of cellulose [56].

The noncentrosymmetry requirement of SFG can be very useful to resolve an ambiguity in XRD analysis. For example, amylose (one of the two components of starch) is another homopolymer of glucose but connected through α-1,4-glycosidic bond, forming helical chains. Amylose can be crystallized naturally into A and B forms; both have the parallel packing of double helices in the unit cell (see schematics in Figure 7.29A). In contrast, the V-types induced with or without association with guest molecules are known to have antiparallel single-helix structures. Although these structures were determined

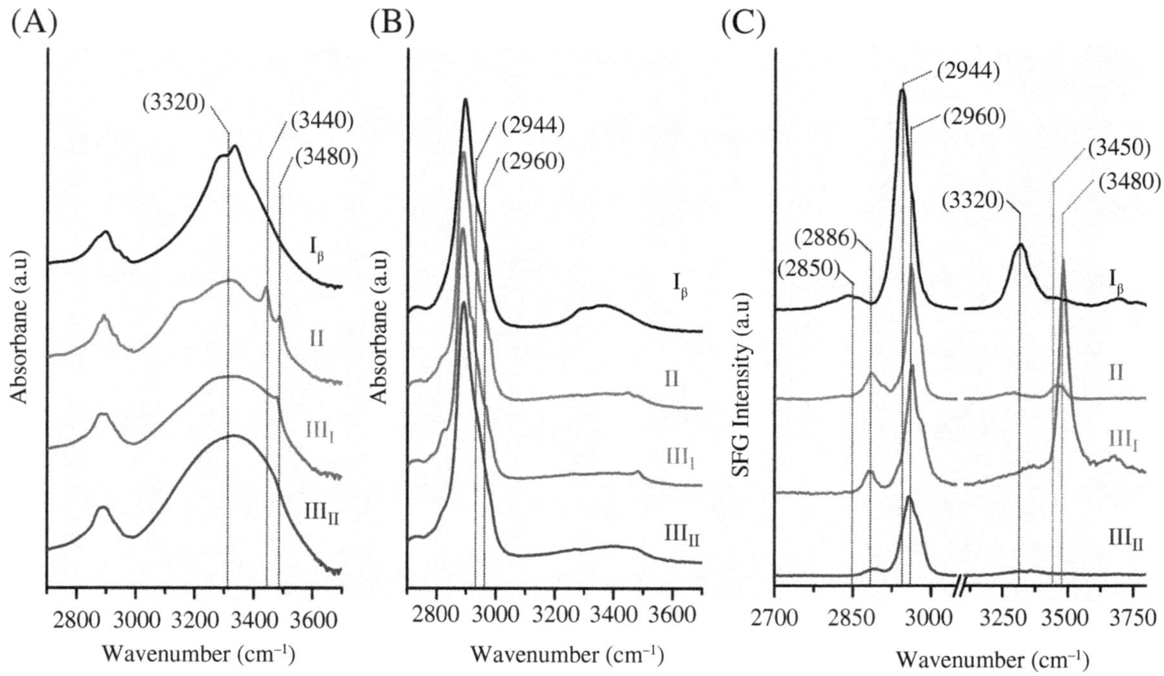

Figure 7.28 (A) IR, (B) Raman, and (C) *ssp*-SFG spectra of cellulose Iβ, II, and III_I, and III_{II}. Reproduced from Lee et al. [54]/with permission of Springer Nature.

based on XRD, there remains an uncertainty regarding the parallel double-helix structure for the B-type. It is known that the V-type with the antiparallel single-helix structure can be easily converted to the B-type at room temperature (far below the melting points) at low humidity; but such a drastic transformation is difficult to imagine without fully disassembling and reassembling the chains. If one could grow a single-crystalline B-polymorph that is large enough for single-crystal XRD analysis, this ambiguity could be resolved easily. But such a crystal could not be grown. Here, SFG can be a useful technique to determine whether amylose helices are parallel or antiparallel in the B-polymorph structure. If they are parallel in the unit cell, the B-type would give a strong SFG signal (similar to the A-type); if the unit cell has an antiparallel structure, it would give a weak signal (similar to the V-type).

Figure 7.29A compares the SFG spectra of amylose samples with different polymorphic structures but similar crystallinities [40]. The B-type sample shows very strong SFG spectral features that are very similar to the A-type, while V-type samples show very weak SFG signals. So, this confirms that B-type amylose also has parallel helices in the unit cell. Although small, there are noticeable differences in the SFG spectral features, which could be attributed to the side chain conformation difference between the A- and B-types [40].

Another example where the noncentrosymmetry requirement of SFG is proven to be useful to differentiate the crystal structure is chitin. In nature, two forms of chitin are produced – α and β. The crystal structure of β-chitin is well established from X-ray and neutron diffraction studies. Both anhydrous and dihydrate structures of β-chitin have a monoclinic one-chain unit cell with $P2_1$ symmetry with all chains arranged in parallel. This is similar to the parallel chain polarity found in the native cellulose microfibrils. On the other hand, the structure of α-chitin is less established involving inherent disorders in the structural model. Based on systematic absence of certain diffraction peaks, an orthorhombic unit cell with $P2_12_12_1$ symmetry is proposed, which requires an antiparallel two-chain arrangement in the unit cell. Naturally occurring polysaccharide crystals have predominantly parallel structures, which are well aligned with their synthesis mechanisms. Thus, the natural occurrence of antiparallel chain arrangement in α-chitin is somewhat counter intuitive, because it would require a very different synthetic machinery or mechanism that have not been discovered yet. Thus, it was critically needed to independently check and confirm the chain polarity in α-chitin unit with a complimentary characterization method.

Figure 7.29B compares the SFG spectra of α-chitin (isolated from *Phaeocystis globosa*) and β-chitin (isolated from *Thalassiosira weissflogii*) with similar crystallinities [37]. While β-chitin shows very strong SFG signals, α-chitin exhibits very weak SFG signals. This supports that the $P2_12_12_1$ symmetry with the antiparallel chains in the unit cell is appropriate

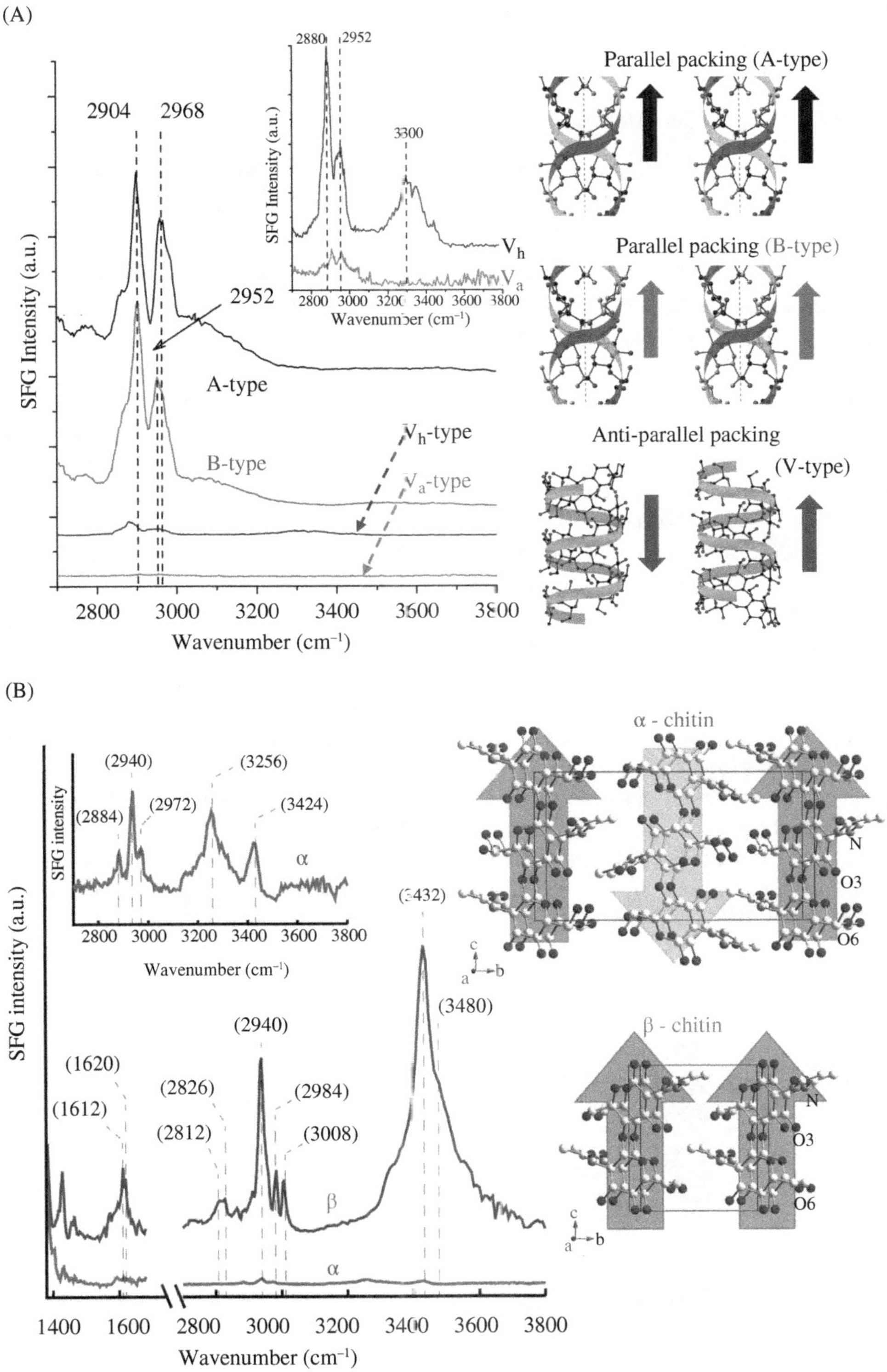

Figure 7.29 (A) *ssp*-SFG vibrational spectra of A-type starch (acid hydrolyzed pea starch), B-type starch (acid hydrolyzed Hylon VII starch), V_h-type amylose–palmitic acid inclusion complex, and V_a-type amylose without guest. Inset shows the magnified spectra of V_h and V_a. Reproduced from Kong et al. [40]/with permission of American Chemical Society. (B) *ssp*-SFG spectra from extracted α-chitin (*P. globose*) and β-chitin (*T. weissf logii*). Inset shows the magnified spectrum of α-chitin. Reproduced from Ogawa et al. [37]/with permission of American Chemical Society.

for the α-chitin crystal structure. Then, the question is shifted from "whether the $P2_12_12_1$ symmetry is reasonable or not" to "how the α-chitin crystal with antiparallel chains in the unit cell is produced in nature."

7.4.4 Sensitivity to Mesoscale Order of Crystalline Domain

SFG signals originating from noncentrosymmetric domains in the bulk are sensitive to not only the crystal structure of each domain but also the arrangement of those domains in space. In analogy to the 2D SFG analysis, the crystal structure dependence would be equivalent to the molecular hyperpolarizability ($\beta_{i',j',k'}$), and the spatial arrangement effect would be to the symmetry cancellation in second-order susceptibility ($\chi^{(2)}_{ijk}$). For example, Figure 7.30 displays the SFG spectra of cellulose Iβ nanocrystals (CNCs) isolated from *Halocynthia roretzi* [36]. Figure 7.30A is the *ssp*-SFG spectrum of cellulose Iβ nanocrystals uniaxially aligned in a polyvinyl alcohol (PVA) matrix. The sample was prepared by stretching the PVA matrix containing cellulose Iβ nanocrystals uniaxially. Thus, nanocrystals can have equal probability of pointing the opposite directions along the alignment axis. For that reason, the OH-stretch intensity is much smaller than the SFG intensity in the CH stretch region. Note that since the SFG spectrum was taken with the *ssp* polarization, the surface of the PVA matrix would also have produced SFG signals of PVA; but the surface signal is so weak compared to the bulk signal, it is buried within the noise of the SFG spectrum of the bulk phase. In the next step, a few drops of water were placed at the spot of the sample where the first SFG spectrum was taken. Since PVA dissolves in water, it was sufficient to disrupt uniaxial alignments and randomize the orientation of nanocrystals. When the SFG spectrum was taken after the sample was dried (Figure 7.30B), the SFG spectral features were drastically changed. The OH stretch bands are stronger than the CH stretch bands, and the CH stretch peak shape is also drastically changed. These changes were solely due to the disruption of the uniaxial alignment of nanocrystals with a net antiparallel polarity in average over the probe volume. In the SFG spectrum of cellulose, the OH signals at ~3320 cm^{-1} as well as the CH peak at 2920 cm^{-1} and 2968 cm^{-1} are suppressed when the cellulose Iβ nanocrystals are aligned uniaxially with the net antiparallel polarity. This is most likely due to the symmetry cancellation. The cancellation effect may vary depending on the degree of antiparallel packing and the types of vibrational modes (especially, their dipole orientation with respect to the crystallographic axis).

Figure 7.31 compares the SFG spectra of cellulose in natural states [36]. Figure 7.31A compiles the SFG spectra of fully developed fibers (flax, ramie, and cotton), grass stalks (*Brachypodium*, maize, and switchgrass), herbaceous plant stem (*Arabidopsis*), and stems of woody plants (poplar and pine). All these samples have cellulose Iβ microfibrils uniaxially aligned along the fiber or stem axis. The spectral features are remarkably similar to the one in Figure 7.30A. Figure 7.31B displays

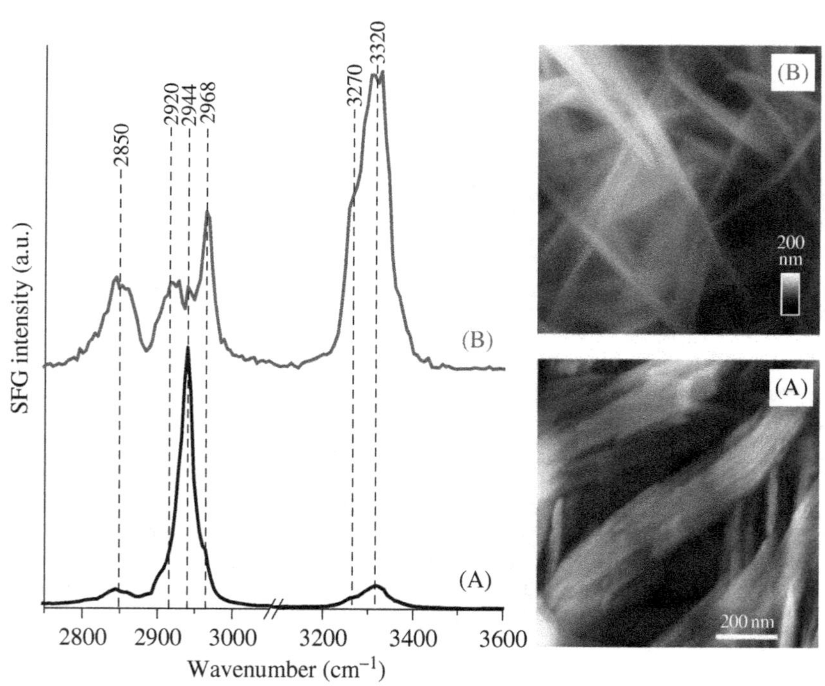

Figure 7.30 Comparison of *ssp*-SFG spectra and AFM images of (A) uniaxially aligned and (B) randomly packed cellulose Iβ nanocrystals. Reproduced from Lee et al. [36]/with permission of Royal Society of Chemistry.

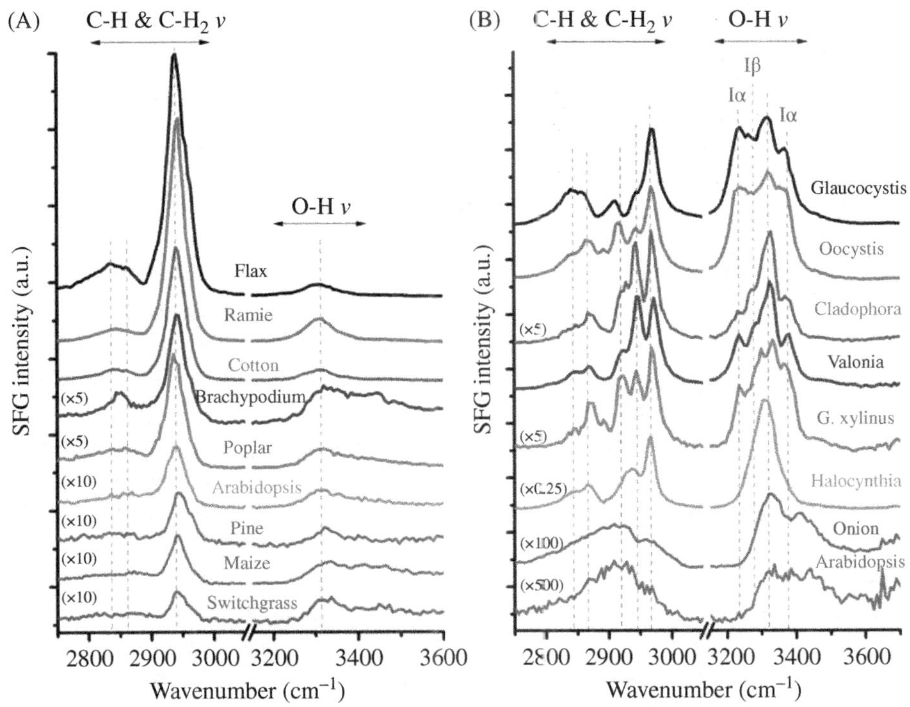

Figure 7.31 SFG spectra of cellulose in biological tissues. (A) Mature secondary cell walls of land plants: flax, ramie, cotton, *Brachypodium*, poplar, *Arabidopsis*, pine, maize, and switchgrass. (B) Algal cell walls (*Glaucocystis*, *Oocystis*, *Valonia*, and *Cladophora*), cellulose biofilm produced from *G. xylinus*, *Halocynthia* mantle, onion epidermis, and Arabidopsis petiole. Reproduced from Lee et al. [36]/with permission of Royal Society of Chemistry.

the SFG spectra of cellulose Iα in the algal cell walls (*Glaucocystis*, *Oocystis*, *Cladophora*, and *Valonia*) and bacterial pellicle (*G. xylinus*) and cellulose Iβ in *Halocynthia* tunic, onion epidermis, and *Arabidopsis* petiole. Note that the SFG spectral feature of *Halocynthia* is very close to the one in Figure 7.30B. In fact, in tunics of tunicates, thick ribbon-like bundles of cellulose Iβ fibrils are randomly weaved in multiple layers, forming hard protective skins. In algal cell walls, ribbon-shaped bundles of cellulose Iα fibrils are arranged in the crossed polylamellate structures. Thus, the CH stretch region of their SFG spectra resembles the one in Figure 7.30B. Since the cellulose chain packing in the unit cells of cellulose Iα and Iβ is slightly different, the OH stretch band (which is composed of multiple peaks) is quite different between these two allomorphs. In bacterial pellicles, the cellulose Iα fibrils are also randomly weaved in multiple layers. Similarly, the onion epidermis and *Arabidopsis* petioles are primary cell walls in which cellulose micrpfibrils are assembled in the crossed polylamellate structures. Thus, the OH signal intensities are relatively strong and even comparable to the CH intensities in SFG spectra.

In Figure 7.27D, we see that the OH/CH intensity ratio is somewhat larger in the top region of the stem of six-week old *Arabidopsis*. This is because the cell walls in the top region of the growing plant are not fully developed; so, the SFG spectral features of cellulose in the top region are different from those of cellulose in the fully developed middle and bottom regions.

We have looked at a few examples of SFG analysis of cellulose in biological samples in their native state as well as CNCs in engineered samples. Certainly, more examples can be found in the literature; but it would be redundant to cover all those examples. Not only SFG but also SHG will be affected by spatial arrangement of noncentrosymmetric domains in polycrystalline samples. In order to more quantitatively understand the relationship between the SFG spectral features and the special arrangement of SFG-active domains in the bulk sample, one needs to theoretically predict the SFG intensity change as a function of nano- and mesoscale structural parameters in 3D.

7.4.5 Calculating SFG Signal Intensity for 3D Material

Considering the existence of the random quasi-phase matching condition (Section 7.4.1), the equation describing the SFG intensity for the 3D material in which SFG-active domains are interspersed can be written as:

$$\frac{I_{SF}}{I_{Vis}I_{IR}} \propto \left| \sum\nolimits_{i,j,k} L_{Fres}\hat{e}_{proj} \sum_{f=0}^{N-1} \chi_{ijk}^{(2)} \left(\frac{e^{-i\Delta k \cdot d} - 1}{\Delta k} \right) \left(e^{-i\Delta k(d+\Delta l)f} \right) \right|^2 \quad (7.84)$$

where $L_{Fres}\hat{e}_{proj}$ is the product of tensorial Fresnel factors and electric field vector projections, which corresponds to $L_i^{SF}L_j^{VIS}L_k^{IR}\hat{e}_{SF}\hat{e}_{VIS}\hat{e}_{IR}$ in eq. (7.39), N and d are the number and size of SFG-active domains within the coherence length ($L_c = \pi/\Delta k$) inside the 3D matrix, respectively, and Δl is the inter-domain distance, the $\left(\frac{e^{-i\Delta k \cdot d}-1}{\Delta k}\right)$ term is the phase synchronization factor (eq. (7.19)), and the $(e^{-i\Delta k(d+\Delta l)f})$ term accounts for the random quasi-phase matching between domains at varying distances [11]. The last two terms are not considered in eq. (7.65) when the SFG signal intensity is calculated for the 2D interface case (Section 7.3.3). Note that the N term in eq. (7.56) is replaced with the $\sum_{f=0}^{N-1}$ sum in eq. (7.84), and the rest $\sum_{i',j',k'}\left\langle R_{ii'}R_{jj'}R_{kk'}\beta_{i',j',k'}\right\rangle$ remain unchanged in eq. (7.84). If the size of crystalline domains has a Gaussian distribution and their orientations is fully random, eq. (7.84) becomes equivalent to $N_{eff}I_{coh}$, where N_{eff} is the effective number of SFG-active domains, and I_{coh} is the SFG intensity of a single coherence length of the material [48], which is shown schematically in Figure 7.25. In experiments, the scattering loss of the signal in the material also needs to be considered if the refractive index contrast between the SFG-active domain and the surrounding medium is large.

In 3D materials, SFG-active domains can be distributed not entirely uniformly; at the mesoscale (between nano- and microscales), they can be distributed with preferential azimuth (ϕ), tilt (θ), and rotation (ψ) angles. The relative polarity among adjacent domains can be anywhere between perfectly parallel and perfectly antiparallel. The degree of net polarity can be quantified using the term "directionality excess (*DE*)." Because too many degrees of freedom are involved in eq. (7.84), it would be difficult to use the analytical integration method described in Section 7.3.3 to predict the SFG intensity for a given peak. An alternative approach of using eq. (7.84) is to simulate the SFG intensity numerically.

To incorporate the mesoscale structural distributions within the coherence length (L_c), one can calculate eq. (7.84) for M_n replicas generated using the Gaussian probability functions with appropriate standard deviations (σ_ϕ, σ_θ, σ_ψ) for the angular distributions and the random choice function with a weighted probability for the polarity bias (DE) and then take the average of these M_n cases [11]:

$$I_{SF}^{caln} = \sum_{m=1}^{M_n} \frac{I_{SF}(m)}{I_{VIS}I_{IR}} / M_n \tag{7.85}$$

Taking the arithmetic average can be justified with the random quasi-phase matching condition. This numerical simulation be done with *Mathematica*, *MATLAB*, or *Python* software. An example of *Mathematica* program can be found in Ref. [11].

In summary, the parameters that we should consider for calculation of the SFG intensity of 3D materials include individual SFG-active domain size (d), inter-domain separation (Δl), averaged azimuth (ϕ), tilt (θ), and rotation (ψ) angles as well as standard deviations in their distribution functions (σ_d, $\sigma_{\Delta l}$, σ_ϕ, σ_θ, σ_ψ), and directional excess (*DE*). Recall that there are also the $\beta_{i',j',k'}$ terms which depends on the crystalline order in the nano-scale. Although bulk SFG signals can be measured in eight different polarization combinations (Table 7.2), the total degree of freedom is larger than the number of measurables if none of these parameters are known. So, it is not possible to determine all these variables independently from SFG. But, if many of them can be predetermined from other characterization techniques (e.g., $\beta_{i',j',k'}$ values from crystal symmetry, d from x-ray diffraction, Δl from small angle x-ray scattering or volumetric concentration, ϕ or θ from polarized IR or Raman, and so on), SFG can be used to determine the remaining variables that cannot be addressed with other experimental methods.

Figure 7.32 demonstrates the use of eq. (7.84) [11]. The sample was a freestanding film of uniaxially aligned cellulose Iβ nanocrystals extracted from tunicates. Figure 7.5 shows the angles that correlate the molecular coordinate to the lab coordinate. Since the crystals are uniaxially aligned, we know the azimuth angle ($\phi = 0°$ for parallel orientation or 90° for perpendicular orientation) with respect to the laser incidence plane, and the tilt angle (θ) is 90°. We can assume a small degree of deviation from the perfect alignment, for example, $\sigma_\phi = \sigma_\theta = 7.5°$. From the XRD analysis, it was found that there was some preferential packing of crystallographic planes of cellulose Iβ with respect to the film surface, namely, $(1\bar{1}0)$ facet (i.e., $\psi = 43.7°$ or 223.7°) with a probability of 0.57, (110) facet ($\psi = 136.3°$ or 316.3°) with 0.27 probability, and (200) facet ($\psi = 0°$ or 180°) with 0.16 probability. From the width of the x-ray diffraction peak, the diameter of nanocrystals was determined to be 20 nm. Since the nanocrystals were tightly packed (without any matrix), the interparticle distance could be set to a minimum (say, $\Delta l = 1$ nm). For the given input beam geometry and the dispersion of refractive index of cellulose, the coherence length (L_c) was calculated to be 210 nm. If we simplify the calculation by modeling as a 1D problem, the number (N) of crystallites within L_c is 10. Since the uniaxial alignment was done without any preferential directionality, DE can be set to zero (i.e., no net polarity in the sample). Lastly, the $\beta_{i',j',k'}$ term is assumed to be governed mostly by the IR transition

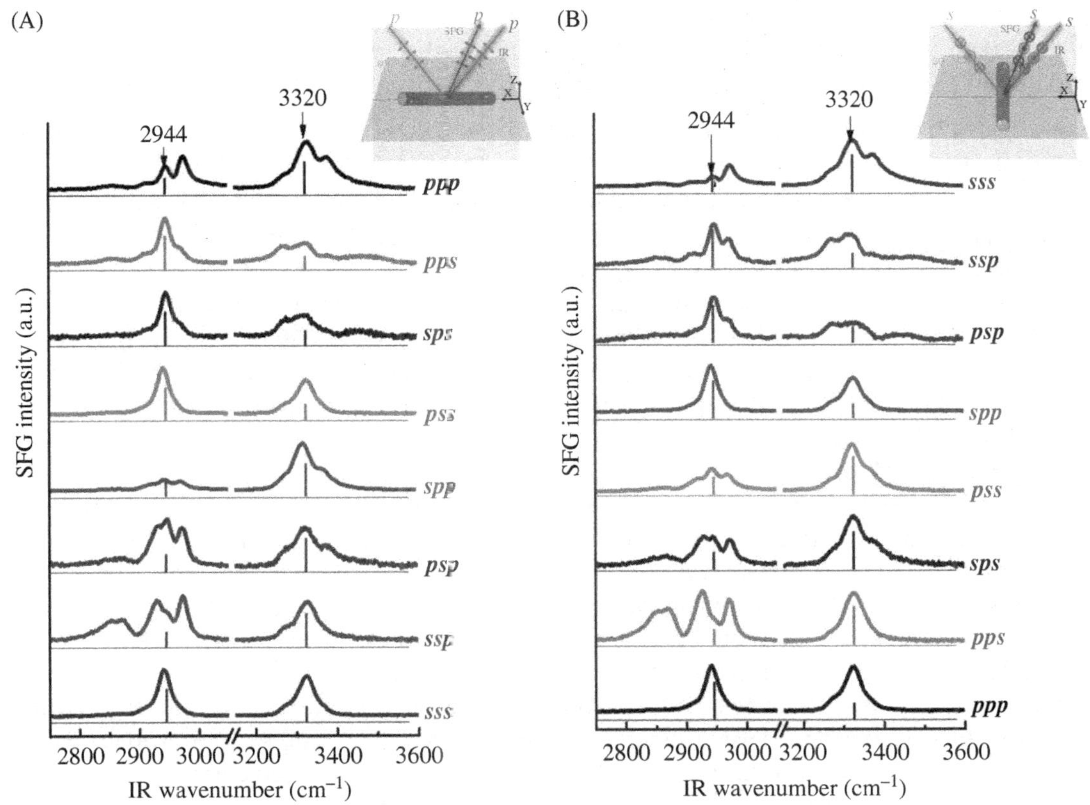

Figure 7.32 SFG spectra of a thin film of uniaxially aligned cellulose Iβ nanocrystals placed in the XY plane with the alignment axis (A) parallel and (B) perpendicular to the laser incidence plane (XZ) collected with all eight possible polarization combinations in the reflection mode with the counter-propagating VIS and IR incidence angles of ±18°. The signal intensity is normalized with the strongest peak. The vertical bars at 2944 cm^{-1} and 3320 cm^{-1} are the relative intensities calculated with eq. (7.85) with the parameters mentioned in the text. Reproduced from Choi et al. [11]/with permission American Chemical Society.

dipole; thus, $\beta_{i'j',k'} \propto \left(\frac{\partial \alpha_{i'j'}}{\partial Q_q}\right) \otimes \left(\frac{\partial \mu_{k'}}{\partial Q_q}\right) \cong \left(\frac{\partial \mu_{k'}}{\partial Q_q}\right)$. In other words, the effect of Raman tensors is insignificant as compared to that of IR tensors. This assumption was made since the Raman tensor was not available and the previous time-dependent density functional theory (TD-DFT) calculation showed that the directionality of $\beta_{i'j',k'}$ in the molecular coordinate is similar to that of $\left(\frac{\partial \mu_{k'}}{\partial Q_q}\right)$ [57]. The dipole angles of the 2944 cm^{-1} (CH stretch) and 3320 cm^{-1} (OH stretch) modes, with respect to the cellulose chain axis, were determined through the dichroic analysis using polarized FT-IR [47]. These two peaks were chosen because they are observed always in the SFG spectra of cellulose, and their dipole angles are above and below magic angle (54.7°) with respect to the cellulose chain axis: 62° for 2944 cm^{-1} and 35° for 3320 cm^{-1}.

Then, numerical calculations were done using eq. (7.84), in which $\chi^{(2)}_{ijk}$ is described with eq. (7.56), for 1000 cases of cellulose nanoparticle arrangements generated with the probability functions constrained with the distributions mentioned above, and then the averages of those 1000 cases were calculated as described by eq. (7.85). The calculation results are shown as bar graphs, along with the experimental SFG spectra, in Figure 7.32. Although several assumptions were made (including $\beta_{i'j',k'} \cong \left(\frac{\partial \mu_{k'}}{\partial Q_q}\right)$), the relative intensities of the 2944 cm^{-1} and 3320 cm^{-1} predicted using eq. (7.85) match quite reasonably well with the experimental trend. It is intriguing to note that the *ppp*-SFG spectra are quite similar to the *sss*-SFG spectra of the sample rotated by 90°. When the uniaxially oriented sample is rotated by 90° with respect to the laser incidence plane, the *y*-component in the tensor probed by *s*-polarization becomes the *x*-component probed by *p*-polarization. At $\phi = 0°$, the *ppp* polarization probes $\chi^{(2)}_{xxx}$, $\chi^{(2)}_{xxz}$, $\chi^{(2)}_{xzx}$, $\chi^{(2)}_{zxx}$, $\chi^{(2)}_{zzx}$, $\chi^{(2)}_{zxz}$, $\chi^{(2)}_{xzz}$, and $\chi^{(2)}_{zzz}$ (see Table 7.2). When the sample is rotated to $\phi = 90°$, the *sss* polarization probes the $\chi^{(2)}_{yyy}$ component. Nonetheless, their spectral features are remarkably similar.

A similar trend is observed for other polarizations – that is, the spectral features at $\phi = 0°$ (parallel to the laser incidence plane) are similar to the ones collected with the orthogonal polarization combination (e.g., *pps* ↔ *ssp*) at $\phi = 90°$ (perpendicular to laser incidence plane). The *pps* polarization at $\phi = 0°$ probes $\chi^{(2)}_{xxy}$, $\chi^{(2)}_{xzy}$, $\chi^{(2)}_{zxy}$, and $\chi^{(2)}_{zzy}$. When the sample is rotated to $\phi = 90°$, the $\chi^{(2)}_{xxy}$ becomes $\chi^{(2)}_{yyx}$ which is probed in the *ssp* polarization. If these terms are the dominant ones compared to other terms probed in those polarizations, these two spectra will have similar spectral features. The second possibility could be the case where the net contribution from $\chi^{(2)}_{xzy}$, $\chi^{(2)}_{zxy}$, and $\chi^{(2)}_{zzy}$ in the *pps* polarization at $\phi = 0°$ is similar to the $\chi^{(2)}_{yyz}$ contribution in the *ssp* polarization at $\phi = 90°$; but this is rather unlikely compared to the first probability.

In Figure 7.32, it is also noted that the 2968 cm^{-1} peak is strong when the 3320 cm^{-1} peak is stronger than the 2944 cm^{-1}. Although the 2968 cm^{-1} peak was not calculated, that trend could be explained with the fact that the dipole tilt angle of the 2968 cm^{-1} peak is 41° with respect to the cellulose chain axis, below the magic angle as in the 3320 cm^{-1} case [47]. That also explains the trend seen in Figure 7.31. When the 2944 cm^{-1} peak is dominant and the 3320 cm^{-1} peak is small, then the 2968 cm^{-1} peak is negligible (Figure 7.31A). When the OH peaks at ~3320 cm^{-1} are strong, the 2968 cm^{-1} peak is stronger than the 2944 cm^{-1} peak (Figure 7.31B).

Figure 7.33 compares the azimuth angle dependence of 2944 cm^{-1} and 3320 cm^{-1} intensity as well as their ratios of CNCs isolated from Whatman No. 1 filter paper and then deposited on a slide glass with a preferential orientation using the Langmuir–Blodgett (LB) deposition method (see Figure 7.19) [11]. In the *psp*-polarization, both 2944 cm^{-1} and 3320 cm^{-1} are peaked at $\phi = 90°$, and only 2944 cm^{-1} has a small peak at $\phi = 0°$. So, the 2944 cm^{-1}/3320 cm^{-1} intensity ratio has four lobes in the polar plot. In the *spp*-polarization, the 3320 cm^{-1} intensity is larger along the $\phi = 0°$ direction than the 2944 cm^{-1} peak and is very small at $\phi = 90°$. So, the 3320 cm^{-1}/2944 cm^{-1} ratio has two lobes in the $\phi = 0°$ direction in the polar plot. TD-DFT calculations show similar patterns in the polar plots of their ratios (Figure 7.33B). The numerical simulation results were done using eq. (7.85) with the following parameters: $d = 4$ nm (thinner because nanocrystals were produced from filter papers), $\Delta l = 1$ nm, $L_c = 50$ nm (data were collected with a different SFG system), $N = 10$, $DE = 0\%$, $\theta = 90°$ with $\sigma_\theta = 0°$ (a monolayer on a slide glass), $\phi =$ incrementally varying with $\sigma_\phi = 7.5°$, $\psi =$ random (assuming no preferential packing in the LB film), and $M_n = 1000$. The simulation results are qualitatively in good agreement with the experimental data and the TD-DFT calculations. Recall that the susceptibility tensor terms were assumed to be governed by the IR tensor only in this simulation. If the full tensor terms are described more accurately (through experimental methods or computationally), the simulation result could be improved further.

7.4.6 Structural Inhomogeneity in Biological Materials

Biological samples are seldomly homogeneous. Different tissues carry different biological and/or physical functions, so they may have very different structures although their chemical compositions are similar. In Figure 7.27, the SFG spectra were collected with a system where two input beams were focused with separate focusing lenses. Since the focused beam spot size was a few hundred microns, the spectra were the average of many cell walls. In the stem of *Arabidopsis*, there are many different tissues, namely, interfascicular fiber providing mechanical support for the plant, xylem transporting water from the root to the leaf, phloem carrying the photosynthesized glucose from the leaf to all other parts of the plant, cortex made up of parenchymatous cells transporting and storing nutrients, and epidermis protecting the plant from the external environment (Figure 7.34) [58]. So, the area-averaged analysis may not provide spectral information relevant to different functions of these individual components. If the SFG spectral contribution comes from one part of the sample which happens to be a minor one, then correlating such data to the volume-averaged property of the sample may lead to a misinterpretation [59].

Figure 7.34 compares the SFG spectra of cellulose in cell walls of different regions inside the stems of six-week and eight-week old *Arabidopsis* [58]. The SFG signal intensities of the phloem and cortex regions are much weaker than those of other regions; this could be because they are thinner (so, less walls in the probe volume) and primary cell walls (i.e., less amount of cellulose in cell wall). The latter point is corroborated by the weak intensity in the cross-polarization microscope images which measures the degree of birefringence that is proportional to the cellulose amount. The SFG spectra as well as hyperspectral imaging show that the OH/CH intensity ratio (3320 cm^{-1}/2944 cm^{-1}) is relatively stronger in the xylem region as compared to the interfascicular fiber region.

Could that difference be due to the random orientation of cellulose fibers in the xylem region and the uniaxial orientation of cellulose fibers in the interfascicular region? If these data were compared with the spectra shown in Figures 7.30 and 7.31, one could make such a conclusion. But, from the x-ray diffraction study, it is known that cellulose fibers are pretty much aligned along a specific angle, which is called microfibril angle (MFA), with respect to the stem axis [60]. So, the random

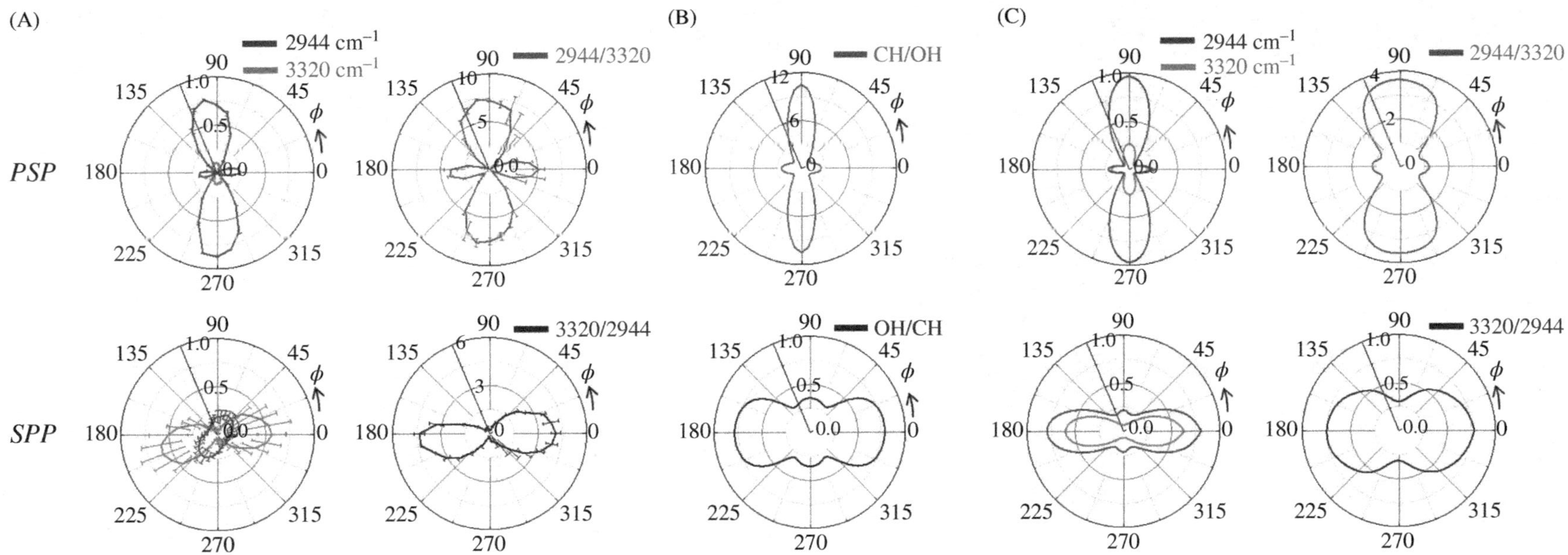

Figure 7.33 (A) Azimuth polar plot of the *psp*- and *spp*-SFG intensity of the 2944 cm^{-1} CH-stretch and 3320 cm^{-1} OH-stretch modes and their ratio of CNCs uniaxially aligned using the LB method. (B) Azimuth polar plot of the *psp*- and *spp*-SFG intensity ratios calculated from TD-DFT calculations for the modes that correspond to the 2944 cm^{-1} and 3320 cm^{-1} peaks of the experimental spectra. (C) Azimuth polar plot numerically calculated using eq. (7.85) with the same approximation used in Figure 7.32. Reproduced from Choi et al. [11]/with permission American Chemical Society.

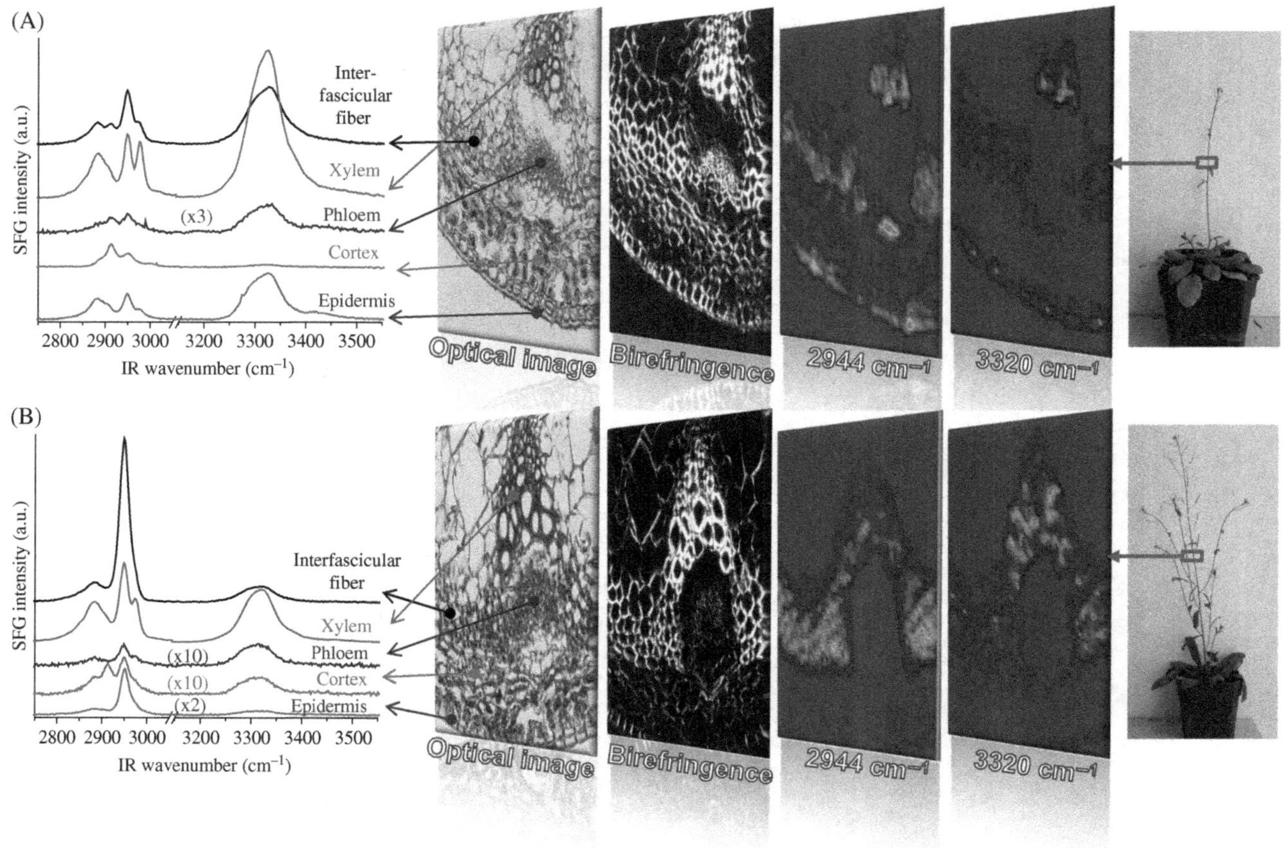

Figure 7.34 Microscopic SFG spectra, optical image, birefringence image, and hyperspectral images at 2944 cm^{-1} and 3320 cm^{-1} of (A) six-week old and (B) eight-week old *Arabidopsis* stems. All images and spectra were taken in D_2O. The polarization combination of probe beams is *pps* and the laser incidence plane is in the vertical direction of the optical image. The SFG hyperspectral image was collected with the 15× objective lens. Reproduced from Huang et al. [58]/with permission of American Chemical Society.

orientation possibility can be ruled out since it is not consistent with the structural model independently determined and confirmed by other experimental techniques. Then, what could be the cause for such a drastic difference in the SFG spectral features of cellulose in the xylem and interfascicular fiber regions?

7.4.7 Relative Polarity Among Noncentrosymmetric Domains

In Section 7.4.3, we have seen the examples where the SFG spectral features are sensitive to the parallel vs. antiparallel directionality of chains in the crystalline unit cell (i.e., the chain polarity in the nanoscale); the same directionality issue pertains to the mesoscale structure. Since all cellulose chains in the microfibril are synthesized by CSCs or their clusters moving in one direction, they have the same directionality; thus, cellulose microfibril itself also has a polarity. If multiple CSCs are moving in two ways along a guiding track and the probabilities to move in two opposite directions are the same, the bundles of cellulose microfibrils deposited along the guiding track direction will have a zero polarity in average. In this case, the directional excess (DE) will be zero. If the CSC movements are biased to one direction over the other, then the bundles of microfibrils will have a net polarity, which leads to DE $\neq$ 0. Since the dipole angles of the 2944 cm^{-1} and 3320 cm^{-1} with respect to the chain axis are above and below the magic angle, respectively, their relative intensities will vary differently with the degree of polarity cancellation among adjacent microfibrils (similar to the effect of chain polarity within the unit cell).

Figure 7.35 shows a high-resolution hyperspectral SFG analysis result of the xylem and interfascicular fiber regions of an eight-week old *Arabidopsis* stem [61]. The full SFG spectra of individual pixels around a single cell were extracted and plotted. From the microscopic images, the azimuth angle of the cellulose in each pixel with respect to the laser incidence plane could be determined (Figure 7.35H). Then, the SFG spectra extracted from different locations around the cell could be used to construct the azimuth angle polar plot (Figures 7.35F,G). While the SFG spectra of cellulose in the interfascicular fiber

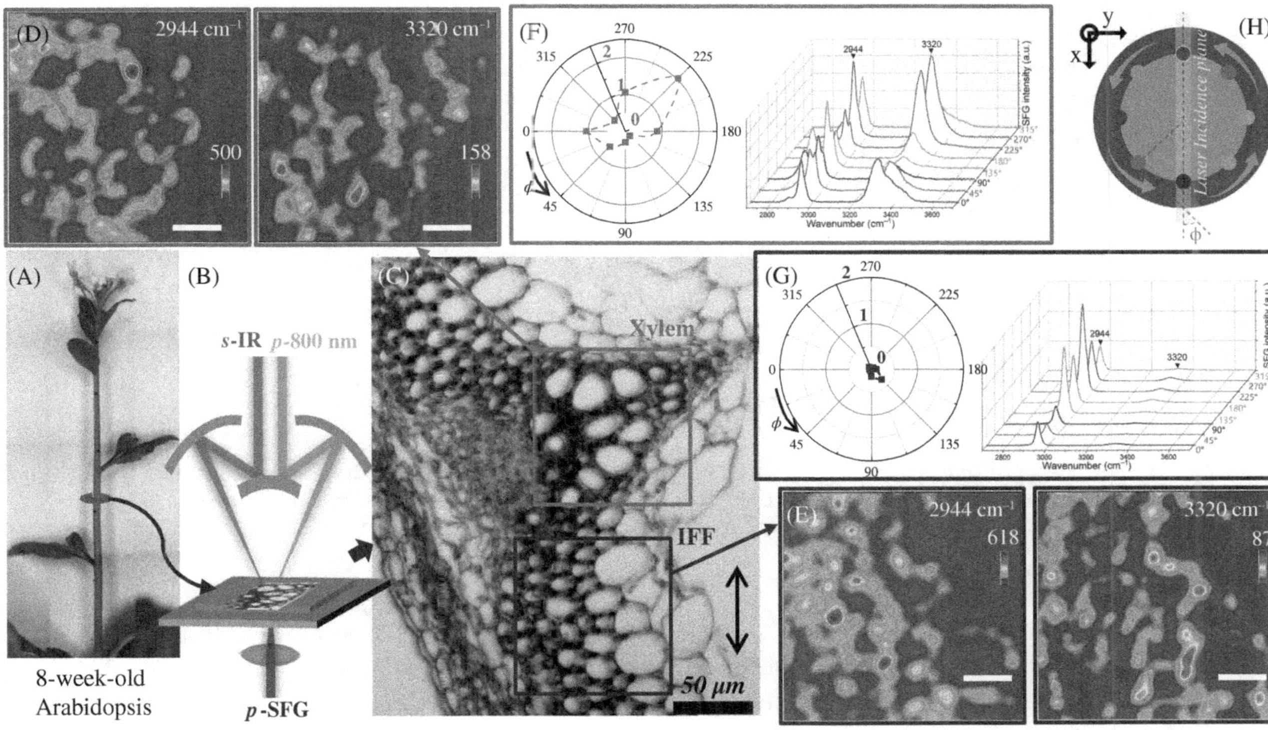

Figure 7.35 (A) Eight-week-old *Arabidopsis* inflorescence stem. (B) A ~10-μm thick sectioned sample was placed on an SFG-microscope. (C) Image of xylem and interfascicular fiber (IFF) regions analyzed. (D,E) Hyperspectral intensity maps of 2944 cm^{-1} and 3320 cm^{-1} of the xylem and IFF regions. (F,G) Full *pps*-SFG spectra extracted from eight locations around one single cell wall that have different azimuth angle with respect to the laser incidence plane (H). Reproduced from Choi et al. [61]/with permission of Springer Nature.

cell wall show a very small OH/CH intensity ratio and relatively small variations from location to location, those of cellulose in the xylem cell wall exhibit large variations in both OH/CH ratio and the CH peak shape around a single cell. Since the cellulose fibrils are highly aligned along the MFA axis and the azimuth angle distribution is very narrow at a given location (since the cell wall is thin), this spectral feature difference must originate from the DE difference, not from the randomness of orientation.

One challenge here is that in addition to the DE of cellulose microfibrils, other parameters can also alter the OH/CH intensity ratio and their peak shapes. The data in Figures 7.32 and 7.33 (which are the case where the tilt angle was 90°) show that the SFG spectral features change as the azimuth angle changes. Similarly, at a given azimuth angle, the SFG spectral features will change as the tilt angle changes. So, one needs to deconvolute the tilt angle dependence and the DE dependence. This could be done by calculating the OH/CH intensity ratio as a function of azimuth angle for various tilt angle and DE cases and compare the experimental data with the calculation results. Since the multiple variables need to be compared simultaneously, it is difficult to do it by looking at one variable effect at a time. This requires a multivariate analysis approach (see Chapter 8). In this specific example, the theoretical calculation results were used as a training dataset for multilayer perceptron (MLP) neural network algorithm, which then was used to estimate the tilt angle (θ) and DE of microfibrils in individual cell walls through a regression.

The analysis results are shown in Figure 7.36 [61]. The cellulose microfibrils in the interfascicular fiber cell walls are found to be highly aligned along the stem axis (MFA < 5°) without any bias in polarity (DE = 0%). In contrast, cellulose microfibrils in the xylem cell walls are found to be tilted away from the stem axis by 20–25° and they are highly uniaxially biased, that is, polar (DE = 80±20 %).

The mesoscale microfibril orientation and directionality in these cell walls with different functions can be compared with engineering principles of composite materials synthesized for specific purposes. In the stem, the interfascicular fibers serve a mechanical function supporting the body weight of the plant. In a fiber-reinforced polymer composite sheet, the in-plane load bearing capacity is higher along the fiber direction. Thus, the fact that cellulose microfibrils in these fibers are highly aligned along the vertical stem axis attests that the fiber cell wall structure is optimally designed to support the compressive stress due to gravity.

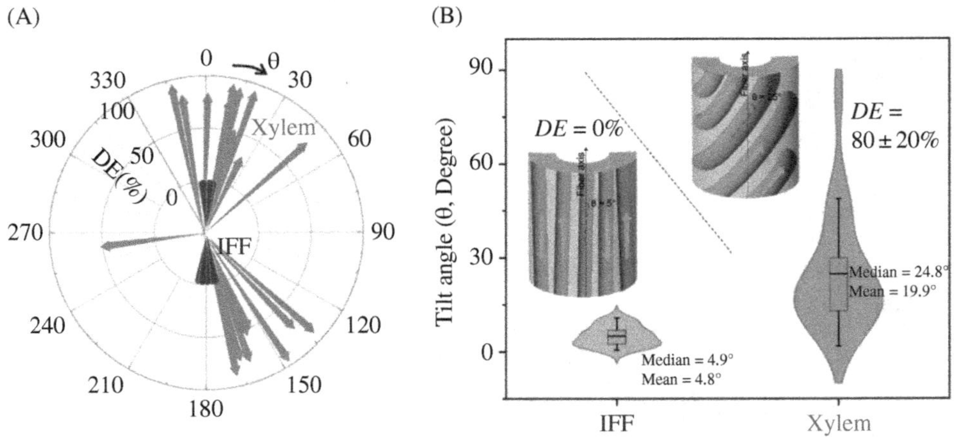

Figure 7.36 (A) Polar plot and (B) distribution of tilt angle (θ) and polarity (DE) of cellulose microfibrils, determined via the MLP analysis of azimuth angle (ϕ) dependence of the 3320 cm^{-1}/2944 cm^{-1} intensity ratio, for 20 IFF cells and 20 xylem cells selected in the SFG hyperspectral images of transverse cross-sections of eight-week-old *Arabidopsis* stem. Reproduced from Choi et al. [61]/with permission of Springer Nature.

The main function of xylem in *Arabidopsis* stem is the water transport from root to leaf via negative pressure produced by evapotranspiration; thus, the cell wall must have good resistance against vessel collapse. If the cellulose microfibril synthesis in the secondary cell wall is suppressed through genetic mutations, then the xylem cells collapse due to the negative pressure gradient generated by evapotranspiration. The engineering design for high resistance to out-of-plane collapse or rupture of a conduit is to have a spiral fiber reinforcement, as can be seen in a coil-reinforced hose. Such a coiled structure in xylem cell walls must be achieved by controlling the orientation of the cellulose microfibrils (i.e., tilting way from the stem axis). The coiling of microfibrils could be facilitated if they are aligned unidirectionally (DE close to 100%). Each cellulose microfibril can be twisted due to the chirality of the cellulose structure. If all microfibrils in a thick bundle are aligned unidirectionally, then their strain energy could add up and eventually cause coiling to occur naturally. Such unidirectional packing of microfibrils requires the unidirectional movement of CSCs during the cellulose synthesis.

In the previous example, whether cellulose microfibrils in the *Arabidopsis* xylem walls are pointing upward or downward (i.e., the phase of microfibrils) cannot be determined. Since there are so many factors that can affect the absolute intensities of SFG signals of the biological sample (see Section 7.4.2), the CH stretch peak was used as an internal reference and the OH/CH intensity ratio change as a function of azimuth angle was analyzed. By doing so, all experimental artifacts can be cancelled out, allowing extraction of the information needed to determine DE (in other words, the relative phase among adjacent microfibrils) and tilt angle (θ; MFA in the case of microfibrils in stem) simultaneously from hyperspectral SFG imaging data.

If the absolute phase of microfibrils is to be determined, we need to have an external reference with a known phase [62]. This is called heterodyne detection. Figure 7.37 demonstrates the phase-sensitive SFG imaging of collagen fibrils in a rat tail tendon tissue [63]. In this experiment, the mid-IR (broadband 3.4 μm), up-conversion beam (776 nm), and phase-reference beam (630 nm) were focused on the sample using one single objective lens. Then, the interference of the SFG signal with this reference beam allowed to determine the phase of the SFG signal of each pixel in the image with respect to the phase of the 630 nm beam. Figure 7.37C shows that collagen fibrils of certain locations have opposite phases from the neighboring fibrils. At this time, the biological or physical function of this mixed directionality of fibrillar protein fibers in animal tissues is not known yet.

As shown in the above examples, there are multiple ways to analyze the phase information of SFG-active domains in 3D bulk matrices from SFG imaging. In the first example reviewed here, the relative phase (i.e., DE) of cellulose microfibrils in the xylem walls of *Arabidopsis* stem was determined by analyzing SFG signal change as a function of azimuth angle. In the second example, it was done by overlapping the reference signal with a known phase to the sample along with the IR and up-conversion beams. Similarly, in the 2D interface analysis, the phase analysis can be done in two ways. One is to analyze interference patterns among adjacent peaks in the intensity spectrum (see Figure 7.24). The other is overlapping the SFG signal with the reference signal with the known phase. The reference signal can be generated internally by using a nonlinear

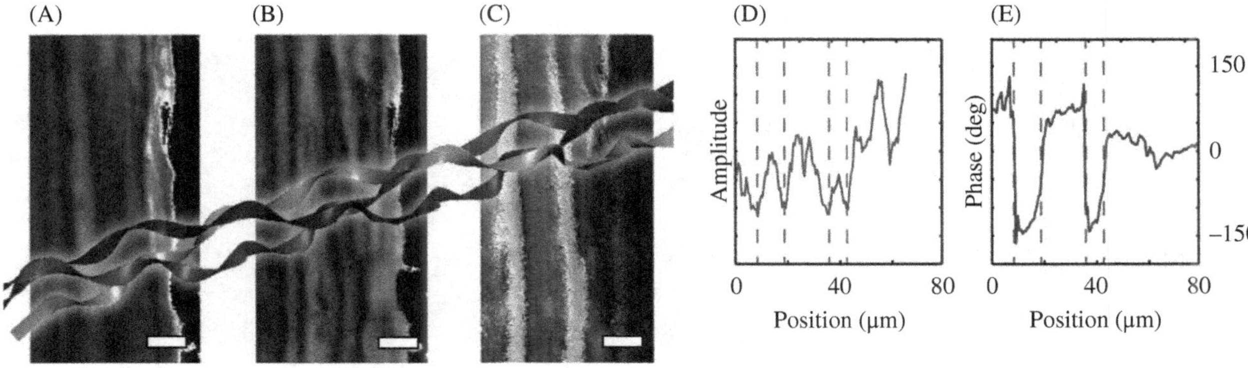

Figure 7.37 SFG images of rat tail tendon t ssue on top of *z*-cut quartz: (A) intensity image from homodyne detection, (B) amplitude, and (C) phase images from heterodyne detection. (D, E) Line profiles across the amplitude images. The scale bar is 20 μm. Reproduced from Han et al. [63]/with permission of American Chemical Society.

crystal (see Figure 7.23) or generated externally and then overlapped with the input beams on the sample (which is heterodyne detection).

In summary, the phase information cannot be read directly from the intensity. It requires an additional reference signal for interference with the SFG signal of interest. In the examples reviewed previously, the reference signal was all optical, and the phase was determined from the measured interference pattern. In some cases, phase information can be deduced even without optical interference. For example, the dipole of a specific functional group can be aligned in response to an electric field gradient that is naturally present or externally applied; then, such information can be used to interpret the SFG data. For example, if we know the isoelectric point of the solid surface and the pH of aqueous solution with which the solid is in contact, then we already know what surface charge of the solid/liquid interface will have, which will govern the orientation of water molecules. The responsiveness of water orientation to the surface charge at the solid/water interface has already been confirmed by the phase-sensitive SFG analysis [64].

A similar concept can be applied to the analysis of SFG-active domains in the bulk phase. Figure 7.38 shows the SFG spectra of poly(vinylidene fluoride-hexafluoropropylene) (PVDF-HFP) [65]. When a PVDF-HFP film is physically stretched, the α-phase (which is naturally formed) is transformed to the β-phase. In the β-phase, the chain backbones are aligned along

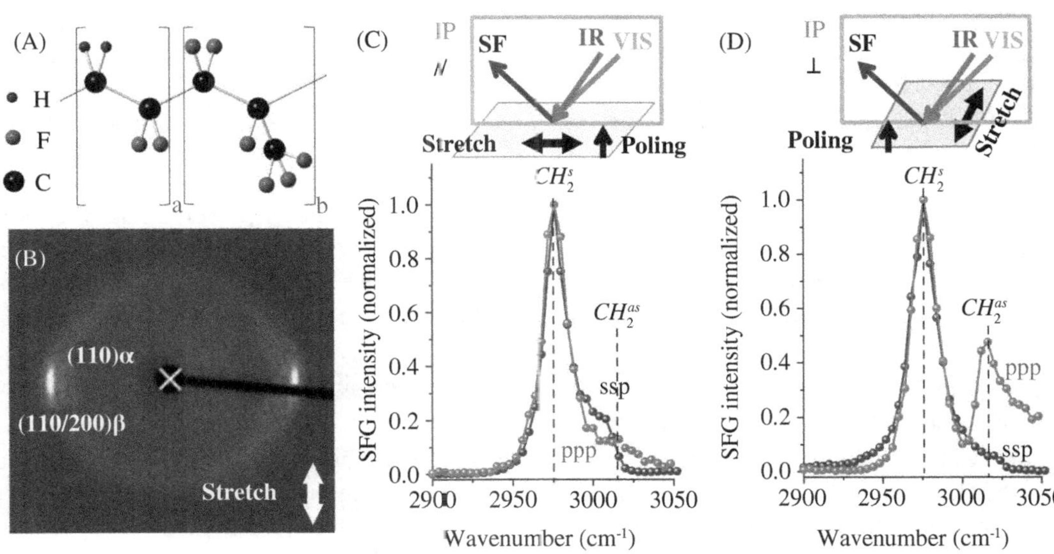

Figure 7.38 (A) Structure of P(VDF-HFP). (B) 2D wide-angle x-ray diffraction image of a stretched P(VDF-HFP) film. (C, D) *ssp*- and *ppp*-SFG spectra of a stretched and poled 8-μm thick P(VDF-HFP) film at two azimuthal angles: the laser incidence plane (IP) is parallel (in C) and perpendicular (in D) to the stretch direction. Reproduced from Chae et al. [65]/with permission American Chemical Society.

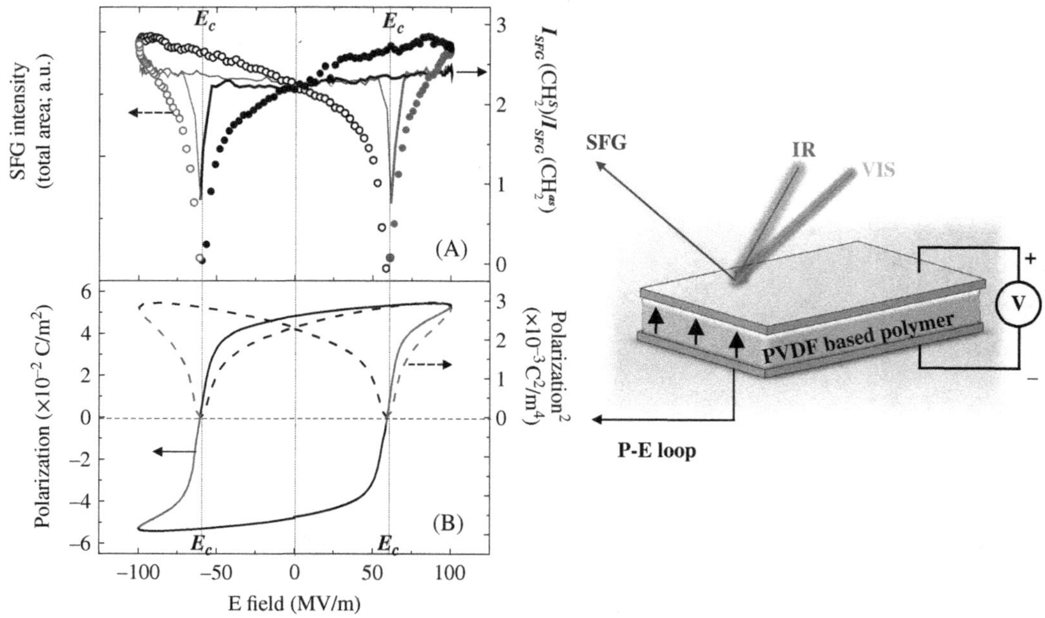

Figure 7.39 Comparison of (A) I_{SFG}-E and (B) P-E hysteresis curves. In the P-E plot, the P^2-E curve is shown for comparison with the I_{SFG}-E curve. The poling cycles are from E_c to the maximum applied bias (±100 MV/m) and the depoling cycles are from the maximum bias to E_c. Reproduced from Chae et al. [65]/with permission American Chemical Society.

the stretch direction and assume the *all-trans* conformation in which the CH_2 and CF_2 groups of the vinylidene unit are faced in the opposite directions (Figure 7.38A). So, a net dipole is produced within the crystalline β-domain. The dipole orientation can be rotated by applying electric field to the sample, which is the ferroelectric behavior. When the poling is done by applying the electric field across the stretched PVDF-HFP film, the CH_2 groups are aligned along the film thickness direction. When the laser incidence plane is parallel to the stretch direction, then the transition dipole of the CH_2 asymmetric stretch mode is perpendicular to the laser incidence plane. Thus, the 3016 cm^{-1} peak is weak in both *ssp*- and *ppp*-SFG spectra (Figure 7.38C). The weak peak at ~3000 cm^{-1} in the *ssp* spectrum must be the Fermi resonance peak of the CH_2 symmetric stretch (see Section 7.3.4). The *ppp* polarization probes surface-parallel components within the laser incidence plane (see Table 7.6). When the laser incidence plane is perpendicular to the stretch direction, the surface-parallel component of the CH_2 asymmetric stretch mode is within the laser incidence plane, thus it becomes stronger in the *ppp* polarization than the *ssp* polarization (Figure 7.38D).

Figure 7.39 compares the total SFG peak area (I_{SFG}) monitored *in situ* during poling (i.e., sweeping the *E*-field) with the polarization (*P-E*) hysteresis curve. The I_{SFG} is proportional to the square of the total polarization, $\left|\chi_R^{(2)} + \chi_{NR}^{(2)}\right|^2$ (see eq. (7.82)). Here, the nonresonant term comes from $\chi^{(3)}E_{\text{applied}}$. Thus, the overall shape of the I_{SFG}-E curve closely follows the P^2-E curve (dashed line in Figure 7.39B). Although we cannot tell the phase of the CH_2 group of the β-domain (whether pointing up or down in the lab coordinate) from the SFG signal intensity, we can deduce it from the hysteresis of the polarization. Regardless of the phase of the β-domain, the SFG intensity is exactly the same if the degree of polarization is the same.

In Figure 7.39, although very weak, the SFG intensity at the coercive electric field (E_c) shows a finite value. At E_c, the net poling order of the β-domain is nulled; thus, the $\chi_R^{(2)}$ term would be zero. Then, this small intensity must be due to a weak contribution from the nonresonant $\chi^{(3)}E_c$ term of the unpoled sample; or it could be incomplete randomization of the polar domains inside the sample. The CH_2^s/CH_2^{as} intensity ratio is also the lowest when the net polarization is zero. The net alignment of the CH_2 groups along the poling direction is negligible at E_c; therefore, the enhancement of the CH_2^s mode due to alignment is also insignificant.

7.5 Further Reading for SFG

In this chapter, we have studied how nonlinear spectroscopy (esp. SFG) can be utilized to probe molecules at the buried interface and noncentrosymmetric domains inside the bulk phase. The theory involved here appears a lot more complicated than the linear spectroscopy theory. Nonetheless, it can still be viewed as an extension of the equations that we already studied in the linear spectroscopy; just the number of terms to consider is increased significantly. This chapter provides the basic theory of SFG and show how they are utilized to determine orientations of molecules at the 2D interface with C_∞ symmetry and the study the nano and mesoscale structures of noncentrosymmetric domains in amorphous bulk matrices. Books by Robert Boyd [66] and Ron Shen [24] would be a good reference for readers with deeper interests or who need to do more advanced SFG characterizations.

Practice Problems

7.1 In typical IR analysis, a thermal lamp (blackbody radiation) is used. For Raman spectroscopy, a low-power CW layer is sufficient. But, for SFG spectroscopy, a high-power pulsed laser is required. Explain the reason.

7.2 In reflection–absorption infrared spectroscopy (RAIRS), the signal is detected at the specular direction. In Raman spectroscopy, the signal can be detected anywhere away from the specular reflection direction. In SFG analysis of air/liquid interfaces, the signal cannot be detected at the specular directions of IR and up-conversion beams or in random positions away from the specular direction. It must be a specific angle with respect to two input beams. Explain the reason and how can you find that angle?

7.3 Why SFG cannot detect amorphous polymers in the bulk phase, but it can detect their functional groups at the polymer/air or polymer/liquid interface?

7.4 An analyte is dissolved in a liquid (e.g., water) which is in equilibrium with the gas phase. Explain the reason that SFG can detect the molecules at the liquid/gas interface without interferences from the gas-phase molecules and the molecules dissolved in the bulk liquid.

7.5 Hypothetical IR and Raman spectra of a molecule are shown below. Sketch the molecular hyperpolarizability β of this molecule.

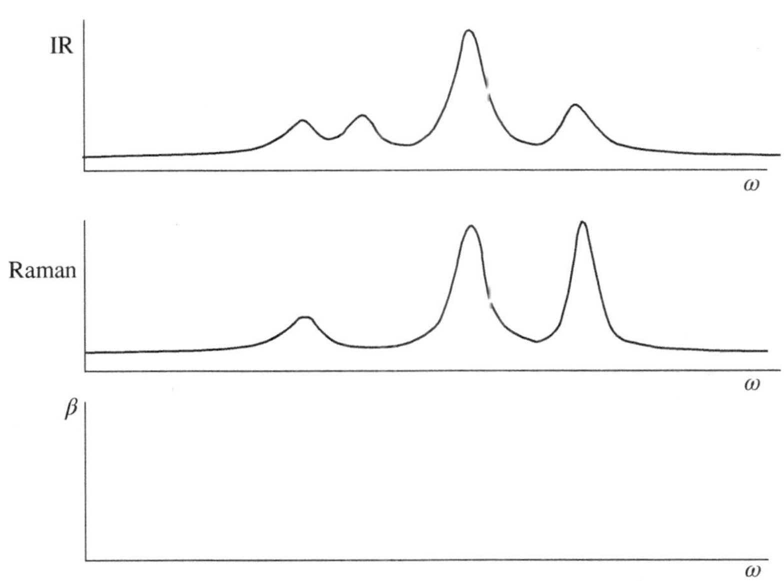

7.6 The incident visible and IR laser beams are shown below. Sketch the reflected and transmitted visible and IR beams and the reflected and transmitted SFG beams as well as visible and IR beams. Justify your answer.

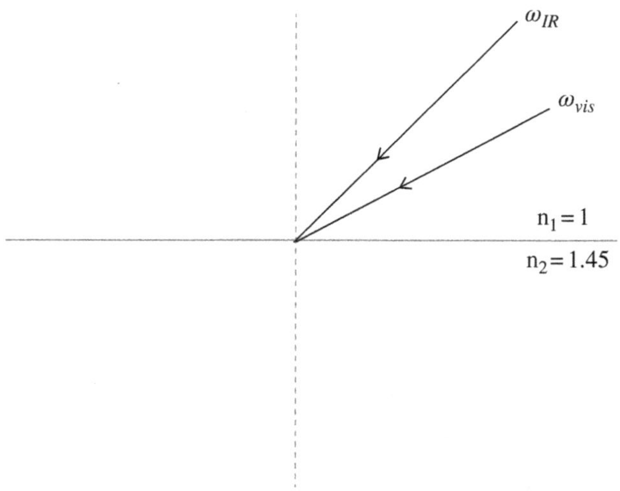

7.7 How is the SFG intensity (I_{SFG}) related to the concentration or density (N) of the adsorbed molecules?

7.8 The second-order susceptibility is expressed as $\chi^{(2)} = Ae^{i\delta}$, where A is the amplitude and δ is the phase angle. The molecule adsorbed on a solid surface has a resonant peak at $\omega_o = 3000\ \text{cm}^{-1}$ with $A_R = 1$ (i.e., $\chi_R^{(2)} = 1e^{i\delta}$). The damping factor of this peak is $\Gamma = 10\ \text{cm}^{-1}$.

a) Sketch the SFG spectrum of this molecule, $I_{SFG}(\omega)$, when it is adsorbed on a dielectric solid without nonresonant susceptibility ($\chi_{NR}^{(2)} = 0$).
b) Sketch the SFG spectrum of the same molecule, $I_{SFG}(\omega)$, adsorbed on a metal surface with a nonresonant background susceptibility, $A_{NR} = 1$ and $\delta_{NR} = -90°$ (i.e., $\chi_{NR}^{(2)} = 1e^{i(-\pi/2)}$).

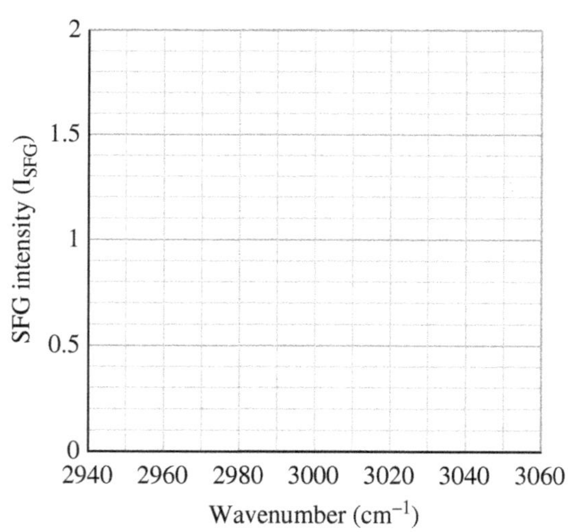

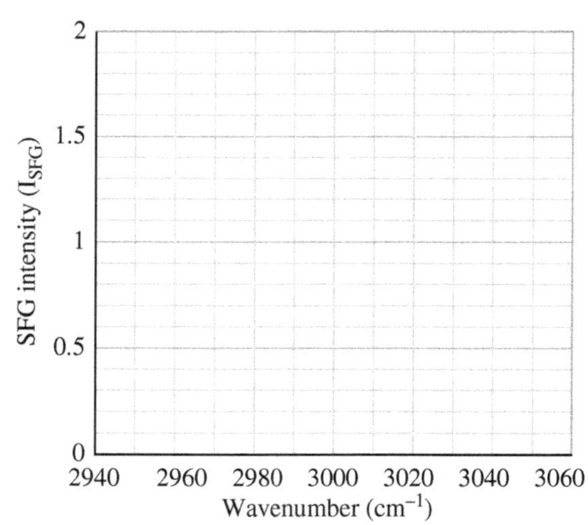

7.9 In SFG analysis of an organic SAM formed on metal surfaces, why can the spectral intensity and shape vary depending on the substrate, even though the areal density and degree of packing of the SAM are the same?

7.10 You are applying the Gibbs adsorption isotherm to estimate the surface excess concentration of surfactants in aqueous solutions (https://doi.org/10.1016/j.cis.2017.07.018) and decide to use SFG, as a complementary tool for the surface tension measurement, to directly measure the surfactant molecules at the air/liquid interface. Can you say that the measured SFG intensity will follow the Beer–Lambert law and be proportional to the concentration of surfactant molecules at the interface?

7.11 Shown here are SFG intensity spectra of water at a charged surface simulated for $\omega_1 = 3200\,\text{cm}^{-1}$, $\varepsilon_1 = 0°$, $\omega_2 = 3400\,\text{cm}^{-1}$, $\varepsilon_2 = 180°$, $A_1 = A_2 = 10$, $\Gamma_1 = \Gamma_2 = 120\,\text{cm}^{-1}$, $\chi_1^{(3)} = 500$, and $\chi_2^{(3)} = 150$ using charge densities of + 0.001, −0.005, −0.025, −0.05 −0.17, and −0.25 C/m^2 and a constant ionic strength of 0.1 M calculated with absorptive–dispersive interactions. From Internet search, you learned that SFG is "surface-sensitive"; then, why do the SFG spectral intensities of water vary with the charge density of the surface in contact with water?

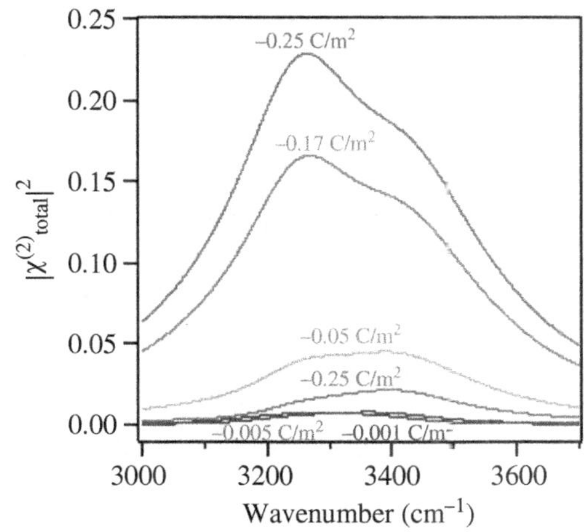

References

1 Williams CT, Beattie DA. Probing buried interfaces with non-linear optical spectroscopy. Surface Science. 2002;500(1):545–76.

2 Chen Z, Shen YR, Somorjai GA. Studies of polymer surfaces by sum frequency generation vibrational spectroscopy. Annual Review of Physical Chemistry. 2002;53(1):437–65.

3 Hosseinpour S, Roeters SJ, Bonn M, Peukert W, Woutersen S, Weidner T. Structure and dynamics of interfacial peptides and proteins from vibrational sum-frequency generation spectroscopy. Chemical Reviews. 2020;120(7):3420–65.

4 Boyd RW. Order-of-magnitude estimates of the nonlinear optical susceptibility. Journal of Modern Optics. 1999;46(3):367–78.

5 Boyd RW. Chapter 1 – The nonlinear optical susceptibility. In: Boyd RW, editor. Nonlinear Optics (Third Edition). Burlington: Academic Press; 2008. p. 1–67.

6 Li Y-S, Jang S, Khan AM, Martin TV, Ogrinc AL, Wang QJ, et al. Possible origin of D- and G-band features in Raman spectra of tribofilms. Tribology Letters. 2023;71(2):57.

7 Boyd RW. Chapter 2 – Wave-equation description of nonlinear optical interactions. In: Boyd RW, editor. Nonlinear Optics (Third Edition). Burlington: Academic Press; 2008. p. 69–133.

8 Shen YR. Basic theory for surface sum-frequency generation. In: Fundamentals of Sum-Frequency Spectroscopy. Cambridge Molecular Science. Cambridge: Cambridge University Press; 2016. p. 23–50.

9 Lee CM, Kafle K, Huang S, Kim SH. Multimodal broadband vibrational sum frequency generation (MM-BB-V-SFG) spectrometer and microscope. The Journal of Physical Chemistry B. 2016;120(1):102–16.

10 Cao J-M, Liu A-A, Yang S-Y, Peng X-X, Wang H-F. Vibrational spectra of liquid interfaces with a 100 kHz Sub-1 cm^{-1} high-resolution broadband sum-frequency generation vibrational spectrometer (HR-BB-SFG-VS). The Journal of Physical Chemistry C. 2023;127(28):14013–22.

11 Choi J, Lee J, Makarem M, Huang S, Kim SH. Numerical simulation of vibrational sum frequency generation intensity for non-centrosymmetric domains interspersed in an amorphous matrix: a case study for cellulose in plant cell wall. The Journal of Physical Chemistry B. 2022;126(35):6629–41.

12 Lambert AG, Davies PB, Neivandt DJ. Implementing the theory of sum frequency generation vibrational spectroscopy: a tutorial review. Applied Spectroscopy Reviews. 2005;40(2):103–45.

13 Hirose C, Akamatsu N, Domen K. Formulas for the analysis of the surface SFG spectrum and transformation coefficients of cartesian SFG tensor components. Applied Spectroscopy 1992;46(6):1051–72.

14 Wang H-F, Gan W, Lu R, Rao Y, Wu B-H. Quantitative spectral and orientational analysis in surface sum frequency generation vibrational spectroscopy (SFG-VS). International Reviews in Physical Chemistry. 2005;24(2):191–256.
15 Moad AJ, Simpson GJ. A unified treatment of selection rules and symmetry relations for sum-frequency and second harmonic spectroscopies. The Journal of Physical Chemistry B. 2004;108(11):3548–62.
16 Choi J, Kwansa AL, Yingling YG, Kim SH. DFT-based calculation of molecular hyperpolarizability and SFG intensity of symmetric and asymmetric stretch modes of alkyl groups. The Journal of Physical Chemistry B. 2023;127(39):8456–67.
17 Ye S, Wei F, Li H, Tian K, Luo Y. Chapter seven – Structure and orientation of interfacial proteins determined by sum frequency generation vibrational spectroscopy: method and application. In: Christov CZ, editor. Advances in Protein Chemistry and Structural Biology (Vol. 93). Cambridge: Academic Press; 2013. p. 213–55.
18 Powell RC. Group theory. In: Symmetry, Group Theory, and the Physical Properties of Crystals. New York, NY: Springer; 2010. p. 25–53.
19 Hu X-H, Wei F, Wang H, Wang H-F. α-quartz crystal as absolute intensity and phase standard in sum-frequency generation vibrational spectroscopy. The Journal of Physical Chemistry C. 2019;123(24):15071–86.
20 Ohno PE, Wang H-f, Geiger FM. Second-order spectral lineshapes from charged interfaces. Nature Communications. 2017;8(1):1032.
21 Ohno PE, Wang H-f, Paesani F, Skinner JL, Geiger FM. Second-order vibrational lineshapes from the air/water interface. The Journal of Physical Chemistry A. 2018;122(18):4457–64.
22 Lu R, Gan W, Wu B-h, Zhang Z, Guo Y, Wang H-f. C—H stretching vibrations of methyl, methylene and methine groups at the vapor/alcohol ($n = 1$–8) interfaces. The Journal of Physical Chemistry B. 2005;109(29):14118–29.
23 Simpson GJ, Rowlen KL. An SHG magic angle: dependence of second harmonic generation orientation measurements on the width of the orientation distribution. Journal of the American Chemical Society. 1999;121(11):2635–6.
24 Shen YR. Fundamentals of Sum-Frequency Spectroscopy. Cambridge: Cambridge University Press; 2016.
25 Weeraman C, Yatawara AK, Bordenyuk AN, Benderskii AV. Effect of nanoscale geometry on molecular conformation: vibrational sum-frequency generation of alkanethiols on gold nanoparticles. Journal of the American Chemical Society. 2006;128(44):14244–5.
26 Barnette AL, Kim SH. Coadsorption of n-propanol and water on SiO_2: study of thickness, composition, and structure of binary adsorbate layer using attenuated total reflection infrared (ATR-IR) and sum frequency generation (SFG) vibration spectroscopy. The Journal of Physical Chemistry C. 2012;116(18):9909–16.
27 Powell RC. Tensor properties of crystals. In: Symmetry, Group Theory, and the Physical Properties of Crystals. New York, NY: Springer; 2010. p. 55–78.
28 Simpson GJ. Molecular origins of the remarkable chiral sensitivity of second-order nonlinear optics. ChemPhysChem. 2004;5(9):1301–10.
29 Chae I, Ngo D, Makarem M, Ounaies Z, Kim SH. Compression-induced topographic corrugation of air/surfactant/water interface: effect of nanoparticles adsorbed beneath the interface. The Journal of Physical Chemistry C. 2019;123(42):25628–34.
30 Yan ECY, Wang Z, Fu L. Proteins at interfaces probed by chiral vibrational sum frequency generation spectroscopy. The Journal of Physical Chemistry B. 2015;119(7):2769–85.
31 Barth A, Zscherp C. What vibrations tell about proteins. Quarterly Reviews of Biophysics. 2002;35(4):369–430.
32 Nihonyanagi S, Mondal JA, Yamaguchi S, Tahara T. Structure and dynamics of interfacial water studied by heterodyne-detected vibrational sum-frequency generation. Annual Review of Physical Chemistry. 2013;64(1):579–603.
33 Shen YR. Phase-sensitive sum-frequency spectroscopy. Annual Review of Physical Chemistry. 2013;64(1):129–50.
34 Wang H-F. Sum frequency generation vibrational spectroscopy (SFG-VS) for complex molecular surfaces and interfaces: spectral lineshape measurement and analysis plus some controversial issues. Progress in Surface Science. 2016;91(4):155–82.
35 Barnette AL, Bradley LC, Veres BD, Schreiner EP, Park YB, Park J, et al. Selective detection of crystalline cellulose in plant cell walls with sum-frequency-generation (SFG) vibration spectroscopy. Biomacromolecules. 2011;12(7):2434–9.
36 Lee CM, Kafle K, Park YB, Kim SH. Probing crystal structure and mesoscale assembly of cellulose microfibrils in plant cell walls, tunicate tests, and bacterial films using vibrational sum frequency generation (SFG) spectroscopy. Physical Chemistry Chemical Physics 2014;16(22):10844–53.
37 Ogawa Y, Lee CM, Nishiyama Y, Kim SH. Absence of sum frequency generation in support of orthorhombic symmetry of α-Chitin. Macromolecules. 2016;49(18):7025–31.
38 Chae I, Zu R, Barhoumi Meddeb A, Ogawa Y, Chen Z, Gopalan V, et al. Electric field-induced polarization responses of noncentrosymmetric crystalline biopolymers in different frequency regimes - a case study on unidirectionally aligned beta-chitin crystals. Biomacromolecules. 2021;22(5):1901–9.

39 Mizutani G, Koyama T, Tomizawa S, Sano H. Distinction between some saccharides in scattered optical sum frequency intensity images. Spectrochimica Acta Part A: Molecular and Biomolecular Spectroscopy. 2005;62(4–5):845–9.

40 Kong L, Lee C, Kim SH, Ziegler GR. Characterization of starch polymorphic structures using vibrational sum frequency generation spectroscopy. The Journal of Physical Chemistry B. 2014;118(7):1775–83.

41 Rocha-Mendoza I, Yankelevich DR, Wang M, Reiser KM, Frank CW, Knoesen A. Sum frequency vibrational spectroscopy: the molecular origins of the optical second-order nonlinearity of collagen. Biophysical Journal. 2007;93(12):4433–44.

42 Han Y, Hsu J, Ge NH, Potma EO. Polarization-sensitive sum-frequency generation microscopy of collagen fibers. The Journal of Physical Chemistry B. 2015;119(8):3356–65.

43 Nguyen KT, King JT, Chen Z. Orientation determination of interfacial β-sheet structures in situ. The Journal of Physical Chemistry B. 2010;114(25):8291–300.

44 Li X, Deng G, Ma L, Lu X. Interchain overlap affects formation of silk fibroin secondary structures on hydrophobic polystyrene surface detected via achiral/chiral sum frequency generation. Langmuir. 2018;34(32):9453–9.

45 Fu L, Wang Z, Psciuk BT, Xiao D, Batista VS, Yan ECY. Characterization of parallel β-sheets at interfaces by chiral sum frequency generation spectroscopy. The Journal of Physical Chemistry Letters. 2015;6(8):1310–5.

46 Tan J, Zhang J, Luo Y, Ye S. Misfolding of a human islet amyloid polypeptide at the lipid membrane populates through β-sheet conformers without involving α-helical intermediates. Journal of the American Chemical Society. 2019;141(5):1941–8.

47 Makarem M, Lee CM, Kafle K, Huang S, Chae I, Yang H, et al. Probing cellulose structures with vibrational spectroscopy. Cellulose. 2019;26(1):35–79.

48 Baudrier-Raybaut M, Haïdar R, Kupecek P, Lemasson P, Rosencher E. Random quasi-phase-matching in bulk polycrystalline isotropic nonlinear materials. Nature. 2004;432(7015):374–6.

49 Huang S, Kiemle SN, Makarem M, Kim SH. Correlation between crystalline cellulose structure and cellulose synthase complex shape: a spectroscopic study with unicellular freshwater alga micrasterias. Cellulose. 2020;27(1):57–69.

50 Barnette AL, Lee C, Bradley LC, Schreiner EP, Park YB, Shin H, et al. Quantification of crystalline cellulose in lignocellulosic biomass using sum frequency generation (SFG) vibration spectroscopy and comparison with other analytical methods. Carbohydrate Polymers. 2012;89(3):802–9.

51 Lee C, Dazen K, Kafle K, Moore A, Johnson DK, Park S, et al. Correlations of apparent cellulose crystallinity determined by XRD, NMR, IR, Raman, and SFG methods. In: Rojas OJ, editor. Cellulose Chemistry and Properties: Fibers, Nanocelluloses and Advanced Materials. Cham: Springer International Publishing; 2016. p. 115–31.

52 Ling Z, Wang T, Makarem M, Santiago Cintrón M, Cheng HN, Kang X, et al. Effects of ball milling on the structure of cotton cellulose. Cellulose. 2019;26(1):305–28.

53 Park YB, Lee CM, Koo B-W, Park S, Cosgrove DJ, Kim SH. Monitoring meso-scale ordering of cellulose in intact plant cell walls using sum frequency generation spectroscopy. Plant Physiology. 2013;163(2):907–13.

54 Lee CM, Mittal A, Barnette AL, Kafle K, Park YB, Shin H, et al. Cellulose polymorphism study with sum-frequency-generation (SFG) vibration spectroscopy: identification of exocyclic CH_2OH conformation and chain orientation. Cellulose. 2013;20(3):991–1000.

55 Ngo D, Liu H, Chen Z, Kaya H, Zimudzi TJ, Gin S, et al. Hydrogen bonding interactions of H_2O and SiOH on a boroaluminosilicate glass corroded in aqueous solution. npj Materials Degradation. 2020;4(1):1.

56 Kafle K, Greeson K, Lee C, Kim SH. Cellulose polymorphs and physical properties of cotton fabrics processed with commercial textile mills for mercerization and liquid ammonia treatments. Textile Research Journal. 2014;84(16):1692–9.

57 Lee CM, Chen X, Weiss PA, Jensen L, Kim SH. Quantum mechanical calculations of vibrational sum-frequency-generation (SFG) spectra of cellulose: dependence of the CH and OH peak intensity on the polarity of cellulose chains within the SFG coherence domain. The Journal of Physical Chemistry Letters. 2017;8(1):55–60.

58 Huang S, Makarem M, Kiemle SN, Hamedi H, Sau M, Cosgrove DJ, et al. Inhomogeneity of cellulose microfibril assembly in plant cell walls revealed with sum frequency generation microscopy. The Journal of Physical Chemistry B. 2018;122(19):5006–19.

59 Lee J, Choi J, Feng L, Yu J, Zheng Y, Zhang Q, et al. Regiospecific cellulose orientation and anisotropic mechanical property in plant cell walls. Biomacromolecules. 2023;24(11):4759–70.

60 Liu J, Inouye H, Venugopalan N, Fischetti RF, Gleber SC, Vogt S, et al. Tissue specific specialization of the nanoscale architecture of Arabidopsis. Journal of Structural Biology. 2013;184(2):103–14.

61 Choi J, Makarem M, Lee C, Lee J, Kiemle S, Cosgrove DJ, et al. Tissue-specific directionality of cellulose synthase complex movement inferred from cellulose microfibril polarity in secondary cell walls of Arabidopsis. Scientific Reports. 2023;13(1):22007.

62 Yamaguchi S, Otosu T. Progress in phase-sensitive sum frequency generation spectroscopy. Physical Chemistry Chemical Physics. 2021;23(34):18253–67.

63 Han Y, Raghunathan V, Feng R-r, Maekawa H, Chung C-Y, Feng Y, et al. Mapping molecular orientation with phase sensitive vibrationally resonant sum-frequency generation microscopy. The Journal of Physical Chemistry B. 2013;117(20):6149–56.

64 Shen YR, Ostroverkhov V. Sum-frequency vibrational spectroscopy on water interfaces: polar orientation of water molecules at interfaces. Chemical Reviews. 2006;106(4):1140–54.

65 Chae I, Ahmed S, Atitallah HB, Luo J, Wang Q, Ounaies Z, et al. Vibrational sum frequency generation (SFG) analysis of ferroelectric response of PVDF-based copolymer and terpolymer. Macromolecules. 2017;50(7):2838–44.

66 Boyd RW. Nonlinear Optics (Fourth Edition). Burlington: Academic Press; 2020.

8

Multivariate Data Analysis

We have studied the IR and SFG principles in Chapters 6 and 7 as an example of linear and nonlinear optical spectroscopy, respectively. In Chapters 2–4, we have learned the elemental analysis principles for XPS, AES, and SIMS analyses, respectively. Once those spectroscopic data are obtained, they are to be interpreted to get physical insights about the sample. When there is only one independent variable (e g., the concentration of one analyte) that affects one spectral feature (which is the dependent variable), then it is straight forward to see the correlation between those two variables. This is the *univariate analysis*, which can be done easily by using the "trendline" function in the Excel spreadsheet. In the spectroscopic analysis, the intensities of more than one peak are often recorded in each measurement. In the case of ToF-SIMS imaging, the entire mass spectrum is recorded at each pixel of the image, and we often need to analyze the full spectra for proper identification of many analyte species simultaneously. The same is true in hyperspectral imaging with vibrational spectroscopy coupled with microscopic techniques. It might be possible to select a few representative or characteristic variables for the data analysis; but, selecting those requires *a priori* knowledge about the system under study, which can be biased due to insufficient knowledge or user's intents. This kind of task can be done more reliably via *multivariate analysis*. It is a statistical method that provides an objective approach using all available information in a data set without needing manual identification and selection of key peaks and features; thus, it eliminates or significantly reduces the potential bias. Multivariate analysis is widely used in chemometrics that relates the measurement data of a chemical system to a certain state of the system via statistical algorithms.

Multivariate analysis employs simultaneous statistical processing of two or more variables in the data set, finding covariances between different variables. By describing the data using a reduced number of statistical variables, the interpretation of complex data sets involving a large number of dependent variables can be done quickly and efficiently. Simply, it can be said that the multivariate analysis starts with pattern recognition within the data set provided by the user. Many commercial software programs are available; for example, *Mathematica* and *MATLAB* also have built-in modules for multivariate analysis. However, unexperienced users may find ambiguities or confusions in utilizing the method or interpreting the results. This chapter will provide basic concepts and practical aspects of multivariate analysis.

This chapter begins with a brief review of univariate analysis and then covers the matrix representation of data and its relation to factor analysis. Matrix calculations involved in factor analysis are tedious to do by hand and become too complicated to do as the dimension of data set increases. So, they are done with math software. Moreover, there are many commercial software that can do various types of multivariate analysis. Without knowing the basic concepts and principles, using such programs is like dealing with a "black box" approach.

In this chapter, we will use some basic examples to build conceptual understanding of principal component analysis (PCA) and principal component regression (PCR). The goal of PCA is to extract factors, or "principal components" (PCs), that capture the largest amount of variance within the multi-dimensional data set. PCA is often the first step for data reduction prior to application of other methods of statistical analysis. PCR produces models that relate multiple sets of variables for quantitative predictions, which can be used for the quantification of unknown samples with the same analytes. There are many other multivariate analyses such as partial least squares (PLS) regression, multivariate curve resolution (MCR), discriminant function analysis (DFA), hierarchical cluster analysis (HCA), artificial neural network (ANN), and machine learning (ML). A good introduction to these methods can be found elsewhere [1]; the basic concept studied in this chapter will be the foundation to understand those review articles and books.

Surface and Interface Analysis: Principles and Applications, First Edition. Seong H. Kim.

Companion website: www.wiley.com/go/Kim

8.1 Least Squares Analysis

8.1.1 Single Component (Univariate) Least Squares Analysis

Figure 8.1 is the ATR-IR spectra of liquid mixtures of n-propanol and water collected with a Si-IRE (see Section 6.1). The alkyl stretch region shows three main peaks at 2875, 2935, and 2962 cm^{-1}. Small peaks identifiable in the 2850–3000 cm^{-1} region of the water spectrum are due to the gasket used to contain the liquid solution in the ATR-cell. Since both n-propanol and water have the OH group, the OH stretch region (3000–3700 cm^{-1}) has contributions from both analytes. In pure water, the OH stretch band extends down to 2700 cm^{-1} due to a broad distribution of hydrogen bond interactions. Thus, the OH stretch mode of water affects the background of the alkyl stretch region. The peak at 1640 cm^{-1} is the H_2O bending vibration. This peak is also affected by the long tail of the n-propanol spectrum extending from the low wavenumber side (black line). The sharp spiky features are due to small changes in the water vapor in the IR beam path during the experiment.

To find the correlation between the peak intensity in the spectrum and the analyte concentration in the solution, one can do the least squares fitting of individual peaks. In typical least squares analysis, we can choose a peak that is characteristic to each analyte and plot its intensity as a function of concentration. Ideally, the peak without any interference should be chosen; but there is no such peak in Figure 8.1. We can still choose the peak that is dominant by the contribution of one component, for example, 2875 cm^{-1} for n-propanol and 3400 cm^{-1} for water. The inset plots the intensity of these peaks vs. water mole fraction, along with the least squares regression (LSR) with a linear fit model, $y = ax + b$. The slope (a) and intercept (b) are calculated using the following equations:

$$a = \frac{\sum_i (x_i - \bar{x})(y_i - \bar{y})}{\sum_i (x_i - \bar{x})^2} \tag{8.1}$$

$$b = \bar{y} - a\bar{x} \tag{8.2}$$

where $\bar{x}$ and $\bar{y}$ are the mean of the analyte concentration (x_i) and the intensity (y_i), respectively. Here, the total variance of the data is $SST = \sum_i (y_i - \bar{y})^2$, the sum of squares explained by regression is $SSR = \sum_i \left(y_{i,fit} - \bar{y}\right)^2$, and the sum of squares

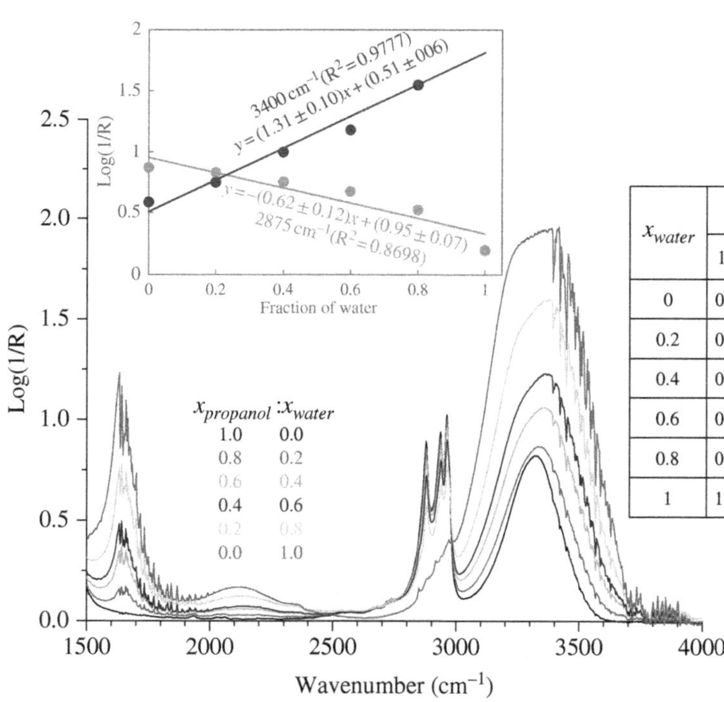

x_{water}	Wavenumber (cm^{-1})					
	1640	2875	2935	2962	3320	3400
0	0.036	0.870	0.935	1.024	0.823	0.585
0.2	0.143	0.828	0.900	1.007	0.862	0.749
0.4	0.337	0.753	0.844	0.957	1.036	0.998
0.6	0.442	0.673	0.777	0.887	1.192	1.180
0.8	0.714	0.523	0.645	0.741	1.554	1.550
1	1.140	0.196	0.310	0.380	1.927	1.901

Figure 8.1 ATR-IR spectra of liquid mixtures of n-propanol and water collected with a Si-IRE. The mole fraction of each component is shown in the figure. The peak intensities at five positions are shown in the table. The inset plots the 2875^{-1} and 3400 cm^{-1} intensities vs. the water fraction in the solution.

due to error is $SSE = \sum_i (y_i - y_{i,\,fit})^2$. Note that $SST = SSR + SSE$. We often check R-squared (R^2) value to evaluate the quality of fit, which is defined as $R^2 = \frac{SSR}{SST} = 1 - \frac{SSE}{SST}$. This value represents the proportion of the variance for a dependent variable that is explained by the independent variable in the regression model used. The calculated values are shown in the inset figure. It also shows the standard error of the fit parameters.

If we interpret this model in the context of the Beer–Lambert law, we can say that the slope is governed by the absorptivity difference of two analytes at the peak position chosen for analysis and the ATR-IR probe depth at the chosen wavenumber, and the intercept is affected by the interference between analytes. The R^2 value of the 3400 cm^{-1} fit is better than that of the 2875 cm^{-1} fit. If this is to be used as the calibration curve of n-propanol/water mixtures with unknown concentrations, one may decide to use the 3400 cm^{-1} fit to calculate x_{water} in the mixture and then the n-propanol concentration is obtained by $x_{propanol} = 1 - x_{water}$.

In a quick glance of the data in Figure 8.1, it appears that there is only one degree of freedom (i.e., concentration of one of the analytes). Then, the above approach should work. But we can see that, even though R^2 is larger than 0.97 (so, close to 1) for the 3400 cm^{-1} data set, the fit is not good. There is a nonlinearity in the data. There are many reasons that the correlation is not linear. One obvious reason is an experimental error. The more important scientific reason is because n-propanol and water mixtures are not an ideal solution. In other words, their molar volume does not change linearly with the composition, and the negative excess volume exists as a result of interactions of water with the nonpolar alkyl (hydrophobic) group and the polar hydroxyl (hydrophobic) group of the alcohol, which vary with concentration [2]. The VLE with the azeotrope point shown in Figure 7.15 is also a nonideal solution behavior. Knowing this, one may fit the experimental data with a nonlinear function to construct the calibration curve. In this example, we have *a priori* knowledge; but for unknown samples, there will be no such information that can justify the use of different regression model.

8.1.2 Multivariate Least Squares Analysis

In the previous example, the negative excess volume could be treated as another degree of freedom. As the degree of freedom increases, the number of peaks to be analyzed should be increased accordingly. If we have two degrees of freedom, we should process at least two peak intensities simultaneously. For that, we can use matrix algebra. For the 2875 cm^{-1} and 3400 cm^{-1} intensities of the data in Figure 8.1, we can write the following equations using the Beer–Lambert law:

$$I^{2875} = a_{PrOH}^{2875} d_p^{2875} c_{PrOH} + a_{water}^{2875} d_p^{2875} c_{water} \tag{8.3}$$

$$I^{3400} = a_{PrOH}^{3400} d_p^{3400} c_{PrOH} + a_{water}^{3400} d_p^{3400} c_{water} \tag{8.4}$$

This can be re-written in the following matrix form:

$$\begin{pmatrix} I^{2875} \\ I^{3400} \end{pmatrix} = \begin{pmatrix} a_{PrOH}^{2875} & a_{water}^{2875} \\ a_{PrOH}^{3400} & a_{water}^{3400} \end{pmatrix} \begin{pmatrix} d_p^{2875} \\ d_p^{3400} \end{pmatrix} \begin{pmatrix} c_{PrOH} \\ c_{water} \end{pmatrix} \tag{8.5}$$

Here, the left-hand side is the intensity matrix, which will be called $\boldsymbol{A}$ hereafter. The first, second, and third terms on the right-hand side is the absorptivity matrix (called $\boldsymbol{E}$), the path length matrix (called $\boldsymbol{L}$), and the concentration matrix (called $\boldsymbol{C}$), respectively. In most absorbance measurement experiments, the pathlength is typically constant. In our example, it is determined by the evanescent wave penetration from the Si-IRE into the liquid medium (see Figure 5.15). It varies with the solution concentration because the refractive index at 2875 cm^{-1} and 3400 cm^{-1} is not constant (because $n_2(\varpi)$ in eq. (6.12) varies with the solution concentration); thus, we just found that there is another degree of freedom that we should consider. But for simplicity, let us assume that it does not vary significantly. Then, the product of the absorptivity and path length matrices can be replaced with one matrix, $\boldsymbol{EL} = \boldsymbol{K}$, which can be called calibration coefficient matrix. This gives:

$$\boldsymbol{A} = \boldsymbol{KC} \tag{8.6}$$

We can solve this equation to get the $\boldsymbol{K}$ matrix. We first multiply the transpose matrix of $\boldsymbol{C}$ to both sides of equation (to make the matrix to be converted into a square form) and then multiply the inverse of $\boldsymbol{CC^T}$; this gives the solution in the following form:

$$\boldsymbol{K} = \boldsymbol{AC^T}\left(\boldsymbol{CC^T}\right)^{-1} \tag{8.7}$$

This is the matrix definition of the multivariate least squares regression. Once we find $\boldsymbol{K}$, it can be used to calculate concentrations of analytes from the intensity matrix of the unknown samples ($\boldsymbol{A}_{ukn}$) using the following equation:

$$\boldsymbol{C}_{ukn} = \left(\boldsymbol{K}^T\boldsymbol{K}\right)^{-1}\boldsymbol{K}^T\boldsymbol{A}_{ukn} \tag{8.8}$$

Here the $\boldsymbol{A}_{ukn}$ matrix can have any arbitrary number of samples (i.e., $2 \times n$ matrix) as long as the peak intensities are measured at the same wavenumbers used for the calibration.

Let us apply this method to the previous example. Since we have two peaks (2875 cm^{-1} and 3400 cm^{-1}) and six concentrations for two analytes, $\boldsymbol{A}$ and $\boldsymbol{C}$ are 2×6 matrices. The absorbance and concentration matrices are written in the following form:

$$\boldsymbol{A} = \begin{pmatrix} 0.870 & 0.828 & 0.753 & 0.673 & 0.523 & 0.196 \\ 0.585 & 0.749 & 0.998 & 1.180 & 1.550 & 1.901 \end{pmatrix} \tag{8.9}$$

$$\boldsymbol{C} = \begin{pmatrix} 1.0 & 0.8 & 0.6 & 0.4 & 0.2 & 0.0 \\ 0.0 & 0.2 & 0.4 & 0.6 & 0.8 & 1.0 \end{pmatrix} \tag{8.10}$$

Then, using eq. (8.7), we find the $\boldsymbol{K}$ matrix to be:

$$\boldsymbol{K} = \begin{pmatrix} 0.9523 & 0.3287 \\ 0.5059 & 1.8151 \end{pmatrix} \tag{8.11}$$

To see how well this two-variate regression works, we can use the same absorbance matrix used for calibration as the absorbance matrix of an unknown sample set with six different concentrations (i.e., $\boldsymbol{A}_{ukn}$ = eq. (8.9)) and calculate the concentration of this unknown sample set ($\boldsymbol{C}_{ukn}$) using the $\boldsymbol{K}$ matrix shown in eq. (8.11). Following eq. (8.8), we get:

$$\boldsymbol{C}_{ukn} = \begin{pmatrix} 0.888 & 0.804 & 0.665 & 0.534 & 0.282 & -0.172 \\ 0.075 & 0.188 & 0.365 & 0.502 & 0.775 & 1.10 \end{pmatrix} \tag{8.12}$$

In an ideal case, we expect to get the answer to be the same as the $\boldsymbol{C}$ matrix shown in eq. (8.10). But, the result is quite different from what we expected. We even get a large negative concentration, which physically is not possible.

Matching the number of peaks to be processed to the degree of freedom is only the minimum requirement. We can use more peaks than the minimum number of peaks required. This can decrease the experimental uncertainty because it can cancel out random noises in specific signals. Also, some peaks may follow the Beer–Lambert law more closely than others. In the above example, there are more than two degrees of freedom (i.e., analyte concentration, concentration dependence of excess volume of mixture, and concentration dependence of probe depth). Considering this, one can try using five peaks – three peaks (2875, 2935, and 2962 cm^{-1}) in the alkyl stretch region and two peaks (3320 and 3400 cm^{-1}) in the OH stretch region. The 1640 cm^{-1} peak in the H_2O bending region can also be included and more peaks can be chosen in the OH stretch region; but it does not improve the result significantly. So, let us just look at the multivariate least squares fit using five peaks. One can construct the $\boldsymbol{A}$ matrix using the table in Figure 8.1. Then, it is written as:

$$\boldsymbol{A} = \begin{pmatrix} 0.870 & 0.828 & 0.753 & 0.673 & 0.523 & 0.196 \\ 0.935 & 0.900 & 0.844 & 0.777 & 0.645 & 0.310 \\ 1.024 & 1.007 & 0.957 & 0.887 & 0.741 & 0.380 \\ 0.823 & 0.862 & 1.036 & 1.192 & 1.554 & 1.927 \\ 0.585 & 0.749 & 0.998 & 1.180 & 1.550 & 1.901 \end{pmatrix} \tag{8.13}$$

The calibration coefficient matrix calculated using eq. (8.7) is as follows:

$$\boldsymbol{K} = \begin{pmatrix} 0.9523 & 0.3287 \\ 1.0178 & 0.4525 \\ 1.1247 & 0.5407 \\ 0.6786 & 1.7861 \\ 0.5059 & 1.8151 \end{pmatrix} \tag{8.14}$$

Again, we use eq. (8.13) as $\boldsymbol{A_{ukn}}$ and calculate $\boldsymbol{C_{ukn}}$ using eq. (8.8) with the $\boldsymbol{K}$ matrix in eq. (8.14), we get the following answer:

$$\boldsymbol{C_{ukn}} = \begin{pmatrix} 0.873 & 0.804 & 0.677 & 0.549 & 0.288 & -0.192 \\ 0.103 & 0.183 & 0.343 & 0.479 & 0.767 & 1.12 \end{pmatrix} \tag{8.15}$$

We see that even if we use more peaks than what we think is sufficient, the accuracy of the calibration does not increase. This is because the drift in baseline or intensity (for whatever reasons) cannot be fully modeled in the least squares regression. Moreover, if there are impurities or interferents whose concentrations do not appear in any of the calibration equations (i.e., beyond *a prior* knowledge), their effects cannot be modeled in regression. These are limitations of the least squares regression. These limitations can be circumvented or overcome by doing PCA and PCR.

8.2 Factor Analysis

8.2.1 Finding Factors (Principal Components)

Factor analysis is a statistical approach based on linear algebra and matrix theory, which analyzes correlations among a large number of variables in a data set and explains these variables in terms of a smaller number of common underlying dimensions. In other words, it involves finding a way of condensing the information contained in the original variables into a smaller set of implicit variables that describe variances in orthogonal dimensions that are called "factors" or "principal components (PCs)".

For example, let us consider the 3400 cm^{-1} intensity graph in the inset of Figure 8.1. The data are plotted two orthogonal axes – x axis for concentration and y axis for peak intensity. But the largest variance in the data is along the trend line. So, the first factor (principal component 1, PC-1) is the unit vector along the trend line. The variance among the data set can be described with a set of "scores" that is multiplied to this unit vector. Then, the line orthogonal to this trendline becomes the second factor (PC-2). The shortest distance of each data point to the trendline (i.e., deviation from PC-1) is the score along this PC-2 axis.

The factor analysis starts from identifying and quantifying variances among the correlated data. We are familiar with the concept of variance (σ^2) which explains the spread of the data. The square root of the variance is the standard deviation (σ). For one-dimensional (1D) data set along the x-axis, the variance is defined as:

$$\sigma_x^2 = \frac{1}{N-1}\sum_{i=1}^{N}(x_i - \overline{x})^2 = \sigma(x,x) \tag{8.16}$$

Let us consider two-dimensional (2D) hypothetical data shown in Figure 8.2. We can calculate variance in the x- and y-directions separately, that is, $\sigma(x, x)$ and $\sigma(y, y)$. However, these variances do not explain the diagonal correlation. This mutual correlation can be captured by extending the notion of variance to the "covariance" of the data, which is defined as:

$$\sigma(x,y) = \frac{1}{N-1}\sum_{i=1}^{N}(x_i - \overline{x})(y_i - \overline{y}) \tag{8.17}$$

For 2D data, total four variances can be obtained, which can be arranged in a matrix form, called the covariance matrix:

$$\boldsymbol{\Sigma} = \begin{bmatrix} \sigma(x,x) & \sigma(x,y) \\ \sigma(y,x) & \sigma(y,y) \end{bmatrix} \tag{8.18}$$

The covariance matrix defines both the spread (variances on its diagonal elements) and orientation (covariances on its off-diagonal elements) of the data. If x is positively correlated with y, y is also positively correlated with x; that is, $\sigma(x, y) = \sigma(y, x)$ and both are positive. If the data are negatively correlated, then both are equal and negative. If there is no correlation, then $\sigma(x, y) = \sigma(y, x) = 0$.

The covariance matrix can be represented with the vectors that point into the directions of the largest spreads of the data, and whose magnitudes equal the spreads (variances) in those directions:

$$\boldsymbol{\Sigma}\hat{\boldsymbol{v}} = \xi\hat{\boldsymbol{v}} \tag{8.19}$$

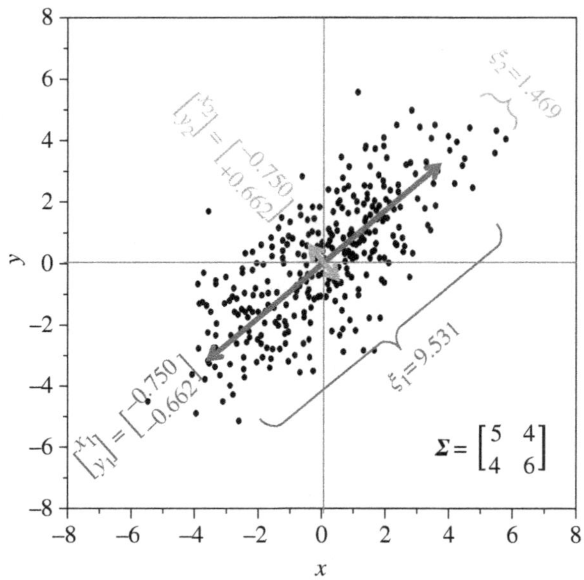

Figure 8.2 Hypothetical 2D data set with corresponding covariance matrix.

Here, $\hat{\boldsymbol{v}}$ and ξ are called eigenvectors and eigenvalues, respectively. We need to find the nullspace of the covariance matrix that defines the set of all n-dimensional unit vectors:

$$(\boldsymbol{\Sigma} - \xi \boldsymbol{I})\hat{\boldsymbol{v}} = \hat{\boldsymbol{0}} \tag{8.20}$$

Here, $\boldsymbol{I}$ is the identity matrix. This can be done by solving for determinant $det(\boldsymbol{\Sigma} - \lambda \boldsymbol{I}) = 0$ to find ξ and then, for each ξ, determine $\hat{\boldsymbol{v}}$ whose magnitude is unity (=1).

Let us practice using the hypothetical data set in Figure 8.2 with $\boldsymbol{\Sigma} = \begin{bmatrix} 5 & 4 \\ 4 & 6 \end{bmatrix}$. The following steps show how to solve $det(\boldsymbol{\Sigma} - \lambda \boldsymbol{I}) = 0$:

$$\begin{bmatrix} 5-\xi & 4 \\ 4 & 6-\xi \end{bmatrix} = (5-\xi)(6-\xi) - 4\cdot 4 = 14 - 11\xi + \xi^2 = 0 \tag{8.21}$$

From this, we get $\xi_1 = 9.531$ and $\xi_2 = 1.469$. For the first eigenvalue (larger one), we find the eigenvector $\hat{\boldsymbol{v}}_1 = \begin{bmatrix} x_1 \\ y_1 \end{bmatrix}$ through the following steps:

$$\begin{bmatrix} 5-9.531 & 4 \\ 4 & 6-9.531 \end{bmatrix} \cdot \begin{bmatrix} x_1 \\ y_1 \end{bmatrix} = \begin{bmatrix} 0 \\ 0 \end{bmatrix} \tag{8.22}$$

$$\begin{cases} -4.531x_1 + 4y_1 = 0 \\ 4x_1 - 3.531y_1 = 0 \end{cases} \tag{8.23}$$

From this, we get:

$$\begin{cases} -4.531x_1 + 4y_1 = 4x_1 - 3.531y_1 \\ x_1^2 + y_1^2 = 1 \end{cases} \tag{8.24}$$

The second condition in eq. (8.24) comes from the requirement that the magnitude of $\hat{\boldsymbol{v}}_1$ should be unity. Solving eq. (8.24), we find the first eigenvector:

$$\hat{\boldsymbol{v}}_1 = \begin{bmatrix} x_1 \\ y_1 \end{bmatrix} = \begin{bmatrix} -0.662 \\ -0.775 \end{bmatrix} \text{ or } \begin{bmatrix} +0.662 \\ +0.750 \end{bmatrix} \tag{8.25}$$

We can do the same for the second eigenvalue, $\xi_2 = 1.469$, and find the second eigenvector:

$$\hat{\boldsymbol{v}}_2 = \begin{bmatrix} x_2 \\ y_2 \end{bmatrix} = \begin{bmatrix} -0.750 \\ +0.775 \end{bmatrix} \text{ or } \begin{bmatrix} +0.750 \\ +0.662 \end{bmatrix} \tag{8.26}$$

These unit vectors are the factors ($\hat{\boldsymbol{v}}_1$ = PC1 and $\hat{\boldsymbol{v}}_2$ = PC2) for the data with $\boldsymbol{\Sigma} = \begin{bmatrix} 5 & 4 \\ 4 & 6 \end{bmatrix}$. The eigenvalue value of each vector is the spread along that PC.

The vectors corresponding to eigenvalue times eigenvector ($\xi_1\hat{\boldsymbol{v}}_1$ and $\xi_2\hat{\boldsymbol{v}}_2$) are marked with long and short arrows in Figure 8.2. If we rotate the principal axes of the graph from (x, y) to $(\hat{\boldsymbol{v}}_1, \hat{\boldsymbol{v}}_2)$, ξ_1 is the variance of the data along the $\hat{\boldsymbol{v}}_1$ axis, $\sigma(\hat{\boldsymbol{v}}_1, \hat{\boldsymbol{v}}_1)$; and ξ_2 is the variance of the data along the $\hat{\boldsymbol{v}}_2$ axis, $\sigma(\hat{\boldsymbol{v}}_2, \hat{\boldsymbol{v}}_2)$. By doing this, the covariance between the data points described along the $\hat{\boldsymbol{v}}_1$ axis and the data described along the $\hat{\boldsymbol{v}}_2$ axis becomes zero, that is, $\sigma(\hat{\boldsymbol{v}}_1, \hat{\boldsymbol{v}}_2) = 0$, because these two axes are orthogonal to each other. In other words, all variances among the data set can be described with these two variances. Note that there are three variances when the data are described with the original (x, y) axes (four, but two are identical as shown in eq. (8.18)); but the variances are reduced to two when the data are described along the eigenvectors. For this reason, factor analysis is viewed as a data reduction technique, that is, describing the data set with a smaller set of implicit variables.

8.2.2 Regression Using Principal Components

In the least squares regression (LSR), we find the relationship between the original independent variables (x_i) and the dependent variables (y_i) at each chosen peak position (k): $y_i = a_k x_i + b_k$. Then, we can use this calibration equation to calculate the independent variables of unknowns from the measured values at each peak position: $x_{unk} = (y_{unk} - b_k)/a_k$. Doing this step for one peak at a time is the univariate regression (Section 8.1.1), and doing the same step simultaneously for multiple peaks is the multivariate regression (Section 8.1.2). Although this is mathematically simple and straightforward, we have witnessed severe limitations when we try to process the data in Figure 8.1 in this way.

We can do the same regression using the factors (principal components). In this case, we first fine the covariance patterns among dependent variables (y_i) of all peaks of our interest. We can choose as many peaks as we want. In fact, all data points in the spectrum can be used. From this, we find all factors ($\boldsymbol{F}$) that we need to describe the variances of the data set sufficiently. Then, we represent all dependent variables with a set of values along each factor dimension. These values are called scores ($\boldsymbol{S}$). This is *principal component analysis* (PCA).

The factors and scores can be organized in the matrix form. Then, we can find the relationship between the independent variables (x_i) and the scores (S). This matrix is the calibration coefficient matrix, which will be called $\boldsymbol{B}$ hereafter. As in the linear regression case, we can use the $\boldsymbol{F}$ and $\boldsymbol{B}$ matrices to calculate the independent variables of unknowns from the measured values. This is *principal component regression* (PCR).

8.3 Matrix Algebra for Principal Component Analysis (PCA) and Principal Component Regression (PCR)

8.3.1 Mathematical Steps of PCA and PCR Process

Let us say n is the number of standard samples used, and p is the number of peak positions to use for regression. In the example shown in Figure 8.1, n is 6 and p is 5 (without including 1640 cm^{-1}). We will try to describe the variances in the measured data (all dependent variables) with factors in a matrix form:

$$\boldsymbol{A} = \boldsymbol{S}\boldsymbol{F} \tag{8.27}$$

where $\boldsymbol{A}$ is the matrix of peak intensities of order $n \times p$, $\boldsymbol{S}$ is the matrix of scores of order $n \times f$, and $\boldsymbol{F}$ is the matrix of factors of order $f \times p$. Note that the order of $\boldsymbol{A}$ in eq. (8.27) is different from the least squares regression case; in eqs. (8.9) and (8.13), $\boldsymbol{A}$ is the $p \times n$ matrix. To completely reproduce the entire set of standard spectra (calibration date set), we would need to find factors that span all the variance in the data.

Our primary interest is to know how the changes in component concentration (independent variable) cause the spectra (dependent variable) to change. First, find the factor (eigenvector of covariance) with the largest eigenvalue (i.e., the maximum possible variance) in the set of standard spectra $\boldsymbol{A}$. This is the step we have studied in Section 8.2.1. Next, the amount of factor in each standard spectrum is calculated, which is its score. This can be done by multiplying $\boldsymbol{A}$ with the transpose of factor matrix and then multiplying the inverse of the squared matrix of factor (the same as in eq. 8.7):

$$\boldsymbol{S} = \boldsymbol{AF}^T \left(\boldsymbol{FF}^T\right)^{-1} \tag{8.28}$$

Next, calculate residuals in the matrix form:

$$\boldsymbol{R} = \boldsymbol{A} - \boldsymbol{SF} \tag{8.29}$$

This step strips the variance spanned by the first factor (PC1) from the data set, leaving behind the amount of variances that are not described by PC1. Go back and repeat eq. (8.27) using $\boldsymbol{R}$ matrix, instead of $\boldsymbol{A}$ matrix. Then, we find the second factor (PC2) with the second largest variance. Use the PC1 and PC2 together in eq. (8.28) to find a set of scores (S1 and S2) along the PC1 and PC2 directions. Then, we can calculate residuals again using eq. (8.29). These are the variances that are not described by PC1 and PC2. We can repeat these processes iteratively until the values in $\boldsymbol{R}$ matrix become negligible (ideally, all zero). Once we reach this point, we have found factor matrix $\boldsymbol{F}$ and score matrix $\boldsymbol{S}$ that fully describe all variances in the given calibration data set. So, the PCA step is completed.

The number of this iteration determines the number of factors, f, which is the number of rows in $\boldsymbol{F}$ matrix. If the degree of freedom is two in a given system, then there will be two factors (so, two rows in $\boldsymbol{F}$ matrix). In $\boldsymbol{F}$, each row has p number of columns. If the spectral intensity is measured at five wavenumbers, then each row will have five columns.

To find the relationships between the dependent variables and the independent variable, we need to find the calibration coefficient matrix $\boldsymbol{B}$ that correlates the scores of factors (which is organized in score matrix $\boldsymbol{S}$) and the independent variable matrix $\boldsymbol{C}$ (which is the concentration matrix in the example shown in Figure 8.1). That relationship can be written as:

$$\boldsymbol{C} = \boldsymbol{BS}^T \tag{8.30}$$

Here, $\boldsymbol{C}$ has an order $m \times n$ where m is the number of analytes in the system, $\boldsymbol{B}$ has an order $m \times f$, and $\boldsymbol{S}^T$ has an order $f \times n$. Each row of $\boldsymbol{B}$ contains the coefficients for a specific analyte and each column of $\boldsymbol{B}$ contains the coefficients for a specific factor. We can solve this equation for $\boldsymbol{B}$ in the same way we did in eq. (8.28):

$$\boldsymbol{B} = \boldsymbol{CS}^{TT} \left(\boldsymbol{S}^T \boldsymbol{S}^{TT}\right)^{-1} \tag{8.31}$$

Since $\boldsymbol{S}^{TT} = \boldsymbol{S}$, we get:

$$\boldsymbol{B} = \boldsymbol{CS}\left(\boldsymbol{S}^T \boldsymbol{S}\right)^{-1} \tag{8.32}$$

The final step is to obtain the prediction equation using $\boldsymbol{F}$, $\boldsymbol{S}$, and $\boldsymbol{B}$ matrices. In eq. (8.28), $\boldsymbol{FF}^T = \boldsymbol{I}$ because the factors in factor matrix are orthogonal. Thus, eq. (8.28) becomes $\boldsymbol{S} = \boldsymbol{AF}^T$. Then, $\boldsymbol{S}^T$ in eq. (8.30) is $\boldsymbol{S}^T = (\boldsymbol{AF}^T)^T = \boldsymbol{FA}^T$. From eq. (8.30), we get the final equation that can be used to find the analyte concentrations from the measured values:

$$\boldsymbol{C}_{unk} = \boldsymbol{BFA}^T \tag{8.33}$$

This is the PCR process. This step is basically the same as inverse least squares regression; the main difference from the classical least squares regression discussed in Section 8.1.2 is that, instead of using the raw measured data, PCR uses the scores along the PC directions for regression. When an unknown spectrum is measured, its intensities are organized in $\boldsymbol{A}^T$ matrix. Since we know $\boldsymbol{B}$ and $\boldsymbol{F}$ from the calibration, we can predict the unknown concentration $\boldsymbol{C}_{unk}$ by simply multiplying these matrices together in the order described in eq. (8.33).

8.3.2 Conceptual Understanding of PCA and PCR Principles

Before we apply the PCA and PCR methods to complicated real problems (for example, Figure 8.1), let us run some practices with simple examples to understand the physical meaning of $\boldsymbol{F}$, $\boldsymbol{S}$, and $\boldsymbol{B}$. There are many commercial programs that can do various types of multivariate analysis including PCA and PCR. Just using those programs following the written procedure can give an answer quickly, but it would be difficult to fully understand the principles behind all those built-in functions.

Table 8.1 is a hypothetical transmission absorption data set obtained with six calibration samples. This example is so simple that one can do least squares regression easily. Comparing the PCA/PCR procedures and results with the LSR method is helpful to build a physical sense of factors and scores. In this practice, *Mathematica* is used; but other math software can be used to do the same.

Table 8.1 Transmission absorbance data of a set of s x concentrations.

Concentration	1	2	3	4	5	6
Abs-1	3	6	9	12	15	18
Abs-2	6	9	12	15	18	21

Case 1: Let us use Abs-1 data only. It looks like there is no experimental error and the data fits perfectly with $y = 3x$. The concentration and absorbance matrices of the Abs-1 set are as follows:

$$\boldsymbol{C} = (1\ \ 2\ \ 3\ \ 4\ \ 5\ \ 6) \tag{8.34}$$

$$\boldsymbol{A1} = \begin{pmatrix} 3 \\ 6 \\ 9 \\ 12 \\ 15 \\ 18 \end{pmatrix} \tag{8.35}$$

We calculate the covariance of **A1** using *Mathematica*:

$$\boldsymbol{CovA} = \text{Covariance}[\boldsymbol{A1}] = 31.5 \tag{8.36}$$

Here, the Covariance[] is the syntax used in *Mathematica* to calculate covariance of argument inside the square brackets, that is, doing eq. (8.17). In the subsequent calculations, *Mathematica* syntaxes like this one will appear. Since there is only one data set in the dependent variable, this is the same as the variance of the measured Abs-1 values from its average that is calculated with eq. (8.16). By doing eq. (8.20), we get the eigenvalue of **CovA1**:

$$\boldsymbol{EV1} = \text{Eigenvalues}[\boldsymbol{CovA1}] = 31.5 \tag{8.37}$$

Again, the value is the same as the variance of **A1** since there is only one data set. The eigenvector of **CovA1** is found to be:

$$\boldsymbol{F1} = \text{Eigenvectors}[\boldsymbol{CovA1}, 1] = 1 \tag{8.38}$$

Since there is only one data set varying linearly with the concentration, there is only one degree of freedom. If we force *Mathematica* to find two eigenvectors by typing "Eigenvectors[**CovA1**, 2]," it gives an error message stating that "*cannot take eigenvectors 1 through 2 out of the total of 1 eigenvectors.*" So, **F1** is a 1×1 matrix. The magnitude of this factor is 1, since it is a unit vector along the variance direction. To find the scores for the data with respect to this principal component, we use eq. (8.28) and get the following result:

$$\boldsymbol{S1} = \boldsymbol{A1} \cdot \text{Transpose}[\boldsymbol{F1}] = \begin{pmatrix} 3 \\ 6 \\ 9 \\ 12 \\ 15 \\ 18 \end{pmatrix} \tag{8.39}$$

Here we can see that all information of the measured data is now in the score matrix. If we calculate the residual matrix using eq. (8.29), we see that all elements in **R1** are zero, which means that **F1** and **S1** describe all the variance in the measured data set:

$$\boldsymbol{R1} = \boldsymbol{A1} - \boldsymbol{S1} \cdot \boldsymbol{F1} = \begin{pmatrix} 0 \\ 0 \\ 0 \\ 0 \\ 0 \\ 0 \end{pmatrix} \tag{8.40}$$

We then calculate the calibration coefficient matrix ***B1*** using eq. (8.32):

$$\boldsymbol{B1} = \boldsymbol{C} \cdot \boldsymbol{S1} \cdot \text{Inverse}[\text{Transpose}[\boldsymbol{S1}] \cdot \boldsymbol{S1}] = 0.3333 \quad (8.41)$$

We see that this is equal to the inverse of the slope that we find in the linear regression of the Abs-1 calibration data. If we use ***A1*** as $\boldsymbol{A1_{unk}}$ and calculate $\boldsymbol{C1_{unk}}$ using eq. (8.33), we get the result that we expect:

$$\boldsymbol{C1_{unk}} = \boldsymbol{B1} \cdot \boldsymbol{F1} \cdot \text{Transpose}[\boldsymbol{A1_{unk}}] = \begin{pmatrix} 1 & 2 & 3 & 4 & 5 & 6 \end{pmatrix} \quad (8.42)$$

Case 2: Now, let us use the Abs-2 data only. We can see the relationship between the dependent and independent variables is $y = 3x + 3$. Here the dependent variable matrix is:

$$\boldsymbol{A2} = \begin{pmatrix} 6 \\ 9 \\ 12 \\ 15 \\ 18 \\ 20 \end{pmatrix} \quad (8.43)$$

If we run covariance calculation, we get:

$$\boldsymbol{CovA2} = \text{Covariance}[\boldsymbol{A2}] = 31.5 \quad (8.44)$$

Still, this is the same as the variance of the Abs-1 values because the only difference between Abs-1 and Abs-2 is the average value. Thus, the eigenvalue and eigenvector of ***CovA2*** are the same as Case 1:

$$\boldsymbol{EV2} = \text{Eigenvalues}[\boldsymbol{CovA2}] = 31.5 \quad (8.45)$$

$$\boldsymbol{F2} = \text{Eigenvectors}[\boldsymbol{CovA2}, 1] = 1 \quad (8.46)$$

Again, all data information is reflected in the score matrix and the residual matrix becomes zero:

$$\boldsymbol{S2} = \boldsymbol{A2} \cdot \text{Transpose}[\boldsymbol{F2}] = \begin{pmatrix} 6 \\ 9 \\ 12 \\ 15 \\ 18 \\ 21 \end{pmatrix} \quad (8.47)$$

$$\boldsymbol{R2} = \boldsymbol{A2} - \boldsymbol{S2} \cdot \boldsymbol{F2} = \begin{pmatrix} 0 \\ 0 \\ 0 \\ 0 \\ 0 \\ 0 \end{pmatrix} \quad (8.48)$$

When we calculate the calibration coefficient matrix ***B2***, we now see that the result is different from Case 1:

$$\boldsymbol{B2} = \boldsymbol{C} \cdot \boldsymbol{S2} \cdot \text{Inverse}[\text{Transpose}[\boldsymbol{S2}] \cdot \boldsymbol{S2}] = 0.2686 \quad (8.49)$$

The ***B2*** value is not the same as the inverse of the linear regression slope. If we use ***A2*** as $\boldsymbol{A2_{unk}}$ and calculate $\boldsymbol{C2_{unk}}$ using eq. (8.33), we get the result that is not accurate:

$$\boldsymbol{C2_{unk}} = \boldsymbol{B2} \cdot \boldsymbol{F2} \cdot \text{Transpose}[\boldsymbol{A2_{unk}}] = \begin{pmatrix} 1.61 & 2.42 & 3.22 & 4.03 & 4.83 & 5.64 \end{pmatrix} \quad (8.50)$$

In this example, the absorbance is perfectly proportional to the concentration (with $R^2 = 1$), but there is an offset in the baseline (probably due to a poor calibration of the zero point of the instrument at that wavelength). So, two variants (slope and intercept) are needed to describe the entire date set. This is *a priori* knowledge that we have; but factor analysis algorithm does not have it. So, the covariance calculation finds only one variance (linearity in the measured data). The baseline offset is not a part of variance because it does not vary at all through the entire data set of Abs-2.

Case 3: We can use as many data sets as possible in factor analysis. Let us use both Abs-1 and Abs-2 data for PCA/PCR. Then, the dependent variable matrix for the entire data in Table 8.1 becomes:

$$\boldsymbol{A3} = \begin{pmatrix} 3 & 6 \\ 6 & 9 \\ 9 & 12 \\ 12 & 15 \\ 15 & 18 \\ 18 & 20 \end{pmatrix} \tag{8.51}$$

If we run covariance calculation, we get:

$$\boldsymbol{CovA3} = \text{Covariance}[\boldsymbol{A3}] = \begin{pmatrix} 31.5 & 31.5 \\ 31.5 & 31.5 \end{pmatrix} \tag{8.52}$$

The diagonal elements are the same as ***CovA1*** and ***CovA2***, and the off-diagonal elements are symmetric and positive (because there is a positive correlation among two sets of data). The eigenvalue and eigenvector of ***CovA3*** are found to be:

$$\boldsymbol{EV3} = \text{Eigenvalues}[\boldsymbol{CovA3}] = \begin{pmatrix} 63 \\ 0 \end{pmatrix} \tag{8.53}$$

$$\boldsymbol{F3} = \text{Eigenvectors}[\boldsymbol{CovA3}, 2] = \begin{pmatrix} 0.7071 & 0.7071 \\ -0.7071 & 0.7071 \end{pmatrix} \tag{8.54}$$

Now, ***F3*** becomes a 2 × 2 matrix. One factor is PC1 = (0.7071, 0.7071) with a variance of 63, which describes the linearity in both Abs-1 and Abs-2 data sets. The other factor is PC2 = (−0.7071, 0.7071) with a variance of 0, which describes the constant difference between two measurements (i.e., baseline offset). PC1 and PC2 are orthogonal to each other. The score matrix calculated with these PC1 and PC2 factors describes the variances within the two data along the PC1 and PC2 directions:

$$\boldsymbol{S3} = \boldsymbol{A3} \cdot \text{Transpose}[\boldsymbol{F3}] = \begin{pmatrix} 6.364 & 2.121 \\ 10.607 & 2.121 \\ 14.849 & 2.121 \\ 19.092 & 2.121 \\ 23.335 & 2.121 \\ 27.577 & 2.121 \end{pmatrix} \tag{8.55}$$

The score matrix has two numbers in each row. Each row corresponds to each calibration sample, and two numbers in each row corresponds positions of Abs-1 and Abs-2 data of that specific calibration sample along the PC1 and PC2 axes. The residual matrix is not perfectly zero, but the values are essentially zero:

$$\boldsymbol{R3} = \boldsymbol{A3} - \boldsymbol{S3} \cdot \boldsymbol{F3} = \begin{pmatrix} -8.9 \times 10^{-16} & -8.9 \times 10^{-16} \\ 0 & -1.8 \times 10^{-15} \\ -1.8 \times 10^{-15} & -1.8 \times 10^{-15} \\ 0 & 0 \\ 0 & 0 \\ -3.5 \times 10^{-15} & -3.5 \times 10^{-15} \end{pmatrix} \tag{8.56}$$

The calibration coefficient matrix is calculated to be:

$$\boldsymbol{B3} = \boldsymbol{C} \cdot \boldsymbol{S3} \cdot \text{Inverse}[\text{Transpose}[\boldsymbol{S3}] \cdot \boldsymbol{S3}] = \begin{pmatrix} 0.2357 & -0.2357 \end{pmatrix} \tag{8.57}$$

If we use ***A3*** as $\boldsymbol{A3_{unk}}$ and calculate $\boldsymbol{C3_{unk}}$ using eq. (8.33), we get the result that is now accurate:

$$\boldsymbol{C3_{unk}} = \boldsymbol{B3} \cdot \boldsymbol{F3} \cdot \text{Transpose}[\boldsymbol{A3_{unk}}] = \begin{pmatrix} 1.0 & 2.0 & 3.0 & 4.0 & 5.0 & 6.0 \end{pmatrix} \tag{8.58}$$

If three unknown samples are measured and their absorbances are recorded to be (21, 24) in Abs-1 and Abs-2 wavelengths for sample #7, (24, 27) for sample #8, and (24, 27) for sample #9, the absorbance matrix of these unknown samples is:

$$\boldsymbol{A3_{unk2}} = \begin{pmatrix} 21 & 24 \\ 24 & 27 \\ 27 & 30 \end{pmatrix} \tag{8.59}$$

If we calculate their concentrations using **B3** and **F3**, we get very accurate results:

$$\boldsymbol{C3_{unk2}} = \boldsymbol{B3} \cdot \boldsymbol{F3} \cdot \text{Transpose}[\boldsymbol{A3_{unk2}}] = (\,7.0 \quad 8.0 \quad 9.0\,) \tag{8.60}$$

Case 4: Let us consider one more case. This time, there is an experimental error in the Abs-1 measurement; so, the measured data in Abs-1 fluctuates slightly. The data in Abs-2 is perfectly linear with a constant baseline offset. For example, say that the obtained data set is:

$$\boldsymbol{A4} = \begin{pmatrix} 2.98 & 6 \\ 6.10 & 9 \\ 8.80 & 12 \\ 12.05 & 15 \\ 14.97 & 18 \\ 18.00 & 20 \end{pmatrix} \tag{8.61}$$

The least squares regression of the Abs-1 data gives $y = 2.999x - 0.013$ with $R^2 = 0.9997$. The covariance, eigenvalues, and eigenvectors of **A4** are different from Case 3 where both measurement data have $R^2 = 1$.

$$\boldsymbol{CovA4} = \text{Covariance}[\boldsymbol{A4}] = \begin{pmatrix} 31.486 & 31.488 \\ 31.488 & 31.5 \end{pmatrix} \tag{8.62}$$

$$\boldsymbol{EV4} = \text{Eigenvalues}[\boldsymbol{CovA4}] = \begin{pmatrix} 62.981 \\ 0.005 \end{pmatrix} \tag{8.63}$$

$$\boldsymbol{F4} = \text{Eigenvectors}[\boldsymbol{CovA4}, 2] = \begin{pmatrix} 0.7070 & 0.7072 \\ -0.7072 & 0.7070 \end{pmatrix} \tag{8.64}$$

Due to the fluctuation in the Abs-1 measurement, the second eigenvalue is not zero anymore. And the directions of PC1 and PC2 in this case are slightly rotated from those of Case 3. The calculated score matrix also reflects the fluctuation of the data in one of the measurements.

$$\boldsymbol{S4} = \boldsymbol{A4} \cdot \text{Transpose}[\boldsymbol{F4}] = \begin{pmatrix} 6.350 & 2.135 \\ 10.678 & 2.049 \\ 14.709 & 2.261 \\ 19.128 & 2.084 \\ 23.313 & 2.140 \\ 27.577 & 2.118 \end{pmatrix} \tag{8.65}$$

$$\boldsymbol{R3} = \boldsymbol{A4} - \boldsymbol{S4} \cdot \boldsymbol{F4} = \begin{pmatrix} -4.4 \times 10^{-16} & 0 \\ 0 & 0 \\ -1.8 \times 10^{-15} & -1.8 \times 10^{-15} \\ 0 & 0 \\ -5.3 \times 10^{-15} & -3.6 \times 10^{-15} \\ -3.6 \times 10^{-15} & -3.6 \times 10^{-15} \end{pmatrix} \tag{8.66}$$

The values in the residual matrix are essentially zero. For this data set, the calibration coefficient matrix is calculated to be:

$$\boldsymbol{B4} = \boldsymbol{C} \cdot \boldsymbol{S4} \cdot \text{Inverse}[\text{Transpose}[\boldsymbol{S4}] \cdot \boldsymbol{S4}] = (\,0.2354 \quad -0.2306\,) \tag{8.67}$$

If we use $\boldsymbol{A4}$ as $\boldsymbol{A4_{unk}}$ and calculate $\boldsymbol{C4_{unk}}$ using eq. (8.33), we get the result that is fairly accurate, but not perfect, because of the fluctuation in the calibration data:

$$\boldsymbol{C4_{unk}} = \boldsymbol{B4} \cdot \boldsymbol{F4} \cdot \text{Transpose}[\boldsymbol{A4_{unk}}] = \begin{pmatrix} 1.00 & 2.04 & 2.94 & 4.02 & 4.99 & 6.00 \end{pmatrix} \tag{8.68}$$

8.3.3 Practicing PCR with Selected Data from ATR-IR Spectra

We have done enough practice. Now, let us process the real data shown in Figure 8.1 with the PCA/PCR method. Note that symbols used in this section look identical to the ones used in the previous section, but their meanings will be different; so, they should not be confused. The concentration matrix $\boldsymbol{C}$ is the same as eq. (8.10):

$$\boldsymbol{C} = \begin{pmatrix} 1.0 & 0.8 & 0.6 & 0.4 & 0.2 & 0.0 \\ 0.0 & 0.2 & 0.4 & 0.6 & 0.8 & 1.0 \end{pmatrix} \tag{8.10}$$

The stepwise calculations shown below follow the procedure described in Section 8.3.1. For the PCA, the absorbance matrix is now changed to (transpose of eq. (8.13)):

$$\boldsymbol{A} = \begin{pmatrix} 0.870 & 0.935 & 1.024 & 0.823 & 0.585 \\ 0.828 & 0.900 & 1.007 & 0.862 & 0.749 \\ 0.753 & 0.844 & 0.957 & 1.036 & 0.998 \\ 0.673 & 0.777 & 0.887 & 1.192 & 1.180 \\ 0.523 & 0.745 & 0.741 & 1.554 & 1.550 \\ 0.196 & 0.310 & 0.380 & 1.927 & 1.901 \end{pmatrix} \tag{8.69}$$

First, let us find the factor describing the largest variance; for that, we can do the following:

$$\boldsymbol{CovA} = \text{Covariance}[\boldsymbol{A}] \text{ (not displaying because it is too big)} \tag{8.70}$$

$$\boldsymbol{EV1} = \text{Eigenvalues}[\boldsymbol{CovA}, 1] = (0\ 5982) \tag{8.71}$$

$$\boldsymbol{F1} = -\text{Eigenvectors}[\boldsymbol{CovA}, 1] = \begin{pmatrix} -0.3209 & -0.2951 & -0.3086 & 0.5570 & 0.6359 \end{pmatrix} \tag{8.72}$$

As shown in eqs. (8.25) and (8.26) which are plotted in Figure 8.2, there are two identical eigenvectors pointing opposite directions. We can choose either one. In eq. (8.72), the −1 value is multiplied to make the values corresponding to the alkyl peaks (first three columns) to be negative (because their intensities decrease as the water content increases) and the values corresponding to the OH peaks (last two column) to be positive. Even if −1 is not multiplied (so choosing the one with opposite sign), the calculation results are the same; the only difference is that the values in score matrix $\boldsymbol{S1}$ will have the opposite sign. Note that $\boldsymbol{F1}$ is the unit vector along the axis describing the largest variance (which is calculated to be 0.5982 in this example) in the covariance of the measured data set. Thus, its magnitude is 1 ($= \sqrt{0.3209^2 + 0.2951^2 + 0.3086^2 + 0.5570^2 + 0.6359^2}$). The scores describing the intensity values of six reference samples along this single factor are:

$$\boldsymbol{S1} = \boldsymbol{A} \cdot \text{Transpose}[\boldsymbol{F1}] = \begin{pmatrix} -0.0407 \\ 0.1144 \\ 0.4257 \\ 0.6954 \\ 1.2645 \\ 2.0106 \end{pmatrix} \tag{8.73}$$

The score values are monotonically increasing, which reflects the relationship between the variable described by this factor and the independent variable (the second row in eq. (8.10)). Note that the first-row value for each sample in $\boldsymbol{C}$ is automatically determined if the second-row value is known, since the sample is a binary mixture (i.e., $x_{propanol} = 1 - x_{water}$). The residual matrix is calculated to be:

$$\boldsymbol{R1} = \boldsymbol{A} - \boldsymbol{S1} \cdot \boldsymbol{F1} = \begin{pmatrix} 0.857 & 0.923 & 1.011 & 0.846 & 0.611 \\ 0.865 & 0.934 & 1.042 & 0.798 & 0.676 \\ 0.890 & 0.970 & 1.088 & 0.799 & 0.727 \\ 0.896 & 0.982 & 1.102 & 0.805 & 0.738 \\ 0.929 & 1.018 & 1.131 & 0.850 & 0.746 \\ 0.841 & 0.903 & 1.001 & 0.807 & 0.622 \end{pmatrix} \tag{8.74}$$

Since the residual values are large, we know that we need more factors. In the second round, we can use **R1**, instead of **A**, and run covariance, eigenvalue, eigenvector to find the second factor:

$$\boldsymbol{CovR1} = \text{Covariance}[\boldsymbol{R1}] \quad \text{(not displaying)} \tag{8.75}$$

$$\boldsymbol{EV2} = \text{Eigenvalues}[\boldsymbol{CovR1}, 1] = (0.0089) \tag{8.76}$$

$$\text{Eigenvectors}\ [\boldsymbol{CovR1}, 1]\ \text{(not displaying)} \tag{8.77}$$

$$\boldsymbol{F2} = \{\boldsymbol{F1}[[\mathbf{1}]], \%[[\mathbf{1}]]\} = \begin{pmatrix} -0.3209 & -0.2951 & -0.3086 & 0.5570 & 0.6359 \\ 0.3265 & 0.4446 & 0.5576 & 0.0250 & 0.6198 \end{pmatrix} \tag{8.78}$$

The eigenvector of **CovR1** is calculated in eq. (8.77), and the result is combined with **F1** in eq. (8.78). The first row is PC1 (same as eq. (8.72)), and the second row is PC2. The score and residual matrix are now calculated using these two factors (PC1 and PC2):

$$\boldsymbol{S2} = \boldsymbol{A} \cdot \text{Transpose}[\boldsymbol{F2}] = \begin{pmatrix} -0.0407 & 1.6539 \\ 0.1144 & 1.7178 \\ 0.4257 & 1.7992 \\ 0.6954 & 1.8210 \\ 1.2645 & 1.8703 \\ 2.0106 & 1.6402 \end{pmatrix} \tag{8.79}$$

$$\boldsymbol{R2} = \boldsymbol{A} - \boldsymbol{S2} \cdot \boldsymbol{F2} = \begin{pmatrix} 0.317 & 0.188 & 0.089 & 0.804 & -0.414 \\ 0.304 & 0.170 & 0.084 & 0.755 & -0.388 \\ 0.302 & 0.170 & 0.085 & 0.754 & -0.388 \\ 0.302 & 0.173 & 0.086 & 0.759 & -0.391 \\ 0.318 & 0.187 & 0.088 & 0.803 & -0.413 \\ 0.306 & 0.174 & 0.087 & 0.765 & -0.394 \end{pmatrix} \tag{8.80}$$

The variation in the PC-2 score (second column in eq. (8.79)) is smaller than that in the PC-1 score, because PC-2 describes the second largest variance within the measured data set. After subtracting the largest variance (**S1** · **F1**), the remaining variance appears to have a nonmonotonic trend; the values in the PC-2 scores show a parabolic dependence on the dependent variable.

Obviously, **R2** is smaller than **R1**, but still far from zero. In the third round, we use **R2** and run covariance, eigenvalue, eigenvector to find the third factor:

$$\boldsymbol{CovR2} = \text{Covariance}[\boldsymbol{R2}] \quad \text{(not displaying)} \tag{8.81}$$

$$\boldsymbol{EV3} = \text{Eigenvalues}[\boldsymbol{CovR2}, 1] = (0.0008) \tag{8.82}$$

$$\text{Eigenvectors}\ [\boldsymbol{CovR2}, 1]\ \text{(not displaying)} \tag{8.83}$$

$$\boldsymbol{F3} = \{\boldsymbol{F1}[[\mathbf{1}]], \boldsymbol{F2}[[\mathbf{2}]], \%[[\mathbf{1}]]\} = \begin{pmatrix} -0.3209 & -0.2951 & -0.3086 & 0.5570 & 0.6359 \\ 0.3265 & 0.4446 & 0.5576 & 0.0250 & 0.6198 \\ -0.2557 & -0.2831 & -0.0616 & -0.8179 & 0.4262 \end{pmatrix} \tag{8.84}$$

In eq. (8.84), the first, second, and third rows show PC1, PC2, and PC3, respectively. The score and residual matrix are now calculated using these three factors:

$$\boldsymbol{S3} = \boldsymbol{A} \cdot \text{Transpose}[\boldsymbol{F3}] = \begin{pmatrix} -0.0407 & 1.6539 & -0.9741 \\ 0.1144 & 1.7178 & -0.9144 \\ 0.4257 & 1.7992 & -0.9125 \\ 0.6954 & 1.8210 & -0.9188 \\ 1.2645 & 1.8703 & -0.9725 \\ 2.0106 & 1.6402 & -0.9728 \end{pmatrix} \tag{8.85}$$

$$\boldsymbol{R3} = \boldsymbol{A} - \boldsymbol{S3} \cdot \boldsymbol{F3} = \begin{pmatrix} 0.068 & -0.088 & 0.029 & 0.008 & 0.001 \\ 0.070 & -0.089 & 0.028 & 0.007 & 0.001 \\ 0.069 & -0.089 & 0.029 & 0.008 & 0.001 \\ 0.067 & -0.088 & 0.030 & 0.008 & 0.001 \\ 0.069 & -0.089 & 0.029 & 0.007 & 0.001 \\ 0.068 & -0.088 & 0.029 & 0.008 & 0.001 \end{pmatrix} \tag{8.86}$$

Now, the magnitudes of residuals are all less than 0.09.

If we do one more round of calculations finding another factor (not shown since this the repeat of the same steps from eq. (8.81)), then we get:

$$\boldsymbol{F4} = \begin{pmatrix} -0.3209 & -0.2951 & -0.3086 & 0.5570 & 0.6359 \\ 0.3265 & 0.4446 & 0.5576 & 0.0250 & 0.6198 \\ -0.2557 & -0.2831 & -0.0616 & -0.8179 & 0.4262 \\ 0.8510 & -0.3424 & -0.3718 & -0.0499 & 0.1337 \end{pmatrix} \tag{8.87}$$

$$\boldsymbol{S4} = \begin{pmatrix} -0.0407 & 1.6539 & -0.9741 & 0.0767 \\ 0.1144 & 1.7178 & -0.9144 & 0.0793 \\ 0.4257 & 1.7992 & -0.9125 & 0.0778 \\ 0.6954 & 1.8210 & -0.9188 & 0.0752 \\ 1.2645 & 1.8703 & -0.9725 & 0.0785 \\ 2.0106 & 1.6402 & -0.9728 & 0.0775 \end{pmatrix} \tag{8.88}$$

And the magnitude of residual values in $\boldsymbol{R4}$ are less than 0.06 (not shown); only small improvement from $\boldsymbol{R3}$. That means the number of factors that we found are sufficient to describe most of the variances in the data set $\boldsymbol{A}$. This means we have found all factors that are significant; the remaining ones are all minor, probably describing noises in the measured data set. So, we can stop the PCA step here. This means that all variances (physical information) of the measured data are now represented in the score matrix.

Now, we can proceed to the PCR step and calculate the calibration coefficient matrix using the score matrix from the PCA step:

$$\boldsymbol{B4} = \boldsymbol{C} \cdot \boldsymbol{S4} \cdot \text{Inverse}[\text{Transpose}[\boldsymbol{S4}] \cdot \boldsymbol{S4}] = \begin{pmatrix} -0.4671 & -0.9071 & -1.3933 & 14.587 \\ 0.4674 & 0.9516 & 1.1409 & -5.7454 \end{pmatrix} \tag{8.89}$$

If we use $\boldsymbol{A}$ in eq. (8.69) as $\boldsymbol{A}_{unk}$ and calculate $\boldsymbol{C}_{unk}$ using eq. (8.33), we get the result that is fairly close to the original $\boldsymbol{C}$ matrix in eq. (8.10):

$$\boldsymbol{C}_{unk} = \boldsymbol{B4} \cdot \boldsymbol{F4} \cdot \text{Transpose}[\boldsymbol{A}_{unk}] = \begin{pmatrix} 0.995 & 0.819 & 0.576 & 0.402 & 0.213 & -0.005 \\ 0.003 & 0.189 & 0.423 & 0.577 & 0.810 & 0.997 \end{pmatrix} \tag{8.90}$$

So far, we have done PCA one at a time (eqs. (8.72), (8.78), (8.84), and (8.87)) to see how the factors and scores evolve from the measured data set. There is no need to do it stepwise. Since we are using matrix algebra, all can be done at once. This time, we will try with five factors.

$$\boldsymbol{EV5} = \text{Eigenvalues}[\boldsymbol{CovA}, 5] = \begin{pmatrix} 0.5982 \\ 0.0089 \\ 0.0008 \\ 2 \times 10^{-6} \\ 2 \times 10^{-8} \end{pmatrix} \tag{8.91}$$

$$\boldsymbol{F5} = \text{Eigenvectors}[\boldsymbol{CovA}, 5] = \begin{pmatrix} 0.3209 & 0.2951 & 0.3086 & -0.5570 & -0.6359 \\ 0.3265 & 0.4446 & 0.5576 & 0.0250 & 0.6198 \\ -0.2557 & -0.2831 & -0.0616 & -0.8179 & 0.4262 \\ -0.8510 & 0.3424 & 0.3718 & 0.0499 & -0.1337 \\ 0.0291 & -0.7196 & 0.6722 & 0.1325 & -0.1092 \end{pmatrix} \tag{8.92}$$

$$\boldsymbol{S5} = \boldsymbol{A} \cdot \text{Transpose}[\boldsymbol{F5}] = \begin{pmatrix} 0.0407 & 1.6539 & -0.9741 & -0.0767 & 0.0859 \\ -0.1144 & 1.7178 & -0.9144 & -0.0793 & 0.0858 \\ -0.4257 & 1.7992 & -0.9125 & -0.0778 & 0.0861 \\ -0.6954 & 1.8210 & -0.9188 & -0.0752 & 0.0858 \\ -1.2645 & 1.8703 & -0.9725 & -0.0785 & 0.0859 \\ -2.0106 & 1.6402 & -0.9728 & -0.0775 & 0.0859 \end{pmatrix} \tag{8.93}$$

$$\boldsymbol{B5} = \boldsymbol{C} \cdot \boldsymbol{S5} \cdot \text{Inverse}[\text{Transpose}[\boldsymbol{S5}] \cdot \boldsymbol{S5}] = \begin{pmatrix} 0.4673 & -0.9311 & -1.2568 & -9.8060 & 6.2939 \\ -0.4673 & 0.9517 & 1.2539 & 9.7048 & 5.2129 \end{pmatrix} \tag{8.94}$$

Here, we can compare the eigenvalues of each factor. From eq. (8.91), we can see that the first-row number is 98.40% of the sum of all values; this means that the first factor describes 98.40% of the total variance within the measured data set. Similarly, we can see that the second factor describes 1.46% of the variance within the data set after subtracting the variance along the first factor direction. The third factor describes 0.13% of the variance in the data set. The variances described by the fourth and fifth factors are negligible.

Note that the values in the first factor (the first row in eq. (8.92) have the sign opposite to the ones in eq. (8.72). That is because *Mathematica* chooses the unit vector pointing the opposite direction. Reflecting this sign change in the factor, the first score values (the first column in eq. (8.93) have the sign opposite to the ones in eq. (8.73). Similarly, the signs of the first column values in $\boldsymbol{B5}$ are switched from eq. (8.89).

As we have done previously, we can use $\boldsymbol{A}$ in eq. (8.69) as $\boldsymbol{A_{unk}}$ and calculate $\boldsymbol{C_{unk}}$ using eq. (8.33) to see how well the PCR works. This leads to:

$$\boldsymbol{C_{unk}} = \boldsymbol{B5} \cdot \boldsymbol{F5} \cdot \text{Transpose}[\boldsymbol{A_{unk}}] = \begin{pmatrix} 0.997 & 0.813 & 0.578 & 0.412 & 0.200 & -0.001 \\ 0.004 & 0.185 & 0.425 & 0.586 & 0.800 & 1.001 \end{pmatrix} \tag{8.95}$$

The results are almost identical to eq. (8.90) where we use only four factors found in stepwise calculations; the differences are negligible (within error range).

8.3.4 Doing PCA with Full ATR-IR Spectra

Going further, we can use the entire spectral data points in all collected spectra for the PCA analysis. Figure 8.3 shows a screenshot of the Excel spreadsheet of the raw data of the ATR-IR spectra shown in Figure 8.1. The first column shows the wavenumber of the data point, and the first row shows the concentration (mole fraction) of water in the n-propanol/water binary mixture solutions. The rest of the table is the intensity data at each wavenumber for each solution. Also shown in Figure 8.3 is the excerpts from the *Mathematica* program. The top part is to import the data from the Excel file and the reconstruct the data table for processing. The middle part is to calculate the first principal component (PC1) with the largest eigenvalue (largest variance in the covariance matrix) and calculate the score and residuals. These calculations are the same as the ones in eqs. (8.70)–(8.74). One additional calculation added here is the factor loading which is the factor (*PC*) times the square root of the eigenvalue ($\sqrt{\xi}$). Then, the processed data are put together in a table form for plotting. The bottom part is the repeat of the middle part using the residual matrix of the above step to find the second principal component (PC2) with

	A	B	C	D	E	F	G
1	WN	0	0.2	0.4	0.6	0.8	1
2	1500.347	0.1651636	0.1821849	0.2200061	0.2462996	0.3179405	0.4223347
3	1502.276	0.1523938	0.1703436	0.2097927	0.2374991	0.3116315	0.4208258
4	1504.204	0.1523938	0.1703436	0.2097927	0.2375039	0.3116315	0.4208258
5	1506.133	0.1455362	0.1627693	0.2014665	0.2288492	0.3023960	0.4075919
6	1508.061	0.1356666	0.1550956	0.1967757	0.2255539	0.3026140	0.4152816
7	1509.990	0.1333122	0.1539740	0.1972213	0.2269687	0.3058262	0.4241754
8	1511.918	0.1292045	0.1510973	0.1956686	0.2263332	0.3068605	0.4291151
9	1513.846	0.1258392	0.1482556	0.1933150	0.2244257	0.3057682	0.4301304
10	1515.775	0.1216468	0.1442348	0.1896814	0.2211585	0.3034050	0.4292399
11	1517.703	0.1171882	0.1405251	0.1866633	0.2187493	0.3021997	0.4303272
12	1519.632	0.1138516	0.1377569	0.1844780	0.2172567	0.3016324	0.4321237
13	1521.560	0.1092352	0.1333936	0.1805243	0.2136798	0.2987930	0.4300142
14	1523.489	0.1050465	0.1303649	0.1789886	0.2133008	0.3006028	0.4371801
15	1525.417	0.1026079	0.1282891	0.1773384	0.2121108	0.3002738	0.4392037
16	1527.346	0.0990117	0.1256611	0.1760394	0.2116112	0.3014470	0.4443038
17	1529.274	0.0963585	0.1238178	0.1751232	0.2112333	0.3022236	0.4481279
18	1531.203	0.0939202	0.1217873	0.1736066	0.2102982	0.3021944	0.4503181
19	1533.131	0.0913204	0.1192713	0.1714438	0.2084372	0.3011952	0.4503776
20	1535.060	0.0887014	0.1176712	0.1711643	0.2091133	0.3038563	0.4574529
21	1536.988	0.0873170	0.1163979	0.1700004	0.2082313	0.3033538	0.4581559
22	1538.917	0.0850671	0.1138491	0.1674889	0.2059387	0.3010421	0.4544851
23	1540.845	0.0817108	0.1106123	0.1648278	0.2035484	0.2998477	0.4545859
24	1542.773	0.0797863	0.1098464	0.1651946	0.2050354	0.3036090	0.4655204
25	1544.702	0.0783393	0.1091306	0.1652842	0.2058134	0.3057678	0.4718721
26	1546.630	0.0769942	0.1086605	0.1661103	0.2073952	0.3092304	0.4801014
27	1548.559	0.0757481	0.1076649	0.1654207	0.2070450	0.3097105	0.4829917
28	1550.487	0.0740457	0.1065481	0.1650146	0.2074066	0.3117142	0.4889942
29	1552.416	0.0729027	0.1057875	0.1646430	0.2076013	0.3130170	0.4930629
30	1554.344	0.0713890	0.1045228	0.1640557	0.2076365	0.3143041	0.4971189
31	1556.273	0.0702178	0.1033373	0.1630457	0.2069474	0.3141133	0.4979478

```
SetDirectory[NotebookDirectory[]];
ATRdata = Import["Raw ATR-IR_propanol-water.xlsx"][[1]]
Tdata = Transpose[ATRdata][[2 ;; -1, 2 ;; -1]];
Tconc = Transpose[ATRdata][[2 ;; -1, 1]];
Twn = Transpose[ATRdata][[1, 2 ;; -1]];

CVfull = Covariance[Tdata];
F1full = -Eigenvectors[CVfull, 1];
F1fullLoadings = F1full * Sqrt[Eigenvalues[CVfull, 1]];
S1full = Tdata.Transpose[F1full];
R1full = Tdata - S1full.F1full;

F1fullPlot = Append[{Twn}, F1full[[1]]];
F1fullLoadingsPlot = Append[{Twn}, F1fullLoadings[[1]]];
S1fullPlot = Append[{Tconc}, Transpose[S1full][[1]]];

CVR1full = Covariance[R1full];
-Eigenvectors[CVR1full, 1];
F2full = {F1full[[1]], %[[1]]};
F2fullLoadings = F2full[[2]] * Sqrt[Eigenvalues[CVR1full, 1]][[1]];
S2full = Tdata.Transpose[F2full];
R2full = Tdata - S2full.F2full;

F2fullPlot = Append[{Twn}, F2full[[2]]];
F2fullLoadingsPlot = Append[{Twn}, F2fullLoadings];
S2fullPlot = Append[{Tconc}, Transpose[S2full][[2]]];
S1vS2Plot = Append[Transpose[S1full], Transpose[S2full][[2]]];
```

Figure 8.3 Screenshot of Excell spreadsheet of the full data of ATR-IR spectra in Figure 8.1 and Mathematica program syntaxes importing the data, finding the first and second factors and their scores, and plotting them for graphical comparison.

the second largest eigenvalue of the covariance matrix of the original data. These calculations are the same as the ones in eqs. (8.75)–(8.80). The sign of the second factor is changed by multiplying −1 to make it consistent with the result from eq. (8.77). The only addition is the table plotting score-1 vs. score-2 of each sample.

Figure 8.4A plots the first and second principal components (PC1 and PC2), that is, the unit vectors describing the directions of the largest and second largest variances at each wavenumber in the covariance matrix of all data points measured. There are 1297 data points in each spectrum (from 1500 to 4000 cm^{-1} with 2 cm^{-1} increments); so, each factor is an array of 1297 values. The sum of squares of all values in each factor is 1, since it is a unit vector. In the PC1 graph, the values at three alkyl peaks at 2875, 2935, and 2962 cm^{-1} are negative and the values at two OH peaks at 3320 and 3400 cm^{-1} are positive, which is consistent with the trend seen in eq. (8.72). In the PC2 graph, the values at 2875, 2935, 2962, and 3400 cm^{-1} are large and the value at 3320 cm^{-1} is small, which is consistent with the trend seen in eq. (8.78). The exact values are different from the previous case, which is simply because the numbers of data points used in PCA are different.

Figure 8.4B plots the factor loadings as a function of wavenumber. The loadings refer to coefficients of the regression of the data matrix onto the factors. In loadings, values close to −1 or + 1 indicate that the factor strongly influences the variable with negative or positive correlations, respectively Values close to zero means that the factor has a weak influence on the variable. In the first factor loading, we can see that the peaks in the alkyl stretch region of n-propanol and the broad band in the OH stretch region of water have opposite values. We see the same trend in eq. (8.72). That is because changes in the n-propanol and water concentrations are opposite in our sample set. Overall, the loading of the second factor loading is smaller than that of the first factor. In the second factor loading plot, the values between 3000 and 3340 cm^{-1} are negative, while the alkyl stretch region and the higher wavenumber region of the OH stretch band are positive. This is probably because the variance in that region is described too much by PC1. Physically, it might be related to the concentration dependence of excess volume in n-propanol and water mixture [2]. But, please note that this kind of interpretation is not always possible since factors are abstract mathematical functions and do not necessarily reflect specific physical properties.

In Figure 8.4C, the score-1 increases nonlinearly but monotonically with water concentration (x_{water}). The score-2 varies nonmonotonically within a relatively smaller range. These trends are the same as the ones found in eq. (8.79). Figure 8.4D plots the scores of PC2 vs. the scores of PC1. This is called score plot. In our example, the score plot shows the trend within the data set – a monotonic change along PC1 and a small and nonmonotonic change along PC2. If the first two PCs account for most of the variance in the data, the score plot can show the data structure, such as trends, clusters, and outliers. If three PCs are needed to describe the data, the score plot becomes a 3D plot. The clustering patterns are especially useful for

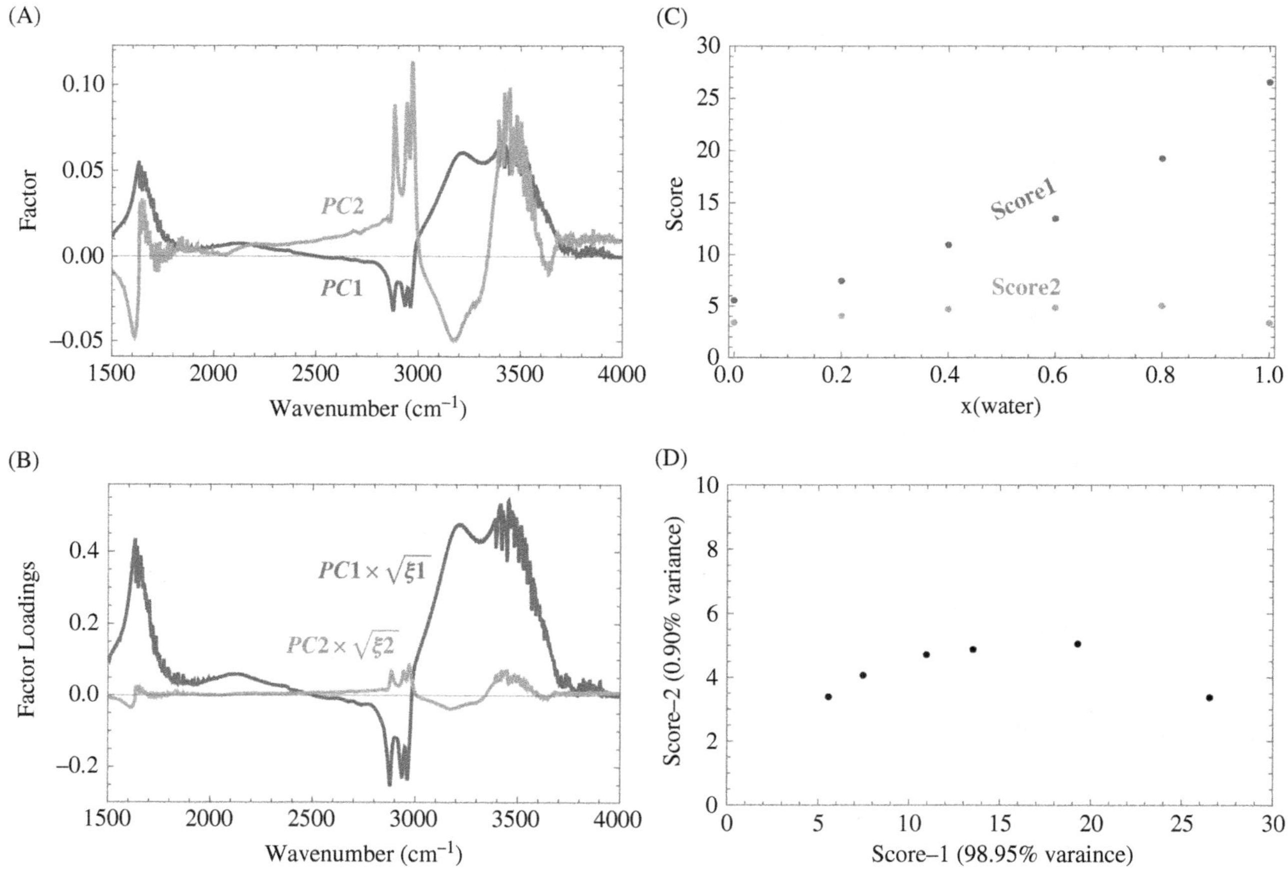

Figure 8.4 Plots of (A) first and second factors (PC1 and PC2), (B) factor loadings, (C) scores vs. water fraction (x_{water}), and (D) score-1 vs. score-2 from calculations shown in Figure 8.3.

identification and classification of specific characteristics within the data set [1]. The clustering patterns in the score plot are frequently used for identification and classification of specific characteristics in ToF-SIMS analyses (see Section 4.2.1) [1].

8.3.5 Comparing PCR with Least Squares Regression (LSR)

We have done two types of regression – LSR and PCR – for the data set presented in Figure 8.1. Now, let us compare the LSR and PCR results. For comparison, those results are listed here again. Equation (8.10) is the concentration of the unknown samples (which we know since we prepared the solutions). Equations (8.12) and (8.15) are the concentrations calculated through the LSR method using two peaks (LSR2) and five peaks (LSR5), respectively. Equations (8.90) and (8.95) are the results calculated through the PCR method using four factors found sequentially (PCR4) and five factors found simultaneously (PCR5), respectively. For comparison, the same two-point data used in LSR2 (intensities at 2875 cm^{-1} and 3400 cm^{-1}; that is, transpose of eq. (8.9)) are processed with PCA and PCR (steps are the same as eqs. (8.51)–(8.58)) and the result are shown as $\boldsymbol{C_{PCR2}}$ in eq. (8.96).

$$\boldsymbol{C_{original}} = \begin{pmatrix} 1.0 & 0.8 & 0.6 & 0.4 & 0.2 & 0.0 \\ 0.0 & 0.2 & 0.4 & 0.6 & 0.8 & 1.0 \end{pmatrix} \tag{8.10}$$

$$\boldsymbol{C_{LSR2}} = \begin{pmatrix} 0.888 & 0.804 & 0.665 & 0.534 & 0.282 & -0.172 \\ 0.075 & 0.188 & 0.365 & 0.502 & 0.775 & 1.10 \end{pmatrix} \tag{8.12}$$

$$\boldsymbol{C_{PCR2}} = \begin{pmatrix} 0.853 & 0.777 & 0.649 & 0.528 & 0.296 & -0.128 \\ 0.096 & 0.205 & 0.374 & 0.505 & 0.767 & 1.07 \end{pmatrix} \tag{8.96}$$

$$C_{LSR5} = \begin{pmatrix} 0.873 & 0.804 & 0.677 & 0.549 & 0.288 & -0.192 \\ 0.103 & 0.183 & 0.343 & 0.479 & 0.767 & 1.12 \end{pmatrix} \tag{8.15}$$

$$C_{PCR4} = \begin{pmatrix} 0.995 & 0.819 & 0.576 & 0.402 & 0.213 & -0.005 \\ 0.003 & 0.189 & 0.423 & 0.577 & 0.810 & 0.997 \end{pmatrix} \tag{8.90}$$

$$C_{PCR5} = \begin{pmatrix} 0.997 & 0.813 & 0.578 & 0.412 & 0.200 & -0.001 \\ 0.004 & 0.185 & 0.425 & 0.586 & 0.800 & 1.00 \end{pmatrix} \tag{8.95}$$

When only two peaks representative of two analytes are selected and used, both C_{LSR2} and C_{PCR2} results are quite inaccurate. This suggests that there are more than two degrees of freedom. In the binary mixture solution of n-propanol and water (Figure 8.1), we know that in addition to the concentration of these two components, there are other factors that affect the peak intensities. Those include excess molar volume due to molecular interactions between n-propanol and water [2] which will affect the molar absorptivity of each absorption band as well as peak width (because damping factors may also be altered), changes in the evanescent penetration depth due to refractive index change, peak overlap, baseline shift, and so on. In the case of LSR2, these could not be modeled with the simple Beer–Lambert law. In the case of PCR2, the effects of these variables could not be fully described either, because insufficient numbers of factors are used.

When five peaks are selected and used, the outcomes are quite different. While the LSR5 results are still inaccurate, the PCR4 and PCR5 results are acceptable within reasonable experimental error ranges. This shows the inefficacy of LSR to model all variables that are specific to the system under consideration, especially describing the deviation from the linear dependence of the peak intensity on the analyte concentration. In PCR, using four or five factors looks like sufficient to model the variances existing in the data set. It means most of intrinsic complications of the system (such as excess molar volume) as well as experimental uncertainties (measurement errors) in the calibration data set are reflected in those four or five factors. But PCA cannot tell what those complications are because factors are arbitrary mathematical functions describing the unit vector along the major variances within the data set. The factors are found without *a priori* knowledge of the system under consideration. If an additional component is added to the system or a different instrument is used, then the calibration coefficient matrix obtained from the previous PCR does not work; so, a new calibration experiment should be done.

8.3.6 How Many Principal Components Are Needed

From the comparison of C_{PCR2}, C_{PCR4}, and C_{PCR5}, we may ask how many factors or principal components are needed and how to determine the minimum number of factors? That can be done by checking relative magnitudes of eigenvalues. The eigenvalue of a specific factor describes the variance described by that principal component. The eigenvalue of each factor can be divided by the sum of all eigenvalues and listed in a plot format. This is called a scree plot. In this plot, each point shows the relative amount of variance described by each PC.

Let us go back to the example shown in Figures 8.3 and 8.4 and check their eigenvalues. Since we have used the full spectrum containing 1297 data points, there can be as many as 1297 eigenvalues and 1297 eigenvectors. This means 1297 factors. We do not care about the factors that describe noises in the data. So, let us look at the first seven eigenvalues in the decreasing order and display them in the scree plot. It is shown in Figure 8.5. This plot shows that PC1 and PC2 describe 98.95% and 0.895% of the total variance. These are the numbers shown in the axis titles in Figure 8.4D. It means 98.95% and 0.895% of the covariance in the raw data are along the PC1 and PC2 directions, respectively. The rest (0.15%) is described by other minor factors.

If PCA is used for identification or classification of patterns in the data set, usually the two main PCs are sufficient [1]. In PCR, it depends on the precision of regression results. This can be seen clearly if we do PCR of the full ATR-IR spectra shown in Figure 8.1 with different numbers of factors. The results are organized in Table 8.2 for comparison.

The PCR result with two factors is not good; in fact, it is even worse than the C_{PCR2} result in eq. (8.96) which is obtained using the intensities of five selected peaks. This is because the spectral regions outside the characteristic bands of the two analytes under consideration (i.e., the spectral regions with mostly noises only) are included in the PCA of the full spectra. The accuracy of the PCR results with three factors is similar to that of C_{PCR4} and C_{PCR5} from PCR of five selected peaks shown in (8.90) and (8.95). From the scree plot, we can see that the sum of three factors describes 99.977% of the variance. The addition of another factor (thus, covering 99.998% of the variance) does not make any significant improvement. If five factors are used, the accuracy of the high x_{water} results become almost perfect. When six factors are used, all calculated values

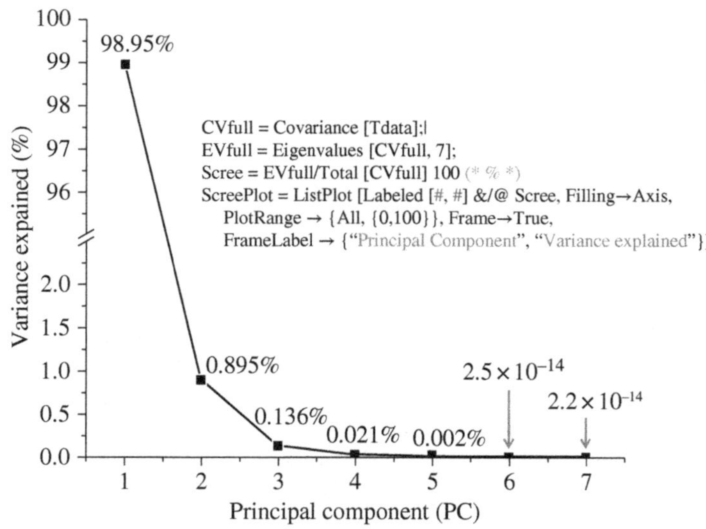

Figure 8.5 Scree plot of the eigenvalues of the first seven factors of the data in Figure 8.3. The script in the figure is the *Mathematica* syntax to get the scree plot.

Table 8.2 Comparison of concentration calculation results from PCR conducted with different number of factors. The full spectra were used for PCA.

	Expected x_{water}					
# of factors used in PCR	**0.0**	**0.2**	**0.4**	**0.6**	**0.8**	**1.0**
2	0.175	0.244	0.381	0.487	0.729	1.07
3	−0.009	0.206	0.442	0.569	0.786	1.01
4	−0.006	0.202	0.436	0.588	0.772	1.01
5	0.003	0.193	0.403	0.600	0.800	0.999
6	0.000	0.200	0.400	0.600	0.800	1.00
7	0.388	0.590	0.794	0.895	1.03	1.21

are exactly the same as the expected values. This means that six factors fully describe all the variances in the system mathematically (recall that there were six samples in the calibration set). When seven factors are used, then the calculation results are even worse than the results obtained with three to six factors. This is because the factor that is completely irrelevant to the correlations within the data set (so, just noise) is included in the calculation. In this case, *Mathematica* program gave an error message stating "the result for inverse of badly conditions matrix may contain significant numerical errors."

In this specific example, the use of three or four factors could be justified as "reasonable." Without knowing the actual concentration of unknown samples, it would be difficult to justify the use of six factors for PCR. Recall that the purpose of factor analysis is to use the minimum number of factors needed.

Nowadays, PCA is frequently used to summarize a large data set of correlated variables into a smaller and more easily interpretable axes of variation (as demonstrated in Figure 8.4B) [1]. It should be noted that PCs need to be distinct from each other to be meaningful; otherwise, they only represent random directions or noises in the data (as seen in the PCR result with seven PCs in Table 8.2). Correlation matrices can be found even if there are no real patterns; their eigenvalues cluster around zero. The loadings of the PCs need to be larger than expected by random data to be useful in the calculation of PC-scores. PC-scores calculated from indistinct PCs have very large standard errors and cannot be used for meaningful interpretations. Keep in mind that PCA can be very useful but great care is needed to avoid spurious results [3].

8.3.7 Multivariate Analysis of Hyperspectral Images

So far, we have studied the use of PCA and PCR for the analysis of one set of spectra. This can be extended to the analysis of multivariate images (for example, the hyperspectral SFG images shown in Figure 7.35 in Chapter 7). Such image data sets contain a spatial raster of $m \times n$ pixels, where each pixel contains a full spectrum consisting of p variables (i.e., intensities as a function of wavenumber). In order to perform PCA, each pixel in the hyperspectral image can be treated as a separate sample. Then, the three-dimensional image data with an order of $m \times n \times p$ is collapsed into a two-dimensional data matrix with an order of $(m \cdot n) \times p$. This has the same data format that we have dealt with in Section 8.3.4, so we can carry out the PCA procedure to obtain scores and loadings as described previously (similar to the ones shown in Figures 8.4). Upon completion of PCA, we get the scores matrix with an order of $(m \cdot n) \times f$, where f is the number of PCs with significant eigenvalues. This output can be redistributed to have a three-dimensional matrix with an order of $m \times n \times f$. The score values of each pixel can be plotted into a two-dimensional image with $m \times n$ pixels for each *PC* for identification of a specific pattern (variance) described by each PC.

Figure 8.6 displays an example of processing the hyperspectral data with PCA. The hyperspectral images collected for the xylem and interfascicular fiber (IFF) regions of an *Arabidopsis* stem have 50 × 50 pixels (100 μm scan with a 2 μm step size) and each pixel has the SFG spectrum from 2800 to 3500 cm^{-1} which contains structural information about cellulose in individual cell walls. The intensity maps for the 2944 and 3320 cm^{-1} peaks extracted from SFG spectra of individual pixels for the same regions are shown in Figure 7.35, and the spectroscopic interpretation of the full hyperspectral data is described in Section 7.4.7.

In the xylem region, the loading plot of PC1 describes 78% of the variance in the SFG spectral feature (Figure 8.6A). The PC-2 loading plot of the xylem region, which describes 16% of the variance, has negative peaks in the CH stretching region and a broad positive peak in the OH stretching region (Figure 8.6A). This reflects the large changes in the OH/CH intensity ratio depending on the location of the cell wall inside the imaged region (see the spectra in Figure 7.35F); so, the (PC-1 × score-1) is subtracted too little in some locations and too much in other locations. The score-1 image reflects the relative amount of cellulose in each location (Figure 8.6C) and the score-2 image shows the locations where OH/CH intensity ratio is larger or lower than the average described by PC-1 (Figure 8.6E). This is the reason that, in Figure 7.36, the polar plot of cellulose microfibril orientations has a large degree of scatter, and the standard deviation of the directionality excess is large for the xylem data.

In contrast, the SFG spectra of cellulose in the IFF region are much more uniform (Figure 7.35G) and 94% of the variance is described by PC-1 (Figure 8.6B). Thus, the score-1 image in this region (Figure 8.6D) looks almost identical to the 2944 cm^{-1} intensity map in Figure 7.35E. This is consistent with the fact the cellulose microfibril orientation is distributed within a fairly small range of angles from the stem axis and the directionality of cellulose microfibrils is also fairly uniform for the IFF region (Figure 7.36) The score-2 plot (Figure 8.6F) shows local variances in cellulose structure. Although the local differences look large in the auto-scale of color plot, it reflects only 4% of total variance. The region that looks like having more differences (toward the right side in Figure 8.6E) is the boundary between the IFF and the pith part inside the stem. Within the IFF region, the difference appears to be small. These are the trends that might be difficult to find if we look at the individual intensity maps (Figure 7.35E); but such trends can be identified easily by processing the hyperspectral image data with PCA.

8.3.8 Partial Least Squares (PLS) Regression

PLS regression (PLSR) is a more generalized technique that combines key features from PCA and multiple regression. It is particularly useful when there are multiple independent variables. In the binary mixture example discussed previously, once the concentration of one analyte is determined, the concentration of the other can be obtained automatically. So, when the full ATR-IR spectra of n-propanol/water mixtures are processed with PCR in Section 8.3.6, only one variable (x_{water}) is sufficient in the concentration matrix (Table 8.2); once it is determined, then the other concentration can be calculated using the constraint equation of the binary mixture ($x_{propanol} = 1 - x_{water}$). We can treat both $x_{propanol}$ and x_{water} as independent variables and run PCR simultaneously with both concentrations in the $\boldsymbol{C}$ matrix to determine both values of unknown samples. This is what we have done in Section 8.3.3; in this case, we can see that $x_{propanol}$ and x_{water} do not add up to 1 (see Section 8.3.5), although their sum must be one since we analyze the binary mixture. One can claim that it is because the original measured data has experimental errors. When the system under consideration is more complex due to multiple variables with measurement errors and we need to find correlations among those variables, PLSR can be more useful. The uncertainty due to measurement errors can exist in both independent and dependent variables. The main difference between PCR and PLSR is that PLSR does not use original data in the independent variable matrix. Instead, it performs PCA of the independent variable matrix and then use their PCs for regression of dependent variable matrix. So, PLSR can handle measurement uncertainty better.

The basic process of PLSR is as follows. We perform PCA for independent variables ($\boldsymbol{C} = \boldsymbol{S_C F_C}$) as well as dependent variables ($\boldsymbol{A} = \boldsymbol{S_A F_A}$) in tandem. By carrying out PCA for the $\boldsymbol{C}$ matrix, we find the factors that capture meaningful variances

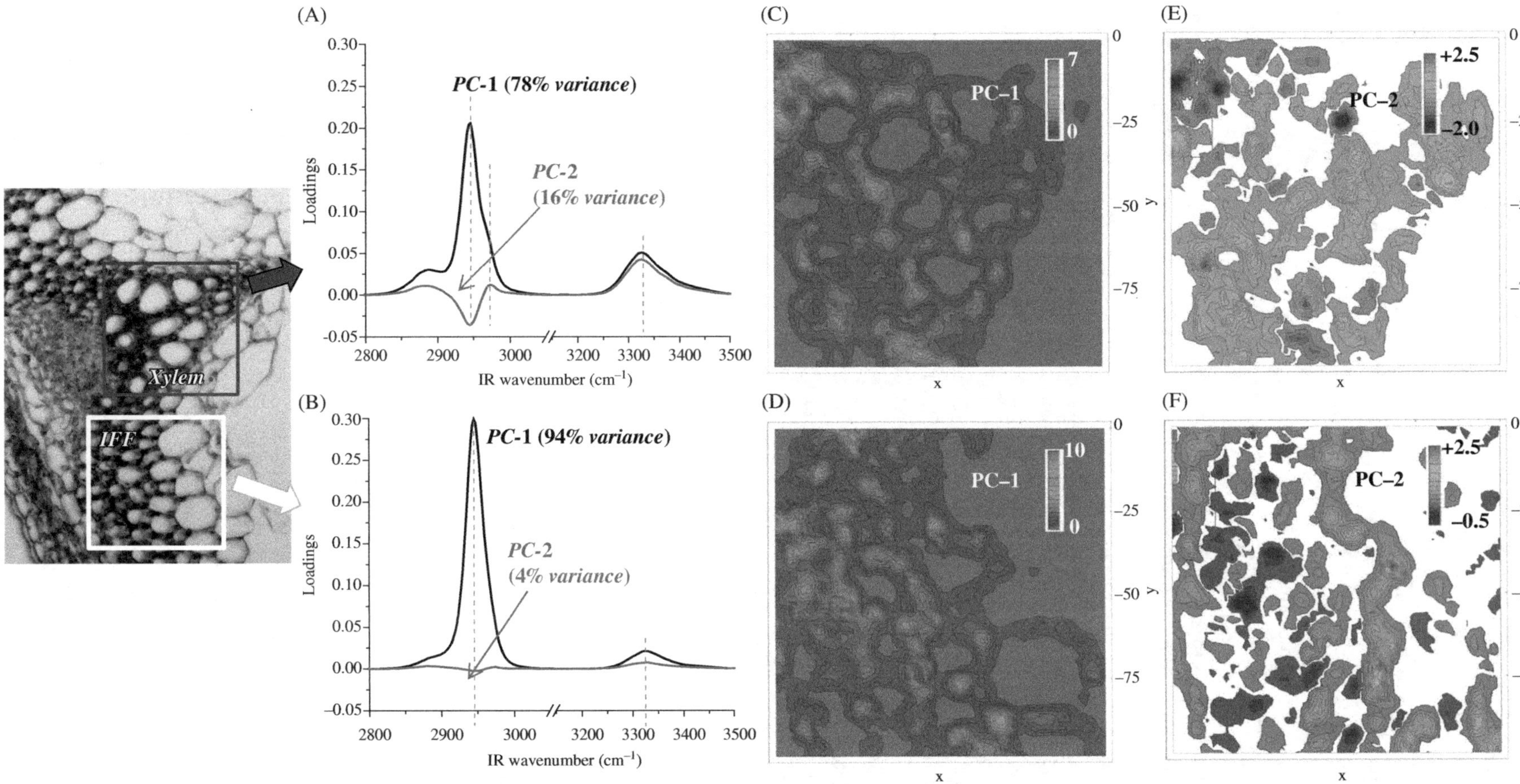

Figure 8.6 (A,B) PC-1 and PC-2 loading plots and score-1 and score-2 maps of the hyperspectral SFG data of (A) xylem and (B) IFF regions of a cross-sectioned stem of 8-week-old Arabidopsis inflorescence. (C,D) Score-1 and (E,F) score-2 images of (C,E) xylem and (D,F) IFF regions. The intensity maps of the 2944 and 3320 cm^{-1} peaks are shown in Figure 7.35.

in the C matrix containing uncertainties (for example, concentration fluctuations). Then, we do least squares regression between the scores of independent variables (S_C) and the scores of dependent variables (S_A) to find the calibration matrix. When this calibration matrix is used to process the measured data of unknown samples (A_{unk}) in the same way described in Section 8.3.1 for the PCR step, we get the $S_{C,unk}$ matrix of the unknown samples; then, by multiplying $S_{C,unk}$ with the F_C matrix, we get the C_{unk} matrix. Note that in PCR, the independent variables (C) are compared with the S_A matrix to find the calibration matrix (see eq. (8.30) in Section 8.3.1); so, C_{unk} is obtained directly from PCR. In ordinary LSR described in Section 8.1.2, the independent variables (C) are directly compared with the dependent variables (A).

Let us practice PLSR with a real example. Figure 8.7 shows the near-infrared (NIR) absorbance spectra of powdered stalk samples of 98 sorghum plants produced with different genetic mutations. The goal of this experiment was to develop a spectroscopic method that can be used for routine analysis of cell wall compositions in plant samples without doing time-consuming and laborious chemical analysis. NIR analysis in the 4000 and 9000 cm^{-1} region is chosen for that purpose. Even though spectroscopic interpretation of peaks in NIR spectra is difficult since they are overtones and combination bands, NIR has an important advantage over the mid-IR analysis. Due to weak absorption cross section, NIR penetrates much deeper (at least, two orders of magnitude deeper) into the sample than mid-IR. Thus, variations or artifacts due to sample preparation (particle size, packing, etc.) as well as surface contamination issues are much less. The small differences among the spectra in Figure 8.7 reflect the compositional differences in the cell wall. Such small differences are to be correlated to the chemical compositions (cellulose, hemicellulose, and lignin) determined using the conventional dietary fiber (DF) analysis method [4].

First, out of 98 samples, 20 samples were randomly chosen and used as a calibration data set. Their chemical compositions (C) and NIR spectra (A) were processed through PLSR to obtain the correlation matrix. Then, the original spectra were assumed to be the unknown sample spectra (A_{ukn}; as we did in the previous examples in Sections 8.3.3 and 8.3.4) and re-processed with the correlation matrix. The resulting outcome was multiplied with the F_C factor matrix from the PCA of the C matrix to get the unknown concentrations (C_{ukn}). The PLSR result is compared with the original compositions of 20 samples in Figure 8.8. Although we are not able to deconvolute or interpret the NIR spectra at all due to overlap of spectral features and the concentration data of the reference samples are expected to have a fair amount of measurement errors, the correlation between the concentrations predicted from PLSR and the original concentrations is quite good. The R^2 value for the lignin component is relatively smaller than others. That is because the variance in the lignin concentration among the sample is small.

Once the PLSR model constructed with 20 samples was verified in the above step by analyzing the training data set again, it can be used to predict the cell wall concentrations of the rest 78 samples from their NIR spectra. This was done by treating 78 NIR spectra as unknowns (A_{ukn}) and calculating the corresponding wall compositions (C_{ukn}) using the correlation matrix determined from the 20 reference samples. Figure 8.9 shows the cell wall compositions determined from the PLSR

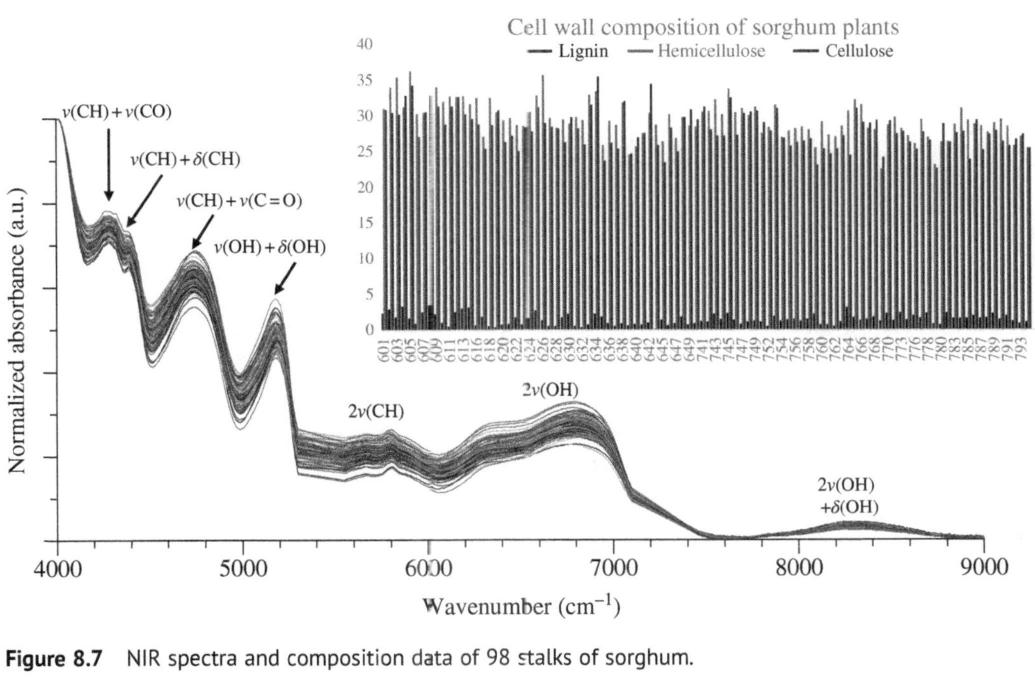

Figure 8.7 NIR spectra and composition data of 98 stalks of sorghum.

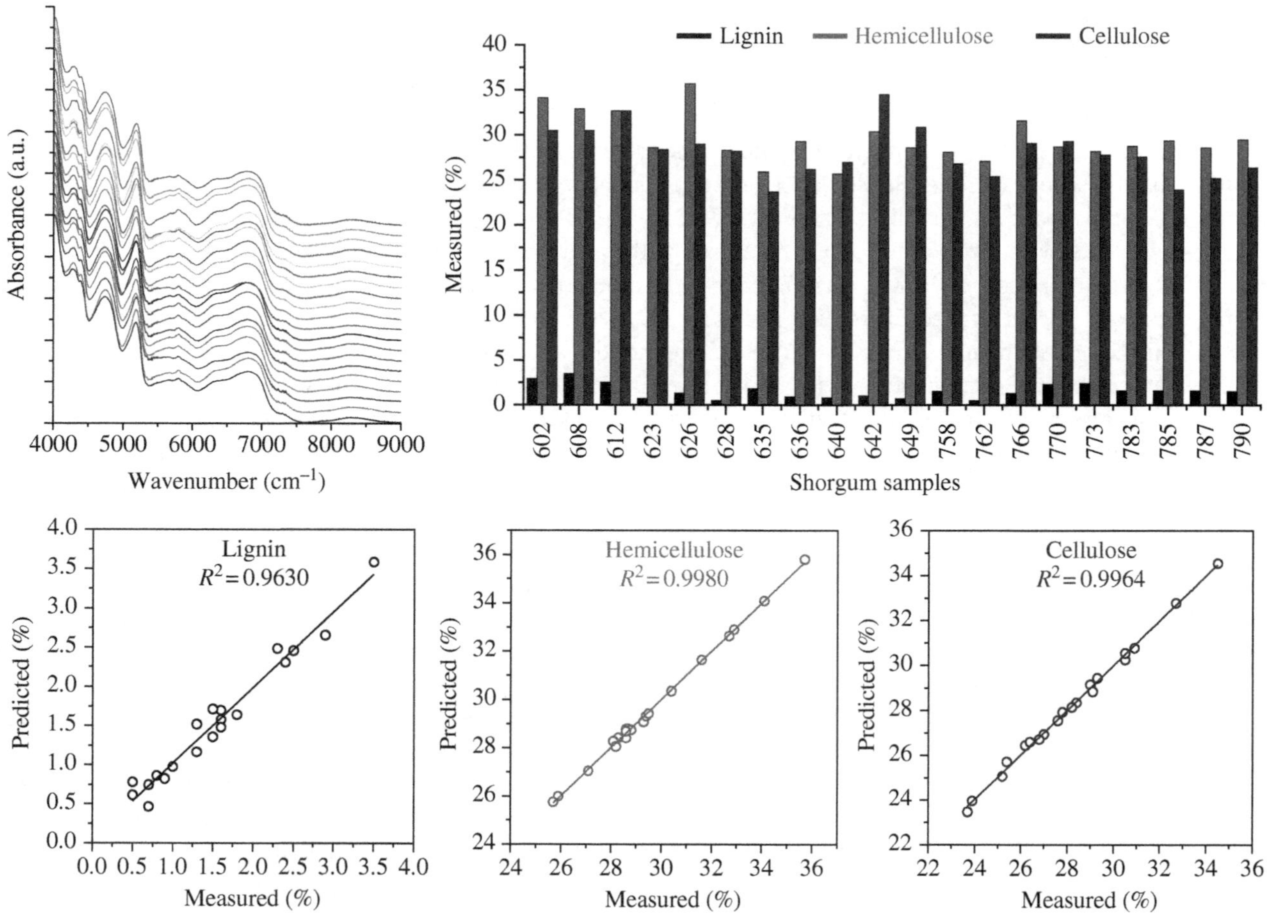

Figure 8.8 NIR spectra of randomly chosen 20 sorghum stalks and comparison of the PLSR results and the measured compositions of cell wall.

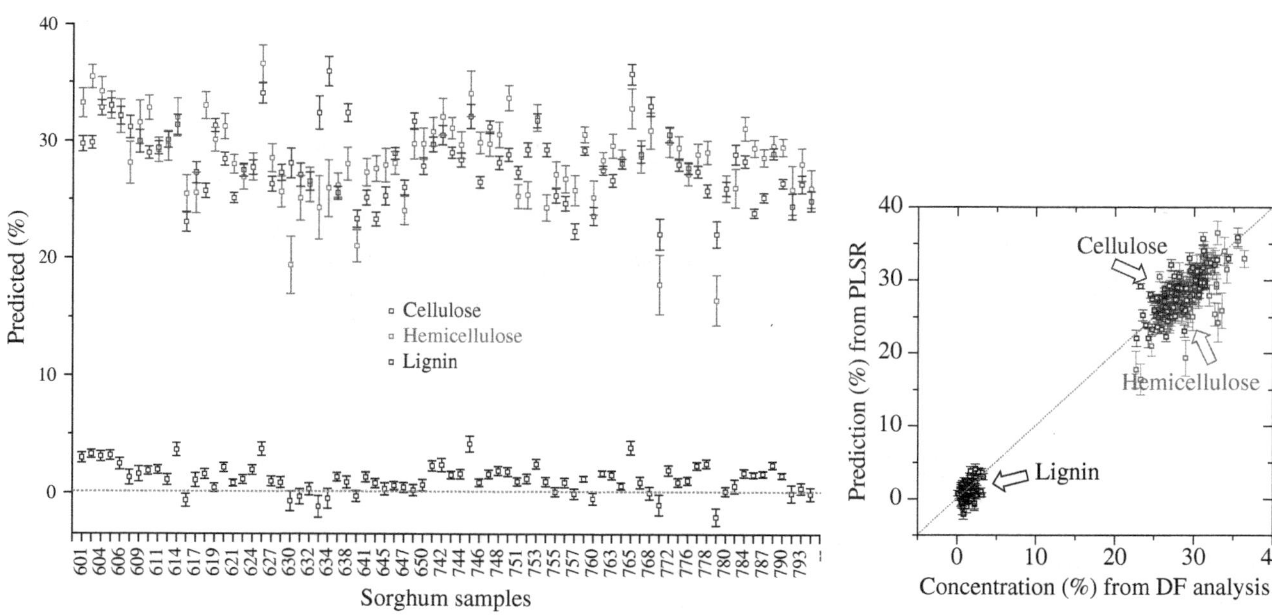

Figure 8.9 Prediction of 78 samples using the PLSR model constructed from 20 samples.

process and compares them with the chemical compositions determined from the DF analysis method. There are a few outliners or bad prediction points (such as negative lignin concentrations in several cases). Overall, the correlation between the predicted values from PLSR and the values determined from DF analysis are not too bad, suggesting the NIR analysis coupled with PLSR could be used as a routine screening tool although its accuracy is somewhat limited. The uncertainty can come from errors in NIR measurements or errors in DF analysis, or both. It is also possible that biological variances existing among 78 test samples produced via different mutations were not modeled by the 20 reference samples selected.

8.4 Further Reading for Multivariate Analysis

In commercial software, many additional data manipulations steps are used. Those include mean centering, normalization, variance scaling, and Poisson scaling, which are not discussed in this chapter and can be studied further as needed [5]. The basic concepts and procedures covered in this chapter are sufficient for processing most spectroscopic and image data. Other multivariate analysis methods are beyond the scope of this book. Readers can find good tutorials and books with exploratory examples on MCR [6], maximum autocorrelation factors (MAFs), and DFA, HCA, and ANN. Studying such methods can be done when the need to use such methods arises.

Practice Problems

8.1 In an analytical lab, a TA provided six reference samples with known concentrations and two unknown samples. A team of students measured absorbances at three wavelengths. The data are shown in the table below. One of the team members has done the trendline analysis, using Excel, of absorbance at each wavelength.

	Absorbance		
Concentration (a.u.)	**at λ_1**	**at λ_2**	**at λ_3**
1	4	6	10
2	6.9	7.9	11
3	10.1	9.9	12
4	12.9	12.1	13
5	16.1	14.1	14
6	19	16	15
Unk-1	20.8	17.7	16.2
Unk-2	23.5	20.2	16.7

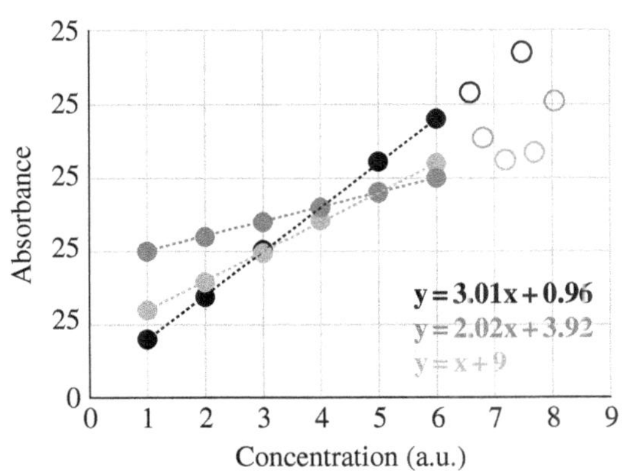

The team was asked to do PCA/PCR for the entire data set collected at three wavelengths.

a) Write the covariance of the absorbance data set (it should be 3 × 3 matrix) and eigenvalue matrix. Plot the eigenvalues in the scree plot format.
b) List the factor matrix, score matrix, and residue matrix for the case where PCA is done with only one factor (using the one with the largest eigenvalue). And run PCR using this one factor and find the correlation matrix and the concentrations of unknown samples.
c) List the factor matrix, score matrix, and residue matrix for the case where PCA is done with two factors (using the largest and the second largest ones). And run PCR using these two factors and find the correlation matrix and the concentrations of unknown samples.
d) List the factor matrix, score matrix, and residue matrix for the case where PCA is done with three factors. And run PCR using all three factors and find the correlation matrix and the concentrations of unknown samples.
e) Why are the three concentrations of unknown samples obtained with different numbers of factors different? Which one would be the most accurate? And why?
f) Look at the scores in the PCA with one factor. The increments of score values reflect the overall trend of the absorbance data set that can be described with the factor consisting of a set of three numbers only. The scree plot shows that it covers >99.98% of the variances in the data set. But the residues are very large. Explain the reason that the residues are so large.
g) When the PCA is done with two factors, then 99.999% of the variance is described; but the residues of the two-factor PCA result are still large. Compare the residue values of the two-factor PCA result with the third scores in the score matrix and the values in the residues matrix of the three-factor PCA. What aspects of the variances that have not described by the first two factors are now described by the third factor?
h) Now, let us compare the PCR result with the linear regression results. The unknown concentration can be calculated from the data at each wavelength; the values are shown as open circles in the chart. From that, the average of three values predicted from the individual trendline equations can be calculated. Compare the PCR result with the averaged value from the linear regression. Are they the same? Explain why they are different.

References

1 Lee JLS, Gilmore IS. The application of multivariate data analysis techniques in surface analysis. In: Vickerman JC, Gilmore IS, editors. Surface Analysis – The Principal Techniques. West Sussex, UK: John Wiley & Sons; 2009. p. 563–612.
2 Dethlefsen C, Sørensen PG, Hvidt A. Excess volumes of propanol-water mixtures at 5, 15, and 25°C. Journal of Solution Chemistry. 1984;13(3):191–202.
3 Björklund M. Be careful with your principal components. Evolution. 2019;73(10):2151–8.
4 Van Soest PJ, Robertson JB, Lewis BA. Methods for dietary fiber, neutral detergent fiber, and nonstarch polysaccharides in relation to animal nutrition. Journal of Dairy Science. 1991;74(10):3583–97.
5 Morais CLM, Lima KMG, Singh M, Martin FL. Tutorial: multivariate classification for vibrational spectroscopy in biological samples. Nature Protocols. 2020;15(7):2143–62.
6 de Juan A, Jaumot J, Tauler R. Multivariate curve resolution (MCR). Solving the mixture analysis problem. Analytical Methods. 2014;6(14):4964–76.

9

Thin Film Analysis via Reflectometry and Ellipsometry

We have studied, in Chapter 5, how the light (electromagnetic wave) propagates through an optical medium and how it is reflected and refracted at the interface of two media with different refractive indexes. From that, Snell's law and Fresnel coefficient equations are derived. Chapter 6 describes how to utilize these equations for linear spectroscopy, especially ATR-IR, SR-IR, RAIRS, PM-RAIRS, BAT-IR, and DRIFT. We also have touched the infrared spectroscopic imaging at the end of Chapter 6, which includes microscopic IR analysis, photothermal AFM-IR, s-SNOM nano-IR, and PiFM. In Chapter 7, Snell's law and Fresnel coefficients are combined with nonlinear polarization responses of the medium and applied to vibrational SFG spectroscopy. In Chapter 8, we have studied how to process the spectroscopic data (especially signal intensities in measured spectra) using multivariate analysis methods for quantification of analytes under consideration. Now, let us revisit Snell's law and Fresnel coefficient equations and see how they are used to analyze the structure of thin films, especially determining the thickness of thin films.

In fact, we already have discussed the relevant principle in Chapter 6. For example, in ATR-IR analysis, the experimentally measured $log(1/R)$ of adsorbate layers on the IRE surface can be compared with the theoretically calculated intensity to determine the effective thickness of the adsorbed layer (see Section 6.1.3). In RAIRS experiment, the experimentally measured $\Delta R/R_o$ of adlayers on dielectric materials or metallic surfaces can be compared with theoretically calculated intensities for a given thickness of thin film with a known refractive index (see Sections 6.3.2 and 6.3.4). For adsorbates on metallic surfaces, it is even possible, using PM-RAIRS, to measure the thickness of adsorbates in the presence of the same molecules in the gas phase (see Section 6.3.5). For thin adsorbate or organic layers on transparent substrates, a slight change in the transmittance of the *p*-polarized beam at the Brewster incidence angle can be measured and compared with the theoretical calculation results to determine the thickness information of adlayers on the surface (see Section 6.4). In SFG, the spectral intensity and shape are influenced by interferences between the signal generated at the top and bottom interfaces of the adlayer or thin film. Although the interference effect on the SFG signal intensity can be theoretically predicted [1], it is difficult to determine the layer thickness from the measured intensity because multiple parameters affect the SFG intensity (see Chapter 7).

All methods described in Chapters 6 and 7 are based on "absorption" of the electromagnetic wave by the medium (molecules at the interface and/or in the subsurface region) along the propagation path (transmission, RAIRS, and PM-RAIRS), within the attenuation depth (SR-IR), within the evanescence decay depth (ATR-IR), or within the coherence length (SFG). So, they are coupled with molecule-specific identification. The multivariate data processing of these absorption signals is discussed in Chapter 8. Film thickness measurements can be done "without absorption," as long as there is a contrast in refractive index at the wavelength of the light used. This chapter covers such techniques that do not rely on resonance absorption – namely, *reflectometry* and *ellipsometry*. Since the resonance absorption is not required, these techniques can be applied to any materials that permit propagation of electromagnetic wave. Although it does not require absorption, it can also work when absorption occurs.

9.1 Recap of Light Reflection and Transmission Principles

We have already derived the theoretical equations in previous chapters. Snell's law is explained in Section 5.2.1:

$$n_i \sin\theta_i = n_t \sin\theta_t \tag{5.61}$$

Surface and Interface Analysis: Principles and Applications, First Edition. Seong H. Kim.

Companion website: www.wiley.com/go/Kim

Here, θ_i and θ_t are the incidence angle of the incoming (i) and outgoing (t) wave at the interface of two media with different refractive indexes, n_i and n_t. In Section 5.2.1, we have considered non-absorbing medium only; but this equation works even for the absorbing medium. So, n_i and n_t becomes complex numbers ($\tilde{n} = n + ik$; eq. (5.92)) when we deal with absorbing medium.

At the interface of two media with different optical densities (i.e., $n_i \neq n_t$), a fraction of incoming beam is reflected back to the incoming medium and refracted into the outgoing medium. Their electric field amplitude ratios of the s- and p-polarized lights are defined by the Fresnel coefficients:

$$r_s = \frac{n_i \cos\theta_i - n_t \cos\theta_t}{n_i \cos\theta_i + n_t \cos\theta_t} = -\frac{\sin(\theta_i - \theta_t)}{\sin(\theta_i + \theta_t)} \tag{5.70}$$

$$t_s = \frac{2n_i \cos\theta_i}{n_i \cos\theta_i + n_t \cos\theta_t} = \frac{2\cos\theta_i \sin\theta_t}{\sin(\theta_i + \theta_t)} \tag{5.71}$$

$$r_p = \frac{n_t \cos\theta_i - n_i \cos\theta_t}{n_t \cos\theta_i + n_i \cos\theta_t} = \frac{\tan(\theta_i - \theta_t)}{\tan(\theta_i + \theta_t)} \tag{5.74}$$

$$t_p = \frac{2n_i \cos\theta_i}{n_t \cos\theta_i + n_i \cos\theta_t} = \frac{2\cos\theta_i \sin\theta_t}{\sin(\theta_i + \theta_t)\cos(\theta_i - \theta_t)} \tag{5.75}$$

For a single interface, these are all equations that we need to use to describe all optical phenomena occurring at that interface. When a layer is present, then we are dealing with two interfaces – reflection and refraction at the top interface of the layer, and reflection and refraction at the bottom interface of the layer. The same five equations are used to describe the reflection and refraction at each interface. The only additional term to consider is interference of the waves reflected and refracted at two interfaces separated by the thickness of the layer (d_2), which is described in Figure 9.1 (which is the same one studied in Section 6.3.1). The final outcomes of these interferences are derived for reflection (r_{123}) in Section 6.3.1 and transmission (t_{123}) in Section 6.4:

$$r_{123} = \frac{r_{12} + r_{23}e^{2i\beta}}{1 + r_{12}r_{23}e^{2i\beta}} \tag{6.33}$$

$$t_{123} = \frac{t_{12}t_{23}e^{i\beta}}{1 + r_{12}r_{23}e^{2i\beta}} \tag{6.47}$$

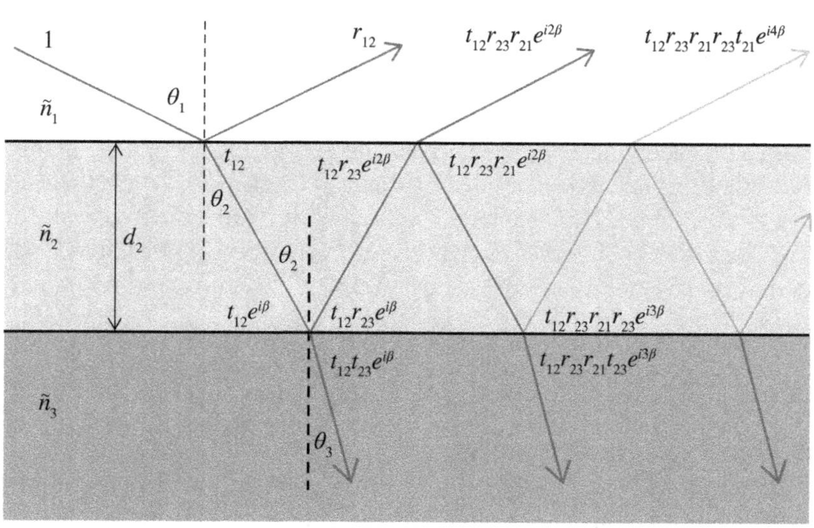

Figure 9.1 Illustration of a three-medium optical system ($\tilde{n}_1/\tilde{n}_2/\tilde{n}_3$) where an incident wave (whose intensity is normalized to 1) is being reflected and transmitted due to the boundaries of separated by d_2.

where subscript numbers indicate the medium through which light is propagating and β is the phase difference of the beam at the medium-1/2 interface and the medium-2/3 interface:

$$\beta = \frac{2\pi d_2}{\lambda}\tilde{n}_2 \cos\theta_2 = \frac{2\pi d_2}{\lambda}\sqrt{\tilde{n}_2^2 - \tilde{n}_1^2 \sin^2(\theta_1)} \tag{6.25}$$

For the non-absorbing medium, the complex refractive index ($\tilde{n}_2$) is replaced with a real number (n_2). The intensities of the reflected and transmitted beams at the 1/2/3 interface are the absolute square of the reflection and transmission coefficients. We should do this calculation for *s*- and *p*-polarizations separately:

$$R_{s,123} = r_{s,123} \cdot r^*_{s,123} \tag{6.34}$$

$$R_{p,123} = r_{p,123} \cdot r^*_{p,123} \tag{6.35}$$

$$T_{s,123} = t_{s,123} \cdot t^*_{s,123} \tag{9.1}$$

$$T_{p,123} = t_{p,123} \cdot t^*_{p,123} \tag{9.2}$$

Here, the terms with subscript *s* are calculated by inserting eqs. (5.70) and (5.71) into eqs. (6.33) and (6.47), and those with subscript *p* are calculated by inserting eqs. (5.74) and (5.75) instead. This kind of calculations can be expanded to multilayer systems (Section 6.4). When unpolarized light is used, then its reflectance and transmittance are the average of the *s*- and *p*-polarizations: $R_{unpol,123} = (R_{s,123} + R_{p,123})/2$ and $T_{unpol,123} = (T_{s,123} + T_{p,123})/2$.

9.2 Reflectometry

The reflectance of a non-absorbing thin film system varies as a function of its refractive index n_2 and thickness d_2 (Figure 9.1). Equations (6.34) and (6.35) can be re-written in a trigonometric form:

$$R = \frac{r_{12}^2 + r_{23}^2 + 2r_{12}r_{23}\cos(2\beta)}{1 + r_{12}^2 r_{23}^2 + 2r_{23}r_{23}\cos(2\beta)} \tag{9.3}$$

Again, r_{12} and r_{23} are the Fresnel reflection coefficients for the air/film and film/substrate interfaces, respectively, and β is the phase different of the wave at these two interfaces. Here, we drop the *s* and *p* in the subscript, for simplicity; but, they should not be dropped when reflectance is calculated. We can see that reflectance will oscillate following the cos(2β) function as the film thickness increases. At the normal incidence ($\theta_i = 0°$), the max and min of reflectance can be calculated to be:

$$R_{\max} = \left(\frac{n_1 - n_3}{n_1 + n_3}\right)^2 \text{ when } 2n_2 d_2 = m\lambda \tag{9.4}$$

$$R_{\min} = \left(\frac{n_2^2 - n_1 n_3}{n_2^2 + n_1 n_3}\right)^2 \text{ when } 2n_2 d_2 = \left(m + \frac{1}{2}\right)\lambda \tag{9.5}$$

where *m* in integer. As the film thickness increases, the maximum intensity will be observed periodically at:

$$d_m = \frac{m\lambda}{2n_2 \cos\theta_2} \tag{9.6}$$

Figure 9.2 shows an example where this principle is used to monitor, *in situ*, changes in thickness of a film growing in reactive conditions [2]. In this specific case, a Ziegler–Natta polymerization catalyst was produced on a gold foil in an ultra-high vacuum (UHV) system and then enclosed in a high-pressure reaction cell to which ethylene gas was supplied. For the polymerization kinetics study, it was required to measure the amount of polyethylene growing on top of the catalyst. If the reaction was done with catalyst powders with a large surface area in a reactor system where gravimetric weighing could be done *in situ*, the polymer growth kinetics could be monitored by measuring the weight change. But, for the high-pressure reaction cell fitted inside the UHV chamber, it was not possible. So, the reflectometry method was applied. Through a small window above the sample, a diode laser beam was irradiated at an incidence angle of 15° and the reflected beam intensity

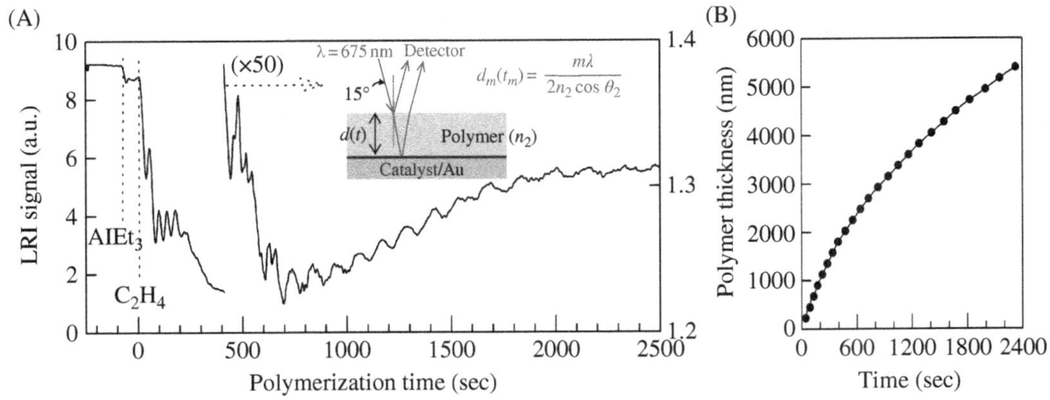

Figure 9.2 Reflectrometric monitoring of thickness of a polyethylene film growing on a Ziegler–Natta polymerization catalyst. (A) Raw signal recorded and (B) calculated thickness of polymer layer as a function of time. Reproduced from Kim et al. [2]/with permission of Elsevier.

was measured in real time. As soon as the ethylene gas was charged into the reactor cell, the reflected beam intensity significantly dropped (Figure 9.2A). This sudden drop indicated that the polymer film growing on the catalyst surface was not uniform. The film surface was not smooth, and the crystallization of polyethylene chains occurred spontaneously; so, these caused scattering loss of the light intensity. In the reflectance data, undulating patterns with small amplitudes could still be seen (Figure 9.2A). The maximum points of the data can be counted as a function of reaction time and then the thicknesses at those points are calculated using eq. (9.6), assuming $n_2 = 1.5$. The resulting plot of polymer thickness as a function of time is shown in Figure 9.2B, from which the polymerization kinetics could be deduced [2].

In the previous example, it is possible to calculate the film thickness by simply counting the R_{max} point because m can be counted sequentially from zero as the film thickness increases. For an unknown sample, the R measurement at one wavelength at one incidence angle cannot give the film thickness. That is because when we solve eq. (9.3) for d_2, we get infinite numbers of solution, that is, $d_2 + \dfrac{m\lambda}{2n_2 \cos\theta_2}$ where $m = 0, 1, 2, 3$, and so on. To get a definitive answer, we need to do additional measurement at different wavelength (λ) or different incident angle (θ_i). Changing incident angle may need realignment of the beam, so it is inconvenient. Instead, we can do the measurement at multiple wavelengths at a fixed incident angle. The latter case is called spectroscopic reflectometry. By doing the measurements at more than the minimum requirement (which is three in this example since we have three unknowns variables: d_2, n_2, and m), we can increase the accuracy of the calculation (we have learned it in Chapter 8).

Figure 9.3 shows the reflectance of the s-polarized light calculated for a hypothetical thin film with $n_2 = 1.5$ (assuming no dispersion for simplicity) on a silicon wafer ($n_3 = 3.4$; again, assuming no dispersion) at two different incident angles ($\theta_i = 0°$ and 45°) in the visible wavelength region (300 nm < λ < 800 nm). At a given angle, R_{max} and R_{min} are fixed, but the periodicity decreases as the film thickness (d_2) increases. When the film is thin, R changes monotonically in this wavelength region and the rate of change with λ is a function of d_2. If R is measured with low noise, then thickness measurement can be done down to 5~10 nm fairly accurately. In this example where $n_2 = 1.5$ and $n_3 = 3.4$, R at normal incidence is 0.279 at 300 nm and 0.295 at 800 nm if $d_2 = 10$ nm. For the s-polarization, reflectance increases monotonically as the incidence angle increases (see Figure 5.13); so, the difference between R_{max} and R_{min} increases as well. When the full range of the maximum variance in R increases, the measurement precision can be improved. The determination of film thickness (d_2) is done by fitting the measured $R(\lambda)$ at a given incident angle using eq. (9.3). Since all commercial instruments come with the automatic fitting function, we can skip the data fitting practice.

Not only the film thickness but the refractive index of the film can also be determined from the spectroscopic reflectance measurement. Figure 9.4 displays the calculation results done for 50 nm and 100 nm thick films with different refractive indexes. For a given film thickness, the lower limit in R (eq. 9.5) and the periodicity (eq. (9.6)) vary as the refractive index changes. So, by doing spectroscopic reflectance measurements, both refractive index (n_2) and thickness (d_2) of a film can be determined simultaneously.

For ultrathin films, the reflectance measurement in the visible wavelength range does not work well. That is because the reflectance chance is so small. For a 3-nm thick organic film on a silicon wafer, we would need to measure reflectance with λ less than 80 nm. This is an extreme ultraviolet (EUV) region. Photons in this wavelength range ($hc/\lambda > 15$ eV) are readily

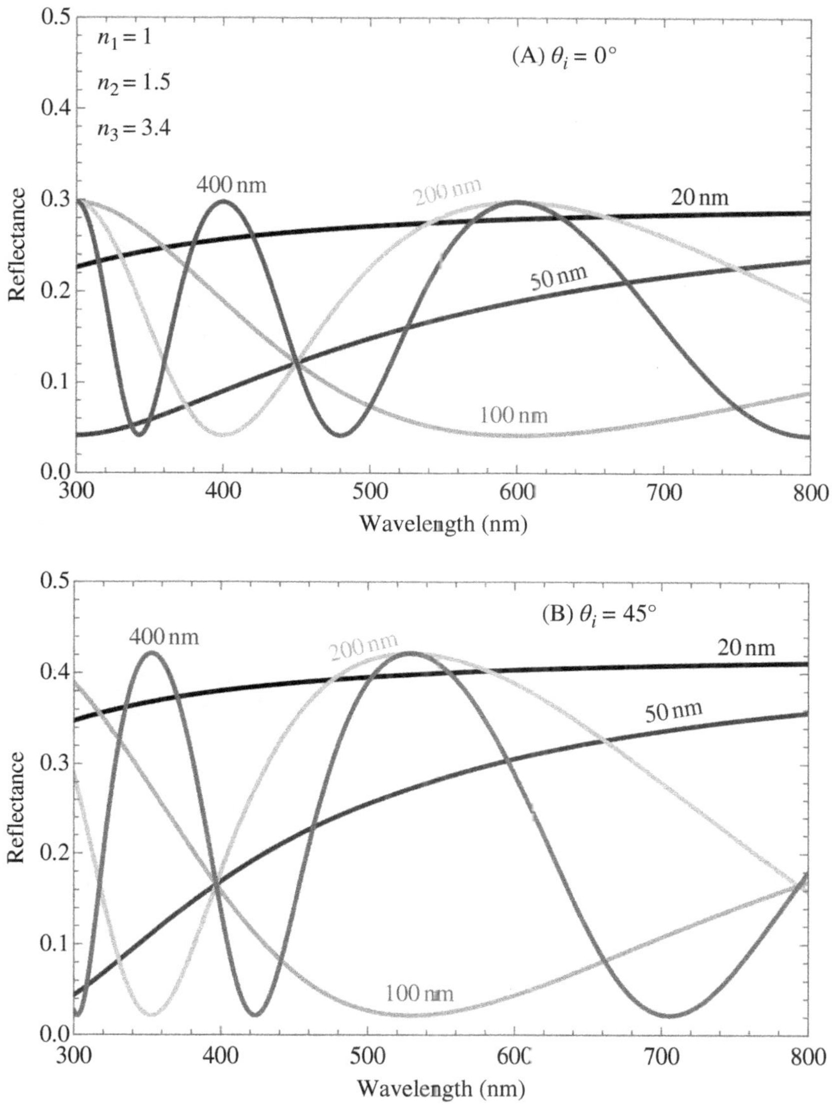

Figure 9.3 Spectroscopic reflectance of *s*-polarized light calculated for a thin film with n_2 = 1.5 with different thickness (d_2) on a substrate with n_2 = 3.4 at incidence angle of (A) 0° and (B) 45°.

absorbed by most materials through electronic transitions (ionization), which often leads to degradation of materials. Thus, reflectometry cannot be employed. For such ultrathin films, x-ray reflectometry (XRR) can be applied [3, 4]. But, XRR does not work well for films thicker than a few hundred nanometers. In Chapter 6, we have studied that the resonance absorption in the mid-IR range can be used to determine the film thickness of organic samples down to a few Angstroms; but this method works only when we know the refractive index of the sample. An alternative technique that can be applied to various types of materials with a wide range of thickness from ultrathin (≲1 nm) to thick (>10 μm) films is ellipsometry [5, 6].

One may have noticed that the basic principle used for film thickness measurement through the reflectance measurement is the same as the one used for RAIRS. In both cases, the measured intensity is calculated by multiplying eq. (6.33) with its conjugate. In other words, they detect the absolute magnitude of the reflected electric fields only. The only difference between RAIRS and reflectometry is whether the probe beam interacts with the film through resonance absorbance or not. In ellipsometry, the same eq. (6.33) is used again; but what is measured is the ellipticity of the polarization, not the intensity, of the reflected beam. Ellipsometry measures the relative magnitude of *p*- and *s*-polarization components and their phase difference. Since more variables are measured in ellipsometry, more information about the film can be extracted.

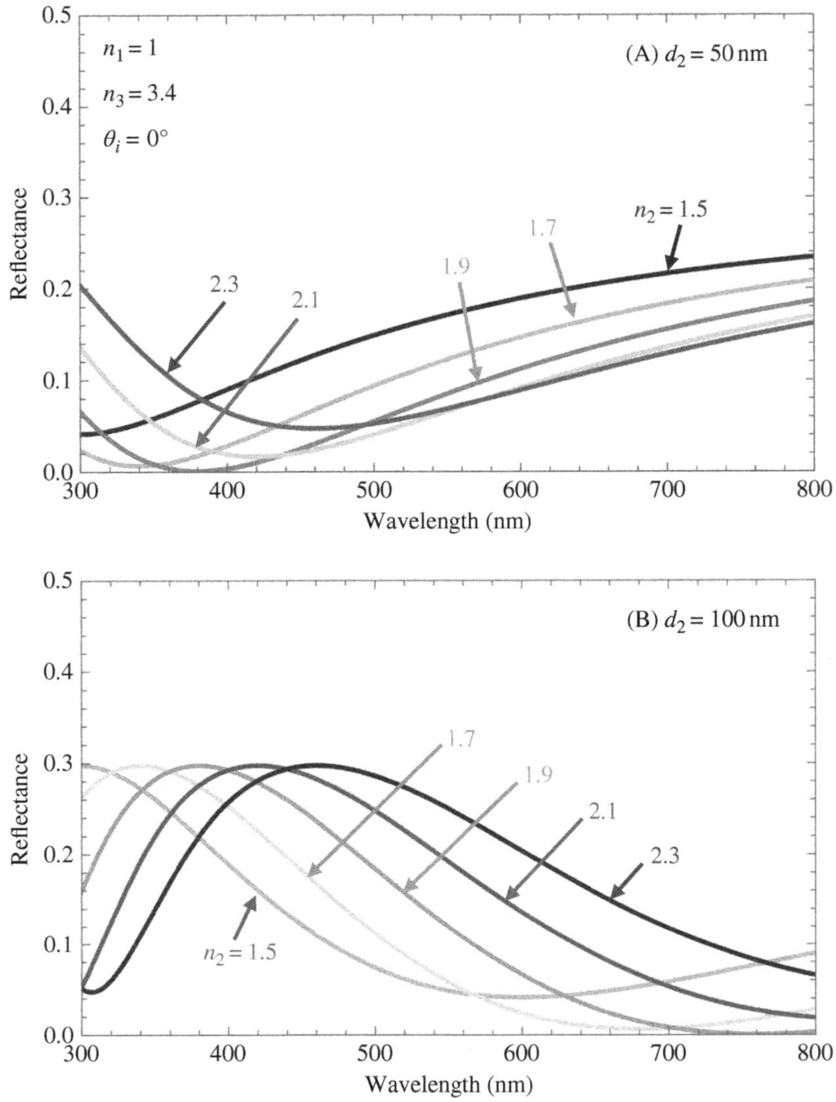

Figure 9.4 Spectroscopic reflectance calculated for a thin film with different refractive index (n_2) and a thickness of (A) 50 nm thick and (B) 100 nm on a substrate with n_3 = 3.4 at the normal incidence.

9.3 Ellipsometry

9.3.1 Polarization Change of Light upon Reflection from a Thin Film

When a linearly polarized light with its polarization angle is not aligned parallel (p) or perpendicular (s) to the incidence plane and it makes a reflection at the interface of two materials with different refractive indexes, there is a phase shift as well as an amplitude change. For example, when $n_1 < n_2$, r_p is positive and r_s is negative below the Brewster angle (θ_p), and both are negative above the Brewster angle (see Figure 5.13). Thus, the phase difference between p and s is either 180° below θ_p or 0° above θ_p. If the incident beam is linearly polarized, then the reflected beam is still linearly polarized. But the electric field amplitudes of the reflected beam along the p and s polarization directions are different unless the incident angle is zero (where the distinction of p and s polarization cannot be made; see Figure 5.13). If the same linearly polarized light is reflected off a thin film, the phase shift and amplitude change in p- and s-polarizations vary further due to interference of the reflected beams from two interfaces (top and bottom of the film; Figure 9.1).

This interference makes the reflected light from a film to become elliptically polarized (unless the incident beam is perfectly p- or s-polarized). In Section 6.3.1, eq. (6.25) describes the single-path phase difference between the top and bottom interfaces, which is the same for both p- and s-polarizations; eq. (6.26) is the same for both p- and s-polarizations. But, the

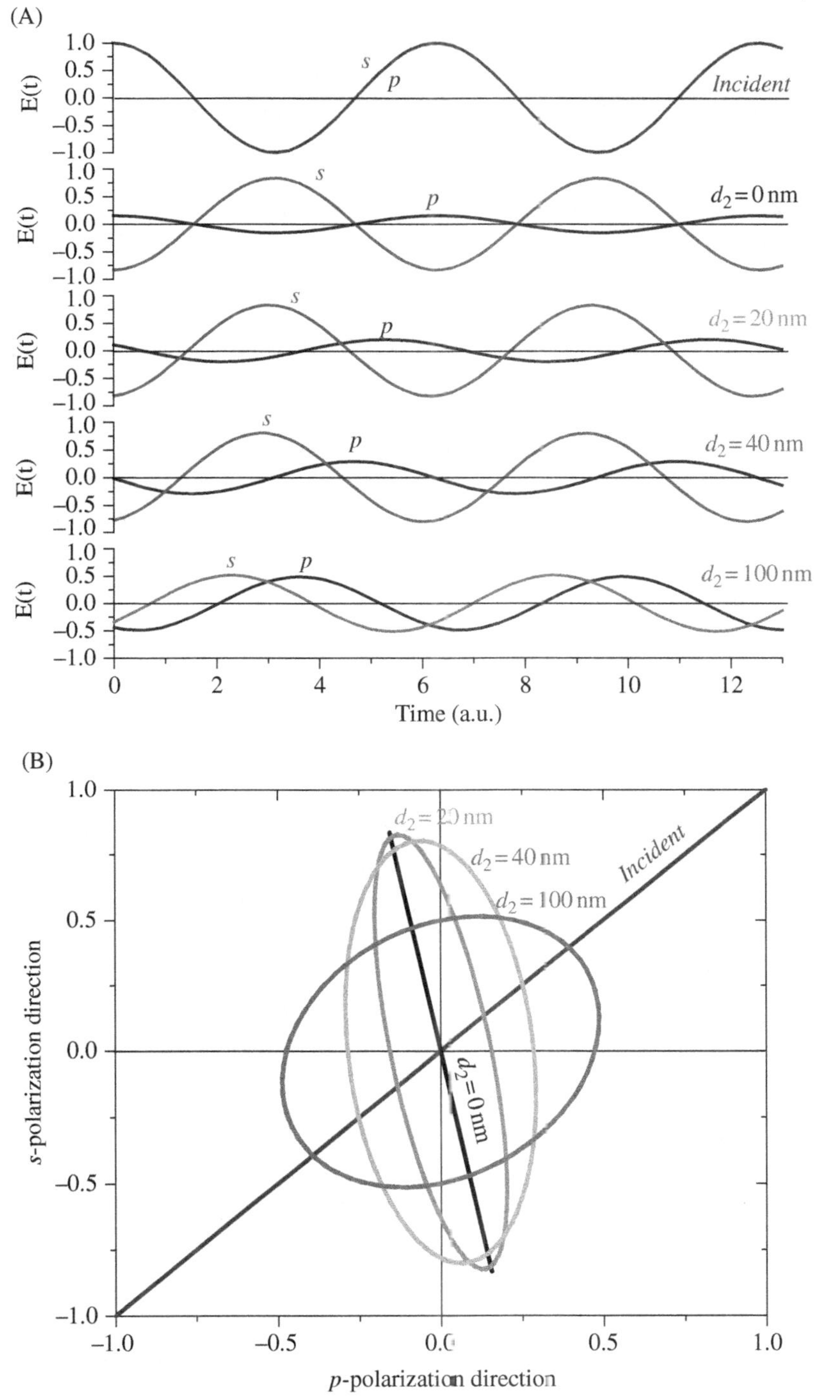

Figure 9.5 (A) Simulated $E_p(t)$ and $E_s(t)$ of a linearly polarized incident beam (632.8 nm) with a 45° polarization angle with respect to the incidence plane and the beam reflected from a thin film (n_2 = 1.5) with different thicknesses (n_2) on a silicon substrate (n_3 = 3.882; θ_i = 70°). (B) Time-averaged projection of E-field to the plane perpendicular to the beam propagation direction.

magnitude of q in eq. (6.27) is different; that is, $r_{p,21}r_{p,23} \neq r_{s,21}r_{s,23}$. In other words, the number of multiple reflections contributing to the interferece is different for p- and s-polarizations. Due to this difference, the phase shift and amplitude change in p- and s-polarizations are different; i.e., $r_{p,123} \neq r_{s,123}$ in eq. (6.33). The extent of ellipticity depends on optical properties of the substrate as well as optical properties and thickness of overlying films.

Figure 9.5 illustrates the electric field of the beam propagating in the z-direction projected on the x-axis (within the incidence plane; so, p-polarization component) and y-axis (perpendicular to the incidence plane; so, s-polarization). The simulation is done using $r_{p,123}$ and $r_{s,123}$ (eq. 6.33) for a system mimicking a SiO_2 film with a given thickness (d_2)

on a Si wafer irradiated with a HeNe laser at a 70° incident angle. The incident beam is linearly polarized with the polarization angle of 45° with respect to the laser incidence plane. So, the $E_p(t)$ and $E_s(t)$ components are perfectly in-phase (overlapped) in time domain and their magnitudes are set to be 1 (Figure 9.5A). If the arrow described by $E_p(t)$ and $E_s(t)$ components of the incident beam is projected to the plane perpendicular to the beam propagation direction (i.e., the detector surface), it oscillates linearly along the diagonal direction (incident line in Figure 9.5B). The Brewster angle of the air/Si interface is 75.55°. Since the incident angle is smaller than this, the p-component of the beam reflected from the air/Si interface (i.e., $d_2 = 0$ nm) is small ($|r_{p,\,13}| = 0.1556$) and its sign remains the same as the incident beam. In contrast, the s-component of the reflected beam is large ($|r_{s,\,13}| = 0.8335$) and its sign is reversed with respect to the incident beam (i.e., 180° phase shift, δ). When the trajectory described by $E_p(t)$ and $E_s(t)$ components of the reflected beam is projected to the plane perpendicular to the beam propagation direction, it oscillates along the linear line marked as $d_2 = 0$ nm in Figure 9.5B.

When a 20-nm thick SiO_2 is present on the Si surface, then the amplitude and phase of the p- and s-components of the reflected beam are calculated to be: $|r_{p,123}| = 0.2031$ with $\delta = 57.64°$ and $|r_{s,123}| = 0.8265$ with $\delta = -182.82°$. When the trajectory described by these components is projected to the plane perpendicular to the beam propagation direction, it rotates along the elliptical orbit described by the line marked as $d_2 = 20$ nm in Figure 9.5B. When the SiO_2 film thickness is changed to 40 nm, $|r_{p,123}|$ increases to 0.2928 with $\delta = 92.70°$ from the incident beam, and $|r_{s,123}|$ decreases to 0.8033 with $\delta = -165.09°$. The trajectory made by time-averaged projection of this field projected onto the plane perpendicular to the propagation direction is given by the elliptical orbit marked as $d_2 = 40$ nm in Figure 9.5B. When the film thickness is changed to 100 nm, the elliptical orbit becomes more rounded and its main axis close to the diagonal direction of the incident beam as shown in Figure 9.5B.

In this simulation, we see that the degree of ellipticity of the reflected beam is a function of film thickness on the substrate. Conversely, it is possible to calculate the thickness of film from the measurement of the ellipticity of the reflected beam. This is the basic principle of ellipsometry.

9.3.2 Polarization Ellipse (Ellipsometric Angles)

In ellipsometry, the ratio of the amplitude of p- and s-polarizations and their phase differences are measured. These are expressed using two angles called Ψ and Δ. The graphical definitions of these ellipsometric angles are shown in Figure 9.6. The p- and s-polarization components can be described as $E_p = E_{0p}e^{i\delta_p}$ and $E_s = E_{0s}e^{i\delta_s}$, respectively, using their amplitudes and phase angles with respect to the incident beam. The relative amplitude ratio of the p- and s-polarization components defines the Ψ angle of the polarization ellipse:

$$\tan\Psi = \frac{E_{0p}}{E_{0s}} \tag{9.7}$$

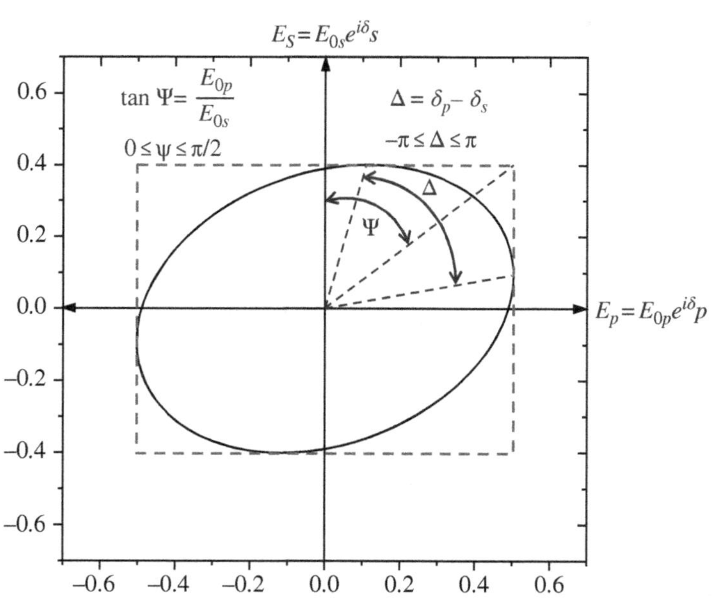

Figure 9.6 Parameters (Ψ and Δ) describing the polarization ellipse.

Ψ varies from 0 to $\pi/2$. The relative difference between their angles defines the Δ angle of the polarization ellipse:

$$\Delta = \delta_p - \delta_s \tag{9.8}$$

This angle tells that the maximum of the p-polarization component is reached after the time interval of Δ/ω from the maximum of the s-polarization component. Δ can vary from $-\pi$ to π and the positive value means clockwise (right-handed) precession and the negative value means counterclockwise (left-handed) precession. In this scheme, the amplitude (Ψ) and phase (Δ) angles of the incident beam in Figure 9.5 are 45° and 0°, respectively. The Ψ and Δ angles of the reflected beam for $d_2 = 0$ nm in Figure 9.5 are 10.57° and $-\pi$, respectively. At the Brewster angle (θ_p), Ψ becomes zero because the amplitude of the reflected p-polarization component is zero. As the incident angle passes through θ_p, Δ shift by π because the phase of p-polarization is flipped.

9.3.3 Singe-Wavelength Ellipsometry (SWE)

The linearly polarized incident beam with the polarization angle of 45° has $\Psi = 45°$ and $\Delta = 0°$ (Figure 9.5). For a thin film system (Figure 9.7A), the individual reflectivity of p- and s-components is calculated with eq. (6.33). As an example, the calculated $r_{p,123}$ and $r_{s,123}$ for a SiO_2 film on a Si wafer are plotted as a function of film thickness in Figure 9.7B in which the incident beam wavelength and angle are 632.8 nm and 70°, respectively. As in the reflectometry case (where the intensity of the reflected beam is measured), the polarization ellipticity of the reflected beam also varies in a cyclic pattern as the film thickness increases. The periodicity of this cyclic pattern is calculated with the following equation (which is essentially the same as eq. (9.6)):

$$d_{2,m} = \frac{\lambda}{2\sqrt{n_2^2 - (n_1 \sin\theta_i)^2}} \tag{9.9}$$

For the example shown in Figure 9.7, this periodicity is 284.16 nm.

The ellipsometric angles of the reflected beam are calculated by taking the ratio (ρ) of the p- and s-polarization reflectivity coefficients:

$$\rho = \frac{r_{p,123}}{r_{s,123}} = \tan\Psi \exp(i\Delta) \tag{9.10}$$

This may look like one question for two unknowns (Ψ and Δ); in fact, there are two relationships in eq. (9.9) since both sides of the equation are complex numbers. So, we get the following equations to solve:

$$\mathrm{Re}\left[\frac{r_{p,123}}{r_{s,123}}\right] = \tan\Psi\cos\Delta \quad \text{and} \quad \mathrm{Im}\left[\frac{r_{p,123}}{r_{s,123}}\right] = \tan\Psi\sin\Delta \tag{9.11}$$

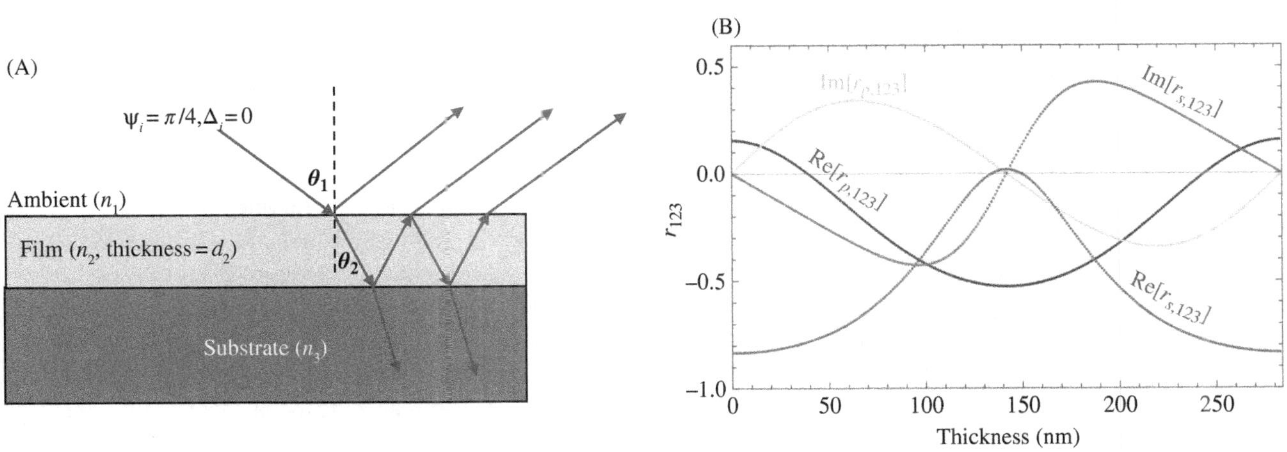

Figure 9.7 (A) Ellipsometry measurement of a thin film systems. (B) Real and imaginary parts of theoretically calculated $r_{p,123}$ and $r_{s,123}$ as a function of thickness (d_2) of a SiO_2 film (n_2 = 1.457) on a Si wafer (n_3 = 3.8812 + 0.0195i) at λ = 632.8 nm and θ_i = 70° in air (n_1 = 1).

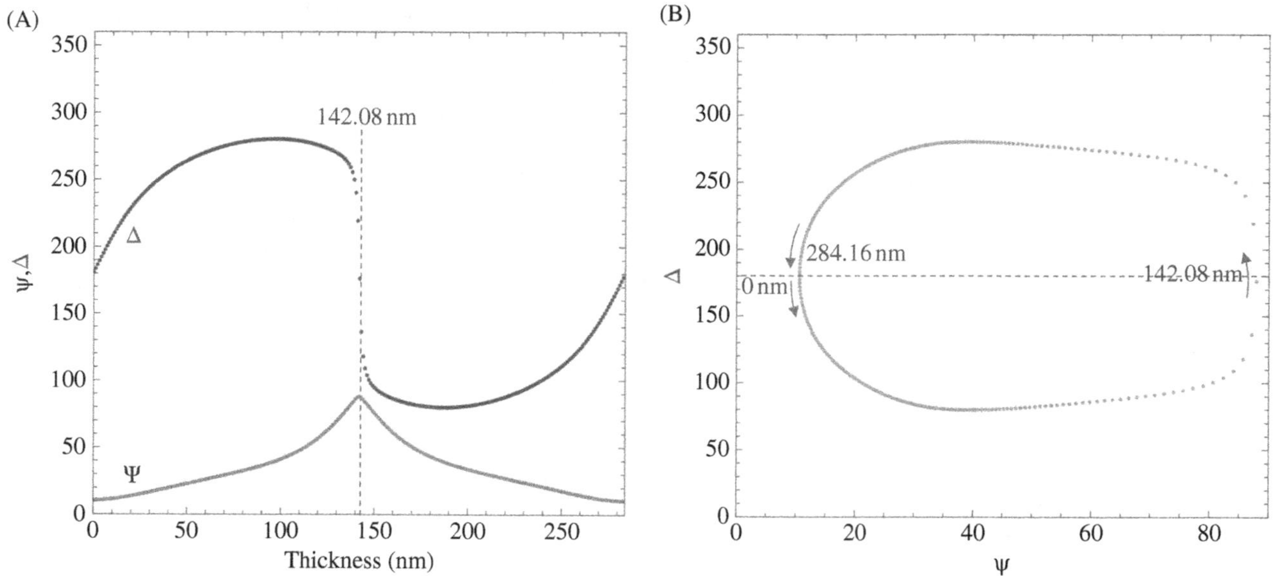

Figure 9.8 (A) Changes in Ψ and Δ as a function of film thickness (d_2) for the system shown in Figure 9.7. (B) Plot of ellipsometric angles in the Ψ – Δ coordinate.

There will be two sets of solutions satisfying eq. (9.11). Since the absolute magnitude ratio in eq. (9.7) will be always positive, we choose the solution with a positive value in Ψ. Most commercial instruments set δ of $r_{p,12}$ at $\theta_i < \theta_p$ to be 360°, instead of 0°; thus, the negative phase value (−180°~0°) is plotted in the 180°~360° domain. Figure 9.8A plots $\Psi(d_2)$ and $\Delta(d_2)$ calculated using the data in Figure 9.7B (from 0 to 284 nm with 1 nm increment). Figure 9.8B plots these solutions in the Ψ-Δ coordinate. The Ψ-Δ trajectory is divided into two half circles: one with increasing Ψ from 0 to 142.08 nm and the other with decreasing Ψ from 142.08 to 284.16 nm. These two branches have opposite phases.

When we try to find the thickness of film (with a known refractive index) from the measured ellipsometric angles (Ψ and Δ) at a single wavelength at a single incidence angle, the unknown variable in eq. (9.11) is d_2. When this equation is solved, then we get multiple solutions, that is, $d_{solved} + m \cdot d_{2,m}$ where $m = 0, 1, 2$, and so on. In order to find the exact solution, we need to find another information to determine the m value. This can be done if we measure the ellipsometric angles at another incidence angle. This is because $d_{2,m}$ is a function of incident angle (see eq. (9.9)). This is called variable-angle (VA) ellipsometry. Figure 9.9 plots $\Psi(d_2)$ and $\Delta(d_2)$ calculated for the SiO_2/Si system at different incident angles ($\theta_i = 50°$, 60°, 70°, 80°) of the 632.8 nm probe beam. For a given thickness (d_2), there is a unique set of $\Psi(\theta_i)$ and $\Delta(\theta_i)$. Thus, if Ψ and Δ are measured at two different incident angles, then we can find a definitive solution for d_2.

In the case of the SiO_2/Si system, the silica film is transparent to the He–Ne laser beam ($\lambda = 632.8$ nm). So, there is no attenuation of the probe beam by absorption. But, if the film absorbs the probe beam, then the ellipsometric angles does not come back to the original value after the thickness is increased to $d_{2,m}$. Figure 9.10 plots the $\Psi(d_2)$ and $\Delta(d_2)$ calculated for a SiGe film on a Si wafer probed with a 496 nm beam at θ_i of 70° and 75° [7]. The spiral patterns in the Ψ-Δ coordinate with increasing d_2 are governed by two factors: (i) the periodicity which would appear if the absorption is negligible and (ii) the gradual loss of the beam intensity reflected at the film/substrate interface due to the attenuation in the absorbing film.

The incident angle can be optimized to improve detection sensitivity for a small change in film thickness. At the Brewster angle of the substrate (i.e., $r_{p,13} = 0$), the phase of the p-polarization changes by 180°, which makes Δ vary over 360°. Thus, by choosing the incident angle close to the Brewster angle of the substrate, we can make $\Delta(d_2)$ vary over the full 360° scale as the film thickness (d_2) changes [8]. This becomes particularly useful when the refractive index difference between the film and substate is small and both are transparent. For the ellipsometric measurement of a polystyrene (PS; $n_2 = 1.5983$) film on a fused silica substrate (SiO_2, $n_2 = 1.4607$) with a green laser ($\lambda = 532$ nm) at $\theta_i = 70°$, the total change in Ψ and Δ is only 6° and 20°, respectively, when d_2 increases from 0 to 205.6 nm (which is $d_{2,m}$). Then, by choosing the incident angle close to the Brewster angle of the substrate, the sensitivity to the film thickness can be increased significantly. This is demonstrated in Figure 9.11. When the incident angle is 55°, Δ varies over 100° when d_2 varies over 194 nm (which is $d_{2,m}$ at $\theta_i = 55°$). When the incident angle is changed to 56°–58°, then Δ varies over 360°, so the sensitivity to small changes in d_2 is increased even

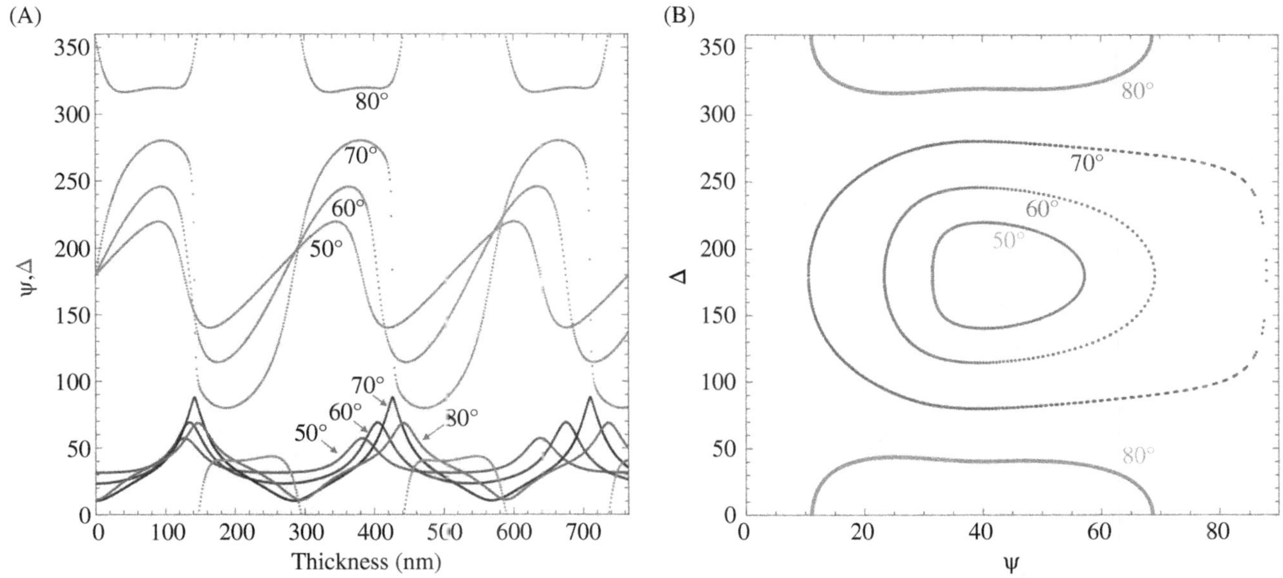

Figure 9.9 (A) Changes in Ψ and Δ, for different incident angles (θ_i) with λ = 632.8 nm, as a function of SiO_2 film thickness (d_2) on a Si wafer. (B) Plot of ellipsometric angles in the Ψ – Δ coordinate.

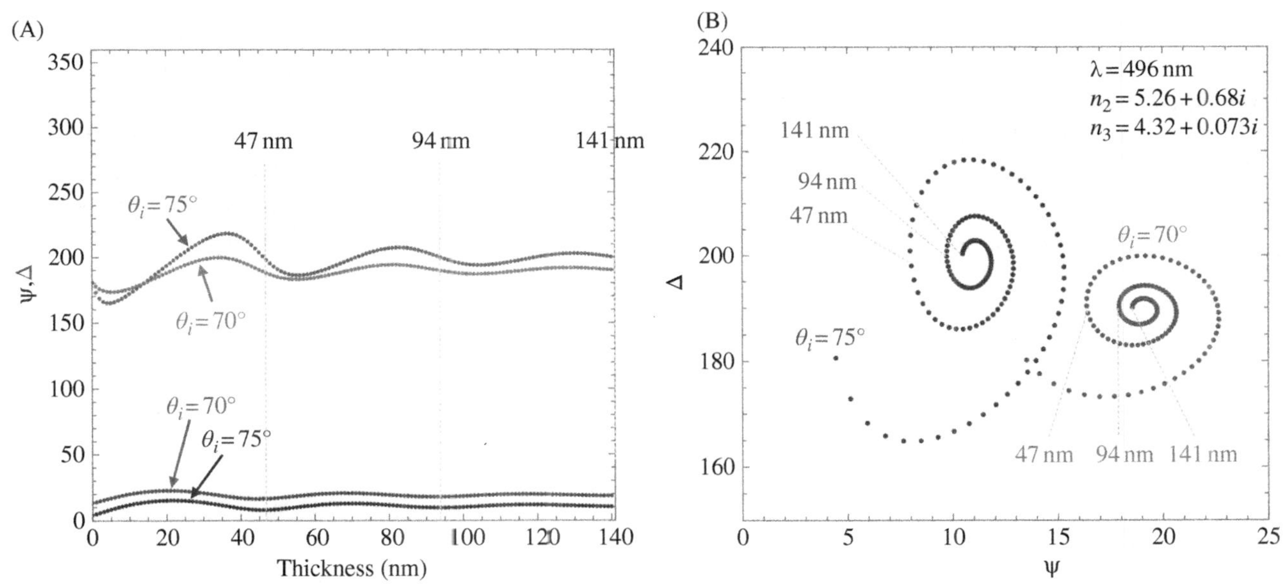

Figure 9.10 (A) Changes in Ψ and Δ as a function of SiGe film thickness (d_2) on a Si wafer probed with a 496 nm beam at θ_i of 70° and 75°. (B) Plot of ellipsometric angles in the Ψ-Δ coordinate.

further (3.6 times compared to the 55° case and 18 times compared to the 70° case). Note that, for the non-absorbing medium, the reflectivity of *p*-polarization becomes zero at the Brewster angle (i.e., $r_{p,13} = 0$; see Figure 5.13). But, for the thin film case, the $r_{p,123}$ value does not go through zero due to the phase difference in the beams reflected at the air/film and film/substrate interfaces.

When a dielectric film is present on a highly reflective medium (such as metal), the change in Ψ is negligible. That is because the reflectivity of the substrate is very high for both *p*- and *s*-polarizations at the incident angle below 80° (see Figure 6.27A). Figure 9.12A compares the ellipsometric angles calculated for the PS film on Si ($n_3 = 4.152 + 0.0518i$) and silver (Ag; $n_3 = 0.054 + 3.429i$) at an incident angle of 70° (λ = 532 nm). On the Si substrate, the presence of PS film

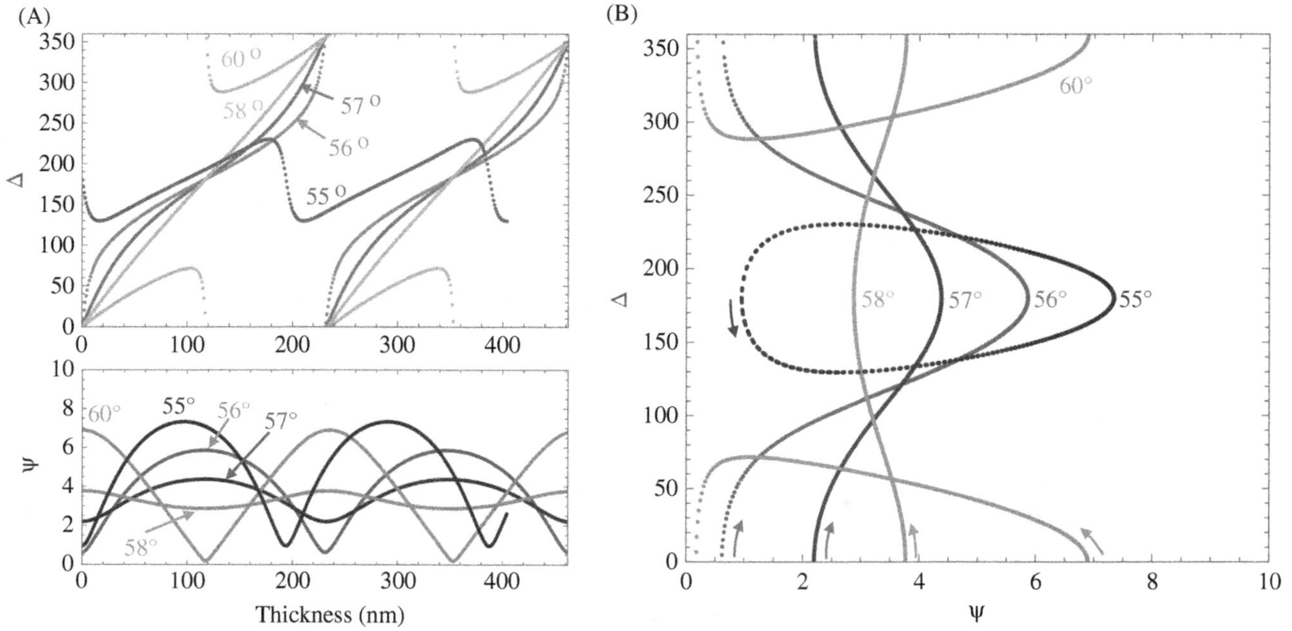

Figure 9.11 (A) Changes in Ψ and Δ as a function of film thickness (d_2) for the PS/silica system at various incident angles of a 532 nm beam. (B) Plot of ellipsometric angles in the Ψ-Δ coordinate. The arrows in (B) show the initial change in Ψ and Δ as d_2 increases from 0 nm.

makes both Ψ and Δ vary over a fairly large range. On the Ag substrate, Ψ is within ±1° from 45° because both $r_{p,23}$ and $r_{s,23}$ are close to 1; in contrast, Δ varies between 110° and 285° due to the large phase difference between $r_{p,23}$ and $r_{s,23}$, making ellipsometric determination of film thickness still feasible.

When a metallic film is deposited on a dielectric substrate, ellipsometric analysis with visible light can be done only when the film thickness is quite small. That is because the metal surface reflects most of the light and the extinction coefficient of metal is quite large for the visible light. Figure 9.12B plots the Ψ and Δ values calculated for an Ag film on Si with varying thicknesses. It shows that ellipsometric analysis can be done for the Ag film reliably up to 10–12 nm only. Beyond this, the

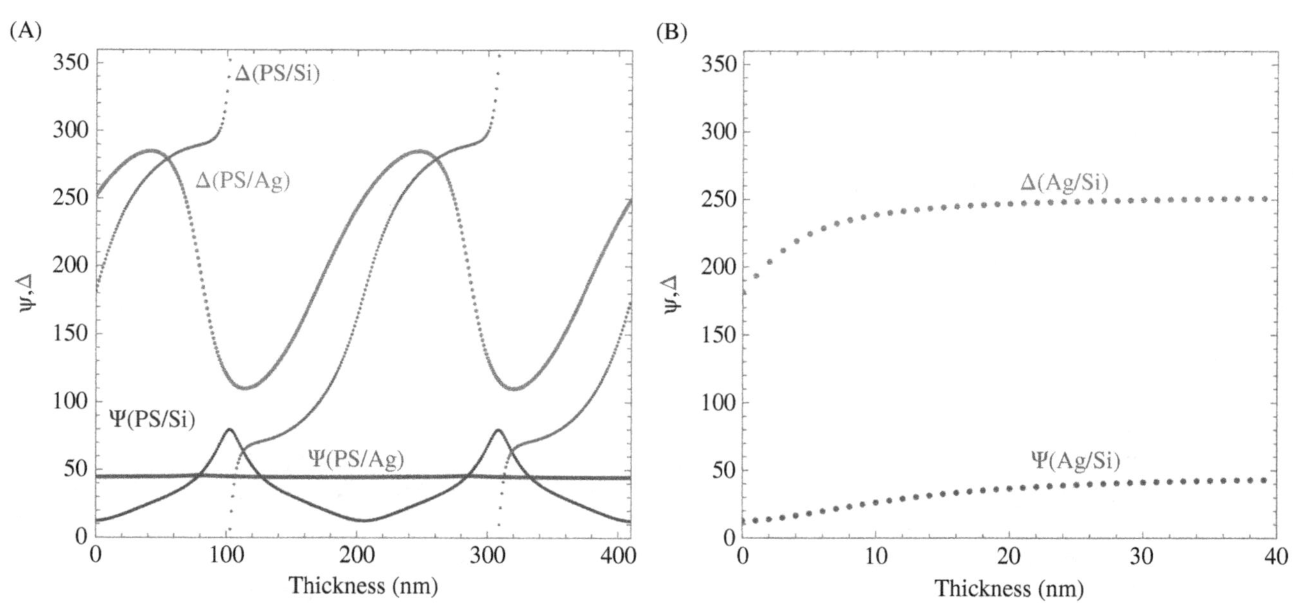

Figure 9.12 (A) Changes in Ψ and Δ as a function of film thickness (d_2) for PS/Si and PS/Ag systems. (B) Changes in Ψ and Δ as a function of film thickness (d_2) for an Ag film on Si. The laser wavelength is 532 nm, and the incident angle is 70°.

thickness sensitivity of Ψ and Δ is lacking. This is understandable if we consider the penetration depth (d_p) of 532 nm into the Ag film. Using eq. (6.20), it is calculated to be ~4 nm at 70° incident angle. After traveling $3d_p$, only 5% of the incident beam reaches the film/substrate interface. So, the interference effect from the beam reflected at the film/surface interface becomes negligible.

Not only the thickness (d_2) but the refractive index (n_2) of film can also be determined if Ψ and Δ are measured using a single wavelength at three different incident angles. More incident angles would give more accurate values of d_2 and n_2. But, this requires physical movement of optics to adjust the incident angle, so it is somewhat inconvenient unless the instrument is fully motorized and the incident angle (and so the angle of detection as well) is automatically controlled. An alternative approach is to change the wavelength of the probe beam because $d_{2,m}$ is a function of λ as well (see eq. (9.9)). If the wavelength is continuously varied over a large range by passing a broadband light through a monochromator, it is called spectroscopic ellipsometry (SE).

9.4 Spectroscopic Ellipsometry (SE)

9.4.1 Calculation of SE Angles

Spectroscopy ellipsometry (SE) can determine not only the film thickness but also the dispersion of refractive index of dielectric materials, the band-gap transition of semiconductor materials, and so on. Commercial instruments typically cover the wavelength from 300 to 900 nm. This can be extended to deep UV region (down to 200 nm) as well as near IR (up to 1.7 μ m). When the mid-IR light source is used, then molecular vibrations of organic materials and phonons of inorganic materials can be studied.

The governing equation for SE is the same as that of single-wavelength ellipsometry (SWE); the only difference is that the n_2 term is now a function of λ in SE, instead of being a constant in SWE. In SWE, we get only two equations at one incident angle (eq. (9.10)). If we know which $\Psi - \Delta$ cycle the film thickness is in (i.e., m value in $d_{solved} + m \cdot d_{2,m}$), we can determine the film thickness from SWE experiment at one incident angle (θ_i). If m needs to be determined, we need to do SWE measurements at two incident angles. If we need to determine the refractive index and thickness of the film simultaneously, we need to do SWE measurements at least three incident angles. In SE, all these parameters can be determined by one measurement at one incident angle since we get hundreds of equations from the Ψ and Δ data measured as a function of λ. Of course, the incident angle can be changed or optimized in SE to maximize the detection sensitivity.

Let us look at the SiO_2 film on a Si wafer, as an example. The refractive index of these materials can be found online [9]. Then, we just have to do the eq. (9.11) calculation at each wavelength to get the spectroscopic SE data, $\Psi(\lambda)$ and $\Delta(\lambda)$. Figure 9.13A,B plots the refractive index of fused silica (SiO_2) and Si downloaded from online refractive index database [9]. Figure 9.13C,D plots $\Psi(\lambda)$ and $\Delta(\lambda)$ for silica films with different thicknesses on a Si wafer. As the film thickness increases from 0 nm, $\Psi(\lambda)$ and $\Delta(\lambda)$ shift gradually with little change in the λ dependence. Up to ~20 nm thick film, the absorption band of the Si substrate can be seen in the $\Psi(\lambda)$ and $\Delta(\lambda)$ spectra. When the film thickness is larger than 50 nm, $\Psi(\lambda)$ and $\Delta(\lambda)$ are dominated by the interference effects of the beam reflected at the top and bottom interfaces of the silica film and show maximum and minimum in the λ dependence. As in the spectroscopic reflectometry, the wavelength difference between two consecutive maxima decreases as the film thickness increases further.

Figure 9.14 plots the $\Psi(\lambda)$ and $\Delta(\lambda)$ calculated for a GaInP film on a GaAs substrate. Figure 9.14A plots their refractive indexes obtained from online database [9]. Without the film, the $\Psi(\lambda)$ and $\Delta(\lambda)$ are obviously governed by the GaAs substrate. When a 50-nm thick GaInP film is present, the short wavelength side is already dominated by the GaInP feature, although a small feature of the GaAs component can still be seen at $\lambda \approx 415$ nm (Figure 9.14B). The interference fringes appear prominently in the long wavelength side only ($\lambda > 660$ nm) where the GaInP absorption is negligible.

9.4.2 Analysis of Measured SE Data

In SE analysis, there are sufficiently large number of equations to solve simultaneously (two equations at each wavelength). Thus, both complex refractive index ($n_2(\lambda) + ik_2(\lambda)$) and thickness ($d_2$) of an unknown film (even multiple layers) can be determined from the SE data. This involves fitting of the measured $\Psi(\lambda)$ and $\Delta(\lambda)$ with an optical model [6, 10]. Typical models used in SE analysis are schematically summarized in Figure 9.15. If the material thickness is larger than five times the penetration depth (d_p) of the probe beam (Figure 9.15A), then it can be considered "bulk." This is because the intensity

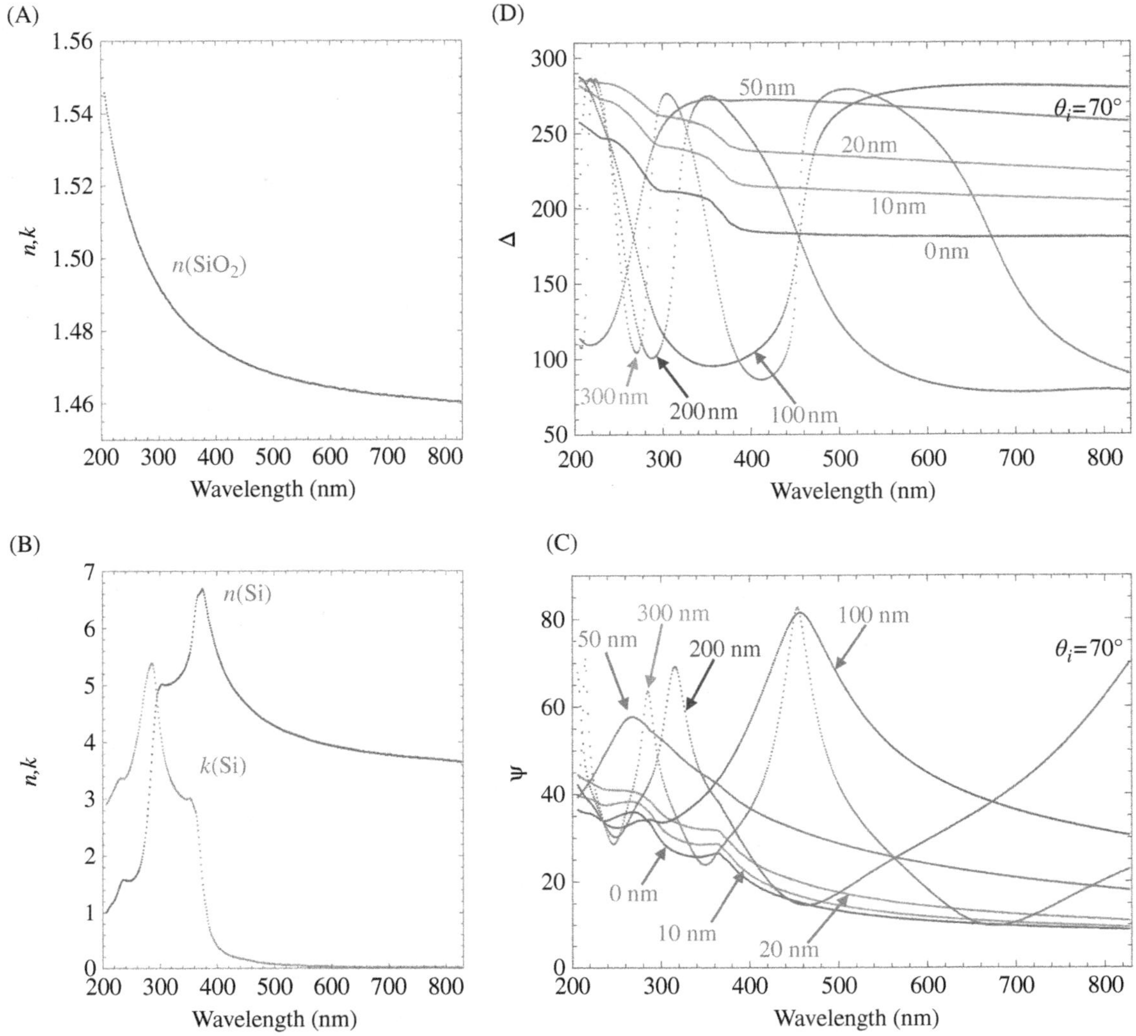

Figure 9.13 Refractive index of (A) amorphous silica and (B) silicon from online database. Simulated (C) $\Psi(\lambda)$ and (D) $\Delta(\lambda)$ as a function of wavelength for silica films with different thicknesses on a Si wafer.

of the beam reaching the bottom of the material is less than 0.7%; so, the interference effect of the beam reflected from the bottom interface on the reflection signal is negligible. If the back side of the material is roughened or in contact with a liquid with the same refractive index, it can also be modeled as bulk. If the sample is thinner than $5d_p$ and the backside is flat enough to reflect the probe beam, then it should be considered as another interface. Often, this backside reflection causes a depolarization issue because the back reflection has 180° phase shift with respect to the front reflection, causing destructive interference especially when the path difference through the transparent bulk is even multiples of $\lambda/2$.

In all simulations shown in the previous sections (which is based on Figure 9.7A), we have assumed that the top film surface is ideally flat (Figure 9.15B). In reality, the top surface may be rough. Then, the surface roughness layer can be considered a separate layer with a distinct refractive index (Figure 9.15C); its thickness (d_{sr}) will be proportional to the root-mean-square (RMS) roughness. The interface between the layer-1 and the bulk may not be flat. Then, the interfacial region can also be modeled as a separate layer (layer-2 in Figure 9.15D). Or layer-1 and bulk may react and form a new compound, which can be modeled as layer-2. If multiple layers of different materials are expected, then the number of layers can be increased until the calculation result of the model fits the experimental data reasonably well. We already have derived the equations for reflection from multiple layers in Section 6.4 (see Figure 6.34). So, a modified version of eq. (6.48) can be used to calculate $\Psi(\lambda)$ and $\Delta(\lambda)$ from multiple layers.

The effective refractive index of the surface roughness layer can be calculated using the effective medium approximation (EMA). Also, using EMA, we can calculate volume fractions in composite or porous materials from SE analysis. Based on the

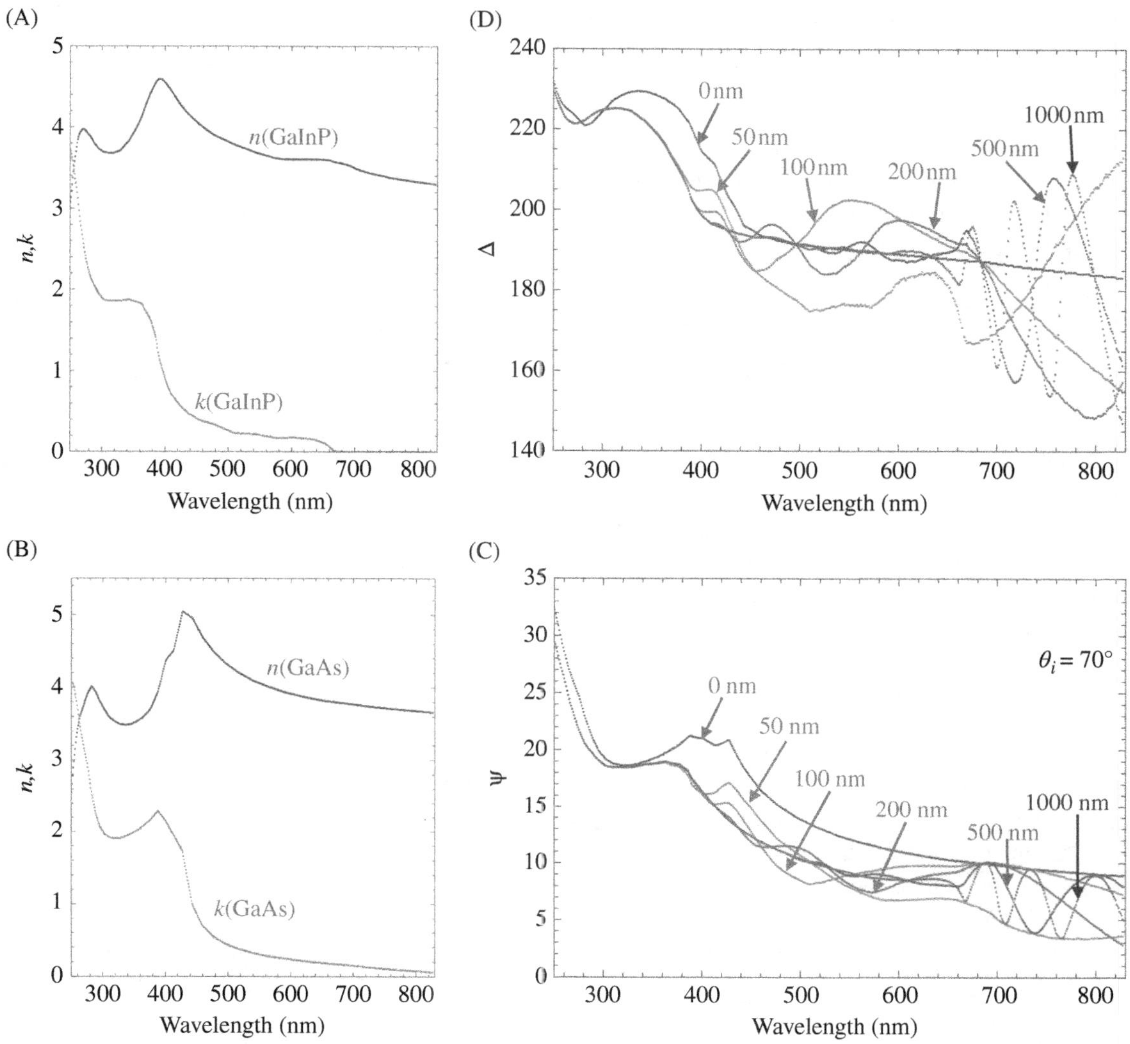

Figure 9.14 Refractive index of (A) GaInP and (B) GaAs from online database. Simulated (C) $\Psi(\lambda)$ and (D) $\Delta(\lambda)$ as a function of wavelength for GaInP films with different thicknesses on a GaAs wafer.

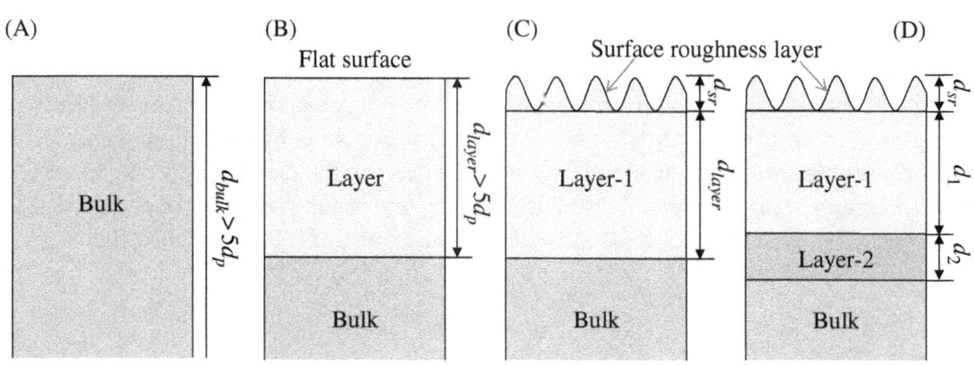

Figure 9.15 Optical models used in SE data analysis. (A) Bulk material. (B) One single layer without surface roughness. (C) A layer with surface roughness. (D) Multiple layers with surface roughness. Layer-2 may model interfacial roughness between layer-1 and bulk or a different material. More layers can be added.

Clausius–Mossotti relation, the relative dielectric permittivity (ε_r) of a material is a function of the polarizability (α) of constituent atoms or molecules and their number density (N):

$$\frac{\varepsilon_r - 1}{\varepsilon_r + 2} = \frac{N\alpha}{3\varepsilon_r} \tag{9.12}$$

Applying this to a material composed of two spherical dielectric components with relative permittivity ε_a and ε_b with volume fractions f_a and $(1-f_a)$, respectively, that are uniformly distributed in a host medium with relative permittivity ε_h, we can express the effective dielectric permittivity of the material as:

$$\frac{\varepsilon_r - \varepsilon_h}{\varepsilon_r + 2\varepsilon_h} = f_a\frac{\varepsilon_a - \varepsilon_h}{\varepsilon_a + 2\varepsilon_h} + (1-f_a)\frac{\varepsilon_b - \varepsilon_h}{\varepsilon_b + 2\varepsilon_h} \tag{9.13}$$

If the host material is air ($\varepsilon_h = 1$), eq. (9.13) is the Lorentz–Lorentz (LL) relation. When the host material is component a and surrounding component b, then $\varepsilon_h = \varepsilon_a$ and eq. (9.13) becomes:

$$\frac{\varepsilon_r - \varepsilon_a}{\varepsilon_r + 2\varepsilon_a} = (1-f_a)\frac{\varepsilon_b - \varepsilon_a}{\varepsilon_b + 2\varepsilon_a} \tag{9.14}$$

This is the Maxell–Garnett (MG) relation, which works well when the fraction of component b is small, that is, $f_a > (1-f_a)$. When the host material is the effective medium itself, then $\varepsilon_h = \varepsilon_r$ and eq. (9.13) becomes:

$$f_a\frac{\varepsilon_a - \varepsilon_r}{\varepsilon_a + 2\varepsilon_r} + (1-f_a)\frac{\varepsilon_b - \varepsilon_r}{\varepsilon_b + 2\varepsilon_r} = 0 \tag{9.15}$$

This is the Bruggeman approximation, which works well when volume fractions of both components are comparable. Recall that $\varepsilon = n^2$ (eq. 5.19). The EMA modeling works reasonably well when the sizes of inclusion domains in a composite material are smaller than $\lambda/10$ and its dielectric function is independent of size and shape.

Using the Bruggeman EMA model, we can estimate the effective refractive index of the surface roughness layer (n_{sr}). Let us assume that the void fraction within the roughness layer is $f_{void} = 0.5$ and n_a is the refractive index of the solid fraction (layer-1 in Figure 9.15C) and $n_b = 1$ (for air). If $n_a = 1.5$, then eq. (9.15) gives $n_{sr} = 1.24$. The void fraction can vary depending on the skewness of surface roughness, and the thickness of the surface roughness layer (d_{rs}) will vary with the RMS roughness and kurtosis. For skewness and kurtosis, see Section 10.2.

Then, the Ψ and Δ of the optical model in Figure 9.15C can be calculated using the reflectivity of three interfaces (see eq. 6.48):

$$\frac{r_{p,1234}}{r_{s,1234}} = \tan\Psi\exp(i\Delta) \tag{9.16}$$

where subscripts 1, 2, 3, and 4 are air, surface roughness layer, layer-1, and bulk, respectively. Each j-th layer will have its own refractive index (n_j) and thickness (d_j). The Ψ and Δ of the optical model in Figure 9.15D can be calculated using the reflectivity of four interfaces (see eq. 6.48):

$$\frac{r_{p,12345}}{r_{s,12345}} = \tan\Psi\exp(i\Delta) \tag{9.17}$$

where subscripts 1, 2, 3, 4, and 5 represent air, surface roughness layer, layer-1, layer-2, and bulk, respectively.

These equations are used to fit the measured $\Psi(\lambda)$ and $\Delta(\lambda)$ with a plausible optical model to find the refractive index (n_j) and thickness (d_j) of each j-th layer in the model. Recall that the refractive index here is a complex number if the layer is an absorbing material. The data fitting process, which is nonlinear parameter estimation, is mathematically intense; fortunately, most commercial SE systems come with various fitting functions where the user can build and modify optical models to fit the data. The goodness of the fitting result, that is, how well the model actually represents the measured system, can be judged by checking mean squared error (MSE) and mean squared weighted deviation (MSWD, also called reduced χ^2) [6, 10]. They are defined as:

$$MSE = \frac{1}{N-m-1}\sum_{j=1}^{N}\left(\rho_{ex}(\lambda_i) - \rho_{fit}(\lambda_i)\right)^2 \tag{9.18}$$

$$\chi^2 = \frac{1}{N-m-1}\sum_{j=1}^{N}\left(\frac{\rho_{ex}(\lambda_i) - \rho_{fit}(\lambda_i)}{\delta(\lambda_i)}\right)^2 \tag{9.19}$$

where the subscripts *ex* and *fit* represent experimental and model fit values at each wavelength, respectively, N and m are the number of measurement points and model parameters, respectively, and $\delta(\lambda_i)$ is the measurement error for ellipsometry instruments.

Note that MSE value is dependent on the representation of the SE data, that is, whether $\rho_{ex}(\lambda_i)$ is expressed with (Ψ, Δ), or $(\tan\Psi\cos\Delta, \tan\Psi\sin\Delta)$, or measured coefficients (α_1, α_2). Moreover, the value of MSE has the dimensions of the chosen parameterization and thus the meaning of absolute magnitude varies. The main use of MSE is to check if there is any other way to reduce the MSE value smaller or not. The rule of sum in data fitting is to reach a small MSE value with the minimum number of parameters (m). A marginal decrease in MSE after adding additional parameters may not mean the model with more parameters is better. The same has already discussed in Section 8.3.6. In the case of MSWD, we can say that the model is good if χ^2 is close to 1. If χ^2 is significantly larger than 1, then it means the model does not fit the data well.

9.4.3 Case Study of Employing SE

SE analysis is not just to determine the film thickness, but it can also provide the information relevant to the composition or structure of the film. From the refractive index determined from SE, the effective dielectric permittivity (ε_r) is calculated, which can be interpreted with the EMA model. It should be noted that such information is obtained with a series of assumptions to build the optical model to fit the SE data. Figure 9.16 shows an example of using SE to determine the thickness and porosity of a surface layer on a glass formed by aqueous corrosion [11]. A boroaluminosilicate glass (called international simple glass, ISG; modeling the glass used for nuclear waste immobilization) was corroded in silica-saturated, pH7 solution with 40 mM KCl at 90 °C for 1625 days (nearly 4.5 years). During the corrosion process, sodium, boron, and calcium species are leached out, leaving a silica-rich layer on the glass. This layer, a.k.a. alternation layer, is usually formed at the beginning of the corrosion process (within a month or so) and protects the bulk glass from further corrosion for the rest of the experimental duration. The pristine ISG glass was analyzed first with SE to get the refractive index of the bulk glass (Figure 9.16A, B). In Figure 9.16C,D, the 1625-day corroded sample shows oscillatory behaviors in $\Psi(\lambda)$ and $\Delta(\lambda)$ due to the presence of the alteration layer with a refractive index different from the bulk. The data were fitted with four layers – a surface roughness

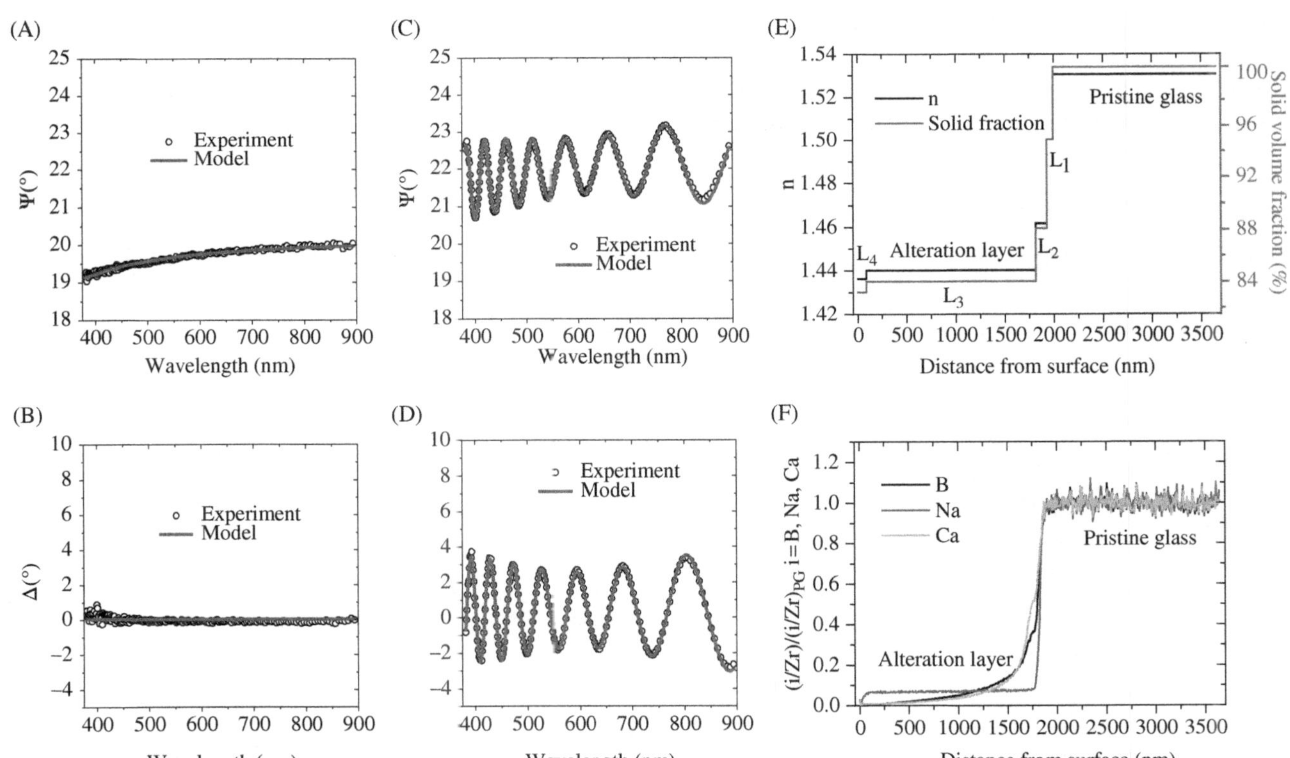

Figure 9.16 (A,B) SE data and fit of bulk ISG glass and (C,D) SE data and fit of ISG glass corroded in an aqueous solution for 1625 days. (E) Optical model constructed from the fitting of data in (C,D). (F) ToF-SIMS depth profile of the 1625-day corroded sample. Ngo et al. [11]/Springer Nature/CC BY 4.0.

layer (L_4), the main alteration layer (L_3), a transition layer (L_2), and a reaction front (L_1) where the corrosion reaction occurred (Figure 9.16E). In the data fitting, the refractive index of each sublayer was modeled using a Sellmeier oscillator model for the wavelength dependence and a Bruggeman EMA accounting for fractions of solid and void. The refractive index of the solid fraction may have changed due to the composition change, but it was assumed to be the same as the pristine glass. This assumption was made due to the difficulty in accurately determining the refractive index of the solid part of a porous leached layer. The optical model could be expanded to more layers, or a continuous gradient in refractive index in the L_2 and L_1 regions could be applied. Although the increased degrees of freedom (m in eq. (9.18)) could improve the overall fitting, it would also increase the uncertainty in the physical structure model applied. Thus, all fittings in SE data analysis were limited to the minimum number of layers needed. When the final optical model is compared with the depth profile obtained with the time-of-flight secondary ion mass spectrometry (ToF-SIMS; Figure 9.16F), the agreement appears to be good.

Due to the loss of network modifier ions (sodium and calcium) and part of the network forming species (boron), the alteration layer becomes porous, and the pore size distribution is an important parameter for the transport of the ions from and to the reaction front. Perhaps, SE is the most efficient technique to characterize the pore size distribution in a thin film on a monolithic sample surface; no other methods can do that easily. The basic principle is to use adsorption isotherm. For example, as relative humidity (RH) in the surrounding air increases, water is adsorbed in the pores, which leads to an increase of the effective refractive index of the alteration layer. This can be measured with SE (Figure 9.17A) and analyzed with the Bruggeman EMA model [11]. Assuming the layer thickness and glass volume fraction in each sublayer remained the same as the 0% RH values, the increase in refractive index can be converted to the amount of water in the pore. The amount of water in the alteration layer (L_3) determined during the adsorption and desorption processes is plotted in Figure 9.17B. The same calculation could be done for the L_2 layer as well. For this calculation, the pore sizes were considered to be small in comparison to the wavelength of the incident light. This assumption was required to validate the use of the EMA in representing the optical response of sublayers consisting of glass, void, and water components. The RH-dependent analysis result indicated that the majority of pores are filled with adsorbed water at ~50% RH due to the capillary condensation, and there is a small hysteresis in the desorption isotherm.

The pore sizes could be estimated using the Kelvin equation describing the relationship between the relative humidity of vapor (p/p_o) and the radius of a meniscus (r_m) of the liquid in the pores [11]. Again, several assumptions were made in the calculation. The surface tension and molar volume of liquid water were used, although those thermodynamic properties of the adsorbed water layer might be different. The pores were assumed to be cylindrical and the water contact angle on the pore wall was assumed to be 0°. Based on those assumptions, the meniscus radius can be set equal to the radius of pore (r_k).

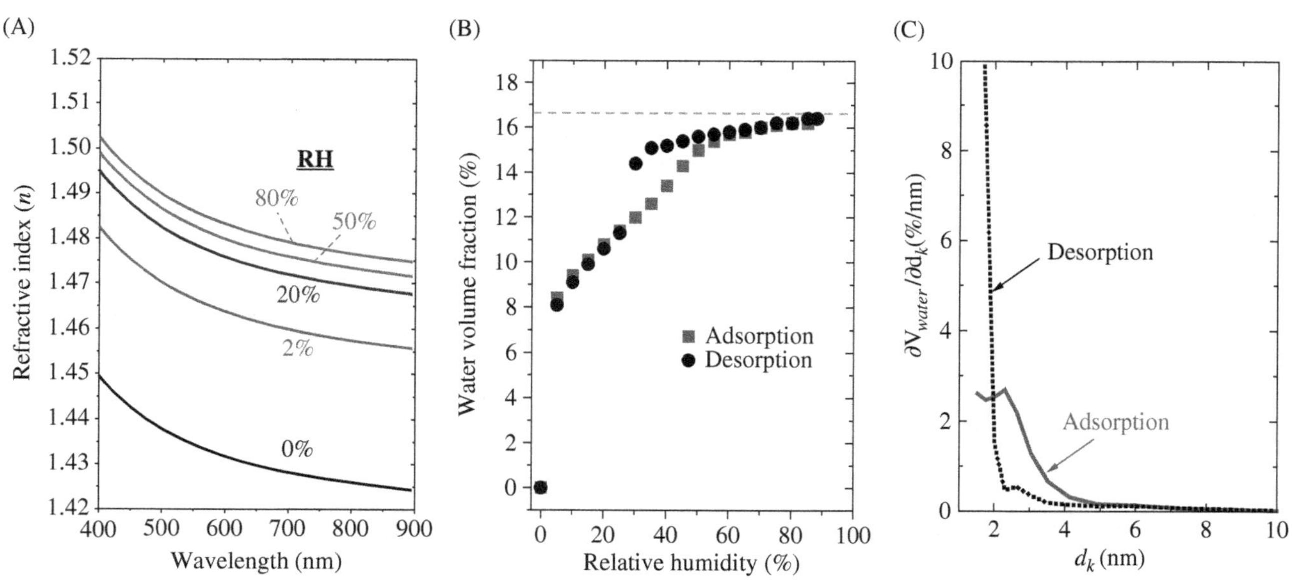

Figure 9.17 (A) Refractive index of the L_3 layer of the 7-day corroded sample. (B) Adsorption and desorption isotherm of water in the L_3 layer of the 1625-day corroded sample. The dashed line is the void fraction determined by fitting the SE data at 0% RH. (C) Pore size distribution in L_3 of the 1625-day corroded sample. Ngo et al. [11]/Springer Nature/CC BY 4.0.

The distribution of pore diameter ($d_k = 2r_k$) was determined by taking the first derivative of water volume fraction with respect to the pore diameter ($\partial V_{water}/\partial d_k$), which is plotted in Figure 9.17C. The thickness of the physiosorbed water layer in the pores cannot be determined. The RH-controlled SE experiment used to obtain adsorption/desorption data for pore-size calculations is not very sensitive to pore diameters less than 2 nm.

In the above SE analysis, the data fitted reasonably well with the assumption that refractive index is homogeneous and isotropic (averaged throughout each layer). So, it was modeled using isotropic Bruggeman EMA consisting of the optical response of solid and void (empty or water-filled) volume fractions. For another ISG sample corroded in a different condition (silica-saturated, pH 7 at 90 °C without KCl electrolyte for 2046 days (nearly 5.6 years)), the fit result of the SE data was not good when the refractive index of the alteration layer was assumed to be isotropic [12]. The MSE value of the fit shown in Figure 9.18A was 2.75. Although the periodicity of the undulating feature was well represented by the model (so, the determined thickness would be reasonably good), the maximum and minimum values of interference fringes did not match the experimental data. To improve the fit, the out-of-plane refractive index (n_z) and the in-plane refractive index ($n_x = n_y$) of the thickest sublayer (L_3) of the alteration layer were allowed to vary independently while keeping the pore volume fraction fixed. The qualitative agreement in the modeled SE data was improved, and the MSE was reduced to 1.0 (Figure 9.18B). Further fitting the thickness and pore fractions while fixing the refractive index from this anisotropic model did not reduce the MSE value and fitting parameters did not change noticeably.

The SE analysis results for the 2046-day corroded ISG sample are summarized in Figure 9.18C,D. For this sample, water ingress into the porous alteration layer was accompanied by a small increase in the layer thickness, implying a swelling of the layer along the out-of-plane direction. Unlike the sample described in Figure 9.17, not all pores are fully filled with condensed water at near-saturation RH condition (Figure 9.18C), implying that some pores are not accessible by gas-phase water molecules. The network structure of the alteration layer can undergo reorganizations during the corrosion process in an aqueous solution, and this can lead to pore closure. The refractive index anisotropy must originate from anisotropic

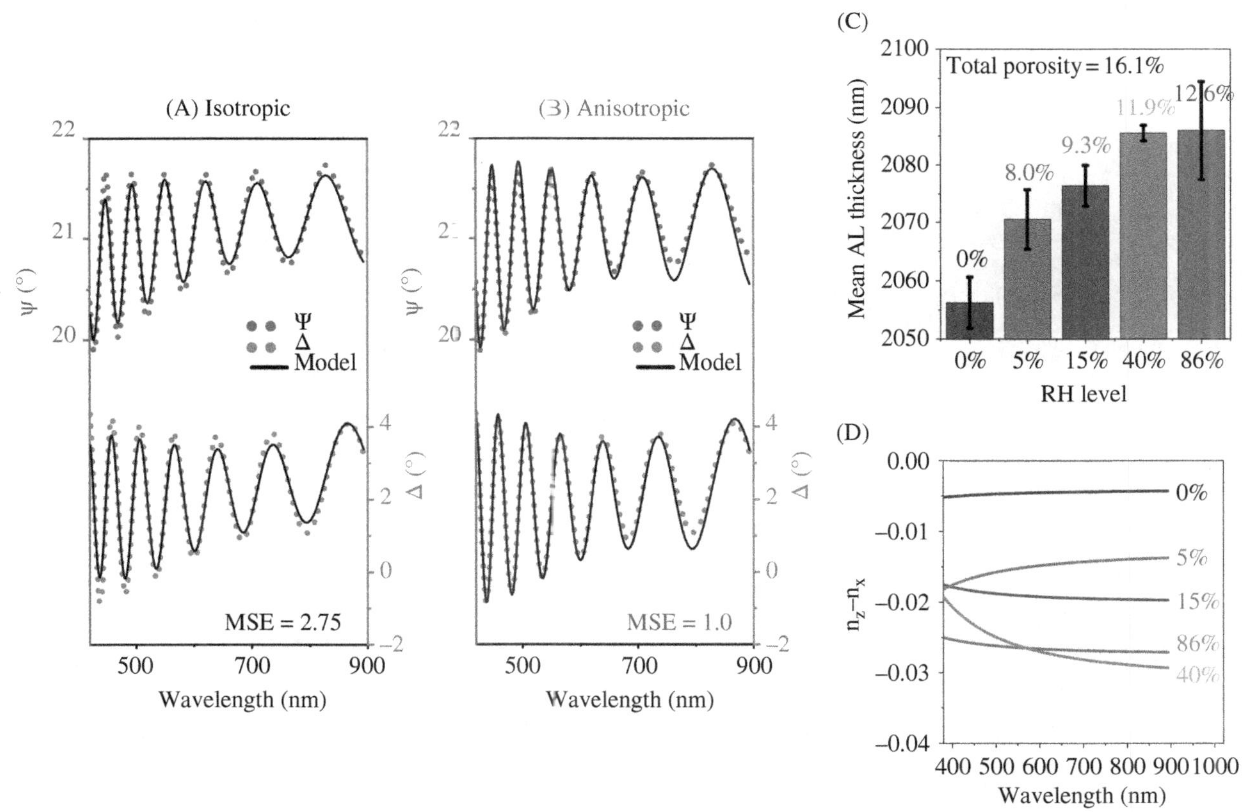

Figure 9.18 Comparison of the data fitting, assuming (A) isotropic refractive index and (B) anisotropic refractive index, of SE angles (Ψ and Δ) measured for an ISG sample corroded for 2046 days in silica-saturated solution in pH7 at 90 °C. (C) Corresponding alteration layer thickness and (D) birefringence ($n_z - n_{x,y}$) as a function of wavelength at different RH levels. In (C), the water content calculated from the EMA model is shown above each bar. Reproduced with permission from Kaya et al. [12]/American Chemical Society.

swelling. Swelling of amorphous material is an isotropic process. However, the alteration layer can swell in the out-of-plane direction only, but not along the in-plane direction. This is because the bottom interface is chemically bonded to the bulk glass and the bulk does not swell. This may be manifested as an internal stress. From the strain in the overall thickness, a biaxial tensile stress was estimated to be ~760 MPa. The changes in the layer thickness and refractive index anisotropy were large when RH was increased from 0% to 40%. This implied that the internal stress develops primarily during water adsorption and condensation in the low-to-intermediate RH regime. This may explain why cracking and delamination of alteration layers is often observed upon full dehydration in dry air or cycling of dehydration and rehydration.

After 2312 days (~6.3 years) in aqueous solution, the thickness of the alteration layer on the ISG glass was found to be ~2000 nm for corrosion at pH7 and ~300 nm for corrosion at pH9 (Figure 9.19A) [13]. These layers were formed within the first 30 days or so and the alteration layer thickness did not grow substantially further. In order to check the hypothesis that this difference in the alteration layer thickness is due to the remnant network structure of the silica-rich layer, specular reflection infrared (SR-IR) analysis was conducted (Figure 9.19B). At the first glance, the Si–O–Si stretch band region (900–1200 cm^{-1}) looks very different between the pH7 and pH9 samples. Also, it did not make sense to detect a prominent band at ~1400 cm^{-1} for the pH7 sample. Based on the peak position, this peak could be attributed to the B-O stretch; but, based on ToF-SIMS depth profile (Figure 9.19A), there was no boron species remaining within the SR-IR probe depth (inset in Figure 9.19B). These features suggested that the spectral features in SR-IR might have been complicated due to interferences of the IR beam reflected from the top and bottom interfaces of the alteration layer (see Figure 6.24).

To circumvent such complications, IR-SE was conducted at multiple incident angles (50°, 55°, 60°, and 65°). While SR-IR detects only the overall intensity of the reflected beam, IR-SE detects the phase change of the reflected beam, allowing to simultaneously calculate the complex refractive index and thickness of the alteration layer. The optical constants obtained for the pristine ISG were used for the bulk phase beneath the alteration layer of the corroded samples. The $\Psi(\lambda)$ and $\Delta(\lambda)$ of the ISG samples corroded at pH7 and pH9 (Figure 9.20A) were fitted with an optical model consisting of a single layer of finite thickness on top of the bulk phase. Since the IR wavelength was much larger than the surface roughness detected with the visible SE (e.g., L_4 in Figure 9.16), it was not necessary to have a separate surface roughness layer. The thickness of the alteration layer was fitted along with the parameters for a new set of Gaussian oscillators that accounted for the absorbance

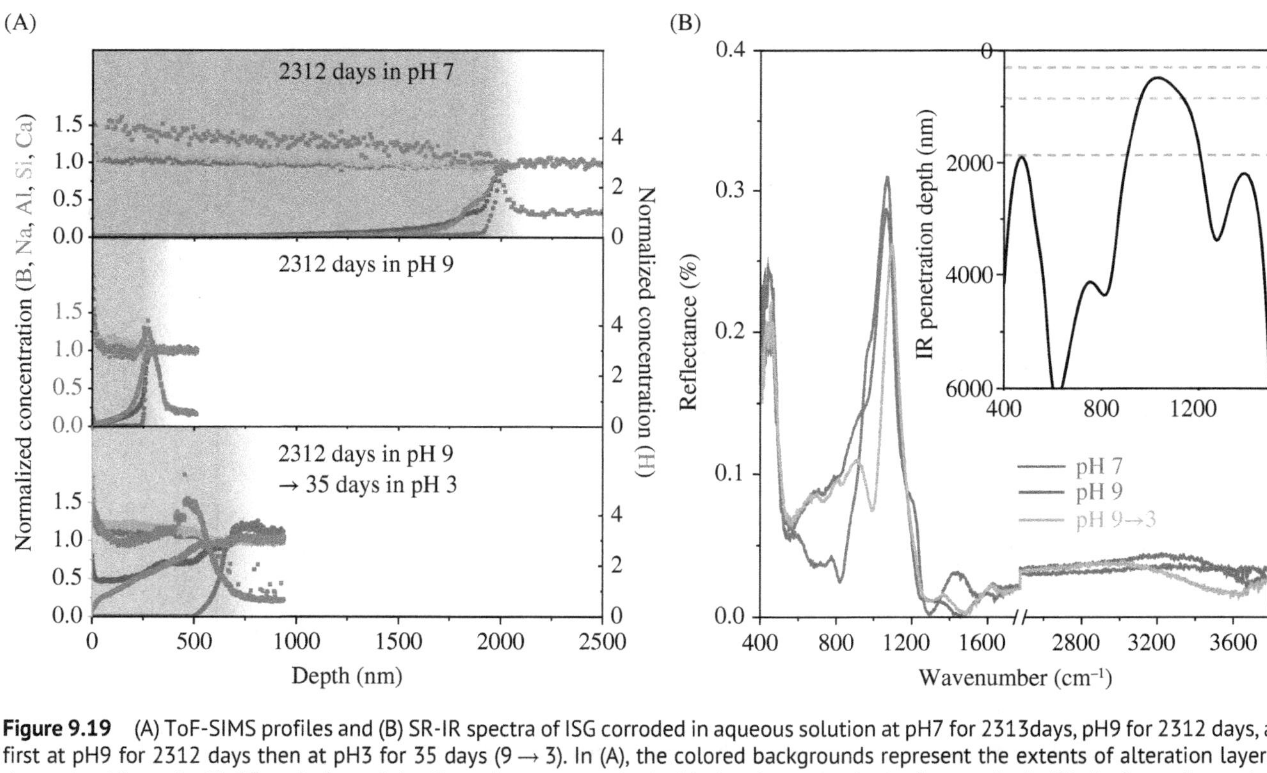

Figure 9.19 (A) ToF-SIMS profiles and (B) SR-IR spectra of ISG corroded in aqueous solution at pH7 for 2313days, pH9 for 2312 days, and first at pH9 for 2312 days then at pH3 for 35 days (9 → 3). In (A), the colored backgrounds represent the extents of alteration layers determined from the IR-SE analysis, and the line colors are matched with the element color in the *y*-scale. In (B), the inset shows the IR attenuation depth calculated with the complex refractive index of the bulk ISG glass, and the colored dashed lines mark the thicknesses of the alteration layers from (A). Reproduced with permission from Kaya et al. [13]/John Wiley & Sons.

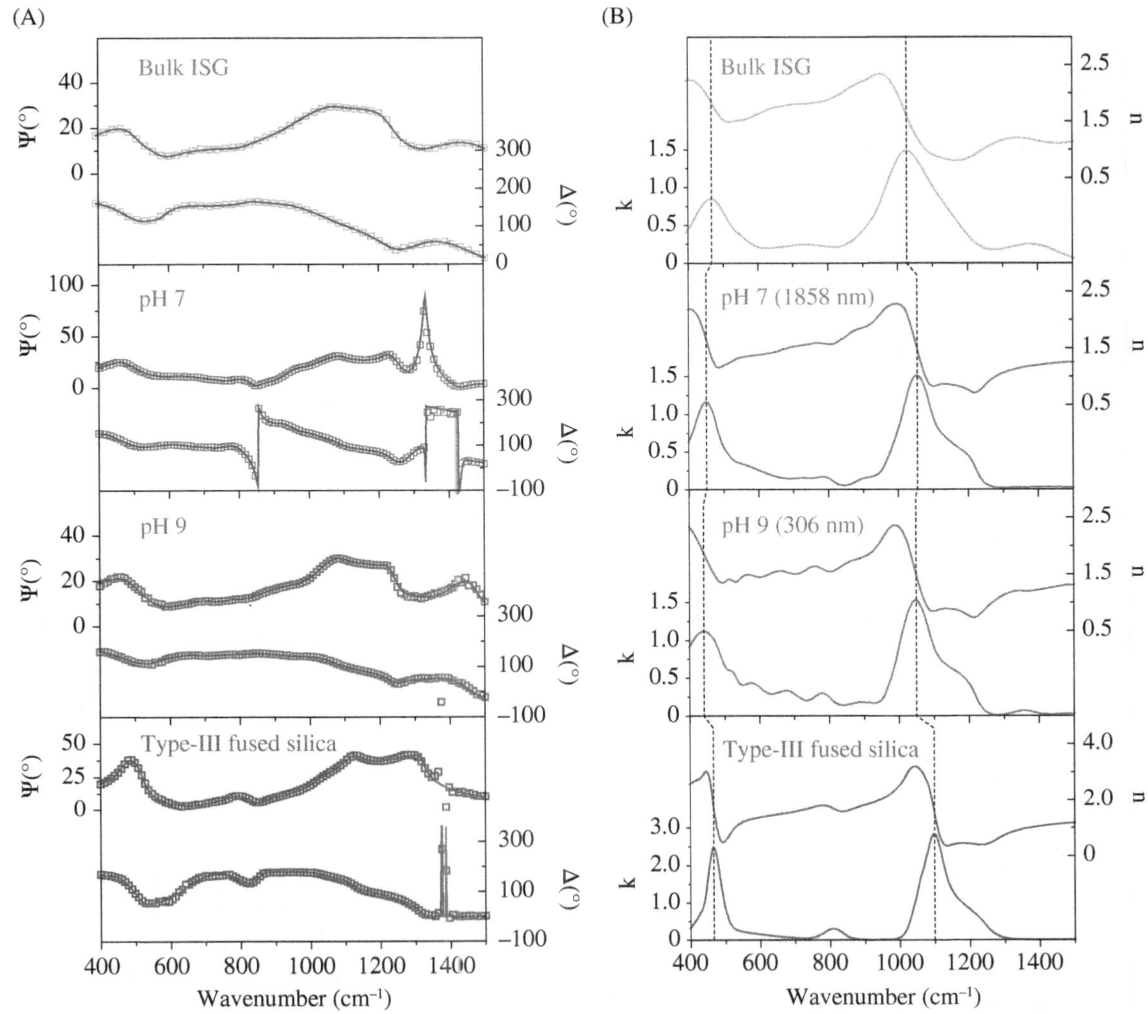

Figure 9.20 (A) $\Psi(\lambda)$ and $\Delta(\lambda)$ data (symbols) measured for bulk ISG, pH7-corroded ISG, pH9-corroded ISG, and type-III silica collected at an incidence angle of 55° and fit results (solid lines). (B) Corresponding complex refractive index ($n + ik$) calculated by simultaneously fitting all $\Psi(\lambda)$ and $\Delta(\lambda)$ data collected. The two dashed lines are guides to eyes to for the Si–O–Si bending and Si–O stretching bands in the k spectra. Reproduced with permission from Kaya et al. [13]/John Wiley & Sons.

features of the alteration layers of each sample. These parameters were simultaneously adjusted to fit the $\Psi(\lambda)$ and $\Delta(\lambda)$ measured at four incident angles.

One challenge in the recursive fit of the IR-SE spectra is that the solution is not unique to a single profile of complex refractive index. Thus, a small MSE value for the fit may not necessarily mean that the result is physically meaningful. An alternative way to check if the fit result is acceptable or not is to calculate the theoretical SR-IR spectrum using the complex refractive index and thickness determined from the IR-SE fit and compare it to the measured SR-IR spectrum (Figure 9.21A,B). This can be done using eqs. (6.33)–(6.35), which is shown in Section 6.3.3.

The refractive index fit results are shown in Figure 9.20B. The data of type-III silica containing ~1000 ppm of OH species are shown for comparison. From the k-spectra (imaginary part of complex refractive index), it is clear that the silica-rich networks in both alteration layers are quite similar to each other although they are formed in different pH conditions. The Si–O stretch band is blue-shifted toward the main peak of type-III silica. The absorption peak at 1400 cm^{-1} band, which is clearly seen in the bulk spectrum, is negligible in both spectra of the alteration layers. This means that the 1400 cm^{-1} peak in the SR-IR spectra in Figure 9.19B is due to the optical interference, not due to the absorption of B-O species in the alteration layer [13]. The difference in the 800–950 cm^{-1} regions of SR-IR spectra of the pH7 and pH9 samples in Figure 9.19B is also due to the difference in optical interference (different thicknesses of the alteration layer), not due to the difference in glass network structure in the alteration layer. Although the alteration layer is expected to have a large amount of hydrous species

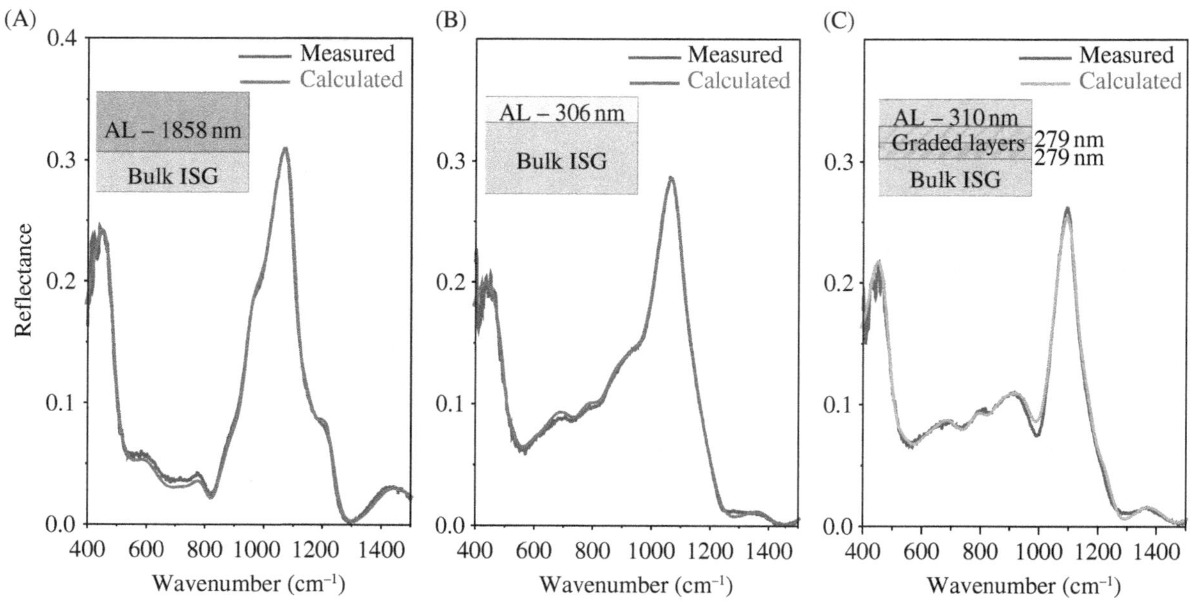

Figure 9.21 (A) Calculated SR-IR spectra of ISG corroded at (A) pH7, (B) pH9, and (C) pH9→3 using the refractive index obtained from IR-SE analyses. Reproduced with permission from Kaya et al. [13]/John Wiley & Sons.

(such as Si–OH), they do not give any definitive band in the 800–900 cm^{-1} region of the SR-IR spectra. It is noted that the *k*-spectrum of type-III silica is also indistinguishable from that of type-I silica which contains less than 5 ppm of OH species [13]. The fine features in the 500–800 cm^{-1} region of the *k*-spectrum of the pH9 sample are interesting, but difficult to interpret.

Since the ISG sample stored in pH9 for ∼6.3 years did not show any measurable increase in the alteration layer thickness more than ∼300 nm which was formed within the first month or so and protected the bulk from further corrosion for ∼6.2 years, it can be considered "fully passivated." When this sample was exposed to pH3 solution, then the alteration layer thickness grew again to ∼800 nm within 35 days (Figure 9.19A) [13]. If the ISG sample had been corroded only at pH3 without the passivation step in pH9 for 2312 days, the B, Ca, and Na depleted alteration layer would have been more than one order of magnitude thicker. Thus, the silica-rich, porous layer formed by pre-corrosion in pH9 must have acted as a "barrier" limiting the transport of B and Ca species from the reaction front to the pH3 solution. Unlike the pH7 and pH9 cases, the IR-SE data fitting needed the inclusion of a composition gradient layer in the interface region between the alteration layer and the bulk glass (Figure 9.21C). This gradient must be associated with the gradual depth profiles of B and Ca species detected by ToF-SIMS as shown in Figure 9.19A. The alteration layer formed after exposure to pH3 for 35 days had a more blue-shift of the Si–O stretching closer to that of silica (Figure 9.22A). This means that the alteration layer network becomes more "silica-like." This blue-shift is also evident in the SR-IR analysis (Figure 9.19B). This means that the B and Ca species detected in ToF-SIMS depth profile are not directly incorporated into the silica-rich network of the alteration layer. Instead, they must form amorphous aggregates trapped in the pores. This was supported by the re-appearance of the broad band at ∼1400 cm^{-1} upon exposure to pH3.

Figure 9.22B compares the *k* spectra obtained by fitting the hydrous species region of the IR-SE data [13]. The sharp band at ∼1640 cm^{-1} is the bending mode of H_2O and the broad band in the 2500–3700 cm^{-1} is ascribed to the stretching mode of H_2O and Si–OH groups. Note that the molecular H_2O species detected in IR-SE are the ones retained or re-adsorbed in the pores of the alteration layer exposed to humid air (Figures 9.17B and 9.18C). The long tail of the OH stretch band far below 3200 cm^{-1} indicates the presence of strongly hydrogen-bonded species (see Sections 6.1.3 and 6.4). In vibrational spectroscopy, it is not possible to distinguish whether such species are Si–OH or H_2O. But knowing that such strong hydrogen-bonding interactions are rare in liquid water, they are likely to be water species interacting with functional groups exposed at the internal pore surface (see Figure 6.37A for IR spectra of water layers adsorbed on a flat glass surface). The difference in the OH stretch band shape (Figure 9.22B) implies that the structure of the internal pore surface is different depending on the pH of the corroding solution. But further details cannot be extracted from the IR-SE study.

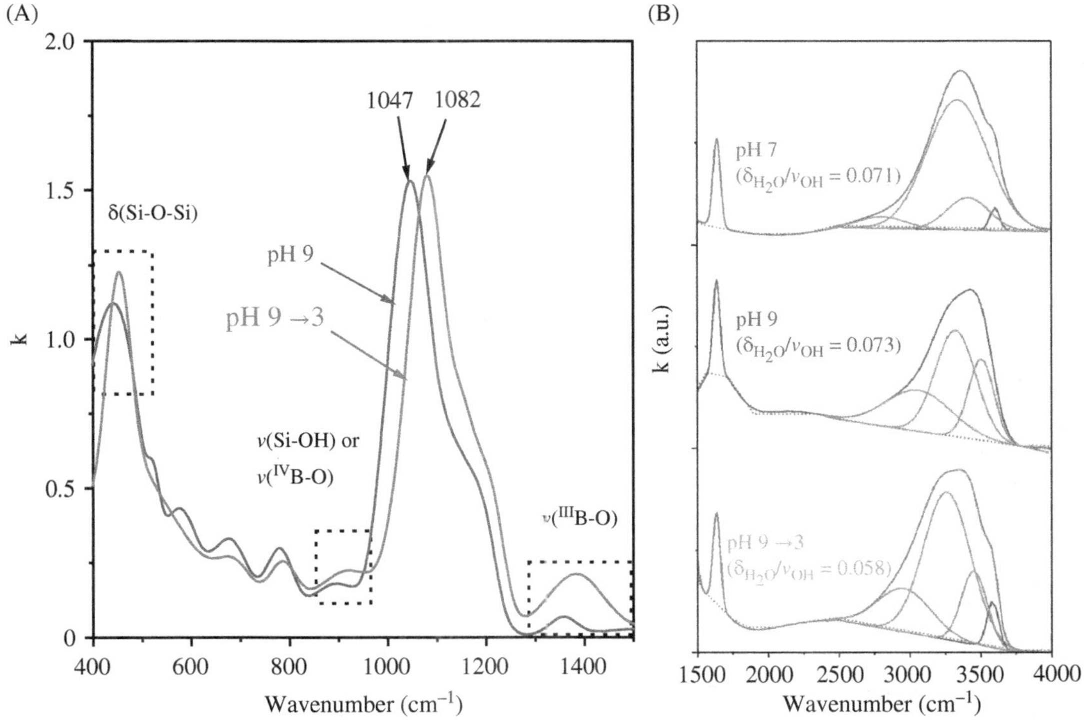

Figure 9.22 (A) Changes in the *k*-spectra of the alteration layer upon exposure of the ISG pre-corroded at pH9 for 2312 days to pH3 solution for 35 days. (B) Spectral region of hydrous species inside the alteration layers on ISG formed by aqueous corrosion at pH7 (2312 days), pH9 (2312 days), and first at pH9 (2312 days) then at pH3 (35 days). Reproduced with permission from Kaya et al. [13]/John Wiley & Sons.

In the above examples, SE and IR-SE were able to find the thickness and refractive index of surface layers. The optical model used for data fitting and the thickness determined from data fitting were compared with the independently measured depth profile. Once the optical model was tested and confirmed to be reasonable and acceptable, the effective refractive index determined from the fitting could be interpreted further to get compositional and structural information about the film or surface layer. It even showed how SR-IR analysis (measuring the intensity only) could be mis-interpreted if the layered structure is not properly taken into account. This kind of detailed analysis was possible because certain assumptions could be made based on *a priori* knowledge about the system under consideration. Without such knowledge, it could be difficult to fit the SE data with complicated optical models with fine details.

Let us re-visit the example shown in Figure 6.22. The sample was a soda lime silica (SLS) glass treated with pH1 solution for 7 days at 90 °C. After the treatment, about 130 nm thick Na-leached layer was formed (Figure 6.22A) and its SR-IR spectrum showed a clearly discernable increase in the 900 cm^{-1} region (Figure 6.22B). Without knowing the refractive index of the Na-leached layer, one may interpret the increase of the 900 cm^{-1} shoulder as an enhancement of the Si–OH stretch mode. But, we know that the Si-OH group in type-III fused quartz (with contains ~1000 ppm of OH species) does not show a shoulder peak at ~900 cm^{-1} (Figure 9.20B); so, interpreting the ~900 cm^{-1} shoulder as the Si–OH stretch inside the Na-leached layer cannot be justified. The refractive index of the Na-leached layer was obtained independently by scattering-based scanning near-field optical microscopy (s-SNOM) coupled with infrared spectroscopy using the data processing method described in Section 6.6.3 [14]. Figure 9.23A compares the experimental SR-IR spectrum of pristine SLS and the theoretical SR-IR spectrum of the Na-leached phase calculated from the refractive index determined from the s-SNOM experiment. The SR-IR spectrum of the Na-leached phase does not have a band at 900 cm^{-1}. Then, where does the 900 cm^{-1} shoulder come from? That is because of the optical interference of the beam reflected at the air/leached layer interface and the leached layer/bulk interface. Figure 9.23A also shows the simulation SR-IR spectra of the Na-leached layer with different thicknesses on the pristine SLS bulk (using eq. (6.34) and (6.35); see Section 6.3.3). We can clearly see that the shoulder feature at ~900 cm^{-1} region becomes more prominent as the Na-leached layer thickness increases.

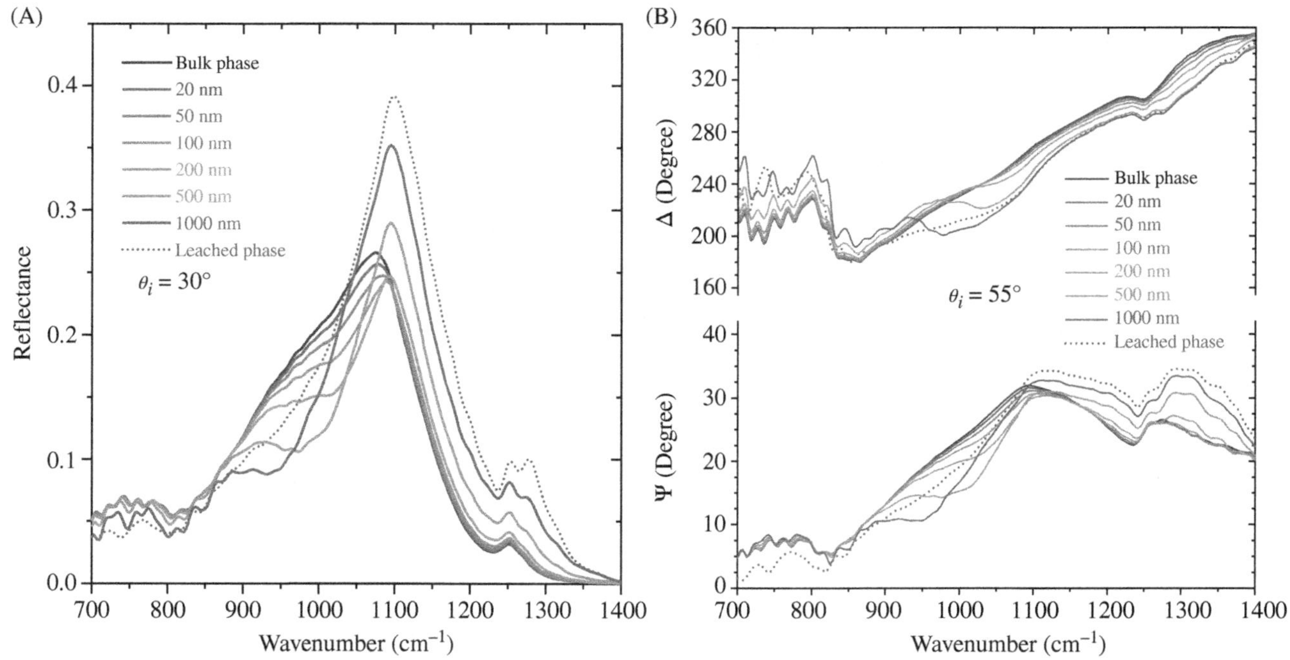

Figure 9.23 (A) Experimental SR-IR spectrum of pristine SLS glass (black line), theoretical SR-IR spectrum of the Na-leached layer calculated from the refractive index obtained from s-SNOM nano-IR analysis (dotted line, from [14]), and simulated SR-IR spectra calculated for the Na-leached layer with different thicknesses on the pristine SLS bulk. (B) $\Psi(\lambda)$ and $\Delta(\lambda)$ spectra simulated for the same materials in (A) using eq. (9.11).

Figure 9.23B plots the simulated $\Psi(\lambda)$ and $\Delta(\lambda)$ spectra for the same systems considered in Figure 9.23A. As long as we know the refractive index of the bulk phase, the IR-SE data for surface layers thicker than 200 nm could be fitted with a model assuming the presence of a surface layer and find the corresponding refractive index of the layer (as in the case of pH9 corroded ISG in Figure 9.20B). But, without *a priori* knowledge of the film structure, the IR-SE data for the leached layer thinner than 100 nm could be fitted with any models assuming the presence of surface roughness layers. In other words, it would be difficult, from IR-SE analysis only, to determine the presence of <100 nm thick surface layer with a slightly different refractive index. So, caution must be taken when optical models are constructed to fit SE data. The theoretical simulations demonstrated in this section are a few examples showing how to construct proper optical models or test the self-consistency level of the fit result.

9.5 Mueller Matrix Ellipsometry (MME)

9.5.1 Jones Parameter and Stokes Parameter

In Figure 9.6, the polarization ellipse is described with the ratio of the amplitudes of the *p*- and *s*-polarization components (E_{0p}/E_{0s}) and their phase difference ($\delta_p - \delta_s$). These two components can be lumped into a matrix form:

$$\begin{pmatrix} E_p \\ E_s \end{pmatrix} = \begin{pmatrix} E_{0p}e^{i\delta_p} \\ E_s e^{i\delta_s} \end{pmatrix} \tag{9.20}$$

This is called Jones vector (***J***). Upon reflection at a sample surface, the Jones vector of the incident beam transforms into another Jones vector through a linear transformation:

$$\begin{pmatrix} E_p^{out} \\ E_s^{out} \end{pmatrix} = \begin{pmatrix} J_{pp} & J_{ps} \\ J_{sp} & J_{ss} \end{pmatrix} \begin{pmatrix} E_p^{in} \\ E_s^{in} \end{pmatrix} \tag{9.21}$$

The 2×2 matrix on the right-hand side of eq. (9.21) is the Jones matrix. For a flat and isotropic sample, the Jones matrix takes on a diagonal form, that is, $J_{ps} = J_{sp} = 0$. For this type of samples, the two nonvanishing elements can be written in terms of the two Fresnel coefficients:

$$\begin{pmatrix} E_p^{out} \\ E_s^{out} \end{pmatrix} = \begin{pmatrix} J_{pp} & 0 \\ 0 & J_{ss} \end{pmatrix} \begin{pmatrix} E_p^{in} \\ E_s^{in} \end{pmatrix} = \begin{pmatrix} r_p & 0 \\ 0 & r_s \end{pmatrix} \begin{pmatrix} E_p^{in} \\ E_s^{in} \end{pmatrix} \tag{9.22}$$

So, we can write the Jones matrix of sample in the following form:

$$\boldsymbol{J}_{sample} = \begin{pmatrix} r_p & 0 \\ 0 & r_s \end{pmatrix} \tag{9.23}$$

The standard ellipsometry discussed in the previous sections can be described with this $\boldsymbol{J}$ matrix.

There is a more generalized way to describe the polarized electric field using a four-dimensional vector, called Stakes vector ($\boldsymbol{S}$). This four-dimensional vector can be derived in the following way. The p- and s-polarization components of the electric field of light at an instant time can be written as:

$$E_p(t) = E_{0p}(t)\cos\left[\omega t + \delta_p(t)\right] \tag{9.24}$$

$$E_s(t) = E_{0s}(t)\cos[\omega t + \delta_s(t)] \tag{9.25}$$

The explicit removal of the frequency term (ωt) between eqs. (9.24) and (9.25) yields the equation describing the polarization ellipse:

$$\frac{E_p^2(t)}{E_{0p}^2} + \frac{E_2^2(t)}{E_{0s}^2} - \frac{2E_p(t)E_s(t)}{E_{0p}E_{0s}}\cos\delta = \sin^2\delta \tag{9.26}$$

where $\delta = \delta_p - \delta_s$. The experimental value measured is the time average of this ellipse. From eq. (9.26), we can derive the time-averaged equation [15]:

$$\left(E_{0p}^2 + E_{0s}^2\right)^2 - \left(E_{0p}^2 - E_{0s}^2\right)^2 - \left(2E_{0p}E_{0s}\cos\delta\right)^2 = \left(2E_{0p}E_{0s}\sin\delta\right)^2 \tag{9.27}$$

Each term in eq. (9.27) is denoted as:

$$S_0 = E_{0p}^2 + E_{0s}^2 \tag{9.28}$$

$$S_1 = E_{0p}^2 - E_{0s}^2 \tag{9.29}$$

$$S_2 = 2E_{0p}E_{0s}\cos\delta \tag{9.30}$$

$$S_3 = 2E_{0p}E_{0s}\sin\delta \tag{9.31}$$

These are the Stokes polarization parameters of the Stokes vector. For a plane wave, eq. (9.27) can be written as $S_0^2 = S_1^2 + S_2^2 + S_3^2$. For a partially polarized light, the sum of three terms is smaller than S_0^2. So, for any state of polarized light, the Stokes parameters can be written as:

$$S_0^2 \geq S_1^2 + S_2^2 + S_3^2 \tag{9.32}$$

Here, the equality is for the completely polarized light, and the inequality is for the partially polarized light.

The time-average Stokes parameters can be related to the real optical amplitudes that can be measured experimentally:

$$\boldsymbol{S} = \begin{pmatrix} S_0 \\ S_1 \\ S_2 \\ S_3 \end{pmatrix} = \begin{pmatrix} I \\ Q \\ U \\ V \end{pmatrix} = \begin{pmatrix} I_p + I_s \\ I_p - I_s \\ I_{+45} + I_{-45} \\ I_L - I_R \end{pmatrix} \tag{9.33}$$

Here, I_p, I_s, I_{+45}, and I_{-45} are the intensities which would be measured through ideal linear polarizers oriented along the p, s, +45°, and −45° in the plane perpendicular to the propagation direction, respectively, and I_L and I_R are the intensities which would be measured with the left and right circular polarizer. We use the p and s polarizations here because we can define them with respect to the laser incidence plane. They can be replaced with x and y to describe the light propagating along the

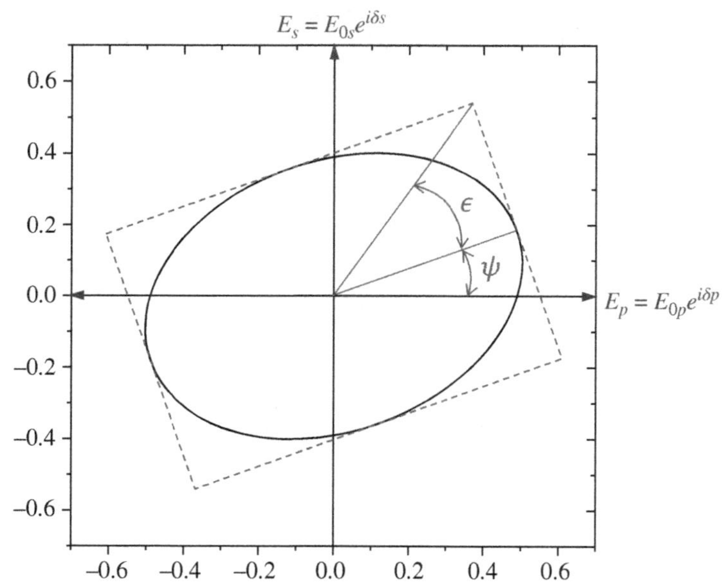

Figure 9.24 Orientation (ψ) and ellipticity (ϵ) angle of the polarization ellipse.

z-direction. Using the Stokes parameters, the linearly p-polarized, linearly s-polarized, linearly +45° polarized, linearly −45° polarized, left circularly polarized, and right circularly polarized states can be expressed in the following form:

$$\boldsymbol{S}_p = \begin{pmatrix} 1 \\ 1 \\ 0 \\ 0 \end{pmatrix}, \ \boldsymbol{S}_s = \begin{pmatrix} 1 \\ -1 \\ 0 \\ 0 \end{pmatrix}, \ \boldsymbol{S}_{+45} = \begin{pmatrix} 1 \\ 0 \\ 1 \\ 0 \end{pmatrix}, \ \boldsymbol{S}_{-45} = \begin{pmatrix} 1 \\ 0 \\ -1 \\ 0 \end{pmatrix}, \ \boldsymbol{S}_L = \begin{pmatrix} 1 \\ 0 \\ 0 \\ 1 \end{pmatrix}, \ \boldsymbol{S}_R = \begin{pmatrix} 1 \\ 0 \\ 0 \\ -1 \end{pmatrix} \tag{9.34}$$

Here, the parameters are normalized with the total intensity.

If we describe the polarization ellipse with the orientation angle (ψ) and ellipticity angle (ϵ) as in Figure 9.24 (note the difference from Figure 9.6), the Stokes vector can be calculated in the following formula:

$$\boldsymbol{S} = \begin{pmatrix} I \\ Q \\ U \\ V \end{pmatrix} = \begin{pmatrix} I \\ I\cos 2\epsilon \cos 2\psi \\ I \cos 2\epsilon \sin 2\psi \\ I \sin 2\psi \end{pmatrix} \tag{9.35}$$

Within the Stokes formalism, the degree of polarization is defined as:

$$P = \frac{\sqrt{Q^2 + U^2 + V^2}}{I} \quad (0 \le P \le 1) \tag{9.36}$$

Here, $P = 0$ and 1 correspond to the fully depolarized and fully polarized state, respectively, and any value between these means a partially polarized light.

9.5.2 MME Using Stokes Vector

As in eq. (9.21), upon reflection at a sample surface, the Stokes vector of the incident beam transforms into another Stokes vector through a linear transformation:

$$\begin{pmatrix} I^{out} \\ Q^{out} \\ U^{out} \\ V^{out} \end{pmatrix} = \begin{pmatrix} M_{11} & M_{12} & M_{13} & M_{14} \\ M_{21} & M_{22} & M_{23} & M_{24} \\ M_{31} & M_{32} & M_{33} & M_{34} \\ M_{41} & M_{42} & M_{43} & M_{44} \end{pmatrix} \begin{pmatrix} I^{in} \\ Q^{in} \\ U^{in} \\ V^{in} \end{pmatrix} \tag{9.37}$$

Here, the 4×4 matrix is called Mueller matrix, $\boldsymbol{M}$. Thus, the ellipsometry process can be expressed as:

$$\boldsymbol{S}^{out} = \boldsymbol{M}_{sample}\, \boldsymbol{S}^{in} \tag{9.38}$$

Since the Stokes vector components are all real numbers, all elements in the Mueller matrix are also real numbers. Since we know $\boldsymbol{S}^{in}(\lambda)$ and measure $\boldsymbol{S}^{out}(\lambda)$, $\boldsymbol{M}_{sample}(\lambda)$ can be calculated through a simple matrix algebra at each wavelength (see Section 8.1.2). This is the Mueller matric ellipsometry (MME).

Due to the capability of the Stokes vector to describe any polarization state (either fully polarized or depolarized), the Mueller matrix can thoroughly describe the polarimetric properties of any sample with real numbers. If the sample is non-depolarizing, the Jones matrix in eq. (9.21) can be transformed to the Muller matrix using the following formula [15]:

$$\boldsymbol{M} = \boldsymbol{T}(\boldsymbol{J} \otimes \boldsymbol{J}^*)\boldsymbol{T}^{-1} \text{ where } \boldsymbol{T} = \begin{pmatrix} 1 & 0 & 0 & 1 \\ 1 & 0 & 0 & -1 \\ 0 & 1 & 1 & 0 \\ 0 & i & -i & 0 \end{pmatrix} \tag{9.39}$$

where $\otimes$ is the Kronecker product. The Mueller matrix does not provide any information on the overall optical phase shift upon reflection. Inversion of this allows the calculation of Jones parameters from experimentally measured Mueller matrix [16].

The standard ellipsometry with a 45° polarized incident beam for an isotropic refractive index sample (thus, no depolarization) can be described with the following Mueller matrix:

$$\boldsymbol{M}_{sample} = \begin{pmatrix} 1 & -N & 0 & 0 \\ -N & 1 & 0 & 0 \\ 0 & 0 & C & S \\ 0 & 0 & -S & C \end{pmatrix} \tag{9.40}$$

and $N^2 + S^2 + C^2 = 1$. Here, the M_{11} element is unity, which means the normalized intensity is used (as in eq. (9.34). The experimentally determined MME parameters can be fitted with an optical model using the following relationship:

$$\rho = \frac{r_p}{r_s} = \tan(\Psi)e^{i\Delta} = \frac{C + iS}{1 + N} \tag{9.41}$$

$$N = \cos 2\Psi \tag{9.42}$$

$$S = \sin 2\Psi \sin\Delta \tag{9.43}$$

$$C = \sin 2\Psi \cos\Delta \tag{9.44}$$

If the sample is isotropic, the following off-block diagonal elements vanish: $M_{13} = M_{14} = M_{23} = M_{24} = M_{31} = M_{32} = M_{41} = M_{42} = 0$. If the sample has a uniaxial alignment, they can have nonzero values, except the case where the uniaxial alignment axis is within the incidence plane. Thus, the orientation angle of that axis with respect to the incidence plane can be determined in MME.

If $N^2 + S^2 + C^2 \neq 1$, then it mean the sample is depolarizing. Depolarization is the reduction of the degree of polarization state. It can occur if the reflected light is composed of several incoherent contributions due to thickness nonuniformity, incoherent superposition of light from dissimilar materials, large surface roughness or edge roughness, as well as limitations of optics used in SE instrument. The degree of depolarization is described with the depolarization index (DI):

$$DI = \frac{\sqrt{\sum_{ij} M_{ij}^2 - M_{11}^2}}{\sqrt{3}M_{11}} \tag{9.45}$$

Here, $D = 1$ corresponds to no depolarization. If $D = 0$, then it means the sample behaves like a perfect depolarizer.

9.5.3 Case Study of Employing MME

Figure 9.25 displays the Mueller matrix data obtained for an atomically smooth Si(100) wafer in the 1.5–6.5 eV (191–827 nm) region [17]. The Si(100) is an isotropic material; we can see that $M_{12} = M_{21}$ (which is $-N$), $M_{33} = M_{44}$ (which is C), and $M_{34} = -M_{43}$ (which is $-s$), and the 2×2 off-block diagonal elements are zero. The M_{11} element is one, so not shown; instead, the degree of polarization $P = \sqrt{N^2 + C^2 + S^2}$ is shown. The P value is slightly higher than 1 in the low-energy region, which could be due to calibration artifacts. Figure 9.26 exhibits the Muller matrix elements as a function of photon energy calculated from the SE data collected for a Si(110) surface with a 45° azimuth angle with respect to the incidence plane [16]. Unlike Si(100), the Si(110) is anisotropic, and its crystallographic axis is not aligned with the incidence plane. Thus, the 2×2 off-block diagonal elements do not vanish, although their values are relatively small. When the sample is rotated so that the angle 0° or 90°, the amplitudes of these matrix elements vanish.

MME can be used to determine the birefringent refractive index of optically anisotropic materials. Figure 9.27 is an example of such application [18]. Cellulose is a birefringent material that has different refractive indexes along and perpendicular to the chain axis. In order to determine these two values, cellulose needs to be aligned. The uniaxially aligned cellulose nanocrystal (CNC) sample was prepared via Langmuir–Blodgett (LB) technique. When cellulose nanocrystals are prepared via H_2SO_4-catalyzed hydrolysis of lignocellulosic biomass, their surfaces are negatively

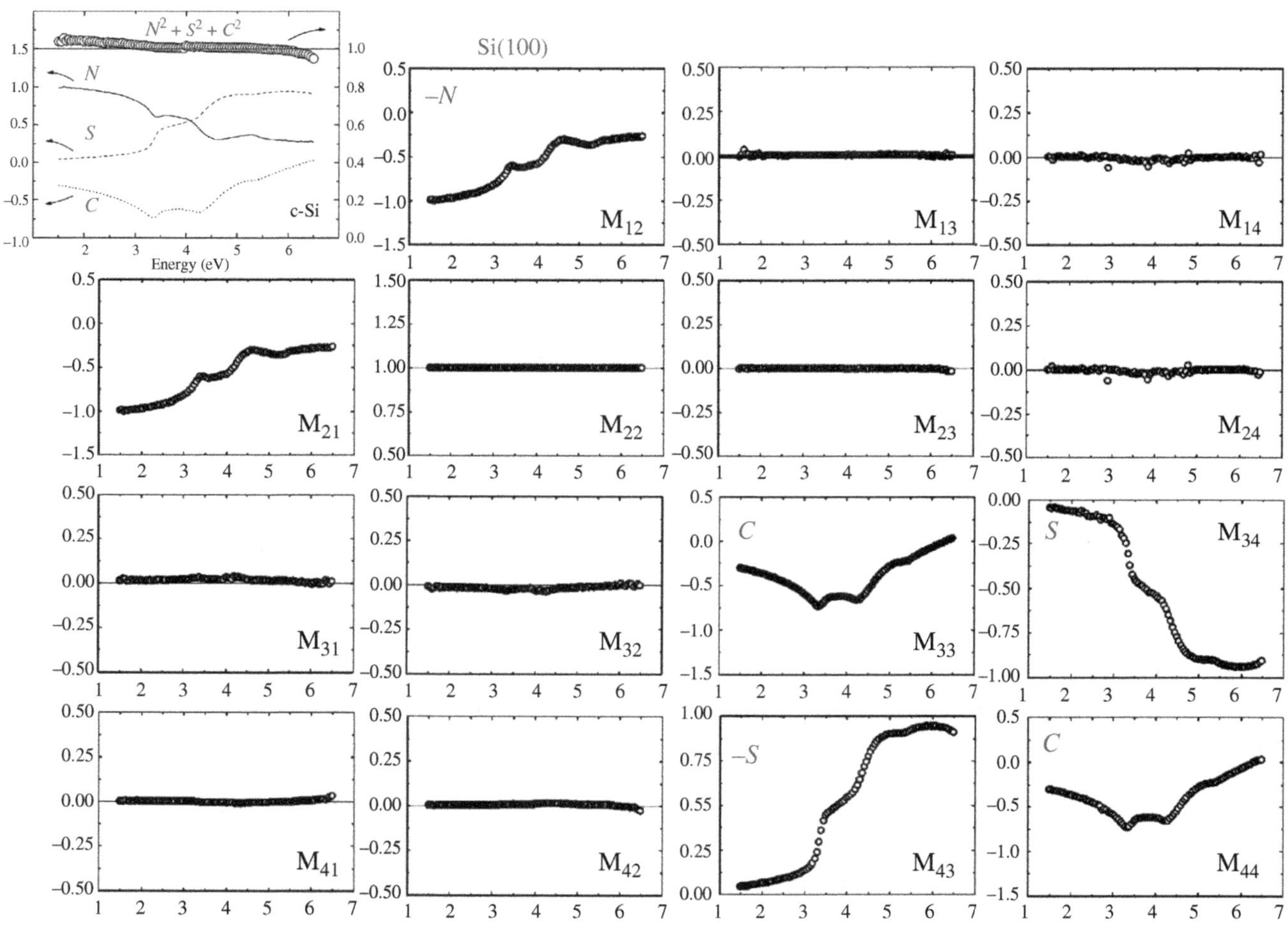

Figure 9.25 Muller matrix calculated from the SE data of atomically smooth Si(100) wafer measured at 70° incident angle. Reproduced from Laskarakis et al. [17]/with permission of Elsevier.

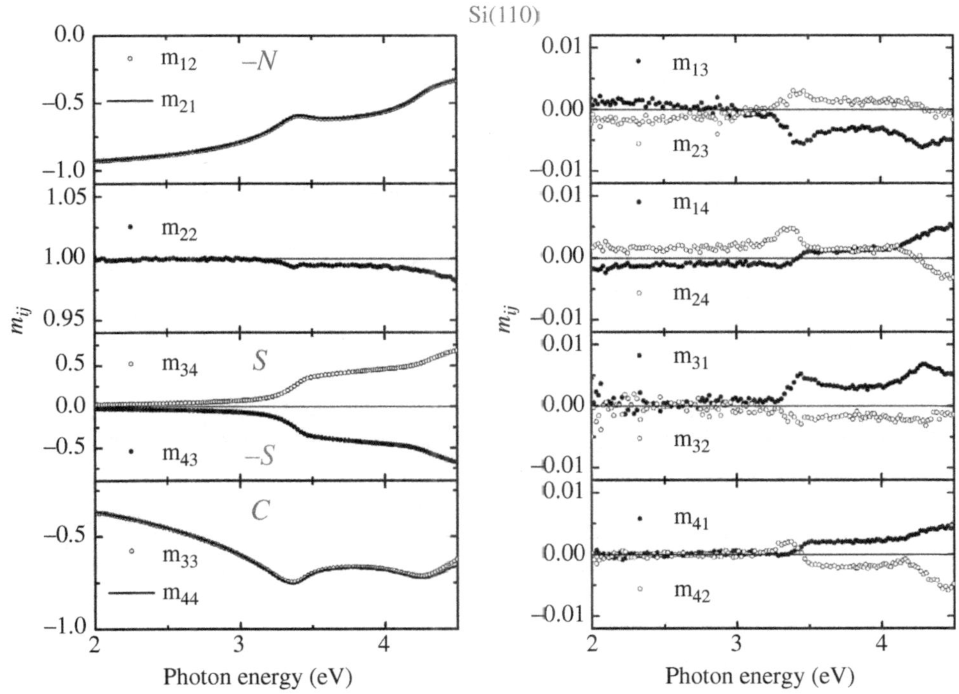

Figure 9.26 Muller matrix calculated from the SE data of a Si(110) wafer with a 45° azimuth angle with respect to the incidence plane collected at 69.9° incident angle. Reproduced from Chen et al. [16]/with permission of Elsevier.

charged due to the presence of sulfate groups. Then, they can be pulled toward the water/air interface which is already covered with a cationic surfactant such as dioctadecyldimethylammonium (DODA). The DODA/CNC Langmuir film at the water/air interface can be compressed to laterally align CNCs underneath the DODA layer (see Figure 7.19). The compressed DODA/CNC film can be transferred to a glass slide for MME measurement. Figure 9.27 shows the Mueller matrix calculated from the SE data measured at two incident angles (50° and 75°) for a single layer of uniaxially-aligned DODA/CNC film [18]. Since the major axis of CNCs were aligned to the incidence plane, the off-block diagonal components are essentially zero. Small deviations were due to imperfect alignment of CNCs in the film. The optical response of CNC films was modeled using Sellmeier expressions to describe the unique optical properties along the material principal axes ($n_x = n_y \neq n_z$). The optical response of the surface roughness was modeled using anisotropic Bruggeman EMA assuming 0.5 DODA/CNC fraction and 0.5 void volume fraction in the film. The structural model consisted of a semi-infinite glass substrate with the DODA/CNC film and surface roughness. A least-squares regression method was used to fit parameters defining the optical and structural model which included Sellmeier amplitude and resonance energy in each direction and layer thicknesses, respectively, until the calculated spectra matched the experimentally measured Mueller matrix spectra. The result is shown in the M_{11} position in Figure 9.27 [18]. The refractive index determined from the MME analysis is in good agreement with the value predicted from time-dependent density functional theory (TD-DFT) calculations.

Many beetles show astonishingly iridescent colors. Such colors are made not with pigments, but through interference of reflected lights. The light reflected off exoskeletons of some beetles is nearly circularly polarized (see pictures of *Cetonia aurata* in Figure 9.28). This near-circular polarization of the reflected light originates from a complex nanostructure in the exoskeleton of the beetles, especially a twisted layered structure of chitin which is birefringent. Determining the refractive index of the materials constituting cuticles in beetles is quite challenging. Cuticle structures are inhomogeneous, which causes scattering of the probe beam. Figure 9.28 shows the color contour maps of Mueller matrix spectra measured at multiple angles in the 400–800 nm spectral region [19]. By fitting these data with a twisted layered structure with chirality, the uniaxial or biaxial refractive indices of different layers could be obtained.

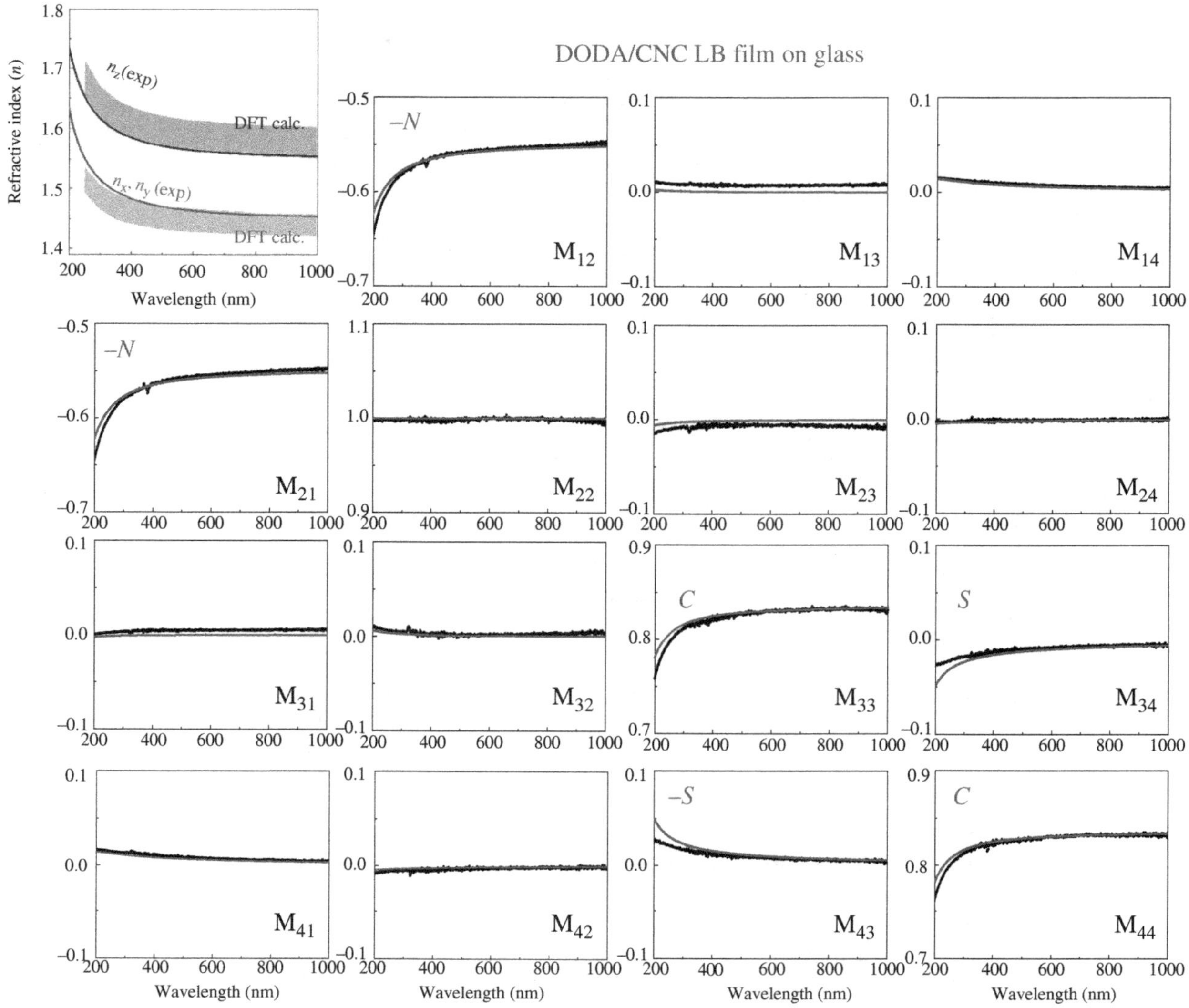

Figure 9.27 Muller matrix measured for a single LB film consisting of DODA/CNC on a glass slide. Reproduced with permission from Chae et al. [18]/John Wiley & Sons.

Not only to find optical properties but MME can also be employed for nanostructure metrology [20]. This is especially useful for diffractive nanostructures (such as grating-like structures). The reconstruction of the structural profile from the measured Mueller matrix is an inverse diffraction problem to find a profile whose theoretical signature can match the measured one through nonlinear regression or library search methods. Figure 9.29 shows an example of employing MME to characterize trenched nanostructures with three grating layers [20]. The collected Mueller matrix was fitted with a structural model constructed based on the TEM cross-section image, and the regression analysis was allowed to vary the width and depth parameters, while keeping the optical constants of Si_3N_4, SiO_2, and Si fixed to the literature values. The comparison of the ellipsometer-measured and calculated best-fit Mueller matrix spectra is shown in Figure 9.29B; the reduced χ^2 was 18.5. The fit parameters shown in Figure 9.29A matched quite well with the values measured with TEM.

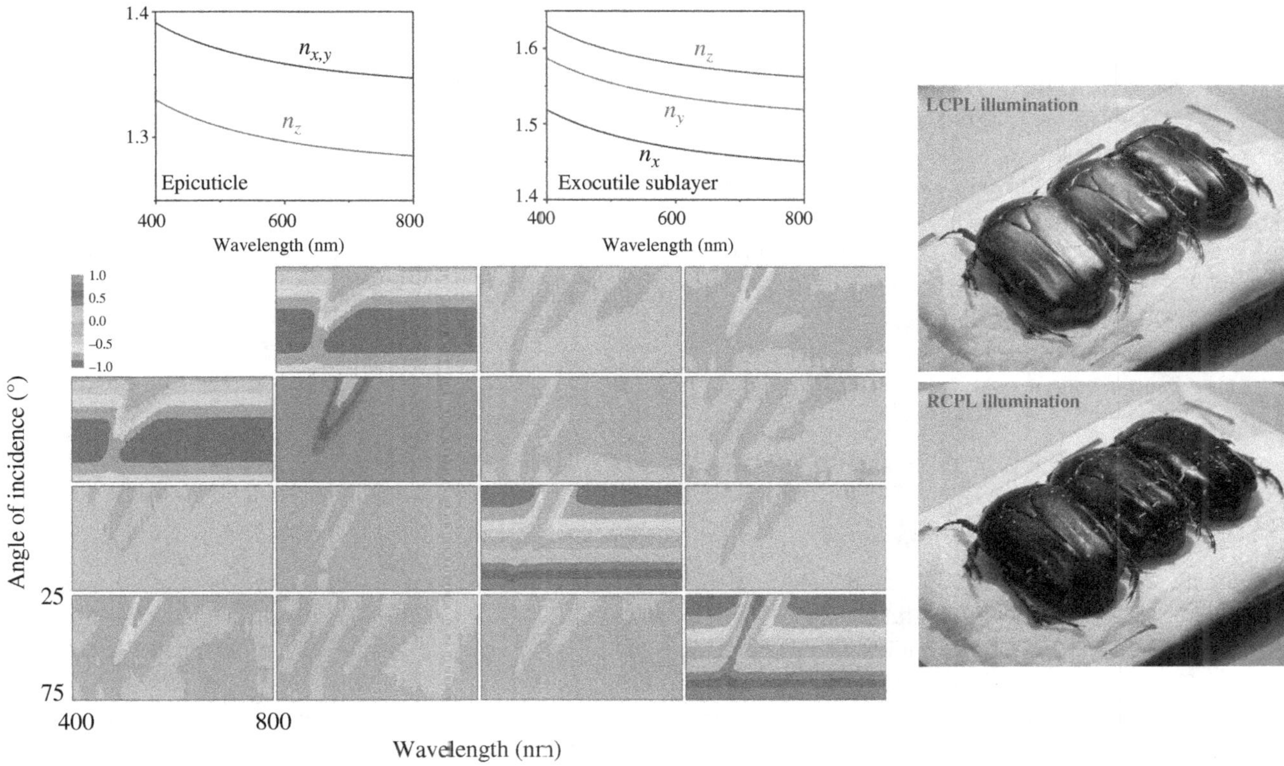

Figure 9.28 Contour plot showing spectral Mueller-matrix data vs. angle of incidence measured on a green-colored *Cetonia aurata*. Arwin et al. [19]/Optical Society of America. Also shown are pictures of *C. aurata* taken with left- and right-circularly polarized light illumination. Reproduced with permission of 2004, Oriol Arteaga Barriel.

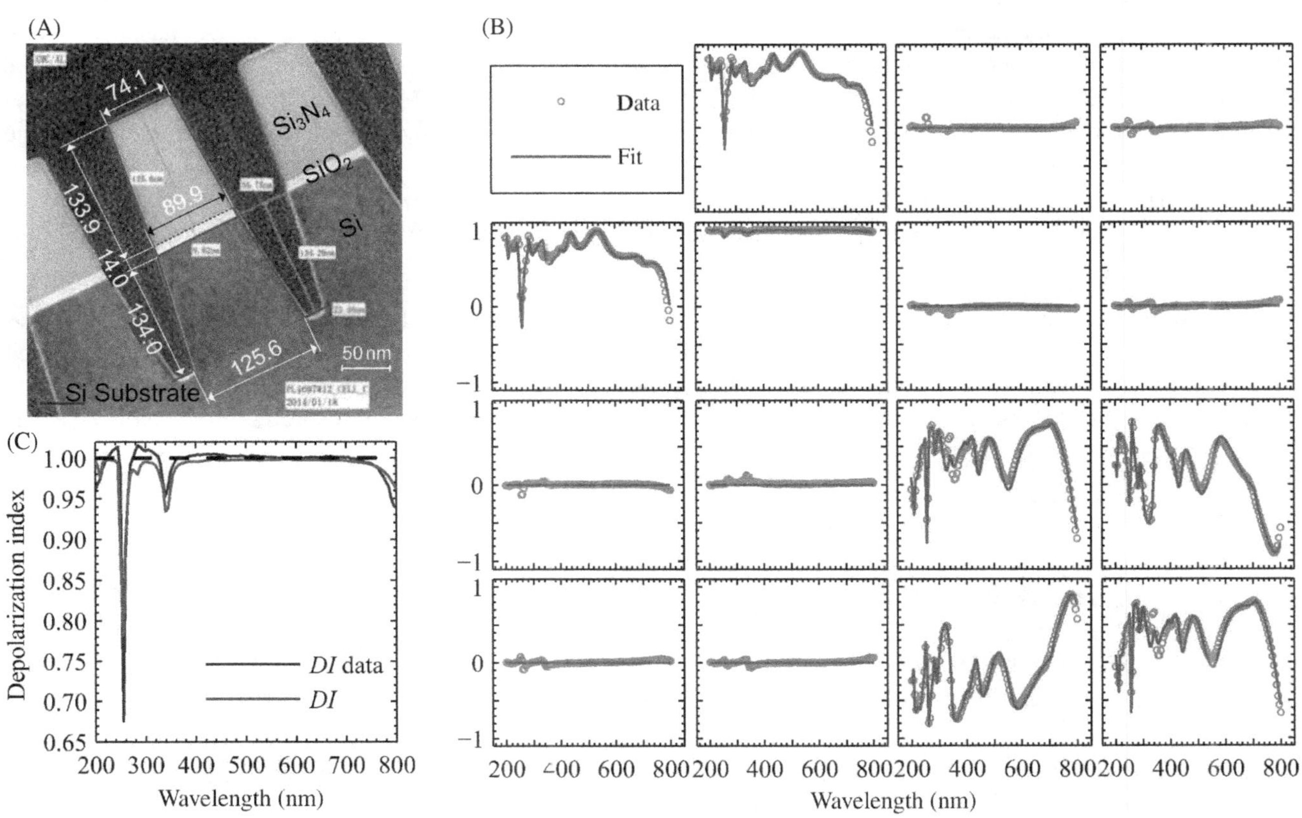

Figure 9.29 (A) TEM micrograph of an etched trench nanostructure with the dimensions determined from MME. (B) Fitting result of the Mueller matrix measured at 55° incident angle and 0° azimuth angle. (C) Depolarization index of the measured and calculated best-fit Mueller matrix spectra. Reproduced from Liu et al. [20]/with permission of Elsevier.

9.6 Further Reading for Ellipsometry

The content covered in this chapter is very basic fundamentals of SWE and SE. Just doing the theoretical calculations shown in this chapter will give you solid conceptual understanding of the key principles. Being able to do such calculations also allows you to not only design SWE and SE experiments properly but also fit the experimental data better and interpret physical meaning of the fit result more accurately. Readers who are interested in more advanced concepts and instrumentations are encouraged to study *Handbook of Ellipsometry* by H. G Tompkins and E. A. Irene [21] or Spectroscopic Ellipsometry: Principles and Applications by Hiroyuki Fujiwara [6]. Also, websites of many SE manufacturers have good tutorial materials [22].

Practice Problems

9.1 Sketch the Fresnel reflection coefficients for *p*- and *s*-polarizations (r_p and r_s) for $n_1 = 1$ and $n_2 = 3.4$. Sketch ellipsometry angles (Ψ, Δ) as a function of incidence angle from 0° to 90°. The relationship between the Fresnel coefficient and ellipsometry angle is given as $\tan\Psi \cdot e^{i\Delta} = \frac{r_p}{r_s}$.

9.2 Calculate the reflectance, $R(\lambda)$, and the ellipsometry angles, $\Psi(\lambda)$ and $\Delta(\lambda)$, for the following five cases: $n_1 = 1$ (air), $n_3 = 3.5$ (substrate), and thin film ($n_2 = 1.5$) with a thickness (d_2) varying from (i) 100 nm, (ii) 125 nm, (iii) 150 nm, (iv) 175 nm, to (v) 200 nm. The incidence angle (θ_i) is 70°. Plot $R(\lambda)$, $\Psi(\lambda)$, and $\Delta(\lambda)$ for λ from 300 nm to 800 nm. Plot the five cases in the same graph for comparison.

9.3 Calculate the reflectance, $R(\lambda)$, and the ellipsometry angles, $\Psi(\lambda)$ and $\Delta(\lambda)$, for the following three cases: $n_1 = 1$ (air), $n_3 = 3.5$ (substrate), and thin film with $n_2 =$ (i) 1.5, (ii) 1.7, and (iii) 1.9. The thickness of film (d_2) is 200 nm and the incidence angle (θ_i) is 70°. Plot $R(\lambda)$, $\Psi(\lambda)$, and $\Delta(\lambda)$ for λ from 300 to 800 nm. Plot the three cases in the same graph for comparison.

9.4 Let us assume that you are doing a SWE analysis in water ($n_{water} = 1.33$). The sample is a 50 nm thick polymer film ($n_{polymer} = 1.5$; non-absorbing) on a Si wafer ($n_{Si} = 3.8812 + 0.0195i$). When the light source is a He–Ne laser ($\lambda = 632.8$ nm) and the incidence angle is 70°, the ψ and Δ are measured to be 11.9653° and 80.9382°, respectively. Now assume that the polymer film absorbs a dye from the aqueous solution; because the dye absorption is irreversible, the color remains after the solution is exchanged back to pure water. When you repeat the ellipsometry measurement at the same SWE condition, the ψ and Δ values are changed to 11.6486° and 104.946°, respectively. The dye absorption does not swell the polymer film nor change the real part (*n*) of the effectively refractive index of the film; it changes the imaginary part (*k*) only.

a) Determine the *k* value of the polymer film with dye molecules.
b) Can you determine the absorptivity (α) of the dye molecule at 632.8 nm from the data provided here? Justify your answer.

9.5 A "sodium-leached, silica-rich" surface layer (SL) is grown on a SLS glass. The refractive index of the layer was determined with scattering scanning near-field optical microscopy (s-SNOM) coupled with Fourier transform infrared spectroscopy. Download the refractive index of the pristine SLS and the leached SL from the supplementary information of *Journal of Non-Crystalline Solids* 600 (2023) 121989), and calculate theoretical spectra of the following cases:

i) air/SLS glass (no SL on SLS)
ii) air/20 nm-thick SL/SLS glass
iii) air/50 nm-thick SL/SLS glass
iv) air/100 nm-thick SL/SLS glass
v) air/200 nm-thick SL/SLS glass
vi) air/500 nm-thick SL/SLS glass
vii) air/1 μm-thick SL/SLS glass
viii) air/SL only (assume the surface layer is the bulk glass)

a) Plot specular reflection infrared (SR-IR) spectra of all 8 cases (together with different color for each case) for an IR incidence angle = 30° from the surface-normal direction.

b) Plot specular reflection infrared (SR-IR) spectra of all 8 cases (together with different color for each case) for an IR incidence angle = 70° from the surface-normal direction.
c) Plot spectroscopic ellipsometry angles (ψ and Δ) of all 8 cases (together with different color for each case) for an IR incidence angle = 30° from the surface-normal direction.
d) Plot spectroscopic ellipsometry angles (ψ and Δ) of all 8 cases (together with different color for each case) for an IR incidence angle = 70° from the surface-normal direction.
e) The SR-IR spectra calculated for the 70° incidence angle has a stronger peak at 1250–1300 cm^{-1}? Is it a "new" absorption band that can be observed only at the oblique incidence angle? Justify your answer. [hint: *J Am Ceram Soc.* **2022**; 105 : 2355–2384].
f) In the IR-SE spectra, the data calculated for the 30° incidence angle look upside down compared to the ones calculated for the 70° incidence angle. In other words, as the wavenumber increases from 700 to 1400 cm^{-1}, the ψ plot varies "from high to low" in the 30° incidence, but it varies "from low to high" in the 70° incidence. Explain the reason for this opposite trend.
g) In the IR-SE spectra, there are some "noise-like" fluctuations above 1450 cm^{-1}. Explain the reason for that artifact.

References

1 McGall SJ, Davies PB, Neivandt DJ. Interference effects in sum frequency vibrational spectra of thin polymer films: an experimental and modeling investigation. The Journal of Physical Chemistry B. 2004;108(41):16030–9.
2 Kim SH, Vurens G, Somorjai GA. Direct measurement of polymer growth rate in a model Ziegler–Natta polymerization system using laser reflection interferometry. Journal of Catalysis. 2000;193(2):171–5.
3 Yasaka M. X-ray thin-film measurement techniques V. X-ray reflectivity measurement. Rigaku Journal. 2010;26(2):1–9.
4 Tanner BK. Grazing incidence X-ray reflectivity and scattering. In: Ida N, Meyendorf N, editors. Handbook of Advanced Non-Destructive Evaluation. Cham: Springer International Publishing; 2018. p. 1–34.
5 Tompkins HG. A User's Guide to Ellipsometry. West Sussex, UK: Elsevier; 1993.
6 Fujiwara H. Spectroscopic Ellipsometry: Principles and Applications. Chichester, UK: Wiley; 2007.
7 Humlíček J. 1 – Polarized light and ellipsometry. In: Tompkins HG, Irene EA, editors. Handbook of Ellipsometry. Norwich, NY: William Andrew Publishing; 2005. p. 3–91.
8 Tompkins HG, Smith S, Convey D. Optimizing the ellipsometric analysis of a transparent layer on glass. Surface and Interface Analysis. 2000;29(12):845–50.
9 Polyanskiy MN. Refractive Index Database [Available from: https://refractiveindex.info].
10 Jellison GE. 3 – Data analysis for spectroscopic ellipsometry. In: Tompkins HG, Irene EA, editors. Handbook of Ellipsometry. Norwich, NY: William Andrew Publishing; 2005. p. 237–96.
11 Ngo D, Liu H, Sheth N, Lopez-Hallman R, Podraza NJ, Collin M, et al. Spectroscopic ellipsometry study of thickness and porosity of the alteration layer formed on international simple glass surface in aqueous corrosion conditions. npj Materials Degradation. 2018;2(1):20.
12 Kaya H, Ngo D, Hahn SH, Li M, He H, Yedikardeş B, et al. Estimating internal stress of an alteration layer formed on corroded boroaluminosilicate glass through spectroscopic ellipsometry analysis. ACS Applied Materials & Interfaces. 2021;13(42):50470–80.
13 Kaya H, Gin S, Vogt BD, Kim SH. Impact of aqueous solution pH on network structure of corrosion-induced surface layers of boroaluminosilicate glass. Journal of the American Ceramic Society. 2022;105(11):6581–92.
14 Ogrinc AL, Zhou Y, Hahn SH, Lin Y-T, Kim SH. Revealing the structure of the sodium-leached layer of soda lime silica glass: a comprehensive spectroscopic analysis. Journal of Non-Crystalline Solids. 2023;600:121989.
15 Goldstein DH. Polarized Light (Third edition). Boca Raton, FL: CRC Press; 2017.
16 Chen C, An I, Ferreira GM, Podraza NJ, Zapien JA, Collins RW. Multichannel Mueller matrix ellipsometer based on the dual rotating compensator principle. Thin Solid Films. 2004;455–456:14–23.
17 Laskarakis A, Logothetidis S, Pavlopoulou E, Gioti M. Mueller matrix spectroscopic ellipsometry: formulation and application. Thin Solid Films. 2004;455–456:43–9.

18 Chae I, Ngo D, Chen Z, Kwansa AL, Chen X, Meddeb AB, et al. Anisotropic optical and frictional properties of Langmuir–Blodgett film consisting of uniaxially-aligned rod-shaped cellulose nanocrystals. Advanced Materials Interfaces. 2020;7 (9):1902169.

19 Arwin H, Berlind T, Johs B, Järrendahl K. Cuticle structure of the scarab beetle *Cetonia aurata* analyzed by regression analysis of Mueller-matrix ellipsometric data. Optics Express. 2013;21(19):22645–56.

20 Liu S, Chen X, Zhang C. Development of a broadband Mueller matrix ellipsometer as a powerful tool for nanostructure metrology. Thin Solid Films. 2015;584:176–85.

21 Tompkins H, Irene EA. Handbook of Ellipsometry. Norwich, NY: William Andrew Publishing; 2005.

22 Woollam JA. Complete EASE Training Series 2024 [Available from: https://www.jawoollam.com/resources/videos/completeease-training-series].

10

Topography Analysis via Light Reflection

So far, we have studied the principles that allow us to get the *compositional or structural information of the surface and subsurface region* of the sample by analyzing the light reflected off the sample. All basic equations are derived in Chapter 5. The application of those equations to infrared spectroscopy is described in Chapter 6. The extension of those principles to nonlinear vibrational spectroscopy is described in Chapter 7. Chapter 8 covers statistical data processing of spectroscopic signals. In Chapter 9, we have studied how the same equations derived in Chapters 5 and 6 are used to analyze *the layered structures and compositions beneath the topmost surface* of the sample. In this chapter, we will study how the same principle is used for topographic metrology, that is, to measure the *three-dimensional (3D) shape of the topmost surface* of the sample. The technique is called optical profilometry. This is a totally noncontact mode of profilometry; in that sense, it contrasts with the stylus profilometry (which is not covered in this book) and atomic force microscopy (AFM; which is covered in Chapter 11).

10.1 White Light Interferometry (WLI)

10.1.1 Conventional Optical Microscopy

With conventional optical microscopes operated in the (normal) bright-field mode, we can see the topographic information in projection to a two-dimensional (2D) image plane. The side walls with vertical height change appear dark in the bright-field image. Figure 10.1A shows the optical microscope images of two gratings used for AFM calibration. The left image is a one-dimensional array of triangular steps; the focal plane is aligned with the apex of the triangular step and other regions look dark. The right one is a checkerboard-like array of square pillars with sharp undercut edges; the top surfaces of the pillars look brightest, the base regions between the pillars look slightly darker, and the side wall regions are completely dark.

When an annual filter is added in the irradiation beam path and the sample is irradiated with annular rays, the beams passing through or reflected from flat regions do not come to the image plane. Instead, the beams scattered from the side walls are imaged; thus, the flat regions are dark, and the side walls are bright in the image. This is called a dark-field mode. In the left image of Figure 10.1B, one side of the triangular step looks brightest, the flat base is darker, and the other side of the triangular step is darkest; this contrast is due to the optics alignment. In the right-side image, the top surfaces of the pillars are dark, the base regions between the pillars are brighter, and the side wall regions of the pillars are brightest.

The comparison of these bright- and dark-field images allows detecting local height changes within the field of view. If the height change is larger than the depth of field of the objective lens used, then one could move the sample stage up and down while tracking the focus in the image plane to tell whether the height change is upward or downward. If the topographic height change is smaller than the depth of field, then we cannot obtain quantitative height information with conventional optical microscopy.

This quandary can be resolved with optical profilometry, which is basically interference microscopy, combining optical microscopy with interferometry. We have already studied the basic principle in the reflectometry section in Chapter 9. When one beam is split into two and two beams travel different paths and then combine again at the detection (image) plane, they interfere with each other, and the measured intensity varies depending on the phase difference between the two beams and the wavelength of light used. The phase difference is a function of the pathlength difference and the refractive index of the material that the beam travels through and is reflected from (see Section 9.2).

Surface and Interface Analysis: Principles and Applications, First Edition. Seong H. Kim.

Companion website: www.wiley.com/go/Kim

(A) (B)

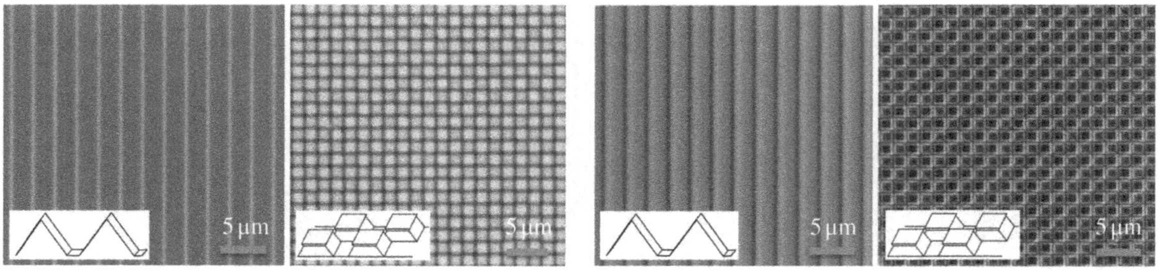

Figure 10.1 (A) Bright- and (B) dark-field images of (left) one-dimensional array of triangular steps and (right) chessboard-like array of square pillars with sharp undercut edges made by lithographic etching of Si wafers. The image is taken with an ×100 objective lens (NA = 0.9). Note that the bright and dark contrasts are changed in (A) and (B) images of each sample.

10.1.2 Interferogram of Light Reflected from Two Surfaces

On a rainy day, we often see a rainbow sheen on a wet floor at a gas station. That is because the spill of a few drops of gasoline forms a thin layer of oil on water covering the road. The spread of oil on the water surface creates two interfaces from which the light can be reflected. We can calculate the optical pathlength difference (OPD) between the light reflected at the air/oil (top) interface and the one passing through this interface and reflected from the oil/water (bottom) interface. From eq. (6.25), we get $OPD = 2d_2\sqrt{n_2^2 - \sin\theta_1}$, where d_2 and n_2 are the thickness of refractive index of oil and θ_1 is the incidence angle (viewing angle from the surface normal direction; note that n_1^2 is not showing since $n_{air} = 1$). If OPD contains an odd number of $\lambda/2$, then the maximum constructive interference will occur; if OPD contains an even number of $\lambda/2$, then the maximum destructive interference will occur. Note that we normally think that constructive interference occurs when OPD is a multiple of the whole wavelength; but the pattern here is opposite. That is because the sign of the electric field of the light reflected from the top interface (where $n_{air} < n_2$) is shifted by 180° from that of the light reflected from the bottom interface (where $n_2 > n_{water}$), see Figures 5.13A and 5.16A. For a given thickness of oil slick on water, we can calculate the wavelengths at which constructive and destructive interferences will occur:

$$\lambda_{constr} = 4d_2\sqrt{n_2^2 - \sin^2\theta_1}/(2m+1) \tag{10.1}$$

$$\lambda_{destr} = 2d_2\sqrt{n_2^2 - \sin^2\theta_1}/m \tag{10.2}$$

where m is the integer. For a given thickness, multiple wavelengths can give constructive interferences. For that reason, rainbow sheen is observed when we look at a thin oil film on water especially when d_2 is larger than ~0.5 μm and less than ~3 μm. This principle is used to estimate the thickness of oil slicks on water when oil is split on the sea [1].

Since we have studied reflectometry, we can do the full spectral calculation over the entire visible wavelength region. Figure 10.2 plots the reflectance spectra, calculated using eq. (9.3), for a 1-μm thick oil film on water as various viewing angles. Note that the incident angle of the light that arrives at our eyes varies as a function of distance; so, the spectral intensity varies for different locations, which is perceived as different colors by our brain. In reality, there might be other factors such as inhomogeneity in thickness or absorption and scattering by impurities.

The above example is applying the optical principle to light reflection from *internal* surfaces. The same principle of interference is applied to the light reflection by two *external* surfaces. Figure 10.3A is the theoretically calculated interference pattern of 600-nm light reflected from a set of two mirrors where one is tilted by 0.014° (0.25 mrad) from the other. Thus, the OPD varies as the position of the beam moves from the left to the right, which causes the interference pattern with undulating bright and dark lines. The OPD between two adjacent bright lines or two adjacent dark lines is exactly 600 nm. Figure 10.3B shows the Newton's ring pattern calculated for a plano-convex lens with a 200 cm radius of curvature placed on a flat mirror under the illumination with a 600-nm light. In both cases, the light reflected at the bottom surface of the top lens and the one reflected at the top surface of the bottom mirror interfere with each other and produce oscillating intensity patterns which are governed by eqs. (10.1) and (10.2). In these cases, n_2 is 1 since the medium is air and θ_1 is zero since the light is illuminated from the surface normal direction. In the case of two nearly parallel mirrors (Figure 10.3A), we can calculate the difference in the gap distance as a function of location. In the case of Newton's rings (Figure 10.3B), we can even

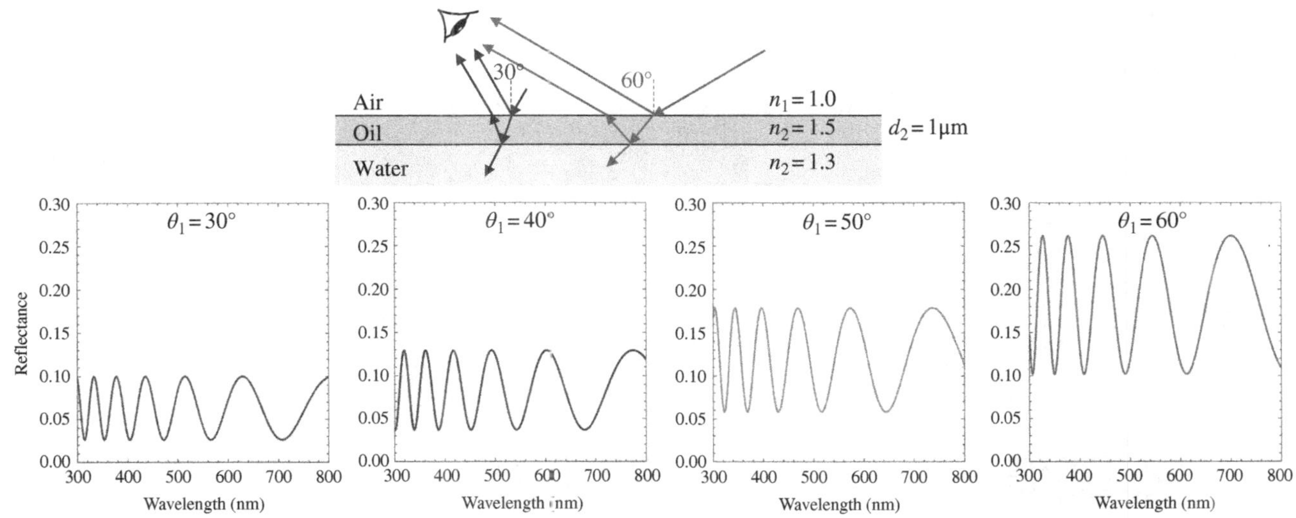

Figure 10.2 Variation in reflectance as a function of wavelength, causing rainbow sheens, calculated for a 1-μm thick oil film on water as a function of viewing angle.

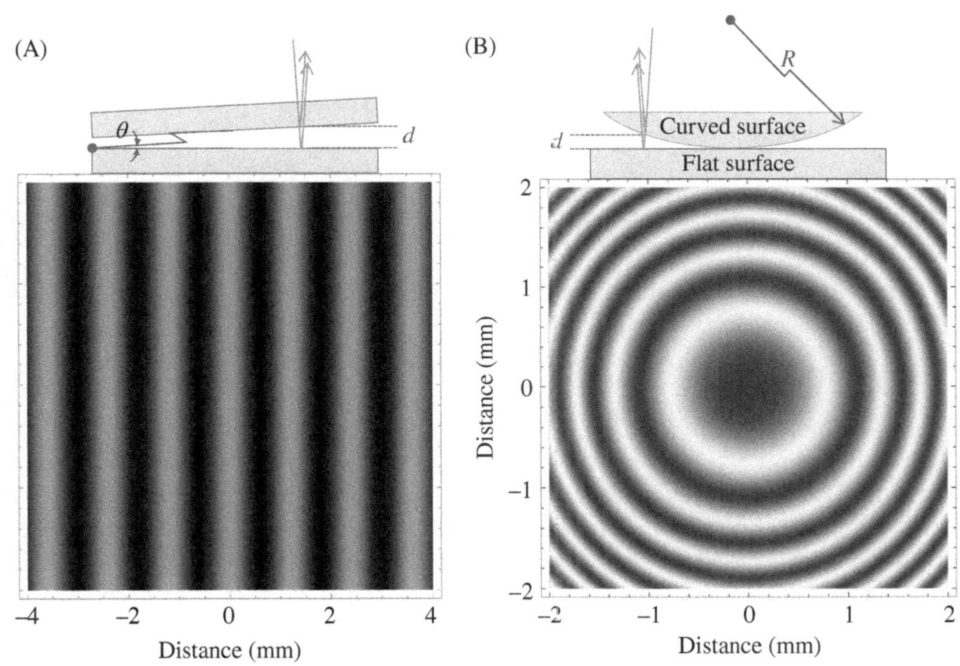

Figure 10.3 (A) Interference pattern created by reflection of 600 nm from two mirrors tilted by 0.014°. (B) Newton's ring pattern created by reflection of 600 nm from a plano-convex lens with a radius 200 cm of curvature on a flat mirror. Both are calculated for the normal incidence.

calculate the exact gap distance as a function of position from the center since we can count m for each dark or bright circles. Going further, we can determine the radius of curvature from Newton's ring patterns measured at a given wavelength.

10.1.3 Objective Lens for Interferometric Microscopy

From these examples, we can see that the distance between the two reflecting planes can be determined by analyzing the interference pattern. To apply this principle to optical microscopy, we need a special objective lens that has an internal reference plane. Two types of such lenses are developed. The first is the Michelson-type objective lens (Figure 10.4A). This

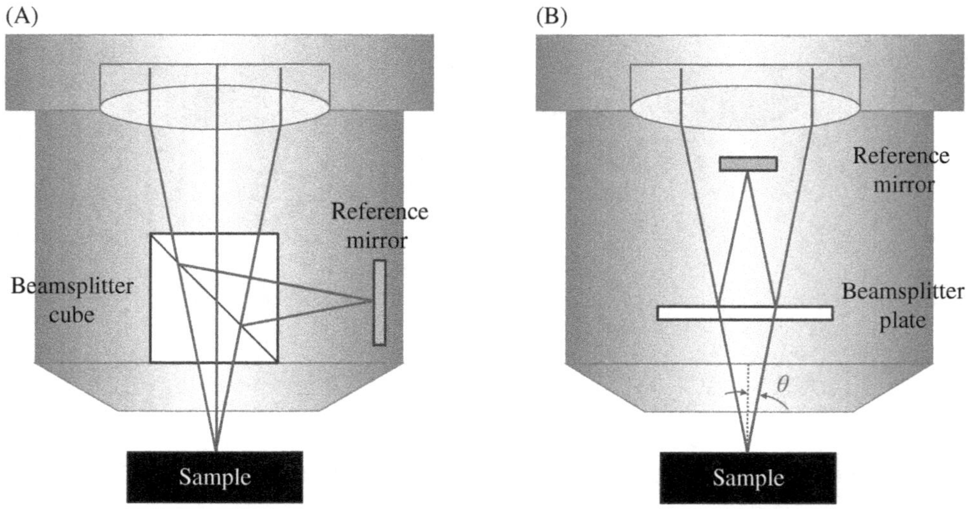

Figure 10.4 (A) Michelson-type and (B) Mirau-type objective lenses used for optical profilometry.

one has a beam splitter cube and a reference mirror. The illuminated light from the objective lens is split equally and one half is sent to the reference mirror and the other is to the sample. The beam reflected from the sample is combined with the beam reflected from the reference mirror at the image plane (optical camera). The second is the Mirau-type objective (Figure 10.4B). In this design, a beam splitter plate is used and the reference mirror is placed at the objective lens axis. The reference mirrors are finely adjusted so that the zero group-velocity OPD position is identical to the position of best focus. The Michelson-type works well in low magnification with a large field of view. The Mirau-type works better for high magnification.

10.1.4 Determining Sample Height in WLI

For a single wavelength (λ) illumination, the interference intensity (I) generated by the recombination of the beams reflected from the reference mirror and the sample surface can be described as:

$$I(z_s, z_o) = I_{DC} + I_{AC}\cos\left(2\pi\frac{OPD}{\lambda} + \varphi\right) \text{where } OPD = 2(z_s - z_o)\cos\theta \tag{10.3}$$

where z_s and z_o are the sample height and the objective position, respectively, I_{DC} and I_{AC} are the intensities of DC background and oscillating fringe, respectively, θ is the illumination angle (see Figure 10.4B), and φ is the phase offset which is determined by the reflection and transmission properties of the interferometer components. Figure 10.5 plots the interference fringes as a function of OPD at four different wavelengths. Fringes formed at different wavelengths have different OPD dependences. The sum of all interference signals produced by white light forms a fringe pattern with quickly decreasing modulation. The bottom panel of Figure 10.5 shows the sum of all interference fringes for wavelengths varying from 350 nm to 750 nm at an increment of 1 nm step, divided by 400 (assuming all wavelengths have the same intensity). The actual white light fringe looks different from this because of the intensity profiles and the divergence of illumination angle. The fringe modulation reaches a maximum intensity when OPDs for all beams are equal, which corresponds to OPD = 0. Thus, by monitoring the maximum fringe intensity as a function of z_o, we can determine the height of the sample surface (z_s).

This is graphically illustrated in Figure 10.6A [2]. The objective lens is lowered stepwise at a fixed increment setting and the white light interference fringe is recorded at each pixel as a function of the objective lens position (z_o). When a certain pixel shows a maximum fringe intensity than others at a given z_o, we know the height of that sample surface position (z_s) is at zero OPD; but we do not know if that position is higher or lower than the other regions. By monitoring the interference fringe intensity as a function of z_o, we can obtain such information at every pixel within the field of view. This is done through computer-controlled data acquisition and processing, and the map of peak contrast offsets at all pixels is digitally reconstructed to make 3D topography image of the entire field of view (Figure 10.6B). This is called vertical scanning interferometry (VSI) or scanning white-light interferometry (SWLI).

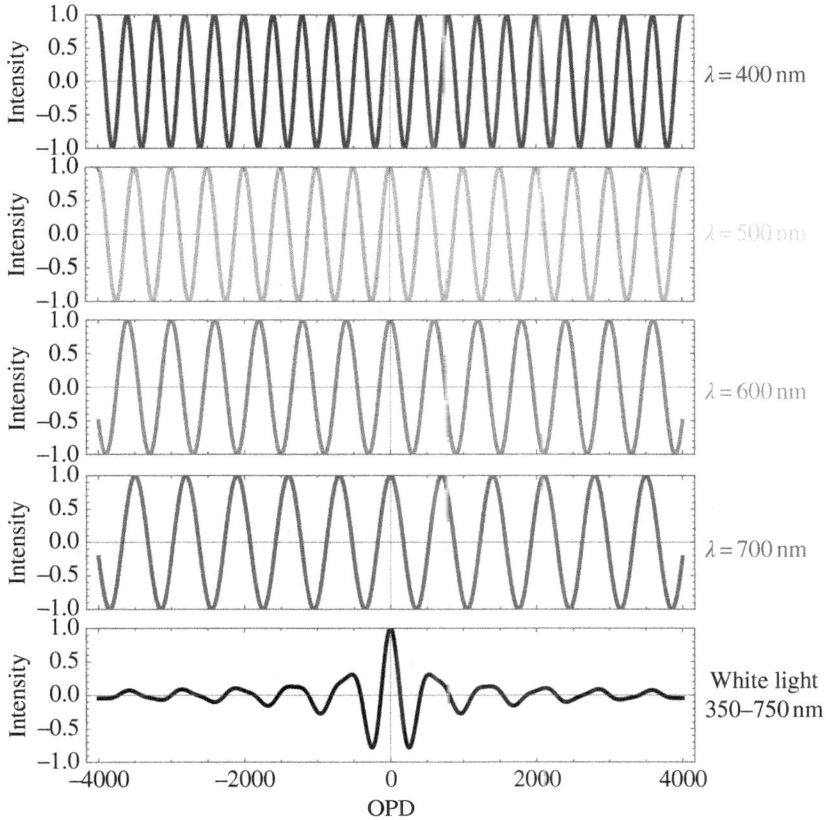

Figure 10.5 Interference fringes made with 400 nm, 500 nm, 600 nm, 700 nm, and 350–750 nm lights.

VSI can measure non-specular surfaces fairly well, and the result is invariant of intensity variations across the field of view. The height resolution is determined by the step scan setting and the digital processing for accurately resolving the maximum fringe intensity. The lateral resolution is determined by the numerical aperture of the objective ($NA = \sin\alpha$, where $\alpha = \arctan(P/f)$ with P = pupil size and f = focal distance) as well as camera pixel spacing and noise filtering of the digitized image.

10.1.5 Case Study of Vertical Scanning Interferometry

Figure 10.7 displays the images of the same gratings shown in Figure 10.1 obtained with an optical profilometer (VSI). The gray-scale 2D images are the interference fringe patterns taken when the objective lens was focused on the highest surface of the sample. In the case of triangular steps, the apex of the triangular step can be seen clearly (i.e., OPD = 0), but the trough regions are all fuzzy. In the case of square pillars, the top surfaces are all bright (i.e., OPD = 0) and the lower surfaces between the pillars are all dark. Due to diffraction of lights, the bottom region looks somewhat rounded. A series of fringe images were taken as a function of the z-step scan of the objective lens (or sample stage) and the collected fringe images were automatically processed at every pixel to construct the 3D height maps of the imaged surface. The perspective views of the processed 3D images are also shown with an artificially added color scale for height in Figure 10.7.

Once the 3D image is constructed, a cross-sectional line profile can be obtained for quantitative analysis of fine details. In Figure 10.7A, we can see that the height change of the slanted surface is well resolved. The lateral distance between adjacent apexes is measured to be 3 μm, which is accurate based on the manufacturer's specification; but the measured depth is slightly larger than the true value. The angle of the slanted surface from the bottom plane is supposed to be 54.7° (angle between Si(111) and Si(100) planes), but the measured angle is 77°. The width of the bottom floor in the valley is much larger than the value calculated from the crystallographic angle and the measured height. This must be an artifact caused by reflection from the highly tilted surface and/or interference of the light reflected from the adjacent slanted surface. Figure 10.7B is

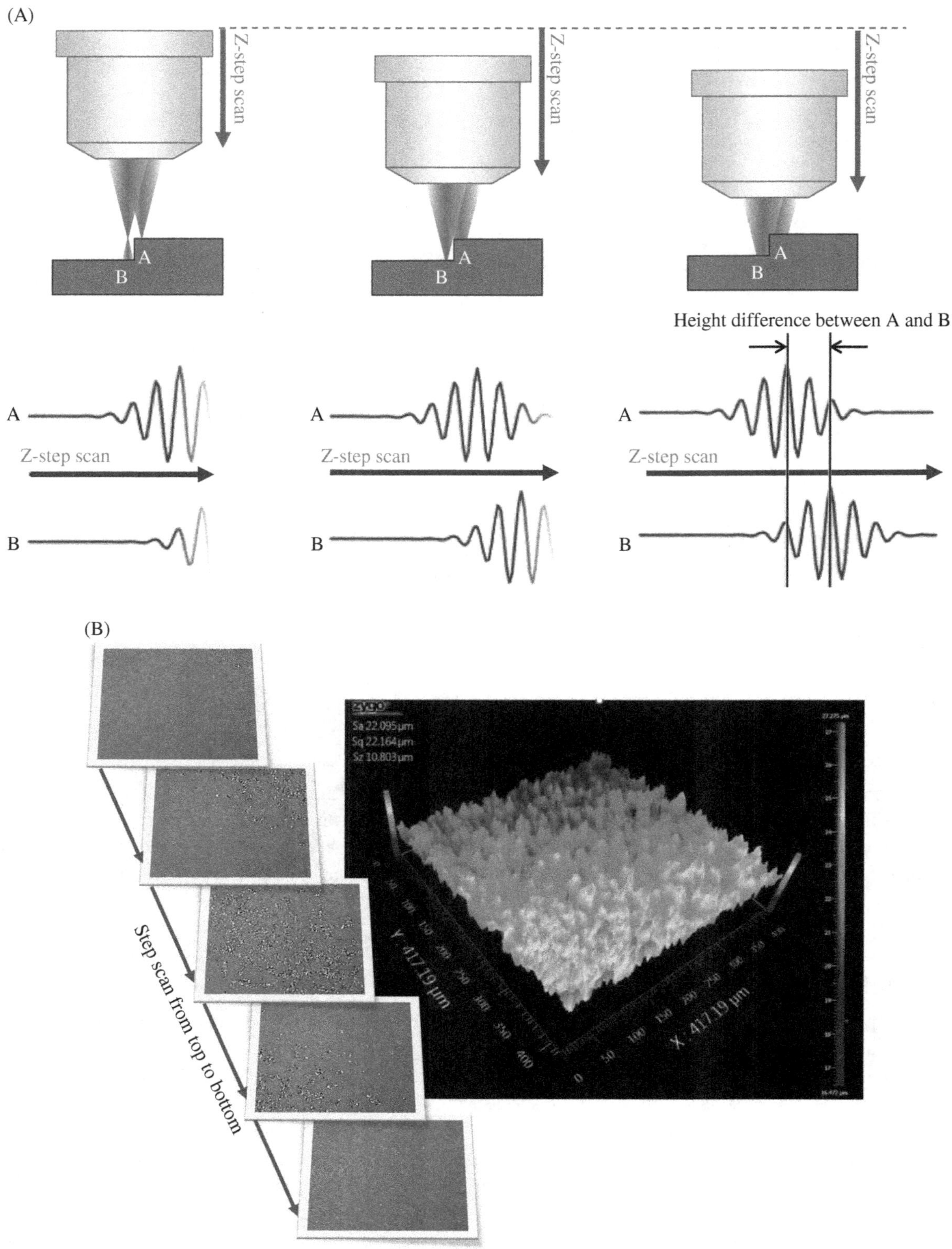

Figure 10.6 (A) Schematic illustration of data acquisition in SWLI (also called VSI). (B) Images of interference fringe patterns collected at different step scan heights and processed 2D topography map. Reproduced from de Groot [2]/with permission of Zygo Corporation.

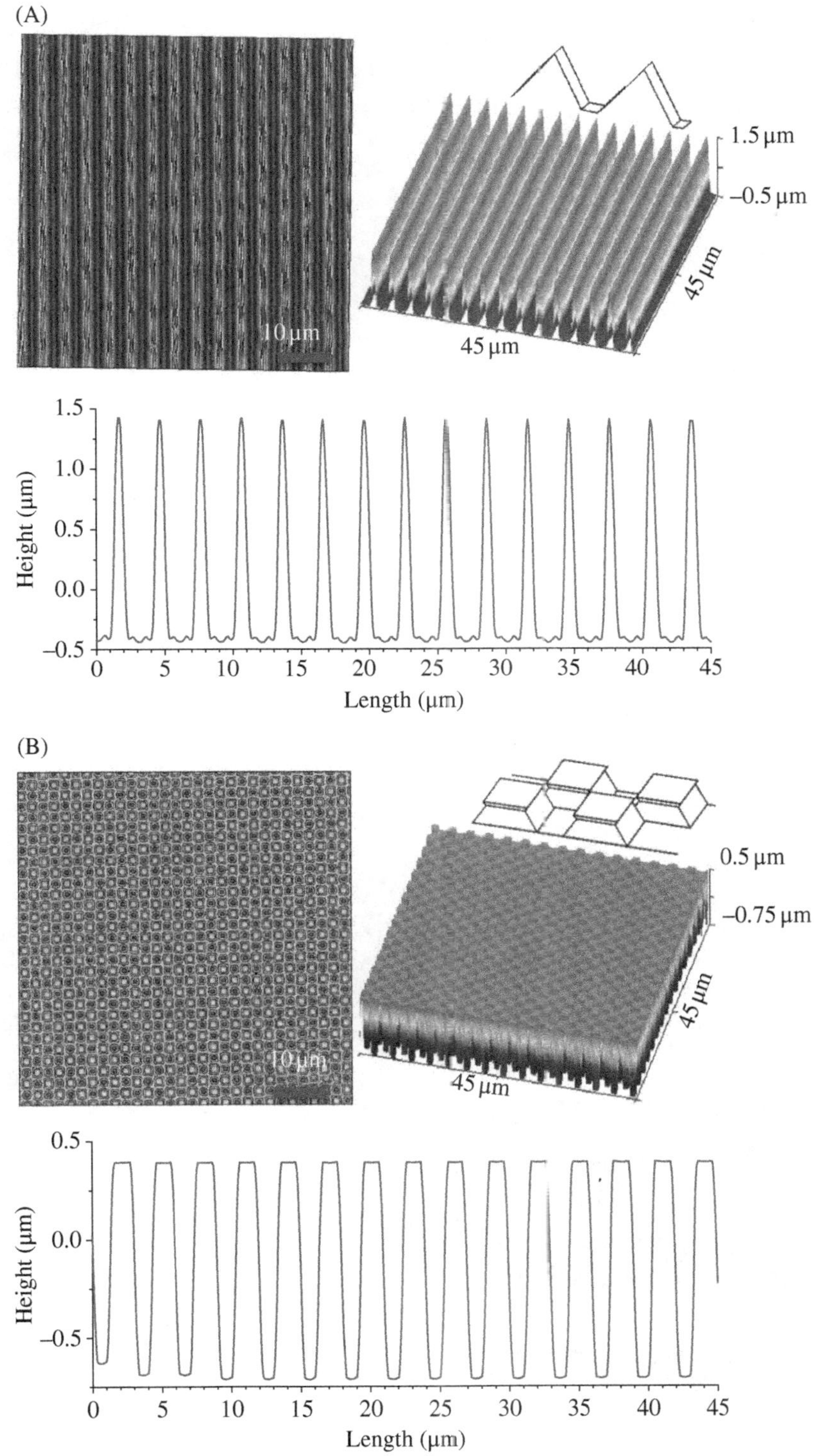

Figure 10.7 Interference fringe images (focused on the highest point), 3D topographic images, and cross-sectional line profiles of AFM calibration gratings shown in Figure 10.1: (A) one-dimensional array of triangular steps and (B) chessboard-like array of square pillars with sharp undercut edges. The image is taken with an ×50 Mirau-type objective lens (NA = 0.55).

the analysis result of the squared pillar sample. The lateral dimension is found to be accurate, that is, in good agreement with the manufacturer's specification. But the measured depth is much larger than the true value. AFM measures the pillar height to be 0.6 μm, which is close to the manufacturer's specification. Again, this discrepancy must be due to interference of lights reflected from the side walls of the surrounding pillars. Obviously, the undercut region cannot be imaged with VSI since that region is not illuminated; instead, the slanted side wall looks smooth in the computer-generated 3D topography. These two examples show the capability and limitation of the VSI in measuring the true height.

Figure 10.8 shows the results from optical profilometric analysis of soda lime silica (SLS) glass surfaces after friction tests with borosilicate spheres (Pyrex balls) in three relative humidity (RH) conditions (20%, 40%, and 90%) [3]. The diameter of

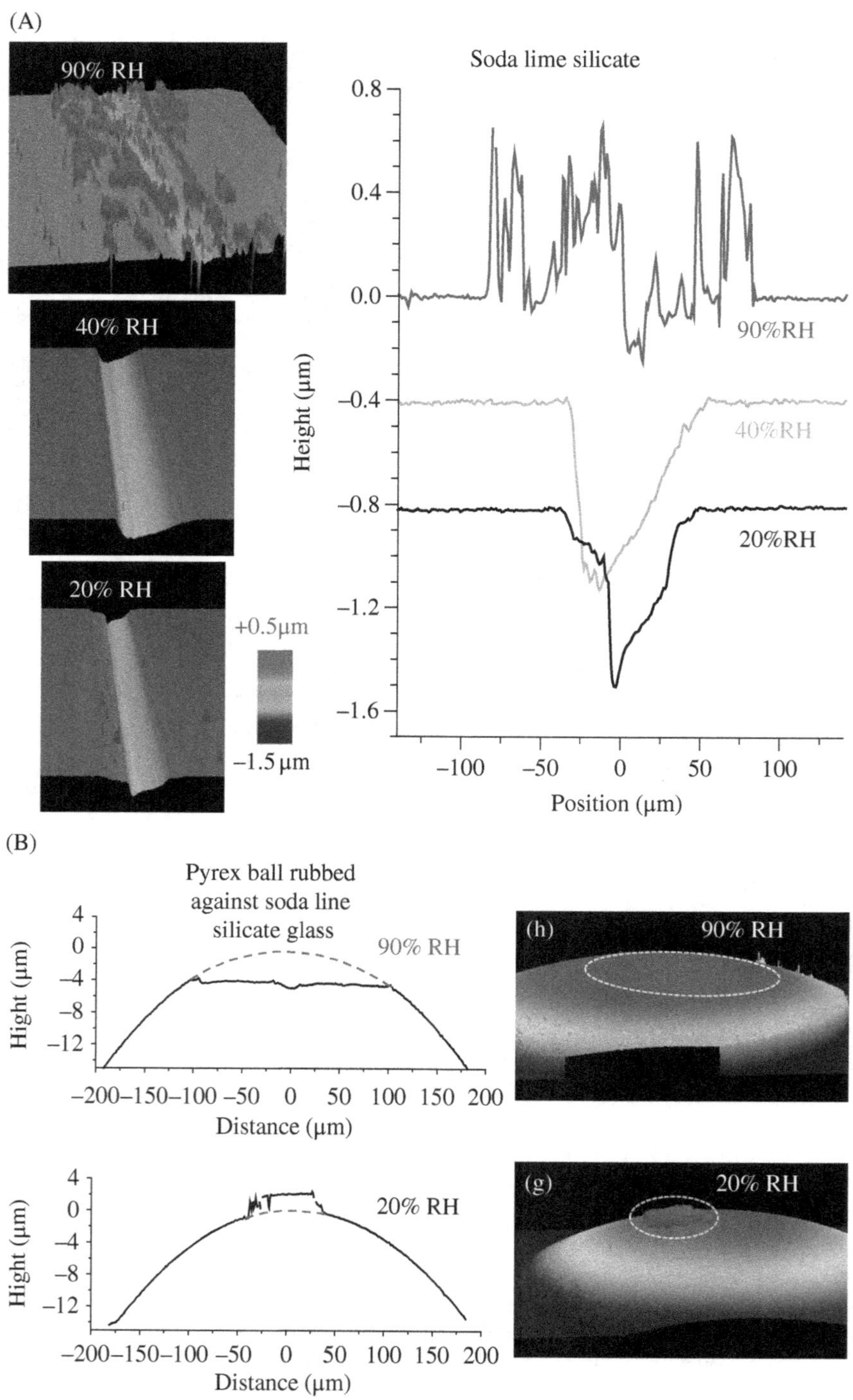

Figure 10.8 Optical profilometry images and cross-sectional line profiles of (A) SLS surface wear tracks and (B) Pyrex ball surfaces after friction test at the same load condition but in different humidity conditions. In (A), the negative volume for 20% and 40% RH is calculated to be −13,000 μm^3 and 18,500 μm^3, respectively, and the positive volume for 90% RH is 25,000 μm^3. In (B), the red-dotted lines show the original ball surface before friction testing. Reproduced from Bradley et al. [3]/with permission of John Wiley & Sons.

Pyrex spheres was 3 mm. When a 20 g load is applied to this sphere on the SLS surface, the maximum Hertzian pressure reaches about 200 MPa at the center of the contact area, which is still far below the hardness of SLS glass (H = ~7 GPa). The sliding speed was only 3 mm/sec, low enough not to cause any frictional heating. The measured friction coefficient was about 0.7–0.8 in all three RH conditions; but the wear pattern was found to be drastically different. Topographic analysis of wear tracks on SLS (Figure 10.8A) revealed that the SLS glass surface wore off, leaving a smooth trench in low RH conditions, while in high RH conditions, the wear of SLS surface was negligible and some debris particles were found in the sliding track. By defining the flat surface outside the wear track as a reference plan, the negative and positive volumes below and above that plane can be calculated, which corresponds to the substrate wear volume and the volume of debris on the surface, respectively. These changes also affected the counter-surface wear (Figure 10.8B). When RH was low, the wear debris of SLS were found to adhere to the counter-surface, forming a protrusion. In the high RH condition, the Pyrex ball was worn off (polished); thus, the debris seen on the substrate must come from the counter-surface wear. Since the exact same materials were used with the exact same load and sliding conditions and the only difference was the RH condition, it could be hypothesized that the difference observed in the wear pattern is caused by the amount and structure of water layers adsorbed on the glass surface (see Figures 6.6 and 6.36).

Not all samples can be analyzed reliably with VSI. Figure 10.9 shows an example where optical profilometric imaging fails to give accurate topographic information. When a stainless steel (SS) surface is friction-tested in an environment containing n-propanol vapor, friction coefficient is observed to be 0.12–0.15 and there is no wear mark left (unless the applied load is too high to produce plastic deformation) [4]. When the same test was conducted in 2-propenol vapor environment, an iridescent layer was observed inside and around the friction-tested track [5]. When the surface was rinsed with an alcohol swab, the iridescent layer was completely removed, and no visible wear mark could be found. This suggests that the iridescent color is some sort of organic films (called tribo-film) produced by rubbing of adsorbed 2-propenol on the SS surface [5]. When such organic films "on" the SS surface were analyzed with VSI, the result showed as "depression" of the surface (Figure 10.9). This is opposite to what we expected. This kind of problem is caused because the tribofilm was semitransparent (thus, causing iridescent colors due to differences in local OPD, as discussed in Section 10.1.2).

The data processing of a series of interference fringe patterns at each pixel is done with the assumption that the interference fringe pattern is produced by recombination of lights reflected from two surfaces (see Figure 10.4). But the presence of transparent or semitransparent film creates two reflection planes at the sample surface (see Figure 10.2); thus, the

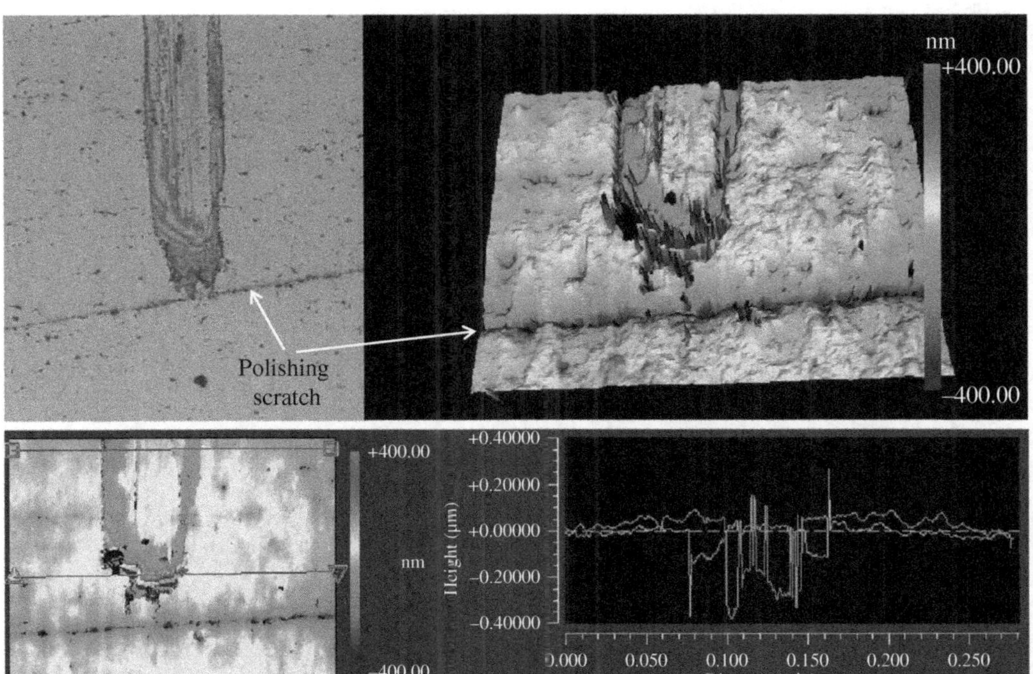

Figure 10.9 Optical image and VSI topography images of an organic deposit (tribofilm) formed on a SS surface. When the deposit was rinsed with alcohol, it was removed, and no physical wear of the SS surface could be detected. Thus, the negative depth of the deposit region from VSI analysis was just an artifact.

interference pattern is generated by recombination of three lights. This problem can be resolved by combining the ellipsometry principle with the white light interferometry [6–8]. Although such a function is available in advanced profilometry instruments, the proper use of that mode requires *a priori* knowledge of the sample since an optical model needs to be built to solve the ellipsometric equations (see Section 9.4.2). If the surface film thickness is uniform across the entire imaged area, model-based ellipsometric or reflectometric VSI analysis can be done reliably. But, if the film thickness varies without any specific patterns and the film property (refractive index) is not known, then such analysis would be quite challenging.

10.2 Surface Roughness

Once the 3D topography of surface is measured, we need to describe it quantitatively. It is done by calculating the moments of topographic height distributions within the imaged area. Figure 10.10 graphically illustrates various parameters used for that purpose. We can simply record the highest and lowest points of the profile; the deviation of those points from the mean value is called maximum peak height (R_p) and minimum valley depth (R_v). The sum of these two is the maximum peak-to-valley height (R_z). These values can be read directly from rather peculiar points and do not have any statistical meaning.

More frequently used one is the arithmetic mean deviation of roughness (R_a) which is the average of profile height deviations from the mean line normalized with the sampling length (L). R_a is typically used to describe the roughness of machined surfaces. Another way is a quadratic mean or root-mean-square (RMS) average of profile height deviations from the mean line, which is denoted as R_q or R_{rms}. R_q is used to describe the finish of polished surfaces. When the roughness of 2D area is calculated, L is replaced with area (A) and the single integral ($\int dx$) is replaced with double integrals ($\int\int dxdy$). The 2D roughness is often denoted as S_a and S_q.

It should be noted that R_a and R_q (for 1D line) and S_a and S_q (for 2D area) are a function of the sampling length or area. As the sampling size increases, they become smaller. So, the comparison of arithmetic mean roughness and RMS roughness of different samples is most meaningful when the sampling dimensions are the same or at least comparable.

It is quite common to describe the degree of surface roughness with R_a and R_q only; but they are not sufficient because they do not give any information on the shape of height profile distribution. For that, a different property is defined. The skewness of roughness (R_{sk}) is a measure of symmetry of the profile about the mean line. If R_{sk} is zero, it depicts symmetric distribution about the mean height line (Figure 10.10). A positive R_{sk} value indicates that a greater percentage of the height

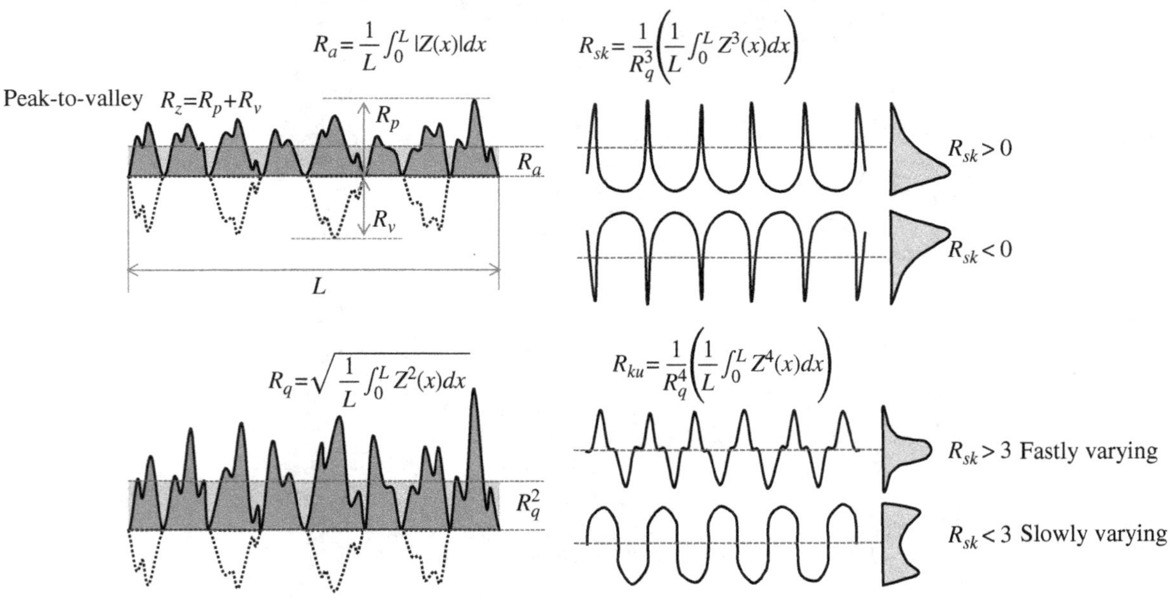

Figure 10.10 Various parameters describing different aspects of surface roughness. In this schematic illustration, the roughness calculated over the length (L) along one-dimensional direction. The dotted contour lines in the left panels are the profiles below the mean value, which are flipped to the positive side for graphical illustration purpose. In the right panels, height distribution profiles are drawn as a representative example.

profile is below the mean line (i.e., having spikes or pillars protruding out of the surface). The opposite is a negative R_{sk} case which means a greater percentage is above the mean line (i.e., having pits or holes penetrating into the surface). The kurtosis of roughness (R_{ku}) is a measure of the sharpness or tailedness of the height profile distribution (Figure 10.10). A perfectly random surface has $R_{ku} = 3$. If R_{ku} is larger than 3, it means a rapidly varying profile; if R_{ku} is less than 3, it means a slowly varying profile. Sometimes, excess kurtosis is used, which is $R_{ku} - 3$. So, more positive excess kurtosis means a sharper distribution having fewer extreme outliers than the normal distribution, and more negative excess kurtosis means a broader or flatter distribution with more outliers than the normal distribution. We can imagine that the surfaces with the same R_q but different R_{sk} and R_{ku} will have quite different mechanical properties or deformation patterns.

We can practice the roughness calculation using the topography profiles shown in Figure 10.7. The analysis data are summarized in Table 10.1. The triangular-shape step grating has a positive skewness and a slightly negative excess kurtosis. The square-pillar checkerboard grating has a negative skewness and a large negative excess kurtosis (i.e., slowly varying, broad distribution).

Another important parameter is the power spectral density (PSD) which provides the distribution of the roughness amplitude as a function of the spatial frequency (= wavenumber or wavevector) of the roughness which is the inverse of the wavelength of the roughness features. Mathematically, it is calculated by taking Fourier transform of the autocorrelation function of the roughness signal, which contains just the power (and not the phase) across a range of wavenumbers. Figure 10.11 shows the PSD analysis of two specific examples [9]. Figure 10.11A is a surface roughness created mathematically by overlapping two sine waves with different frequencies. The PSD function of this surface shows two spikes that corresponds to the frequency of two sine waves. Figure 10.11B is a non-self-affine random surface. For most isotropic surfaces, PSD decreases from low to high wavenumbers. The way the PSD decreases depends on the distribution of

Table 10.1 Roughness of calibration gratings shown in Figure 10.7.

	One-dimensional array of triangular steps	Checkerboard-like array of square pillars
Mean roughness	552 nm	392 nm
RMS roughness	652 nm	434 nm
Skewness	1.2	−0.60
Excess kurtosis	−0.25	−1.3

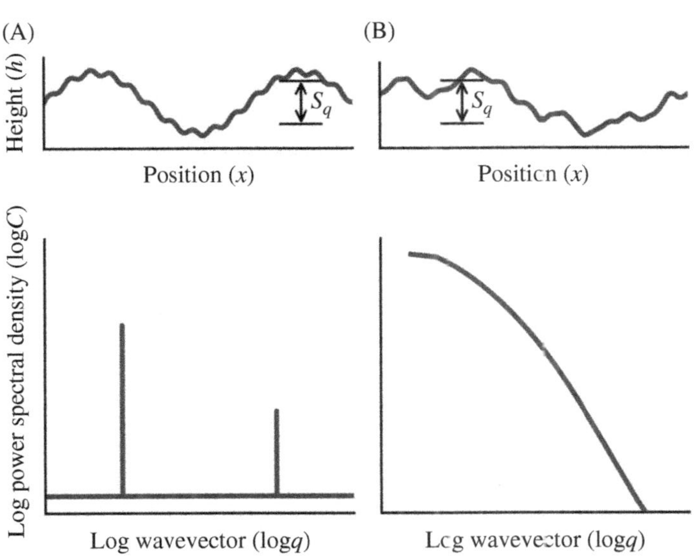

Figure 10.11 Power spectral density (PSD) of (A) a rough surface created by superposition of two sine waves and (B) a frozen capillary wave surface. These two have the same RMS roughness. Reproduced with permission from [9]/IOP Publishing.

roughness features which are a function of process by which the surface roughness is produced [10]. The PSD analysis is particularly useful to find periodic surface features inherent to or characteristic of the system such as grains or fibrillar patterns that might otherwise appear random [11]. It is also useful to compare the roughness determined from different instruments over different size scales [9, 12].

10.3 Further Reading of Optical Profilometry

To understand the optical imaging of surface topographic features, it is important to understand basic concepts and formulas in microscopy. Websites of many optical microscope manufacturers provide quite useful information and interactive tutorials with graphical illustrations and animations [8, 13–15]. The tutorial article written by P. de Groot in *Advances in Optics and Photonics* **7**, 1–65 (2015) is a good resource for optical profilometry [2].

Practice Problems

10.1 Prove that there is 180° phase shift between the light reflected from the bottom of the top glass plate and the one reflected from the top of the bottom glass plate.

10.2 Shown below is the intensity line profile of Newton's ring pattern imaged for a plano-convex lens of a flat mirror with 500 nm in the reflection mode. The incident angle is 0°. Calculate the radius of curvature of the lens surface.

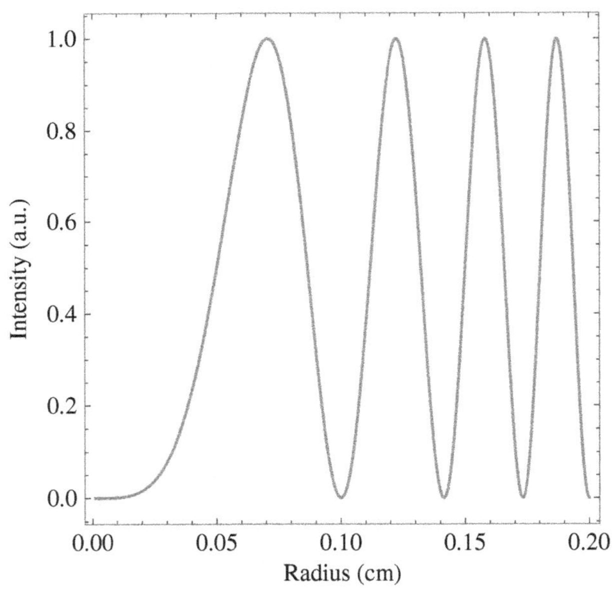

10.3 Compare the characteristic features of skewness and kurtosis by drawing the roughness distributions for the following five cases:

a) Zero skewness with zero excess kurtosis
b) Positive skewness with positive excess kurtosis
c) Positive skewness with negative excess kurtosis
d) Negative skewness with positive excess kurtosis
e) Negative skewness with negative excess kurtosis

References

1 Fingas M. Visual appearance of oil on the sea. Journal of Marine Science and Engineering. 2021;9(1):97.

2 de Groot P. Principles of interference microscopy for the measurement of surface topography. Advanced Optical Materials 2015;7(1):1–65.

3 Bradley LC, Dilworth ZR, Barnette AL, Hsiao E, Barthel AJ, Pantano CG, et al. Hydronium Ions in soda-lime silicate glass surfaces. Journal of the American Ceramic Society. 2013;96(2):458–63.

4 Barthel AJ, Kim SH. Lubrication by physiosorbed molecules in equilibrium with vapor at ambient condition: effects of molecular structure and substrate chemistry. Langmuir. 2014;30(22):6469–78.

5 Barthel AJ, Combs DR, Kim SH. Synthesis of polymeric lubricating films directly at the sliding interface via mechanochemical reactions of allyl alcohols adsorbed from the vapor phase. RSC Advances. 2014;4(50):26081–6.

6 Colonna de Lega X, de Groot P. Characterization of Materials and Film Stacks for Accurate Surface Topography Measurement Using a White-light Optical Profiler: SPIE; 2008.

7 Fay M, Dresel T. Applications of model-based transparent surface films analysis using coherence-scanning interferometry. Optical Engineering. 2017;56(11):111709.

8 Knowledge Center: Edmund Optics; [Available from: https://www.edmundoptics.com/knowledge-center/]

9 Jacobs TDB, Junge T, Pastewka L. Quantitative characterization of surface topography using spectral analysis. Surface Topography: Metrology and Properties. 2017;5(1):013001.

10 Duparré A, Ferre-Borrull J, Gliech S, Notni G, Steinert J, Bennett JM. Surface characterization techniques for determining the root-mean-square roughness and power spectral densities of optical components. Applied Optics. 2002;41(1):154–71.

11 Cutler C, Thackeray J, Trefonas P, Millward D, Lee CB, Mack C. Pattern roughness analysis using power spectral density: application and impact in photoresist formulation. Journal of Micro/Nanopatterning, Materials, and Metrology. 2021;20(1):010901.

12 González Martínez JF, Nieto-Carvajal I, Abad J, Colchero J. Nanoscale measurement of the power spectral density of surface roughness: how to solve a difficult experimental challenge. Nanoscale Research Letters. 2012;7(1):174.

13 Basic Concepts and Formulas in Microscopy: Nikon Instruments [Available from: https://www.microscopyu.com/microscopy-basics].

14 Tutorials and Education Resources: Olympus [Available from: https://www.olympus-lifescience.com/en/microscope-resource/primer/resources/tutorials/].

15 Olympus E-Books: Olympus [Available from: https://www.olympus-ims.com/en/resources/books/].

11

Topography Analysis via Scanning Probe

All surface characterization techniques we have studied so far in this book are spectroscopic and spectrometric methods, which cover a large portion of the methods listed in Table 1.1. When the sample is irradiated with x-ray (high-energy electromagnetic wave), core electrons can be excited and emitted; by analyzing their binding energies, we can do elemental analysis (Chapter 2). Alternatively, high-energy electrons or ions can be used to cause Auger electron emission or secondary ion ejection, respectively, which can provide elemental compositions and concentration information (Chapters 3 and 4). By irradiating the sample with ultraviolet and visible wavelength photons, we can induce excitation of valence band electrons within the sample, from which we can do chemical identification and concentration determination. Upon interaction with infrared light, electronic configuration of the constituent ions or molecules in the sample does not change, but their vibrational excitation can occur affecting the reflection, absorption, and transmission of IR at the sample surface; by measuring IR reflection or transmission, molecular configuration and conformation of the interface can be deduced (Chapters 5–8). By using the same optical principles associated with light reflection, we can analyze the internal structure of layered samples (Chapter 9) or obtain topographic information of the exterior surface of samples (Chapter 10). All analysis methods studied thus far in this book are conducted without any physical contact with the sample; the only exceptions are the surface energy analysis where the sample is touched with probing liquids (Section 1.5) and ATR-IR where solid samples are compressed against the internal reflection element (Section 6.1).

In this chapter, we will study the topographic analysis of the topmost surface of sample through physical contact (Figure 11.1) [1]. It is called scanning probe microscopy (SPM). There are two types of SPM – scanning tunneling microscopy (STM) [2] and atomic force microscopy (AFM) [3]. In this chapter, we will focus on AFM which can be applied to various types of samples in many different operation modes (see Table 11.1). STM can be used for electrically conducting samples only and its applications are often limited to atomically flat surfaces with atomic or molecular features; readers interested in those STM studies can find relevant books and review articles [4]. The basic principle of AFM is the same as stylus profilometry. Obviously, the sharper the physical contact probe is used, the better the spatial resolution of the obtained image can be. So, it is important to use a probe that is sharper than the topographic feature to be analyzed. But there is a limit on how accurately the image can be measured, which in turn is a function of how accurately the force (or interaction) between the probing tip and the sample can be controlled. There are many books that are written for beginners with a comprehensive coverage of many different aspects of technical details [5]; here, we will review only the basic principles that enables contact, noncontact, and tapping mode operations of AFM. All other modes are extensions of these basic operation modes with some additional controls and electronics. We already have studied how AFM technique can be combined with IR spectroscopy to get vibrational spectroscopic imaging of samples with spatial resolutions much smaller than the diffraction limit (see Section 6.6 for AFM-IR, Nano-FTIR, and PiFM).

11.1 Tip–Sample Interactions in Atomic Force Microscopy (AFM)

Figure 11.2A schematically illustrates the basic components of AFM. Obviously, the essential part of AFM is the probe which consists of a rectangular or triangular-shape cantilever with a sharp tip protruding out of the bottom side of the cantilever. An example of a rectangular shape probe is shown in Figure 11.2B. The probes are typically made of silicon using photolithography and etching processes. Sometimes, cantilevers made of silicon nitride are also used, which are fabricated

Surface and Interface Analysis: Principles and Applications, First Edition. Seong H. Kim.

Companion website: www.wiley.com/go/Kim

Figure 11.1 Conceptual illustration of SPM principle. Reproduced from Binnig and Rohrer [1]/with permission of American Physical Society.

Table 11.1 Various types of techniques based on AFM.

AFM	Contact mode AFM	PFM	Piezoelectric force microscopy
	Noncontact mode AFM	C-AFM	Conductive AFM
	Tapping mode AFM	SICM	Scanning ion-conductance microscopy
	Peak-force tapping AFM	SECM	Scanning electrochemical microscopy
MFM	Magnetic force microscopy	SThM	Scanning thermal microscopy
CFM	Chemical force microscopy	AFM-IR	Photothermal AFM coupled with IR
FMM	Force-modulation microscopy	Nano-FTIR	Scattering-type scanning near-field optical microscopy coupled with FTIR
EFM	Electrostatic force microscopy		
KPFM	Kelvin-probe force microscopy	PiFM	Polarization-induced force microscopy
SPFM	Scanning polarization force microscopy	TERS	Tip-enhanced Raman spectroscopy

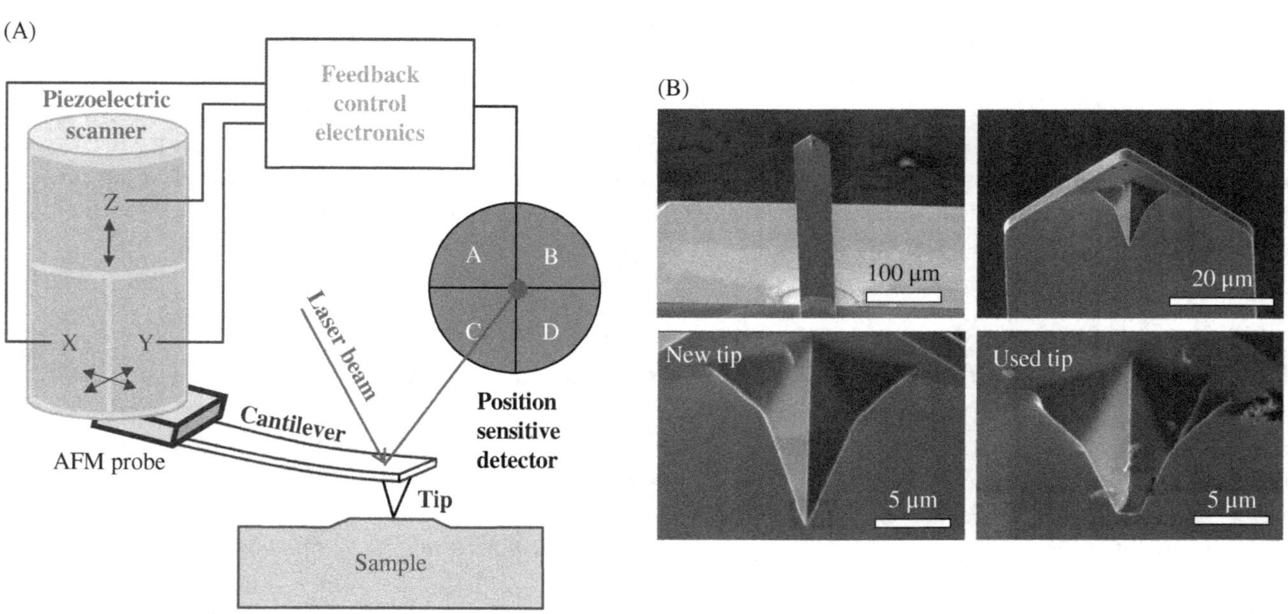

Figure 11.2 (A) Basic constituents of AFM parts (not drawn in scale). (B) SEM images of a typical AFM probe consisting of a rectangular cantilever and a sharp tip.

through additional fabrication processes. The AFM probe is attached to a piezoelectric scanner that can move in the X, Y, and Z directions over the sample surface with precise controls. Alternatively, the tip can be mounted on a stationary stage and the sample is mounted on a scanner. When the tip is in contact (or interacting) with the surface, the cantilever will bend following attractive or repulsive interactions. The degree of cantilever bending is measured typically through the reflection of a laser beam off the top side of the cantilever using a so-called position sensitive detector (PSD). To increase the reflectivity, a thin metal coating is applied to the top side of the cantilever (opposite side of the tip). When electrical interactions (such as electrostatic forces or current flow) are to be probed, the bottom side of the cantilever is also coated with a thin conducting layer (such as Au or Pt). If a magnetic interaction is of interest, a magnetic alloy (such as CoCr) is coated. The PSD signal is fed to a feedback-control electronics unit which adjusts the movement of the scanner. The feedback control keeps the cantilever deflection or oscillation constant while the tip is scanned over the sample. The degree of this deflection or oscillation is related to the force applied to the sample by the tip or vice versa, which can vary from <1 nN to tens of μN depending on the cantilever used. The extent of scanner movement in the Z direction (which is the voltage signal applied to the scanner by the controller) is recorded as a "height" while the tip (or sample) is moving along the X and Y directions. Then, a set of X, Y, and Z voltages representing the position of the scanner are converted to the lateral distances and the vertical height using a set of calibration factors which are predetermined before the sample scanning. This gives a 3D topography image of the sample surface. All other techniques listed in Table 11.1 measure additional forces or other properties simultaneously along with the topography map.

11.1.1 Forces Involved in AFM

What kinds of forces does AFM measure? We can do a back-of-the-envelope estimation for the force between the two atoms bonded together. At the atomic energy scale, typical bond energy (E_b) is about 1–5 eV, which corresponds to 1.6–8 × 10^{-19} J. Typical bond length (l_b) is about 0.2 nm. Since force is energy divided by distance ($F = E_b/l_b$), one could estimate the force between two atoms bonded to each other to be around 1–4 nN. This is within the force range that can be controlled with AFM. Then, is it really, as allured by the name of the technique, the "atomic force" that AFM is measuring? Not really. Let us delve deeper into basic physics.

The most universal force relevant to AFM would be the van der Waals (vdW) force. If the AFM tip end is modeled as a sphere, the force (F_{vdW}) originating from vdW interactions with a flat surface can be estimated as [6, 7]:

$$F_{vdW} = \frac{AR}{6D^2} \tag{11.1}$$

where A is the Hamaker constant which depends on the polarizability of materials and surrounding medium, R is the tip radius, and D is the tip–surface separation distance. Typical values for Hamaker constant between solid materials in air is 4∼20 × 10^{-20} J [6, 7]. If the tip radius is 20 nm and the tip–sample separation distance is 0.2 nm (i.e., in direct contact), then F_{vdW} is calculated to be 3–17 nN. If the materials interacting each other is immersed in a dielectric medium (such as liquid), then Hamaker constant is much smaller (by an order of magnitude) and so is F_{vdW}. Note that this is the attractive force. If the cantilever is pressed against the sample surface, additional repulsive force is applied to the tip–surface contact region. These are the primary contributors to the interaction force between the tip and the sample [8].

If the sample surface is charged or electrically biased, then there would be an electrostatic interaction between the tip and the surface. The electrostatic force (F_E) between a sphere and a flat surface in air can be calculated to be [7]:

$$F_E = \frac{\pi \varepsilon_o R^2 U^2}{D(D+R)} \tag{11.2}$$

where U is the electrical potential difference (or applied bias voltage) between the tip and the sample and ε_o is the dielectric permittivity of air. When $U = 1$ V, $R = 20$ nm, and $D = 0.2$ nm, we get $F_E = 3$ nN. This force is added to F_{vdW}. If the tip is held above the sample at a constant distance (e.g., $D = 2$ nm), then F_E is ∼0.3 nN. Note that this magnitude is comparable or even larger than the attractive vdW force at that distance ($F_{vdW} = 0.2$ nN assuming $A = 20 \times 10^{-20}$ J and $R = 20$ nm).

If both tip and sample are magnetic, their interaction produces a magnetic force (F_M) [8]:

$$F_M = \mu_o \int \nabla \left(M_{tip} \cdot H_{sample}\right) dV_{tip} \tag{11.3}$$

where μ_o is the magnetic permittivity (1.257×10^{-6} N/A^2), M_{tip} is the magnetic moment of tip, H_{sample} is the stray magnetic field of sample, and integration is performed over the volume of the magnetic coating over the tip. The magnitude of F_M is inversely proportional to the tip–sample distance (D) and in the order of 10^{-1} nN when $D = 10$ nm [9].

When the AFM probe surface has specific functional groups that can have chemical interactions with the sample surface (such as hydrogen bonding interactions), transient chemical bond formation can alter the adhesion force (F_{adh}) between the tip and the surface [10]. Since such chemical interactions occur only when the tip is in direct contact with the surface, their effects are observed mostly in the contact-mode scan (manifested as changes in friction force during the lateral scan) or in tapping-mode scan (manifested as changes in vertical adhesion force) [11, 12].

In humid air, water can adsorb and condense at the annular region of the tip in contact with the surface and the negative curvature of the condensed liquid meniscus can exert a capillary force. The capillary force (F_{cap}) is the sum of surface tension contribution (F_{st}) and Laplace pressure contribution (F_{Lap}) of the liquid meniscus [13]:

$$F_{cap} = F_{st} + F_{Lap} = 2\pi\gamma R\left[\sin(\phi)\sin(\theta_1 + \phi) - R\sin^2(\phi)/(2r_K)\right] \quad (11.4)$$

where γ is the surface tension of liquid forming the meniscus via capillary condensation, ϕ is the angle from the center of the sphere (modelling the tip end) to the meniscus-sphere contact point (so-called filling angle), θ_1 is the contact angle of the liquid meniscus edge with the tip surface (which can be assumed to be the same as the macroscopic contact angle, see Section 1.5), and r_K is the Kelvin radius (negative value) which is a function of the molar volume and relative partial pressure (p/p_{sat}) of the vapor that is in equilibrium with the condensed liquid phase at a given temperature. If the liquid completely wets the solid surface (as in the case of water or alcohol adsorption on a clean silica surface), then the boundary condition of the system defined by the liquid contact angle (θ_1) is replaced by the adsorption isotherm (i.e., thickness of the adsorbed liquid layer, see Sections 6.1.3, 6.3.5, and 6.4) [14]. The capillary force is often normalized with $4\pi\gamma R$ for comparison of effects of different tip radii and liquid surface tensions. In very dry condition (i.e., $p/p_{sat} \leq 1{\sim}2\%$), $F_{cap}/4\pi\gamma R \approx 1$. Note that, for liquid water, $4\pi\gamma R$ is around 18 nN when $R = 20$ nm, which is already larger than F_{vdW}. As p/p_{sat} increases, $F_{cap}/4\pi\gamma R$ increases to a maximum and then decreases. The maximum value is a function of the molar volume of the condensed liquid; in the case of ethanol, it can be as high as $F_{cap}/4\pi\gamma R \approx 4$ at $p/p_{sat} = 10$–20% [15]. In the case of water adsorption in humid air, if the ice-like structure is formed on the clean silicon oxide surface of the tip and the substrate (see Section 6.1.3), this force can be even higher due to strong hydrogen-bonding interactions [16].

From these calculations, we can see that the governing force measured or controlled in AFM analysis is not certainly atomic force. It varies depending on the system and the AFM operation mode. In the contact scan mode, F_{vdW}, F_{cap}, and forces due to chemical interactions will contribute to attractive interactions. Among these, F_{cap} can be the most significant factor if the measurement is done for hydrophilic surfaces in humid air. If an external force is applied to the surface in contact with the tip by the cantilever, the total force that the surface will feel is the sum of the applied force and all these attractive forces originating from the tip–surface interactions. Note that hydrophilic surfaces typically have a high surface energy and tend to be contaminated readily by adventitious organic vapor adsorption from ambient air. Thus, keeping or controlling the sample cleanliness is important. In the noncontact mode, F_{vdW}, F_E, and F_M play important roles depending on the tip–sample separation distance (D). In the tapping mode (which is also called "intermittent contact" mode), the effects of all these forces are intertwined and their effects vary depending on the surface properties of the sample and the environment condition of the surrounding medium.

In aqueous solution, various types of solid–liquid(water)–ion interactions play critical roles, which is a function of the surface charge of the solid and the pH and ionic concentration of the solution. Such interactions include repulsive forces due the presence of electrical double layers near the surfaces charged with the same polarity, attractive forces that can be described by the DLVO (Derjaguin–Landau–Verwey–Overbook) theory, solvation forces arising from ordering of solvent molecules (which varies oscillatory with decreasing D to zero), hydration forces due to formation of strongly hydrogen-bonded water network in the presence of chemisorbed cations at the hydrophilic surfaces, attractive hydrophobic forces between two hydrophobic surfaces, and so on [7, 17]. In organic solvent, solid-like packing of molecules can occur when they are sandwiched by the flat substrate and the probe tip end, which is called nanoconfinement effects [18]. Such effects are observed readily in surface force apparatus (SFA) where the sandwiched area is large [19, 20]; the same effects can also be observed with a relatively sharp tips with AFM [21, 22]. These variations may sound sophisticated; but, such variability makes AFM unique and versatile. If one can control the main governing force, then it is possible to probe different physical and chemical properties of various materials under a wide range of environmental conditions.

11.1.2 Atomic Resolution in AFM

Can AFM really get images with an atomic resolution, that is, be able to detect individual atoms? The answer is yes and no. Figure 11.3A is probably the first AFM image with the atomic resolution showing the (7 × 7) reconstructed Si(111) surface [23, 24]. In the region marked with dotted lines, individual atoms arranged in the (7 × 7) reconstruction pattern can be seen clearly. In some regions, the atomic contrast is poor probably due to changes in the tip condition during the scan. Figure 11.3B compares the STM and AFM images of the graphite basal plane [24, 25]. STM detects the electronic corrugation at the Fermi level and not all surface carbon atoms conduct electricity; so, it only sees every other atom of the graphite surface. In contract, every surface atom exerts attractive or repulsive forces and AFM can detect the corrugation of such forces over every carbon atom at the graphite surface.

These examples demonstrate the capability of AFM to obtain "atomic" images; but it does not mean that AFM detects forces of individual atoms. To understand this, let us compare the detection principle of STM and AFM. In STM, the tip is not in contact with the surface; instead, it is held a few Angstroms above the surface and the quantum-mechanical tunneling current between the tip and the surface is measured. The tunneling probability through the open gap decays exponentially as the tip–surface separation distance increases [4]. So, the current from the atom extruded farthest from the tip to the closest surface atom with the allowed electronic state for electron tunneling will have the dominant contribution to the measured intensity (Figure 11.4A). We do not know the exact location of the apex atom, but as long as its position does not change during the scanning, one can get surface images with atomic resolution.

In a noncontact mode of AFM operation (note that the images shown in Figure 11.3 are taken in this mode), the forces involved in imaging are long-range interactions (Section 11.1.1). Thus, several atoms on the tip interact simultaneously with several atoms on the sample surface. That effect is schematically illustrated in Figure 11.4B. Let us consider a hypothetical tip with three atoms at the tip end scanning over a hypothetical periodic array of atoms of the surface. Because the atoms of

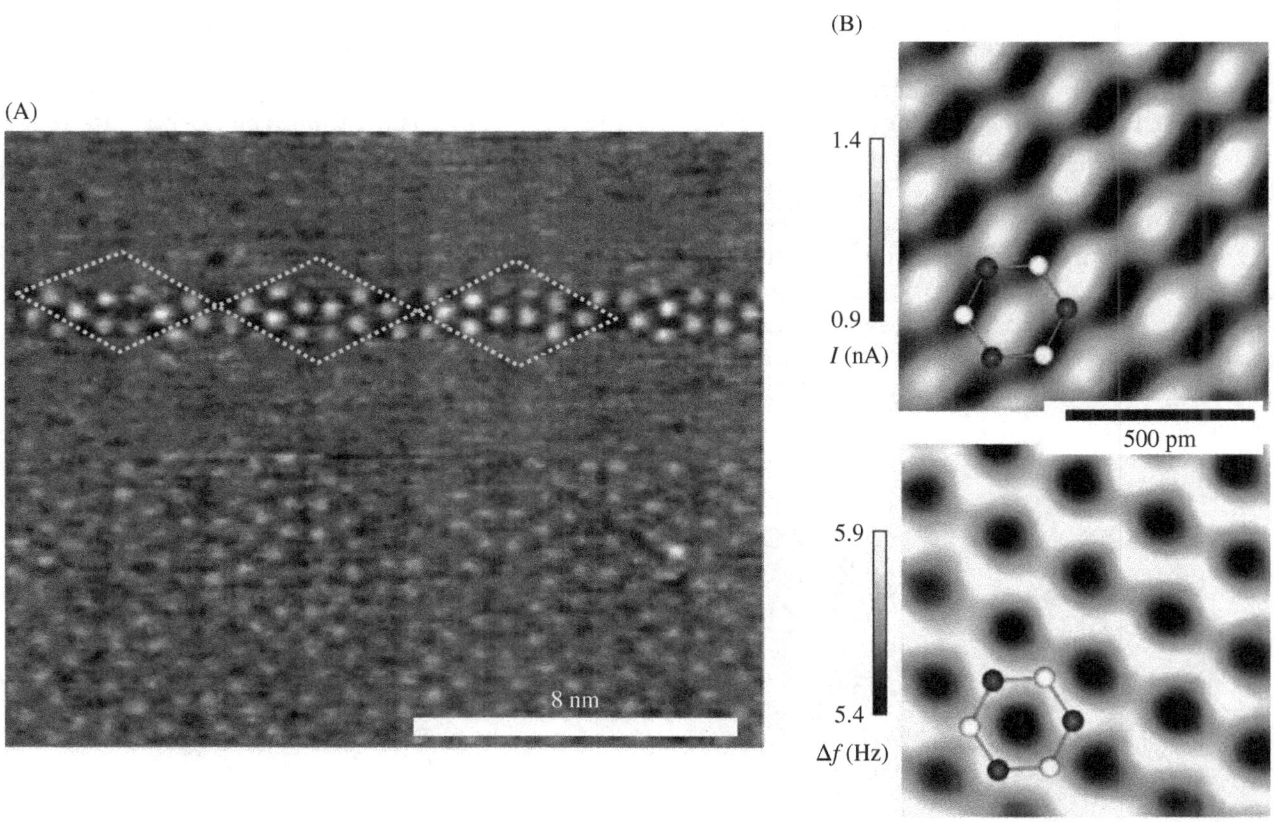

Figure 11.3 (A) AFM image of the Si(111)-(7 × 7) reconstruction with atomic resolution in ultra-high vacuum at room temperature. The (7 × 7) reconstruction pattern is marked with dotted lines. (B) STM (top) and AFM (bottom) images of graphite basal plane in ultrahigh vacuum at liquid temperature. Reproduced from Giessibl [24]/with permission of Elsevier.

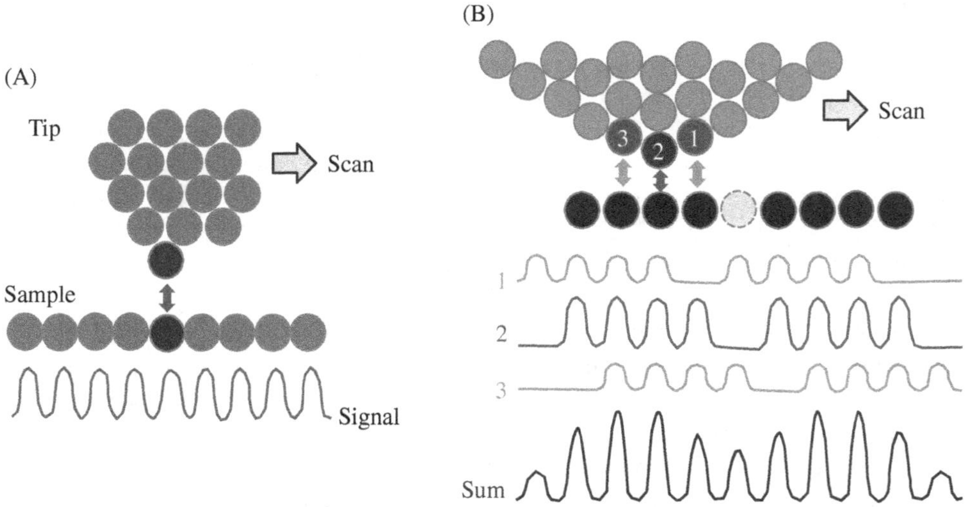

Figure 11.4 Schematic illustration of (A) tunneling current signal between an STM tip and an atomically flat surface and (B) interaction forces between an AFM tip and an atomically flat surface with one single atomic defect in the middle, as the tip moves above the sample from left to right.

the tip are in different positions, the surface atoms that each tip atom at a given scanner position sees are shifted laterally. Also, the tip atoms are at different heights from the surface atoms. The interaction force felt by each tip atom varies with the distance from the surface. The measured signal is the sum of all interactions. The final image may have the correct periodicity; but, vertical corrugation may not look like a single atom defect in the middle. In summary, the atomic features in the recorded image originate from the interactions of ensembles of atoms at the tip and the surface. The height signal recorded is not the true height difference, but just local contrasts in the detected or controlled signal.

Figure 11.5A shows images generated through computer simulations for a single-atom tip and a hexagonally packed seven-atom flake tip scanning over the (111) surface of a face-centered cubic (fcc) lattice with a point defect in the center [26]. The force between the tip and the surface are calculated by adding the pairwise interactions between the tip atoms and the surface atoms which are modeled by a Lennard–Jones potential. Obviously, the single-atom tip can image every individual site in the surface and find the point vacancy accurately. On the other hand, when the same surface is imaged with the flake tip with all seven atoms in registry with the surface lattice, the vacancy is not detected (Figure 11.5B); this is because of the coherent superposition of the forces between all atoms in the flake with all surface atoms under the flake. If the flake tip is rotated, then the tip atoms are placed out of registry with respect to the surface atoms. When the tip is rotated by 14°, the vacancy site is barely visible (Figure 11.5C). When the tip is rotated by 23°, the distortion around the vacancy site is very strong and gives an impression of an extended defect (Figure 11.5D). In reality, such control is difficult to achieve, if not impossible.

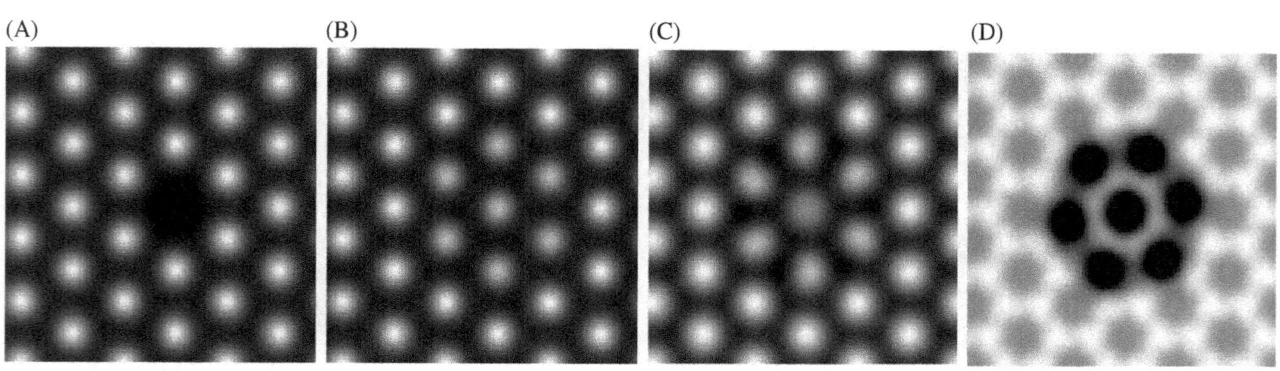

Figure 11.5 Simulated AFM images of a fcc (111) surface with a point vacancy for cases scanning with (A) a single-atom tip, (B) a seven-atom flake tip with its atoms in registry with the surface atoms, (C) the same flake tip rotated by 14°, and (D) the same flake tip rotated by 23°. Reproduced with permission from Koutsos et al. [26]/IOP Publishing Ltd.

11.1.3 Change in Tip Shape Upon Direct Contact with the Surface

In the contact mode scanning, the situation is even worse. The radius (R) of tip curvature is several nanometers at best or often tens of nanometers. Ignoring the adhesion effect, we can calculate the contact radius and pressure using the Hertzian contact mechanics [27]. Let us assume that the tip end is silica and radius of curvature is 10 nm, and the surface is a crystalline silicon wafer. If the applied force is 1 nN, then the diameter of the tip–surface contact area is 1 nm (which is already larger than the size of individual atoms) and the contact pressure is 1.1 GPa. If the applied force is 10 nN, then the contact diameter increases to 2.3 nm and the contact pressure is about 2.5 GPa (see Section 11.4.2). This simple calculation shows that even though the applied load is small, the contact pressure is extremely large due to the small contact area. So, the tip can undergo plastic deformation, becoming blunter and then increasing the contact area even larger. As the tip is scanned over a periodic lattice, the periodic corrugation of the interaction force may be detected as a corrugation in the lateral dragging force (i.e., friction force), even though such corrugations may not be recorded in the topographic image [28, 29]. In this case, if the scan distance is 1 μm, for example, and the data recording is 1024 pixels per line, then each pixel corresponds to 1 nm; so, the lateral signal corrugation corresponding to the lattice dimension cannot be recorded. To record the lateral signal corrugation due to the lattice atoms, the scan size should be reduced to ~10 nm, which gives 0.01 nm/pixel. Again, atomic defects cannot be detected here because the total corrugation of the lateral signal will be dominated by the contribution with periodic atoms within the contact area.

In principle, the number of atoms at the tip end can be counted and their orientations with the surface can be manipulated and measured. Figure 11.6A shows the field ion microscopy (FIM) images of a W(111) tip manipulated by field ion emission processes and the same tip after multiple cycles of contact with and detachment from an Au(111) surface [30]. The forces measured as a function of tip–sample separation distance during the approach and retraction cycles are shown in Figure 11.6B [31]. When the tip movement is controlled such that the tip–surface interaction force does not go into the repulsive region (i.e., compression), the approach and retraction curves are almost on top of each other (top panel in Figure 11.6B). This implies that the tip surface has not been altered by physical contact. There is a small attractive force before the compressive force is applied; this is due to the vdW interaction between the tip and the sample surface. But, when

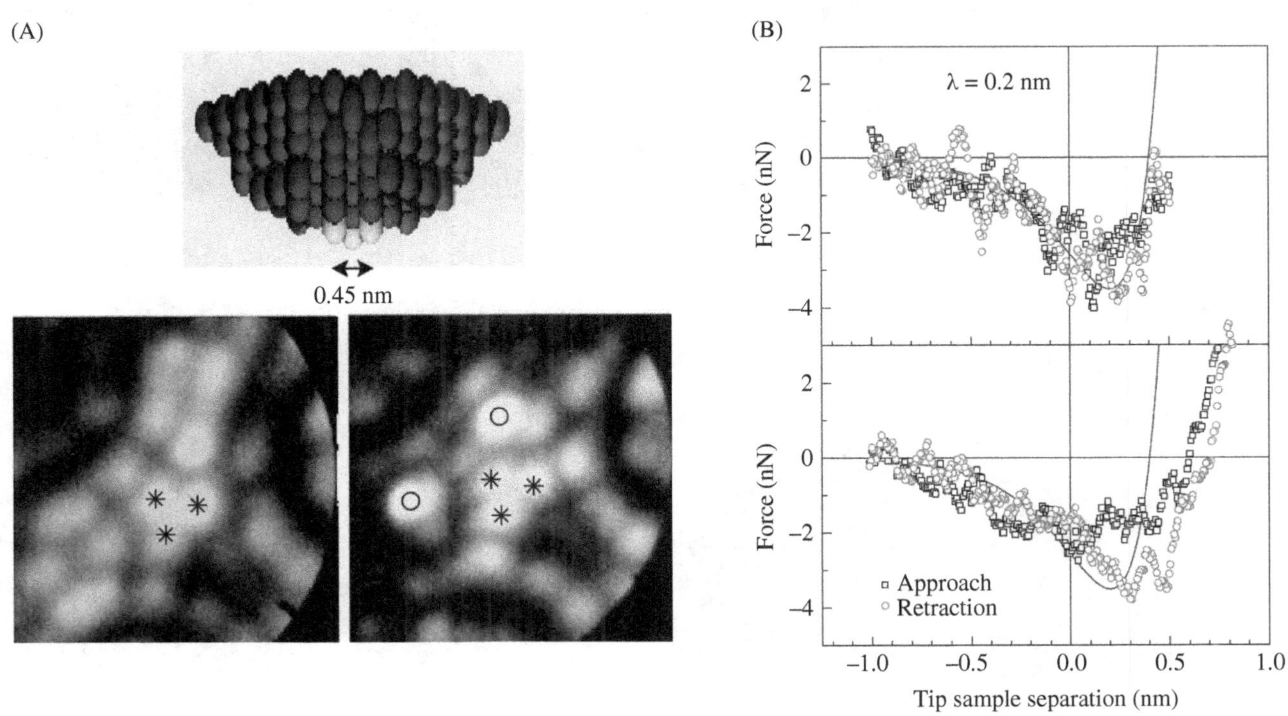

Figure 11.6 (A) FIM images of a W(111) tip before (left) and after (right) the contact with Au(111) surface. Three W atoms at the end are marked with asterisks. The two adatoms are the ones picked up from the surface after the contact. The top is a hard-sphere model of the tip apex reconstructed from FIM images. Reproduced from Cross et al. [30]/with permission of American Physical Society. (B) Force–distance curve for a W(111) tip, which was examined by FIM approaching and reacting from an atomically flat Au(111) surface. Reproduced with permission from Schirmeisen et al. [31]/IOP Publishing.

the tip is compressed against the surface exerting a positive force, then a small hysteresis is observed (bottom panel in Figure 11.6B). This must be due to a change in the tip condition. The FIM image of the tip used for several cycles of compression (Figure 11.6A) shows that some surface atoms were picked by the tip although the overall shape of the W tip did not change much.

The transfer of materials between the tip and the sample surface is also predicted in molecular dynamics (MD) simulations with proper interatomic potentials. Figure 11.7A shows the force vs. tip–surface distance (*F-D*) plot for a Ni tip approaching to and retracting from an Au surface as well as the energy-minimized configurations of atoms at the jump-to-contact and near the separation steps [32]. When the Ni tip is in contact with the Au surface, the atoms at the periphery of the contact are rearranged to optimize their embedding energies, while maintaining their individual material cohesive binding (so, some gold atoms are attached to the Ni tip surface in the middle panel of Figure 11.7A). During the retraction of the tip, a hysteresis occurs in the *F-D* plot due to the adhesive bonding between the two materials and the separation is accompanied by inelastic processes in which the topmost layer of the Au sample adheres to the Ni tip [32].

Even for the materials that are not intrinsically adhesive, indentation-induced material transfer is possible in AFM experiments if the applied load is high enough. Figure 11.7B shows the snapshots from MD simulations when the hydrogen (H)-terminated diamond (111) tip is compressed against the H-terminated diamond (111) surface [33]. Here the apex carbon atoms of the diamond tip are not in perfect registry with the H-terminated carbon atoms in the top layer of diamond lattice of the surface. When the simulation was done for a maximum force of 200 nN, the approach and retraction curves were identical without any hysteresis; only elastic deformation occurred (not shown here). When the maximum load was increased to 250 nN, the carbon atoms along the periphery of the tip move slightly to minimize the C–H repulsion, causing other atoms to be out of alignment. This leads to damages to the periodic (111) structure and the bond formation between tip atoms and substrate carbon atoms. Thus, upon separation, atomic transfer occurs which is accompanied by a hysteresis in the *F-D* curve [33].

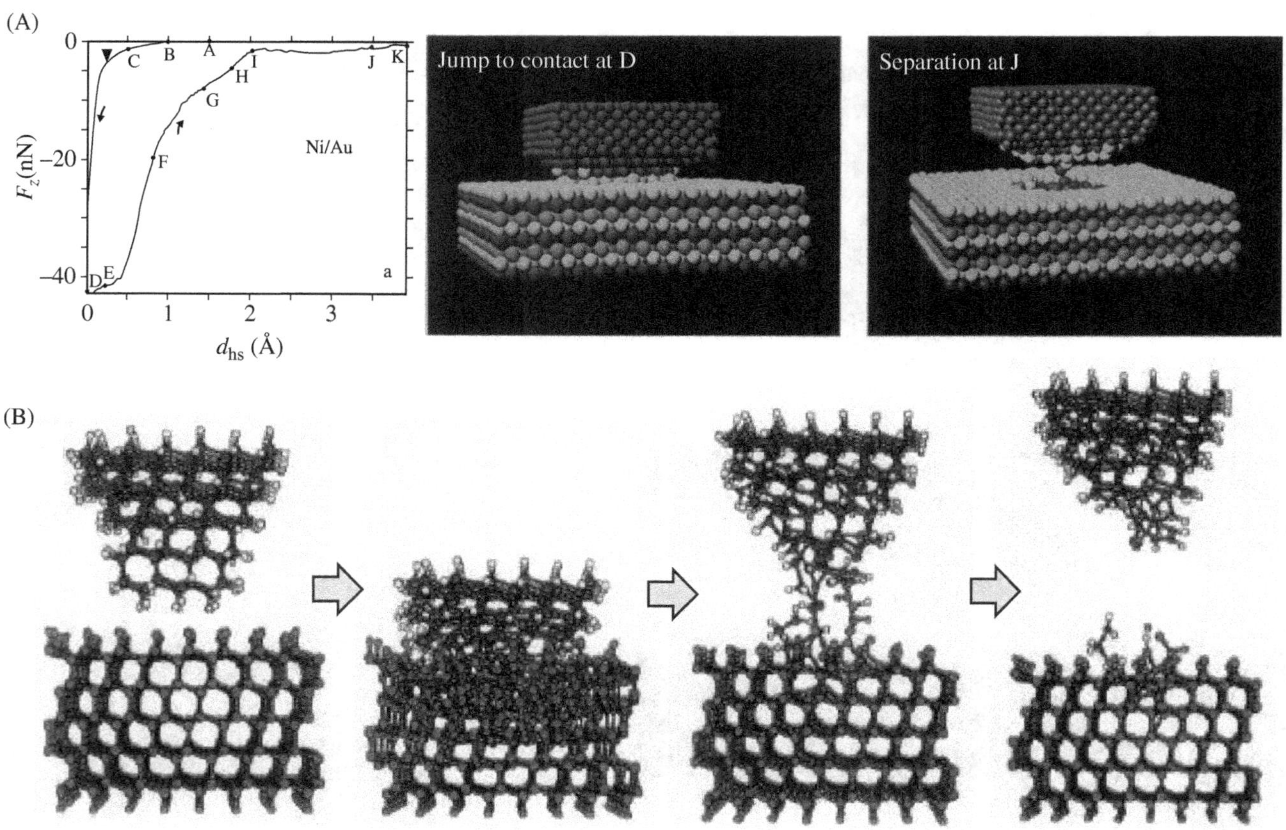

Figure 11.7 (A) Calculated force vs. tip -surface distance curve for a Ni tip approaching to and retracting from an Au surface from MD simulations and snapshots at the tip -surface contact and the separation. In the snapshop images, the top box with a faceted bottom is the Ni tip, and the bottom box with a flat top surface is the Au substrate. The Au atoms in the first, second, and third layers are presented with different shades. Reproduced from Landman et al. [32]/with permission of American Association for the Advancement of Science – AAAS. (B) MD simulation snapshots during the indentation of the H-terminated diamond (111) surface with a H-terminated diamond (111) tip. Reproduced from Harrison et al. [33]/with permission of Elsevier.

These experimental and computational data are a few selected examples showing that the AFM tip surface can be changed upon contact with the sample surface. How readily that change will occur depends on how much force is applied as well as the chemistry of the tip and surface materials. Other than atomic reconfiguration due to adhesive contacts discussed above, other mechanisms such as plastic deformation or frictional wear of the tip are also possible. If the contact stress becomes comparable to the yield strength of the tip material, then plastic deformation will occur during the normal indentation. During the contact scan, the tip can also undergo physical and chemical wear (see the SEM image of the used tip in Figure 11.2B). If the substrate is soft polymeric materials, tip wear is less likely, but contamination is very likely due to material transfer from the substrate. Thus, it is a bad assumption that the AFM tip will keep its original shape and chemistry, while the contact scan is being made or during the repetitive scanning of the same area. Such tip damages can occur not only during the contact-mode scan but also during the tapping-mode scan although the latter is often called "gentle touch" mode. Contact-induced blunting of the tip can be avoided in the noncontact mode scanning [34]; but, the noncontact scanning is sometimes challenging when the surface has large and abrupt topographic changes.

11.2 Force Measurement Through Cantilever Deflection

11.2.1 Cantilever Deflection vs. Force Applied to/by Tip

Figure 11.8A displays a typical force vs. tip–sample distance (*F-D*) curve that plays a pivotal role in the tip–surface interactions. But, it is not known at firsthand or directly controlled; instead, it is "deduced" from the bending of the cantilever (Δz) during the expansion or contraction (Z) of the piezo-scanner to which the cantilever base is mounted (Figure 11.8B). When the tip is far away from the surface, the tip will not feel any force and the cantilever will be straight without any bending. This is called the "free standing" state (which corresponds to point 1 in both Figures 11.8A,B). As the piezo-scanner expands, the cantilever base is lowered toward the sample surface and the tip–sample distance decreases. The tip starts feeling the long-range attractive force ($F < 0$ and $dF/dD > 0$) at point 2. Note that the schematics are drawn for the case doing experiments in air; in aqueous solution, the long-range hydration force can be attractive or repulsive depending on the surface charge. As D decreases further, the tip is pulled toward the surface, causing the cantilever bent toward the surface ($\Delta z < 0$). The attractive force to the tip (F) is balanced with the elastic recovery force of the cantilever, which can be calculated by Hooke's law, $F = k\Delta z$, where k is the spring constant of the cantilever and Δz is the extent of cantilever bending. At point 3, the tangent of the *F-D* curve (dF/dD) becomes the same as the spring constant (k) of the cantilever and

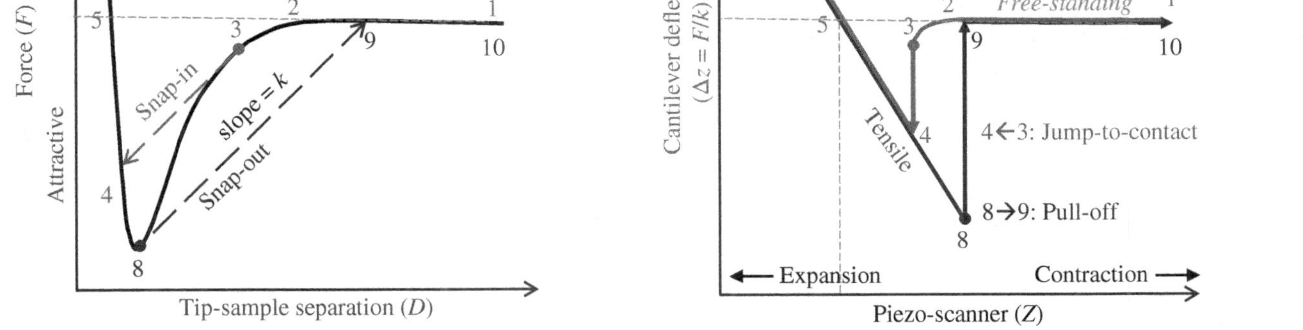

Figure 11.8 (A) Force vs. tip–sample distance (*F-D*) curve and (B) corresponding cantilever defection ($\Delta z = F/k$) vs. Piezo-scanner position (Z) curve during the tip approach and retraction in AFM. The inset in (A) shows the dependent and independent variables in *F-D* and Δz-Z curves. The inset in (B) depicts the principle of measuring Δz in the laser beam reflection method.

the tip snaps to the contact (point 4). The region from point 3 to point 4 of the *F-D* curve cannot be probed due to the instability in force balance. If that region is to be probed, then the spring constant of the cantilever should be higher than the tangent of the attractive force region of the *F-D* curve (i.e., using an extremely stiff cantilever, $k > dF/dD$).

As the base of the cantilever is brought closer to the surface (i.e., decreasing Z further), the tip–surface interaction climbs up the negative-slope side of the *F-D* curve ($dF/dD < 0$), and the contact region undergoes elastic deformation. The degree of elastic deformation would be negligible if the sample is a stiff material (i.e., with high elastic modulus) or can be large if the sample is soft and compliant. Between points 4 and 5, the contact region experiences the tensile stress originating from the convex curvature of the cantilever. At point 5, the cantilever bending is straightened up again ($\Delta z = 0$) and the net force applied to the surface by the tip is zero. In other words, the long-range attractive force between the tip and the surface is balanced by the short-range repulsive force from the elastic deformation of the contact region. Upon further lowering of the cantilever base toward the sample surface (i.e., moving toward point 6), the cantilever vends upward ($\Delta z > 0$), and its concave curvature creates compressive stress to the contact.

Upon retraction from point 7, the concave curvature of the cantilever decreases. Passing point 5 where net force is zero, the tip remains in contact with the surface and the cantilever bends down (i.e., convex curvature again); thus, in the region between points 5 and 8, the contact is under tensile stress again. At point 8, the elastic force of the cantilever becomes equal to the sum of all adhesive forces of the contact, i.e., $k\Delta z = (F_{vdW} + F_{adh} + F_{cap} + \ldots)$; thus, the tip snaps off the surface and the cantilever suddenly springs back to the free-standing state (point 9). Upon further contraction of the piezo-scanner, both D and Z increase at the same rate since Δz remains zero.

11.2.2 Converting Cantilever Deflection to Force

In order to know the force (F) exerted by the surface to the tip or applied by the tip to the surface, we need to measure the cantilever deflection (Δz) with a high precision and know the cantilever spring constant (k) accurately. The most widely used method to measure Δz in commercial instruments is to reflect off a laser beam from the back of the cantilever and detect the position of the reflected beam using the position sensitive detector (Figure 11.2A). The basic principle can be understood using the schematic diagram shown in Figure 11.8B. The position sensitive detector compares the laser beam intensities in the top-half and bottom-half regions, and this difference is converted, using a proper calibration factor which is predetermined independently, to the position or displacement of the reflected beam (Δz_{PSD}) at the detector plane. Let us assume that the piezo-scanner height is fixed, and the sample height changes by Δh, causing the cantilever bending by Δz (assuming the elastic deformation is negligible). Then, the tilt angle of the cantilever will change by $\Delta\theta = \frac{\Delta z}{L}$, where L is the cantilever length. If the PSD plane is located at a distance L_D from the cantilever (i.e., the optical path of the reflected beam from the cantilever to the PSD plane $= L_D$), then the reflected beam spot moves by $\Delta z_{PSD} = 2\Delta\theta \times L_D$. Thus, the cantilever deflection can be calculated to be $\Delta z = \Delta z_{PSD} \times (L/2L_D)$. If the cantilever is 100 μm long and the optical path is 5 cm, a displacement of 1 μm in Δz_{PSD} corresponds to a 1-nm height in Δz.

The Δz resolution is mainly limited by the thermal noise of the deflection detection system, which can be estimated to be $\sqrt{\frac{4k_BT}{3k}}$, where k_B is the Boltzmann constant. At room temperature, the thermal noise of a rectangular cantilever is $0.074\ \text{nm}/\sqrt{k\,(\text{N/m})}$. So, if k is 1 N/m, then the vertical resolution of AFM can be as small as 0.074 nm.

The spring constant (k) of the cantilever can be calculated, using the beam bending theory, from the known elastic modulus (E) of the material and the dimensions of the cantilever. For a rectangular cantilever (shown in Figure 11.2B), the spring constant can be calculated to be [7, 17]:

$$k = \frac{EWT^3}{4L^3} \tag{11.5}$$

where L, W, and T are the length, width, and thickness of the cantilever. The spring constant of a V-shaped triangular cantilever can be calculated to be [7, 17]:

$$k = \frac{EWT^3}{2L^3} \frac{\cos\alpha}{1 + 4(W/b)^3(3\cos\alpha - 2)} \tag{11.6}$$

where α is the half-angle between the two arms, b is the overall width of two arms at the base of the V-shaped cantilever, and other variables are the same as the rectangular cantilever. Typical values of spring constants are 0.01–0.03 N/m for contact mode scanning of soft materials and 20–100 N/m for tapping or noncontact mode scanning.

The length and width of the cantilever can be measured accurately using a simple optical microscope, but the thickness is difficult to measure because it is too small (thin) and requires vertical mounting of the probe. Since the spring constant is proportional to T^3, a 20% uncertainty in T from the manufacturer's specification leads to a large deviation of the spring constant from the nominal value provided by the manufacture, that is, as small as 50% or as large as 170% of the manufacturer's specification. For most cantilevers, a thin metallic coating is applied to the back side to increase reflectivity of the laser beam. The presence of this coating also adds a significant source of error in the force constant calculation from the measured dimensions of the cantilever.

Thus, for every cantilever to be used, the spring constant must be calibrated. One should not assume that the cantilevers in the same batch of the purchased package would have the same or similar k values. Many calibration methods have been tested and published in the literature, which include measuring resonant thermal vibration frequencies, measuring resonance frequency change upon adding a known mass, deflecting with a reference spring with a known stiffness, and so on. Based on easiness and convenience, the method developed by J. E. Sader is most widely used [35]. It relies solely on the measurement of the vibrational amplitude of cantilever as a function of driving frequency, which can be deduced from the detector signal as a function of frequency, $\Delta z_{PSD}(\nu)$, monitored with an oscilloscope or a lock-in amplifier. The frequency at the maximum amplitude is the resonant frequency (ν_o), and the ratio of this to the full width ($\Delta\nu$) at which the amplitude decreases to $1/\sqrt{2}$ of the maximum amplitude defines the quality factor, $Q = \nu_o/\Delta\nu$. Note that we use the amplitude decrease to $1/\sqrt{2}$ because it corresponds to 1/2 in the energy scale. Then, the spring constant of a rectangular cantilever is calculated using [35]:

$$k = 0.1906\, \rho_f L W^2 Q \cdot \Gamma_i(Re) \cdot (2\pi\nu_o)^2 \tag{11.7}$$

where ρ_f is the fluid density (1.18 kg/m^3 for air) and $\Gamma_i(Re)$ is the imaginary part of the hydrodynamic function in which $Re = 2\pi\nu_o\rho_f W^2/(4\eta)$, where η is the fluid viscosity (1.86×10^{-5} kg/ms for air). An online calculator is available at [36].

11.3 Cantilever Oscillation in Noncontact and Tapping Mode AFM

11.3.1 Dynamic Oscillation of Freestanding Cantilever

In the noncontact or tapping mode of AFM operation, the cantilever is dynamically oscillated and the interactions between the tip and the surface is monitored through changes in the amplitude at a given frequency or changes in the driving frequency to obtain a certain amplitude. If we know the *F-D* relationship (Figure 11.8A) at firsthand, we could in principle predict the change in the cantilever vibration and compare it with experimental results (at least qualitatively).

Newton's equation of motion that governs the oscillatory vibration behavior of the cantilever has basically the same form as the Lorentzian damped oscillator model that we have studied for the oscillation of bound electrons under electromagnetic field (i.e., light) (see eq. (5.28) in Section 5.1.3):

$$m^* \frac{d^2 z(t)}{dt^2} + m^* \gamma \frac{dz(t)}{dt} + kz(t) = F(t) \tag{11.8}$$

Here, m^* is the effective mass of cantilever which is given as $m^* = 0.24(\rho LWT) + m_{tip}$, where ρ is the density of the cantilever material and m_{tip} is the mass of the tip portion at the end of the cantilever, and γ is the damping factor ($=2\pi\nu_o/Q$). In the left-hand side of eq. (11.8), the first term is the kinetic force, the second is the damping term, and the last is Hooke's law for the restorative force. The right-hand side of eq. (11.8) is the force driving the cantilever to vibrate, which can be expressed with a sine function, $F(t) = F_o \sin(2\pi\nu_{dr}\, t)$ where ν_{dr} is the driving frequency. It comes from oscillatory shaking of the cantilever base which is mounted to the piezo-scanner. Although the exact value of F_o is not known, its magnitude is proportional to the amplitude of the oscillatory driving voltage applied to the piezo-scanner.

The analytical solution for the cantilever vibration amplitude as a function of driving frequency, $z(\nu)$, can be derived from eq. (11.8); but, let us first look at how the time-dependent vibration amplitude, $z(t)$, evolves since it is directly related to the detector signal, $\Delta z_{PSD}(t)$, displayed in an oscilloscope. For that, we solve eq. (11.8) numerically with a set of realistic parameters. Let us consider a cantilever whose length (L), width (W), and thickness (T) are 225 μm, 30 μm, and 3 μm, respectively. Assume that the tip has a conical shape with the base radius (r) is 4 μm and the height (h) is 8 μm. AFM probes are typically made of silicon, and the density (ρ) and modulus (E) of silicon are 2.3 kg/m^3 and 130 GPa, respectively. Then, the total mass of the cantilever and the tip is the following:

$$m_{cantilever} + m_{tip} = \rho\left(LWT + \pi r^2 h\right) = 4.75 \times 10^{-11}\ \text{kg} \tag{11.9}$$

The spring constant and resonant frequency of the cantilever are calculated from the dimension and modulus of the cantilever [7, 17]:

$$k = \frac{EWT^3}{4L^3} = 2.31\ \text{N/m} \tag{11.10}$$

$$\nu_o = 0.162\,\frac{T}{L^2}\sqrt{\frac{E}{\rho}} = 72.174\ \text{kHz} \tag{11.11}$$

In reality, ν_o is determined from the $\Delta z_{PSD}(\nu)$ measurement and k is calibrated using with eq. (11.7). From the relationship between ν_o and k (see eq. (5.24)), we can calculate the effective mass of the probe in the freestanding position:

$$m^* = \frac{k}{(2\pi\nu_o)^2} = 1.12 \times 10^{-11}\ \text{kg} \tag{11.12}$$

The damping affects the resonance frequency of the cantilever, which makes the maximum of the oscillation amplitude to occur at a frequency slightly lower than ν_o [37]. If the quality factor is $Q = 50$, then the effective resonance frequency becomes:

$$\nu_{eff} = \frac{1}{2\pi}\sqrt{\frac{k}{m^*}}\left(1 - \frac{1}{2Q^2}\right)^{1/2} = 72.167\ \text{kHz} \tag{11.13}$$

If the damping is larger (as in the case where AFM is done in aqueous solution), then ν_{eff} becomes significantly smaller than the ν_o calculated from the physical dimension of the cantilever. Putting all these terms together, eq. (11.8) can be modified to:

$$\frac{d^2 z(t)}{dt^2} + \left(\frac{\omega_{eff}}{Q}\right)\frac{dz(t)}{dt} + \omega_{eff}^2 z(t) = \frac{F_o}{m^*}\sin(\omega_{dr}\,t) \tag{11.14}$$

where $\omega_{eff} = 2\pi\nu_{eff}$ and $\omega_{dr} = 2\pi\nu_{dr}$.

This ordinary differential equation can be solved numerically with initial conditions; we can choose $z(t = 0) = 0$ nm and $\left.\frac{dz(t)}{dt}\right|_{t=0} = 0$ nm/s. Figure 11.9A,B plots the numerical solution of eq. (11.14), $z(t)$, for the case where the driving force (F_o) is 1 nN and the driving frequency (ν_{dr}) is the same as ν_{eff}. Figure 11.9A shows the evolution of $z(t)$ for the initial 1 msec; the system appears to reach the steady state at ~1 msec. Figure 11.9B shows the steady-state oscillation of the cantilever vibration amplitude for the next 200 μsec. If the $\Delta z_{PSD}(t)$ signal is monitored with an oscilloscope, the detected waveform will be exactly the same as Figure 11.9B, except the y-axis which will be displayed in a voltage scale. Once the detector sensitivity is determined, the detected voltage value can be converted to the actual vibration amplitude in the nm unit. If the data in Figure 11.9B are fitted with a sine function, $z(t) = A\sin(2\pi\nu\ t + \varphi)$, then we can find the frequency is $\nu = 72.167$ kHz, the amplitude is $A = 21.5$ nm, and the phase is $\varphi = -90°$. Since the driving frequency (ν_{dr}) is the same as ν_{eff}, the phase of the cantilever vibration amplitude is 90° with respect to the driving force. This behavior is the same as the polarization of bound electrons in the optical frequency that we have studied in Section 5.1.3 (see Figure 5.5). The sign of phase is changed to positive if the data is fitted with $z(t) = A\sin(2\pi\nu\ t - \varphi)$.

Figure 11.9C,D plots the $z(t)$ data when the driving frequency is off by 5 kHz from the resonance frequency. Obviously, the steady-state amplitude is much smaller; only 3.2 nm when $\nu_{dr} = \nu_{eff} - 5$ kHz and 3.0 nm $\nu_{dr} = \nu_{eff} + 5$ kHz. The phase difference between these two cases (i.e., below and above ν_{eff}) is 180°.

While tuning the AFM system for noncontact or tapping mode measurements, we first measure this vibration amplitude at the free-standing position as a function of driving frequency (i.e., monitoring $z(t)$ while sweeping ν_{dr}) to find the resonance frequency (ν_{eff}). Mathematically, this can be done by solving eq. (11.14) numerically while changing ν_{dr} stepwise with small increments; at each ν_{dr}, the numerical solution can be fitted with $z(t) = A(\nu_{dr})\sin(2\pi\nu_{dr}\,t + \varphi(\nu_{dr}))$ to get $A(\nu_{dr})$ and $\varphi(\nu_{dr})$. Figure 11.10 plots the calculation results for the cantilever described in Figure 11.9 with three different quality factors. The quality factor is determined by various damping mechanisms of the cantilever itself and how well it is connected to the

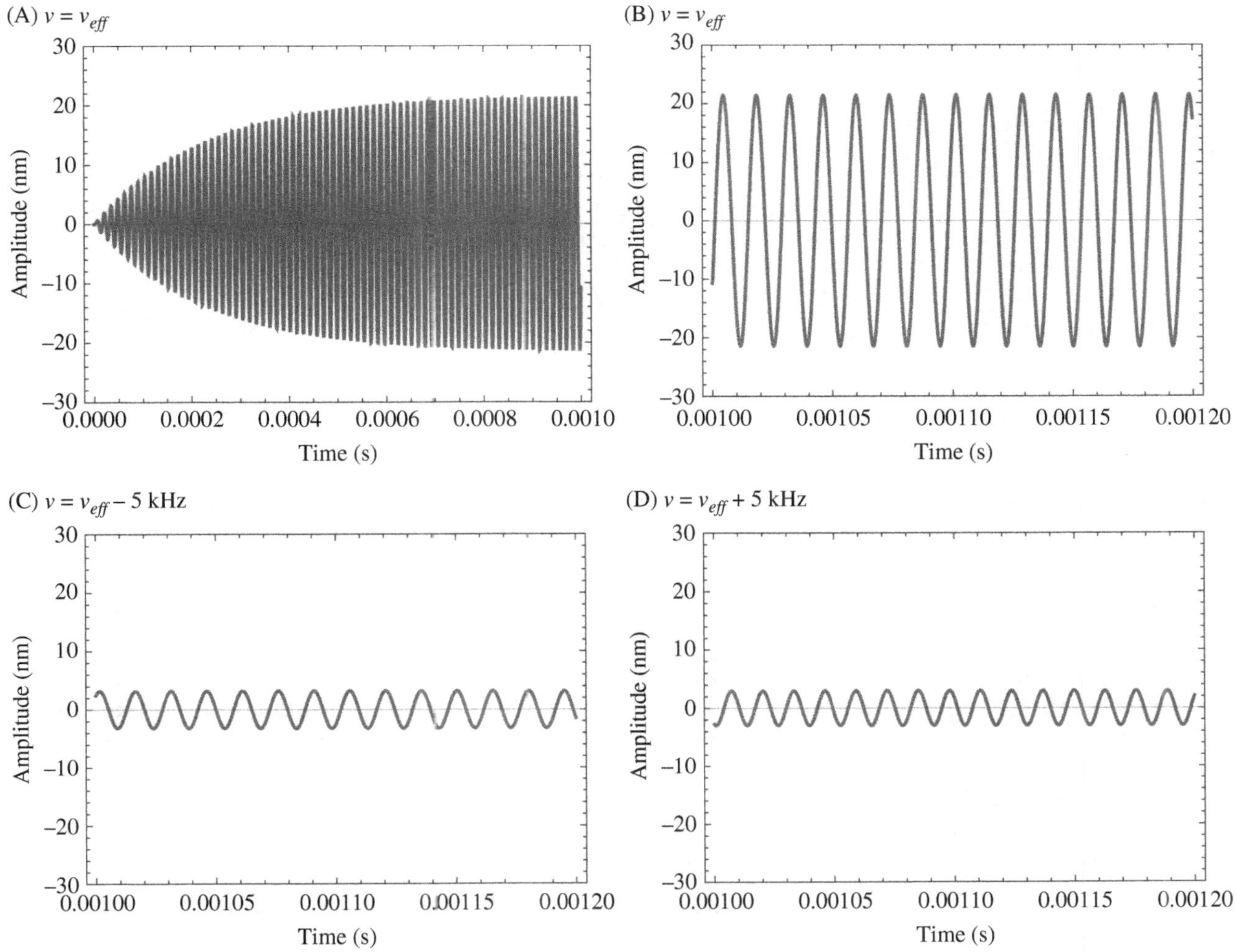

Figure 11.9 (A) Evolution of the cantilever vibration amplitude $z(t)$ for the first 1 msec after the driving force at ν_{eff} is applied. (B) Steady-state vibration at ν_{eff} over a 200 μsec period after the initial transient period. (C,D) Steady-state oscillation at $\nu_{eff} \pm 5$ kHz. The cantilever dimension and the driving force are described in the main text; the quality factor is set to $Q = 50$.

actuator (piezo scanner). For a larger Q, a larger fraction of the energy delivered to the cantilever is channeled into the elastic response of the cantilever, so the vibration amplitude is larger at the resonance frequency (i.e., $A(\nu_{dr} = \nu_{eff})$) and the peak width is smaller in Figure 11.10A. The same behavior is also observed in optical spectroscopy (see Figure 5.8). In Figure 11.10B, we can see that the phase changes over a larger frequency range as the damping increases (i.e., Q decreases). For a better control of the vibration amplitude, we need to use the cantilever that has a high Q value.

For the steady state of cantilever vibration, the analytical solution of eq. (11.4) can be expressed as [37, 38]:

$$A(\nu_{dr}) = \frac{F_o/m^*}{(2\pi)^2\sqrt{\left(\nu_{dr}^2 - \nu_{eff}^2\right)^2 + \left(\nu_{dr}\nu_{eff}/Q\right)^2}} \tag{11.15}$$

$$\varphi(\nu_{dr}) = \tan^{-1}\frac{\nu_{dr}\nu_{eff}/Q}{\nu_{dr}^2 - \nu_{eff}^2} \tag{11.16}$$

Note that these equations are the same as eqs. (5.31) and (5.32) in Section 5.1.3. When eqs. (11.15) and (11.16) are plotted for the cantilever used for the numerical calculations, we get the same results as Figures 11.10A,B. From eq. (11.15), we can see $A\left(\nu_{eff}\right) = \frac{QF_o}{\left(2\pi\nu_{eff}\right)^2 m^*} = \frac{QF_o}{k}$, which is exactly what we see in Figure 11.10A. Eq. (11.16) is plotted from −90° to +90°, which is equivalent to 0–180° with a constant phase shift.

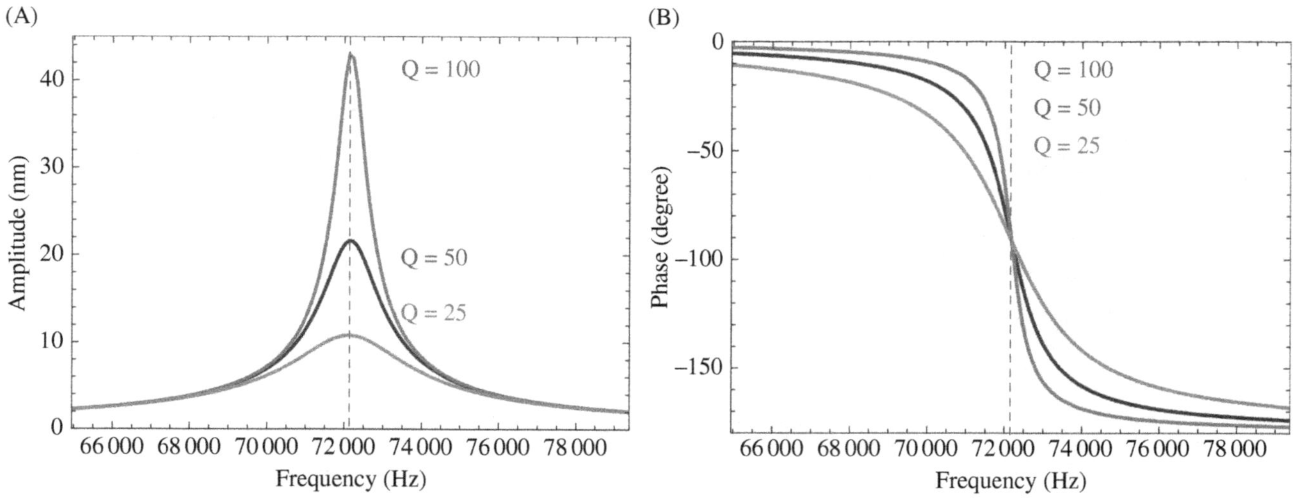

Figure 11.10 Variation of (A) oscillation amplitude $A(\nu_{dr})$ and (B) phase $\varphi(\nu_{dr})$ of the cantilever with different quality factors (Q). The cantilever dimension and the driving force are the same as Figure 11.9. The dotted line is drawn at ν_{eff} for Q = 50 as a guide to eye.

The calculations demonstrated so far are all for the oscillations near the first harmonic of the resonance frequency. The same equations can be used to calculate overtones, such as second and third harmonics, in higher frequency regions; all we have to do is to set the driving frequency on the right-hand side of eq. (11.14) to $2\nu_{dr}$, $3\nu_{dr}$, and so on, and calculate the $A(\nu_{dr})$ solution near those higher harmonic frequencies. The qualitative behavior $A(\nu_{dr})$ at these higher harmonics is basically the same as the first harmonic behavior; just the amplitude becomes significantly smaller as we go to higher harmonics.

11.3.2 Dynamic Oscillation of Cantilever with Tip–Surface Interactions

When the tip is brought close to the surface, the tip–surface interaction (Figure 11.8A) alters the cantilever vibration. As the cantilever bends due to this interaction, it feels an additional force which is equal to $F'_{ts}\Delta z$, where $F'_{ts} = \left(\frac{dF_{ts}}{dz}\right)_{z=0}$; this term can be added to the right-hand side of eq. (11.8), changing F_o to $F_o + F'_{ts}\Delta z$. Alternatively, we can keep the driving force term only in the right-hand side of eq. (11.8) and modify the effective spring constant to $(k - F'_{ts})$ in the left-hand side. This changes eq. (11.13) to:

$$\nu_{eff} = \frac{1}{2\pi}\sqrt{\frac{k - F'_{ts}}{m^*}}\left(1 - \frac{1}{2Q^2}\right)^{1/2} \tag{11.17}$$

Here, F'_{ts} is positive in the attractive interaction region and negative in the repulsive interaction region (Figure 11.8A). Thus, ν_{eff} decreases slightly when the tip–surface distance is in the attractive interaction region and increases in the repulsive interaction region.

Figure 11.11A,B compares the $A(\nu_{dr})$ and $\varphi(\nu_{dr})$ plots calculated for the cases with a small attractive force (F'_{ts} = 0.05 N/m) and two small repulsive forces (F'_{ts} = −0.05 and −0.10 N/m) with those for the freestanding case (F'_{ts} = 0). Here, the tip–surface interaction dynamics is assumed to be conserved so that Q does not change from the free-standing case. The calculation results show that the resonance shifts to a lower frequency when the tip–surface interaction is attractive and a higher frequency when the interaction is repulsive. Thus, when the driving frequency is fixed at the freestanding ν_{eff}, then the amplitude decreases once the tip starts interacting with the surface; and the phase changes from −90° to a larger negative value if the interaction is attractive and a smaller negative value if the interaction is repulsive. In the case of dissipative tip–surface interactions, the quality factor of the cantilever oscillation increases as the tip gets closer to the surface; then, the magnitudes of the amplitude and phase changes are even larger (see Figure 11.10).

A minute change, due to the tip–surface interaction, in dynamic oscillation of the cantilever is used for noncontact and tapping mode imaging. For that, we need to know how the amplitude and phase, $A(\nu_{dr})$ and $\varphi(\nu_{dr})$, change with the piezo-scanner position (Z) at a chosen ν_{dr} [39]. Let us do a thought experiment with the data shown in Figure 11.11. We can choose the cantilever driving frequency first and then see how the amplitude and phase at that frequency increase or decrease as the

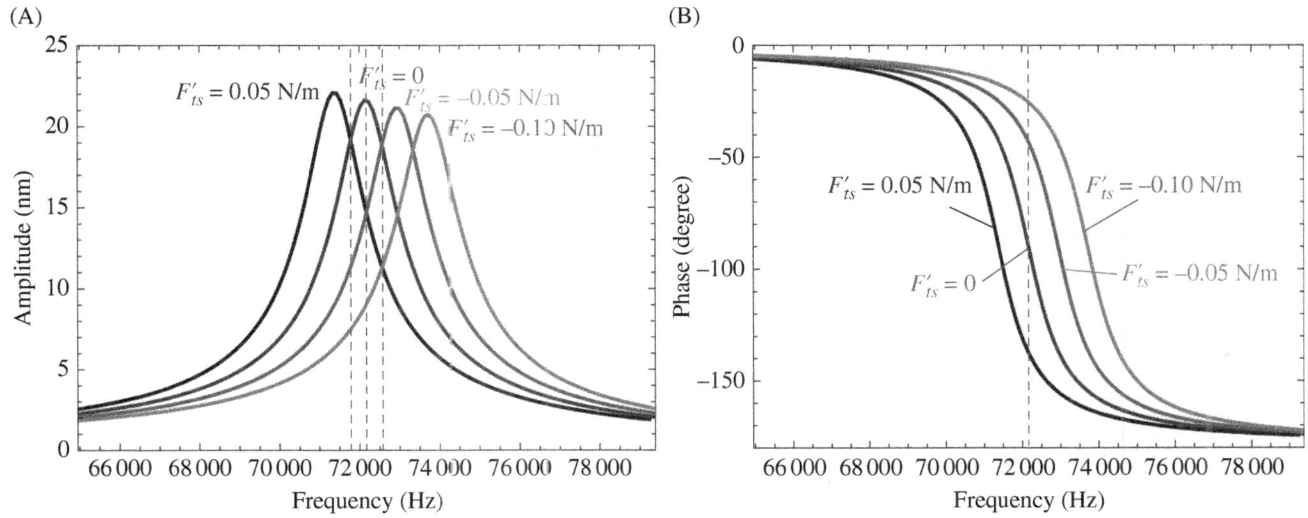

Figure 11.11 Driving frequency dependence of (A) oscillation amplitude $A(\nu_{dr})$ and (B) phase $\varphi(\nu_{dr})$ of the cantilever with Q = 50 (conserved) experiencing different tip–surface interaction forces. The cantilever dimension and the driving force are the same as Figure 11.10. As a guide to eye, a vertical dotted line is drawn at ν_{dr} = "free-standing ν_{eff}."

effective resonance frequency shifts to lower or higher regions as the tip–surface interaction varies following the *F-D* curve (shown in Figure 11.8A). Let us assume that the cantilever is excited at the free-standing ν_{eff} (i.e., $\nu_{dr} = \nu_{eff}^{fs}$, marked with the vertical line in Figure 11.11). The approaching of the tip toward the surface will modify $\nu_{eff}(Z)$ in such a way that the resonance peak position shifts to a lower frequency side first when Z is in the attractive interaction region (since $F'_{ts} > 0$) and then suddenly shifts to a higher frequency side upon jump-to-contact (since $F'_{ts} < 0$). If we monitor the amplitude at ν_{eff}^{fs} as a function of Z, we will see that it gradually decreases while Z decreases in the attractive interaction region, suddenly drops at the jump-to-contact point of the tip to the surface, and then continues decreasing upon further lowering the piezo-scanner.

The amplitude change is accompanied by the phase change. Initially, the phase slightly shifts to the negative side due to the attractive interaction (since $F'_{ts} > 0$); then it suddenly switches to the opposite side of the resonance upon the jump-to-contact of the tip to the surface (since $F'_{ts} < 0$), and continue increasing further lowering of the piezo-scanner. Based on the phase shift with respect to the free-standing state, one can tell if the AFM tip is interacting with the sample in the attractive or repulsive force regime. This can be used to set up the noncontact and tapping mode operations, see Section 11.5.2 for more details.

Note that the above discussion was based on the damped harmonic oscillator model. This model implicitly assumes that the tip–surface interaction induces the frequency shift only as described with eq. (11.17) without affecting the energy transfer (so Q is conserved). It also assumes that the force gradient is much smaller than the cantilever spring constant ($F'_{ts} \ll k$) and invariant with the tip–surface separation (D). Although these assumptions simplify the situation and facilitate the conceptual understanding for noncontact and tapping mode experiments, they limit the validity of the model. In fact, this model fails to provide quantitative agreement with the experiment. For more precise prediction, an anharmonicity of the tip–surface interaction should be included. Readers interested in deeper understanding of the *A-Z* and φ-*Z* behaviors can find more advanced literatures [38, 40].

11.4 Surface Deformation in Contact Mode AFM

11.4.1 Contact Stiffness

In the contact mode AFM experiment, the sample surface as well as the tip apex can undergo elastic and plastic deformations. Figure 11.8B is drawn for the case where there is no deformation of the tip–sample contact region and no hysteresis of the piezo-scanner (see Section 11.5.1) during the approach and retraction. When the elastic deformation occurs, not only the cantilever bends upwards but the sample surface also sinks downward as the distance between the sample surface and the

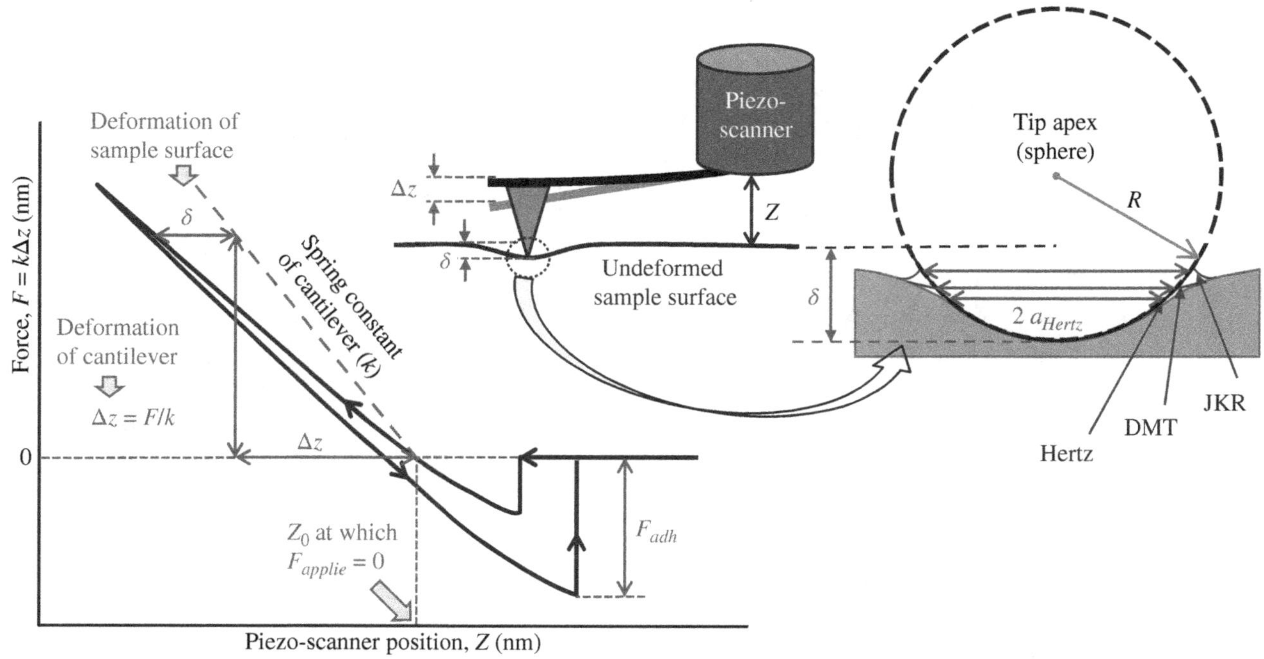

Figure 11.12 Force (F) – piezo-scanner position (Z) curve for the case with elastic deformation of the sample surface. The deformation depth and the contact diameter are marked as δ and $2a$, respectively.

piezo-scanner decreases. Thus, the slope of the contact part of the F-Z curve is smaller than the spring constant of the cantilever (k), which is schematically illustrated in Figure 11.12. If the plastic deformation occurs during the compression by the tip, then the sample surface does not fully recover during the retraction. So, the retraction part in the contact region will be significantly lower than the approach part at the same Z position. The plastic deformation case will be discussed further in Chapter 12. Here, let us consider the case where the sample surface deforms elastically, and the piezo-scanner hysteresis is negligible. Then, the approach and retraction curves in the contact region of the F-Z curve will overlap on top of each other.

The AFM system does not know where the sample surface is at first hands, and it simply measures the cantilever deflection (Δz) as a function of the piezo-scanner expansion and contraction (Z) with respect to an arbitrary reference voltage. Then, the user should define the height of the "undeformed" sample surface as the cross-point (Z_o) where the zero-force line (extension of the horizontal line from the free-standing region) and the slope of the approach curve in the positive force region meet (Figure 11.12). When the sample surface deforms elastically, the displacement of the piezo-scanner (Z) results in changes in both cantilever deflection (Δz) and sample deformation (δ). In equilibrium, the force applied to the sample ($k\Delta z$) is the same as the elastic force pushing back by the sample ($k_s\delta$). Thus, we can write:

$$Z = \Delta z + \delta = (1 + k/k_s)\Delta z \tag{11.18}$$

Rearranging this, we get the following equation:

$$F = k\Delta z = \frac{k\,k_s}{k + k_s} Z = k_{eff} Z \tag{11.19}$$

Equation (11.19) shows that the slope (k_{eff}) of the F-Z curve is a measure of the sample surface stiffness. If the sample is much stiffer than the cantilever (i.e., $k \ll k_s$), then k_{eff} is the same as the spring constant (k) of the cantilever. If the cantilever stiffness is larger than the sample (i.e., $k \gg k_s$), then k_{eff} essentially probes the sample stiffness (k_s). By choosing the cantilever with a proper spring constant (k), the sample stiffness can be determined from the slope (k_{eff}) of the F-Z curve [41]. And, the area of the retraction curve below the zero-force line is the work of adhesion between the tip and the surface, which can be calculated through $W_{ts} = F_{adh}^2/(2k_{eff})$.

The stiffness of the sample can be related to reduced Young's modulus (E_r) through:

$$k_s = 2aE_r \qquad \text{with } \frac{1}{E_r} = \frac{1-\nu_s^2}{E_s} + \frac{1-\nu_t^2}{E_t} \tag{11.20}$$

where a is the radius of the contact area between the surface and the tip, and ν_s, E_s, ν_t, and E_t are Poisson's ratio and Young's modulus of the sample and the tip, respectively. The Poisson ratio and Young's modulus of silicon are typically 0.26–0.28 and 140–180 GPa, respectively, depending on the crystallographic orientation. If the tip is much stiffer than the sample, we can neglect the deformation of the tip and reduce eq. (11.20) to:

$$k_s \approx 2a\frac{E_s}{1-\nu_s^2} \tag{11.21}$$

11.4.2 Contact Area and Adhesion

The radius (a) of contact area can be estimated using elastic contact mechanics theories. Table 11.2 summarizes the three most widely used theories that relate the contact radius (a) to the applied force by the cantilever (F) using a sphere-on-flat geometry [7, 17]. It also shows the relationships for the deformation depth (δ) and the contact pressure distribution ($P(x)$, where $0 \le x \le a$) [7, 17]. In the Hertz model, the adhesion force (F_{adh}) between the tip and the sample is neglected. Hence, the Hertz theory works only when F_{adh} is negligible or much smaller than the applied force (as in macroscale experiments). The other two theories, namely, Derjaguin–Muller–Toporov (DMT) and Johnson–Kendall–Roberts (JKR), take adhesion into account. For a sharp and hard tip on stiff materials, the contact area is small- and the long-range attractive interactions outside the contact area is not negligible. And, since the contacting materials are hard and stiff, adhesive deformation is relatively small (see the contour lines marked as DMT in Figure 11.12). This is the case where the DMT theory is mostly applicable. For a large radius tip in contact with soft and compliant materials, the adhesive interactions between the tip and the surface are large and can deform the contact line in the annular region (see the contour lines marked as JKR in Figure 11.12). In this case, the long-range interaction outside the contact region is negligible compared to the adhesion inside the contact region. This is the case where the JKR theory works well. Note that the DMT and JKR theories can be viewed as the opposite limits of the possible contact responses of materials, and the transition between these two theories and the intermediate cases can be predicted by the Maugis theory [42].

Knowing the contact radius between the tip and the surface (Table 11.2), we can compare the experimental F-Z curves with the elastic deformation of the sample predicted with the Hertz, DMT, or JKR theories and calculate the elastic modulus (E_s) of the sample surface [43–46]. For that, the compressive region of the F-Z curve can be fitted using eqs. (11.19) and (11.20) with the appropriate formula for the contact area (a), or by calculating the deformation depth (δ) first based on the description in Figure 11.13 and then using its relationship to the applied (F) and adhesion (F_{adh}) forces.

Of course, this calculation requires accurately knowing the tip radius (R) and the cantilever spring constant (k) as well as the conversion of Δz_{PSD} (deflection detector signal, which is in volt) to Δz (cantilever bending). The latter can be done by performing the F-Z measurement on a stiff sample surface (so not deforming at all). In this case, since the extend of cantilever deflection (Δz) is the same as the movement of the piezo-scanner (ΔZ), we can obtain the calibration factor converting Δz_{PSD} to Δz. The spring constant can be calibrated using Sader's method (eq. 11.7). The tip radius can be estimated from the adhesion force if the work of adhesion (W in Table 11.2) is calculated from eq. (1.9) for identical materials or eq. (1.10) for

Table 11.2 Contact mechanics equations calculating the contract radius (a), the deformation depth (δ), and contact pressure distribution ($P(x)$, where $0 \le \le a$) for the measured adhesion force (F_{adh}) and the applied load (F) for a sphere (modeling the tip apex) on a flat surface, according to Hertz, DMT, and JKR theories [7, 17]. R is the tip radius and W is the work of adhesion per unit area.

	Hertz	DMT	JKR
F_{adh}	0	$2\pi RW$	$\frac{3}{2}\pi RW$
a	$\sqrt[3]{\frac{3}{4}\frac{R}{E_r}F}$	$\sqrt[3]{\frac{3}{4}\frac{R}{E_r}(F+F_{adh})}$	$\sqrt[3]{\frac{3}{4}\frac{R}{E_r}\left(F+2F_{adh}+\sqrt{4F\,F_{adh}+(2F_{adh})^2}\right)}$
δ	$\frac{a^2}{R}=\left(\frac{3}{4}\frac{F}{R^{1/2}E_r}\right)^{2/3}$	$\frac{a^2}{R}=\left(\frac{3}{4}\frac{F+F_{adh}}{R^{1/2}E_r}\right)^{2/3}$	$\frac{a^2}{R}-\left(\frac{4}{3}\frac{F_{adh}\,a}{RE_r}\right)^{\frac{1}{2}}$
$P(x)$	$\frac{3F\sqrt{1-(x/a)^2}}{2\pi a^2}$	$\frac{3(F+F_{adh})\sqrt{1-(x/a)^2}}{2\pi a^2}$	$\frac{2E_r a}{\pi R}\sqrt{1-(x/a)^2}-\sqrt{\frac{4}{3}\frac{E_r F_{adh}}{\pi^2 Ra}}\frac{1}{\sqrt{1-(x/a)^2}}$

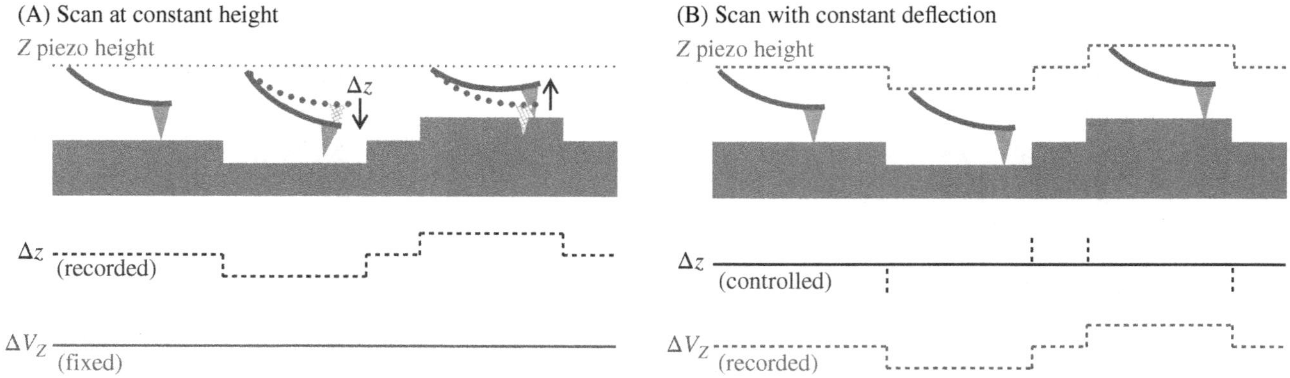

Figure 11.13 Two scan modes in AFM: (A) constant height scan vs. (B) constant set-point scan.

dissimilar materials [16]. If the tip and the surface are the same material, then the work of adhesion per unit area is twice the surface energy ($W = 2\gamma$). Alternatively, we can determine the tip radius by using high-resolution electron microscopy or deconvoluting the tip artifact effect from the image of surface features that are sharper than the tip (see Section 11.5.4).

11.5 Scanning AFM Probe

11.5.1 Piezoelectric Scanner

To achieve precise control of nanoscale motions or displacements during the approach, scan, and retraction in AFM, piezoelectric actuators are normally used. Piezoelectric ceramics have a noncentrosymmetric crystal structure. When such crystals are subjected to external stress, mechanical strain is accompanied by generation of electric charge across the crystal. This is the direct piezoelectric effect. Conversely, when such crystals are subjected to electric fields, then the electrical potential is converted to mechanical strain. This phenomenon is used to produce precise movements of scanners. The most commonly used material for this purpose is a PZT ceramics (lead zirconate titanate; $PbZr_xTi_{1-x}O_3$) [47].

In order to position the AFM probe (tip at the end of the cantilever) with an accuracy of Å or less in all three dimensions, a tube-shape piezo element is widely used as the actuator (Figure 11.2A). This kind of scanner consists of a hollow tube of properly poled PZT ceramics with several electrodes to apply voltage biases across three orthogonal directions. The vertical displacement (expansion or contraction) of the tube element can be achieved by applying a voltage bias (ΔV_Z) across the thickness of the tube (i.e., inner and outer surfaces of the tube):

$$\Delta Z = \left(d_{31}\frac{L_t}{t_t}\right)\Delta V_Z \tag{11.22}$$

where d_{31} is the transverse piezoelectric coefficient of the piezoelectric ceramic material of the tube scanner (typically a few Å/V [47]), and L_t and t_t are the length and thickness of the hollow tube, respectively. The lateral displacement (ΔX or ΔY) can be achieved by applying a voltage bias (ΔV_X or ΔV_Y) across the left and the right side of the external tube surface:

$$\Delta X = \left(d_{31}\frac{L_t^2}{D_t t_t}\right)\Delta V_X \tag{11.23}$$

where D_t is the mean diameter of the tube.

It should be noted that AFM does not record ΔZ as a function of ΔX and ΔY. It records ΔV_Z (instead of ΔZ) as a function of ΔV_X and ΔV_Y applied for the X and Y positions, and then convert the recorded voltage signals to height or distance using the predetermined conversion factors. So, accurately calibrating the vertical height and lateral displacement of piezo-scanner is very critical. If these calibrations are not accurate, obviously the measured topographic dimensions are not correct. *It is a bad assumption that the AFM instrument is well calibrated by the manufacturer or by the technician in charge of the instrument; it is the user's responsibility to check or confirm the accuracy of the vertical and lateral calibration factors by scanning proper reference samples.* The calibration factors will vary as the scanner gets old and heavily used; it is not surprising to see that more voltage bias is applied to the scanner to get the same displacement as the scanner gets old and used more, compared to the brand-new condition.

In principle, the X, Y, and Z displacements can be achieved with a single tube scanner. Since the X and Y motions are achieved through bending of the tube, the X and Y displacements result in a slight displacement in the Z-direction. This kind of coupling can be compensated by adjusting bias functions to the piezo unit, or using multiple piezo units. Also, it should be noted that the positioning performance of piezoelectric actuators is not always linear as described in eqs. (11.22) and (11.23). The degree of actuator displacement becomes nonlinear as the applied voltage becomes large. Also, it is a function of the rate at which the applied voltage is swept. Moreover, it can have hysteresis and creep depending on the history of bias voltage (magnitude and sweeping rate) [48]. A simple nonlinearity can be corrected by programming the vias voltage with nonlinear mathematic functions [48]. The hysteresis and creep effects can be controlled by using a real-time closed-loop control unit to adjust the applied bias in response to unwanted lagging or drifting of the scanner position.

11.5.2 Scanning with Feedback Control

In principle, the topographic scanning can be done in two different modes. One is to translate the cantilever base laterally at a constant height plane (i.e., ΔV_Z applied to the piezo-scanner is fixed) and then recording the change in Δz_{PSD} which can be converted to Δz. If the scan is done in the contact mode and the sample is stiff, then $\Delta z = \Delta h$ (topographic height change of the sample). This is called "constant height" mode (Figure 11.13A). It may work if the sample is mounted perfectly horizontal and we already know the distance from the surface; however, before we scan the sample surface to get the topography, we do not know if the sample surface is perfectly aligned with the piezo scan plane.

A more realistic way is to adjust the piezo-scanner voltage (ΔV_Z) to keep Δz constant whenever the tip is translated to a new position (i.e., step change in ΔV_X and ΔV_Y applied to the scanner) and record the adjusted ΔV_Z amount (in V); then, the recorded ΔV_Z can be converted to Δz (in nm), which is displayed as sample height in the topography image. This mode is schematically illustrated in Figure 11.13B. The target Δz used for scanning is called "set point." In the contact mode scan, it is the extent of the upward bending of the cantilever; since Δz is directly related to the compressive force applied to the surface, this is called a constant-force mode. In the noncontact and tapping modes with the oscillation amplitude modulation (AM), the set point is the ratio of the amplitude at the tip–surface interaction distance to that at the free-standing distance (i.e., A_{ts}/A_{fs}); this is called a constant-amplitude mode. If the noncontact and tapping modes are done with the frequency modulation mode (FM), the set point is the shift in the driving frequency ($\Delta \nu_{dr}$) needed to keep the oscillation amplitude constant.

In the constant set-point scan mode, the topographic height is not the true height of the sample, but the vertical displacement of the piezo-scanner made by the AFM controller unit to keep the detector signal (cantilever bending or oscillation amplitude) constant at the set point. This is called "feedback control." Figure 11.14 schematically shows the basic

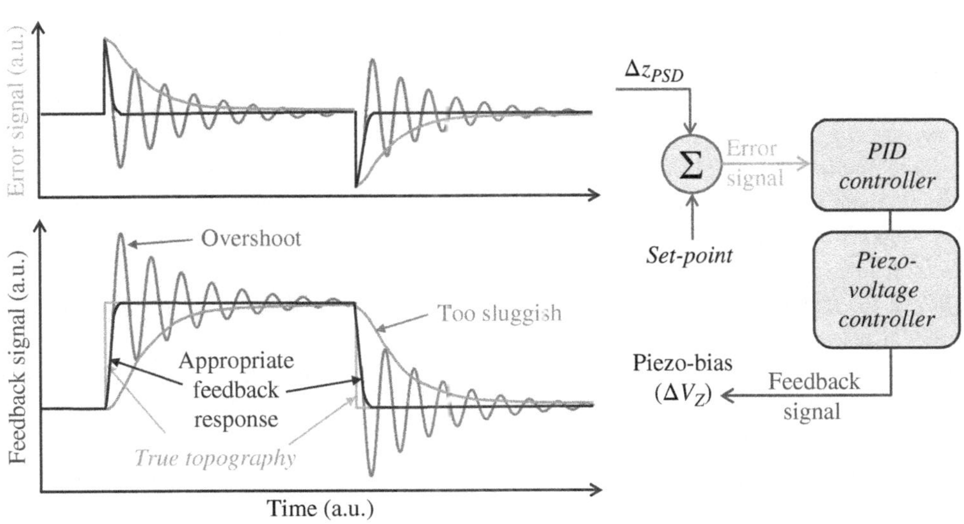

Figure 11.14 Changes in feedback signals adjusted by the controller to keep the error signal zero for the cases with properly and poorly tuned PID control parameters.

components of the feedback control loop. First, it compares the detector signal (Δz_{PSD}) with the set point and calculate the error signal. When the error signal deviates from zero (e.g., because the topographic height in the next pixel position is different from the previous one), the PID (proportional-integral-derivative) controller adjust the bias voltage fed to the piezo-scanner until the error signal returns back to zero.

In an ideal situation, the PID controller response should be instantaneous (Figure 11.13B); in reality, the perfectly instantaneous response is not possible and there is always a time delay. How promptly and efficiently the feedback controller responds are a function of the proportional band (or gain, K), the integral time (τ_I), and the differential time (τ_D). The P-control involves correcting the feedback signal in proportion to the magnitude of the current error signal. The P-gain (K) is typically defined as the inverse of proportional band. If the P-gain is too small, the feedback response could be too small and sluggish (Figure 11.14). As the P-gain is increased, the feedback response becomes faster; but if it is too large, then the feedback signal may fluctuate in response to small noises. In the P-control only, the error signal may not reach zero because the response signal becomes smaller as the error becomes smaller; this may leave an offset from the set point. The I-control attempts to fix this offset by effectively integrating the past error signal from the P-action and increase the correction factor. How long this integration is to be performed is determined by τ_I. Since the PI control tries to make the cumulative error to zero, it may not stop changing the feedback signal right at the moment when the error signal becomes zero, resulting in an overshoot (Figure 11.14). The D-control attempts to minimize this overshoot by slowing the correction factor to be used as the target value is approached, which is a function on τ_D. By adjusting K, τ_I, and τ_D, one can get the proper feedback control. This means that the topography image obtained by AFM can be affected depending on how well the user tunes the PID control parameters [48].

Let us take a look at an example to see how the constant set-point scan mode works. Figure 11.15 compares the amplitude and phase changes as a function of piezo-scanner voltage, that is, $A(\nu_{dr})$-V_Z and $\Delta\varphi(\nu_{dr})$-V_Z plots, in the tapping-mode AFM operation and the images of a reference sample. The sample was a commercial grating used to calibrate scan direction length and Z-height of the scanner (step periodicity $= 3 \pm 0.05$ μm; step height $= 110 \pm 10$ nm). The driving frequency (ν_{dr}) was set at the effective resonance frequency at the free-standing state $\left(\nu_{eff}^{fs}\right)$. The $A(\nu_{dr})$ and V_Z values are shown in the V unit without conversion to the nm unit. Note that $A(\nu_{dr})$ is the root-mean-square (rms) value. Increasing V_Z means the expansion of the piezo-scanner, so lowering the cantilever toward the sample. The phase signal was auto-zeroed while the tip is in the free-standing distance, so only the shift from the resonance phase (−90°) is shown.

Figure 11.15A shows the data collected with a small free-standing oscillation amplitude. From the calibration of the Δz_{PSD}, the 1 V_{rms} signal in this operation condition was found to correspond to 6.4 nm oscillation. As the tip approaches the surface, the tip starts feeling the attractive force (i.e., $F'_{ts} > 0$); so, $A(\nu_{dr})$ decreases and $\Delta\varphi(\nu_{dr})$ shifts to the negative side (see Figure 11.11). At some point, the tip jumps to contact with the surface and the adhesion force of the surface completely damps the cantilever vibration, and $\Delta\varphi(\nu_{dr})$ becomes positive. Note that the details of the magnitude of the sudden change and the snap-in point as well as the slope and hysteresis in the A-Z and $\Delta\varphi$-Z curves can vary drastically depending on the spring constant and quality factor of the cantilever (k and Q), the driving force (F_o, which is controlled by the voltage applied to shake the piezo-scanner), the adhesion force between the tip and the surface ($F_{vdW} + F_{chem} + F_{cap} + \ldots$). Upon the retraction of the cantilever, the tip remains in contact with the surface until the elastic strain force becomes larger than the adhesion force. Initially, the oscillation amplitude is gradually regained and then suddenly jumps back to the free-standing position; it means the tip is snapped off the surface.

From the $A(\nu_{dr})$-V_Z curve, we can see that imaging can be done with the set point (A_{ts}/A_{fs}) of 0.7. The tip is in the attractive force regime and the amplitude has not completely damped. So, this is basically the noncontact mode scanning. The topography image shown in Figure 11.15A looks reasonable. The error signal (deviation from the set point) is zero and not very noisy when the tip is moving over a flat surface. When the tip descends the step edge of the grating, the amplitude increases momentarily from the set point (since the average distance from the cantilever position and the sample at the next pixel position increases). As the error signal increases, the feedback control unit expands the piezo-scanner, so the tip moves closer to the sample. The adjusted ΔV_Z amount is converted to the nm scale using the predetermined calibration factor and recorded as the surface height. When the tip ascends the grating step, the error signal decreases, and the feedback control retracts the cantilever. Again, the ΔV_Z adjustment is converted to the nm unit and recorded as height at that position. The percentage of the error signal in the height change region is quite large (400 mV/760 mV = 52%), and $\Delta\varphi(\nu_{dr})$ becomes noisy whenever the error signal deviates from zero. Based on the magnitude of the phase change in the height change region, we can tell that the tip momentarily touches the surface.

Figure 11.15B shows the data collected with a large free-standing oscillation amplitude (50 nm, instead of 7 nm). As the tip approaches the surface, the $A(\nu_{dr})$ decreases continuously without showing the snap-in behavior. In the $\Delta\varphi(\nu_{dr})$ plot, we can

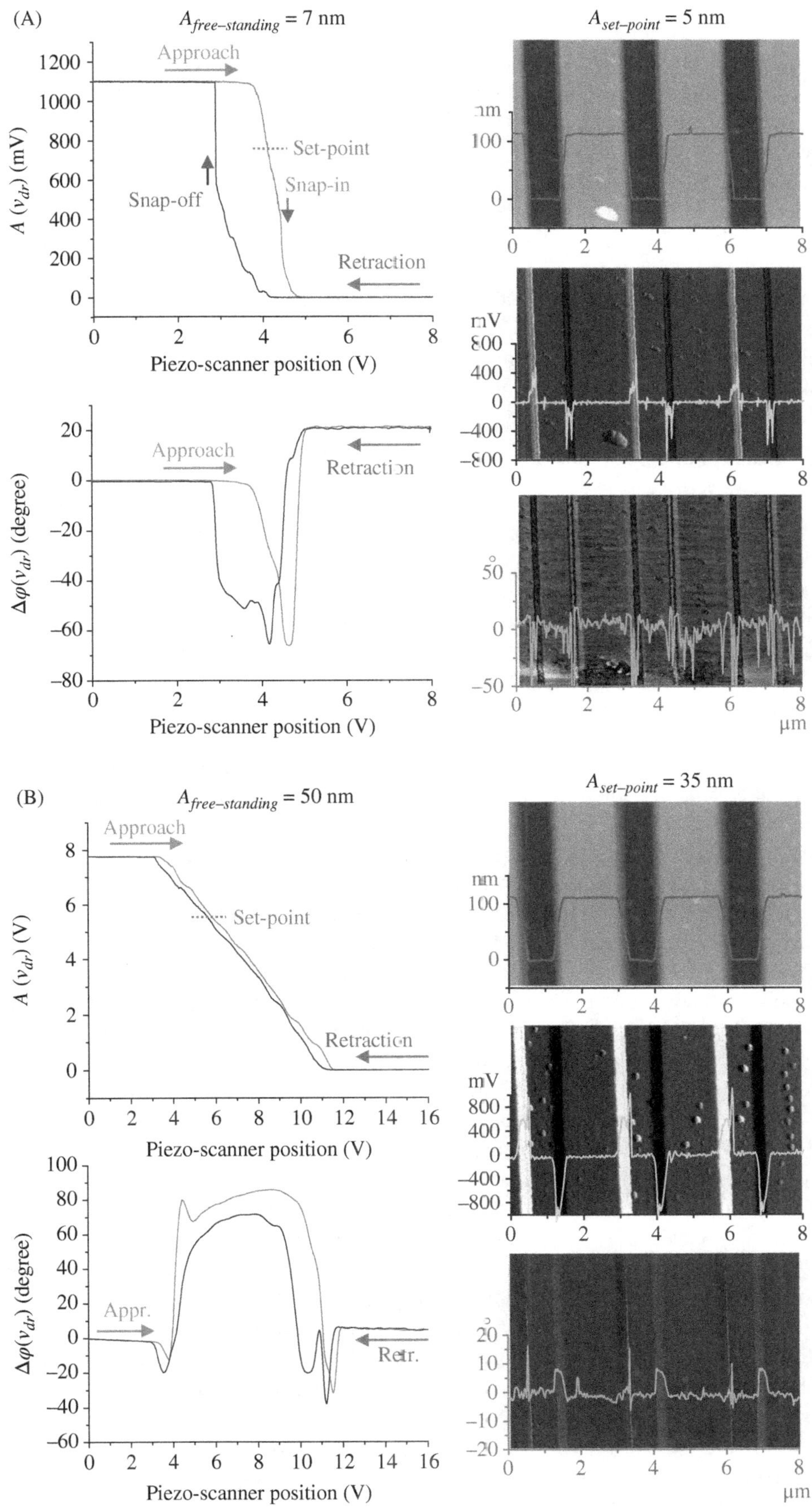

Figure 11.15 $A(\nu_{dr})$-V_Z and $\Delta\varphi(\nu_{dr})$-V_Z plots as well as topography (nm), error signal (mV), and phase (°) images of a calibration grating with periodic steps obtained with two different free-standing amplitudes: (A) 7 nm and (B) 50 nm. The spring constant of the cantilever was 48 N/m, and the driving frequency was 261 kHz. The left-to-right scan speed was 20 µm/s. The P- and I-gains of the feedback control were 1 and 0.44, respectively. The $\Delta\varphi(\nu_{dr})$ signal was auto-zeroed before starting measurement.

see that the phase shift is initially negative (i.e., probing the attractive force; $F'_{ts} > 0$) and then suddenly shifts to the positive side. This is the moment that the tip–surface interaction is switched to the repulsive regime ($F'_{ts} < 0$; see Figure 11.11); so, the tip is now "tapping" the surface. Since the oscillation amplitude is large, the cantilever carries very large kinetic energy. In this specific example, the kinetic force of oscillation was larger than the adhesion force, so the cantilever oscillation amplitude did not drop to zero upon contact with the surface. As the tip is pressed harder by lowering the cantilever, the oscillation amplitude continues decreasing to zero. At the complete damping, $\Delta\varphi(\nu_{dr})$ also drops to zero (no oscillation at all). As the cantilever is pulled away from the surface, the oscillation amplitude grows linearly, and the phase shift follows the reverse path of the approach curve.

In the tapping mode imaging with the large free-standing amplitude, the error signal appeared to be larger, but the percentage error from the set point was smaller (800 mV/5.6 V = 14%). The $\Delta\varphi(\nu_{dr})$ signal change in the step-up region is different from that in the step-down region. Overall, the phase signal image is more stable than the noncontact imaging case. This is because the slope of the $\Delta\varphi(\nu_{dr})$ vs. V_z plot near the set-point region is smaller.

11.5.3 Artifacts Due to Improper Feedback and Scan Control

Since topography images taken with AFM originate from the feedback control of physical interactions that are totally different from those used for imaging in conventional optical and electron microscopy, imaging artifacts may not be readily recognizable by users who are not accustomed to the AFM operation principles [48]. Here, let us look at some of those examples that unexperienced users may not pay enough attention during the data collection.

A good rule of thumb is to collect the topography image with the largest proportionality gain (K) and the smallest integral time (τ_I) at which the collected image does not have any oscillatory instability (overshoot and undershoot, or periodic noises along the scan direction). A small overshoot could be remedied by increasing the differential time (τ_D), or simply by lowering K or increasing τ_I slightly. By doing so, the topography images (which is obtained by multiplying the ΔV_Z feedback signal at each measurement pixel with the predetermined calibration factor) becomes sharper and closer to the true topography. *A good practice is to record the error signal along with the topography signal and adjust the PID control parameters until the absolute magnitude in the error signal becomes the smallest.* Note that the tuning parameter may need to be adjusted when the scanning condition (such as scan speed or scan width) is changed.

Figure 11.16 compares the tapping mode images collected with different feedback control settings. The AFM instrument used to collect these data had a PI control only (not PID). In this instrument, the I-control was done by adjusting the I-gain which is essentially the inverse of τ_I. Some other instruments directly adjust the integral time constant (τ_I), instead of the I-gain. Figure 11.16A is the images collected with the optimum PI control. Compared to Figure 11.16A, the phase image collected with a much smaller P-gain has ringing noises (Figure 11.16B), which can also be seen in the error signal image. When we see the topography image only, we may miss tiny noise signals overlaid on top of the height signal. But, after seeing such noises in the error signal and phase shift images, we can see that the same noises are present in the topography image. The noise components in the topography image look much smaller just because the topography full scale is larger; although small, such noises make the topography image blurry. When the P-gain is increased too much, the error signal and phase shift images have periodic noises; the same periodic noise can be seen in the topography image as well (Figure 11.16C). This means that the feedback loop was too sensitive that the feedback kept overshooting near the set point.

Figure 11.16D is the data taken when I-gain was too small. The error signal in the flat region looks even better than Figure 11.16A, but the error during the height change is significantly larger. It means the feedback did not respond fast enough, which is the consequence of a large τ_I. Since the feedback response became too slow, the error signal was increased. Although not obvious in the image displaced in the full scale, the sharpness of the step corner region was reduced. Since there was a large time delay in the control loop, the transition region of the stepping-down edge appears to be slanted (see the line profile overlaid in the image). This is called a "fly-over" error; in other words, the feedback signal did not decrease fast enough to tightly keep up with the topographic height change. Figure 11.16E is the data taken with a slightly larger I-gain. In this case, the feedback signal kept overshooting with a certain periodicity because τ_I was too small, and the topography image was grossly distorted.

In AFM, the image resolution is limited by the number of pixels recorded along the scan direction. Figure 11.17 compares the same region images collected at the same scan speed (10 μm/sec) in two different settings. In Figure 11.17A, a 10-μm full scan range was imaged with a 0.5 Hz scanning frequency (i.e., 20 μm round trip over 2 sec) and only 5 μm region is cropped and shown. In Figure 11.17B, a 5-μm full scan range was imaged with a scanning frequency of 1 Hz. In both cases, the image resolution was fixed at 512 pixels per line along the scan direction. It means that the tip moved 20 nm per pixel in

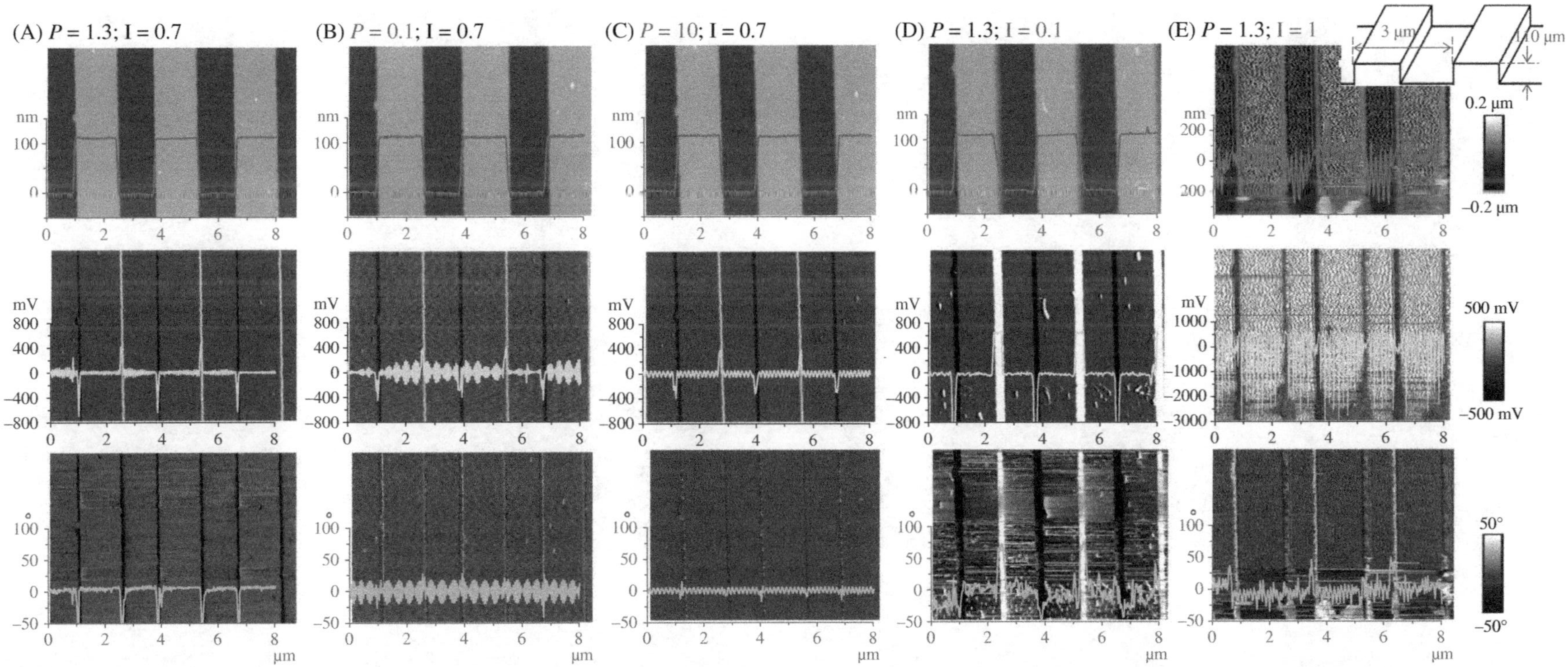

Figure 11.16 Tapping-mode AFM images (top = topography, middle = error signal, bottom = phase) of the same calibration grating used in Figure 11.15, obtained with various P- and I-gains: (A) P = 1.3, I = 0.7; (B) P = 0.1, I = 0.7; (C) P = 10, I = 0.7; (D) P = 1.3, I = 0.1; (E) P = 1.3, I = 1.0. The free-standing oscillation amplitude and set point were 7 nm and 0.7, respectively, and the scan speed was 20 μm/s.

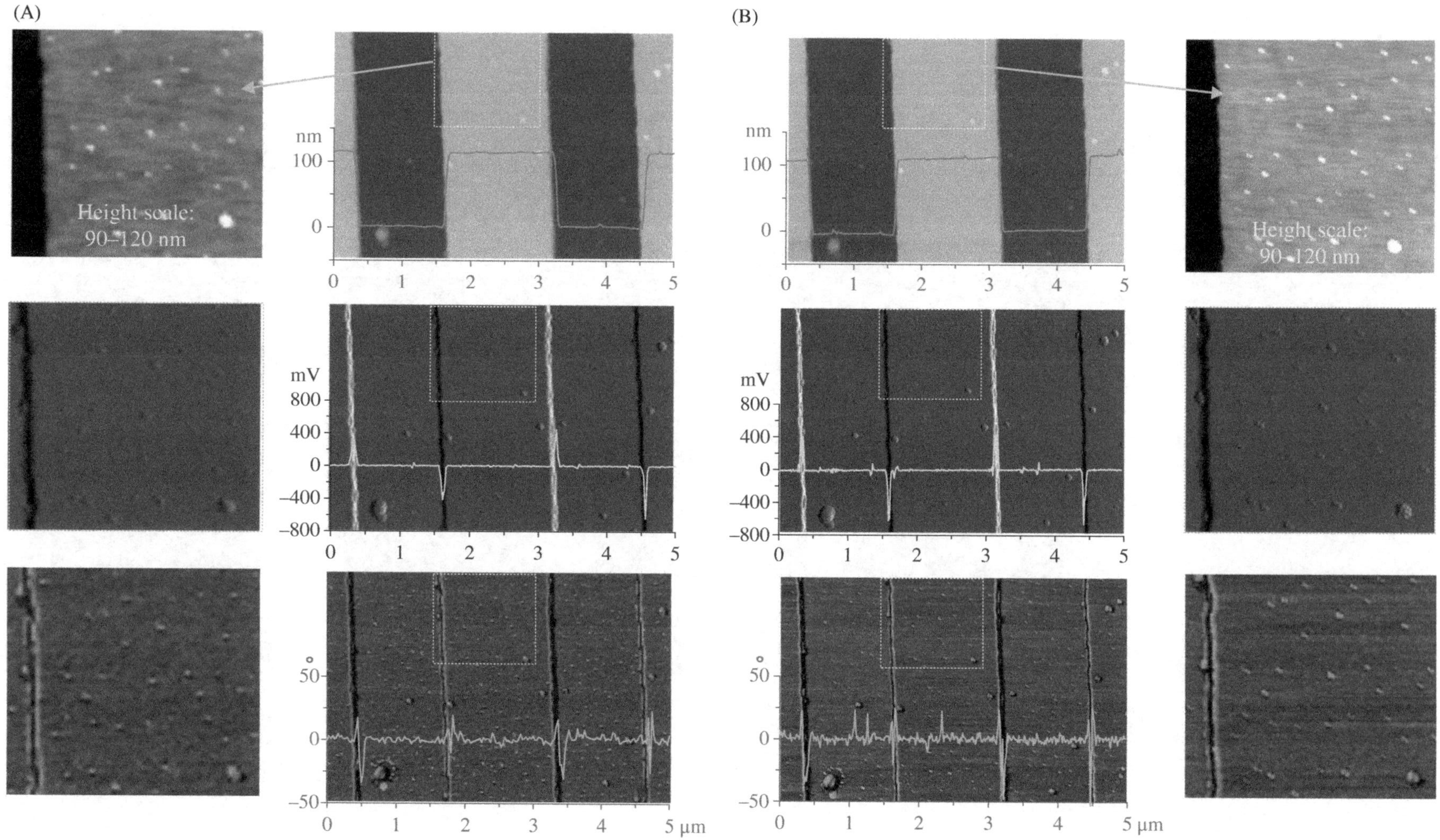

Figure 11.17 Tapping-mode AFM images of the same calibration grating used in Figure 11.15. The P- and I- gains were fixed at 1 and 0.44, respectively. The freestanding oscillation amplitude and set point were 7 nm and 0.7, respectively; the scan speed was set to 10 µm/s in two different ways: (A) 0.5 Hz cycle over 10 µm range and (B) 1 Hz cycle over 5 µm range. The image resolution was 512 pixels per line.

Figure 11.17A and 10 nm per pixel in Figure 11.17B. Although the difference is not easily noticeable in the topography images, we can see that tiny speckles in the error signal and phase shift images collected with the small scan size look like "twins" aligned in the same diagonal direction. That is because the probe had "double tips" (two apexes) separated by ~10 nm. Such fine features cannot be detected when the image is collected with a lower resolution. Also, we can see that it is much easier to recognize such fine features in the error signal or phase image than the topography image. Because the topography image is displayed in the full height scale to cover the large features, tiny fine features may not be easily noticed. When the topography image is zoomed in, we can see the same features in the magnified height scale.

11.5.4 Artifacts Due to Tip Shape

In AFM, the topography image is not a true reflection of the sample image; it is always convoluted by the tip image. For example, the steps in the calibration grating have vertical walls; but the step edges in the topography images in Figures 11.15 and 11.17 are all slanted. This is because the step edge was in contact with the different regions of the tip end, not always with the apex point of the tip. In Figure 11.15A, we can see that all small fine debris (probably contaminants) are slightly elongated in the 45° diagonal direction. In the error signal, they all look like "ants" pointing the same direction. The probability of getting all contaminant particles aligned along the same direction is essentially zero. This is an artifact caused by the tip shape (same as Figure 11.17B); from the image, we can tell that the tip end has two apexes. For that reason, the step edges in the topography image look like a double-vision image. In Figure 11.15B which was collected with the same tip, all particulates look very much rounded. That indicates that the tip end shape was changed (probably due to micro-cracking) during the tapping-mode imaging with a large amplitude (so, hammering with a large force). This means the AFM tip got blunt during the use.

Often, the object to be imaged with AFM is sharper than the AFM tip used. Then, the topography image is dominated by the AFM tip shape. Figure 11.18 exhibits SEM images of an AFM tip and a commercial razor blade (retrieved from an

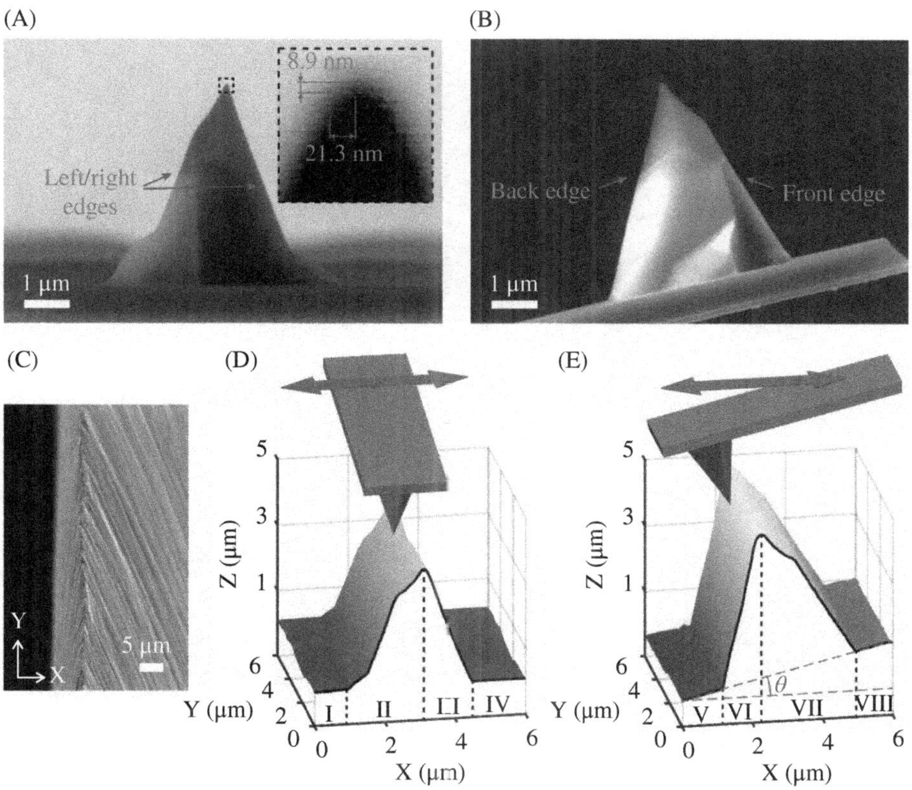

Figure 11.18 (A) Front- and (B) side-view SEM images of a Si tip on a silicon nitride cantilever ($k = 0.4$ N/m). (C) SEM image of a razor blade mounted with a 4° tilt angle. (D,E) Topographic images from the peak-force-tapping mode scan of the razor blade with the AFM shown in (A,B) in two scan directions. Reproduced with permission from Chen et al. [49]/Oxford University Press.

off-the-shelf shaver) and the AFM images obtained in two different scan directions [49]. The cross-sectional cone angle (~20°) of the razor blade edge is smaller than that of the AFM tip. Thus, when the razor blade sample was imaged with the AFM tip, the topography image essentially shows the shape of the AFM tip rather than the sample. In other words, the razor blade imaged the AFM probe, although our intention was the other way around. When the AFM cantilever axis was perpendicular to the razor blade, the base region of the image was tilted by 13.5°. This is because the AFM probe is intentionally mounted at a certain angle so that the main body of the probe does not touch the sample.

If we know the exact shape of the AFM tip apex (as shown in the inset of Figure 11.18A), we can still determine the radius of curvature of the razor blade edge. Figure 11.19 explains that principle. When a rounded sample is imaged with an ideally sharp tip (Figure 11.19A), the height profile of the image consists of two tilted lines and an arc. The arc region defined by points B, C, and D is the height profile imaged by the tip apex. When the blunt tip is used (Figure 11.19B), we can see that at point G, the right end of the circular segment of the tip is in contact with the sample (G′); at point J, the lowest point of the tip is in contact with the sample (J′); and at point K, the left end of the tip circular segment is in contact with the sample (K′). Using the dimensions of the tip end, we can find points G′, J′, and K′ from the experimentally found points G, J, and K. Then,

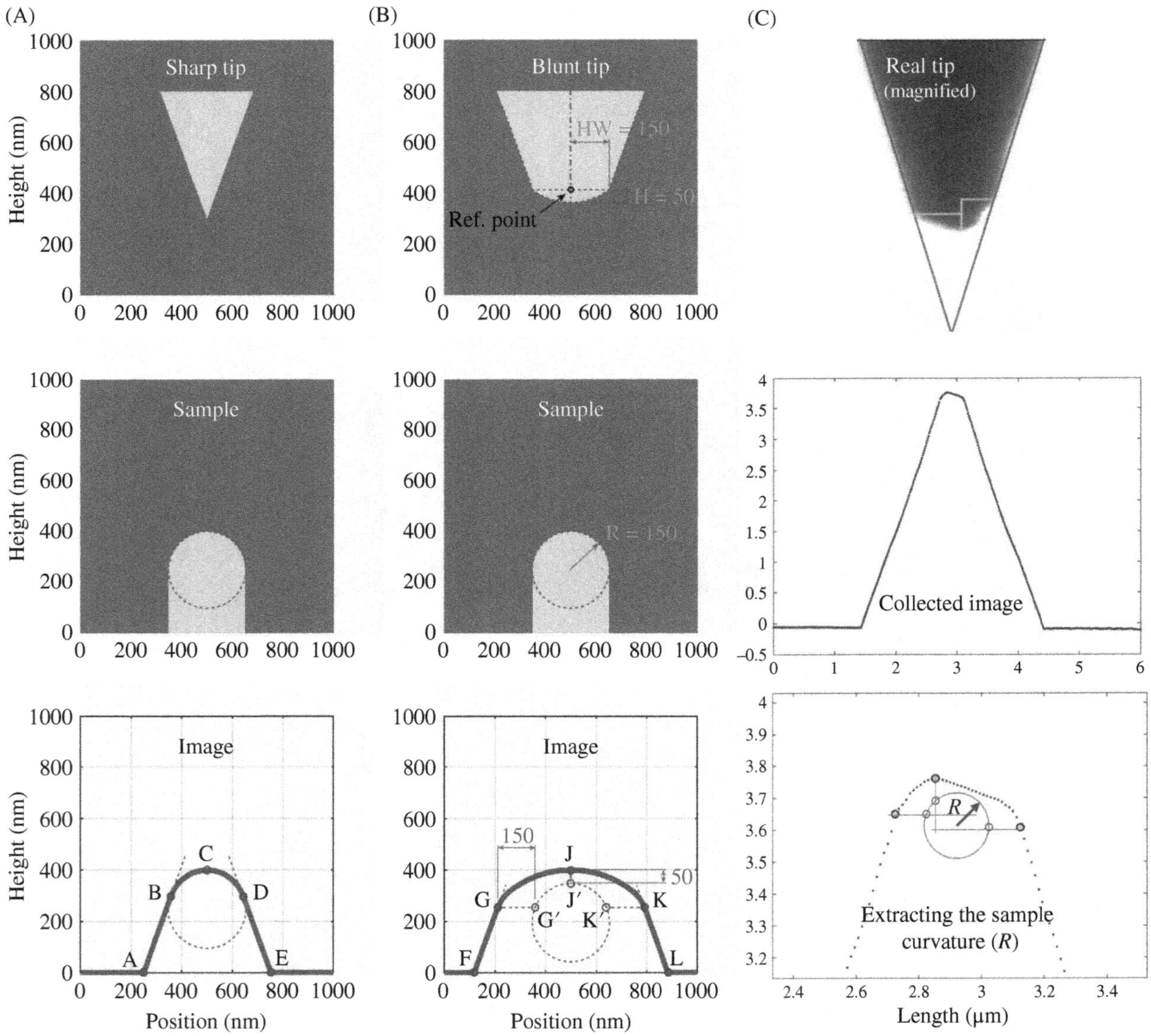

Figure 11.19 Simulation of (A) sharp and (B) blunt tip scanning across over a rounded sample and the resulting height profile. (C) Example of determining the radius of sample curvature by deconvoluting the tip shape. Reproduced with permission from Chen et al. [49]/Oxford University Press.

a circle going through these G′, J′, and K′ points can be drawn mathematically, from which the radius (R) of sample curvature can be calculated. A real example of this approach is shown in Figure 11.19C. Conversely, if we know the dimension of the sample imaged with a tip, we can determine the radius of the tip curvature [50].

As a rule of thumb, it should be remembered that all topographic features having a radius of curvature smaller than the radius of curvature of the scanning tip are not imaged properly. If many features in the AFM image have the same shape or elongated in the same direction, it is the shape of the blunt tip imaged by the topographic feature at the sample surface. The tip shape may change (due to fracture or wear) during imaging or over repeated usages. There are dead zones in the surface topography that cannot be reached by the AFM tip. For example, as we can see in Figures 11.16–11.19, imaging the side walls of tall topographic features is not possible unless a special instrument is built to tilt the AFM head such that the AFM tip apex can touch the side walls [51]. The trenches whose width is smaller than the tip radius cannot be imaged with AFM.

In addition to the tip-shape artifacts caused by the finite size and shape of the AFM tip, there are other artifacts that can influence or alter the surface image. The effects of poor feedback control are shown in Section 11.5.3. The nonlinearity and hysteresis of the scanner as well as thermal drift of the sample can affect the image quality. Of course, the bad calibration of the scanner and detector signal will limit the precision of the dimensions in the AFM image. Optical interference or electronic noises can create periodic noises in AFM imaging. In the tapping mode imaging, the root-mean-square amplitude of the cantilever oscillation is extracted with a lock-in amplifier; then the time constant of the lock-in should be adjusted and chosen properly for a given scan speed (i.e., a dwell time of the tip at one-pixel location). Recognizing all artifact effects in AFM experiments is important but sometimes not obvious. It requires a lot of experiences; in that sense, it would be very helpful to practice identifying artifacts in the examples shown in various review papers [52, 53].

11.6 Material Properties Measured Along with Topography

11.6.1 Nanoscale Friction

In addition to topographic imaging of the surface, AFM can also be used to probe various material properties of the surface. In the contact mode scan, friction between the tip and the surface will make the scanning probe tilt sideway which causes the cantilever to twist. If the degree of cantilever twisting during the lateral scan is calibrated, then nanoscale friction force can be measured and recorded simultaneously along with the topography image of the sample. In AFM, friction force is defined as half of the lateral signal hysteresis between the trace and retrace scans (see Figure 11.20). In this way, the offset of zero force drops out from the lateral signal measurement. Many theoretical models have been developed and used for calibration and conversion of the lateral signal from the cantilever twisting to the frictional force at the tip [54, 55]. It can be done by compressing the tip over a slanted surface, scanning the tip over a wedge, scanning the tip across a vertical step, and so on. All these methods rely on the force balance at the contact point. A simple alternative method could be to scan over a flat sample with a known coefficient of friction (COF) and compare the lateral signal with the expected friction force [56]. All methods have advantages and limitations, which can be found in the literature [54]. Once the lateral signal is converted to the friction force, the COF of the surface can be determined by dividing it with the normal force which is the sum of the applied force (which is governed by the set point for the contact imaging) and the adhesion force of the contact (see Section 11.4.2). Alternatively, the friction force can be described as the shear stress (τ) times the contact area (A). Since A can be calculated with the relationships shown in Table 11.2, one can find the effect of adhesion force on friction of the single-asperity contact and determine the shear stress of the contact [57]. In many cases, friction force correlates with adhesion force, but not always especially when topographic height change contributes to friction even if the height change is a single atomic layer [58].

On atomically smooth crystalline surfaces, the AFM friction signal may vary in a stick-slip fashion with the periodicity coinciding with the crystalline lattice parameter [29]. This behavior is often explained with the Prandtl–Tomlinson model which is based on the elastic strain energy balance on a corrugated potential energy surface. Note that it is not always guaranteed to observe the stick-slip behavior commensurate with the lattice periodicity of crystalline samples. In repeated scans with a single tip on the same sample, the stick-slip behavior could disappear and reappear as the tip shape changes. If the test environment changes, the stick-slip behavior can also change [59]. This behavior is still under investigation.

Although the effect of a single atomic defect on friction cannot be measured with AFM (for the reason discussed in Section 11.1.2), the effect of an atomic step on friction can be measured with AFM. Figure 11.20 displays friction signals measured over edges of single-layer thick graphene layers on the basal plane of graphite [56]. The friction force measured

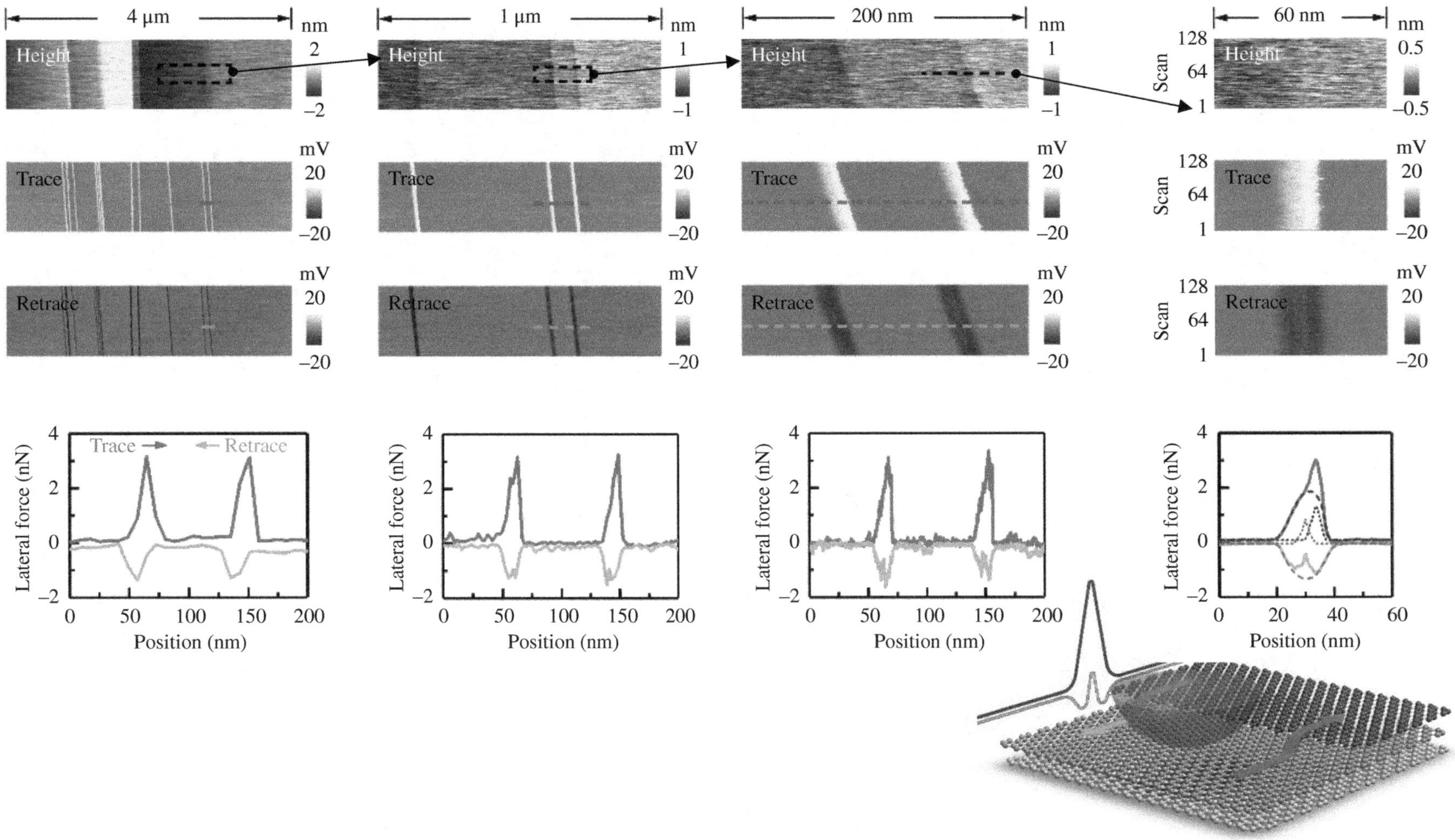

Figure 11.20 Topographic images and trace/retrace lateral signals at the step edge of a single-layer graphene on the basal plane of graphite collected at different scan sizes marked at the top of each panel. Reproduced from Chen and Kim [56]/with permission of Springer Nature.

during the step-up process is significantly larger than that during the step-down process. If the scan resolution is increased by reducing the scan size, further fine details can be revealed. The broad component with the opposite sign in the trace and retrace scans is due to the chemical interaction between the tip and the functional group at the step edge. The sharp component with the same sign in the trace and retrace scans is due to the change in the scan direction across the step (although the single graphene layer step is only 0.34 nm tall) [11].

11.6.2 Indentation Modulus and Adhesion

In the amplitude-modulation (AM) tapping mode scan, the phase shift can be sensitive to mechanical properties of the sample surface. From eq. (11.16) and using $\delta\nu = \nu_{dr} - \nu_{eff}$ and $\tan^{-1}x \approx -\frac{\pi}{2} - \frac{1}{x}$ at $x \gg 1$, we can derive the following approximation form [60]:

$$\varphi(\nu_{dr}) = \tan^{-1}\frac{\nu_{dr}\nu_{eff}}{Q\left(\nu_{dr}^2 - \nu_{eff}^2\right)} \approx \tan^{-1}\frac{\nu_{eff}}{2Q\delta\nu} \approx -\frac{\pi}{2} - 2Q\frac{\delta\nu}{\nu_{eff}} \tag{11.24}$$

And from eq. (11.17) and using $\sqrt{1+x} = 1 + \frac{x}{2}$ at $x < 1$, we can approximate:

$$\delta\nu = \nu_{dr} - \nu_{eff} = \nu_{dr} - \nu_0\sqrt{1 - \frac{F'_{ts}}{k}} \approx \nu_{dr} - \nu_0\left(1 - \frac{F'_{ts}}{2k}\right) \approx \nu_0\frac{F'_{ts}}{2k} \tag{11.25}$$

So, the phase shift $\delta\varphi$ from the $-90°$ resonance phase can be expressed as:

$$\delta\varphi \approx -Q\frac{F'_{ts}}{k} \tag{11.26}$$

Here, we can see that the phase shift depends linearly on both the effective quality factor and the force gradient of the tip–sample interaction. The magnitude of Q is related to the energy dissipated by the sample, and the F'_{ts} term is a function of the surface stiffness which can be related to elastic modulus of the sample (see Section 11.4.1). So, in principle, by monitoring the phase shift during AM-tapping mode imaging, one can monitor changes in the tip–surface interactions. In practice, the AM-tapping mode imaging is performed at a fixed set-point amplitude; in this condition, $\delta\varphi$ is relatively insensitive to the elastic modulus [61]. It means changes in $\delta\varphi$ can be interpreted as changes in the energy dissipated by the tip–sample interactions [62]. Note that the phase shift can also be associated with the topographic height change (see Figures 11.15–11.17), which are caused by the feedback control and not by the material property change. Thus, the correlation between $\delta\varphi$ and Q cannot be applied to the region where topographic height changes.

The mechanical properties such as elastic modulus and adhesion force can be extracted from the force–distance (*F-Z*) curve, as discussed in Sections 11.4.1 and 11.4.2. By repeatedly measuring the *F-Z* curves at user-defined locations or in an array pattern over a certain area, local variations in those mechanical properties can be monitored. But performing *F-Z* measurements at multiple locations could be time consuming. As a means to extract such information simultaneously while taking topography images, force-modulation mode was developed [63, 64]. In this mode, the surface is imaged in the contact mode at a constant force (i.e., feedback controlling the average deflection of the cantilever); at the same time, a small sinusoidal signal is added to the Z-voltage of the piezo-scanner. This applies a small oscillatory force with a high frequency to the sample surface. Then, depending on the elastic modulus or viscoelastic properties, the magnitude of the oscillatory cantilever deflection at the driving frequency varies (Figure 11.21). The amplitude and phase shift of this cantilever oscillation can be extracted and quantified with a lock-in amplifier. The amplitude signal is proportional to the elastic properties of the sample, and the phase signal is sensitive to the viscous properties [63].

A more advanced method that can provide quantitative nanomechanical properties is so-called peak-force tapping mode [65, 66]. The operation of this mode is schematically illustrated in Figure 11.22. It drives the cantilever up and down sinusoidally at a rate in the range of a few hundred Hz up to 1–2 kHz, $Z_{piezo}(t)$, and measures the cantilever deflection in time, $\Delta z_{PSD}(t)$. Then, using a synchronization algorithm, it separates the parasitic cantilever motion not directly related to the tip–sample interaction and extracts the actual vertical deflection caused by the tip–sample interaction, which is used for feedback control and topographic imaging. From the $\Delta z_{PSD}(t)$ portion corresponding to the tip–sample interaction, the

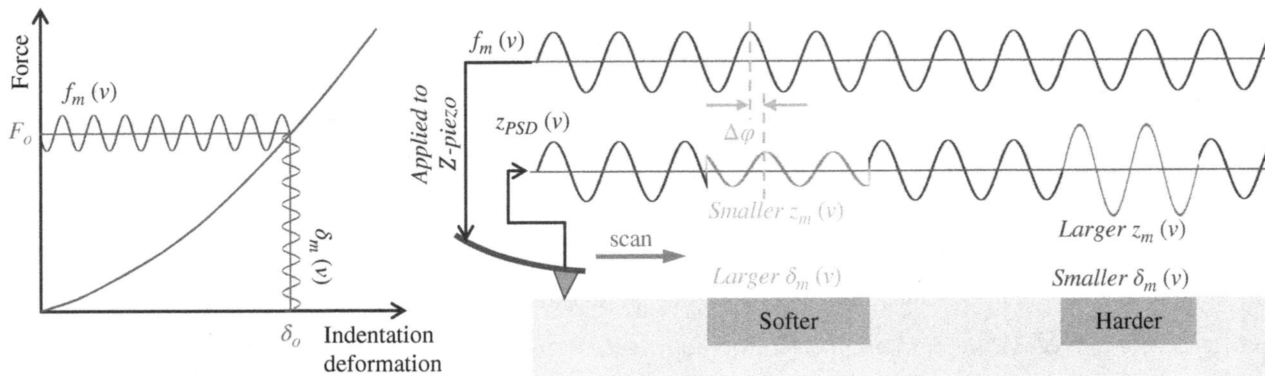

Figure 11.21 Schematic illustration of force-modulation AFM which can detect small changes in surface mechanical properties.

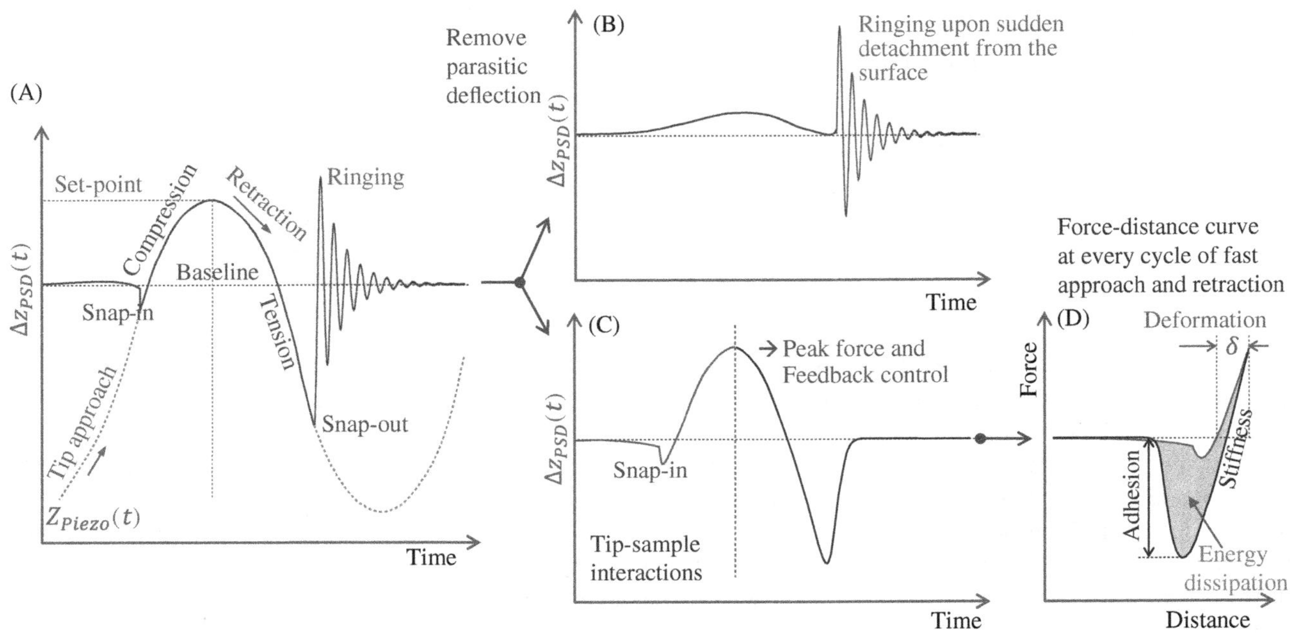

Figure 11.22 Schematic illustration of the peak-force tapping AFM. (A) Raw deflection signal in a period of cantilever modulation. (B) Parasitic deflection signal removed electronically by the system. (C) Deflection signal proportion to the tip–sample interaction force. (D) *F-D* curve of one single cycle of cantilever modulation obtained from (C).

F-D curve can be constructed for every cycle of the tapping (Figure 11.22D). The adhesion force (F_{adh}) and deformation depth (δ) can be determined at every pixel and the reduced modulus can be calculated using the DMT model (see Table 11.2):

$$E_r = \frac{3}{4}\frac{\left(F_{tip} + F_{adh}\right)}{\sqrt{R\delta^3}} \tag{11.27}$$

where F_{tip} is the peak force applied (which is the set point of the feedback control) and R is the radius of the tip curvature. The hysteresis area between the approach and retraction curve is the energy dissipated during the tip–surface interaction. Being able to measure and differentiate the adhesion force, DMT-modulus, and energy dissipation of the surface along with the topography image, peak-force tapping AFM can provide rich mechanical property information of the heterogeneous samples which could not be obtained or are difficult to obtain with other AFM techniques [67].

Figure 11.23 shows an example of peak-force tapping AFM. The sample was polystyrene (PS) nanoparticles of mixed sizes deposited on a Si wafer. The nanoparticles were suspended in a solution and then drop-casted on the wafer. In the topography image, most particles are isolated, but some particles are aggregated with each other. The base plane of the topography image appears slanted, which is because the wafer could not be mounted perfectly horizontally. The slope of the base plane

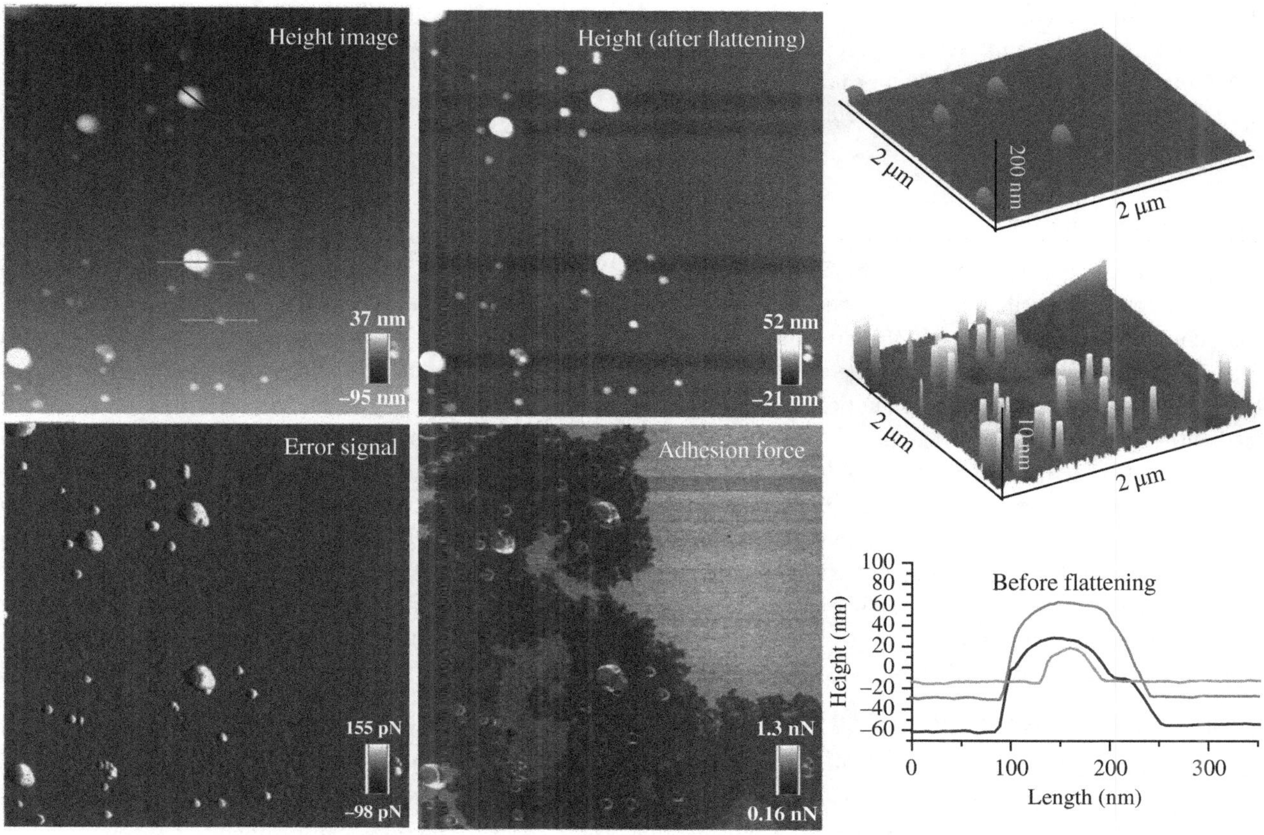

Figure 11.23 Peak-force tapping images of PS nanoparticles on a Si wafer. The set point (peak force) was 0.75 nN and the scan size was 2 μm × 2 μm. Topographic images after flattening to remove the base plane slope, 3D images, and line profiles are also shown.

can be removed by running image flattening with the built-in function of the AFM software. The flattened image shows some dark lines in the left and right sides of large particles. This is an artifact caused by the flattening algorithm. In this case, the program took the average height along the scan direction (left-to-right) and moved it to zero at each horizontal line. By doing so, the overall slope of the sample was removed. But, since the average value was higher in the line with a tall particle, its background along the scan direction appeared lower than other regions without particles in the flattened image. In the error signal, the left sides of the particles are darker (smaller errors in the feedback control, which means ascending a slope) and the right sides of the particles are brighter (larger errors; so, descending).

Around the PS particle region, the error signal shows some fine features. If we reduce the full scale range of the topography image (so, magnify along the height direction), we can see some thin deposits over a large area. The adhesion force map obtained from the peak-force tapping shows that the deposits have a smaller adhesion force than the bare silicon surface (which is top right side of the image). The deposits must come from the organic residues that were present in the PS-suspension solution. In the adhesion map, the small particles appear to have halos; but that is an artifact caused by the topography change at those locations. Inside those halos, the adhesion of the PS particle appears similar to that of organic residues in the surrounding area and significantly smaller than that of the bare substrate area. In this specific example, the DMT modulus of PS particles could not be obtained. Because a soft cantilever was used, the deformation (δ) of the sample surface was too small to be measured reliably; so, the modulus map automatically calculated by the AFM software was very noisy. The dissipated energy map (not shown here) was simply the same as the adhesion map because it was governed by the adhesion process.

The line profiles of three particle locations (marked in the topography image) are also shown in Figure 11.23. All three show that the left side of the particle image is steeper than the right side. One possible reason is the tip shape. The other possibility is the tilt angle in the AFM cantilever mounting (see Figure 11.18). The particle in the red profile appears to be 90 nm tall and 150 nm wide. But, this is a PS nanoparticle, so it should be sphere. The reason that the width in the image is larger than the height is due to the convolution of the tip shape. For the same reason, the particle in the green profile appears to be 35 nm tall and 60 nm wide. The true particle diameter can be read out directly from the height, assuming that the applied peak force was small enough that elastic deformation of the particle was negligible (which was supported by the fact that the DMT-modulus image detected just the noise signals). Also, from these line profiles, one can calculate the effective radius of the tip curvature [50]. In Figure 11.23, the blue line profile shows that two particles are aggregated together. The diameter of these particles could not be determined from the height because the line profile was taken along the direction where the sample slope changed.

11.6.3 Other Properties

All imaging modes described above can be done with regular AFM probes made of silicon or silicon nitride materials. One just has to choose the cantilever with a proper spring constant. The contact mode scan is typically done with soft cantilevers. The noncontact and tapping mode scans are done with stiff cantilevers with high resonance frequencies. Cantilevers with spring constants in the intermediate range are typically chosen for the force-modulation and peak-force tapping modes.

If the tip is coated with functional materials, then other surface properties can also be measured. If the tip surface is derivatized with organic self-assembled monolayers, it is possible to distinguish specific chemical interactions through friction force changes. This is the basis of chemical force microscopy (CFM) [10]. If a magnetic alloy is coated on the tip, local variance in magnetization (magnetic field) can be obtained along with the topography using MFM [68]. If the tip is coated with a conductive metal layer and the sample substrate is conductive, one can measure local changes in electrical conductance (or resistance) along with the topography in the contact mode imaging. This is conductive AFM [69].

When the conductive tip is electrically biased with an AC voltage, the surface potential (more accurately, contact potential difference between the tip and the surface, ΔV_{CPD}) can be imaged. Such techniques are Kelvin probe force microscopy (KPFM) [70], scanning polarization force microscopy (SPFM) [71], and electrostatic force microscopy (EFM) [72]. The basic principles of these methods are all same; they detect the cantilever bending due to the electrostatic force, $F = \frac{1}{2}\frac{dC}{dz}\Delta V_{CPD}^2$. If charges are involved instead of voltages, then the interaction measured is Coulomb force arising between the static charge of the sample and the AC charge induced on the tip. The main differences among KPFM, SPFM, and EFM are how the cantilever vibrations induced by these forces are deconvoluted and recorded. The same can be done in aqueous solution as well [73, 74].

When the sample is piezoelectric, then the converse piezoelectric effect (i.e., strain under electrical bias) can be measured using piezoelectric force microscopy (PFM) [75]. The direct piezoelectric effect (charge displacement under strain) is difficult to measure reliably since the electric signal is so small. AFM can also be used for nanoscale thermal mapping [76]. That can be done by using a tip consisting of thermocouple materials or by measuring electrical resistance change due to heat loss or gain.

As discussed in Section 6.6, AFM can be coupled with infrared spectroscopy for high-resolution vibrational spectroscopic mapping. Those methods are called photothermal AFM-IR, s-SNOM nano-FTIR, and PiFM. The near-field enhancement at the metal-coated AFM tip can also enable nanoscale Raman mapping. This is called tip-enhanced Raman spectroscopy (TERS) [77].

11.7 Further Reading for AFM

Going through all advanced AFM-based nanoscale analyses is beyond the scope of this book. But all those methods listed in Table 11.1 utilize one of the AFM operation principles discussed in this chapter; thus, mastering the AFM principles and understanding the subtle but critical details in AFM operation and data acquisition described in this chapter are very helpful in practice. More thorough coverage of fine details and examples can be seen at "Atomic Force Microscopy" by P. Eaton and P. West [5] and B. Voigtländer [78]. A tutorial animation at Park Systems website would be helpful to visualize various operation modes [79].

Practice Problems

11.1 Use the cantilever dimensions shown in Table 1 in Sader et al.'s paper in *Review of Scientific Instruments* **1999**, *70*, 3967 [35] and calculate the spring constants of the cantilever. The hydrodynamic function, $\Gamma(\mathrm{Re})$, can be approximated from Figure 1 of that paper.

11.2 Look up a commercial AFM probe website and obtain the cantilever dimensions. Then, calculate the resonance frequency for a given cantilever. Assume the quality factor (Q) is 100. Simulate the cantilever oscillation and how it changes when the tip is in attractive and repulsive interactions.

11.3 Atomic force microscopy (AFM) controls or records the voltage signals and then converts them to distances or forces. Below is the typical force – piezo displacement curve from an AFM measurement with a sharp tip on a hard surface. Convert it to the force as a function of the distance between the tip and surface. *Mark the zero points for the force and distance in the right-side plot (your answer).*

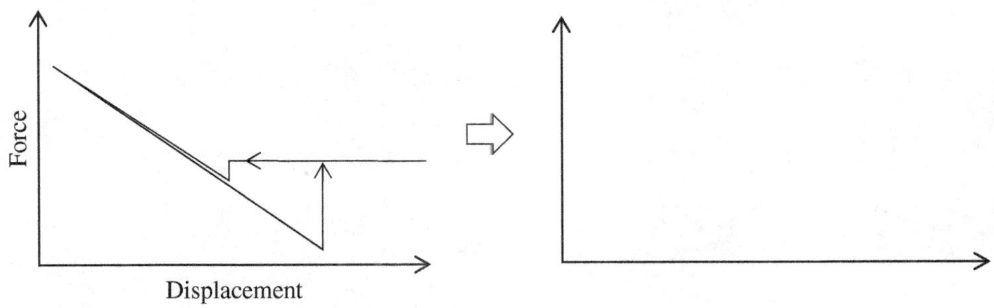

11.4 Calculate the Hertzian contact radius, deformation depth, and contact pressure distribution for the following cases. The modulus and surface energy of materials can be found in the literature.

a) Si AFM tip with a 10-nm radius of curvature pressing a soda lime silica glass surface at an applied load of 10 nN.
b) AFM probe with a 1-μm diameter silica sphere pressing a polystyrene surface at an applied load of 100 nN.
c) Estimate the adhesion force from the surface energy of the above materials and calculate the contact radius, deformation depth, and contact pressure distrubtion using the DMT and JKR theories.

11.5 a) How can AFM produce images of atomically resolved crystalline lattices? Can AFM detect a single atomic defect in a crystalline lattice?
b) How is it possible to distinguish mechanical properties (adhesive vs. repulsive interactions) in tapping mode AFM? Explain the principle briefly.
c) Sketch the AFM topography image of the following three cases:

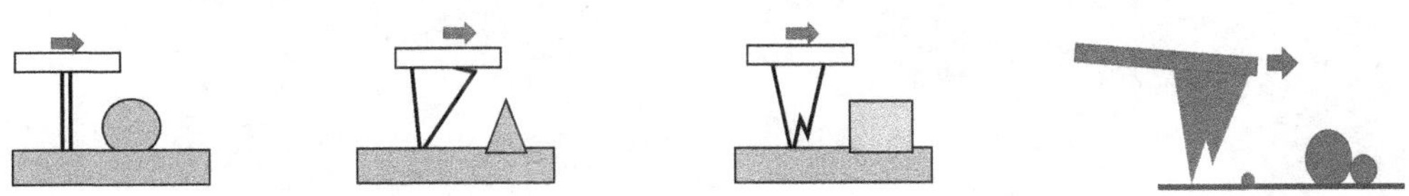

11.6 Nine spherical nanoparticles are arranged in a 3 × 3 matrix. Three nanoparticles in the same horizontal row are made of the same material. The AFM imaging was done over a 14 × 14 nm area with two AFM probes while keeping the cantilever deflection (D_{sp}) constant during the contact scan. The cantilever spring constant (k_{sp}) and the tip radius (R_{tip}) are given below. The height profiles along the x and y directions taken at the + cursor position are also shown.

Cantilever dimension: length = 225 μm, width = 40 μm, thickness = 3 μm
Position of tip = 200 μm from the base (chip); elastic modulus = 110 GPa
Density of material = 2.3 g/cm^3; spring constant (k_{sp}) = 3.71 N/m

Tip #1: Radius (R_{tip}) = 300 pm	***Tip #2***: Radius (R_{tip}) = 3 nm
D_{sp} = 1 pm; contact force = 3.71 pN	D_{sp} = 1 nm; contact force = 3.71 nN

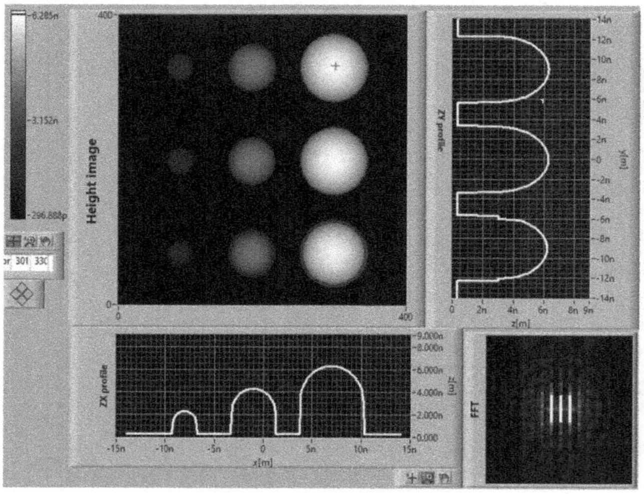

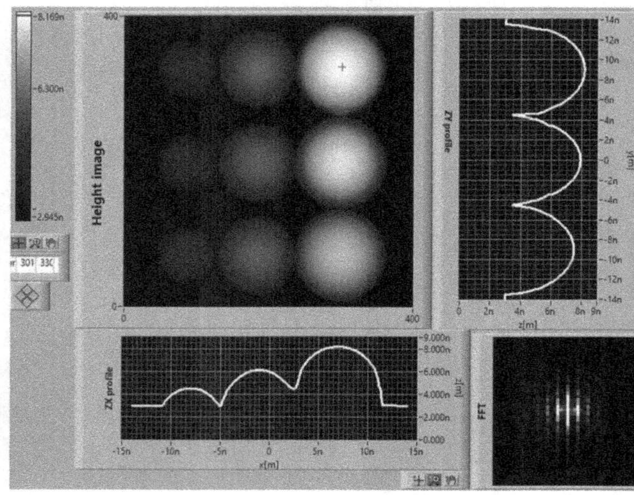

a) The particles in the left image look like an arch-shaped door, but they look like an Eskimo igloo in the right image. Explain the reason.
b) The heights of three particles at $x = +7$ nm look identical in the left image, but their heights vary in the right image. Explain the possible reason.
c) What are the true diameters of the particles at $y = +9$ nm, 0 nm, and -9 nm in the $x = +7$ nm column of the right image?
d) A contact-mode AFM scan image of triangles on a flat surface is displayed here:

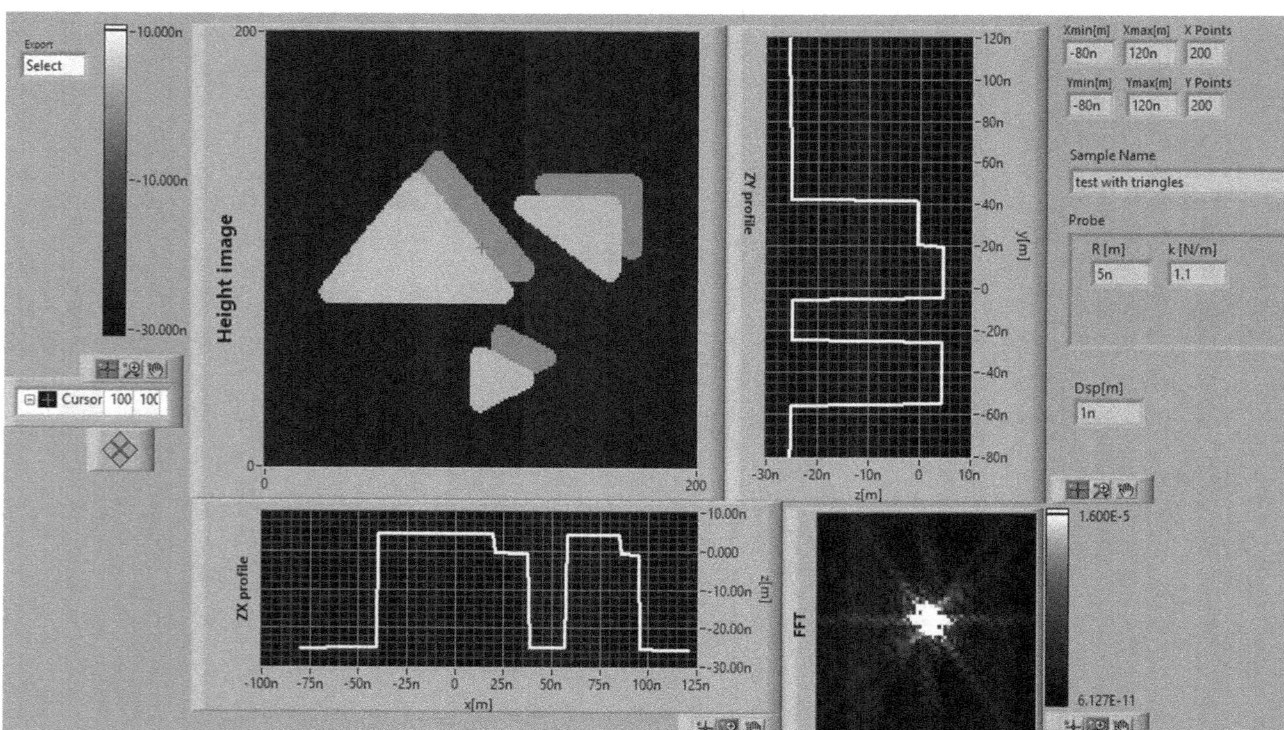

How many triangles are on the surface? What is the true thickness of triangles? Based on the image, what can you say about the tip?

References

1 Binnig G, Rohrer H. In touch with atoms. Reviews of Modern Physics. 1999;71(2):S324–S30.
2 Binnig G, Rohrer H. Scanning tunneling microscopy---from birth to adolescence. Reviews of Modern Physics. 1987;59 (3):615–25.
3 Binnig G, Quate CF, Gerber C. Atomic force microscope. Physical Review Letters. 1986;56(9):930–3.
4 Chen CJ. Introduction to Scanning Tunneling Microscopy. Oxford: Oxford University Press; 2007.
5 Eaton P, West P. Atomic Force Microscopy. Oxford: Oxford University Press; 2010.
6 Israelachvili JN. Chapter 13 – Van der Waals forces between particles and surfaces. In: Israelachvili JN, editor. Intermolecular and Surface Forces (Third Edition). San Diego: Academic Press; 2011. p. 253–89.
7 Butt H-J, Cappella B, Kappl M. Force measurements with the atomic force microscope: technique, interpretation and applications. Surface Science Reports. 2005;59(1):1–152.
8 Rugar D, Mamin HJ, Guethner P, Lambert SE, Stern JE, McFadyen I, et al. Magnetic force microscopy: general principles and application to longitudinal recording media. Journal of Applied Physics. 1990;68(3):1169–83.
9 Zheng X-X, Sun W-F. Magnetic force probe characterizations of nanoscaled ferromagnetic domains: finite-element magnetostatic simulations. Nanomaterials. 2022;12(13):2212.
10 Noy A, Vezenov DV, Lieber CM. Chemical force microscopy. Annual Review of Materials Research. 1997;27:381–421.
11 Chen Z, Khajeh A, Martini A, Kim SH. Chemical and physical origins of friction on surfaces with atomic steps. Science Advances. 2019;5(8):eaaw0513.
12 Yang J, Petitjean SJL, Koehler M, Zhang Q, Dumitru AC, Chen W, et al. Molecular interaction and inhibition of SARS-CoV-2 binding to the ACE2 receptor. Nature Communications. 2020;11(1):4541.
13 Hsiao E, Marino MJ, Kim SH. Effects of gas adsorption isotherm and liquid contact angle on capillary force for sphere-on-flat and cone-on-flat geometries. Journal of Colloid and Interface Science. 2010;352(2):549–57.
14 Asay DB, de Boer MP, Kim SH. Equilibrium vapor adsorption and capillary force: exact laplace–young equation solution and circular approximation approaches. Journal of Adhesion Science and Technology. 2010;24(15–16):2363–82.
15 Asay DB, Kim SH. Molar volume and adsorption isotherm dependence of capillary forces in nanoasperity contacts. Langmuir. 2007;23(24):12174–8.
16 Asay DB, Kim SH. Effects of adsorbed water layer structure on adhesion force of silicon oxide nanoasperity contact in humid ambient. The Journal of Chemical Physics. 2006;124(17):174712
17 Cappella B, Dietler G. Force-distance curves by atomic force microscopy. Surface Science Reports. 1999;34(1):1–104.
18 Israelachvili JN, Pashley RM. Molecular layering of water at surfaces and origin of repulsive hydration forces. Nature. 1983;306 (5940):249–50.
19 Israelachvili JN, Tabor D. The measurement of van der Waals dispersion forces in the range 1.5 to 130 nm. Proceedings of the Royal Society of London A Mathematical and Physical Sciences. 1972;331(1584):19–38.
20 Israelachvili J, Min Y, Akbulut M, Alig A, Carver G, Greene W, et al. Recent advances in the surface forces apparatus (SFA) technique. Reports on Progress in Physics. 2010;73(3):036601.
21 Krämer G, Bennewitz R. Molecular rheology of a nanometer-confined ionic liquid. The Journal of Physical Chemistry C. 2019;123(46):28284–90.
22 Khan SH, Matei G, Patil S, Hoffmann FM. Dynamic solidification in nanoconfined water films. Physical Review Letters. 2010;105(10):106101.
23 Giessibl FJ. Atomic resolution of the silicon (111)-(7×7) surface by atomic force microscopy. Science. 1995;267(5194):68–71.
24 Giessibl FJ. AFM's path to atomic resolution. Materials Today. 2005;8(5):32–41.
25 Hembacher S, Giessibl FJ, Mannhart J, Quate CF. Revealing the hidden atom in graphite by low-temperature atomic force microscopy. Proceedings of the National Academy of Sciences. 2003;100(22):12539–42.
26 Koutsos V, Manias E, Brinke GT, Hadziioannou G. Atomic force microscopy and real atomic resolution. Simple computer simulations. Europhysics Letters. 1994;26(2):103.
27 Fischer-Cripps AC. The Hertzian contact surface. Journal of Materials Science. 1999;34(1):129–37.
28 Mate CM, McClelland GM, Erlandsson R, Chiang S. Atomic-scale friction of a tungsten tip on a graphite surface. Physical Review Letters. 1987;59(17):1942–5.
29 Carpick RW, Martini A, Cannara RJ. Atomic-level stick-slip. In: Wang QJ, Chung Y-W, editors. Encyclopedia of Tribology. Boston, MA: Springer US; 2013. p. 140–8.

30 Cross G, Schirmeisen A, Stalder A, Grütter P, Tschudy M, Dürig U. Adhesion interaction between atomically defined tip and sample. Physical Review Letters. 1998;80(21):4685–8.

31 Schirmeisen A, Cross G, Stalder A, Grütter P, Dürig U. Metallic adhesion and tunnelling at the atomic scale. New Journal of Physics. 2000;2(1):329.

32 Landman U, Luedtke WD, Burnham NA, Colton RJ. Atomistic mechanisms and dynamics of adhesion, nanoindentation, and fracture. Science. 1990;248(4954):454–61.

33 Harrison JA, White CT, Colton RJ, Brenner DW. Nanoscale investigation of indentation, adhesion and fracture of diamond (111) surfaces. Surface Science. 1992;271(1):57–67.

34 True Non-Contact™ Mode Park Systems [Available from: https://www.parksystems.com/index.php/park-spm-modes/217-true-non-contact-mode].

35 Sader JE, Chon JWM, Mulvaney P. Calibration of rectangular atomic force microscope cantilevers. Review of Scientific Instruments. 1999;70(10):3967–9.

36 Sader JE. Calibration of Atomic Force Microscope Cantilevers: Sader Method – Global Calibration Initiative [Available from: https://sadermethod.org/].

37 Voigtländer B. Harmonic oscillator. In: Atomic Force Microscopy. Cham: Springer International Publishing; 2019. p. 15–33.

38 García R, Pérez R. Dynamic atomic force microscopy methods. Surface Science Reports. 2002;47(6):197–301.

39 Anczykowski B, Cleveland JP, Krüger D, Elings V, Fuchs H. Analysis of the interaction mechanisms in dynamic mode SFM by means of experimental data and computer simulation. Applied Physics A. 1998;66(1):S885–S9.

40 Voigtländer B. Intermittent contact mode/tapping mode. In: Voigtländer B, editor. Atomic Force Microscopy. Chapter 14. Cham: Springer International Publishing; 2019. p. 231–53.

41 Opdahl A, Kim SH, Koffas TS, Marmo C, Somorjai GA. Surface mechanical properties of pHEMA contact lenses: viscoelastic and adhesive property changes on exposure to controlled humidity. Journal of Biomedical Materials Research Part A. 2003;67A(1):350–6.

42 Maugis D. Adhesion of spheres: the JKR-DMT transition using a dugdale model. Journal of Colloid and Interface Science. 1992;150(1):243–69.

43 Xi X, Kim SH, Tittmann B. Atomic force microscopy based nanoindentation study of onion abaxial epidermis walls in aqueous environment. Journal of Applied Physics. 2015;117(2):024703.

44 Li G, Gao Y, He X, Huang Q, Chen S, Kim SH, et al. Organosulfide-plasticized solid-electrolyte interphase layer enables stable lithium metal anodes for long-cycle lithium-sulfur batteries. Nature Communications. 2017;8(1):850.

45 He X, Barthel AJ, Kim SH. Tribochemical synthesis of nano-lubricant films from adsorbed molecules at sliding solid interface: tribo-polymers from α-pinene, pinane, and n-decane. Surface Science. 2016;648:352–9.

46 Gao Y, Yan Z, Gray JL, He X, Wang D, Chen T, et al. Polymer–inorganic solid–electrolyte interphase for stable lithium metal batteries under lean electrolyte conditions. Nature Materials. 2019;18(4):384–9.

47 Voigtländer B. Technical aspects of force microscopy. In: Voigtländer B, editor. Atomic Force Microscopy. Chapter 3. Cham: Springer International Publishing; 2019. p. 35–67.

48 Eaton P, West P. Chapter 6: AFM image artefacts. In: Atomic Force Microscopy. Oxford: Oxford University Press; 2010. p. 0.

49 Chen Z, Luo J, Doudevski I, Erten S, Kim SH. Atomic force microscopy (AFM) analysis of an object larger and sharper than the AFM tip. Microscopy and Microanalysis. 2019;25(5):1106–11.

50 Ramirez-Aguilar KA, Rowlen KL. Tip characterization from AFM images of nanometric spherical particles. Langmuir. 1998;14(9):2562–6.

51 Cho S-J, Ahn B-W, Kim J, Lee J-M, Hua Y, Yoo YK, et al. Three-dimensional imaging of undercut and sidewall structures by atomic force microscopy. Review of Scientific Instruments. 2011;82(2):023707.

52 Voigtländer B. Artifacts in AFM. In: Atomic Force Microscopy. Chapter 8. Cham: Springer International Publishing; 2019. p. 137–47.

53 Gołek F, Mazur P, Ryszka Z, Zuber S. AFM image artifacts. Applied Surface Science. 2014;304:11–9.

54 Munz M. Force calibration in lateral force microscopy: a review of the experimental methods. Journal of Physics D: Applied Physics. 2010;43(6):063001.

55 Asay DB, Hsiao E, Kim SH. Corrected direct force balance method for atomic force microscopy lateral force calibration. Review of Scientific Instruments. 2009;80(6):066101.

56 Chen Z, Kim SH. Measuring nanoscale friction at graphene step edges. Friction. 2020;8(4):802–11.

57 Carpick RW, Salmeron M. Scratching the Surface: fundamental investigations of tribology with atomic force microscopy. Chemical Reviews. 1997;97(4):1163–94.

58 Chen Z, Vazirisereshk MR, Khajeh A, Martini A, Kim SH. Effect of atomic corrugation on adhesion and friction: a model study with graphene step edges. The Journal of Physical Chemistry Letters. 2019;10(21):6455–61.

59 Chen Z, Kim SH. Tuning super-lubricity via molecular adsorption. Applied Materials Today. 2022;29:101615.

60 Voigtländer B. Amplitude modulation (AM) mode in dynamic atomic force microscopy. In: Voigtländer B, editor. Atomic Force Microscopy. Chapter 13. Cham: Springer International Publishing; 2019. p. 209–29.

61 Tamayo J, García R. Effects of elastic and inelastic interactions on phase contrast images in tapping-mode scanning force microscopy. Applied Physics Letters. 1997;71(16):2394–6.

62 Tamayo J, García R. Relationship between phase shift and energy dissipation in tapping-mode scanning force microscopy. Applied Physics Letters. 1998;73(20):2926–8.

63 Radmacher M, Tillmann RW, Gaub HE. Imaging viscoelasticity by force modulation with the atomic force microscope. Biophysical Journal. 1993;64(3):735–42.

64 Maivald P, Butt HJ, Gould SAC, Prater CB, Drake B, Gurley JA, et al. Using force modulation to image surface elasticities with the atomic force microscope. Nanotechnology. 1991;2(2):103.

65 Shuiqing H, Lars M, Yan H, Natalia E, Johannes K, Chanmin S, editors. High-speed atomic force microscopy and peak force tapping control. ProcSPIE; 2012.

66 Young TJ, Monclus MA, Burnett TL, Broughton WR, Ogin SL, Smith PA. The use of the PeakForceTM quantitative nanomechanical mapping AFM-based method for high-resolution Young's modulus measurement of polymers. Measurement Science and Technology. 2011;22(12):125703.

67 Xu K, Sun W, Shao Y, Wei F, Zhang X, Wang W, et al. Recent development of PeakForce tapping mode atomic force microscopy and its applications on nanoscience. Nanotechnology Reviews. 2018;7(6):605–21.

68 Kazakova O, Puttock R, Barton C, Corte-León H, Jaafar M, Neu V, et al. Frontiers of magnetic force microscopy. Journal of Applied Physics. 2019;125(6):060901.

69 Hui F, Lanza M. Scanning probe microscopy for advanced nanoelectronics. Nature Electronics. 2019;2(6):221–9.

70 Melitz W, Shen J, Kummel AC, Lee S. Kelvin probe force microscopy and its application. Surface Science Reports. 2011;66(1):1–27.

71 Salmeron M. Scanning polarization force microscopy: a technique for studies of wetting phenomena at the nanometer scale. Oil & Gas Science and Technology – Rev IFP. 2001;56(1):63–75.

72 Paul G. Electrostatic force microscopy: principles and some applications to semiconductors. Nanotechnology. 2001;12(4):485.

73 Marchand DJ, Hsiao E, Kim SH. Non-contact AFM imaging in water using electrically driven cantilever vibration. Langmuir. 2013;29(22):6762–9.

74 Collins L, Jesse S, Kilpatrick JI, Tselev A, Varenyk O, Okatan MB, et al. Probing charge screening dynamics and electrochemical processes at the solid–liquid interface with electrochemical force microscopy. Nature Communications. 2014;5(1):3871.

75 Gruverman A, Alexe M, Meier D. Piezoresponse force microscopy and nanoferroic phenomena. Nature Communications. 2019;10(1):1661.

76 Gomès S, Assy A, Chapuis P-O. Scanning thermal microscopy: a review. physica status solidi (a). 2015;212(3):477–94.

77 Shao F, Zenobi R. Tip-enhanced Raman spectroscopy: principles, practice, and applications to nanospectroscopic imaging of 2D materials. Analytical and Bioanalytical Chemistry. 2019;411(1):37–61.

78 Voigtländer B. Introduction. In: Voigtländer B, editor. Atomic Force Microscopy. Cham: Springer International Publishing; 2019. p. 1–13.

79 How AFM Works: Park Systems [Available from: https://www.parksystems.com/learning-center/nano-academy/how-afm-works.]

12

Mechanical Analysis via Indentation

The scanning probe methods described in Chapter 11 are mainly for the topographic analysis with a high spatial resolution through physical interactions of the probe with the sample surface. In the force-displacement curve analysis or peak-force tapping mode of atomic force microscopy (AFM), we could estimate the elastic modulus of the surface (see Section 11.4.1). If the applied force is increased high enough to cause plastic deformation of the surface, we could measure the hardness of the sample as well. For that, a special cantilever with an extremely high stiffness and an extremely hard tip material is used. In this chapter, we will study nanoindentation and micro-indentation, which are more widely utilized methods for characterization of mechanical properties of solid surfaces.

12.1 Modulus, Hardness, and Toughness

12.1.1 Types of Indenter

Figure 12.1 shows the electron micrographs of typical indenter tips used in nanoindentation and micro-indentation tests [1]. These indenters are directly mounted in the force application (load) axis of the actuator. A sphero-conical indenter (Figure 12.1A) is widely used for measuring soft materials; the cone shape is made for ease of mounting and only the spherical region is used to penetrate the sample surface. This is considered a blunt indenter. A conical indenter (image not shown) could be viewed as an opposite end – sharp indenter. The axial symmetry of these indenters can simplify the data analysis. But, an indenter with the perfect conical shape is not used since the sharp apex can be blunted or damaged easily. Other indenters have square or triangular pyramid shapes. A widely used square pyramid shape is Vickers indenter with a semi-angle (θ) of the faces of 68° (image not shown). It is widely used for micro-indentation. Knoop indenter (Figure 12.1B) is another four-sided pyramidal indenter with two different face angles. It is particularly useful for very hard materials which will have a very small residual indent; the long diagonal line in one direction makes it easy to measure the dimension of residual impression. Three-sided pyramidal types are used for nanoindentation because it is more convenient to produce the tip with a single point, rather than a line due to inaccuracy in aligning four facets. A Berkovich indenter (Figure 12.1C) has a face angle (θ) of 65.27°, which gives the same projected area-to-depth ratio as the Vickers indenter. Another three-sided pyramidal indenter is the cube-corner tip with a face angle $\theta = 35.26°$ (Figure 12.1D); it is a sharp indenter used to study material responses to a deep penetration.

12.1.2 Elastic and Plastic Responses Upon Indentation

The indentation experiment is done by measuring the response of materials during or after indentation with an indenter that is much harder than the sample. Micro-indentation measures the residual impression made in the sample surface, using an optical microscope, *after* indentation at a given load with Vickers or Knoop indenter. Nanoindentation method records the load (force) *during* the loading and unloading of Berkovich, cube-corner, or sphero-conical indenter. Since the residual indentation marks made in nanoindentation experiments are too small to analyze with an optical microscope, it relies on analyzing the load–displacement curve with a theoretical model to calculate the modulus and hardness of the sample. Nowadays, the AFM technique is combined with the nanoindentation instrument and measures the imprints (typically a few hundred nm's or less), which allows calculation of the hardness with the same calculation method used in micro-indentation experiments.

Surface and Interface Analysis: Principles and Applications, First Edition. Seong H. Kim.

Companion website: www.wiley.com/go/Kim

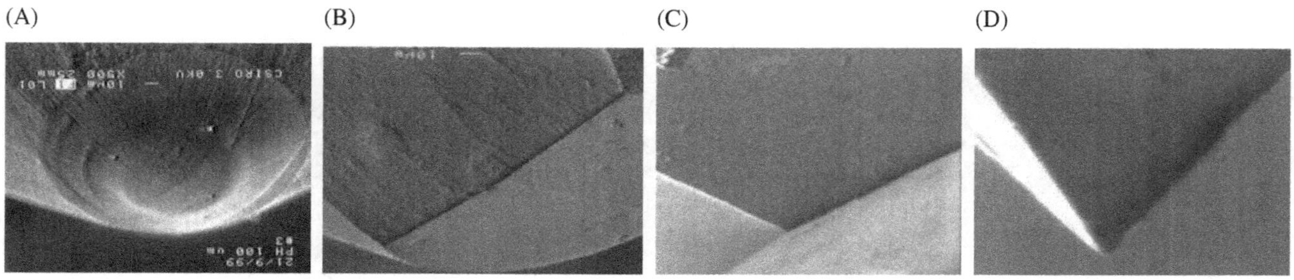

Figure 12.1 Images of (A) sphero-conical tip, (B) Knoop tip, (C) Berkovich tip, and (D) cube-corner tip. Reproduced from Fischer-Cripps [1]/with permission of Springer Nature.

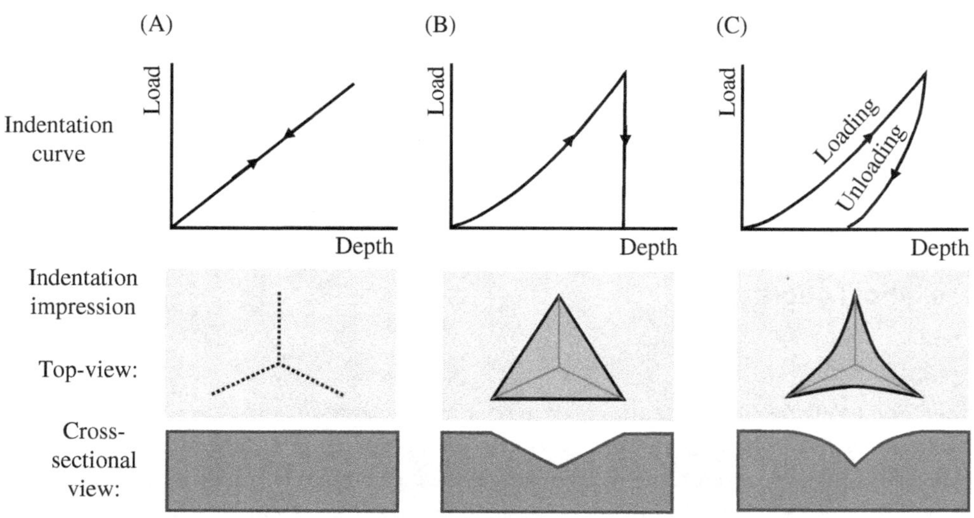

Figure 12.2 Indentation behaviors of (A) perfectly elastic, (B) perfectly plastic, and (C) elastic–plastic behavior of materials.

Figure 12.2 shows three different types of indentation curves and the top and cross-sectional views of the deformed surfaces after the removal of indenter. For an ideally elastic solid, the indented region deforms elastically according to Young's modulus. The forces measured as a function of depth of penetration (also called displacement) during loading and unloading are identical, and there is no impression of the indentation after unloading (Figure 12.2A). For a rigid perfectly plastic solid, permanent deformation occurs once the contact stress reaches the yield strength of the material. There is no elastic recovery during unloading, and the residual impression is the same as the tip shape and its volume is the same as the volume of indenter penetrated the sample at the maximum load (Figure 12.2B). In the case of elastic–plastic solids, the surface deforms elastically according to Young's modulus and then deforms plastically. The elastic deformation is recovered during unloading (Figure 12.2C). In most solid surfaces, plastic deformation is accompanied by some degree of work hardening.

12.1.3 Elastic Modulus, Hardness, and Fracture Toughness

From the indentation behaviors, elastic modulus and/or hardness of the solid surface can be calculated. Elastic modulus is a quantity measuring the resistance to being deformed *elastically* (i.e., without any permanent deformation) when a stress is applied to it. There are three types of elastic modulus. The first is a Young's modulus (E), which pertains to tensile elasticity; it is defined as the ratio of tensile stress to tensile strain. It can be applied to the compressive stress if the stress is applied uniaxially. The second is a bulk modulus (K), which pertains to volumetric elasticity under hydrostatic compression. It is defined as volumetric stress over volumetric strain; the inverse of K is the compressibility. The last is a shear modulus (G), which relates to the tendency to deform the shape at a constant volume under shear force. It is defined as shear (deviatoric) stress over shear strain. For homogeneous isotropic materials, these three moduli are connected to each other through the following relationship:

$$E = 3K(1-2\nu) = 2G(1+\nu) \tag{12.1}$$

where $\nu = \varepsilon_{\perp}/\varepsilon_{\parallel}$ is the Poisson ratio which is the strain of transversal contraction (expansion) divided by the strain of longitudinal elongation (contraction) under an axial tension (compression). The maximum possible value of ν is 0.5, which is observed for incompressible fluids that freely flow.

Hardness (H) is a measure of the resistance to *localized plastic* deformation induced by mechanical indentation. It is defined as the applied load (F_{max}) divided by the residual impression area (A); thus, it can be viewed as mean contact pressure (P_m). There are two ways to calculate the area: one is the "actual" area of the residual impression below the pristine surface (A_c; i.e., the contact area at the detachment of the indenter from the surface) and the other is the "flat" area projected to the two-dimensional reference plane (A_{pr}). It should be noted that different indentation methods define the area term differently. The hardness calculated with A_{pr} is defined as Meyer's hardness. Moreover, the plastic deformation pattern in the subsurface region varies depending on the indenter shape. So, hardness values determined from different methods can vary even for the same material [2]. Hardness can be related to yield strength (σ_Y) or ultimate tensile strength (σ_{UTS}). A widely used empirical relationship is $H \approx 3\sigma_Y$ or $H \approx 3\sigma_{UTS}$ for metallic materials; but, this is a very rough estimation and the actual proportionality can deviate a lot from this ratio [3]. For brittle ceramic materials, H is always significantly larger than $3\sigma_{UTS}$ [3].

In micro-indentation experiments of brittle samples, cracks can be found at or near the corners of the indenter. Then, indentation fracture toughness (K_c) can be calculated from residual indent size and crack length as well as the modulus and hardness. It can be compared with the conventional mode-I fracture toughness (K_{Ic}) that is defined as the critical stress intensity factor at which a sharp crack suddenly propagates rapidly under tensile stress. The indentation fracture toughness can be viewed as "equivalent to" the K_{Ic} determined from the crack opening experiment of bulk samples; but, there are some reservations in that view.

12.2 Nanoindentation

12.2.1 Load–Displacement Curve

In nanoindentation, the load–displacement curve is measured and the size of the projected area (A_{pr}) at the maximum load is *estimated* from the depth of penetration with the known geometry of the indenter. The depth of penetration varies typically from a few tens of nm up to 1~2 μm, depending on the indenter shape and the applied load. The data analysis model was originally derived from the contact between two *elastic* spheres, and it was extended to other shapes, that is, cones and pyramids on a flat surface, with some geometry correction factors. The generalized model is derived by Oliver and Pharr, and thus called Oliver–Pharr method [4]. As mentioned before, the main reasons to rely on the mathematical model analyzing the load–displacement curve are because the nanoindentation method was developed before AFM became popular. *Ex situ* scanning electron microscope (SEM) analysis could have been done to measure the indentation size; but, translating the sample from the nanoindentation instrument to SEM back and forth is inconvenient. Nowadays, an *in situ* nanoindentation unit mountable inside SEM instruments is available for real time monitoring of the indentation process and measuring the area of the impressed region. Alternatively, the nanoindenter tip can be used in the contact-scan mode, similar to the AFM operation, to measure the area of the impressed region projected to the pristine surface plane.

Figure 12.3A shows a typical indentation curve measured for solid surfaces undergoing elastic–plastic deformations. In an experiment, either the maximum load (F_{max}) or the maximum penetration depth (h_{max}) is preset as a control parameter and the data is collected while ramping up the load or penetration depth from zero to the set value and then back to zero. When the indenter is retracted after reaching the preset maximum value, the solid surface attempts to regain its original shape through the relaxation of elastic strains in the subsurface region; but, full recovery does not occur because plastic deformation has already occurred during the loading, leaving residual impression. The analysis of the initial portion of the unloading response gives an estimate of the elastic modulus of the indented material. During unloading, the elastic recovery force exerted by the surface to the indenter becomes zero before the indenter is fully retracted to the original zero displacement. Again, this is because the sample surface is permanently deformed, and this depth of imprint is called residual depth (h_r).

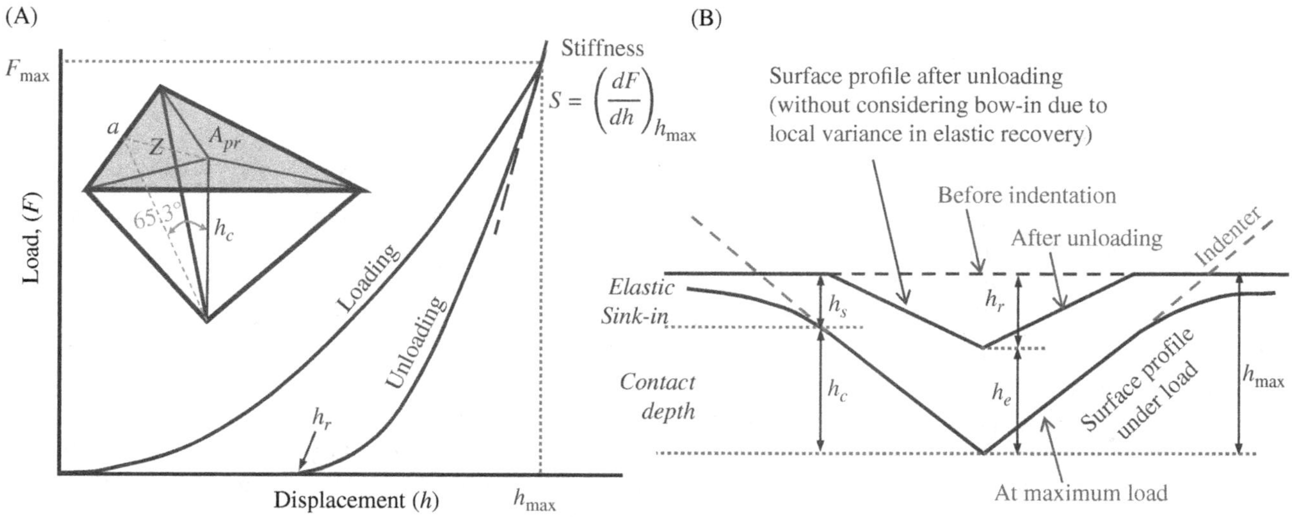

Figure 12.3 (A) Typical load–displacement curve measured during loading and unloading cycles and (B) schematic illustration of the depression of the sample surface during the indentation and after unloading. The inset in (A) shows the imprint geometry made by a Berkovich tip.

The difference between h_{max} and h_r is the elastic recovery depth (h_e). The initial slope of the unloading part at the h_{max} depth is called stiffness of the indented surface, $S = \frac{dF}{dh}$.

Figure 12.3B schematically illustrates the deformation pattern of the surface at the maximum load and after unloading. The solid surface around the indented region sinks from its original plane, following its elastic property, as the indenter penetrates deeper into the sample. The depth from the original surface at which the indenter tip and the sample surface meets is called elastic sink-in depth (h_s). Then, the difference between h_{max} and h_s is the contact depth (h_c). The Oliver–Pharr method calculates the projected area (A_{pr}) at the onset of unloading from h_c and S.

12.2.2 Modulus and Hardness Calculation from Indentation Curve

Based on the elastic contact mechanics for a cone geometry, the reduced modulus (E_r) of the contact is determined from the contact stiffness (S) at the onset of unloading from the maximum load and the contact area (A_{pr}) at that condition using the following relationship [1, 5]:

$$E_r = \frac{1}{2\beta} S \sqrt{\frac{\pi}{A_{pr}}} \quad \text{with} \quad \frac{1}{E_r} = \frac{1-\nu_s^2}{E_s} + \frac{1-\nu_t^2}{E_t} \tag{12.2}$$

where E_s and ν_s are the elastic modulus and Poisson ratio of the sample, respectively, and E_t and ν_t are those of the tip. For diamond, E_t is 1140 GPa and ν_t is 0.07. This is basically the same as eq.(11.20), except the β term. The β term is a geometrical correction factor for the indenter shape ($\beta = 1$ for the rigid-axisymmetric indenter such as a conical tip) [5]. The S value can be determined by either (i) fitting the upper portion of the unloading curve with a straight line or (ii) fitting the entire unloading data with $P = b(h - h_r)^m$ and then analytically evaluating the derivative at F_{max} and h_{max}. Since the hardness is defined as the mean contact pressure, the nanoindentation hardness can be calculated as:

$$H_N = \frac{F_{max}}{A_{pr}} \tag{12.3}$$

Then, all we need for calculation of E_r and H_N of the sample is the relationship between A_{pr} at the maximum indentation depth (h_{max}) under the maximum load (F_{max}). E_s can be calculated from E_r if we know (or can guess) ν_s.

The mathematical expression for A_{pr} can be derived as a function of the contact height (h_c), i.e., $A_{pr} = f(h_c)$. The elastic model predicts the sink-in depth (h_s) at the full load is F_{max}/S times an intercept correction factor (ε) [5]. Then, the contact height at the full load can be estimated to be:

$$h_c = h_{max} - \varepsilon \frac{F_{max}}{S} \tag{12.4}$$

Table 12.1 Equations and parameters used to calculate A_{pr}, h_c, and S in nanoindentation.

Indenter type	Projected area A_{pr}	Semi-angle, θ (deg)	Effective cone angle, α (deg)	Intercept correction factor, ε	Geometry correction factor, β
Sphere	$2\pi R_{tip}h_c$	n/a	n/a	0.75	1
Cone	$\pi h_c^2 \tan^2\alpha$	α	α	0.727	1
Berkovich	$3\sqrt{3}h_c^2 \tan^2\theta$	65.27°	70.3°	0.75	1.034
Vickers	$4h_c^2 \tan^2\theta$	68°	70.3°	0.75	1.012
Knoop	$2h_c^2 \tan\theta_1 \tan\theta_2$	65°, 86.25°	77.64°	0.75	1.012
Cube corner	$3\sqrt{3}h_c^2 \tan^2\theta$	35.26°	42.28°	0.75	1.034

Source: Fischer-Cripps [1]/Springer Nature.

The ε for a given shape of the indenter can be found in Table 12.1. Once h_c is estimated, then A_{pr} can be calculated from the tip geometry. For a spherical tip with a radius of R_{tip}, the contact radius (a) and the contact height (h_c) follows $a^2 + (R_{tip} - h_c)^2 = R_{tip}^2$. Thus, the projected area becomes:

$$A_{pr} = \pi a^2 = \pi(2R_{tip}h_c - h_c^2) \approx 2\pi R_{tip}h_c \tag{12.5}$$

For a conical tip with a half-angle of α, the projected area at a given h_c is calculated to be:

$$A_{pr} = \pi a^2 = \pi(h_c^2 \tan^2\alpha) \tag{12.6}$$

For a Berkovich tip with a semi-angle of the face of 65.27°, the imprint has the equilateral triangular shape. If the side length is a, its area is $\frac{\sqrt{3}}{4}a^2$. In the inset of Figure 12.3A, $\tan 65.27^o = \frac{z}{h_c}$ and $\tan 30^o = \frac{z}{2/a}$. Thus, we get the following relationship:

$$A_{pr} = \frac{\sqrt{3}}{4}a^2 = 3\sqrt{3}h_c^2 \tan^2 65.27^o \approx 24.5h_c^2 \tag{12.7}$$

Table 12.1 summarizes the projected areas, the semi-angle of the indenter face (θ), the intercept correction factor (ε) used in eq. (12.4), and the geometry correction factor (β) used in eq. (12.2). It also shows the effective cone angle (α) for pyramidal indenters, which allows comparison with the axially symmetric conical tip. Using the parameters and equations shown here, the elastic modulus (E_s) and hardness (H_N) of the solid surface can be calculated from the measured load-displacement curve.

12.2.3 Factors Affecting Nanoindentation

Note that the equations shown in Table 12.1 for the calculation of A_{pr} from the estimated h_c are obtained assuming that the indenter shape follows the ideal geometry. In reality, the tip shape is rarely ideal. As it is used, the tip end gets blunter (see the inset in Figure 12.4) [6]. To account for nonideal indenter shape, it is necessary to apply a correction factor, that is, the ratio of actual area to ideal area (A_a/A_i), to the equations shown in Table 12.1. For that, the indenter shape should be the images with AFM or SEM. Alternatively, the area correction can be done by measuring the load-displacement curves for a set of varying h_{max} (so, varying h_c) against a reference material with a known hardness (e.g., fused quartz which has $H = 9.25$ GPa and $E = 70$ GPa) and calculate the A_{pr} and h_c using eqs. (12.3) and (12.4) [7, 8]. Then, we can fit the data with a polynomial function:

$$A_{pr}(h_c) = C_1h_c^2 + C_2h_c + C_3h_c^{1/2} + \cdots \tag{12.8}$$

where the first term represents the ideal area function of the indenter (e.g., $C_1 = 24.5$ for a Berkovich indenter).

In addition to the actual area correction, there are various corrections to be made in practice [9]. The indenter penetration is ideally measured from the load-free surface. But, as in the AFM case, the exact location of the surface is not known until

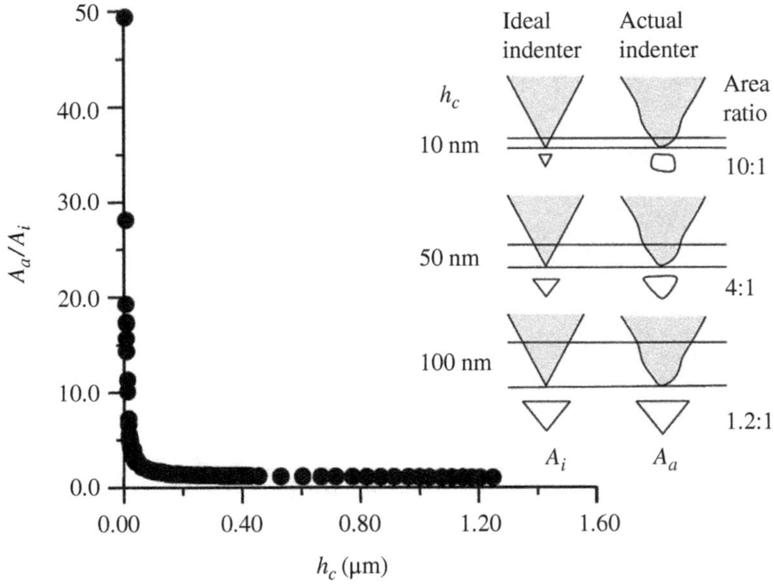

Figure 12.4 Schematic illustration of the projected area of an ideal Berkovich tip (A_i) and the actual area (A_a) of a hypothetical nonideal shape tip, and variance of area correction (A_a/A_i) as a function of h_c. Reproduced with permission from Fischer-Cripps [6]/Springer Nature.

the tip starts penetrating the sample. This initial penetration depth needs to be estimated and added to all the displacement data. When the sample surface is quite stiff, the application of load to the sample may cause deflection of load frame. The instrumental compliance needs to be measured and taken into account for an accurate measurement of the displacement. Even for a homogenous isotropic material, the calculated hardness may vary with indentation depth due to the surface roughness or the presence of residual stress, native oxide, or polishing defects in the subsurface region. This is called indentation size effect (ISE). Often, it is difficult to distinguish ISE from the nonideal indenter shape effect. Plastic flow of material during indentation can cause a pile-up around the indenter above the original surface, and work hardening can lead to a sink-in around the indenter (Figure 12.5) [10]. Then, it may increase or decrease the actual contact area. Based on simulations, the pile-up effect is predicted to be significant when h_r/h_{max} is larger than 0.7 [5, 10]. Also, the elastic modulus

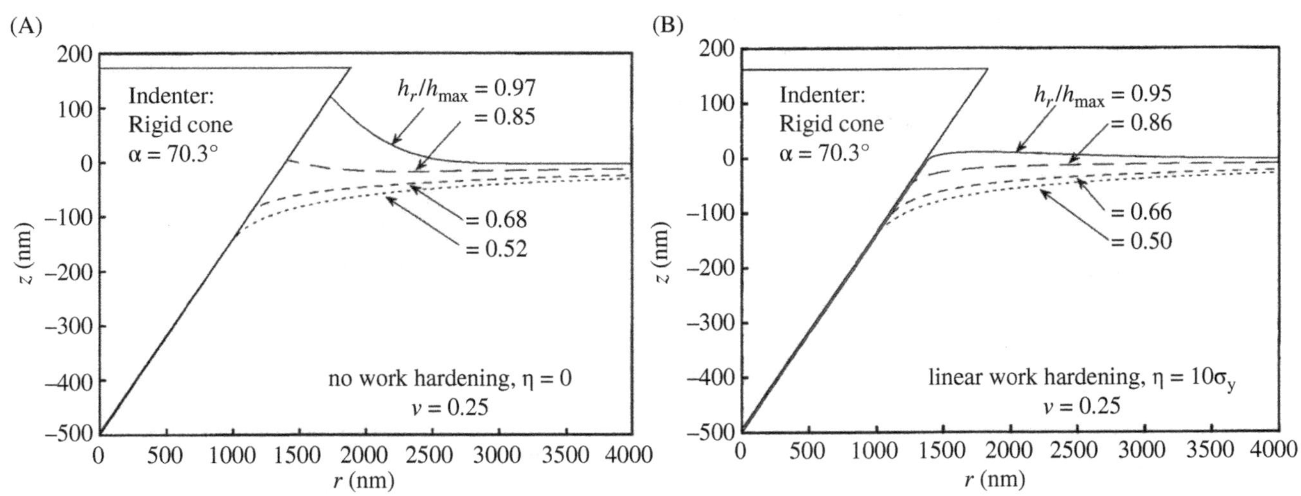

Figure 12.5 Finite-element simulation of contact profiles under load for conical indentation of (A) nonwork hardening materials and (B) linear work hardening materials with work hardening rate, $\eta = 10\,\sigma_y$. Reproduced with permission from Bolshakov and Pharr [10]/Springer Nature.

determined from the nanoindentation may be different from the true elastic modulus of the material, if work hardening occurs during the load cycle.

The data processing method described in Section 12.2.2 assumes that plastic deformation occurs instantaneously. It also assumes that solid materials behave in an elastic–plastic manner and do not exhibit any time-dependent behavior or load-rate dependence. However, solid materials may exhibit a creep behavior. Creep occurs (usually slowly) over time as a result of the diffusion of atoms or irreversible movement of dislocations under the indentation stress field. Then, the modulus and hardness values determined from nanoindentation will vary with the indentation rate or the hold time at the maximum load before unloading starts [11–13].

Since its penetration depth is shallow, nanoindentation can be used to determine mechanical properties of thin films. In this case, the indentation scales with the ratio of contact radius to film thickness (a/t) as well as the ratio of penetration depth to film thickness (h/t). For hard films on soft substrates, the former plays an important role since a is approximately equal to the radius of the hydrostatic core beneath the indenter. For soft films on hard substrates, the latter is more important because it gauges how closely the indenter approaches the substrate. In any case, nanoindentation measurements of the layered materials will have contributions from both film (top layer) and substrate.

There are several models to predict the effective modulus and hardness of the layered materials. The combined modulus (E_{eff}) of the film, substrate, indenter system can be expressed as [14, 15]:

$$E_{eff} = \left[\frac{1-\nu_f^2}{E_f}\left(1-e^{-\alpha t/\sqrt{A}}\right) + \frac{1-\nu_s^2}{E_s}\left(e^{-\alpha t/\sqrt{A}}\right) + \frac{1-\nu_i^2}{E_i}\right]^{-1} \tag{12.9}$$

where subscripts f, s, and i refer to the film, substrate, and indenter, respectively, t is the film thickness, A is the contact area, and α is an empirical constant which is a function of the contact radius to film thickness ratio (a/t). Another empirical relationship used widely is [16]:

$$E_{eff} = E_s + \left(\frac{E_f}{E_s}\right)^L \text{ with } L = \frac{1}{1+B(h/t)^C} \tag{12.10}$$

where B and C are adjustable coefficients obtained by calibration on reference samples of known elastic modulus.

For hardness of layered materials, the following equations could be used [17]:

$$\text{Soft film on hard substrate: } H_{eff} = H_s + \left(H_f - H_s\right)\exp\left(-\frac{\sigma_{Y,f}/\sigma_{Y,s}}{E_f/E_s}\left(\frac{h}{t}\right)^2\right) \tag{12.11}$$

$$\text{Hard film on soft substrate: } H_{eff} = H_s + \left(H_f - H_s\right)\exp\left(-\frac{H_f/H_s}{\left(\sigma_{Y,f}/\sigma_{Y,s}\right)\sqrt{E_f/E_s}}\left(\frac{h}{t}\right)\right) \tag{12.12}$$

where $\sigma_{Y,f}$ and $\sigma_{Y,s}$ are the yield strength of film and substrate, respectively. Another widely used model is in the following form [18]:

$$H_{eff} = H_s + \frac{H_f - H_s}{1+k(h/t)^2} \tag{12.13}$$

where k is an empirical parameter that can be adjusted for data fitting. Since eq. (12.13) has only two unknowns (H_f and k), one can determine H_f by fitting experimental data obtained with varying h/t over a wide range (from ~0.1 to >1) [19]. There are other models proposed to describe the modulus and hardness of layered samples. In general, it is accepted that if the penetration depth is less than 10% of the film thickness (i.e., $h_{max}/t < 0.1$), the contribution from the substrate is negligible in the modulus and hardness. Keep in mind that this is just a rule of thumb and not always valid.

12.3 Micro-indentation

Micro-indentation is generally performed to quantify local variations in hardness over a length scale on the order of 1–10 μm. The applied load typically varies from a few gram-force (gf) to 1 kgf [20], and the indenter penetration depth is on the order of 1–10 μm. The dimension of the residual indent is measured with an optical microscope. Two widely used indenters are

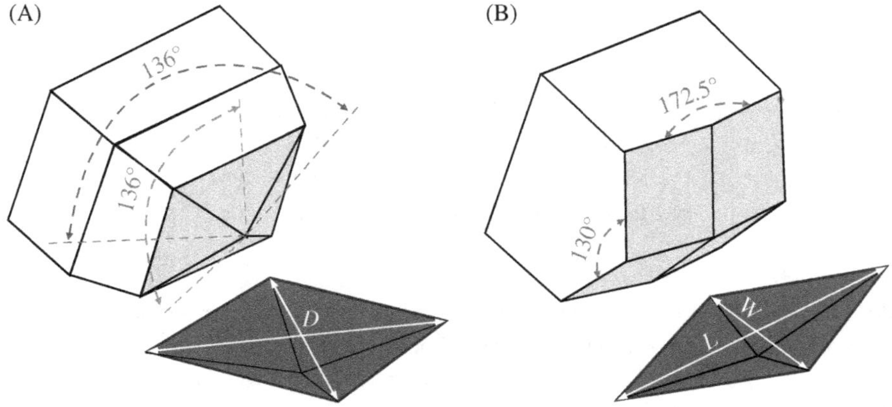

Figure 12.6 Geometry of (A) Vickers and (B) Knoop indenters and the residual impression.

Vickers with a diamond pyramid shape and Knoop with a rhombic pyramid shape (Figure 12.6). In general, the Vickers indenter is better suited for determining bulk properties because Vickers hardness is less sensitive to the test force than Knoop hardness. The Knoop indenter is useful for the study of hard materials because the long diagonal length of the residual indent of a small depth can be measured easily. However, the Knoop indentation is not geometrically identical as a function of depth, and thus the result may vary at test forces less than 200 gf [20]. It should be noted that both methods are susceptible to imprecision at forces less than 25 gf due to the difficulty in measuring extremely small indents (<10 μm width) using optical microscopes with high precision and the deviation of the indenter geometry from the ideal shape. And the ISE may appear at low test forces (see Section 12.4.4).

12.3.1 Vickers Indentation

Vickers hardness (H_V) is defined as the indenter loading (F) divided by the actual area of the residual imprint (A_c) which is calculated as:

$$A_c = \frac{D^2}{2} \frac{1}{\sin\left(\frac{136^o}{2}\right)} \tag{12.14}$$

where D is the diagonal length of the residual imprint and 136° is the angle of two opposite faces. If F is measured in kgf and D is measured in mm, H_V in the unit of kgf/mm^2 is obtained using:

$$H_V = \frac{F}{A_c} = 1.8544 \frac{F(\text{kgf})}{D^2(\text{mm})^2} \tag{12.15}$$

This can be converted to the SI unit (N/m^2) by simply multiplying the standard gravity: H (MPa) = 9.8 H_V. Vickers hardness can also be converted to the mean contact pressure (which is the definition of Meyer's hardness) based on the projected area of the indent ($A_{pr} = D^2/2$) for direct comparison with other hardness tests.

Although the A_c calculated through eq. (12.14) is called "actual" area of the residual imprint, it is not true surface area of the imprint topography. Figure 12.7A shows the optical profilometry image of a Vickers indent made on soda lime silica (SLS) glass and the line profiles along the two main axes of the image [21]. From these data, we can see that, due to elastic recovery upon unloading of the indenter, the residual indent volume is smaller than the geometric volume used for the calculation of A_c assuming a perfectly plastic indent (as shown in Figure 12.2B). Also, there is a pileup due to shear flow during the plastic deformation, which also affects the indenter-surface contact area but is not considered in the calculation of A_c.

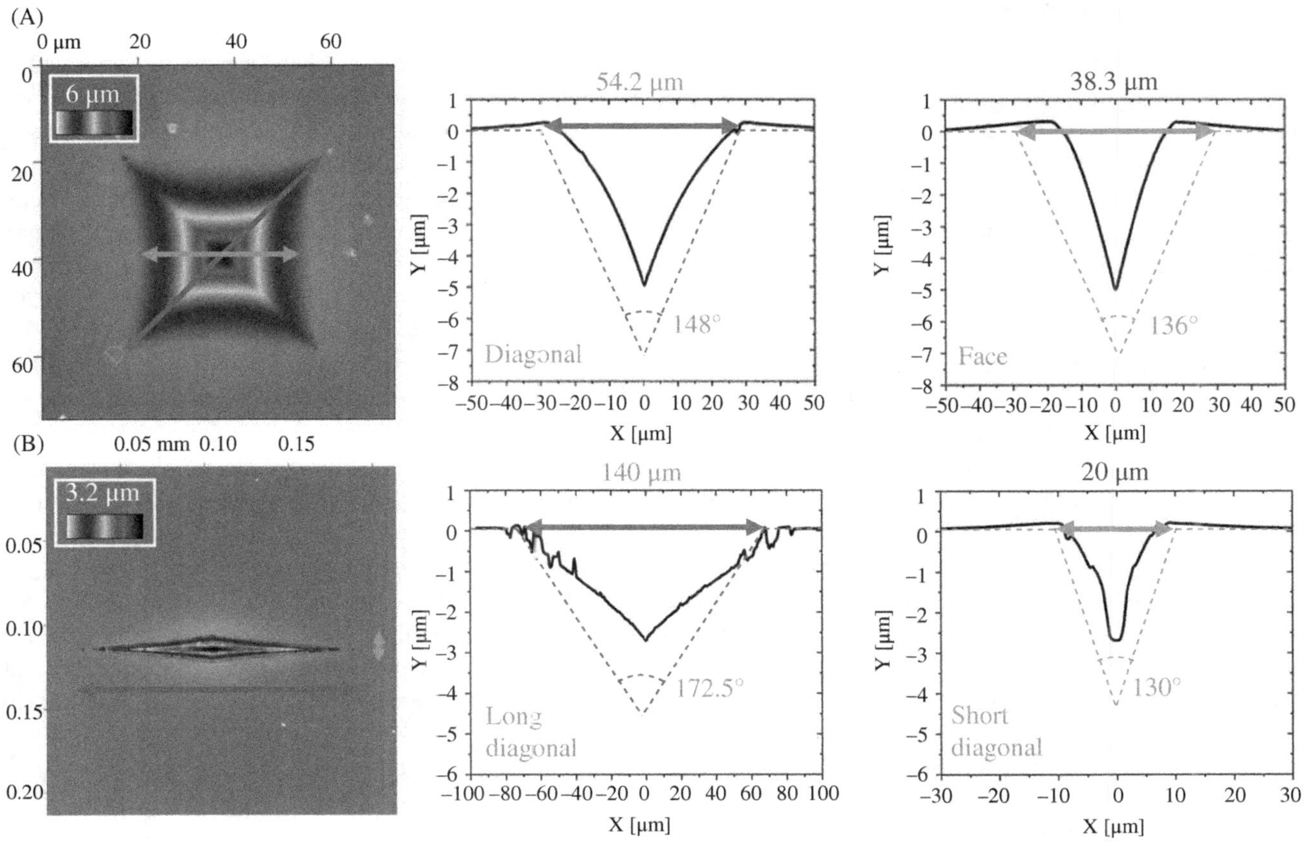

Figure 12.7 Optical profilometry image and line profiles along the two main axes marked in the image of (A) Vickers. (B) Knoop indents. The color scale in the top left corner of the profilometry image gives the depth of the indented surface. The dotted lines show the hypothetical indentation profile without any elastic recovery (i.e., perfectly mirroring the shape of the indenter). Reproduced with permission from Varshneya et al. [21]/John Wiley & Sons.

12.3.2 Knoop Indentation

Knoop hardness (H_K) is defined as the indenter loading (F) divided by the projected area of the residual imprint (A_{pr}) which is calculated from the length of the longer diagonal (L) and the two angles of the opposite faces (172.5° and 130°):

$$A_{pr} = \frac{LW}{2} = \frac{L^2}{2} \frac{\tan\left(\frac{130^o}{2}\right)}{\tan\left(\frac{172.5^o}{2}\right)} \tag{12.16}$$

If F is measured in kgf and D is measured in mm, H_K in the unit of kgf/mm^2 is obtained using:

$$H_K = \frac{F}{A_{pr}} = 14.229 \frac{F(\text{kgf})}{L^2(\text{mm})^2} \tag{12.17}$$

The advantage of Knoop hardness over Vickers hardness is that it makes much shallower indent so more suitable for measurements of brittle materials. The residual imprint of Knoop indention also shows the elastic recovery and pile-up behaviors, so the actual topography of the residual imprint is different from the geometry of the indenter (Figure 12.7B).

12.3.3 Indentation Fracture Toughness

Brittle materials such as ceramics or oxide glasses can undergo local fractures, forming cracks, during indentation with a pyramidal-shape indenter. Upon loading, lateral tensile stresses are induced inside the sample as the radius of plastic zone increases. Upon unloading, additional stresses arise as the elastically strained region outside the plastic zone attempts to recover its original shape but is blocked by the permanent deformation of the plastic zone. The cracking pattern varies depending on not only the material property but also the indenter shape, loading rate, and test environment. Figure 12.8 schematically illustrates general shapes of indentation cracks made by a pyramidal shape indenter. Median crack (Figure 12.8A) is the one formed along the load axis beneath the plastic zone due to the wedging effect of a pyramidal indenter. Median cracks are typically formed during the loading. Radial cracks are vertical cracks propagating from the edge of the plastic impression driven by tensile hoop stress (Figure 12.8B). They extend downward into the subsurface region but are usually quite shallow. Radial cracks can form during loading or unloading, depending on the load condition and material type. Half-penny cracks are usually formed upon unloading and start from the median crack running upwards, which usually join with the radial cracks, but not always. By simple visual inspection of the crack pattern appearing at the top surface of the sample after the removal of the indenter, it is difficult to distinguish radial and half-penny cracks. Lateral cracks are generated beneath the plastic zone and run almost parallel to the surface (Figure 12.8D). If lateral cracks propagate upward to the surface, they form chips. For most brittle materials, lateral cracks are usually formed during the unloading cycle.

When indentation cracks are observed from the corners of pyramidal imprints, one can estimate fracture toughness (K_c) from the indentation crack length. Typically, fracture toughness is calculated from the bending test of a pre-notched or pre-cracked beam, but such tests are very difficult to undertake with brittle materials. Attempts to make test specimens of brittle materials into a specific shape usually end up with catastrophic failures. Instead, indentation cracks can easily be made and take only a small surface area for testing. For that reason, several empirical methods have developed and been used widely with some reservations.

Lawn, Evans, and Marshall found a correlation, for ceramics and glasses, between the fracture toughness and $F/c^{3/2}$, where c is the length of crack measured from the center of crack (Figure 12.8E) and constructed the following empirical equation for cases where $c > 2a$:

$$K_c = k\left(\frac{E}{H}\right)^{1/2} \frac{F}{c^{3/2}} \tag{12.18}$$

where K_c is in the $\mathrm{Pa \cdot m^{1/2}}$ unit, k is the geometry-related empirical constant (0.016 for Vickers or Berkovich when F and c are entered in the N and m scales, respectively). Note that hardness used in eq. (12.19) is Meyer's hardness calculated with the projected area (i.e., F/A_{pr}) [22, 23]; thus, Vickers hardness needs to be converted to the mean contact pressure

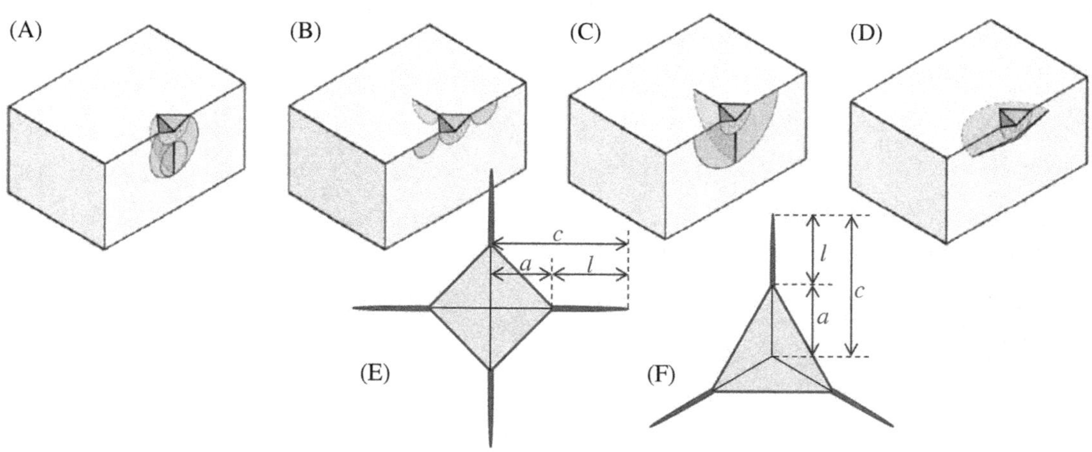

Figure 12.8 Top and cross-sectional view of (A) median, (B) Palmqvist radial, (C) halfpenny, and (D) lateral cracks generated with Vickers indent. The lightly shaded region directly below the indent mark shows the plastic zone. Reproduced with permission from Varshneya et al. [21]/John Wiley & Sons. (E,F) Characteristic dimensions of cracks made with (E) square and (F) triangular pyramid indenters.

($=2F/D^2$; see Section 12.3.1). Laugier analyzed a series of data sets available in the literature and derived the following equation for the Vickers indent [24]:

$$K_c = 0.015\left(\frac{a}{l}\right)^{1/2}\left(\frac{E}{H}\right)^{2/3}\frac{F}{c^{3/2}} \tag{12.19}$$

For the Berkovich indent, Dukino and Swain proposed the following modification [25]:

$$K_c = 1.073 \times 0.015\left(\frac{a}{l}\right)^{1/2}\left(\frac{E}{H}\right)^{2/3}\frac{F}{c^{3/2}} \tag{12.20}$$

For the cube corner (Figure 12.8F), the equation proposed by Harding, Oliver, and Pharr can be used [26]:

$$K_c = 0.036\left(\frac{E}{H}\right)^{1/2}\frac{F}{c^{3/2}} \tag{12.21}$$

It should be noted that all these equations are for rough estimation of fracture toughness. There are many sources of error that limit the precision of the measurement [27]. The empirical correction factors in eqs. (12.18) through (12.21) were the values selected as the best compromise to match the mode-I fracture toughness (K_{Ic}) data for the materials chosen for comparison. It means the correlation implicitly assumes that the tested materials deform and fracture in a self-similar manner, which cannot be justified. Also, all models assume that cracks have semicircular shape in the sample and originate from the corners of the residual imprint; but often, there are many nonideal cracks that run off the indenter main axes or from the side of the imprint (see Section 12.4.5). Also, determining the crack end to measure the crack length (c) is often not trivial. If the sample surface has residual stress, it can also affect the crack length [28]. Even though these uncertainties complicate the calculation, indentation fracture toughness (K_c) is still widely used because of the expediency of the method. As mentioned before, it is difficult to prepare a test specimen with a specific shape and dimension using brittle samples, and indentation can be done even with a small amount of sample (as long as a flat top surface can be prepared) [27].

12.4 Hidden Factors Affecting Indentation Measurement

12.4.1 Residual Stress

Conceptually, the indentation method itself and the data analysis protocols described in Sections 12.2 and 12.3 are relatively straightforward. However, indentation analyses are prone to unexpected errors or limitations beyond what are described in Section 12.2.3 for nanoindentation, shown in Figure 12.7 for uncertainty in area calculation in micro-indentation, and mentioned in Section 12.3.3 for indentation fracture toughness calculation. It would not be possible to cover all those examples, but it would still be useful to go over some of those examples.

SLS float glass (such as a slide glass) is often chosen as a reference material for indentation tests since flat samples with a sub-nm RMS roughness are readily available [27]. Thermal tempering is an industrial process widely used to make SLS glass panels (such as automobile windshields and architectural window panels) stronger and tougher. It involves heating the glass panel above the glass transition temperature followed by fast cooling with a high-speed air stream to room temperature. Then, the cooling rate difference between the surface region and the core region creates a compressive-stress casing. Since the air flow below the panel (facing the conveyer roller) is different from that above the panel (facing the open space), the top and bottom surfaces may have different properties. Figure 12.9 compares the load–displacement curves obtained through nanoindentation with the same Berkovich tip at the air-side of SLS panel that were fully annealed (AN), tempered with facing up (T/up), and tempered with facing down (T/down) [29]. Note that perfectly flat SLS glass panels are produced via the float process in which one side of the glass melt is in contact with the molten tin bath. So, the SLS float glass panel always has two sides that have different chemical compositions and thus different mechanical and chemical properties. The tin-side data are not shown here but can be found in [29]. One may think that if the glass surface becomes stronger due to compressive stress from the thermal tempering process, it would be more difficult to penetrate and would leave smaller residual imprint (so exhibiting a higher hardness) compared to the annealed surface which does not have any compressive stress.

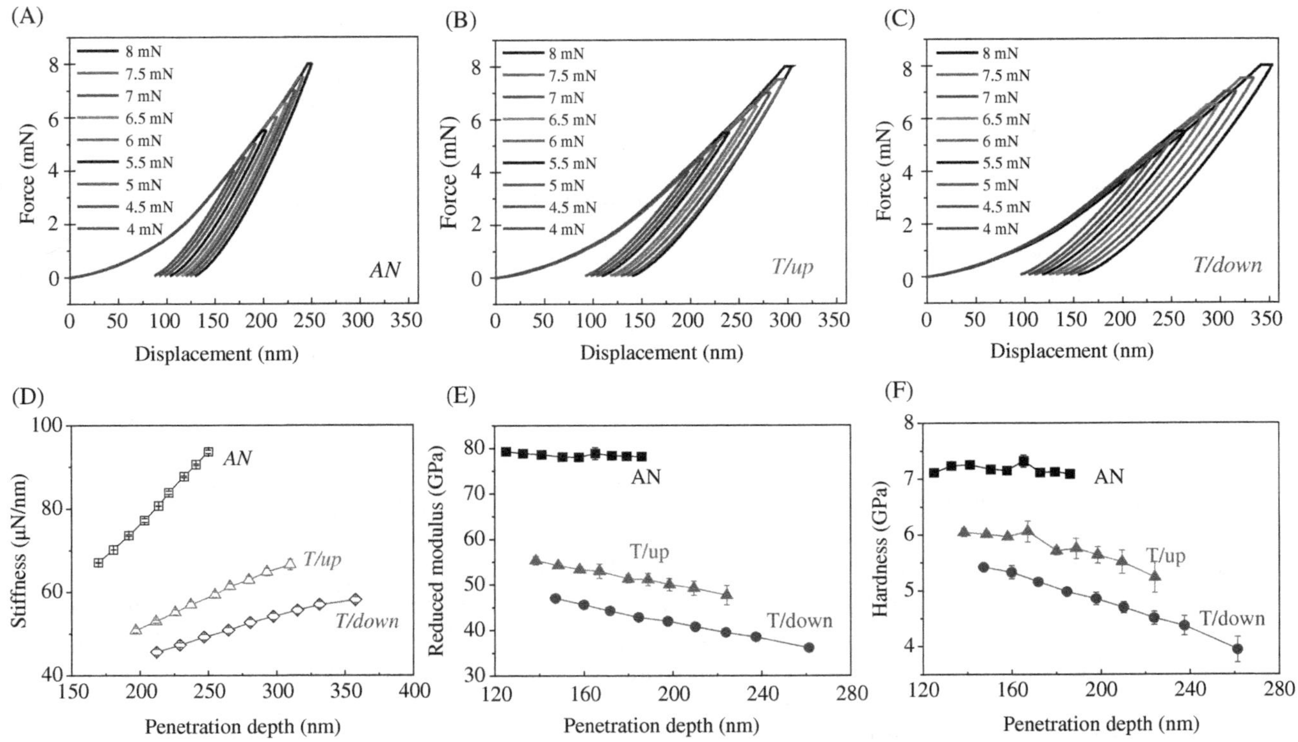

Figure 12.9 Load–displacement curves obtained from nanoindentation test with a Berkovich tip on air-side of SLS float glass panels (A) fully annealed, (B) thermally tempered with facing up, and (C) tempered with facing down. The maximum applied loads used in the test are shown in each graph. (D) Stiffness, (E) reduced modulus, and (F) hardness calculated with the Oliver–Pharr method are also shown. Reproduced with permission from He et al. [29]/John Wiley & Sons.

But, the $F - h$ curves in Figure 12.9 show the opposite. At a given maximum load (F_{max}), the indenter penetrates deeper (so larger h_{max}) into the surface tempered with facing up, even deeper into the surface tempered with facing down. The residual depth (h_r) is also larger, suggesting the imprint depth is larger. The calculated stiffness (S) upon at the onset of unloading from the F_{max} point is obviously smaller for the surface where the indenter penetrates deeper (Figure 12.9D). The reduced modulus (Figure 12.9E) and hardness (Figure 12.9F) calculated using eqs. (12.2)–(12.4), and (12.7) are surprisingly lower for the tempered surfaces than the annealed surface. But, when the Vickers hardness was measured, the differences in H_V was statistically insignificant.

Glass is a nonequilibrium material; thus, its properties (such as density) are highly dependent on thermal history. Since the annealed and tempered surfaces have experienced different cooling rates, one possible explanation for this stark difference might be the density difference caused by this difference in thermal history [30]. But, the picture is not that simple. Figure 12.10 shows the nanoindentation data for SLS glass under flexural tensile and compressive stresses [31]. The samples were fully annealed, so they have the same thermal history. The samples were slightly bent at room temperature using a four-point bending rig (Figures 12.10A, B), so that nanoindenter could access the bent surface. The bending curvature was kept very small so that the sample would not catastrophically break during nanoindentation. The flexural stress was typically 40 MPa, which is >100 times smaller than the hardness of SLS, thus far below its tensile strength [3]. Under the flexural stress condition, the indenter tip was found to penetrate deeper again, even for the compressive stress side (Figure 12.10C). When the tip–sample contact area at the maximum load was calculated from the load-displacement curve using the Oliver–Pharr method (i.e., eq. (12.7)), it was found that $A(h_c)$ is larger for the compressive stress side than the stress-free surface and the largest for the tensile stress side (Figure 12.10D). However, when the residual indent was imaged with AFM after removal of the indenter (Figure 12.10E), there was no discernable difference in the contact area (A_{pr}) or the pileup amount (Figure 12.10D). In the AFM images, the only minor difference is the residual penetration depth (h_r); it appears ~5 nm deeper for the compressive and tensile stress surfaces compared to the stress-free surface (Figure 12.10E). If we calculate the effective modulus and hardness using eqs. (12.2) and (12.3), respectively, the values are significantly lower for the tensile stress surface than others (Figure 12.10F). For the compressive stress surface, the

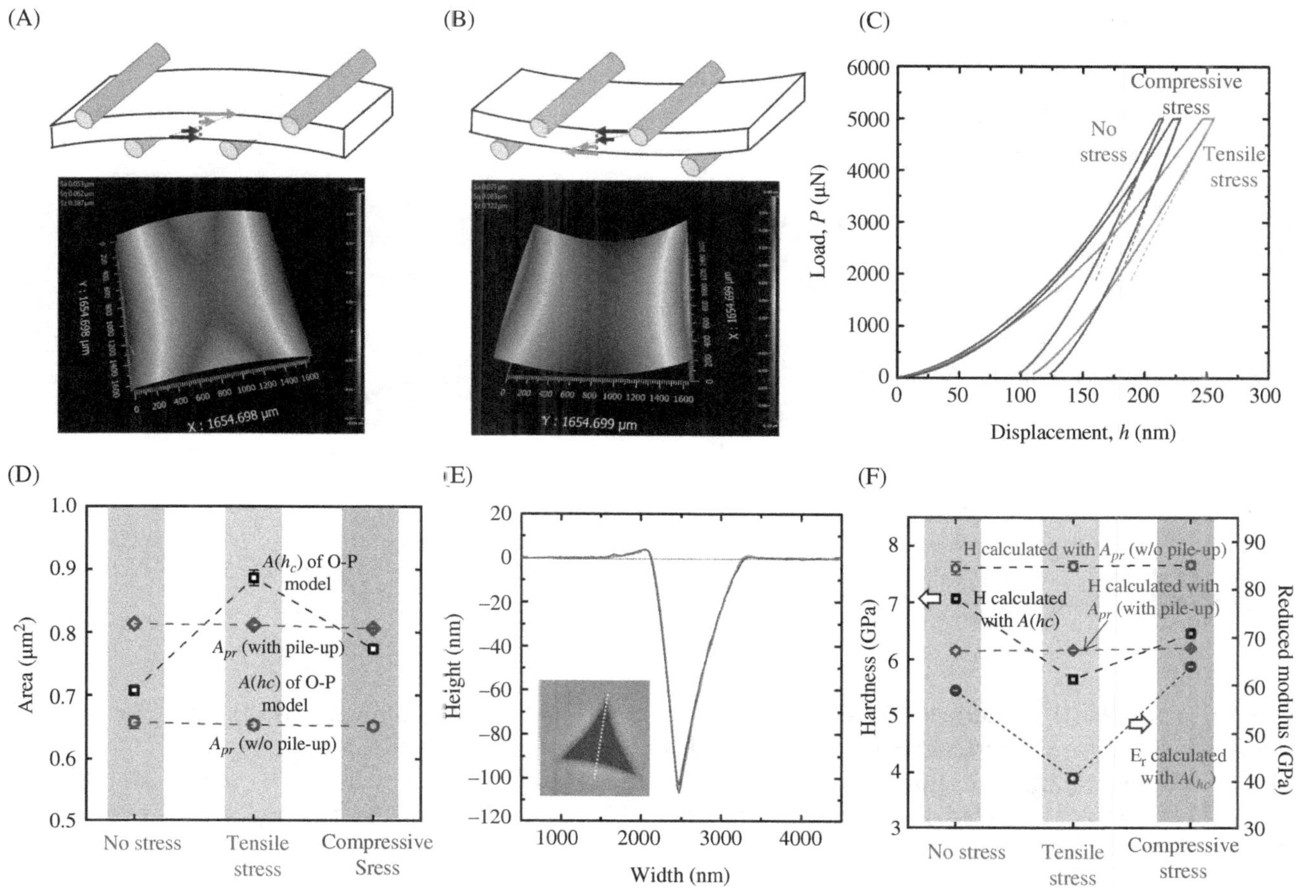

Figure 12.10 Flexural stress generation at the SLS glass surface using a four-point bending method and corresponding optical profilometry images: (A) convex surface with tensile stress. (B) Concave surface with compressive stress. The color scale bar for height is 380 nm in (A) and 320 nm in (B). (C) Load–displacement curves obtained with Berkovich-nanoindentation on stress-free, tensile-stress, and compressive-stress SLS surfaces (all air-side). (D) Projected contact area at h_c calculated Oliver–Pharr method and the area calculated from post-indentation analysis, (E) AFM image and line profile of the imprints, and (F) hardness and reduced modulus calculated with the Oliver–Pharr method and using the actual imprint area. Reproduced from Liu et al. [31]/with permission of John Wiley & Sons.

modulus appears to be slightly higher than the stress-free surface, but the hardness is lower. However, if Meyer's hardness is calculated using the projected area of the actual residual imprint (A_{pr}), the value does not vary with the flexural stress (Figure 12.10F), which makes sense because the curvature was too small to affect the contact area and the flexural stress was two orders of magnitude smaller than the hardness.

Then, what caused such a large error in hardness calculation using the Oliver–Pharr method? Recall that the Oliver–Pharr method was originally derived with the assumption that the material under investigation is homogeneous and stress-free. But, the samples analyzed in Figure 12.10 have stress gradient along the depth direction. Thus, the assumption made for derivation is not valid for these samples. It means the Oliver–Pharr method does not work well for the stressed sample, which was warned by Oliver and Pharr [32]. The same applies to the samples analyzed in Figure 12.9. For the same reasons, the reduced modulus calculated using eq. (12.2) is not accurate either. Similar effects were observed for nanoindentation in the area near the Vickers indent on SLS glass [33]. It is important to know the limitations of the Oliver–Pharr method and check its applicability in nanoindentation analysis.

12.4.2 Environmental Effects

Plastic deformation is the permanent changes occurring in the subsurface region; thus, one may think that nanoindentation measurements would be insensitive to humidity of ambient air. If the applied indentation load is extremely small and

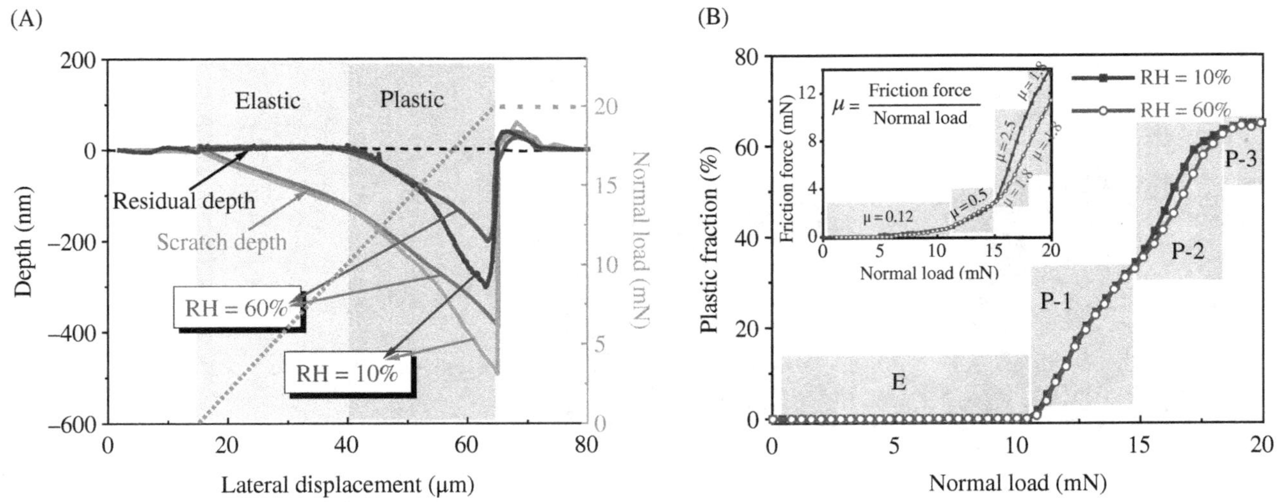

Figure 12.11 (A) Penetration depth and residual depth of nanoscratch with linearly ramping the applied load from 0 to 20 mN on SLS glass at 10% and 60% RH conditions. (B) Plastic fraction of nanoscratch as a function of applied load. The inset shows friction force measured during nanoscratch. Reproduced with permission from He et al. [35]/John Wiley & Sons.

becomes comparable to the adhesion force, the contact area as well as the contact stiffness may vary [34]; then, ambient humidity may influence the measurement since it can alter the adhesion force (see Section 11.1.1).

Nanoindentation systems can also perform nanoscratch. In that case, plastic deformation can be significantly influenced by relative humidity (RH). Figure 12.11A displays the depth during nanoscratch with a ramp load and the residual depth obtained for SLS surface in relatively dry (RH = 10%) and humid (RH = 60%) conditions [35]. Both scratch and residual depths are larger in the dry condition than the humid condition. The plastic deformation fraction can be calculated by dividing the residual depth with the scratch depth, which is plotted in Figure 12.11B. The onset of plastic deformation does not seem to be a function of RH, and the plastic fraction is not sensitive to RH in the small plastic deformation region (marked as P-1) regime. But, in the intermediate regime (P-2), it is larger in the dry condition than the humid condition, and then in the severe regime (P-3), it becomes the same again. The difference in the P-2 regime appears to correlate with the friction coefficient (inset in Figure 12.11B). It may imply that frictional energy dissipation mode at the sliding interface, which is a function of adsorbed water molecules, influences the subsurface deformation mode [35].

12.4.3 Subsurface Structural Changes in Topographically Fully Recovered Region

So far, we have defined that indentation is fully elastic if the surface topography is fully recovered to the original surface. It means there is no permanent change in the surface. But, this definition of "fully elastic" based on topography may not be sufficient. In the case of glass, which is nonequilibrium material, there could be residual structural changes in the subsurface region even if the surface is fully recovered topographically. Figure 12.12 displays the s-SNOM nano-IR analysis data of a Berkovich nanoindent area made on a silica glass surface [36]. The principle of the s-SNOM nano-IR technique is described in Section 6.6.3. In the fresh topography and scattering intensity images, triangular marks are added to show the plastic deformation region. It is noted that right outside the plastic deformation region (marked with "elastic" in Figure 12.12C), the scattering intensities at 1070 cm^{-1} and 1120 cm^{-1} are slightly lower than the pristine surface region, although the topography is fully recovered to the original surface level. The variations in IR scattering in the plastic region are understandable because the density of the silica network in the subsurface is increased (so, the topography is depressed) during the indentation. But, even in the elastic region where the surface topography is fully recovered, the IR scattering intensity is not fully recovered. This means there are residual structural changes in the subsurface beneath the elastic contact region; in other words, even though the density is the same (so, the topography is fully recovered), the glass network structure is different in the region outside the plastic deformation. This is possible because the structure of nonequilibrium material can vary depending on the history – in this case, subjected to indentation compression history. The same pattern is

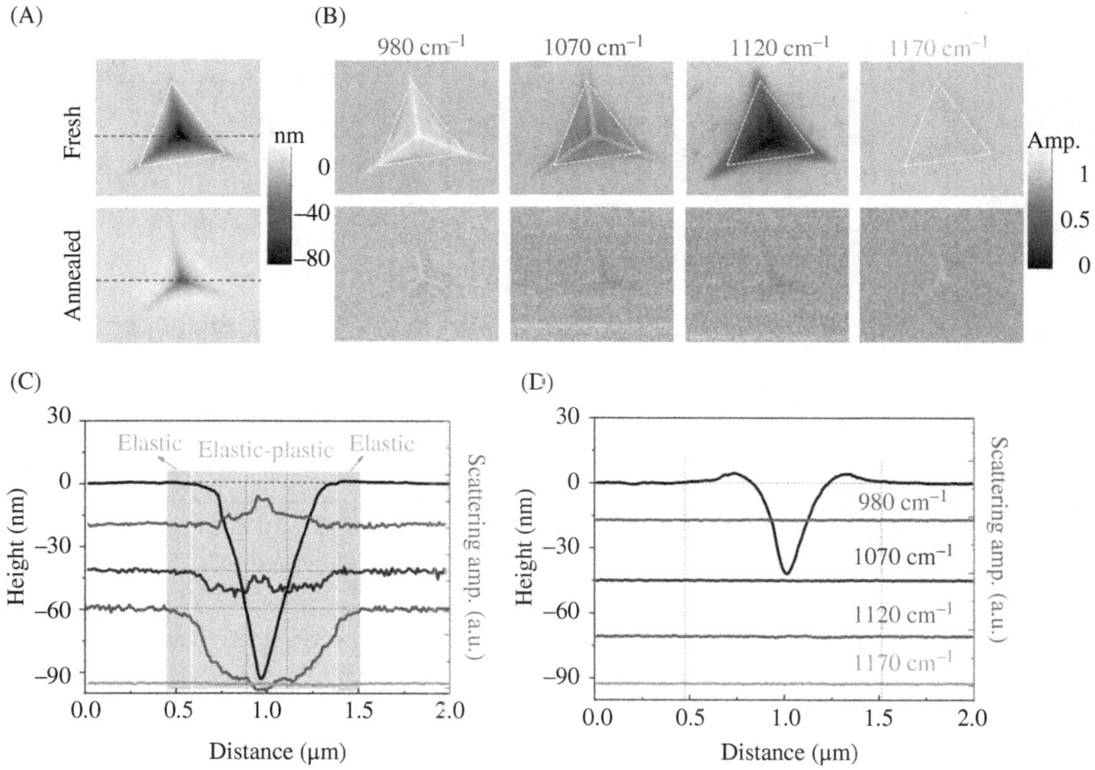

Figure 12.12 (A) Topographic images (2 μm × 2 μm) of fresh and annealed nanoindentation mark on silica glass with a Berkovich tip at an applied load of 5 mN. (B) s-SNOM S_2-amplitude images of the same region at four different IR wavenumbers. (C,D) Cross-sectional line profiles of the topography and S_2 images of (C) fresh and (D) annealed nanoindentation mark along the lines marked in (A). Reproduced from He et al. [36] with permission of Elsevier.

observed for the nanoscratch case [36]. When these surfaces are annealed at a temperature just below the glass transition temperature (T_g), all structural alterations are reverted. In that case, although some residual imprints still remain because plastic flow has occurred during indentation, the subsurface structure is fully recovered and the IR scattering behavior is the same inside and outside the annealed indent (Figure 12.12D).

It is also possible that even though the glass surface has never been touched, the externally applied stress can cause irreversible alteration of the network structure. Figure 12.13 shows the AFM-IR analysis result, along with theoretical calculation of the stress field distribution, for the Hertzian indentation region on a SLS glass surface [37]. When the glass is indented with a tungsten carbide (WC) sphere at 40 N, the center of the contact region is pressed by ~5 μm and the contact diameter (2a) is around 160 μm (Figure 12.13A). The region below the contact area is under compressive stress, but the surrounding area is under tensile stress (Figure 12.13B). When the applied Hertzian contact stress is 3.5 GPa, the maximum principal normal stress is −1.68 GPa just beneath the center of the WC sphere, and +330 MPa at the boundary of the contact area. When the indented area is analyzed with a regular optical microscope, it is not possible to see any residual indent. When it is analyzed with an optical profilometry (see Chapter 10), a tiny indent can be seen. The depth of the residual impression is only a few nanometers, so the topographic recovery of the contact region is more than 99.95% (nearly elastic). When the contact region is analyzed with photothermal AFM-IR (see Section 6.6.3), the Si–O–Si stretch band is found to be red-shifted (center and 50 μm points in Figure 12.13D) compared to the pristine surface (far away from center). This indicates the glass network in the nearly elastically indented area has a slightly longer Si–O bond length distribution which is a consequence of a slightly narrower Si–O–Si bond angle distribution [38]. About 200 μm from the center, the surface has never been touched by the WC sphere but is subjected to tensile stress (about +70 MPa) due to elastic deformation. In that region, the AFM-IR analysis reveals that the silicate network is still deformed from its original state [37]. This finding can be used to explain the variation of the nanoindentation modulus and hardness around the Vickers indent [33]; in other words, the stress gradient in the surrounding area is associated with the structural changes in the silicate network.

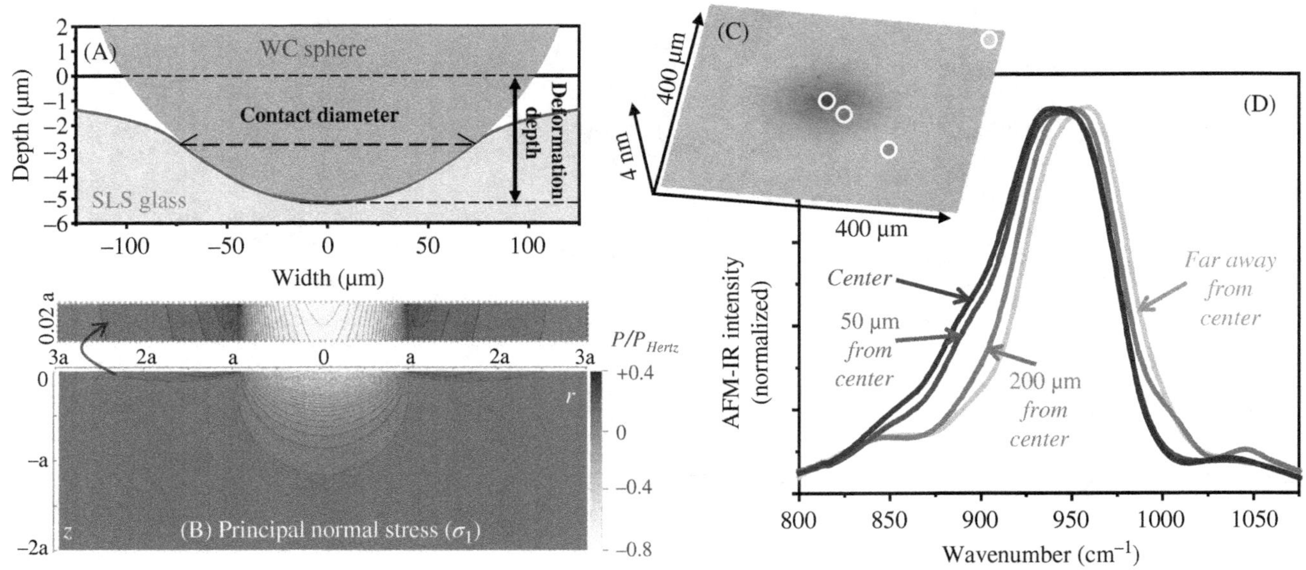

Figure 12.13 (A) Topographic deformation and (B) maximum principal stress distribution of the SLS surface upon indentation with a WC sphere at 40 N. (C) Topography of SLS surface after removal of the indenter and (D) AFM-IR spectra obtained at various locations in and around the residual indent. Reproduced from Lin et al. [37]/with permission of Elsevier.

12.4.4 Indentation Size Effect

ISE can occur not only in nanoindentation but also in micro-indentation. Figure 12.14A plots the Vickers hardness measured at different loads for a series of aluminosilicate glass with different alkali or alkaline earth modifier ions [39]. The experimental data continued above the 1 kgf load, but the load dependence was negligible in that regime. Thus, the changes in hardness value with the applied load at ≤1 kgf are the ISE. One way of defining the magnitude of ISE is to calculate the difference of H_V at the lowest force used and the maximum force at which ISE stops. For the data in Figure 12.14A, it would

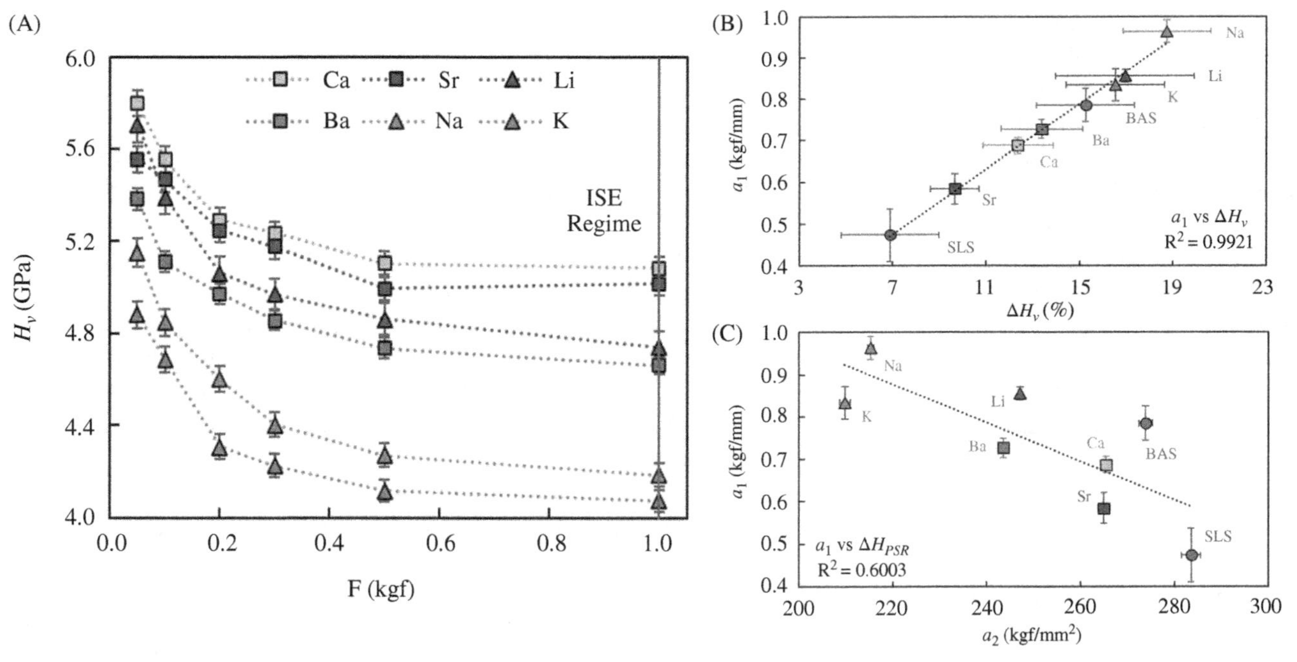

Figure 12.14 (A) Load (F) dependence of Vickers hardness (H_V) for a series of aluminosilicate glasses with different modifier ions. (B) Magnitude of ISE (a_1) from fitting vs. ΔH_V from two-point calculation. (C) Plot of a_1 vs. a_2. Reproduced with permission from Fry et al. [39]/John Wiley & Sons.

be $\Delta H_V = \left(H_V^{0.05kgf} - H_V^{1kgf}\right)/H_V^{0.05kgf}$. Another way is to linearize the data and use a linear regression. Li and Bradt introduced a normalized Meyer's relationship, which assumes that the measured hardness is the sum of the ISE component (a_1) whose effect decreases inversely proportional to the average diagonal distance (D) and the size-independent component (a_2) [40, 41]. It gives the following empirical relation: $H_V = a_1/D + a_2$. Figure 12.14B plots the a_1 from data fitting vs. the ΔH_V calculated from two data points; they correlate to each other very well. There seems a modest correlation between a_1 and a_2 (Figure 12.14C), which implies that a harder material may exhibit a smaller ISE [39]. But it should be noted that whether this correlation can be generalized to other types of materials remains to be tested.

It also should be noted that the deformation mode of the surface under indentation varies with the indentation load. The elastic deformation fraction (V_E) can be recovered immediately upon removal of the load. The plastic deformation fraction (V_P) can be viewed as the sum of isochoric densification fraction (V_D) and non-isochoric shear-flow fraction (V_S). These individual contributions can be estimated in the following method. Figure 12.15A, B compares the Vickers indent made on a aluminosilicate glass before and after sub-T_g annealing [42]. The total geometric volume penetrated by the indent (V_G) can be calculated from the diagonal length (D) and the indenter shape (Figure 12.7). The initial imprint (V_i^-) and pile-up (V_i^+) volumes below and above the original surface plane can be calculated from the 3D image of the indent obtained with optical profilometry or AFM. The densified fraction due to the compressive stress during the indentation can be reverted by annealing. The negative impression (V_a^-) and pile-up (V_a^+) volumes of the residual indent after sub-T_g annealing are measured in the same way. Then, the volume recovery ratio (V_{Rec}) can be defined as [43]:

$$V_{Rec} = \frac{(V_i^- - V_i^+) - (V_a^- - V_a^+)}{V_i^-} \tag{12.22}$$

Then, V_E, V_D, and V_S can be expressed as:

$$V_E = 1 - (V_i^- / V_G) \tag{12.23}$$

$$V_D = V_{Rec}(V_i^- / V_G) \tag{12.24}$$

$$V_S = 1 - (V_E + V_D) \tag{12.25}$$

For the aluminosilicate glass series studied in Figure 12.14, the elastic deformation fraction gradually increases and levels off in the force regime showing the ISE (Figure 12.15C). In the same load range, the densification fraction decreases initially and then levels off (Figure 12.15D), while the shear flow fraction exhibits a trend similar to the elastic fraction over a smaller magnitude (Figure 12.15E). Overall, the plastic deformation fraction ($V_P = V_D + V_S$) decreases in the very low force regime and then become insensitive to the applied load. The initial decrease in VP with the load might seem counterintuitive; but this is somewhat expected because the elastic property of the plastically deformed region would be different from that of the pristine bulk.

12.4.5 Humidity Effect and Surface Layer on Indentation Fracture Measurement

Estimating fracture toughness from indentation experiments is quite sensitive to various extrinsic conditions. Although median crack is formed under the indenter, whether it propagates to the surface in the half-penny crack pattern (which is necessary to use eq. (12.18)) can vary with the test environment and the surface condition. Figure 12.16A shows crack events monitored by taking a video after removal of the Vickers indenter (load = 1.96 N, holding time = 15 s) in the respective environment marked at the top of each panel [44]. When the indentation experiment was carried out in a dry condition, then the radial or half-penny cracks are rarely observed. For the SLS sample leached in pH 1 solution for 72 hours (so, forming ~147-nm thick sodium-depleted layer; determined with ellipsometry), no crack was observed over the 7-min observation time in dry condition. But, cracks were observed immediately for the annealed SLS surface as soon as the indenter was removed. The cracking behavior was found to be drastically different in the humid condition. For the acid-leached samples, cracks gradually popped to the surface over time after retrieval of the indenter, while all cracks were observed for the annealed samples as soon as the indenter was retrieved. The humidity-induced cracking event could be confirmed by performing indentation in a dry condition and then switching to a humid condition while continuously monitoring the residual impression.

When the median cracks propagate to the surface, they do not always behave ideally (i.e., meeting the indent corner, as schematically drawn in Figure 12.8). There are many nonideal cracks formed and, in the case of SLS glass, their abundances are very sensitive to the presence of sodium-leached layer. For acid-leached samples, the number of nonideal cracks and the

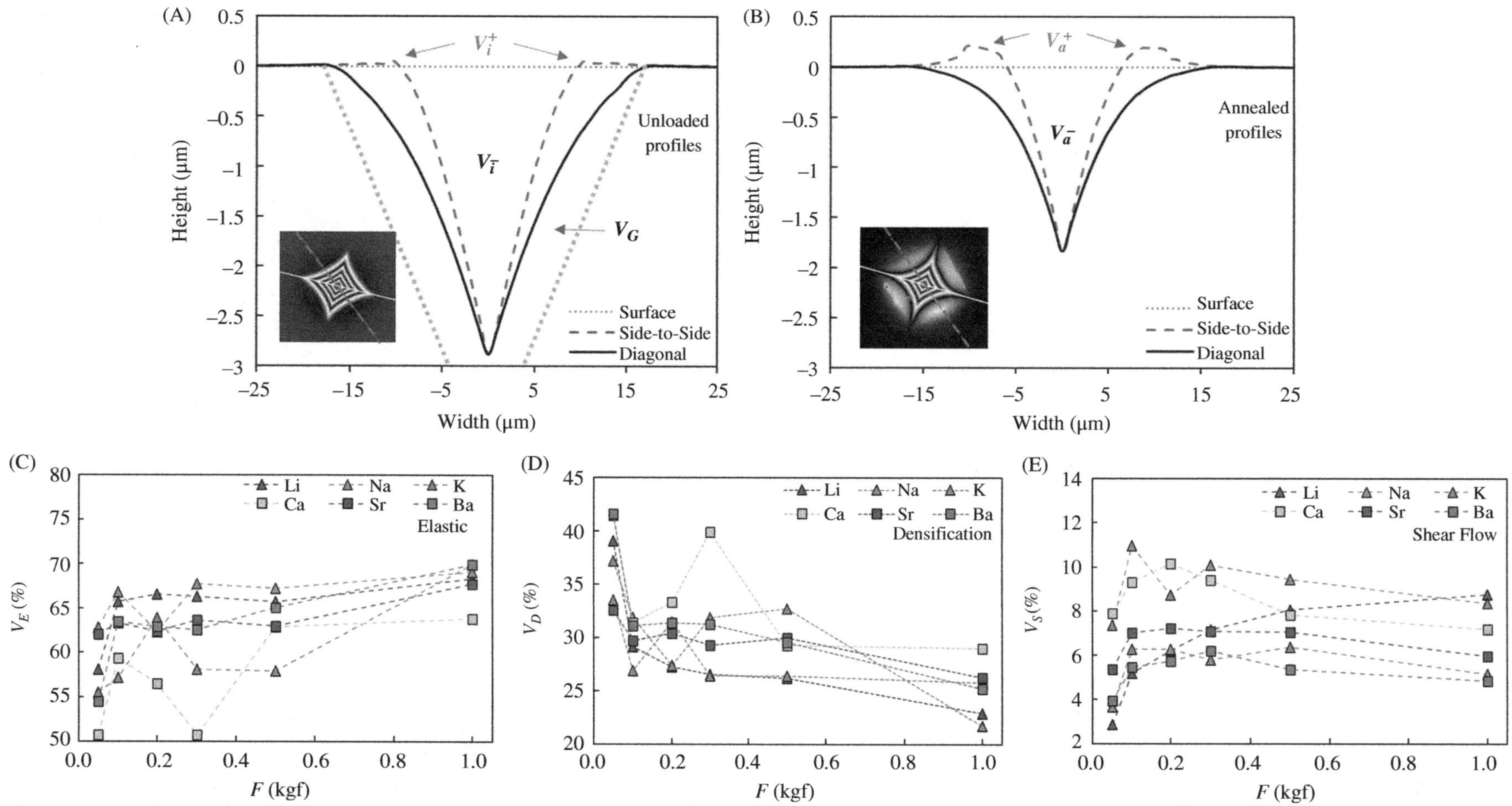

Figure 12.15 Optical profilometry images and line profiles of a Vickers indent (A) before and (B) indent after sub-T_g annealing. Volume fractions of (C) elastic deformation, (D) densification, and (E) shear flow for a series of aluminosilicate glasses. Reproduced with permission from Fry et al. [42]/John Wiley & Sons.

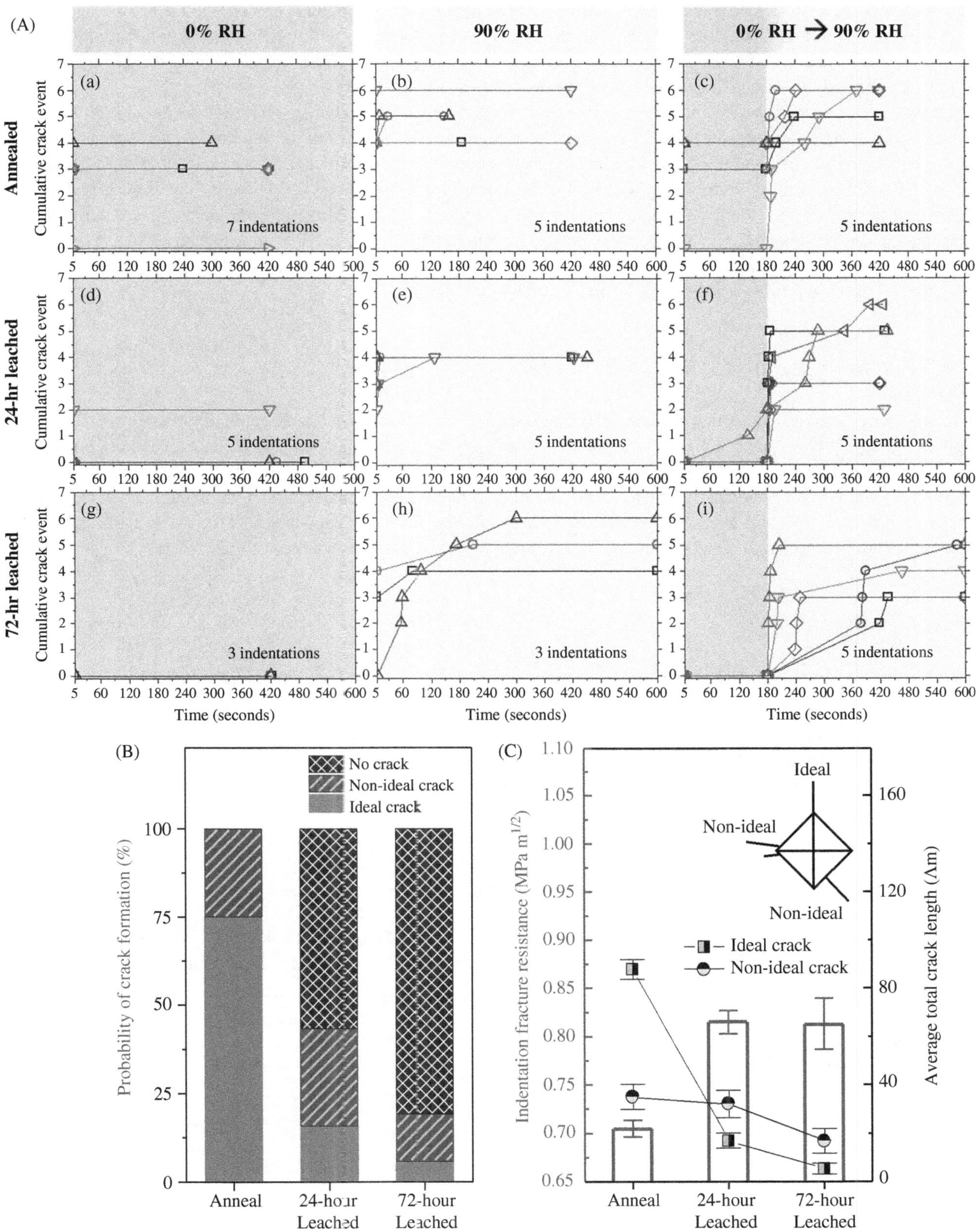

Figure 12.16 (A) Crack events as a function of time for (a–c) annealed, (d–f) 24-hour acid-leached, (g–i) 72-hour acid-leached SLS bottle glass. (B) Vickers indentation at 1.96 N for 15 second hold time at 30% RH on annealed, 24-hour acid-leached, and 72-hour acid-leached SLS bottle glass. (B) Probability of ideal, nonideal, and no crack formation in 30% RH. (C) Total crack length per indent for ideal (square) and nonideal (circle) cracks, and calculated indentation fracture toughness (open bars). Reproduced from Sheth et al. [44]/with permission of Elsevier.

number of corners without any cracks were found to be even larger than that of ideal cracks (Figure 12.16B). The indentation fracture toughness calculation using eqs. (12.18)–(12.21) can be done only with ideal cracks. When those ideal cracks (which are minor fractions) are used for calculation, the indentation fracture toughness (K_C) is estimated to be higher for the sample with the sodium-leached layer than the annealed sample (Figure 12.16C).

When the residual imprints were analyzed with optical profilometry, it was found that the residual depth is larger for the leached samples than the annealed sample. This suggested that during indentation, the sodium-leached layer is densified and does not elastically recover upon the withdrawal of the indenter. Such densification can occur through the condensation of SiOH groups forming bridging oxygen (BO) under the compressive stress [44]. Then, it may exist on the imprint surface as a densified, silica-rich crust, which may act as a barrier for water diffusing from air to the crack tip. So, the water-mediated crack opening may not occur readily, which may explain the trend seen in Figure 12.16A. The delay in crack length growth in oil compared to the humid air condition was also reported by Anstis *et al.*, who proposed the original model for fracture toughness calculation from indentation cracking [23].

12.5 Further Reading for Nanoindentation

There are many excellent books that students can use to study nanoindentation. For example, the book by A. C. Fischer-Cripps is a good one covering from basic principles to applications and limitations [45]. The contact mechanics book by the same author would be the next one to study for deeper-level theories [46]. It would be also very helpful to review potential uncertainties and errors before performing indentation experiments [9]. Analyzing the indentation data and drawing physical conclusions from the indentation tests need to be done carefully with proper knowledge of all possible artifacts and their consequences in experimental results; some of those examples are discussed in Section 12.4.

Practice Problems

12.1 Figure 12.7 shows the optical profilometry analysis results of Vickers and Knoop indents made on a SLS glass. The applied load was 1.0 kgf for both Vickers and Knoop indentations.

a) Calculate Vickers hardness (H_V) based on the indenter geometry using eq. (12.15).
b) Recalculate Vickers hardness based on the "accrual" imprint shape; a rough estimate of the imprint area can be made from the profilometry image (ignoring the bow-in curvature of the surface). How much error is produced in H_V by ignoring the elastic recovery of the imprinted region?
c) Calculate Meyer's hardness for the Vickers indent.
d) Calculate Knoop hardness (H_K) using eq. (12.17).
e) Would Knoop hardness be affected if we consider the elastic recovery of the imprint region?

12.2 Below is a set of nanoindentation data for a silica sample ($\nu = 0.17$ using four diamond indenters with different shapes [47]. Young's modulus and Poisson ratio of diamond are 1220 GPa and 0.2, respectively.

Indenter → Load F (mN)	Sphere R = 5 μm h (mm)	Berkovich θ = 65.3 h (mm)	Vickers θ = 68.0° h (mm)	Cube corner θ = 35.3° h (mm)
0.031	0.0028	0.0169	0.0169	0.0413
0.125	0.0071	0.0338	0.0339	0.0825
0.281	0.0122	0.0508	0.0508	0.1238
0.500	0.0179	0.0677	0.0677	0.1651
0.781	0.0241	0.0846	0.0847	0.2063
1.125	0.0307	0.1015	0.1016	0.2476
1.531	0.0377	0.1184	0.1185	0.2889
2.000	0.0451	0.1354	0.1355	0.3302
2.531	0.0527	0.1523	0.1524	0.3714
3.125	0.0607	0.1692	0.1693	0.4127

Indenter → Load F (mN)	Sphere R = 5 μm h (mm)	Berkovich θ = 65.3 h (mm)	Vickers θ = 68.0° h (mm)	Cube corner θ = 35.3° h (mm)
3.781	0.0689	0.1861	0.1863	0.4540
4.500	0.0774	0.2030	0.2032	0.4952
5.281	0.0861	0.2199	0.2201	0.5365
6.125	0.0951	0.2369	0.2371	0.5778
7.031	0.1042	0.2538	0.2540	0.6190
8.000	0.1136	0.2707	0.2709	0.6603
9.031	0.1232	0.2876	0.2879	0.7016
10.125	0.1329	0.3045	0.3048	0.7429
11.281	0.1428	0.3215	0.3217	0.7841
12.500	0.1530	0.3384	0.3387	0.8254
13.781	0.1632	0.3553	0.3556	0.8667
15.125	0.1737	0.3722	0.3725	0.9079
16.531	0.1843	0.3891	0.3895	0.9492
18.000	0.1950	0.4061	0.4064	0.9905
19.531	0.2060	0.4230	0.4233	1.0317
21.125	0.2170	0.4399	0.4403	1.0730
22.781	0.2282	0.4568	0.4572	1.1143
24.500	0.2395	0.4737	0.4741	1.1556
26.281	0.2510	0.4906	0.4911	1.1968
28.125	0.2626	0.5076	0.5080	1.2381
30.031	0.2744	0.5245	0.5249	1.2794
32.000	0.2862	0.5414	0.5419	1.3206
34.031	0.2982	0.5583	0.5588	1.3619
36.125	0.3103	0.5752	0.5757	1.4032
38.281	0.3226	0.5922	0.5927	1.4444
40.500	0.3349	0.6091	0.6096	1.4857
42.781	0.3474	0.6260	0.6265	1.5270
45.125	0.3599	0.6429	0.6435	1.5682
47.531	0.3726	0.6598	0.6604	1.6095
50.000	0.3854	0.6768	0.6773	1.6508
48.775	0.3791	0.6704	0.6708	1.6444
47.550	0.3727	0.6639	0.6642	1.6379
46.325	0.3663	0.6573	0.6575	1.6314
45.100	0.3598	0.6507	0.6507	1.6247
43.875	0.3533	0.6439	0.6438	1.6180
42.650	0.3467	0.6371	0.6368	1.6111
41.425	0.3400	0.6302	0.6297	1.6042
40.200	0.3332	0.6231	0.6225	1.5972
38.975	0.3264	0.6160	0.6152	1.5900
37.750	0.3196	0.6087	0.6078	1.5828
36.525	0.3126	0.6014	0.6003	1.5754
35.300	0.3056	0.5939	0.5926	1.5679
34.075	0.2985	0.5862	0.5848	1.5603
32.850	0.2913	0.5785	0.5769	1.5525

(*Continued*)

Indenter → Load F (mN)	Sphere R = 5 μm h (mm)	Berkovich θ = 65.3 h (mm)	Vickers θ = 68.0° h (mm)	Cube corner θ = 35.3° h (mm)
31.625	0.2840	0.5706	0.5688	1.5446
30.400	0.2766	0.5625	0.5606	1.5365
29.175	0.2691	0.5543	0.5521	1.5283
27.950	0.2615	0.5458	0.5436	1.5199
26.725	0.2538	0.5372	0.5348	1.5113
25.500	0.2460	0.5285	0.5258	1.5025
24.275	0.2381	0.5194	0.5166	1.4935
23.050	0.2300	0.5102	0.5071	1.4842
21.825	0.2218	0.5007	0.4974	1.4747
20.600	0.2134	0.4910	0.4875	1.4650
19.375	0.2049	0.4809	0.4772	1.4549
18.150	0.1961	0.4705	0.4666	1.4446
16.925	0.1872	0.4598	0.4556	1.4338
15.700	0.1781	0.4487	0.4443	1.4227
14.475	0.1687	0.4371	0.4324	1.4111
13.250	0.1590	0.4250	0.4201	1.3991
12.025	0.1491	0.4124	0.4072	1.3864
10.800	0.1387	0.3991	0.3936	1.3731
9.575	0.1280	0.3850	0.3792	1.3590
8.350	0.1169	0.3700	0.3639	1.3440
7.125	0.1051	0.3538	0.3473	1.3278
5.900	0.0927	0.3362	0.3293	1.3102
4.675	0.0794	0.3166	0.3093	1.2906
3.450	0.0648	0.2942	0.2865	1.2683
2.225	0.0484	0.2674	0.2591	1.2414
1.000	0.0284	0.2313	0.2222	1.2054

a) Calculate the elastic modulus and hardness of silica using the Oliver–Pharr method. Organize the calculation steps (S, h_c, A_c, E_r, E_{silica}, and H_{silica}) and compare the results obtained with different indenters in a table.

Use ε = 0.75	Sphere R = 5 μm	Berkovich θ = 65.3°	Vickers θ = 68°	Cube corner θ = 35.3°
h_r / h_{max}				
S (N/m)				
h_c (μm)				
A_p (μm^2)				
E_r (GPa)				
E_{silica} (GPa)				
H_{silica} (GPa)				

b) Are the calculated modulus and hardness of silica the same regardless of the indenter shape or different depending on the indenter shape? If different, why?
c) Among these four data, which case would have the largest error in the calculated hardness? Why?
d) In which case, do you expect the largest pileup around the indent mark?

References

1 Fischer-Cripps AC. Nanoindentation testing. In: Nanoindentation. New York, NY: Springer; 2011. p. 21–37.
2 Broitman E. Indentation hardness measurements at macro-, micro-, and nanoscale: a critical overview. Tribology Letters. 2016;65(1):23.
3 Zhang P, Li SX, Zhang ZF. General relationship between strength and hardness. Materials Science and Engineering A. 2011;529:62–73.
4 Oliver WC, Pharr GM. An improved technique for determining hardness and elastic modulus using load and displacement sensing indentation experiments. Journal of Materials Research. 1992;7(6):1564–83.
5 Oliver WC, Pharr GM. Measurement of hardness and elastic modulus by instrumented indentation: advances in understanding and refinements to methodology. Journal of Materials Research. 2004;19(1):3–20.
6 Fischer-Cripps AC. Factors affecting nanoindentation test data. Nanoindentation. New York, NY: Springer; 2011. p. 77–104.
7 Schiffmann KI. Area function calibration in nanoindentation using the hardness instead of Young's modulus of fused silica as a reference value. International Journal of Materials Research. 2007;98(5):424–9.
8 Jakes JE. Improved methods for nanoindentation Berkovich probe calibrations using fused silica. Journal of Materials Science. 2018;53(7):4814–27.
9 Jaroslav M. Uncertainties and errors in nanoindentation. In: Jiri N, editor. Nanoindentation in Materials Science. Rijeka: IntechOpen; 2012. Ch. 3.
10 Bolshakov A, Pharr GM. Influences of pileup on the measurement of mechanical properties by load and depth sensing indentation techniques. Journal of Materials Research. 1998;13(4):1049–58.
11 Hill R, Storåkers B, Zdunek AB. A theoretical study of the Brinell hardness test. Proceedings of the Royal Society of London A Mathematical and Physical Sciences. 1989;423(1865):301–30.
12 Sakai M. Time-dependent viscoelastic relation between load and penetration for an axisymmetric indenter. Philosophical Magazine A. 2002;82(10):1841–9.
13 Storåkers B, Larsson P-L. On Brinell and Boussinesq indentation of creeping solids. Journal of the Mechanics and Physics of Solids. 1994;42(2):307–32.
14 King RB. Elastic analysis of some punch problems for a layered medium. International Journal of Solids and Structures. 1987;23 (12):1657–64.
15 Doerner MF, Nix WD. A method for interpreting the data from depth-sensing indentation instruments. Journal of Materials Research. 1986;1(4):601–9.
16 Jung Y-G, Lawn BR, Martyniuk M, Huang H, Hu XZ. Evaluation of elastic modulus and hardness of thin films by nanoindentation. Journal of Materials Research. 2004;19(10):3076–80.
17 Bhattacharya AK, Nix WD. Analysis of elastic and plastic deformation associated with indentation testing of thin films on substrates. International Journal of Solids and Structures. 1988;24(12):1287–98.
18 Korsunsky AM, McGurk MR, Bull SJ, Page TF. On the hardness of coated systems. Surface and Coatings Technology. 1998;99 (1):171–83.
19 Amma S-i, Luo J, Kim SH, Pantano CG. Effect of glass composition on the hardness of surface layers on aluminosilicate glasses formed through reaction with strong acid. Journal of the American Ceramic Society. 2018;101(2):657–65.
20 ASTM. ASTM E384-22: Standard Test Method for Microindentation Hardness of Materials 2022 [Available from: https://www.astm.org/e0384-22.html].
21 Varshneya AK, Macrelli G, Yoshida S, Kim SH, Ogrinc AL, Mauro JC. Indentation and abrasion in glass products: lessons learned and yet to be learned. International Journal of Applied Glass Science. 2022;13(3):308–37.
22 Lawn BR, Evans AG, Marshall DB. Elastic/plastic indentation damage in ceramics: the median/radial crack system. Journal of the American Ceramic Society. 1980;63(9–10):574–81.
23 Anstis GR, Chantikul P, Lawn BR, Marshall DB. A critical evaluation of indentation techniques for measuring fracture toughness: I, direct crack measurements. Journal of the American Ceramic Society. 1981;64(9):533–8.
24 Laugier MT. New formula for indentation toughness in ceramics. Journal of Materials Science Letters. 1987;6(3):355–6.
25 Dukino RD, Swain MV. Comparative measurement of indentation fracture toughness with Berkovich and Vickers indenters. Journal of the American Ceramic Society. 1992;75(12):3299–304.
26 Harding DS, Oliver WC, Pharr GM. Cracking during nanoindentation and its use in the measurement of fracture toughness. MRS Proceedings. 1994;356:663.

27 Quinn GD. Fracture Toughness of Ceramics by the Vickers Indentation Crack Length Method: A Critical Review. Mechanical Properties and Performance of Engineering Ceramics II: Ceramic Engineering and Science Proceedings 2006. p. 45–62.

28 Anunmana C, Anusavice KJ, Mecholsky JJ. Residual stress in glass: indentation crack and fractography approaches. Dental Materials. 2009;25(11):1453–8.

29 He H, Liu H, Lin Y-T, Qu C, Yu J, Kim SH. Differences in indentation and wear behaviors between the two sides of thermally tempered soda lime silica glass. Journal of the American Ceramic Society. 2021;104(9):4718–27.

30 Gross TM, Tomozawa M. Indentation-induced microhardness changes in glasses: possible fictive temperature increase caused by plastic deformation. Journal of Non-Crystalline Solids. 2008;354(34):4056–62.

31 Liu H, He H, Chen Z, Kim SH. Flexural stress effect on mechanical and mechanochemical properties of soda lime silicate glass surface. Journal of the American Ceramic Society. 2022;105(4):2847–57.

32 Tsui TY, Oliver WC, Pharr GM. Influences of stress on the measurement of mechanical properties using nanoindentation: part I. Experimental studies in an aluminum alloy. Journal of Materials Research. 1996;11(3):752–9.

33 Kese KO, Li ZC, Bergman B. Influence of residual stress on elastic modulus and hardness of soda-lime glass measured by nanoindentation. Journal of Materials Research. 2004;19(10):3109–19.

34 Yang F. Effect of adhesion energy on the contact stiffness in nanoindentation. Journal of Materials Research. 2006;21 (10):2683–8.

35 He H, Qiao Q, Xiao T, Yu J, Kim SH. Effect of humidity on friction, wear, and plastic deformation during nanoscratch of soda lime silica glass. Journal of the American Ceramic Society. 2022;105(2):1367–74.

36 He H, Chen Z, Lin Y-T, Hahn SH, Yu J, van Duin ACT, et al. Subsurface structural change of silica upon nanoscale physical contact: chemical plasticity beyond topographic elasticity. Acta Materialia. 2021;208:116694.

37 Lin Y-T, Ogrinc AL, Zoba AN, Lee J, Jang S, Smith NJ, et al. Revealing 'invisible' subsurface structural change/damage in silicate glass made by 'nearly-elastic contact' with a spherical smooth surface. Acta Materialia. 2024;264:119571.

38 Liu H, Kaya H, Lin Y-T, Ogrinc A, Kim SH. Vibrational spectroscopy analysis of silica and silicate glass networks. Journal of the American Ceramic Society. 2022;105(4):2355–84.

39 Fry AL, Ogrinc AL, Kim SH, Mauro JC. Field strength effect on elastoplastic behavior of aluminoborosilicate glass: I. Elastic moduli and indentation size effect. Journal of the American Ceramic Society. 2023;106(8):4664–77.

40 Li H, Bradt RC. The indentation load/size effect and the measurement of the hardness of vitreous silica. Journal of Non-Crystalline Solids. 1992;146:197–212.

41 Li H, Bradt RC. The microhardness indentation load/size effect in rutile and cassiterite single crystals. Journal of Materials Science. 1993;28(4):917–26.

42 Fry AL, Ogrinc AL, Kim SH, Mauro JC. Field strength effect on elastoplastic behavior of aluminoborosilicate glass: II. Volumetric recovery. Journal of the American Ceramic Society. 2023;106(9):5213–32.

43 Yoshida S, Sangleboeuf J-C, Rouxel T. Quantitative evaluation of indentation-induced densification in glass. Journal of Materials Research. 2005;20(12):3404–12.

44 Sheth N, Hahn SH, Ngo D, Howzen A, Bermejo R, van Duin ACT, et al. Influence of acid leaching surface treatment on indentation cracking of soda lime silicate glass. Journal of Non-Crystalline Solids. 2020;543:120144.

45 Fischer-Cripps AC. Nanoindentation. New York, NY: Springer; 2011.

46 Fischer-Cripps AC. Introduction to Contact Mechanics. Boston, MA: Springer US; 2007.

47 Fischer-Cripps AC. Simulation of nanoindentation test data. In: Fischer-Cripps AC, editor. Nanoindentation. New York, NY: Springer; 2011. p. 105–17.

Index

t